Proceedings of the Institute of Industrial Engineers
Asian Conference 2013

Yi-Kuei Lin · Yu-Chung Tsao
Shi-Woei Lin
Editors

Proceedings of the Institute of Industrial Engineers Asian Conference 2013

Volume I

Springer

Editors
Yi-Kuei Lin
Yu-Chung Tsao
Shi-Woei Lin
Department of Industrial Management
National Taiwan University of Science and Technology
Taipei
Taiwan, R.O.C.

ISBN 978-981-4451-97-0   ISBN 978-981-4451-98-7   (eBook)
DOI 10.1007/978-981-4451-98-7
Springer Singapore Heidelberg New York Dordrecht London

Library of Congress Control Number: 2013943711

© Springer Science+Business Media Singapore 2013
This work is subject to copyright. All rights are reserved by the Publisher, whether the whole or part of the material is concerned, specifically the rights of translation, reprinting, reuse of illustrations, recitation, broadcasting, reproduction on microfilms or in any other physical way, and transmission or information storage and retrieval, electronic adaptation, computer software, or by similar or dissimilar methodology now known or hereafter developed. Exempted from this legal reservation are brief excerpts in connection with reviews or scholarly analysis or material supplied specifically for the purpose of being entered and executed on a computer system, for exclusive use by the purchaser of the work. Duplication of this publication or parts thereof is permitted only under the provisions of the Copyright Law of the Publisher's location, in its current version, and permission for use must always be obtained from Springer. Permissions for use may be obtained through RightsLink at the Copyright Clearance Center. Violations are liable to prosecution under the respective Copyright Law. The use of general descriptive names, registered names, trademarks, service marks, etc. in this publication does not imply, even in the absence of a specific statement, that such names are exempt from the relevant protective laws and regulations and therefore free for general use.
While the advice and information in this book are believed to be true and accurate at the date of publication, neither the authors nor the editors nor the publisher can accept any legal responsibility for any errors or omissions that may be made. The publisher makes no warranty, express or implied, with respect to the material contained herein.

Printed on acid-free paper

Springer is part of Springer Science+Business Media (www.springer.com)

Preface

The organizers of the IIE Asian Conference 2013, held July 19–22, 2013 in Taipei, Taiwan, are pleased to present the proceedings of the conference, which includes about 180 papers accepted for presentation at the conference. The papers in the proceedings were selected from more than 270 paper submissions. Based on input from peer reviews, the committee selected these papers for their perceived quality, originality, and appropriateness to the theme of the IIE Asian Conference. We would like to thank all contributors who submitted papers. We would also like to thank the track chairs and colleagues who provided reviews of the submitted papers.

The main objective of IIE Asian Conference 2013 is to provide a forum for exchange of ideas on the latest developments in the field of industrial engineering and management. This series of conferences were started and supported by the Institute of Industrial Engineers (IIE) in 2011 as a meeting of organizations striving to establish a common focus for research and development in the field of industrial engineering. Now after just 2 years of the first event, the IIE Asian Conference has become recognized as an international focal point annually attended by researchers and developers from dozens of countries around the Asian.

Whether the papers included in the proceedings are work-in-progress or finished products, the conference and proceedings offer all authors opportunities to disseminate the results of their research and receive timely feedback from colleagues, without long wait associated with publication in peer-reviewed journals. We hope that the proceedings will represent a worthwhile contribution to the theoretical and applied body of knowledge in industrial engineering and management. We also hope that it will attract some other researchers and practitioners to join future IIE Asian Conferences.

Finally, the organizers are indebted to a number of people who gave freely their time to make the conference a reality. We would like to acknowledge Dr. Ching-Jong Liao, the president of the National Taiwan University of Science and Technology (Taiwan Tech), and all faculties and staffs in the Department of Industrial

Management at the Taiwan Tech, who have contributed tremendous amount of time and efforts to this conference. The Program Committee would also like to express its appreciation to all who participate to make the IIE Asian Conference 2013 a successful and enriching experience.

Program Committee of IIE Asian Conference 2013

Contents

An Optimal Ordering Policy of the Retailers Under Partial Trade Credit Financing and Restricted Cycle Time in Supply Chain 1
Shin'ichi Yoshikawa

An Immunized Ant Colony System Algorithm to Solve Unequal Area Facility Layout Problems Using Flexible Bay Structure 9
Mei-Shiang Chang and Hsin-Yi Lin

Characterizing the Trade-Off Between Queue Time and Utilization of a Manufacturing System 19
Kan Wu

Using Latent Variable to Estimate Parameters of Inverse Gaussian Distribution Based on Time-Censored Wiener Degradation Data 29
Ming-Yung Lee and Cheng-Hung Hu

Interpolation Approximations for the Performance of Two Single Servers in Series 37
Kan Wu

On Reliability Evaluation of Flow Networks with Time-Variant Quality Components 45
Shin-Guang Chen

Defect Detection of Solar Cells Using EL Imaging and Fourier Image Reconstruction 53
Ya-Hui Tsai, Du-Ming Tsai, Wei-Chen Li and Shih-Chieh Wu

Teaching Industrial Engineering: Developing a Conjoint Support System Catered for Non-Majors 63
Yoshiki Nakamura

An Automatic Image Enhancement Framework for Industrial Products Inspection 73
Chien-Cheng Chu, Chien-Chih Wang and Bernard C. Jiang

Ant Colony Optimization Algorithms for Unrelated Parallel Machine Scheduling with Controllable Processing Times and Eligibility Constraints 79
Chinyao Low, Rong-Kwei Li and Guan-He Wu

General Model for Cross-Docking Distribution Planning Problem with Time Window Constraints 89
Parida Jewpanya and Voratas Kachitvichyanukul

A New Solution Representation for Solving Location Routing Problem via Particle Swarm Optimization 103
Jie Liu and Voratas Kachitvichyanukul

An Efficient Multiple Object Tracking Method with Mobile RFID Readers 111
Chieh-Yuan Tsai and Chen-Yi Huang

A New Bounded Intensity Function for Repairable Systems 119
Fu-Kwun Wang and Yi-Chen Lu

Value Creation Through 3PL for Automotive Logistical Excellence 127
Chin Lin Wen, Schnell Jeng, Danang Kisworo, Paul K. P. Wee and H. M. Wee

Dynamics of Food and Beverage Subsector Industry in East Java Province: The Effect of Investment on Total Labor Absorption 133
Putri Amelia, Budisantoso Wirjodirdjo and Niniet Indah Arvitrida

Solving Two-Sided Assembly Line Balancing Problems Using an Integrated Evolution and Swarm Intelligence 141
Hindriyanto Dwi Purnomo, Hui-Ming Wee and Yugowati Praharsi

Genetic Algorithm Approach for Multi-Objective Optimization of Closed-Loop Supply Chain Network 149
Li-Chih Wang, Tzu-Li Chen, Yin-Yann Chen, Hsin-Yuan Miao, Sheng-Chieh Lin and Shuo-Tsung Chen

Replacement Policies with a Random Threshold Number of Faults . . . 157
Xufeng Zhao, Mingchih Chen, Kazunori Iwata, Syouji Nakamura
and Toshio Nakagawa

A Multi-Agent Model of Consumer Behavior Considering
Social Networks: Simulations for an Effective Movie
Advertising Strategy 165
Yudai Arai, Tomoko Kajiyama and Noritomo Ouchi

Government Subsidy Impacts on a Decentralized Reverse Supply
Chain Using a Multitiered Network Equilibrium Model 173
Pin-Chun Chen and I-Hsuan Hong

A Capacity Planning Method for the Demand-to-Supply
Management in the Pharmaceutical Industry 181
Nobuaki Ishii and Tsunehiro Togashi

Storage Assignment Methods Based on Dependence of Items 189
Po-Hsun Kuo and Che-Wei Kuo

Selection of Approximation Model on Total Perceived Discomfort
Function for the Upper Limb Based on Joint Moment 197
Takanori Chihara, Taiki Izumi and Akihiko Seo

Waiting as a Signal of Quality When Multiple Extrinsic
Cues are Presented 205
Shi-Woei Lin and Hao-Yuan Chan

Effect of Relationship Types on the Behaviors of Health
Care Professionals. 211
Shi-Woei Lin and Yi-Tseng Lin

A Simulation with PSO Approach for Semiconductor
Back-End Assembly. 219
James T. Lin, Chien-Ming Chen and Chun-Chih Chiu

Effect of Grasp Conditions on Upper Limb Load During
Visual Inspection of Objects in One Hand 229
Takuya Hida and Akihiko Seo

A Process-Oriented Mechanism Combining Fuzzy Decision
Analysis for Supplier Selection in New Product Development 239
Jiun-Shiung Lin, Jen-Huei Chang and Min-Che Kao

**Reliability-Based Performance Evaluation for a Stochastic
Project Network Under Time and Budget Thresholds** 249
Yi-Kuei Lin, Ping-Chen Chang and Shin-Ying Li

**System Reliability and Decision Making for a Production
System with Intersectional Lines** 257
Yi-Kuei Lin, Ping-Chen Chang and Kai-Jen Hsueh

Customer Perceptions of Bowing with Different Trunk Flexions 265
Yi-Lang Chen, Chiao-Ying Yu, Lan-Shin Huang, Ling-Wei Peng
and Liang-Jie Shi

**A Pilot Study Determining Optimal Protruding Node Length
of Bicycle Seats Using Subjective Ratings** 271
Yi-Lang Chen, Yi-Nan Liu and Che-Feng Cheng

**Variable Neighborhood Search with Path-Relinking
for the Capacitated Location Routing Problem** 279
Meilinda F. N. Maghfiroh, A. A. N. Perwira Redi and Vincent F. Yu

**Improving Optimization of Tool Path Planning in 5-Axis
Flank Milling by Integrating Statistical Techniques** 287
Chih-Hsing Chu and Chi-Lung Kuo

**A Multiple Objectives Based DEA Model to Explore the Efficiency
of Airports in Asia–Pacific** 295
James J. H. Liou, Hsin-Yi Lee and Wen-Chein Yeh

**A Distributed Constraint Satisfaction Approach for Supply
Chain Capable-to-Promise Coordination** 303
Yeh-Chun Juan and Jyun-Rong Syu

**Design and Selection of Plant Layout by Mean of Analytic
Hierarchy Process: A Case Study of Medical Device Producer** 311
Arthit Chaklang, Arnon Srisom and Chirakiat Saithong

Using Taguchi Method for Coffee Cup Sleeve Design 319
Yiyo Kuo, Hsin-Yu Lin, Ying Chen Wu, Po-Hsi Kuo,
Zhi-He Liang and Si Yong Wen

**Utilizing QFD and TRIZ Techniques to Design a Helmet
Combined with the Wireless Camcorder** 327
Shu-Jen Hu, Ling-Huey Su and Jhih-Hao Laio

South East Asia Work Measurement Practices Challenges and Case Study 335
Thong Sze Yee, Zuraidah Mohd Zain and Bhuvenesh Rajamony

Decision Support System: Real-Time Dispatch of Manufacturing Processes 345
Chung-Wei Kan and An-Pin Chen

The Application of MFCA Analysis in Process Improvement: A Case Study of Plastics Packaging Factory in Thailand 353
Chompoonoot Kasemset, Suchon Sasiopars and Sugun Suwiphat

Discussion of Water Footprint in Industrial Applications 363
Chung Chia Chiu, Wei-Jung Shiang and Chiuhsiang Joe Lin

Mitigating Uncertainty Risks Through Inventory Management: A Case Study for an Automobile Company 371
Amy Chen, H. M. Wee, Chih-Ying Hsieh and Paul Wee

Service Quality for the YouBike System in Taipei 381
Jung-Wei Chang, Xin-Yi Jiang, Xiu-Ru Chen, Chia-Chen Lin and Shih-Che Lo

Replenishment Strategies for the YouBike System in Taipei 389
Chia-Chen Lin, Xiu-Ru Chen, Jung-Wei Chang, Xin-Yi Jiang and Shih-Che Lo

A Tagging Mechanism for Solving the Capacitated Vehicle Routing Problem 397
Calvin K. Yu and Tsung-Chun Hsu

Two-Stage Multi-Project Scheduling with Minimum Makespan Under Limited Resource 405
Calvin K. Yu and Ching-Chin Liao

A Weighting Approach for Scheduling Multi-Product Assembly Line with Multiple Objectives 415
Calvin K. Yu and Pei-Fang Lee

Exploring Technology Feature with Patent Analysis 423
Ping Yu Hsu, Ming Shien Cheng, Kuo Yen Lu and Chen Yao Chung

Making the MOST® Out of Economical Key-Tabbing Automation . . . 433
P. A. Brenda Yap, S. L. Serene Choo and Thong Sze Yee

Integer Program Modeling of Portfolio Optimization with Mental Accounts Using Simulated Tail Distribution ... 441
Kuo-Hwa Chang, Yi Shou Shu and Michael Nayat Young

Simulated Annealing Algorithm for Berth Allocation Problems ... 449
Shih-Wei Line and Ching-Jung Ting

Using Hyperbolic Tangent Function for Nonlinear Profile Monitoring ... 457
Shu-Kai S. Fan and Tzu-Yi Lee

Full Fault Detection for Semiconductor Processes Using Independent Component Analysis ... 465
Shu-Kai S. Fan and Shih-Han Huang

Multi-Objective Optimal Placement of Automatic Line Switches in Power Distribution Networks ... 471
Diego Orlando Logrono, Wen-Fang Wu and Yi-An Lu

The Design of Combing Hair Assistive Device to Increase the Upper Limb Activities for Female Hemiplegia ... 479
Jo-Han Chang

Surgical Suites Scheduling with Integrating Upstream and Downstream Operations ... 487
Huang Kwei-Long, Lin Yu-Chien and Chen Hao-Huai

Research on Culture Supply Chain Intension and Its Operation Models ... 497
Xiaojing Li and Qian Zhang

Power System by Variable Scaling Hybrid Differential Evolution ... 505
Ji-Pyng Chiou, Chong-Wei Lo and Chung-Fu Chang

Investigating the Replenishment Policy for Retail Industries Under VMI Strategic Alliance Using Simulation ... 513
Ping-Yu Chang

Applying RFID in Picker's Positioning in a Warehouse ... 521
Kai Ying Chen, Mei Xiu Wu and Shih Min Chen

An Innovation Planning Approach Based on Combining Technology Progress Trends and Market Price Trends ... 531
Wen-Chieh Chuang and Guan-Ling Lin

Contents

Preemptive Two-Agent Scheduling in Open Shops Subject to Machine Availability and Eligibility Constraints 539
Ming-Chih Hsiao and Ling-Huey Su

Supply Risk Management via Social Capital Theory and Its Impact on Buyer's Performance Improvement and Innovation 549
Yugowati Praharsi, Maffie Linda Araos Dioquino and Hui-Ming Wee

Variable Time Windows-Based Three-Phase Combined Algorithm for On-Line Batch Processing Machine Scheduling with Limited Waiting Time Constraints 559
Dongwei Yang, Wenyou Jia, Zhibin Jiang and You Li

Optimal Organic Rankine Cycle Installation Planning for Factory Waste Heat Recovery 569
Yu-Lin Chen and Chun-Wei Lin

Evaluation of Risky Driving Performance in Lighting Transition Zones Near Tunnel Portals 577
Ying-Yin Huang and Marino Menozzi

Application of Maple on Solving Some Differential Problems 585
Chii-Huei Yu

Six Sigma Approach Applied to LCD Photolithography Process Improvement 593
Yung-Tsan Jou and Yih-Chuan Wu

A Study of the Integrals of Trigonometric Functions with Maple 603
Chii-Huei Yu

A Study of Optimization on Mainland Tourist Souvenir Shops Service Reliability 611
Kang-Hung Yang, Li-Peng Fang and Z-John Liu

The Effects of Background Music Style on Study Performance 619
An-Che Chen and Chen-Shun Wen

Using Maple to Study the Multiple Improper Integral Problem 625
Chii-Huei Yu

On Reformulation of a Berth Allocation Model 633
Yun-Chia Liang, Angela Hsiang-Ling Chen
and Horacio Yamil Lovo Gutierrezmil

Forecast of Development Trends in Cloud Computing Industry 641
Wei-Hsiu Weng, Woo-Tsong Lin and Wei-Tai Weng

**Self-Organizing Maps with Support Vector Regression
for Sales Forecasting: A Case Study in Fresh Food Data** 649
Annisa Uswatun Khasanah, Wan-Hsien Lin and Ren-Jieh Kuo

**State of Charge Estimation for Lithium-Ion Batteries Using
a Temperature-Based Equivalent Circuit Model** 657
Yinjiao Xing and Kwok-Leung Tsui

**Linking Individual Investors' Preferences to a Portfolio
Optimization Model** 665
Angela Hsiang-Ling Chen, Yun-Chia Liang and Chieh Chiang

**Models and Partial Re-Optimization Heuristics for Dynamic
Hub-and-Spoke Transferring Route Problems** 673
Ming-Der May

**Hazards and Risks Associated with Warehouse Workers:
A Field Study** 681
Ren-Liu Jang and An-Che Chen

**Green Supply Chain Management (GSCM) in an Industrial Estate:
A Case Study of Karawang Industrial Estate, Indonesia** 687
Katlea Fitriani

**Limits The Insured Amount to Reduce Loss?: Use the Group
Accident Insurance as an Example** 695
Hsu-Hua Lee, Ming-Yuan Hsu and Chen-Ying Lee

**Manipulation Errors in Blindfold Pointing Operation
for Visual Acuity Screenings** 705
Ying-Yin Huang and Marino Menozzi

3-Rainbow Domination Number in Graphs 713
Kung-Jui Pai and Wei-Jai Chiu

Contents

A Semi-Fuzzy AHP Approach to Weigh the Customer Requirements in QFD for Customer-Oriented Product Design 721
Jiangming Zhou and Nan Tu

An Optimization Approach to Integrated Aircraft and Passenger Recovery 729
F. T. S. Chan, S. H. Chung, J. C. L. Chow and C. S. Wong

Minimizing Setup Time from Mold-Lifting Crane in Mold Maintenance Schedule 739
C. S. Wong, F. T. S. Chan, S. H. Chung and B. Niu

Differential Evolution Algorithm for Generalized Multi-Depot Vehicle Routing Problem with Pickup and Delivery Requests 749
Siwaporn Kunnapapdeelert and Voratas Kachitvichyanukul

A Robust Policy for the Integrated Single-Vendor Single-Buyer Inventory System in a Supply Chain 757
Jia-Shian Hu, Pei-Fang Tsai and Ming-Feng Yang

Cost-Based Design of a Heat Sink Using SVR, Taguchi Quality Loss, and ACO 765
Chih-Ming Hsu

Particle Swarm Optimization Based Nurses' Shift Scheduling 775
Shiou-Ching Gao and Chun-Wei Lin

Applying KANO Model to Exploit Service Quality for the Real Estate Brokering Industry 783
Pao-Tiao Chuang and Yi-Ping Chen

Automated Plastic Cap Defect Inspection Using Machine Vision 793
Fang-Chin Tien, Jhih-Syuan Dai, Shih-Ting Wang and Fang-Cheng Tien

Coordination of Long-Term, Short-Term Supply Contract and Capacity Investment Strategy 801
Chiao Fu and Cheng-Hung Wu

An Analysis of Energy Prices and Economic Indicators Under the Uncertainties: Evidence from South East Asian Markets 809
Shunsuke Sato, Deddy P. Koesrindartoto and Shunsuke Mori

Effects of Cooling and Sex on the Relationship Between Estimation and Actual Grip Strength 819
Chih-Chan Cheng, Yuh-Chuan Shih and Chia-Fen Chi

Data Clustering on Taiwan Crop Sales Under Hadoop Platform 827
Chao-Lung Yang and Mohammad Riza Nurtam

Control with Hand Gestures in Home Environment: A Review 837
Sheau-Farn Max Liang

An Integrated Method for Customer-Oriented Product Design 845
Jiangming Zhou, Nan Tu, Bin Lu, Yanchao Li and Yixiao Yuan

Discrete Particle Swarm Optimization with Path-Relinking for Solving the Open Vehicle Routing Problem with Time Windows 853
A. A. N. Perwira Redi, Meilinda F. N. Maghfiroh and Vincent F. Yu

Application of Economic Order Quantity on Production Scheduling and Control System for a Small Company 861
Kuo En Fu and Pitchanan Apichotwasurat

CUSUM Residual Charts for Monitoring Enterovirus Infections 871
Huifen Chen and Yu Chen

A Study on the Operation Model of the R&D Center for the Man-Made Fiber Processing Industry Headquarter 879
Ming-Kuen Chen, Shiue-Lung Yang and Tsu-Yi Hung

Planning Logistics by Algorithm with VRPTWBD for Rice Distribution: A Case BULOG Agency in the Nganjuk District Indonesia 889
Kung-Jeng Wang, Farikhah Farkhani and I. Nyoman Pujawan

A Systematic and Innovative Approach to Universal Design Based on TRIZ Theories 899
Chun-Ming Yang, Ching-Han Kao, Thu-Hua Liu, Hsin-Chun Pei and Yan-Lin Lee

Wireless LAN Access Point Location Planning 907
Sung-Lien Kang, Gary Yu-Hsin Chen and Jamie Rogers

Contents   xvii

**The Parametric Design of Adhesive Dispensing Process
with Multiple Quality Characteristics** 915
Carlo Palacios, Osman Gradiz and Chien-Yi Huang

The Shortage Study for the EOQ Model with Imperfect Items 925
Chiang-Sheng Lee, Shiaau-Er Huarng, Hsine-Jen Tsai and Bau-Ding Lee

**Power, Relationship Commitment and Supplier
Integration in Taiwan** 935
Jen-Ying Shih and Sheng-Jie Lu

**A Search Mechanism for Geographic Information
Processing System** 945
Hsine-Jen Tsai, Chiang-Sheng Lee and Les Miller

Maximum Acceptable Weight Limit on Carrying a Food Tray 953
Ren-Liu Jang

**Fatigue Life and Reliability Analysis of Electronic Packages
Under Thermal Cycling and Moisture Conditions** 957
Yao Hsu, Wen-Fang Wu and Chih-Min Hsu

**Clustering-Locating-Routing Algorithm for Vehicle Routing
Problem: An Application in Medical Equipment Maintenance** 965
Kanokwan Supakdee, Natthapong Nanthasamroeng
and Rapeepan Pitakaso

Whole-Body Vibration Exposure in Urban Motorcycle Riders 975
Hsieh-Ching Chen and Yi-Tsong Pan

Analysis of Sales Strategy with Lead-Time Sensitive Demand 985
Chi-Yang Tsai, Wei-Fan Chu and Cheng-Yu Tu

Order and Pricing Decisions with Return and Buyback Policies 993
Chi-Yang Tsai, Pei-Yu Pai and Qiao-Kai Huang

**Investigation of Safety Compliance and Safety Participation
as Well as Cultural Influences Using Selenginsk Pulp
and Cardboard Mill in Russia as an Example** 1001
Ekaterina Nomokonova, Shu-Chiang Lin and Guanhuah Chen

**Identifying Process Status Changes via Integration
of Independent Component Analysis and Support Vector Machine** ... 1009
Chuen-Sheng Cheng and Kuo-Ko Huang

A Naïve Bayes Based Machine Learning Approach and Application Tools Comparison Based on Telephone Conversations 1017
Shu-Chiang Lin, Murman Dwi Prasetio, Satria Fadil Persada and Reny Nadlifatin

Evaluating the Development of the Renewable Energy Industry 1025
Hung-Yu Huang, Chung-Shou Liao and Amy J. C. Trappey

On-Line Quality Inspection System for Automotive Component Manufacturing Process 1031
Chun-Tai Yen, Hung-An Kao, Shih-Ming Wang and Wen-Bin Wang

Using Six Sigma to Improve Design Quality: A Case of Mechanical Development of the Notebook PC in Wistron 1039
Kun-Shan Lee and Kung-Jeng Wang

Estimating Product Development Project Duration for the Concurrent Execution of Multiple Activities 1047
Gyesik Oh and Yoo S. Hong

Modeling of Community-Based Mangrove Cultivation Policy in Sidoarjo Mudflow Area by Implementing Green Economy Concept 1055
Diesta Iva Maftuhah, Budisantoso Wirjodirdjo and Erwin Widodo

Three Approaches to Find Optimal Production Run Time of an Imperfect Production System 1065
Jin Ai, Ririn Diar Astanti, Agustinus Gatot Bintoro and Thomas Indarto Wibowo

Rice Fulfillment Analysis in System Dynamics Framework (Study Case: East Java, Indonesia) 1071
Nieko Haryo Pradhito, Shuo-Yan Chou, Anindhita Dewabharata and Budisantoso Wirdjodirdjo

Activity Modeling Using Semantic-Based Reasoning to Provide Meaningful Context in Human Activity Recognizing 1081
AnisRahmawati Amnal, Anindhita Dewabharata, Shou-Yan Chou and Mahendrawathi Erawan

An Integrated Systems Approach to Long-Term Energy Security Planning 1091
Ying Wang and Kim Leng Poh

An EPQ with Shortage Backorders Model on Imperfect
Production System Subject to Two Key Production Systems 1101
Baju Bawono, The Jin Ai, Ririn Diar Astanti
and Thomas Indarto Wibowo

Reducing Medication Dispensing Process Time
in a Multi-Hospital Health System 1109
Jun-Ing Ker, Yichuan Wang and Cappi W. Ker

A Pareto-Based Differential Evolution Algorithm
for Multi-Objective Job Shop Scheduling Problems 1117
Warisa Wisittipanich and Voratas Kachitvichyanukul

Smart Grid and Emergency Power Supply on Systems
with Renewable Energy and Batteries: An Recovery
Planning for EAST JAPAN Disaster Area 1127
Takuya Taguchi and Kenji Tanaka

Establishing Interaction Specifications for Online-to-Offline (O2O)
Service Systems. 1137
Cheng-Jhe Lin, Tsai-Ting Lee, Chiuhsiang Lin, Yu-Chieh Huang
and Jing-Ming Chiu

Investigation of Learning Remission in Manual Work Given
that Similar Work is Performed During the Work
Contract Break 1147
Josefa Angelie D. Revilla and Iris Ann G. Martinez

A Hidden Markov Model for Tool Wear Management 1157
Chen-Ju Lin and Chun-Hung Chien

Energy Management Using Storage Batteries in Large
Commercial Facilities Based on Projection of Power Demand 1165
Kentaro Kaji, Jing Zhang and Kenji Tanaka

The Optimal Parameters Design of Multiple Quality Characteristics
for the Welding Thick Plate of Aerospace Aluminum Alloy 1173
Jhy-Ping Jhang

Synergizing Both Universal Design Principles and Su-Field Analysis
to an Innovative Product Design Process 1183
Chun-Ming Yang, Ching-Han Kao, Thu-Hua Liu, Ting Lin
and Yi-Wun Chen

The Joint Determination of Optimum Process Mean, Economic Order Quantity, and Production Run Length 1191

Chung-Ho Chen

Developing Customer Information System Using Fuzzy Query and Cluster Analysis 1199

Chui-Yu Chiu, Ho-Chun Ku, I-Ting Kuo and Po-Chou Shih

Automatic Clustering Combining Differential Evolution Algorithmand k-Means Algorithm 1207

R. J. Kuo, Erma Suryani and Achmad Yasid

Application of Two-Stage Clustering on the Attitude and Behavioral of the Nursing Staff: A Case Study of Medical Center in Taiwan ... 1217

Farn-Shing Chen, Shih-Wei Hsu, Chia-An Tu and Wen-Tsann Lin

The Effects of Music Training on the Cognitive Ability and Auditory Memory 1225

Min-Sheng Chen, Chan-Ming Hsu and Tien-Ju Chiang

Control Scheme for the Service Quality 1233

Ling Yang

Particle Swam Optimization for Multi-Level Location Allocation Problem Under Supplier Evaluation 1237

Anurak Chaiwichian and Rapeepan Pitakaso

Evaluation Model for Residual Performance of Lithium-Ion Battery 1251

Takuya Shimamoto, Ryuta Tanaka and Kenji Tanaka

A Simulated Annealing Heuristic for the Green Vehicle Routing Problem 1261

Moch Yasin and Vincent F. Yu

Designing an Urban Sustainable Water Supply System Using System Dynamics 1271

S. Zhao, J. Liu and X. Liu

An Evaluation of LED Ceiling Lighting Design with Bi-CCT Layouts 1279

Chinmei Chou, Jui-Feng Lin, Tsu-Yu Chen, Li-Chen Chen and YaHui Chiang

Postponement Strategies in a Supply Chain Under the MTO Production Environment 1289
Hsin Rau and Ching-Kuo Liu

Consumer Value Assessment with Consideration of Environmental Impact................................. 1297
Hsin Rau, Sing-Ni Siang and Yi-Tse Fang

A Study of Bi-Criteria Flexible Flow Lines Scheduling Problems with Queue Time Constraints...................... 1307
Chun-Lung Chen

Modeling the Dual-Domain Performance of a Large Infrastructure Project: The Case of Desalination 1315
Vivek Sakhrani, Adnan AlSaati and Olivier de Weck

Flexibility in Natural Resource Recovery Systems: A Practical Approach to the "Tragedy of the Commons"......... 1325
S. B. von Helfenstein

The Workload Assessment and Learning Effective Associated with Truck Driving Training Courses 1335
Yuh-Chuan Shih, I-Sheng Sun and Chia-Fen Chi

Prognostics Based Design for Reliability Technique for Electronic Product Design 1343
Yingche Chien, Yu-Xiu Huang and James Yu-Che Wang

A Case Study on Optimal Maintenance Interval and Spare Part Inventory Based on Reliability................. 1353
Nani Kurniati, Ruey-Huei Yeh and Haridinuto

Developing Decision Models with Varying Machine Ratios in a Semiconductor Company 1361
Rex Aurelius C. Robielos

New/Advanced Industrial Engineering Perspective: Leading Growth Through Customer Centricity 1371
Suresh Kumar Babbar

Scheduling a Hybrid Flow-Shop Problem via Artificial Immune System............................. 1377
Tsui-Ping Chung and Ching-Jong Liao

Modeling and Simulation on a Resilient Water Supply System Under Disruptions 1385
X. Liu, J. Liu, S. Zhao and Loon Ching Tang

A Hybrid ANP-DEA Approach for Vulnerability Assessment in Water Supply System 1395
C. Zhang and X. Liu

An Integrated BOM Evaluation and Supplier Selection Model for a Design for Supply Chain System 1405
Yuan-Jye Tseng, Li-Jong Su, Yi-Shiuan Chen and Yi-Ju Liao

Estimation Biases in Construction Projects: Further Evidence 1413
Budi Hartono, Sinta R. Sulistyo and Nezar Alfian

Exploring Management Issues in Spare Parts Forecast 1421
Kuo-Hsing Wu, Hsin Rau and Ying-Che Chien

Artificial Particle Swarm Optimization with Heuristic Procedure to Solve Multi-Line Facility Layout Problem 1431
Chao Ou-Yang, Budi Santosa and Achmad Mustakim

Applying a Hybrid Data Preprocessing Methods in Stroke Prediction 1441
Chao Ou-Yang, Muhammad Rieza, Han-Cheng Wang, Yeh-Chun Juan and Cheng-Tao Huang

Applying a Hybrid Data Mining Approach to Develop Carotid Artery Prediction Models 1451
Chao Ou-Yang, Inggi Rengganing Herani, Han-Cheng Wang, Yeh-Chun Juan, Erma Suryani and Cheng-Tao Huang

Comparing Two Methods of Analysis and Design Modelling Techniques: Unified Modelling Language and Agent Modelling Language. Study Case: A Virtual Bubble Tea Vending Machine System Development 1461
Immah Inayati, Shu-Chiang Lin and Widya Dwi Aryani

Persuasive Technology on User Interface Energy Display: Case Study on Intelligent Bathroom 1471
Widya Dwi Aryani, Shu-Chiang Lin and Immah Inayati

Investigating the Relationship Between Electronic Image of Online Business on Smartphone and Users' Purchase Intention 1479
Chorng-Guang Wu and Yu-Han Kao

Forecast of Development Trends in Big Data Industry 1487
Wei-Hsiu Weng and Wei-Tai Weng

Reliability Analysis of Smartphones Based on the Field Return Data 1495
Fu-Kwun Wang, Chen-I Huang and Tao-Peng Chu

The Impact of Commercial Banking Performance on Economic Growth. 1503
Xiaofeng Hui and Suvita Jha

Data and Information Fusion for Bio-Medical Design and Bio-Manufacturing Systems 1513
Yuan-Shin Lee, Xiaofeng Qin, Peter Prim and Yi Cai

Evaluating the Profit Efficiency of Commercial Banks: Empirical Evidence from Nepal 1521
Suvita Jha, Xiaofeng Hui and Baiqing Sun

Explore the Inventory Problem in a System Point of View: A Lot Sizing Policy 1529
Tsung-shin Hsu and Yu-Lun Su

Global Industrial Teamwork Dynamics in China and Southeast Asia: Influence on Production Tact Time and Management Cumulative Effect to Teamwork Awareness-1/2 . . . 1539
Masa-Hiro Nowatari

Global Industrial Teamwork Dynamics in Malaysia—Effects of Social Culture and Corporate Culture to Teamwork Awareness—2/2. 1551
Masa-Hiro Nowatari

Relaxed Flexible Bay Structure in the Unequal Area Facility Layout Problem 1563
Sadan Kulturel-Konak

Comparisons of Different Mutation and Recombination Processes of the DEA for SALB-1 1571

Rapeepan Pitakaso, Panupan Parawech and Ganokgarn Jirasirierd

An Exploration of GA, DE and PSO in Assignment Problems 1581

Tassin Srivarapongse and Rapeepan Pitakaso

Author Index 1589

An Optimal Ordering Policy of the Retailers Under Partial Trade Credit Financing and Restricted Cycle Time in Supply Chain

Shin'ichi Yoshikawa

Abstract The traditional EOQ (Economic Order Quantity) model assumes that retailers' capitals are unrestricting and the retailer must pay for items as soon as the retailer receives them from suppliers. However, this may not be true. In practice, the supplier will offer the retailer a delay period. This period is known as the trade credit period. Previously published papers assumed that the supplier would offer the retailer a delay period and the retailer could sell goods and earn interest or investment within the trade credit period. They assumed that the supplier would offer the retailer a delay period but the retailer would not offer the trade credit period to his/her customer. We extend their model and construct new ordering policy. In this paper, the retailer will also adopt the partial trade credit policy to his/her customer. We assume that the retailer's trade credit period offered by the supplier is not shorter than his/her customer's trade credit period offered by the retailer. In addition, they assumed the relationship between the supplier and the retailer is one-to-one. One thing we want to emphasize here is that the supplier has cooperative relations with many retailers. Furthermore, we assume that the total of the cycle time is restricted. Under these conditions, we model the retailers' inventory system to determine the optimal cycle times for n retailers.

Keywords EOQ model · Partial trade credit · Supply chain

1 Introduction

Inventory management is to decide appropriate times and quantities to produce goods. It has a significant impact on the costs and profitability of many organizations. In general, the EOQ model is still a widely used model to guide the

---

S. Yoshikawa (✉)
Nagoya Keizai University, Uchikubo 61-1, Inuyama, Aichi Prefecture 484-8504, Japan
e-mail: greatriver-1@nagoya-ku.ac.jp

Y.-K. Lin et al. (eds.), *Proceedings of the Institute of Industrial Engineers Asian Conference 2013*, DOI: 10.1007/978-981-4451-98-7_1,
© Springer Science+Business Media Singapore 2013

management of inventories in many industrial enterprises and service organizations. The EOQ captures the trade-off between inventory carrying cost and ordering cost for each item with accuracy.

Until now, the EOQ model assumes that the retailer's capitals are unrestricting and the retailer must pay for the items as soon as the retailer receives the items from the supplier. However, this may not be completely true. In practice, the supplier will offer the retailer a delay period. This period is called the trade credit period. During this period, the retailer can sell goods and earn interest or investment (Chang et al. 2001; Chen and Chuang 1999; Kim et al. 1995; Hwang and Shinn 1997).

We extend their model and construct new ordering policy. We reconstruct an ordering policy to stimulate his/her customer demand to develop the retailers' replenishment model. The retailer will adopt the partial trade credit policy to his/her customer (Huang and Hsu 2008). We assume that the retailer's trade credit period M offered by the supplier is not shorter than his/her customer's trade credit period N offered by the retailer ($M \geq N$). Moreover, the problem we want to emphasize here is that the supplier has cooperative relations with n retailers. Furthermore, we assume that the total of the cycle time offered for n retailers is restricted. Under these conditions, we model an optimal ordering policy to determine the optimal cycle times for n retailers. This model can be formulated as a mathematical problem.

2 Model Formulations

2.1 Notation

The notations used in this paper are as follows: D: demand rate per year; A: ordering cost per order; c: unit purchasing price per item; h: unit stock holding cost per item per year excluding interest charges; α: customer's fraction of the total amount owed payable at the time of placing an order offered by the retailer, $0 \leq \alpha \leq 1$, I_e: interest earned per \$ per year; I_k: interest charged per \$ in stocks per year by the supplier; M: the retailer's trade credit period offered by supplier in years; N: the customer's trade credit period offered by retailer in years; T: the cycle time in years; $TCV(T)$: the annual total inventory cost, which is a function of T; T^*: the optimal cycle time of $TCV(T)$.

2.2 Assumptions

(1) Demand rate is known and constant; (2) shortages are not allowed; (3) time period is finite; (4) the lead time is zero; (5) $I_k \geq I_e$, $M \geq N$; (6) when $T \geq M$, the account is settled at $T = M$ and the retailer starts paying for the interest charges on the items in stock with rate I_k. When $T \leq M$, the account is settled at $T = M$ and

An Optimal Ordering Policy of the Retailers 3

the retailer does not need to pay any interest charges; (7) the retailer can accumulate revenue and earn interest after his/her customer pays for the amount of purchasing cost to the retailer until the end of the trade credit period offered by the supplier. That is, the retailer can accumulate revenue and earn interest during the period N to M with rate I_e under the condition of the trade credit.

2.3 Model Formulation

From the above notation and assumptions, we construct a model. First, we must consider all the inventory costs. That is, the annual total inventory cost consists of the following four elements.

1. Annual ordering cost is as follow: $\frac{A}{T}$.
2. Annual stock holding cost (excluding interest charges) is as follow: $\frac{DTh}{2}$.
3. According to Assumption 6, there are two cases to occur in costs of interest charges for the items kept in stock per year. Annual interest payable is as follows

$$\frac{cI_k D(T-M)^2}{2T}, \ M \leq T, \quad 0, \ N \leq T < M, \ T \leq N$$

4. According to Assumption 7, there are three cases to occur in interest earned per year. Annual interest earned is as follows:

$$\frac{sI_e D[M^2 - (1-\alpha)N^2]}{2T}, \ M \leq T, \quad \frac{sI_e D[2MT - (1-\alpha)N^2 - T^2]}{2T}, \ N \leq T < M$$

$$sI_e D\left[M - (1-\alpha)N - \frac{\alpha T}{2}\right], \quad 0 \leq T < N.$$

From the above arguments, the annual total inventory cost for the retailer can be expressed as $TCV(T) =$ ordering cost + stock-holding cost + interest payable cost − interest earned. That is, it is formulated as follows:

$$TCV(T) = \begin{cases} TCV_1(T) & T \geq M, \\ TCV_2(T), & N \leq T \leq M, \\ TCV_3(T), & 0 < T \leq N, \end{cases}$$

$$TCV_1(T) = \frac{A}{T} + \frac{DTh}{2} + \frac{cI_k D(T-M)^2}{2T} - \frac{sI_e D[M^2 - (1-\alpha)N^2]}{2T},$$

$$TCV_2(T) = \frac{A}{T} + \frac{DTh}{2} - \frac{sI_e D[2MT - (1-\alpha)N^2 - T^2]}{2T},$$

$$TCV_3(T) = \frac{A}{T} + \frac{DTh}{2} - sI_e D\left[M - (1-\alpha)N\frac{\alpha T}{2}\right].$$

Since $TCV_1(M) = TCV_2(M)$ and $TCV_2(N) = TCV_3(N)$, $TCV(T)$ is continuous and well-defined. All $TCV_1(T), TCV_2(T), TCV_3(T)$ and $TCV(T)$ are defined on $T > 0$. Also, $TCV_1(T), TCV_2(T)$ and $TCV_3(T)$ are convex on $T > 0$. Furthermore, we have $TCV_1'(M) = TCV_2'(M)$ and $TCV_2'(N) = TCV_3'(N)$. Therefore, $TCV(T)$ is convex on $T > 0$.

5. According to Assumption 7, there are three cases to occur in interest earned per year. Annual interest earned is as follows:

$$\frac{sI_eD[M^2 - (1 - \alpha)N^2]}{2T}, \ M \leq T, \quad \frac{sI_eD[2MT - (1 - \alpha)N^2 - T^2)}{2T}, \ N \leq T < M,$$

$$sI_eD\left[M - (1 - \alpha)N - \frac{\alpha T}{2}\right], \quad 0 \leq T < N.$$

From the above arguments, the annual total inventory cost for the retailer can be expressed as $TCV(T)$ = ordering cost + stock-holding cost + interest payable cost − interest earned. That is, it is formulated as follows:

$$TCV(T) = \begin{cases} TCV_1(T), & T \geq M, \\ TCV_2(T), & N \leq T \leq M, \\ TCV_3(T), & 0 < T \leq N, \end{cases}$$

$$TCV_1(T) = \frac{A}{T} + \frac{DTh}{2} + \frac{cI_kD(T - M)^2}{2T} - \frac{sI_eD[M^2 - (1 - \alpha)N^2)}{2T},$$

$$TCV_2(T) = \frac{A}{T} + \frac{DTh}{2} - \frac{sI_eD[2MT - (1 - \alpha)N^2 - T^2]}{2T},$$

$$TCV_3(T) = \frac{A}{T} + \frac{DTh}{2} - sI_eD\left[M - (1 - \alpha)N\frac{\alpha T}{2}\right].$$

Since $TCV_1(M) = TCV_2(M)$ and $TCV_2(N) = TCV_3(N)$, $TCV(T)$ is continuous and well-defined. All $TCV_1(T), TCV_2(T), TCV_3(T)$ and $TCV(T)$ are defined on $T > 0$. Also, $TCV_1(T), TCV_2(T)$ and $TCV_3(T)$ are convex on $T > 0$. Furthermore, we have $TCV_1'(M) = TCV_2'(M)$ and $TCV_2'(N) = TCV_3'(N)$. Therefore, $TCV(T)$ is convex on $T > 0$.

3 An Optimal Ordering Policy of the Retailers Under Partial Trade Credit Financing and Restricted Cycle Time

We consider an optimal ordering policy of the retailers for the following problem: (1) inventory problem we propose here is in the EOQ model; (2) the number of the retailers is n; (3) the summation of each cycle time t_i $(i = 1, 2, \ldots, n)$ is restricted to T_0; (4) we allocate the restricted cycle time T_0 to n retailers to minimize the summation of total inventory cost.

To solve the problem, we use the inventory model we have already discussed in previous section. First, we formulate the cost function for each retailer. Next, we lead to all retailers' total cost function.

Let t_i be a cycle time for retailer i $(i = 1, 2, \ldots, n)$. Then we obtain the following total inventory cost function for retailer i:

$$TCV_i(t_i) = \begin{cases} TCV_{1i}(t_i), & t_i \geq M_i, \\ TCV_{2i}(t_i), & N_i \leq t_i \leq M_i, \\ TCV_{3i}(t_i), & 0 < t_i \leq N_i. \end{cases}$$

Here, new subscript i represents the retailer's number i $(i = 1, 2, \ldots, n)$. From the above expression, the summation of total inventory cost is shown as follows:

$$\sum_{i=1}^{n} TCV_i(t_i), \quad i = 1, 2, \ldots, n, \tag{1}$$

Therefore, an optimal ordering policy of the retailers under trade credit financing and restricted cycle time we propose here is represented as follows:

$$Minimize \quad \sum_{i=1}^{n} TCV_i(t_i), \tag{2}$$

$$subject\ to \quad \sum_{i=1}^{n} t_i \leq T_0, \quad t_i > 0 \quad (1 = 1, 2, \ldots, n), \tag{3}$$

Now, our aim is to minimize the objective function [Eq. (2)] under the constraints [Eq. (3)]. We differentiate Eq. (2) with respect to t_i. We obtain as follows:

$$TCV'(t_i) = \begin{cases} TCV'_{1i}(t_i) = -\left[\frac{2A_i + c_i D_i M_i^2 I_{ki} - s_i D_i I_{ei}(M_i^2 - (1-\alpha_i)N_i^2)}{2t_i^2}\right] + D_i\left(\frac{h_i + c_i I_{ki}}{2}\right), & t_i \geq M_i, \\ TCV'_{2i}(t_i) = -\left[\frac{2A_i + s_i D_i(1-\alpha_i)N_i^2 I_{ei}}{2t_i^2}\right] + D_i\left(\frac{h_i + s_i I_{ei}}{2}\right), & N_i \leq t_i < M_i, \\ TCV'_{3i}(t_i) = -\frac{A_i}{t_i^2} + D_i\left(\frac{h_i + s_i \alpha_i I_{ei}}{2}\right), & 0 \leq t_i < N_i \end{cases} \tag{4}$$

Here, we define Δ_{1i} and Δ_{2i} respectively.

Table 1 The retailer's optimal cycle time $\tilde{t}_i^*$ for Lagrange's multiplier λ (Case 1)

The range of λ	Optimal cycle time $(\tilde{t}_i^*)$
$\lambda \geq 0$	$\sqrt{\dfrac{2A_i}{D_i(h_i+s_i\alpha_i I_{ei})}}$
$-\infty < \lambda < 0$	$\sqrt{\dfrac{2A_i}{D_i(h_i+s_i\alpha_i I_{ei})-2\lambda}}$

Table 2 The retailer's optimal cycle time $\tilde{t}_i^*$ for Lagrange's multiplier λ (Case 2)

The range of λ	Optimal cycle time $(\tilde{t}_i^*)$
$\lambda \geq 0$	$\sqrt{\dfrac{2A_i+s_iD_i(1-\alpha_i)N_i^2 I_{ei}}{D_i(h_i+s_i I_{ei})}}$
$\Delta_{2i} \leq \lambda < 0$	$\sqrt{\dfrac{2A_i+s_iD_i(1-\alpha_i)N_i^2 I_{ei}}{D_i(h_i+s_i I_{ei})-2\lambda}}$
$-\infty < \lambda < \Delta_{2i}$	$\sqrt{\dfrac{2A_i}{D_i(h_i+s_i\alpha_i I_{ei})-2\lambda}}$

Table 3 The retailer's optimal cycle time $\tilde{t}_i^*$ for Lagrange's multiplier λ (Case 3)

The range of λ	Optimal cycle time $(\tilde{t}_i^*)$
$\lambda \geq 0$	$\sqrt{\dfrac{2A_i+c_iD_iM_i^2 I_{ki}-s_iD_i I_{ei}[M_i^2-(1-\alpha_i)N_i^2]}{D_i(h_i+c_i I_{ki})}}$
$\Delta_{1i} \leq \lambda < 0$	$\sqrt{\dfrac{2A_i+c_iD_iM_i^2 I_{ki}-s_iD_i I_{ei}[M_i^2-(1-\alpha_i)N_i^2]}{D_i(h_i+c_i I_{ki})-2\lambda}}$
$\Delta_{2i} \leq \lambda < \Delta_{1i}$	$\sqrt{\dfrac{2A_i+s_iD_i(1-\alpha_i)N_i^2 I_{ei}}{D_i(h_i+s_i I_{ei})-2\lambda}}$
$-\infty < \lambda < \Delta_{2i}$	$\sqrt{\dfrac{2A_i}{D_i(h_i+s_i\alpha_i I_{ei})-2\lambda}}$

$$\Delta_{1i} = \frac{-2A_i + D_iM_i^2(h_i + s_i I_{ei}) - s_iD_i(1-\alpha_i)N_i^2 I_{ei}}{2M_i^2}, \quad \Delta_{2i}$$
$$= \frac{-2A_i + D_iN_i^2(h_i + s_i\alpha_i I_{ei})}{2N_i^2}$$

where $\Delta_{1i} \geq \Delta_{2i}$ $(i = 1, 2, \ldots, n)$. To solve the problem, we use Lagrange's multiplier λ. The optimal solution $\tilde{t}_i^*$ for our problem is obtained from Tables 1, 2, 3 using $\sum_{i=1}^n t_i \leq T_0$.

For each retailer, we rearrange $\{\Delta_{1i}|1 \leq i \leq n\}$, $\{\Delta_{2i}|1 \leq i \leq n\}$, and $\{0\}$ small order, $B_1 \leq B_2 \leq \cdots \leq B_n \leq \cdots \leq B_{2n} \leq B_{2n+1}$. If $B_k \neq B_{k+1}$, the optimal Lagrange's multiplier λ^* only exists in the interval $B_k \leq \lambda^* < B_{k+1}$ $(0 \leq k \leq 2n + 1)$ because the objective function is convex, where $B_0 = -\infty$ and $B_{2n+2} = \infty$. In this interval, we allocate the restricted cycle time T_0 to n retailers. Then, the summation of each cycle time has to be equal to T_0. That is to say, using $\sum_{i=1}^n t_i = T_0$, we can find the optimal Lagrange's multiplier λ^*.

An Optimal Ordering Policy of the Retailers 7

Table 4 The parameters for each retailer

Retailer no.	D_i	A_i	c_i	h_i	α_i	I_{ei}	I_{ki}	M_i	N_i
1	6,100	160	100	6	0.3	0.13	0.15	0.11	0.08
2	2,600	100	90	4	0.5	0.12	0.15	0.10	0.06
3	1,500	80	80	3	0.2	0.10	0.15	0.10	0.04
4	6,300	100	110	7	0.3	0.12	0.15	0.11	0.08
5	2,650	75	92	4.5	0.5	0.12	0.15	0.10	0.06
6	1,300	130	79	3	0.25	0.10	0.15	0.10	0.045
7	5,000	100	140	7	0.4	0.09	0.15	0.10	0.08
8	2,650	98	94	4	0.5	0.12	0.15	0.10	0.065
9	1,480	75	80	3.5	0.5	0.10	0.15	0.10	0.06
10	5,950	190	98	6.5	0.5	0.13	0.15	0.11	0.08

Table 5 The optimal cycle time $\tilde{t}_i^*$ for each retailer and the total inventory cost $TCV_i(\tilde{t}_i^*)$

Retailer No.	$\tilde{t}_i^*$	$TCV_i(\tilde{t}_i^*)$
1	0.065281	139.9007
2	0.063961	381.1834
3	0.057087	903.6561
4	0.069594	785.8357
5	0.062986	377.0067
6	0.056952	913.8913
7	0.052168	211.1396
8	0.063189	340.6989
9	0.058609	862.7761
10	0.070176	88.1847
Total	0.620000	5,004.2731

4 Numerical Example

In this section, we show the numerical examples. Here, there are 10 retailers in our model. The parameters for each retailer are shown in Table 4.

4.1 The Optimal Retailers' Ordering Policy with the Restricted Cycle

We consider the case of the given T_0 is equal or less than the total of optimal cycle time (0.942389). For example, we set $T_0 = 0.62$. In Table 5, the optimal cycle time $\tilde{t}_i^*$ and the total inventory cost $TCV_i(\tilde{t}_i^*)$ for each retailer are shown. The optimal Lagrange's multiplier $\lambda^* = -19,244.2$. The summation of the total inventory costs are 5,004.2731.

5 Conclusion

This note is a modification of the assumption of the trade credit policy in previously published results to reflect realistic business situations. We assumed that the retailer also adopts the trade credit policy to stimulate his/her customer demand to develop the retailer's replenishment model. In addition, their model assumed the relationship between the supplier and the retailer is one-to-one. In this paper, we assumed the supplier has cooperative relations with many retailers for more suiting and satisfying the real world problems and assumed that the summation of each retailer's cycle time is restricted. Under these conditions, we have constructed the retailers' inventory system as a cost minimizing problem to determine the n retailers' optimal ordering policy.

A future study will further incorporate the proposed model more realistic assumptions such as probabilistic demand, allowable shortages and a finite rate of replenishment.

References

Chang HJ, Hung CH, Dye CY (2001) An inventory model for deteriorating items with linear trend demand under the condition of permissible delay in payment. Prod Plan Cont 12:274–282

Chen MS, Chuang CC (1999) An analysis of light buyer's economic order model under trade credit. Asia Pacific J Opns Res 16:23–24

Huang YF, Hsu KF (2008) An EOQ model under retailer partial trade credit policy in supply chain. Inter J Prod Econ 112:655–864

Hwang H, Shinn SW (1997) Retailer's pricing and lot sizing policy for exponentially deteriorating products under the condition of permissible delay in payment. Comp Opns Res 24:539–547

Kim JS, Hwang H, Shinn SW (1995) An optimal credit policy to increase wholesaler's profits with price dependent demand functions. Prod Plan Cont 6:45–50

An Immunized Ant Colony System Algorithm to Solve Unequal Area Facility Layout Problems Using Flexible Bay Structure

Mei-Shiang Chang and Hsin-Yi Lin

Abstract The Facility Layout Problem (FLP) is a typical combinational optimization problem. In this research, clonal selection algorithm (CSA) and ant colony system (ACS) are combined and an immunized ant colony system algorithm (IACS) is proposed to solve unequal-area facility layout problems using a flexible bay structure (FBS) representation. Four operations of CSA, clone, mutation, memory cells, and suppressor cells, are introduced in the ACS to improve the solution quality of initial ant solutions and to increase differences among ant solutions, so search capability of the IACO is enhanced. Datasets of well-known benchmark problems are used to evaluate the effectiveness of this approach. Compared with preview researches, the IACS can obtain the close or better solutions for some benchmark problems.

Keywords Unequal-area facility layout · Ant colony optimization · Clonal selection algorithm · Flexible bay structure · Constrained combinatorial optimization

1 Introduction

Facility layout problems (FLPs) aim to find the optimal arrangement of a given number of non-overlapping departments with unequal area requirements within a facility and certain ratio constraints or minimum length constraints. The common objective is to minimize the total material handling costs among departments.

M.-S. Chang (✉) · H.-Y. Lin
Department of Civil Engineering, Chung Yuan Christian University, 200, Chung Pei Road, Chung Li 32023, Taiwan, Republic of China
e-mail: mschang@cycu.edu.tw

H.-Y. Lin
e-mail: ac-1722@hotmail.com

Recently, different ACO approaches have been used to solve various versions of FLP problems. Most of them formulate FLP as a quadratic assignment problem (QAP) and obtain promising solutions to several test problems (Baykasoglu et al. 2006; Mckendall and Shang 2006; Nourelfath et al. 2007; Hani et al. 2007). Such approaches need modification in solving FLP. In addition, an ant system approach was first presented to solve the FLP (Wong and Komarudin 2010; Komarudin and Wong 2010). These algorithms use a FBS and a slicing tree structure to represent the FLP respectively. The former also presents an improvement to the FBS representation by using free or empty space. The algorithm can improve the best known solution for several problem instances. The latter one integrates nine types of local search to improve the algorithm performance. No doubt this heuristic shows encouraging results in solving FLP. Moreover, an ACO is proposed to solve the FLP with FBS (Kulturel-Konak and Konak 2011a, b). Compared with metaheuristics such as GA, TS, AS, and exact methods, this ACO approach is shown to be very effective in finding previously known best solutions and making notable improvements. Then an ACS is used to solve the FLP with FBS (Chang and Lin 2012). Compared with the previously best known solutions, the ACS can obtain the same or better solutions to some benchmark problems. Such interesting results inspire us to further explore the capability of applying ACS to solve the FLP.

Generally speaking, fusion of the computational intelligence methodologies can usually provide higher performances over employing them separately. This study proposes an immunized ant colony system (IACS) approach to solve the FLP with the flexible bay structure (FBS). It is based on clonal selection algorithm (CSA) and ACS.

2 Immunized Ant Colony System Algorithm

2.1 Solution Representation

We adopt the ant solution representation proposed by Komarudin (2009) for solving FLPs. Each ant solution consists of two parts: the department sequence codes and the bay break codes, such as (1-4-5-7-2-3-6)–(0-0-1-0-0-1). The former represents the order of n departments, which will be placed into the facility. The latter is n binary numbers. Here, 1 represents a bay break and 0 otherwise. We assume that bays run vertically and the departments are placed from left to right and bottom to top.

Komarudin (2009) presented this intuitive rule: "A department with higher material flow should be located nearer to the center of the facility." The heuristic information function was defined by Eq. (3).

$$\eta_{ij} = \left(\sum\nolimits_{k=1}^{N} f_{ki} + \sum\nolimits_{k=1}^{N} f_{ik} \right) \left(\frac{W}{2} - \left| x_j - \frac{W}{2} \right| + \frac{H}{2} - \left| y_j - \frac{H}{2} \right| \right) \qquad (1)$$

where f_{ij} is the workflow from i and j; x_j is the x-coordinate of the centroid of the department j; and y_j is the y-coordinate of the centroid of the department j. The rectilinear distance between the centroid of the candidate department and the facility boundary is measured.

2.2 Procedure Steps

Based on the mechanisms of ACS and CSA, we propose a hybrid optimization algorithm and name it immunized ant colony system (IACS) algorithm. The overall procedure of the IACS-FBS is given below. It includes standard procedures of ACS, i.e. parts of Step 0 (except Step 0.2), Step 2 (only N initial ant solutions are needed), Step 3, Step 9, parts of Step 10 (except Steps 10.1, 10.4, and 10.10), Step 13, and Step 14. Basically, we extend the study of ACS-FBS proposed by Chang and Lin (2012) except for several minor modifications. In Step 2, ant solutions are constructed by the space filling heuristic with having the most proper bay number. Such a modification is made for achieving better initial solutions.

The rest of the IACS algorithm is developed according to clone selection algorithm. First, a temporary pool is generated in Steps 1–7. The size of the temporary pool is two times the number of the ant colony. In Step 4, certain ants are reselected because of its diversity with the current best solution in order to maintain the ant diversity. For the same consideration, mutated ants are generated in Step 5. Two mutation operations are performed: swap between a department sequence, which exchanges the positions of two departments in the department sequence and switch of a bay break, which conditional changes the value of a bay break code from 0 to 1 or 1 to 0. The first and the last bay break codes are fixed. Sum of three successive values of bay break codes must be less than or equal to 1.

Next, all solutions in the temporary pool are selected for the ant colony in Step 8. Then the ant colony is further improved by optimization searching in Step 9 and by local searching in Step 10. It is different to the standard ACS in this step. We don't perform a local search to all ant solutions. We regard a local search as a mutation operation. The mutation rate of each ant is inversely proportional to its fitness. After the ant colony is put back a mutated ant pool. According to the mutated ants pool, a memory pool and a candidate pool are updated in Steps 12 and 13 respectively. Note that we don't allow identical ants in the memory pool and the candidate pool in order to increase the ant diversity.

The detailed steps of the IACS-FBS are listed herein.
Step 0: Parameter Setting and Initialization

Step 0.1: Set algorithm parameters of ACS, maximum number of iterations (NI), number of ants (N), pheromone information parameter (α), heuristic information parameter (β), and evaporation rate (ρ).

Step 0.2: Set algorithm parameters of CAS, size of memory pool ($r = N \times b\%$), clone number of the best ant-solutions ($s_1 = (N - r) \times d\%$), and clone number of the diverse ant-solutions ($s_2 = (N - r) \times (1 - d\%)$).
Step 0.3: Initialize iteration number counter. Set $I := 0$.
Step 0.4: Initialize pheromone information $\tau_{ij}^0, \forall i, j$.
Step 0.5: Initialize the fitness value of the global best solution. Set $z^* = \infty$.

Step 1: Generate an empty memory pool M
Step 2: Generate initial candidate pool P of ant colony (2 N ants) by performing the modified space filling heuristic proposed by Chang and Lin (2012)

Step 2.1: Initialize ant number counter. Set $p = 0$.
Step 2.2: Initialize the fitness value of the iteration best solution. Set $z_I^* = \infty$.
Step 2.3: Update ant number counter $p = p + 1$.
Step 2.4: Perform a procedure of ant solutions construction to create ant p.
Step 2.5: If the number of ants is less than 2 N, then go to Step 2.3; otherwise continue.

Step 3: Evaluate the fitness of the ant colony in candidate pool P

$$
z = \sum_i \sum_j f_{ij} c_{ij} \left(d_{ij}^x + d_{ij}^y \right) + \lambda \sum_i \left[Ub_i^w - w_i \right]^+ + \left[Lb_i^w - w_i \right]^+ \\
+ \lambda \sum_i \left[Ub_i^h - h_i \right]^+ + \left[Lb_i^h - h_i \right]^+
\tag{2}
$$

where c_{ij} is the cost per unit distance from i and j; d_{ij}^x is the rectilinear distance of the centroids from departments i and j on the x-axis; d_{ij}^y is the rectilinear distance of the centroids from i and j on the y-axis; λ is the relative importance of penalty costs and $\lambda = \sum_i \sum_j 10 f_{ij} c_{ij} WH$; $[\,]^+$ denotes returning a positive value of a subtraction expression or zero, i.e. $[\,]^+ = \max\{0, a - b\}$; Lb_i^h is the lower height limit of department i; Lb_i^w is the lower width limit of i; Ub_i^h is the upper height limit of i and Ub_i^w; and Ub_i^w is the upper width limit of i and Ub_i^h.
Step 4: Generate a temporary pool C from the memory pool M and the candidate pool P

Step 4.1: Clone ants in memory pool M (r ants) into the temporary pool C.
Step 4.2: Clone the best ants in candidate pool P (s_1 ants) into the temporary pool C.
Step 4.3: According to Eq. (3), evaluate the diversity measurement of each ant between the best ant in the candidate pool P.

$$
\delta = \sum_l \left| b_l - b_l^* \right|
\tag{3}
$$

where b_l is the current bay widths; b_l^* is the best bay widths.
Step 4.4: Clone the diverse ants in candidate pool P (s_2 ants) into the temporary pool C.

Step 5: Generate a mutated ants pool $C1$ from the temporary pool C

Perform mutation operations of a department sequence and/or of a bay break to all ants in the temporary pool C.

Step 6: According to Eq. (1), evaluate the fitness of all ants in the mutated ant pool $C1$

Step 7: Update the temporary pool C

If the mutated ant is better than the original one, the original ant is replaced.

Step 8: Select ant colony (n ants) from the temporary pool C

Step 9: Optimization searching of ant colony

Step 9.1: Exploit the selected regions by sending the ants for local search by performing a state transition rule.

$$j = \begin{cases} \arg \max \left[\tau_{ij} \right]^{\alpha} \cdot \left[\eta_{ij} \right]^{\beta}, & \text{if } q \leq q_0 \text{ (exploitation)} \\ S, & \text{otherwise} \quad \text{(exploration)} \end{cases} \qquad (4)$$

$$S = P_{ij}^k = \begin{cases} \left[\tau_{ij} \right]^{\alpha} \cdot \left[\eta_{ij} \right]^{\beta} \Big/ \sum_{q \in N_i} \left[\tau_{iq} \right]^{\alpha} \cdot \left[\eta_{iq} \right]^{\beta}, & \text{if } j \in N_i \\ 0, & \text{otherwise} \end{cases} \qquad (5)$$

where s is a probability to locate department j after department i in the positioning order of departments; τ_{ij} is the pheromone value defined as the relative desirability of assigning department j after department i in the department sequence; η_{ij} is the heuristic information related to assigning department j after department i in the department sequence; q is a random number uniformly distributed in $[0, 1]$; q_0 is a fixed parameter $(0 \leq q_0 \leq 1)$; α is a parameter which determines the relative weight of pheromone information; and β is a parameter which determines the relative weight of heuristic information; P_{ij}^k is a probability of the department j of the department sequence to be chosen by an ant k located in department i and N_i is available alternatives of the department sequence to be chosen by the corresponding ant located in department i.

Step 9.2: Update the local pheromone all the ants according to Eq. (6).

$$\tau_{ij} := (1 - \rho) \cdot \tau_{ij} + \rho \cdot \tau_0 \qquad (6)$$

where $0 < \rho < 1$ is the evaporation parameter; τ_0 represents the initial level of pheromone.

Step 10: Mutate the current ant solutions of this iteration by performing the proposed local searching. The mutation rate of each ant is inversely proportional to its fitness, that is, the mutation rate is proportional to the affinity

Step 10.1: Determine the threshold of mutation rate ϕ by Eq. (7).

$$\phi = N \Big/ \sum_p z_p \tag{7}$$

Step 10.2: Initialize ant number counter. Set $p = 0$.

Step 10.3: Update ant number counter $p = p + 1$.

Step 10.4: Calculate the mutation rate of ant p, $\phi_p = 1/z_p$. If the value ϕ_p is less than the threshold ϕ, go to Step 10.5; otherwise, go to Step 10.10.

Step 10.5: Perform local search operations of a department sequence (swap, one-insert, or two-opt, Chang and Lin 2012) and of a bay break to the ant solution p.

Step 10.6: Calculate the fitness value $\widehat{z}_I$ of the ant p after local search.

Step 10.7: Update the best solution of this iteration z_I^*, once a new best solution is found $(\widehat{z}_I < z_I^*)$.

Step 10.8: Update the local pheromone of the mutated ant p according to Eq. (6), if its fitness value is improved.

Step 10.9: Update the global pheromone of the mutated ant p according to Eqs. (8) and (9), if its fitness value is not improved.

$$\tau_{ij} := (1 - \rho) \cdot \tau_{ij} + \rho \cdot \Delta\tau_{p \in P} \tag{8}$$

$$\Delta\tau_{p \in P} = \sum_{ij \in P} d_{ij}^x + d_{ij}^y \tag{9}$$

Step 10.10: Add ant p or mutated ant p to the mutated ants pool C1.

Step 10.11: If the number of ants is less than N, then go to Step 10.3; otherwise continue.

Step 11: Update the memory pool M

Step 11.1: Clone the best r ants in the mutated ants pool C1 into the memory pool M.

Step 11.2: Delete identical ants in the memory pool M

Step 12: Update the candidate pool P

Step 12.1: Delete identical ants in the candidate pool P to maintain the ant diversity.

Step 12.2: Replace those ant solutions in the candidate pool P by the rest ants with better fitness in the mutated ants pool C1.

Step 13: Update the global best solution

If z_I^* is less than z^*, update the fitness value of the global best solution $z^* := z_I^*$.

Step 14: Stopping criteria

If the maximum number of iterations is realized, then output the global best solution and stop; otherwise, go to Step 4.

3 Computational Experiments

The proposed algorithm was tested using several problem sets, as listed in Table 1. Note that M11a and M15a were modified to allow the use of FBS representation. The location of the last department is fixed by assigning it to the last position of the facility, but the department fixed size constraint is relaxed. The algorithm was coded with C++ and tested using an Intel(R) Core(TM) i7 CPU processor.

Table 2 provides the previous best-known results of the test problems. The IACS-FBS results are compared to other FBS solutions. The comparative results show that the ACS-FBS approach is very promising. For problem Nug15a5, the IACS-FBS found a new best FBS solution.

4 Conclusions

To prevent the premature convergence problem and to escape from a local optimal solution, clone with affinity-related mutation of the CSA is utilized and combined with the ACS in this algorithm. In this study, an IACS-FBS algorithm is proposed to solve unequal area FLP. We regard local searching as a mutation operation and the mutation rate is inversely proportional to its fitness. In addition, the diversities between the current best solution are measured to help choose clone candidates. Identical ants in a memory pool and a candidate pool are deleted to maintain the diverseness among the ant colony. Furthermore, we revise the ACS-FBS to provide more efficient and comprehensive local exploitation, such as the construction of initial solutions and the local search methods. Compared with existing ACO algorithms, the proposed algorithm obtain better or at least the same solution quality, except for problem Nug15a4. For problem Nug15a5, a new best FBS solution is found.

Table 1 Problem set data

Prob. set	No. of Dpt.	Facility size		Maximum aspect ratio
		Width	Height	
O7	7	8.54	13.00	$\alpha^{max} = 4$
O8	8	11.31	13.00	$\alpha^{max} = 4$
FO7	7	8.54	13.00	$\alpha^{max} = 5$
FO8	8	11.31	13.00	$\alpha^{max} = 5$
O9	9	12.00	13.00	$\alpha^{max} = 4, 5$
vC10a	10	25.00	51.00	$\alpha^{max} = 5$
M11a	11	3.00	2.00	$\alpha^{max} = 4, 5$
M15a	15	15.00	15.00	$\alpha^{max} = 5$
Nug12	12	3.00	4.00	$\alpha^{max} = 4, 5$
Nug15	15	3.00	4.00	$\alpha^{max} = 4, 5$

Table 2 Comparisons of the best FBS solutions with other approaches

Problem	Shape cons.	Konak et al. (2006)	Wong and Komarudin (2010)	Kulturel-Konak and Konak (2011a)	Kulturel-Konak and Konak (2011b)	Chang and Lin (2012)	This study
O7	$\alpha^{max} = 4$	–	136.58	–	–	134.19*	134.19*
O8	$\alpha^{max} = 4$	–	–	–	–	245.51*	245.51*
O9a4	$\alpha^{max} = 4$	–	–	–	–	241.06*	241.06*
O9a5	$\alpha^{max} = 5$	241.06	241.06	–	–	238.12*	238.12*
FO7	$\alpha^{max} = 5$	23.12	23.12	–	–	18.88*	18.88*
FO8	$\alpha^{max} = 5$	22.39*	22.39*	–	–	22.39*	22.39*
vC10a	$\alpha^{max} = 5$	21,463.07	21,463.1	21,463.07	20,142.13*	20,142.13*	20,142.13*
M11a4	$\alpha^{max} = 4$	–	–	1,268.55	–	–	1268.55* 1271.56 SingYi
M11a5	$\alpha^{max} = 5$	1,225.00	1,204.15	–	1,201.12*	1,201.12*	1204.15 black
Nug12a4	$\alpha^{max} = 4$	265.5	–	262	257.5*	262	262i7
Nug12a5	$\alpha^{max} = 5$	265.6	262*	–	–	262*	262*
M15a	$\alpha^{max} = 5$	31,779.09	27,545.30	–	27,545.27	–	27701.01*hm
Nug15a4	$\alpha^{max} = 4$	526.72	–	524.75	524.75	–	542.8i7 532.77sil
Nug15a5	$\alpha^{max} = 5$	526.75	536.75	–	–	–	523.67
Approach	–	MIP + FBS	AS + FBS	ACO + FBS	PSO + FBS	ACS + FBS	IACS + FBS

References

Baykasoglu A, Dereli T, Sabuncu I (2006) An ant colony algorithm for solving budget constrained and unconstrained dynamic facility layout problems. Omega 34(4):385–396

Chang MS, Lin HY (2012) A flexible bay structure representation and ant colony system for unequal area facility layout problems. Lecture Notes Eng Comp Sci 2199(1):1346–1351

Hani Y, Amodeo L, Yalaoui F, Chen H (2007) Ant colony optimization for solving an industrial layout problem. Eur J Oper Res 183(2):633–642

Komarudin (2009) An improved ant system algorithm unequal area facility layout problems. Master Thesis, University of Teknologi, Malaysia

Komarudin, Wong KY (2010) Applying ant system for solving unequal area facility layout problems. Eur J Oper Res 202(3):730–746

Konak A, Kulturel-Konak S, Norman BA, Smith AE (2006) A new mixed integer formulation for optimal facility layout design. Oper Res Lett 34:660–672

Kulturel-Konak S, Konak A (2011a) Unequal area flexible bay facility layout using ant colony optimization. Inter J Prod Res 49(7):1877–1902

Kulturel-Konak S, Konak A (2011b) A new relaxed flexible bay structure representation and particle swarm optimization for the unequal area facility layout problem. Eng Optimiz 43:1–25

Mckendall AR Jr, Shang J (2006) Hybrid ant systems for the dynamic facility layout problem. Comp Oper Res 33(3):790–803

Nourelfath M, Nahas N, Montreuil B (2007) Coupling ant colony optimization and the extended great deluge algorithm for the discrete facility layout problem. Eng Optimiz 39(8):953–968

Wong KY, Komarudin (2010) Solving facility layout problems using flexible bay structure representation and ant system algorithm. Exp Syst Appl 37:5523–5527

Characterizing the Trade-Off Between Queue Time and Utilization of a Manufacturing System

Kan Wu

Abstract Characterizing system performance is essential for productivity improvement. Inspired by the underlying structure of tandem queues, an approximate model has been derived to characterize the system performance. The model decomposes system queue time and variability into bottleneck and non-bottleneck parts while capturing the dependence among workstations.

Keywords Queueing systems · Manufacturing systems modeling

1 Introduction

An objective evaluation of the system performance is an essential input for manufacturing system design and productivity improvement. To quantify system performance, performance curves (and their associated variabilities) play a key role since they characterize the trade-off between queue time and utilization. The performance curve is also called a characteristic curve, trade-off curve, operation curve or queueing curve. Bitran and Tirupati (1989) used performance curves to describe the relationship between work-in-process (WIP), cycle time and capacity. Sattler (1996) used performance curves to determine productivity improvements of a semiconductor fab. She assumed variability is independent of utilization and approximated the curve by using a constant to replace the variability term in Kingman's G/G/1 approximation (Kingman 1965).

Since performance curves can be used to quantify the trade-off between queue time and utilization, it would be nice if we can generate the curves analytically. However, due to the complexity of practical manufacturing systems, it is difficult

---

K. Wu (✉)
Georgia Tech, Atlanta, GA 30332, USA
e-mail: kanwu@gatech.edu

Y.-K. Lin et al. (eds.), *Proceedings of the Institute of Industrial Engineers Asian Conference 2013*, DOI: 10.1007/978-981-4451-98-7_3,
© Springer Science+Business Media Singapore 2013

to take all the details into account. Hence, people sometimes model a manufacturing system by the aggregation approaches with a macroscopic view. Rather than creating the curve analytically, Rose (2001), Nazzal and Mollaghasemi (2001) and Park et al. (2001) developed the fab performance curve through simulations. Because simulation studies are time consuming, Sattler (1996) attempted to describe the performance curve of a fab based on Kingman's approximation and demonstrated how to enhance productivity through variability reduction. However, rather than a single server, a fab is composed of a sequence of operations executed by a series of workstations. Predicting the performance curve simply based on the G/G/1 queue is not adequate.

Motivated by Kingman's approximation, Yang, et al. (2007) proposed empirical algorithms to generate performance curves through simulation. Two unknown scalars and one unknown vector have to be determined through complex procedures. The algorithm performs well for M/M/1 systems with various dispatching rules, and is extended to predict the performance curve of manufacturing systems by ignoring the dependence among workstations.

It is difficult to compute the performance curve of a practical manufacturing system exactly. Our goal is to have a good approximation of the performance curve by capturing the main underlying structure of a manufacturing system. Rather than assuming stochastic independence, we can capture the dependence among workstations by the intrinsic ratio discovered by Wu and McGinnis (2013). Based on the intrinsic ratio, an approximate model is derived, and it performs very well in the examined cases with complex routing, scheduling, and batching. The objective is to find a model which can describe factory behavior through regression analysis so that we can use only a few known throughput rates and mean queue times to fit the model parameters. After the parameters are determined, the model can be used to predict mean cycle time at any other utilization. Queue time approximation for tandem queues and manufacturing systems are given in Sects. 2 and 3, respectively. Model validation is given in Sect. 4. Conclusion is given at the end.

2 Multiple Single-Server Stations in Series

Wu and McGinnis (2013) identified the nice property of intrinsic ratio in tandem queues, and extended the results to n single-server stations in series with infinite buffers. Their model captures the dependence among tandem queues through so called intrinsic ratios and considerably outperforms the previous approximate models based on parametric-decomposition or diffusion approximations. Based on the concept of intrinsic ratios, system mean queue time can be approximated by

$$\sum_{i=1}^{n} QT_i = f_1\alpha_1\left(\frac{\rho_1}{1-\rho_1}\right)\frac{1}{\mu_1} + f_2\alpha_2\left(\frac{\rho_2}{1-\rho_2}\right)\frac{1}{\mu_2} + \cdots + f_n\alpha_n\left(\frac{\rho_n}{1-\rho_n}\right)\frac{1}{\mu_n}, \quad (1)$$

where QT_i is the mean queue time of station i in steady states, ρ_i is the utilization of station i, μ_i is the service rate of station i, and α_i is the variability of station i in the ASIA system (Wu and McGinnis 2013). f_i is called contribution factor, since it represents the percentage of station i's ASIA system queue time contributing to the overall system queue time as shown in Eq. (1).

Based on the property of intrinsic ratio, Procedure 1 explains how to determine f_i and approximate the mean queue time of n single-server stations in series. It consists of two stages: decomposition and computation. Stage 1 identifies the main system bottleneck first, and then identifies the next bottleneck within each subsystem. A subsystem is composed of the stations from the first station to the newest identified bottleneck (not included). At the beginning, when no bottleneck has been identified, the subsystem is the same as the original system. The subsystem then gradually becomes smaller until it is composed of solely one single station, which is the first station of the tandem queue.

Procedure 1 (System Queue Time for Single Queues in Series):
Stage I: Decomposition by bottlenecks

1. Identify system bottleneck (BN_1), where $\mu_{BN_1} = \min \mu_i$, for $i = 1$ to n. Let $k = 1$.

   - If more than one station has the same minimum service rate, $BN_1 = \min i$, where $=\mu_i = \mu_{BN_1}$.
2. Identify the next bottleneck (i.e., BN_{k+1}) in front of the previous one (i.e., BN_k), where $\mu_{BN_{k+1}} = \min \mu_i$, for $i = 1$ to $BN_k - 1$.

   - If more than one station has the same minimum service rate, $BN_{k+1} = \min i$, where $\mu_i = \mu_{BN_{k+1}}$.
3. If $BN_{k+1} = 1$, stop. Otherwise, let $k = k + 1$, go to 2.

   Stage II: Determining the parameters

4. Let $k = n$, $f_i = 1$ for $i = 1$ to k.
5. If station k is marked as a bottleneck, $f_i \leftarrow y_k * f_i$ for $i = 1$ to $k - 1$. Otherwise, $f_k \leftarrow x_k * f_k$. Stop if $k = 2$.
6. Otherwise, let $k = k - 1$, go to 5.

Figure 1 shows the bottleneck decomposition. In Eq. (1), the value of f_i always equals 1 for the system bottleneck, which implies that a unit weight is always given to the system bottleneck. However, there will be a (non-unit) weight on all other stations' ASIA system mean queue times. When both x_k and y_k are smaller than 1, the contribution factor will be smaller than 1 and behaves like a discount factor. In this situation, reducing bottleneck service time SCV brings greater improvement on queue time than reducing non-bottleneck service time SCV.

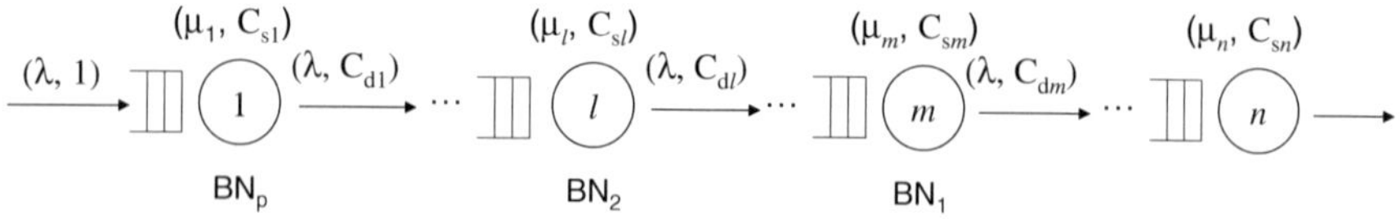

Fig. 1 n single queues in series

3 Approximate Models for Manufacturing Systems

A practical factory is much more complex than single-server queues in series. Each workstation may have multiple servers with different capabilities and each server may have complicated configuration and suffer different types of interruptions. Dispatching rules other than work-conserving policies may be used. Modeling the behavior of a factory through analyzing all details is a formidable task.

Since the intrinsic ratio captures the dependence among stations (Wu and McGinnis 2013), our approach is to abstract a simple form from Eq. (1) to describe the behavior of practical manufacturing systems. We hope this simple form can capture the underlying structure of system performance. We assume system stability can be achieved when the utilization of each station is smaller than one.

Because f_i is 1 at the system bottleneck, system mean cycle time can be expressed as

$$
\begin{aligned}
CT &= \sum_{i=1}^{n} QT_i + PT_f = \sum_{i=1}^{n} f_i \alpha_i \left(\frac{\rho_i}{1 - \rho_i} \right) \frac{1}{\mu_i} + PT_f \\
&= \alpha_{BN} \left(\frac{\rho_{BN}}{1 - \rho_{BN}} \right) \frac{1}{\mu_{BN}} + \sum_{i \neq BN} f_i \alpha_i \left(\frac{\rho_i}{1 - \rho_i} \right) \frac{1}{\mu_i} + PT_f,
\end{aligned} \tag{2}
$$

where CT is mean cycle time, BN stands for bottleneck. Hence, ρ_{BN} is the bottleneck utilization (or system utilization). PT_f is mean total processing time (of a job in a factory). Total processing time can be approximated by the cycle time of a job in light traffic when the station has no cascading, and there is no batch and assembly. The first term of Eq. (2) is the ASIA system mean queue time of the bottleneck, and the second term is the gross mean queue time of the non-bottlenecks.

Queue time of tandem queues is contributed by all stations and so does its variability. From Eq. (2), system variability is composed of two parts: the bottleneck and non-bottlenecks, with the bottleneck part playing a key role in heavy traffic. The coefficient of the bottleneck (α_{BN}) is the same as the variability in its ASIA system. The coefficients of the non-bottlenecks ($f_i \alpha_i$) are the weighted (by the contribution factors) variabilities in their ASIA systems. When all service times are exponential and the original external arrival process is Poisson, all intrinsic ratios are one. Hence, f_i is 1 and Eq. (2) reduces to the cycle time of a Jackson network. When all service times are deterministic, the intrinsic ratios are zero for the system bottleneck and the stations behind it. Hence, f_i is zero for all non-bottlenecks and Eq. (2) reduces to the cycle time of a Friedman's tandem queue (Friedman 1965).

Characterizing the Trade-Off Between Queue Time and Utilization

In Eq. (2), since queue time is dominated by the bottleneck in heavy traffic, we can replace the $(n-1)$ non-bottleneck performance curves by $(n-1)$ identical composite non-bottleneck performance curves. Furthermore, within the non-bottleneck performance curves, system queue time is dominated by the performance curve of the second bottleneck (i.e., the one with the highest utilization among all non-bottlenecks) in heavy traffic. Therefore, a reasonable choice is to pick up the second bottleneck performance curve to represent the composite non-bottleneck performance curves. Based on the above observations, Eq. (2) can be simplified as

$$
\begin{aligned}
CT &= \alpha_{BN}\left(\frac{\rho_{BN}}{1-\rho_{BN}}\right)\frac{1}{\mu_{BN}} + \sum_{i\neq BN} f_i\alpha_i\left(\frac{\rho_i}{1-\rho_i}\right)\frac{1}{\mu_i} + PT_f \\
&\cong k_1\left(\frac{\rho_{BN}}{1-\rho_{BN}}\right)\frac{1}{\mu_{BN}} + (n-1)k_2'\left(\frac{\lambda/k_3}{1-\lambda/k_3}\right)\frac{1}{k_3} + PT_f \\
&= k_1\left(\frac{\rho_{BN}}{1-\rho_{BN}}\right)\frac{1}{\mu_{BN}} + k_2\left(\frac{\lambda}{k_3-\lambda}\right)\frac{1}{k_3} + PT_f,
\end{aligned}
\tag{3}
$$

where λ is the external arrival rate at the first station. In Eq. (3), the first term on the right-hand side corresponds to the bottleneck mean queue time where k_1 is the bottleneck variability. The second term corresponds to the mean queue time of the non-bottleneck stations where k_2 approximates the variability of the composite station (representing the $(n-1)$ non-bottleneck stations), and k_3 represents the capacity of the composite station. Note that k_1 is the same as α_{BN}, the bottleneck variability in the ASIA system if the system is a tandem queue. Since there are three parameters in Eq. (3), we call it the 3-parameter model. When there is reentry or rework, capacity can be computed as follows,

$$
1/\mu_i = \sum_{j=1}^{l} w_j \times ST_j,
\tag{4}
$$

where l is the total reentry or rework frequency at station i, w_j is the rework rate.

In addition to its simplicity, Eq. (3) considerably reduces the burden of data collection in Eq. (2). In order to apply Eq. (2), we assume the service time SCV is available. However, in a setting of complex machine configurations, robot scheduling, interruptions, resource contention, reentry, and product mix, finding out the variance of service time, where service time is the reciprocal of capacity (Wu et al. 2011), can be a formidable task. Wu and Hui (2008) discussed the potential issue when measuring process time in practice. Newell (1979) stated, "In fact, in most applications, one is lucky if one has a good estimate of the service rates (to within 5 % say); the variance rates are often known only to within a factor of 2, seldom to within an accuracy of 20 %." Although SCV of service time is well defined in theory, it may not be accessible in practice.

An easy way to estimate service time SCV is to analyze historical data. Although it is difficult to have a reliable estimate of service time SCV, it is relatively easy if only the mean queue time is needed. In practice, observable mean queue time (first moment estimator) is much more accessible than the intangible service times SCV (second moment estimator). In order to make queueing models more accessible in practice, it is important to have an approach which does not rely on the service time SCV explicitly. The historical mean queue time is a good alternative. Indeed, except for when we construct a brand new factory, the historical performance is commonly accessible in practical applications of queueing theory. Hence, it is practical to develop an approximate model which can predict future performance based on few reliable historical data estimates.

Compared with the approximate model by Wu (2005), Eq. (3) gauges the variability of a manufacturing system from the viewpoint of the bottleneck while adding a correction term to consider the impact from non-bottlenecks.

4 Model Validation

Planning and managing major defense acquisition programs (DAP) requires balancing and synchronizing design, and manufacturing across a network of distributed activities performed by independent commercial entities. To achieve this goal, it is important to have the capability to describe the performance curve of each independent entity accurately. Hence, a simulation model is constructed by one of the manufacturers in a DAP using ARENA®. The model (illustrated in Fig. 2) describes the behavior of a manufacturing facility for a specific product.

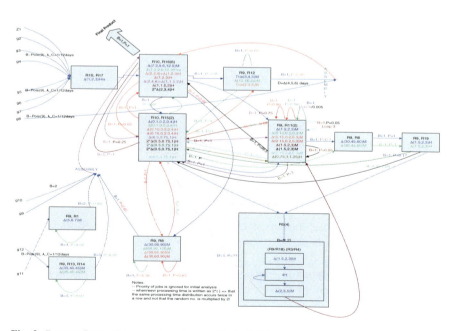

Fig. 2 Process flows of a manufacturing facility in DAP

The output of this facility supplies critical parts to downstream assembly lines in the supply chain. There are 14 workstations (R1, R3, R4, R5, R6, R8, R11, R12, R13, R14, R15, R16, R17, R19) arranged in 11 main process groups (i.e., 9 process groups use 1 workstation, 1 group needs 2 workstations, and 1 group uses 3 workstations). Four workstations have multiple servers (R5 = 4, R11 = 2, R15 = 2, R16 = 6), and the remaining workstations have only a single server. In addition to workstations, each process group requires operators from one of the three operator types (R9, R10 and R18). While the operators have their own shift schedules (i.e., 8 h a day), machines work 24 h a day as long as a job has been loaded by operators. Service times follow triangular distributions. Dispatching rules at critical workstations follow the shortest remaining processing time instead of First-Come-First-Serve (FCFS). Reentry and rework in the system are shown in Fig. 2. 12 different raw parts arrive every 12 days with random batch sizes following a Poisson distribution. System utilization is determined by the mean batch size rather than the arrival intervals. Before the process can be started, raw parts (Z1 and g2–g7) need to be assembled in front of the first process step. Subsequently, an incomplete job has to be assembled again with some other raw parts (g8–g12) in the middle of the process flow.

Due to shift schedule, operator availability, batch arrivals and assembly lines, finding a reliable estimate of inter arrival and service time SCVs is difficult, not to mention computing the cycle time. To get a reliable estimate of cycle time, we resort to simulation. Through experimentation, the system bottleneck has been identified as R8, which is composed of a single server, and system capacity is about 13.7 jobs per 12 days. In total, system performance at 14 different utilizations has been simulated. The performance curve with 99 % confidence intervals is shown in Fig. 3. The cycle time at each utilization is the average of 30 batches

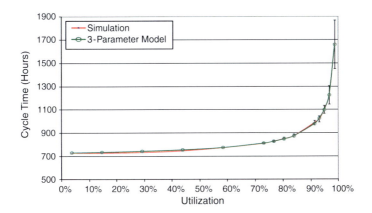

Fig. 3 Performance curve of a manufacturing facility in DAP

from one simulation run. In each simulation run, the first 1,000–10,000 data points are discarded for warm-up, and each batch consists of 1,500–500,000 data points.

The values of k_1, k_2 and k_3 in Eq. (3) (i.e., 3-Parameter model curve in Fig. 3) are 0.356, 422.951 and 14.636, and the largest error is 1.38 % at 29 % utilization. Note that the value of k_3, 14.636, which represents the non-bottleneck capacity, is greater than the bottleneck capacity 13.7 as expected. The results exhibit the accurate prediction capability of Eq. (3) in practical manufacturing systems.

5 Conclusion

Through the nice property of intrinsic ratios, we developed analytical models to quantify the performance of manufacturing systems. From the simulation results, we find the intrinsic ratios discovered in tandem queues can be applied to quantify the performance of manufacturing systems.

Although our models perform well in the examined case, the practical situation can be much more complicated. In real production lines, dispatching rules can be complicated especial when batching exist. Furthermore, there can be multiple products in a production system. They may have different cycle times but share the same resources. Those situations have not been considered in our models and will be left for future research.

References

Bitran GR, Tirupati D (1989) Tradeoff curves, targeting and balancing in manufacturing queueing networks. Oper Res 37:547–564

Friedman HD (1965) Reduction methods for tandem queuing systems. Oper Res 13:121–131

Kingman JFC (1965) The heavy traffic approximation in the theory of queues. In: Proceedings of the symposium on congestion Theory, pp 137–159

Nazzal D, Mollaghasemi M (2001) Critical tools identification and characteristics curves construction in a wafer fabrication facility. In: Proceedings of the winter simulation conference, pp 1194–1199

Newell G (1979) Approximate behavior of tandem queues, Springer

Park S, Mackulak GT, Fowler JW (2001) An overall framework for generating simulation-based cycle time-throughput curves. In: Proceedings of the winter simulation conference, pp 1178–1187

Rose O (2001) The shortest processing time first (SPTF) dispatch rule and some variants in semiconductor manufacturing. In: Proceedings of the winter simulation conference, pp 1220–1224

Sattler L (1996) Using queueing curve approximations in a fab to determine productivity improvements. In: Proceedings of the advanced semiconductor manufacturing conference, 140–145

Wu K (2005) An examination of variability and its basic properties for a factory. IEEE Trans Semicond Manuf 18:214–221

Wu K, Hui K (2008) The determination and indetermination of service times in manufacturing systems. IEEE Trans Semicond Manuf 21:72–82

Wu K, McGinnis L (2013) Interpolation approximations for queues in series. IIE Trans 45:273–290

Wu K, McGinnis L, Zwart B (2011) Queueing models for a single machine subject to multiple types of interruptions. IIE Trans 43:753–759

Yang F, Ankenman B, Nelson B (2007) Efficient generation of cycle time-throughput curves through simulation and metamodeling. Naval Res Logistics 54:78–93s

Using Latent Variable to Estimate Parameters of Inverse Gaussian Distribution Based on Time-Censored Wiener Degradation Data

Ming-Yung Lee and Cheng-Hung Hu

Abstract To effectively assess the lifetime distribution of a highly reliability product, a degradation test is used if the product's lifetime is highly related to a critical product characteristic degrading over time. The failure times, as well as the degradation values, provide useful information for estimating the lifetime in a short test duration. The Wiener process model has been successfully used for describing degradation paths of many modern products such as LED light lamps. Based on this model, the lifetime of a product would follow the Inverse Gaussian (IG) distribution with two parameters. To estimate the parameters, we propose a method using the latent variables to obtain Latent Variable Estimates (LVE) of the parameters of the IG lifetime distribution. The proposed LVEs have simple closed functional form and thus they are easy to interpret and implement. Moreover, we prove the LVEs are consistent estimates. Via simulation studies, we show that the LVEs have smaller bias and mean square error than existing estimates in the literature.

Keywords Latent variable · Time-censored · Wiener process · Inverse Gaussian distribution

---

M.-Y. Lee (✉)
Statistic and Informatics Department, Providence University,
Taichung 43301, Taiwan
e-mail: mylee@pu.edu.tw

C.-H. Hu
Department of Industrial Engineering and Management, Yuan Ze University, No. 135,
Yuan-Tung Road, Chungli, Taoyuan, Taiwan
e-mail: chhu@saturn.yzu.edu.tw

Y.-K. Lin et al. (eds.), *Proceedings of the Institute of Industrial Engineers Asian Conference 2013*, DOI: 10.1007/978-981-4451-98-7_4,
© Springer Science+Business Media Singapore 2013

1 Introduction

To increase the likelihood of observing failure in a short duration, a degradation test is used if we can associate the lifetime of a product to certain Quality Characteristic (QC). For example, Tseng et al. (2003) presented an example of Light Emitting Diode (LED) lamps, whose life is related to the light intensity.

Suppose the value of the QC of the products degrades over time, the product lifetime can then be defined as the time when the QC crosses a pre-specified critical level. In this paper, we consider the Wiener process to model the degradation path. Under this assumption, it can be shown that the product life follows the Inverse Gaussian (IG) distribution. The Wiener/IG distribution model has found many applications in certain studies. For example, see Doksum and Hoyland (1992), Meeker and Escobar (1998), Tseng and Peng (2004), Park and Padgett (2005), and Wang (2010).

When a degradation test is time-constrained, the failure times and the degradation values of the censored units at time both contain valuable information regarding the lifetime distribution. In this paper, we propose an estimation procedure by using Latent Variables. Closed-form Latent Variable Estimates (LVE) of the parameters of the IG distribution are available and the LVE are proved being consistent in Sect. 3. In addition, to compare the performance of the LVE to some existing estimates, a simulation study is proposed in Sect. 4. We numerically evaluate the performances of the LVE, modified E-M estimates (MEME), and Modified MLE (MMLE), for both IG parameters with different sample sizes (For detailed introduction of MEME and MMLE, one can refer to Lee and Tang (2007)). Results show that the resulting LVEs can reduce the bias and variances than that from other two estimation methods. In Sect. 5, we use our LVE for estimating parameters of the IG distribution to existing degradation data proposed in Lee and Tang (2007). Concluding remarks and future research topics are given in Sect. 6.

2 Model Assumption

We assume that the transformed degradation path of the QC at time t, is $W(t) = \eta t + \sigma B(t)$, $t \geq 0$, where η is the drift parameter, $\sigma > 0$ is the diffusion coefficient, and $B(\cdot)$ is a standard Brownian motion. Under this model, the product's failure time (or lifetime), denoted by T, is the defined as the first-passage time over a constant threshold, a. It is well-known that T follows an inverse Gaussian distribution, denoted by IG (μ, λ), with the location and scale parameters $\mu = a/\eta$ and $\lambda = a^2/\sigma^2$. For detailed introduction and systematic overviews of the IG distribution, one can refer to (Seshadri 1999).

For a given random sample $x_1, x_2, \ldots, x_n$ from the IG(μ, λ) distribution, the maximum likelihood estimates of parameters are:

Using Latent Variable to Estimate Parameters

$$\hat{\mu}_{\mathrm{MLE}} = \sum_{i=1}^{n} \left(\frac{x_i}{n}\right) \text{ and } \hat{\lambda}_{\mathrm{MLE}} = n \bigg/ \sum_{i=1}^{n} \left(1/x_i - 1/\hat{\mu}_{\mathrm{MLE}}\right).$$

The discussion above works on a premise that the failure time of all test units are observable. On the other hand, if part of the random sample is censored, the previous MLE has no closed functional form and thus numerical procedures are required for obtaining the estimates. To overcome this problem, we introduce the latent variable and propose LVE in the next section to obtain simple closed-form estimate of IG parameters.

3 Latent Variable Estimates

Assume that n independent units are tested, and, before a pre-specified test censoring time τ, m of them have failed with failure times $t_1, \ldots, t_m$. The remaining units were censored with degradation values, $w_{m+1}(\tau), \ldots, w_n(\tau)$, at time τ. Using the observed sample of these random values from a time-constrained test, we propose LVE for both parameters by using latent variables.

3.1 The LVE of μ

Suppose one can observed the degradation value of all test units at the censoring time, then by the fact that the degradation path is a Wiener process, the observed degradation value is a random sample of $N(\eta\tau, \sigma^2\tau^2)$. However, for those failed units, the degradation value at the censoring time is unknown. Hence, we would add a latent variable to each of the failed units and obtain a pseudo random sample of degradation data. The following discussion illustrates how we proceed.

For the i-th failed unit ($i = 1, 2, \ldots, m$) with failure time t_i, its degradation value at the failure time is a. If one had measured the degradation value of this failed unit at time τ, the value is a random variable $w_i(\tau)$ and the difference between $w_i(\tau)$ and a (the *latent variable*, $\Delta w_i(\tau)$) follows a normal distribution. That is,

$$\Delta w_i(\tau) \sim N(\eta(\tau - t_i), \sigma^2(\tau - t_i)), \quad i = 1, 2, \ldots, m$$

The latent variable is a random variable independent from failure time, T_i because of the independent increment property of Brownian motion. By adding the latent variable to each failed unit, we then obtain a pseudo random sample of degradation data, $(\tilde{w}_1(\tau), \ldots, \tilde{w}_m(\tau), w_{m+1}(\tau), \ldots, w_n(\tau))$ where $\tilde{w}_i(\tau) = a + \Delta w_i(\tau)$. This pseudo random sample are identical and independently distributed as the normal distribution with mean $\tau\eta$ (also $\tau a/\mu$) and variance $\tau\sigma^2$ (also $\tau a^2/\sigma^2$) by Wiener process property.

However, different from the real random degradation values, the latent variables are not actually observable and thus we would replace them by their conditional expected values (i.e., $\eta(\tau - t_i)$ for $\Delta w_i(\tau)$) and obtain the following closed-form LVE for parameter μ:

$$\hat{\mu}_{\text{LVE}} = \frac{(n-m)\tau + \sum_{i=1}^{m} t_i}{m + \sum_{j=m+1}^{n} w_j(\tau)/a} \tag{1}$$

3.2 The LVE of λ

Notice that since $\tilde{w}_1(\tau), \ldots \tilde{w}_m(\tau), w_{m+1}(\tau), \ldots w_n(\tau)$ are assumed to be identical and independent normally distributed random variables. The MLE for $\tau\sigma^2$ is

$$\tau\hat{\sigma}^2 = \frac{1}{n}\left(\sum_{i=1}^{m} (\tilde{w}_i - \tau\eta)^2 + \sum_{j=m+1}^{n} (w_j - \tau\eta)^2 \right).$$

As discussed in Sect. 3.1, each latent variable $\Delta w_i(\tau)$ follows the normal distribution. In other words, we can decompose each $\Delta w_i(\tau)$ as $\Delta w_i(\tau) = \eta(\tau - t_i) + \Delta B_i(\tau)$, where $\Delta B_i(\tau) \sim N(0, \sigma^2(\tau - t_i))$, $i = 1, 2, \ldots, m$. Thus, the original MLE for $\tau\sigma^2$ becomes

$$\tau\hat{\sigma}^2 = \frac{1}{n}\left(\sum_{i=1}^{m} (a - \eta t_i)^2 + \sum_{i=1}^{m} 2(a - \eta t_i)\Delta B_i(\tau) + \sum_{i=1}^{m} \Delta B_i(\tau)^2 + \sum_{j=m+1}^{n} (w_j - \tau\eta)^2 \right). \tag{2}$$

The random variables, $\Delta B_i(\tau)$, is also independent from failure time, T_i, by independent increment prosperity of Brownian motion. Similar to the derivation process for $\hat{\mu}_{\text{LVE}}$, since the latent variables $\Delta w_i(\tau)$ (and thus $\Delta B_i(\tau)$) are not observable, we substitute them with their expected values (i.e., $E(\Delta B_i(\tau)) = 0$ and $E\left(\Delta B_i(\tau)^2\right) = \sigma^2(\tau - t_i)$) into the (2) and obtain an estimate for the parameter σ

$$\hat{\sigma}^2 = \left(\sum_{i=1}^{m} (a - \hat{\eta} t_i)^2 + \sum_{j=m+1}^{n} (w_j - \tau\hat{\eta})^2 \right) \Big/ \left(\sum_{i=1}^{m} t_i + \tau(n-m) \right).$$

Finally, using the relationships that $\mu = a/\eta$, $\lambda = a^2/\sigma^2$, and the estimates of μ and σ, we have an estimate for λ as

$$\hat{\lambda}_{\text{LVE}} = \left(\sum_{i=1}^{m} t_i + \tau(n-m) \right) \Big/ \left(\sum_{i=1}^{m} (1 - t_i/\hat{\mu}_{\text{LVE}})^2 + \sum_{j=m+1}^{n} (w_j/a - \tau/\hat{\mu}_{\text{LVE}})^2 \right). \tag{3}$$

Compare to the estimates in Lee and Tang (2007), advantages of using the LVEs include closed form estimates for both IG parameters are available and thus

Using Latent Variable to Estimate Parameters

easier to interpret and implement. In the next section, the comparisons of performance between the LVE and some other estimates are given.

4 Latent Variable Estimates

For comparing the performance of the proposed LVE to some existing parameter estimates in the literature, a simulation study is conducted and presented in this section. Following the parameters used in Lee and Tang (2007), we fix the true parameter values at $\mu = 100$, $\lambda = 2{,}500$ for simulating random IG distributed failure times, as well as the degradation values. Meanwhile, instead of the actual censoring time, we specify the failure probability (denoted as p) before the censoring time for each simulated sample. Three values of p (0.2, 0.5, and 0.8) are used in this simulation. The actual censoring time can be easily calculated by using the CDF of IG distribution. Different sample sizes ($n = 16$, 32, and 64, and 128) are considered for studying the performance of the estimates under small and large sample size scenarios. For each combination of experimental setting, two hundred samples are simulated by using R Software. Each simulated sample contains both failure times and degradation values at the censoring time. The estimates of the IG distribution parameters using three different methods (MEME, MMLE, and LVE) are obtained for each sample of simulated data. We calculate the sample mean and the variance (and thus the Square roots of Mean Square Error (SMSE)) of the 200 obtained estimates for each method. Results are presented in Table 1.

Based on Table 1, both $\hat{\mu}_{\text{LVE}}$ and $\hat{\lambda}_{\text{LVE}}$ estimates are reasonably close to the true parameter values, even for small sample sizes. Moreover, the SMSE of $\hat{\mu}_{\text{LVE}}$ and $\hat{\lambda}_{\text{LVE}}$ decrease as the sample size increases, which provide numerical evidence that the LVEs are consistent. Comparing to the other estimates, the LVE performs better than MMLE and MEME in all cases with respect to both bias and SMSE criteria. In addition, the bias and SMSE of $\hat{\lambda}_{\text{MEME}}$ are very sensitive to the failure probability, p; both bias and SMSE of $\hat{\lambda}_{\text{MEME}}$ are large when p is small. Comparing to others, the LVE would be a better estimate.

5 LED Example

Tseng et al. (2003) considered the problem of determining the termination time for a burn-in test for a LED lamp. The lifetime of a LED lamp is highly related to its light intensity. Hence, one may define a LED lamp *failed* when its light intensity decreases and becomes smaller than certain threshold (e.g., 50 % of its original brightness).

As an illustrative example, we implement the proposed LVE for estimating the lifetime distribution of a LED lamp. The following data was obtained from one of

Table 1 The mean and square root of mean square error of 200 estimates of by using LVE, MMLE, and MEME when $\mu = 100$ and $\lambda = 2,500$ with different sample sizes

p	LVE (SMSE)		MMLE(SMSE)	MEME(SMSE)	Observed p
	μ	λ	λ	λ	
$n = 16$					
0.2	100.49	2,494.91	3,063.46	2,866.14	0.19
	(5.46)	(1,002.10)	(1,341.26)	(1,225.07)	
0.5	100.63	2,483.93	3,029.97	2,585.47	0.49
	(5.13)	(959.73)	(1,274.11)	(977.06)	
0.8	100.51	2,529.44	3,119.25	2,533.98	0.79
	(4.94)	(979.39)	(1,337.53)	(966.71)	
$n = 32$					
0.2	100.13	2,503.51	2,771.65	2,874.41	0.19
	(3.69)	(691.32)	(804.84)	(883.87)	
0.5	100.23	2,506.81	2,757.69	2,620.75	0.50
	(3.59)	(641.61)	(733.29)	(675.96)	
0.8	100.25	2,516.02	2,787.99	2,536.86	0.79
	(3.51)	(679.42)	(800.22)	(680.96)	
$n = 64$					
0.2	100.14	2,482.12	2,619.05	2,858.00	0.20
	(2.67)	(459.03)	(494.25)	(646.08)	
0.5	100.22	2,479.82	2,612.54	2,610.57	0.49
	(2.62)	(4,35.83)	(474.44)	(475.06)	
0.8	100.25	2,473.34	2,611.85	2,513.43	0.80
	(2.54)	(448.51)	(487.18)	(462.88)	
$n = 128$					
0.2	100.24	2,496.96	2,567.57	2,869.60	0.19
	(1.93)	(325.53)	(339.71)	(537.98)	
0.5	100.27	2,500.38	2,569.16	2,625.25	0.49
	(1.85)	(300.44)	(320.41)	(345.60)	
0.8	100.34	2,493.22	2,563.30	2,520.05	0.79
	(1.80)	(311.53)	(326.83)	(315.73)	

the leading LED manufacturers in Taiwan. The data contains the failure times and degradation values of LED lamps from an accelerated degradation test using electric current $= 10$ amperes and temperature $= 105\ °C$. The test censoring time is 6,480 h. The sample size is $n = 24$ and there were $m = 18$ boundary-crossing times (the defined failure times): 6,274.826, 6,164.547, 6,144.000, 6,102.000, 5,430.000, 6,291.087, 6,259.672, 5,261.236, 3,963.600, 6,034.026, 4,866.947, 3,508.613, 5,008.976, 2,893.333, 6,172.000, 6,158.170, 3,494.400, and 4,801.878. The brightness of the 6 censored units at censoring time are: 0.5027, 0.5438, 0.5768, 0.5516, 0.5267, and 0.5639. The original degradation path (denoted as $L(t)$) did not follow a Wiener process. However, as demonstrated in Tseng et al. (2003), the transformed path, $W(t) = -\ell n(L(t^{0.6}))$, can be modeled by the Wiener process. The threshold is then $a = -\ell n(0.5) = 0.6932$. To apply our results, the boundary-crossing times, the censoring time, as well as the degradation values at the censoring time of censored units, are transformed. The resulting LVEs for the IG parameter are $\hat{\mu}_{\mathrm{LVE}}= 181.39$ and $\hat{\lambda}_{\mathrm{LVE}} = 16,569$. Note that distribution of T can be estimated by $F(t|\hat{\mu}_{\mathrm{LVE}}, \hat{\lambda}_{\mathrm{LVE}})$.

6 Summary

In this paper, a latent variable method for estimating both mean and the scale parameters of the IG distribution is proposed. Comparing to the existing modified maximum likelihood estimate, advantages of our method includes: (1) our estimates has clear closed form solutions and thus easier to compute. (2) Simulation results show that, with the help of the degradation values of the censored units, the performance of LVE is better than MMLE because it has smaller asymptotic variances of the estimators for both IG parameters. A Numerical example is also provided for illustration of our method. A possible direction for future research is to see whether it is better to collect first-passage times of the degradation sample paths over certain non-failure thresholds, instead of collecting degradation values at prescribed time points as we considered in this paper.

Acknowledgments This research was supported by the National Science Council of ROC grand NSC 99-2118-M-126 -002. The authors also gratefully acknowledge the helpful comments and suggestions of the reviewers, which have improved the presentation.

References

Doksum KA, Hoyland A (1992) Models for variable-stress accelerated life testing experiments based on Wiener processes and the inverse Gaussian distribution. Technometrics 34:74–82

Lee MY, Tang J (2007) Modified EM-algorithm estimates of parameters of inverse Gaussian distribution based on time-censored Wiener degradation data. Stat Sinica 17:873–893

Meeker WQ, Escobar LA (1998) Statistical methods for reliability data. Wiley, New York

Padgett WJ, Tomlinson MA Inference from accelerated degradation and failure data based on Gaussian process models. Lifetime Data Anal 10: 191–206

Park C, Padgett WJ (2005) Accelerated degradation models for failure based on geometric Brownian motion and Gamma process. Lifetime Data Anal 11:511–527

Seshadri V (1999) The inverse gaussian distribution: statistical theory and applications. Springer, New York

Tseng ST, Peng CY (2004) Optimal burn-in policy by using integrated Wiener process. IIE Tran 36:1161–1170

Tseng ST, Tang J, Ku IH (2003) Determination of optimal burn-in parameters and residual life for highly reliable products. Naval Res Logistics 50:1–14

Wang X (2010) Wiener process with random effects for degradation data. J Multivariate Anal 101:340–351

Interpolation Approximations for the Performance of Two Single Servers in Series

Kan Wu

Abstract Dependence among servers is the root cause of the analytic intractability of general queueing networks. A tandem queue is the smallest unit possessing the dependence. Understand its behavior is the key to understand the behavior of general queueing networks. Based on observed properties, such as intrinsic gap and intrinsic ratio, a new approximation approach for tandem queues is proposed. Across a broad range of examined cases, this new approach outperforms prior methods based on stochastic independence or diffusion approximations.

Keywords Queueing systems · Manufacturing systems modeling

1 Introduction

In manufacturing systems, the states of the upstream and downstream workstations are mutually dependent in general queueing networks. The dependence plays a critical role in modeling the behavior of manufacturing, computer and communication queueing networks. To understand the impact of dependence, we study the behavior of tandem queues which consists of two single servers in series. It is the simplest queueing system which exhibits dependence among servers.

Virtually all existing tractable results in first-come-first-serve (FCFS) queueing networks only hold exactly under memoryless assumptions, which can be restrictive. An important exception is formed by two early papers that seem forgotten but form a cornerstone for our modeling approach. Specifically, Avi-Itzhak (1965) and Friedman (1965) investigated the behavior of tandem queues with constant service times. In particular, Friedman showed the following important properties. If

---

K. Wu (✉)
Georgia Technology, Atlanta GA 30332, USA
e-mail: kanwu@gatech.edu

Y.-K. Lin et al. (eds.), *Proceedings of the Institute of Industrial Engineers Asian Conference 2013*, DOI: 10.1007/978-981-4451-98-7_5,
© Springer Science+Business Media Singapore 2013

customers arrive at the first stage and proceed though the stages in FCFS order with infinite buffers, then: (a) for any sequence of customer arrival times, the time spent in the system by each customer is independent of the order of the stages; and (b) under certain conditions, a tandem queueing system can be reduced to a corresponding system with fewer stages, possibly a single stage. This procedure is called a reduction method. Consequently, the total system queue time is determined solely by the bottleneck workstation, i.e. the queue time of any customer equals the time the same customer would have been waiting in the queue of a single workstation, which is the bottleneck workstation. Therefore, we can analyze such a queueing system exactly even though there is dependence among workstations.

Jackson (1957) and Friedman's results seem totally different at first sight, but indeed share a very important structure: in both, each server sees the initial arrival process directly. In other words, in a Jackson network, all servers see Poisson arrivals, and in a tandem queue with constant service times, the bottleneck sees the initial arrival process. Based on this key insight, we develop new approximate models without directly dealing with the non-renewal departure process. Based on extensive simulations, we argue that this new approach outperforms earlier methods by the parametric-decomposition and diffusion approximation approaches.

The rest of this paper is organized as follows. An introduction to the prior approximate models is given in Sect. 2 and definitions of some critical terms are given in Sect. 3. Section 4 presents the intrinsic gap and intrinsic ratio, along with some important properties and observations of simple tandem queues. The new approximate model is introduced in Sect. 5. In Sect. 6, we draw conclusions and discuss future work.

2 Prior Approach

The parametric-decomposition method (Kuehn 1979) analyzes the nodes in the queueing networks separately by assuming each node is stochastically independent. This dates back to Kleinrocks' independence assumption (Kleinrock 1976). Whitt (1983) developed Queueing Network Analyzer (QNA) based on this assumption, Kingman's G/G/1 approximation (Heyman 1975), and Marshall's equation (Marshall 1968). The essential model for QNA is:

$$E(QT) = \left(\frac{c_a^2 + c_s^2}{2}\right)\left(\frac{\rho}{1-\rho}\right)\frac{1}{\mu}, \tag{1}$$

$$c_d^2 = c_a^2 + 2\rho^2 c_s^2 - (2\rho)(1-\rho)\mu E(QT) = \rho^2 c_s^2 + (1-\rho^2)c_a^2, \tag{2}$$

where ρ is utilization, μ is service rate (or capacity), σ_a is the standard deviation of inter-arrival time, σ_s is the standard deviation of service time, c_d is the coefficient of variation of departure intervals, c_s is the coefficient of variation of service time and QT is queue time.

Interpolation Approximations for the Performance

Reiser and Kobayashi (1974) use the diffusion process approximation to develop analytical models of computing systems by considering service time distributions of a general form. By using multidimensional reflected Brownian motion, Harrison and Nguyen (1990) proposed QNET to approximate queueing networks. Dai and Harrison (1992) developed QNET further to obtain numerical results. QNET is based on a central limit theorem, involving a sequence of networks in which all nodes are assumed to be in heavy traffic. Such a regime may not always be realistic in practice and there is no guarantee on the speed of convergence.

3 Definitions

Throughout this paper, the term bottleneck refers to a throughput bottleneck, which is the server with the highest utilization. Thus the utilization of a tandem queue is its bottleneck server utilization.The following two idealized tandem queue systems are the foundation for the proposed interpolation approach.

Definition 1 *In a fully coupled system (FCS), the system queue time is determined solely by its bottleneck workstation, and is the same as the bottleneck workstation queue time would be if the bottleneck workstation sees the initial arrival process directly.*

For a given tandem queue, a fully coupled system results from assuming the service times are deterministic. Since Friedman's results are applicable to tandem queues with any specified arrival process, it should be noted that an FCS may have either renewal or non-renewal arrival processes.

Definition 2 *An ASIA system is one in which it is assumed that for the workstations in the system, "all see initial arrivals" (ASIA) directly.*

If the initial arrival process is renewal, the ASIA system queue time can be computed by Kingman's approximation. It should be noted that, for a tandem queue, a Jackson network is a special case of an ASIA system, since all servers in Jackson networks see exponential inter-arrival times.

Definition 3 *For a given tandem queue, the intrinsic gap (IG) is the queue time difference between its ASIA and fully coupled system models, i.e.,*

$$Intrinsic\ Gap = QT\ in\ ASIA\ System - QT\ in\ Fully\ Coupled\ System. \qquad (4)$$

Based on the intrinsic gap, two important ratios are defined.

Definition 4 *For a given tandem queue, the intrinsic ratio (IR) is defined as*

$$Intrinsic\ Ratio = \frac{Actual\ QT - QT\ in\ Fully\ Coupled\ System}{Intrinsic\ Gap}. \qquad (5)$$

Definition 5 *For a given tandem queue, the intrinsic gap ratio (IGR) is the ratio of its IG to its AISA system queue time, i.e.,*

$$Intrinsic\ Gap\ Ratio = \frac{Intrinsic\ Gap}{Queue\ Time\ in\ the\ ASIA\ System}. \tag{6}$$

4 Properties of Simple Tandem Queues

In a simple tandem queue, the first queue time can be approximated by Kingman's equation if the arrival process is renewal. The challenge comes from the second queue time, since its arrival process may not be renewal. There two possible cases: (1) simple tandem queue with backend bottleneck (STQB), where the second server has higher utilization, (2) simple tandem queue with front-end bottleneck (STQF), where the first server utilization is at least as large as the second.

Accurately predicting the bottleneck queue time in STQB is important but difficult, since system queue time is dominated by the bottleneck queue time while the bottleneck facing a non-renewal departure process. Therefore, we will study the behavior of the second queue time in STQB in detail.

For simple tandem queues, the ASIA system queue time of the second server (QT_2^A) is

$$QT_2^A \cong \left(\frac{c_{a1}^2 + c_{s2}^2}{2} \right) \left(\frac{\rho_2}{1 - \rho_2} \right) \frac{1}{\mu_2}, \tag{7}$$

where c_{a1}^2 is the inter-arrival time squared coefficient of variation (SCV) of the first server, c_{s2}^2 is the service time SCV of the second server, μ_2 is the mean service rate of the second server, and ρ_2 is the utilization of the second server.

While the fully coupled system queue time of the second server in STQF is zero, the fully coupled system queue time of the second server (QT_2^A) in STQB is given by

$$QT_2^C = \left(\frac{c_{a1}^2 + c_{s2}^2}{2} \right) \left(\frac{\rho_2}{1 - \rho_2} \right) \frac{1}{\mu_2} - \left(\frac{c_{a1}^2 + c_{s1}^2}{2} \right) \left(\frac{\rho_1}{1 - \rho_1} \right) \frac{1}{\mu_1}, \tag{8}$$

where c_{s1}^2 is the service time SCV of the first server, μ_1 is the mean service rate of the first server, and ρ_1 is the utilization of the first server, and other notation is the same as for Eq. (7).

The intrinsic gap of the second server in STQB is

$$IG = QT_2^A - QT_2^C = QT_1 = \left(\frac{c_{a1}^2 + c_{s1}^2}{2} \right) \left(\frac{\rho_1}{1 - \rho_1} \right) \frac{1}{\mu_1}. \tag{9}$$

Interpolation Approximations for the Performance

In STQB, the intrinsic gap of the second server is exactly the same as the queue time of the first server. Therefore, this intrinsic gap possesses the following nice property.

Property 1 (Heavy Traffic Property of the Intrinsic Gap Ratio in STQB) (Heavy Traffic Property of the Intrinsic Gap Ratio in STQB) *In simple tandem queues with backend bottlenecks, the intrinsic gap ratio of the second server goes to zero as the traffic intensity approaches 1.*

In STQB, when randomness exists, the second queue time as well as its ASIA system and FCS queue times approach infinity when the traffic intensity (ρ_2) approaches 1. However, as traffic intensity approaches 1, the intrinsic gap remains finite. Thus, the IGR of the second server goes to zero in heavy traffic.

To gain further insight into Property 1, we conducted a number of simulation experiments for STQB. In these experiments, we assumed Poisson arrivals (i.e. SCV $= 1$) and gamma-distributed service times. In a series of experiments for "small" SCV, values for the two servers are chosen from {0.1 (low), 0.5 (medium), 0.9 (high)}, resulting in nine experiment settings, i.e. (1, 0.1, 0.1), (1, 0.1, 0.5), (1, 0.1, 0.9), (1, 0.5, 0.1), (1, 0.5, 0.5), (1, 0.5, 0.9), (1, 0.9, 0.1), (1, 0.9, 0.5) and (1, 0.9, 0.9). In each experiment, the service time of the second server is 30, but the first service time is chosen from {10, 20, 25, 29}. Each observation in the tables is the average of 100–200 replications. Depending on the utilization and service times, each replication is the average of 200,000–50,000,000 data points after a warm up period of 50 years or longer. The intrinsic ratios of the second server in the STQB cases, when service time SCV is smaller than 1, are shown in Fig. 1. The results come from 36 experiments. All intrinsic ratios are between 0 and 1 and show regular patterns.

When the service time variability is greater than 1, we examine the cases of STQB. As before, we conduct 36 experiments with Poisson arrivals and gamma distributed service times. The service times are as in the first series of experiments. However, in this series, the service time SCV is chosen from 2, 5 or 8. Each observation is the average of 100 replications. Each replication is the average of 200,000–1,000,000 data points after a warm up period of 50 years or longer. Based on Fig.1, we have the following observation.

Observation 1 (Nearly-Linear Relationship) *The intrinsic ratio of the second server is approximately linear across most traffic intensities.*

Because the intrinsic ratio is developed based on the ASIA system and FCS, we can obtain the intrinsic ratio exactly when the arrival process is Poisson and the service times are exponential or constant. When the arrival process is Poisson, the intrinsic ratio is 1 when service times are exponential, and is 0 when the service times are constant at all utilizations. The intrinsic ratio slopes are 0 in these two cases. From the simulation results, the slope of the intrinsic ratios is also close to 0 when the service time SCV is between 0 and 1. The slopes become positive when the service time SCV is larger than 1. Therefore, we have the following conjecture:

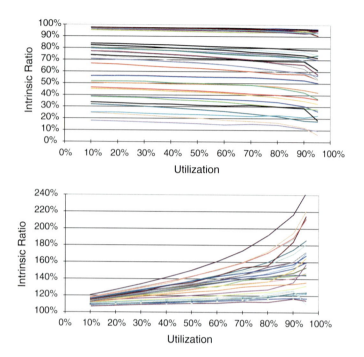

Fig. 1 Intrinsic ratio when service time SCV < 1 (*top*) and SCV > 1 (*below*)

Conjecture 1 (Upper and Lower Bounds for the Second Queue Time in Simple Tandem Queues) *In a simple tandem queue with Poisson arrivals, if the service time SCV of the first server is smaller than 1, the ASIA system queue time is an upper bound and the FCS queue time is a lower bound of the second queue time. If the service time SCV of the first server is greater than 1, the ASIA system queue time becomes a lower bound of the second server queue time.*

Because of Conjecture 1 and Property 1, the second queue time for STQB with Poisson arrivals is expected to be bounded within the intrinsic gap, and the IGR of the second server goes to zero in heavy traffic. These nice properties give us a reliable way to estimate the second queue time in STQB.

5 Simple Tandem Queues with Backend Bottlenecks

Based on the heavy traffic property and nearly-linear relationship of the intrinsic ratios, we propose to approximate the bottleneck queue time in STQB by

$$QT_2 = QT_2^A - (1 - y_2) \times IG$$
$$\cong \left(\frac{c_{a1}^2 + c_{s2}^2}{2}\right)\left(\frac{\rho_2}{1-\rho_2}\right)\frac{1}{\mu_2} - (1-y_2)\left(\frac{c_{a1}^2 + c_{s1}^2}{2}\right)\left(\frac{\rho_1}{1-\rho_1}\right)\frac{1}{\mu_1}, \quad (10)$$

where y_2 is the intrinsic ratio between the first and second server. When y_2 is 1, Eq. (10) is the second queue time in the ASIA system. When y_2 is 0, Eq. (10) is the second queue time in the fully coupled system. The heavy traffic property of the intrinsic gap (Property 1) plays an important role in approximating the second queue time of STQB. Since, based on Observation 1, the intrinsic ratio is finite, when the intrinsic gap in Eq. (5) becomes small, the approximate error of queue time, caused by the imprecise estimate of the intrinsic ratio, also becomes small.

The value of the intrinsic ratio is determined by four factors: initial arrival process, service time ratio (first service time/second service time), and service time SCVs of the first and second servers. If the intrinsic ratio can be approximated by the first and second moments of the inter-arrival and service times, it can be presented as a function of those parameters:

$$\text{Intrinsic Ratio}: \ y_2 \cong f(\lambda, c_{a1}^2, ST_1/ST_2, c_{s1}^2, c_{s2}^2), \tag{11}$$

where λ is the arrival rate, ST_i is service time of the i-th server. The challenge is to determine a good estimate of y_2. The following heuristics are proposed for estimating y_2: make it equal to the coefficient of variation of the first station service time. This heuristic is the direct result of the previous observation: When y_2 is either 0 or 1, Eq. (10) is either the second queue time in the fully coupled system or the ASIA system, respectively. Using cs1 to approximate y_2 seems to be a reasonable choice, since c_{s1} is 0 in the fully coupled system and is 1 in the ASIA system.

When the arrival process is Poisson and service time variability is smaller than 1, the average error of the heuristic is 3.3 % (cf. 11.0 % for QNA and 13.1 % for QNET). Since the first queue time can be computed exactly in this case, it is important to see the impact of the errors on system queue time. If we give a weight to the approximate errors by "Second QT/System QT", the weighted average error becomes 1.8 % for our heuristic, 6.3 % for QNA and 10.3 % for QNET. When the service time SCV is greater than 1, the heuristic does not perform well. In this situation, we may simply use QNA. We call the heuristic the intrinsic ratio (IR) methods, since they are based on the properties of the intrinsic ratio.

6 Conclusion

We have presented a new approximation approach for estimating queue time for simple tandem queues based on newly observed properties, namely the intrinsic gap and intrinsic ratio. Through those underlying properties, the computation is much simpler than the prior methods. Due to the heavy traffic property, the resulting model performs well for simple tandem queues with backend bottlenecks. This is very important, since the system queue time is dominated by the second server in STQB where the second server faces a non-renewal departure process.

In practical manufacturing system, the situation can be more complicated: parallel batching may exist and determination of service time may not be trivial (Wu et al. 2011). All the above will complicate the determination of the intrinsic ratio.

Our approach to approximating queue time does not depend on assumptions about the behavior of mathematical models. Instead, we have identified a fundamental property of tandem queues and exploited this property to deal directly with dependence among workstations. We have achieved notable improvement in the approximation errors, relatively to the prior approaches. The extension of the current approximate models to practical manufacturing systems has been studied by Wu and McGinnis (2012) with promising results.

References

Avi-Itzhak B (1965) A sequence of service stations with arbitrary input and regular service times. Manage Sci 11(5):565–571

Dai J (1992) Performance Analysis of queueing networks using reflecting brownian motions. http://www2.isye.gatech.edu/people/faculty/dai/Software.html

Dai JG, Harrison JM (1992) Reflected brownian motion in an orthant: numerical methods for steady-state analysis. Ann Appl Probab 2:65–86

Friedman HD (1965) Reduction methods for tandem queueing systems. Oper Res 13(1):121–131

Harrison JM, Nguyen A (1990) The QNET method for two-moment analysis of open queueing networks. Queueing Syst 6:1–32

Heyman DP (1975) A diffusion model approximation for the GI/G/1 queue in heavy traffic. Bell Syst Tech J 54(9):1637–1646

Jackson JR (1957) Networks of waiting Lines. Oper Res 5(4):518–521

Kleinrock L (1976) Queueing systems: computer applications. Wiley-Interscience, New York

Kuehn PJ (1979) Approximate analysis of general queuing networks by decomposition. IEEE Trans Commun 27(1):113–126

Marshall KT (1968) Some inequalities in queueing. Oper Res 16(3):651–665

Reiser M, Kobayashi H (1974) Accuracy of the diffusion approximation for some queueing Systems. IBM J Res Dev 18(2):110–124

Whitt W (1983) The queueing network analyzer. Bell Syst Tech J 62(9):2779–2815

Wu K, McGinnis L (2012) Performance evaluation for general queueing networks in manufacturing systems: characterizing the trade-off between queue time and utilization. Eur J Oper Res 221(2):328–339

Wu K, McGinnis L, Zwart B (2011) Queueing models for a single machine subject to multiple types of interruptions. IIE Trans 43(10):753–759

On Reliability Evaluation of Flow Networks with Time-Variant Quality Components

Shin-Guang Chen

Abstract In general, the reliability evaluation of a stochastic-flow network is with time-invariant quality components (including links or vertices). However, in practice, the quality of components for a stochastic-flow network may be variant due to deterioration or improvement by incomplete renewal. This paper presents a method to evaluate the two-terminal network reliability (2TNR) with time-variant quality components. A numerical example is presented to show the application of this method.

Keywords Time-variant quality component · Two-terminal network · Reliability · Minimal path

1 Introduction

The network reliability with time-invariant quality components has been deeply investigated for decades. For example, Aggarwal et al. (1975) first presented the reliability evaluation method for a binary-state network (no flow happens). Lee (1980) used lexicographic ordering and labeling scheme (Ford and Fulkerson 1962) to calculate the system reliability for a binary-state flow network (0 or a positive integer flow exists). Aggarwal et al. (1982) solved such a reliability problem in terms of minimal paths (MPs). Without MPs, Rueger (1986) extended to the case that nodes as well as arcs all have a positive-integer capacity and may fail. Considering that each arc has several states/capacities, such a network is called a stochastic-flow network (Xue 1985, for perfect node cases; Lin 2001, for imperfect node cases). Given the demand d, the system reliability is the probability

---

S.-G. Chen (✉)
Department of Industrial Engineering and Management, Tungnan University,
New Taipei City, Taiwan ROC
e-mail: Bobchen@mail.tnu.edu.tw

Y.-K. Lin et al. (eds.), *Proceedings of the Institute of Industrial Engineers Asian Conference 2013*, DOI: 10.1007/978-981-4451-98-7_6,
© Springer Science+Business Media Singapore 2013

that the maximum flow of the network is not less than d. However, in practice, the quality of components for a stochastic-flow network may be variant due to deterioration or improvement by incomplete renewal.

This paper mainly aims at the reliability evaluation of a two-terminal and stochastic-flow network (2TNR) with time-variant quality component (nodes only) in terms of MPs. A path is a set of nodes and arcs whose existence results in the connection of source node and sink node. A MP is a path whose proper subset is not a path. When the network is live, there are several MP vectors respect to system demand d, called d-MP, can be found. Then, the network reliability is the union probability of all those d-MPs.

The remainder of the work is described as follows: The assumptions are addressed in Sect. 2. The network model is described in Sect. 3. Then, a numerical example is demonstrated in Sect. 4. Section 5 draws the conclusion for this paper.

2 Assumptions

Let $G = (A, B, M)$ be a stochastic-flow network where A is the set of arcs, $B = \{b_i | 1 \leq i \leq s\}$ is the set of nodes, and $M = (m_1, m_2, \ldots, m_s)$ is a vector with m_i (an integer) being the maximum capacity of b_i. Such a G is assumed to satisfy the following assumptions.

1. The capacity of b_i is an integer-valued random variable which takes values from the set $\{0, 1, 2, \ldots, m_i\}$ according to an empirical probability mass function (p.m.f.) μ_{it}, which can be obtained by a statistical observation in a moving time frame at time t. Note that the capacity 0 often means a failure or unavailability of this node at time t.
2. The arcs are perfect. That is, they are excluded from the reliability calculation.
3. Flow in G satisfies the flow-conservation law (Ford and Fulkerson 1962).
4. The states of each node are independent from each other.

3 The Network Model

Let $mp_1, mp_2, \ldots, mp_z$ be the MPs. Thus, the network model can be described in terms of two vectors: the capacity vector $X_t = (x_{1t}, x_{2t}, \ldots, x_{st})$, and the flow vector $F_t = (f_{1t}, f_{2t}, \ldots, f_{zt})$, where x_{it} denotes the current capacity of b_i at time t and f_{jt} denotes the current flow on mp_j at time t. Then, such a vector F_t is feasible iff

$$\sum_{j=1}^{z} \{f_{jt} | b_i \in mp_j\} \leq m_i \quad \text{for } i = 1, 2, \ldots, s. \tag{1}$$

Constraint (1) describes that the total flow through b_i can not exceed the maximum capacity of b_i at time t. We denote such set of F_t as $U_{Mt} = \{F_t|F_t$ is feasible under $M\}$. Similarly, F_t is feasible under $X_t = (x_{1t}, x_{2t}, ..., x_{st})$ iff

$$\sum_{j=1}^{z} \{f_{jt}|b_i \in mp_j\} \leq x_{it} \quad \text{for } i = 1, 2, ..., s \tag{2}$$

For clarity, let $U_{Xt} = \{F_t|F_t$ is feasible under $X_t\}$. The maximum flow under X_t is defined as $V(X_t) = \max\{\sum_{j=1}^{z} \{f_{jt}|F_t \in U_{Xt}\}$.

3.1 Reliability Evaluation

Given a demand d_t, the reliability R_{dt} is the probability that the maximum flow is not less than d_t, i.e., $R_{dt} = \Pr\{X_t|V(X_t) \geq d_t\}$ at time t. To calculate R_{dt}, it is advantageously to find the minimum capacity vector in $\{X_t|V(X_t) \geq d_t\}$. A minimum capacity vector X is said to be a d-MP iff (1) $V(X) \geq d$ and (2) $V(Y) < d$, for any other vector Y such that $Y < X$, in which $Y \leq X$ iff $y_j \leq x_j$, for $j = 1, 2, ...,s$ and $Y < X$ iff $Y \leq X$ and $y_j < x_j$, for at least one j. Suppose there are totally w_t d-MPs at time t: $X_{1t}, X_{2t}, ..., X_{wt}$, and $E_{it} = \{X_t|X_t \geq X_{it}\}$, the probability R_{dt} can be equivalently calculated via the well-known inclusion–exclusion principle or the RSDP algorithm (Zuo et al. 2007)

$$R_{dt} = \Pr\{X_t|V(X_t) \geq d_t\}$$
$$= \Pr(\bigcup_{i=1}^{w_t} E_{it}) \tag{3}$$
$$= \sum_{k=1}^{w_t} (-1)^{k-1} \sum_{I \subset \{1,2,...,w_t\}, |I|=k} \Pr\{\bigcap_{i\in I} E_{it}\},$$

where $\Pr\{\bigcap_{i\in I} E_{it}\} = \prod_{j=1}^{s} \sum_{l=\max\{x_{ij}|\forall i \in I\}}^{m_j} \mu_{jt}(l)$.

3.2 Generation of d-MPs

At first, we find the flow vector $F_t \in U_{Mt}$ such that the total flow of F_t equals d_t. It is defined as in the following equation

$$\sum_{j=1}^{z} f_{jt} = d_t. \tag{4}$$

Then, let $\mathbf{F}_t = \{F_t | F_t \in U_{Mt}$ and satisfies Eq. (4)$\}$. We show that a d-MP X_t exists if there is a $F_t \in \mathbf{F}_t$ by the following lemma.

Lemma 3.1 *Let X_t be a d-MP, then there is a $F_t \in \mathbf{F}_t$ such that*

$$x_{it} = \sum_{j=1}^{z} \{f_{jt} | a_i \in mp_j\} \quad \text{for } i = 1, 2, \ldots, s \tag{5}$$

Proof If X_t is a d-MP, then there is a F_t such that $F_t \in U_{X_t}$ and $F_t \in \mathbf{F}_t$. Suppose there is a k such that $x_{kt} > \sum_{j=1}^{z} \{f_{jt} | a_i \in mp_j\}$. Set $Y_t = (y_{1t}, y_{2t}, \ldots, y_{(k-1)t}, y_{kt}, y_{(k+1)t}, \ldots, y_{st}) = (x_{1t}, x_{2t}, \ldots, x_{(k-1)t}, x_{kt} - 1, x_{(k+1)t}, \ldots, x_{st})$. Hence $Y_t < X_t$ and $F_t \in U_{Y_t}$ (since $\sum_{j=1}^{z} \{f_{jt} | a_i \in mp_j\} \leq y_{it}$, $\forall i$), which indicates that $V(Y_t) \geq d_t$ and contradicts to that X_t is a d-MP. □

Given any $F_t \in \mathbf{F}_t$, we generate a capacity vector $X_{Ft} = (x_{1t}, x_{2t}, \ldots, x_{st})$ via Eq. (5). Then, the set $\Omega = \{X_{Ft} \mid F_t \in \mathbf{F}_t\}$ is built. Let $\Omega_{min} = \{X_t | X_t$ is a minimal vector in $\Omega\}$. Lemma 3.1 indicates that the set Ω includes all d-MPs. The following lemma further proves that Ω_{min} is the set of d-MPs.

Lemma 3.2 Ω_{min} *is the set of d-MPs.*

Proof Firstly, suppose $X_t \in \Omega_{min}$ (note that $V(X_t) \geq d_t$) but it is not a d-MP. Then there is a d-MP Y_t such that $Y_t < X_t$, which implies $Y_t \in \Omega$ and thus contradicts to that $X_t \in \Omega_{min}$. Hence, X_t is a d-MP. Conversely, suppose X_t is a d-MP (note that $X_t \in \Omega$) but $X_t \notin \Omega_{min}$ i.e., there is a $Y_t \in \Omega$ such that $Y_t < X_t$. Then, $V(Y_t) \geq d_t$ which contradicts to that X_t is a d-MP. Hence, $X_t \in \Omega_{min}$. □

4 A Numerical Example

Figure 1 shows a 5-nodes network. The demand is 4. Table 1 gives the results of sampling from the throughputs of all 5 nodes in a period $t = 1$–20. Table 2 shows the changes of b_1 quality from $t = 10$–20 by 10 units moving time frame.

There are 4 MPs found: $mp_1 = \{b_1, b_2, b_4, b_5\}$, $mp_2 = \{b_1, b_2, b_5\}$, $mp_3 = \{b_1, b_3, b_4, b_5\}$, $mp_4 = \{b_1, b_3, b_5\}$. All 4-MPs are generated step-by-step as follows:

Fig. 1 A 5-node network

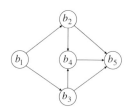

On Reliability Evaluation of Flow Networks

Table 1 The raw data

t	b_1	b_2	b_3	b_4	b_5	R_{dt}
1	2	1	1	1	2	–
2	4	0*	3	2	3	–
3	0	1	0	1	0	–
4	1	0	1	0	2	–
5	6	2	2	3	4	–
6	2	1	2	1	3	–
7	3	2	1	1	3	–
8	5	2	2	2	4	–
9	5	2	3	3	5	–
10	4	2	2	1	4	0.0720
11	6	0	3	2	5	0.1230
12	5	3	2	3	4	0.1728
13	5	3	3	3	5	0.3087
14	0	3	2	3	5	0.4368
15	6	3	0	3	3	0.3479
16	5	3	3	3	6	0.4928
17	5	3	3	3	6	0.6561
18	4	2	1	3	0	0.5616
19	5	3	2	0	6	0.5688
20	4	3	3	3	6	0.5760

*node failure

Table 2 The changes of b_1 quality from $t = 10$–20

t	The p.m.f of b_1						
	0	1	2	3	4	5	6
10	0.10	0.10	0.20	0.10	0.20	0.20	0.10
11	0.10	0.10	0.10	0.10	0.20	0.20	0.20
12	0.10	0.10	0.10	0.10	0.10	0.30	0.20
13	0.00	0.10	0.10	0.10	0.10	0.40	0.20
14	0.10	0.00	0.10	0.10	0.10	0.40	0.20
15	0.10	0.00	0.10	0.10	0.10	0.40	0.20
16	0.10	0.00	0.00	0.10	0.10	0.50	0.20
17	0.10	0.00	0.00	0.00	0.10	0.60	0.20
18	0.10	0.00	0.00	0.00	0.20	0.50	0.20
19	0.10	0.00	0.00	0.00	0.20	0.50	0.20
20	0.10	0.00	0.00	0.00	0.20	0.50	0.20

Step 1. Find the feasible vector $F_t = (f_{1t}, f_{2t}, \ldots, f_{4t})$ satisfying both capacity and demand constraints.

    a. Enumerate f_{jt} for $0 \le f_{jt} \le 6$, $1 \le j \le 4$ do

    b. If f_{jt} satisfies the following equations
$$f_{1t} + f_{2t} + f_{3t} + f_{4t} \le 6, f_{1t} + f_{2t} \le 6, f_{3t} + f_{4t} \le 6, f_{1t} + f_{3t} \le 6,$$
$$f_{1t} + f_{2t} + f_{3t} + f_{4t} = 4,$$
then $\mathbf{F}_t = \mathbf{F}_t \cup \{F_t\}$.

End enumeration.

The result is $\mathbf{F}_{t=10} = \{(0, 0, 0, 4)\ (0, 0, 1, 3),\ (0, 0, 2, 2),\ \ldots, (4, 0, 0, 0)\}$.

Step 2. Generate the set $\Omega = \{X_{Ft} | F_t \in \mathbf{F}_t\}$.

    a. For F_t in $\mathbf{F}_t$ do

    b. $x_{1t} = f_{1t} + f_{2t} + f_{3t} + f_{4t},$ $x_{2t} = f_{1t} + f_{2t},$ $x_{3t} = f_{3t} + f_{4t},$
$x_{4t} = f_{1t} + f_{3t},$
$x_{5t} = f_{1t} + f_{2t} + f_{3t} + f_{4t}.$

    c. $U_{Xt} = U_{Xt} \cup \{X_{Ft}\}$.

End for-loop.

    d. For X_t in U_{Xt} do

    e. If $X_t \notin \Omega$, then $\Omega = \Omega \cup \{X_t\}$.

End for-loop.

At the end of the loop: $\Omega = \{X_{1t=10} = (4, 0, 4, 0, 4), X_{2t=10} = (4,0, 4, 1, 4), \ldots, X_{25t=10} = (4, 4, 0, 4, 4)\}$.

Step 3. Find the set $\Omega_{min} = \{X_t | X_t$ is a minimum vector in $\Omega\}$ via pairwise comparison.

The result is $\Omega_{min} = \{X_{1t=10} = (4, 0, 4, 0, 4), X_{6t=10} = (4, 1, 3, 0, 4), X_{10t=10} = (4, 2, 2, 0, 4), X_{13t=10} = (4, 3, 1, 0, 4), X_{15t=10} = (4, 4, 0, 0, 4)\}$.

Finally, the probability $R_{4t=10}$ can be calculated in terms of 5 4-MPs. Let $E_{1t} = \{X_t | X_t \ge X_{1t}\}$, $E_{2t} = \{X_t | X_t \ge X_{6t}\}$, $E_{3t} = \{X_t | X_t \ge X_{10t}\}$, $E_{4t} = \{X_t | X_t \ge X_{13t}\}$ and $E_{5t} = \{X_t | X_t \ge X_{15t}\}$. From Eq. (3), we get $R_{4t=10} = \Pr\{\bigcup_{i=1}^{5} E_{it}\} = 0.0720$.

5 Conclusion

This paper proposed a reliability evaluation method for two-terminal and stochastic-flow networks (2TNR) with time-variant quality component (node only) in terms of MPs. The network reliability with time-invariant quality components has been deeply investigated for decades, such as the reliability evaluation of bi-state networks, bi-state flow networks, and stochastic-flow networks. However, in practice, the quality of components for a stochastic-flow network may be variant due to deterioration or improvement by incomplete renewal. A numerical example is demonstrated to explain the proposed method.

Future researches are suggested to investigate the reliability of imperfect double-resource networks with time-variant quality components.

Acknowledgments This work was supported in part by the National Science Council, Taiwan, Republic of China, under Grant No. NSC 101-2221-E-236-006.

References

Aggarwal KK, Gupta JS, Misra KB (1975) A simple method for reliability evaluation of a communication system. IEEE T Commun 23:563–565

Aggarwal KK, Chopra YC, Bajwa JS (1982) Capacity consideration in reliability analysis of communication systems. IEEE T Reliab 31:177–180

Ford LR, Fulkerson DR (1962) Flows in networks. Princeton University Press, NJ

Lee SH (1980) Reliability evaluation of a flow network. IEEE T Reliab 29:24–26

Lin YK (2001) A simple algorithm for reliability evaluation of a stochastic-flow network with node failure. Comput Oper Res 28:1277–1285

Rueger WJ (1986) Reliability analysis of networks with capacity-constraints and failures at branches and nodes. IEEE T Reliab 35:523–528

Xue J (1985) On multistate system analysis. IEEE T Reliab 34:329–337

Zuo MJ, Tian Z, Huang HZ (2007) An efficient method for reliability evaluation of multistate networks given all minimal path vectors. IIE Trans 39:811–817

Defect Detection of Solar Cells Using EL Imaging and Fourier Image Reconstruction

Ya-Hui Tsai, Du-Ming Tsai, Wei-Chen Li and Shih-Chieh Wu

Abstract Solar power is an attractive alternative source of electricity nowadays. Solar cells, which form the basis of a solar power system, are mainly based on crystalline silicon. Many defects cannot be visually observed with the conventional CCD imaging system. This paper presents defect inspection of multi-crystalline solar cells in electroluminescence (EL) images. A solar cell charged with electrical current emits infrared light. The intrinsic crystal grain boundaries and extrinsic defects of small cracks, breaks, and finger interruptions hardly reflect the infrared light. The EL image can thus distinctly highlight barely visible defects as dark objects. However, it also shows random dark regions in the background, which makes automatic inspection in EL images very difficult. A self-reference scheme based on the Fourier image reconstruction technique is proposed for defect detection of solar cells in EL images. The target defects appear as line- or bar-shaped objects in the EL image. The Fourier image reconstruction process is applied to remove the possible defects by setting the frequency components associated with the line- and bar-shaped defects to zero and then back-transforming the spectral image into a spatial image. The defect region can then be easily identified by evaluating the gray-level differences between the original image and its reconstructed image. The reference image is generated from the inspection image itself and, thus, can accommodate random inhomogeneous backgrounds. Experimental results on a set of various solar cells have shown that the proposed method performs effectively for detecting small cracks, breaks, and

---

Y.-H. Tsai · W.-C. Li
Mechanical Industry Research Laboratories, Industrial Technology Research Institute, 195, Sec. 4, Chung-Hsing Road, Chutung, Hsinchu, Taiwan, Republic of China
e-mail: yahuitsai@itri.org.tw

W.-C. Li
e-mail: jeecool@itri.org.tw

D.-M. Tsai (✉) · S.-C. Wu
Department of Industrial Engineering and Management, Yuan-Ze University,
135 Yuan-Tung Road, Nei-Li, Tao-Yuan, Taiwan, Republic of China
e-mail: iedmtsai@saturn.yzu.edu.tw

Y.-K. Lin et al. (eds.), *Proceedings of the Institute of Industrial Engineers Asian Conference 2013*, DOI: 10.1007/978-981-4451-98-7_7,
© Springer Science+Business Media Singapore 2013

finger interruptions. The computation time of the proposed method is also fast, making it suitable for practical implementation. It takes only 0.29 s to inspect a whole solar cell image with a size of 550 × 550 pixels.

Keywords Defect detection · Surface inspection · Solar cell · Fourier transform

1 Introduction

In recent years, the demand for solar cells has increased significantly due to growing environmental concerns and the global oil shortage. The demand could even be potentially boosted after the nuclear disaster in Fukushima, Japan. Solar cells, which convert the photons from the sun to electricity, are largely based on crystalline silicon in the currently installed solar power systems because of the competitive conversion efficiency and usable lifespan.

Since defects in solar cells critically reduce their conversion efficiency and usable lifetime, the inspection of solar cells is very important in the manufacturing process. Some fatal defects, such as small cracks lying within the wafer surface and subtle finger interruptions, may not be visually observed in the image captured by a typical CCD camera. In order to highlight the intrinsic and extrinsic deficiencies that degrade the conversion efficiency of a solar cell, the electroluminescence (EL) imaging technique (Fuyuki et al. 2005; Fuyuki and Kitiyanan 2009) has been proposed in recent years. In the EL imaging system, the solar cell is excited with voltage, and then a cooled Si-CCD camera or a more advanced InGaAs camera is used to capture the infrared light emitting from the excited solar cell. Areas of silicon with higher conversion efficiency present brighter luminescence in the sensed image. Figure 1 shows the configuration of the EL imaging system. Figure 2a shows the EL image of a defect-free solar cell. Figure 2b–d presents three EL images of defective solar cells that contain small cracks, breaks, and finger interruptions, respectively. The defect areas are inactive, resulting in dark regions that are visually observable in the sensed EL image. The EL imaging system not only shows the extrinsic defects as dark objects but also presents the dislocation and grain boundaries with dark gray levels in the sensed image. The background shows inhomogeneous patterns of dark blobs and clouds due to the random crystal grains in the multi-crystalline silicon wafer. This characteristic makes the automatic defect detection in EL images very difficult.

Recently, Li and Tsai (2011) presented a machine vision algorithm to detect saw-mark defects in multi-crystalline solar wafers in the wafer cutting process. The Fourier transform is used to smooth the crystal grain background as a non-textured surface. A line detection process is then carried out for each individual scan line of the filtered image to detect possible defect points that are distinctly apart from the line sought.

In this study, we further the development of an automatic detection method for critical surface defects of both exterior and interior small cracks, breaks, and finger

Fig. 1 Configuration of an EL imaging system

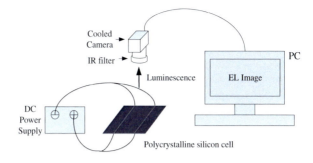

Fig. 2 EL images of solar cells **a** defect-free sample, **b** small crack, **c** break, and **d** finger-interruption

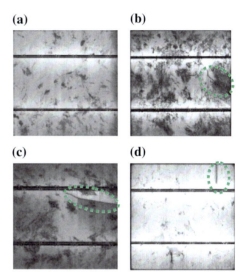

interruptions of a multi-crystalline solar cell in an EL image. The small crack appears as a thin line segment, and the break divides the infected area into a dark and a bright region in the EL image. The disconnected finger affects the area beneath the finger and its neighborhood and appears as a dark bar-shaped object along the finger direction in the EL image. In order to detect such defects against an inhomogeneous background, a self-reference scheme based on the Fourier image reconstruction is proposed for solar cell inspection in EL images.

2 Defect Detection in Inhomogeneous Surfaces

Self-reference approaches (Guan et al. 2003) that generate a golden template from the inspection image itself and image reconstruction approaches (Tsai and Huang 2003) that remove the background pattern have been used for defect detection in textured surfaces. The multi-crystalline solar cell in the EL image falls in the category of inhomogeneous textures. Conventional self-reference or image reconstruction methods cannot be directly extended for defect detection in randomly textured surfaces. The Fourier transform (FT) is a global representation of an image and, thus, gives no spatial information of the defect in the original EL image. In order to locate a defect in the EL image, we can eliminate the frequency components along a line with a specific slope angle in the spectrum image and then back-transform the image using the inverse FT. The reconstructed image can then be used as a defect-free reference image for comparison.

2.1 Fourier Image Reconstruction

Let $f(x, y)$ be the gray level at pixel coordinates (x, y) in an EL image of size $M \times N$. The two-dimensional discrete Fourier transform (DFT) of $f(x, y)$ is given by

$$F(u, v) = \sum_{y=0}^{N-1} \sum_{x=0}^{M-1} f(x, y) \cdot \exp\left[-j2\pi\left(\frac{ux}{M} + \frac{vy}{N}\right)\right] \quad (1)$$

for spectral variables $u = 0, 1, 2,..., M-1$ and $v = 0, 1, 2,..., N-1$. The DFT is generally complex; that is $F(u, v) = R(u, v) + j.I(u, v)$, where $R(u, v)$ and $I(u, v)$ are the real and imaginary parts of $F(x, y)$, i.e.,

$$R(u, v) = \sum_{y=0}^{N-1} \sum_{x=0}^{M-1} f(x, y) \cdot \cos\left[2\pi\left(\frac{ux}{M} + \frac{vy}{N}\right)\right] \quad (2a)$$

$$I(u, v) = \sum_{y=0}^{N-1} \sum_{x=0}^{M-1} f(x, y) \cdot \sin\left[2\pi\left(\frac{ux}{M} + \frac{vy}{N}\right)\right] \quad (2b)$$

The power spectrum $P(x, y)$ of $F(x, y)$ is defined as

$$P(u, v) = |F(u, v)|^2 = R^2(u, v) + I^2(u, v) \quad (3)$$

Since the input image $f(x, y)$ is real, the FT exhibits conjugate symmetry. The magnitude of the transform is symmetric with respect to the DC center in the spectrum image. Figure 3a1 shows the EL image of a defective solar cell that contains a small line crack at an angle of $37°$, and Fig. 3a2 is the corresponding Fourier spectrum image. It is expected to find a $127°$ high-energy straight line passing through the DC center in the spectrum image. However, the resulting spectrum image cannot sufficiently display a visible line due to the extremely short line segment of the small crack with respect to the whole EL image area.

In order to intensify the high-energy line for a defect with either a long or a short line segment in the Fourier spectrum image, the original solar cell image is equally divided into many non-overlapping subimages of smaller size. Figure 3b1

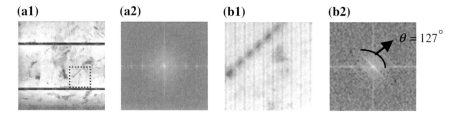

Fig. 3 Fourier spectrum images of solar cells: **a1** full-sized defective EL image of 550 × 550 pixels; **a2** spectrum image of (**a1**) where no lines can be visually observed; **b1** small crack in a 75 × 75 EL subimage; **b2** spectrum image of (**b1**), where a 127°-line can be distinctly observed

shows the crack defect in a 75 × 75 subimage. Figure 3b2 presents the corresponding Fourier spectrum, where the high-energy line associated with the 37° crack is now more distinctly present at 127° in the spectrum image.

2.2 Line Detection in Spectrum Images

To automatically detect high-energy lines associated with line-shaped defects in the Fourier spectrum image, we propose a line detection process for the task. The Hough transform (HT) is an effective technique for line detection under noisy and discontinuous conditions. It involves a voting process that transforms the set of points in the image space into a set of accumulated votes in the parameter space. A local peak larger than some threshold in the parameter space then indicates a line present in the image space. The line equation is then solely defined by the parameter of slope angle. The HT process scans every pixel (u, v) in the spectrum image, and calculates the slope angle

$$\theta = \tan^{-1}((v - v_0)/(u - u_0)) \tag{4}$$

where (u_0, v_0) is the central coordinates of the spectrum image, and is given by $(M/2, N/2)$ for an image of size $M \times N$. The voting weight for the 1-D accumulator $A(\theta)$ at slope angle θ is given by the Fourier spectrum $|F(u, v)|$; i.e.,

$$A(\theta) \leftarrow A(\theta) + |F(u, v)|$$

A local peak with significantly large accumulated magnitude indicates a possible line in the spectrum image. Due to the inherited structure of a solar cell in the EL image, the accumulator will show extremely high magnitudes in horizontal, vertical, and diagonal directions, i.e., angles at 0°, 45°, 90° and 135°. Figure 4a1, b1 show a defect-free and a defective EL subimage of solar cells, respectively. Figure 4a2, b2 are the corresponding accumulators $A(\theta)$ for the EL subimages in Fig. 4a1, b1, excluding the accumulated values at 0°, 45°, 90° and 135°. It can be seen that there are no significant peaks for the Fourier spectrum of the defect-free

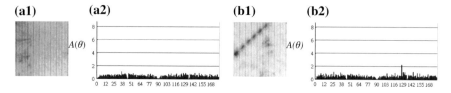

Fig. 4 The accumulator $A(\theta)$ of the Hough transform; **a1** defect-free EL subimage, **b1** defective EL image with a 37°-crack; **a2** accumulator $A(\theta)$ as a function of θ for defect-free sample (**a1**); **b2** accumulator $A(\theta)$ for defective sample (**b1**)

subimage. However, there is a distinct peak at $\theta = 127°$ corresponding to the 37° crack in the EL subimage.

In this study, the adaptive threshold T_θ for θ selection is given by the upper bound of a simple statistical control limit, which is defined as Eq. (5):

$$T_\theta = \mu_{A(\theta)} + K_\theta \cdot \sigma_{A(\theta)} \qquad (5)$$

where K_θ is user specified control constant; it is given by 2.5 in this study. Note that we exclude 0°, 45°, 90° and 135° for the computation of the mean $\mu_{A(\theta)}$ and standard deviation $\sigma_{A(\theta)}$ of $A(\theta)$. All local maximum peaks with $A(\theta)$ greater than the angle threshold T_θ are recognized as suspected slope angles of defects, and are collected in a set:

$$\Theta = \left\{ \theta | A(\theta) > \mu_{A(\theta)} + K_\theta \cdot \sigma_{A(\theta)} \right\} \qquad (6)$$

Since a true line crack generally results in distinctly high-energy frequency components in the spectrum image, setting $K_\theta = 2.5$ in this study is sufficiently small to ensure all true crack lines in the associated spectrum image can be selected.

2.3 Defect Detection

To tackle the problem of defect detection in an inhomogeneous surface, we remove all the possible defects and create a near defect-free reference image from each individual test image under inspection. This is done by assigning zero values to all frequency components in the vicinity of each detected line with slope angle θ^* in the collection Θ, and then back-transforming the revised Fourier spectrum using the inverse DFT.

Let Δw be the band-rejection width. For a given slope angle θ^*, $\theta^* \in \Theta$, the band-rejection region is bounded by two lines $L_{\theta^*}^+$ and $L_{\theta^*}^-$ that are parallel to $L_{\theta^*,0}$ with a width of Δw between these two lines, as shown in Fig. 5. The bounded lines $L_{\theta^*}^+$ and $L_{\theta^*}^-$ are given by

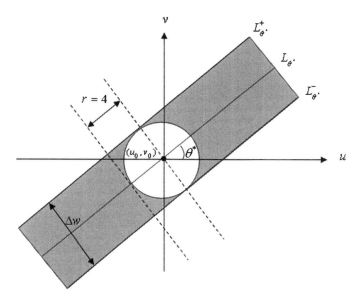

Fig. 5 Band-rejection region in the Fourier spectrum image for defect removal

$$L_{\theta^*}^+ : (u - u_0)\tan\theta^* - (v - v_0) - b = 0 \quad (7a)$$

$$L_{\theta^*}^- : (u - u_0)\tan\theta^* - (v - v_0) - b = 0 \quad (7b)$$

where $b = (\Delta w/2)/\cos\theta^*$.

For any pixel (u, v) in the Fourier spectrum image that lies within the band-rejection region, i.e. $L_{\theta^*}^+ \geq 0$ and $L_{\theta^*}^- \leq 0$, the associated $F(u, v)$ is set to zero. To preserve the background pattern in the reconstructed image, all frequency components lying within a small circle with the DC as the center must be retained without changing their spectrum values. Based on the removal process discussed above, the revised Fourier spectrum is given by

$$\hat{F}(u, v) = \begin{cases} 0, & \text{if}(L_{\theta^*}^+ \geq 0 \text{ and } L_{\theta^*}^- \leq 0) \text{ \& } (u-u_0)^2 + (v-v_0)^2 > r^2 \\ F(u, v), & \text{otherwise} \end{cases}$$

In this study, the circular radius r is set to 4 for a 75×75 EL subimage. The self-reference image of an inspection image $f(u, v)$ of size $M \times N$ can now be reconstructed by the inverse discrete Fourier transform. Hence,

$$\hat{f}(x, y) = \frac{1}{M \cdot N} \sum_{v=0}^{N-1} \sum_{u=0}^{M-1} \hat{F}(u, v) \cdot \exp\left[j2\pi(ux/M + vy/N)\right] \quad (8)$$

Defects in the EL subimage can be easily identified by subtracting the original image $f(u, v)$ from the reconstructed image $\hat{f}(x, y)$. The difference between $f(u, v)$ and $\hat{f}(x, y)$ is given by

$$\Delta f(x, y) = |f(x, y) - \hat{f}(x, y)| \tag{9}$$

The upper bound of a simple statistical control limit is thus used to set the threshold for segmenting defects from the background in the difference image $\Delta f(x, y)$. The threshold for $\Delta f(x, y)$ is given by

$$T_{\Delta f} = \mu_{\Delta f} + K_{\Delta f} \cdot \sigma_{\Delta f} \tag{10}$$

where $\mu_{\Delta f}$ and $\sigma_{\Delta f}$ are the mean and standard deviations of $\Delta f(x, y)$ in the whole image, and $K_{\Delta f}$ is a pre-determined control constant. The detection results can be represented by a binary image, wherein the pixel (u, v) with $\Delta f(x, y) > T_{\Delta f}$ is a defect point and is marked in black. It is otherwise a defect-free point and is marked in white in the binary image. Figure 6a1, b1 show, respectively, a defect-free and a small crack subimage. Their reconstructed images are presented in Fig. 6a2, b2. Figure 6a3, b3 are the corresponding difference images $\Delta f(x, y)$. The results show that the reconstructed image of the defect-free sample is similar to its original one, and no significant differences are found in the difference image.

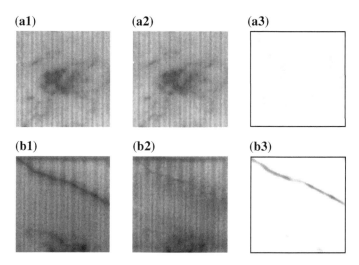

Fig. 6 Image reconstruction and image difference: **a1** defect-free EL subimage; **b1** defective EL subimage; **a2, b2** respective reconstructed images; **a3, b3** respective difference images

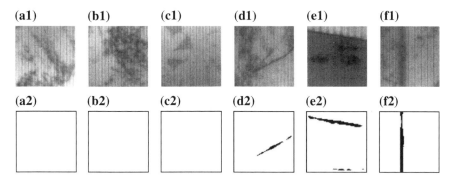

Fig. 7 Detection results on defect-free and defect solar cells: **a1–c1** EL subimages **d1** EL subimages with small cracks; **e1** EL subimages with breaks; **f1** EL subimage with finger interruption; **a2–f2** respective segmentation results

3 Experimental Results

This subsection demonstrates the detection results of various multi-crystalline solar cells in EL images. The parameter values are fixed with band-rejection width $\Delta w = 5$ and control constant $K_{\Delta f} = 1$ for all test samples in the experiment. Figure 7a1–c1 shows a set of three defect-free EL subimages of various solar cell surfaces, where random dark regions are present in the background. The detection results of Fig. 7a1–c1 with the same parameter setting are present in Fig. 7a2–c2. The results show uniformly white images and indicate no defects in the original EL images. Figure 7d1–f1 present, respectively, small cracks, breaks and finger interruption in the EL subimages. All three defects appear with cloud-shaped backgrounds in the EL images. The detection results are displayed as binary images, and are shown in Fig. 7d2–f2. The results indicate that the proposed self-reference method with the Fourier image reconstruction can well detect the small, local defects in EL images with inhomogeneous backgrounds.

In an additional experiment, we have also evaluated a total of 323 EL sub-images of solar cells, of which 308 are defect-free samples and 15 are defective samples containing various defects of small cracks, breaks, and finger interruptions. The proposed method, with the same parameter settings of $\Delta w = 5$ and $K_{\Delta f} = 1$, identifies correctly the defects in all 15 defective samples, and declares no false alarms for all 308 defect-free samples.

4 Conclusions

In this paper, we have proposed a self-reference scheme based on the Fourier image reconstruction to detect various defects in multicrystalline solar cells. Micro-cracks, breaks, and finger interruptions are severe defects found in solar cells. The EL imaging technique is thus used to highlight the defects in the sensed image. Experimental results show that the proposed method can effectively detect various defects and performs stably for defect-free images with random dark regions in the background. The proposed method is efficiently applied to a small subimage and achieves an average computation time of 0.29 s for a whole solar cell image of size 550×550 pixels.

References

Fuyuki T, Kondo H, Yamazaki T, Takahashi Y, Uraoka Y (2005) Photographic surveying of minority carrier diffusion length in polycrystalline silicon solar cells by electroluminescence. Appl Phys Lett 86:262108

Fuyuki T, Kitiyanan A (2009) Photographic diagnosis of crystalline silicon solar cells utilizing electroluminescence. Appl Phys A96:189–196

Guan SU, Xie P, Li H (2003) A golden-block-based self-refining scheme for repetitive patterned wafer inspections. Mach Vis Appl 13:314–321

Li WC, Tsai DM (2011) Automatic saw-mark detection in multicrystalline solar wafer images. Sol Energy Mater Sol Cells 95:2206–2220

Tsai DM, Huang TY (2003) Automated surface inspection for statistical textures. Image Vis Comput 21:307–323

Teaching Industrial Engineering: Developing a Conjoint Support System Catered for Non-Majors

Yoshiki Nakamura

Abstract Previously, business managers and college students seem to have not given enough thought to the study of Industrial Engineering (IE). Increasingly, however, they have become conscious of the importance of IE. In fact, many have started to consider the topic to be useful and critical for their future career. This being the case, it seems highly valuable to develop an educational program which deals specifically with both operation and concept of IE. The program so developed will help improve those who have already studied IE; at the same time, the system would likewise enhance and broaden the knowledge of those whose focus has been confined only to business management. This study tries to create an educational program which conjoins two different faces of business management. On one hand, the program targets on those who have an extensive experience in business management. On the other hand, the system likewise centers on those who know little about business management but have studied IE in the past. By using this cross cutting support method, two different will equally enhance their total knowledge of business administration.

Keywords Industrial and systems engineering education · The support system

1 Introduction

Both business administration students and company presidents should study industrial engineering (IE) because it is basic and essential to all business activity. "Cost management," "work study," and "KAIZEN," are among important IE

Y. Nakamura (✉)
Department of Business Administration, Aoyama Gakuin University, 4-4-25 Shibuya, Shibuya-ku, Tokyo 150-8366, Japan
e-mail: nakamura@busi.aoyama.ac.jp

Y.-K. Lin et al. (eds.), *Proceedings of the Institute of Industrial Engineers Asian Conference 2013*, DOI: 10.1007/978-981-4451-98-7_8,
© Springer Science+Business Media Singapore 2013

activities. Furthermore, effective IE study should include an understanding of the interrelationship between this course and student's or manager's prior knowledge.

In a typical Japanese college, three major methods are applied to teaching business management. First is the teacher–student lecture, wherein teacher's lectures are generally one-sided and the contents, even of the same subject, differ depending on the lecturer. Also, unfortunately, knowledge learned is applied only after graduation.

Second is the "case study," in which teachers present students with a theme based on a real business incident or situation (Alavi and Gallupe 2003; Gorman et al. 1997; Lambert et al. 1990; Mallick and Chaudhury 2000). Typically, students study the case in groups. The students discuss the given case, and present the answer and problem-solving by their own opinions in the case. The case study differs from lecture-style teaching in that students can simulate a company's real work situation. They conduct business meetings and also learn practical business management science in a simulated setting. The problem in this teaching method, however, is that the teachers' expectations and students' problem-solving outcomes do not always coincide.

The third approach is the educational business game (Graham and Gray 1969; Nonoyama et al. 2002; Riis 1995; Tseng 2009). This is a management game simulator that is based on actual business activities. It teaches such business aspects as management and the flow of funds through a game. As the teacher makes the computer program for the students, they learn what the teacher's wants them to. The downside of this teaching method is its strong "game" element. It also begs the question as to whether the game allows students to actually use their own information, knowledge, and reasoning in their decision-making.

Regarding the above-mentioned lecture method and under current educational conditions, we consider that there are two techniques in ideal lecturing: the first is students thinking and problem-solving using their own hands. The second is the student can feel and understand the relationship between the business and the contents the student learned through the support system. That system makes to similar to a real business situation wherein students learn business practical content and the study the student studied before.

This study proposes an educational program wherein students can study IE knowledge through the support system. The objective is to understand the interrelationship between new IE learning and prior knowledge and the influence both these factors have on each other. Concretely, students will form teams and use the mini-belt conveyor and miniature cars for their study. In preparing for the simulation, they must study the major IE fields of production management, cost management, work study, and KAIZEN activity. Because the students' learning process is hands on, I will build the support system wherein the relationship between students' prior knowledge and the knowledge they acquire during the program is observed. For example, how the students understand the influence of "hands-on work" on "business activities and the financial statement" will be observed.

2 The Outline of the Program

2.1 Set the Situation and the Contents of the Methodology

The subject in this educational program is an automobile industry. The company is the mid-sized car company and with a low profit margin. We chose the automobile industry because it is a manufacturing industry and its basic business operations include procurement, manufacturing, and sales (Fig. 1).

As a result, it is easy for the learner to devise a business plan and marketing analysis and understand the company's competitor. The methodology chosen involves making a team wherein students decide who plays the role of COE, CTO, etc.

In this instance, "the learner" is the student who studies the program and "the educator" is the teacher who manages and operates the educational program.

At the beginning of the program, the learner will be given the market and in-house information, a miniature automobile block, a belt conveyor and a support system file made by MS Office Excel. The activity flow will proceed as follows:

(1) Information analysis and forming a decision on the goal rate for the profit margin: learners determine how many cars are to be produced, i.e., the planned production amount. To arrive at this decision, the learners analyze the economic trends, competitor information, and in-house information about the company's finances from the financial statements.
(2) Simulate production experience.
(3) The proposal and the execution of KAIZEN.
(4) Examination of the miniature car's material and environmental elements.
(5) Discussion of the rate of the goal rate whether approaching the learner's set.

In the "marketing" class, for example, students study the importance of the market, competitor information, and the need for business analysis. It is, however,

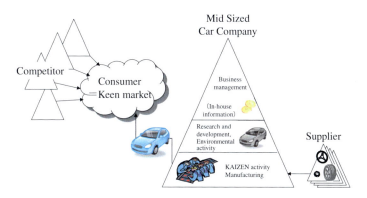

Fig. 1 Program situation

difficult to see the linkage between these studies and decision-making in the actual business activity. As a result, in Step 1, the learner will analyze this information and arrive at a decision on "the amount of planned production" and "the goal for the profit rate." (Details follow).

In Steps 2 and 3, the learners actually make the miniature car and measure the standard work and standard time. They also implement the KAIZEN process and address the environmental problem.

For the cost reduction at Step 4, they must decrease the automobile's materials cost. At the same time, they have to sell the new vehicle to attain the desired goal profit rate. Thus, the learners propose a sellable and popular automobile.

In Step 5, through activities 2, 3, and 4, the learners determine how close they are to the target. If they cannot achieve the set goal, they will determine what the problem is. By carrying out activities outlined in these steps and by having an access to the support system, students will visually understand the relationship among the goal profit rate, production, and the financial statements.

By following these steps in sequence, the learner can simulate business management, and especially the analysis, planning, and implementation phases as well as KAIZEN and the feedback. In this study, it is possible to conduct business management through the experiential and synthetic study, even in a university class.

2.2 The Program's Plan and Flow

The following table shows the flow of this educational program in a university class. The schedule was modeled on the typical Japanese university business class and is comprised of fifteen weekly 90-min sessions (Table 1).

Week 1:

The educator lectures on the educational program, its purpose, goals, and meaning. The learners form a team of three to four members. For the purposes of the team's function, its collegial decision-making nature is understood.

The team decides on the company name, the business philosophy, and each member's role. For unity among the team members in terms of awareness of the program's direction, these decision items will be set.

Weeks 2 and 3:

In the in-house analysis, the learner will examine the financial statements and propose the problem and plan for improvement. The financial analysis framework is used for the business analysis. To calculate the profit rate, this program uses business indicators and "the standard comparison method" with the industry average. If the rate is calculated to be lower than the industry average, it presents a problem for the learners' company. Each team calculates the indicator and determines and presents its own problems.

Teaching Industrial Engineering

Table 1 Program schedule

Week	Contents	Notes
1	Lectures on the educational program	
2	In-house analysis 1	Financial analysis
3	In-house analysis 2	
4	Market and competitor analysis 1	Using information sheet
5	Market and competitor analysis 2	
6	Decides the amount of the production and the goal profit rate	Using support system
7	Midterm presentation	
8	Simulate production experience 1	Study standard time and cost
9	Simulate production experience 2	KAIZEN process
10	Simulate production experience 3	
11	Sell, cost and environmental acuities 1	
12	Sell, cost and environmental acuities 2	
13	Sell, cost and environmental acuities 3	
14	Sell, cost and environmental acuities 4	
15	Final presentation	

Weeks 4 and 5:

As for the market and competitor analysis, the learner is given Information sheet and examines it in relation to how it affects the company. This sheet is composed of macro information and the information on market conditions and the competitor environment. The macro environmental information includes the political environment, the economy, and social and technological factors. Market information consists of the purchase environment, the production situation, and sales factors. Competitor information relates to market share, profitability, and the trends in competitor company factors.

Week 6:

Using week 5 analysis, the team decides the amount of the production and the goal profit rate. After deciding on the production figures, they determine the sales data set and calculate the profit rate with the assistance from the support system.

Week 7:

To clarify what they have learned and accomplished, the team presents the results of the first six weeks' exercises. After the learner's presentation, the educator gives some answers example on the study that the learners might have. Every learner is in agreement with the direction of the study. For example, the educator might point out two things: lower sales and higher cost. Subsequently, this educator challenges them to achieve "more than 4 % of the rate of ordinary profit divided by capital" and "more than 2.8 % of the rate of ordinary profit divided by sales."

Weeks 8–10:

The learner actually produces the miniature automobile in the standard time and work set by the educator. After production, they discuss "unreasonable, unfruitful, and uneven production" and propose the KAIZEN plan. The support system provides information on the KAIZEN plan and reduction of labor cost. In week 10, the learner must make a presentation on the KAIZEN plan.

Weeks 11–14:

During this period, the learners consider two things: lowering the cost of materials and increasing sales. Lowering the material cost is studied according to the figures on the overall automobile. These costs are subject to an environmental index; however, recently in manufacturing, a product's environmental impact must be considered. This program, therefore, presents the indicators of the automobile's effect on the environment. The more environmentally friendly the car, the higher the environmental indicators. Keeping this in mind, the learner understands the relationship between the cost and the environment.

In this program, more than 80 indicators are required for an automobile to be certified as a low-fuel consumption and low-smog emission vehicle. The student needs these indicators to sell the car they develop. Moreover, they need to repeatedly meet the standard of workmanship, the production time, and the KAIZEN until the goal profit rate is accomplished.

Week 15:

The learners make their final presentation.

3 Details of the Support System

In this program, the learner can visually check the support system's influence when they propose the KAIZEN and accordingly change the automobile materials. Figure 2 shows the system flow.

The answers from the educator are "more than 4 % of the rate of the ordinary profit divided by the entire capital amount" and "more than 2.8 % of the rate of the ordinary profit divided by the sales." This problem has to be solved in two ways: "lower the cost" and "increase the sales."

In the "lower the cost" criteria, using the scroll bar of the material cost and the labor cost adjusts the "more than 2.8 % of the rate of the ordinary profit divided by the sales" through the support system (1). The former method does not achieve the "more than 4 % of the rate of the ordinary profit divided by the entire capital amount." Through the support system (2), the scroll bar, cash deposits, accounts receivable, the money due from accounts, products, semi-finished products, materials in process, and raw materials, all have to be reduced to meet the objectives (Fig. 3).

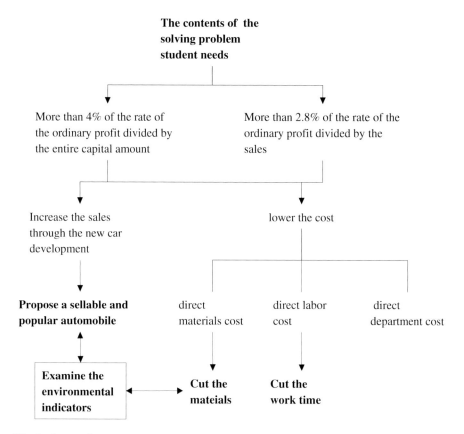

Fig. 2 System flow

In addition, the support system alone (2) does not improve the rate of the ordinary profit divided by the all capital. After scrolling the bar (2), the support system (3) has to be displayed. From screen (3), the learner recognizes that, above all else, they must make an effort to increase sales. In the "saving labor costs sheet," the support system shows the time and width required to cut this cost. On the basis of this sheet, the learners discuss work improvements and attempt to achieve labor cost savings. The "saving materials cost sheet" shows the structure and price of parts. It needs to reach the width reduction from the support system (1).

Learners also have environmental indicators. As a result, in developing the new automobile they need to achieve an environmental index above 80 in addition to material cost savings and higher sales at same time.

Through using the support system, the learner can understand their IE knowledge.

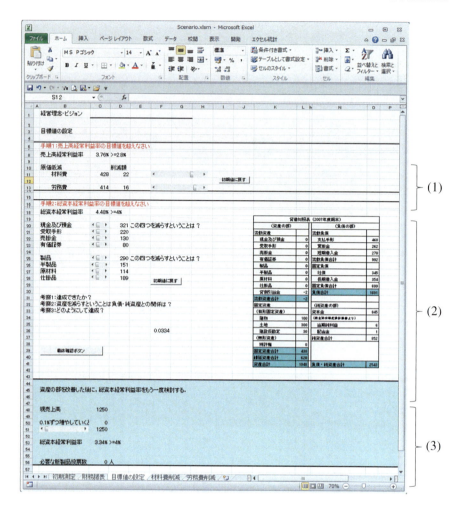

Fig. 3 Support system

4 Conclusion and Future Prospect

This paper presents a proposal for the educational program that addresses the IE study not only for the major's. Especially, IE's basic studies, KAIZEN, standard time and work, cost management, environmental activity and new product development, are able to exercise by their own hand. Through this program, the learners are is to understand not only IE concept, but also the relationship between new IE learning and prior knowledge which they have already.

We implemented this program with university students, a total of 23 students in 4 teams. A questionnaire survey was conducted after the program had ended to evaluate learning effectiveness. From the free comment, "I acquired the knowledge about IE and the relationship with the finance," "the importance of team work and discussion," etc. We believe that this program proposed in this paper is useful to study IE's elements and knowledge the learners studied before.

Future issues to be addressed include (1) implementation of the program for large numbers of learners, (2) the addition of the case, not only automobile industry but also the other manufacturing industries, and (3) enhancement of the support system.

Acknowledgments This research was supported by Grant-in-Aid for Young Scientists (B), (23710179) in Japan.

References

Alavi M, Gallupe R (2003) Using information technology in learning: case studies in business and management education programs. Acad Manage Learn Educ 2(2):139–153

Gorman G, Hanlon D, King W (1997) Some research perspectives on entrepreneurship education, enterprise education and education for small business management: a ten-year literature review. Int Small Bus J 15(3):56–77. doi:10.1177/0266242697153004

Graham R, Gray C (1969) Business games handbook. AMA, New York

Lambert D, Cooper M, Pagh J (1990) Supply chain management: implementation issues and research opportunities. Int J of Logs Manag 9(2):1–20

Mallick D, Chaudhury A (2000) Technology management education in MBA programs: a comparative study of knowledge and skill requirements. J Eng Technol Manage 17(2):153–173

Nonoyama T, Yanagida Y, Takahashi T, Narikawa T (2002) Business game practice (Japanese). Pearson Education Japan, Tokyo

Riis J (1995) Simulation games and learning in production management. Springer, Berlin

Tseng C (2009) Development of business game simulator for supporting collaborative problem-based learning. Paper presented at ICALT 2009, Riga Technical University, Latvia, 15–17 July 2009

An Automatic Image Enhancement Framework for Industrial Products Inspection

Chien-Cheng Chu, Chien-Chih Wang and Bernard C. Jiang

Abstract Image enhancement methods play a key role in image preprocessing. In practical, to obtain product characteristics, image enhancement methods are usually selected by trial-and-error or by experience. In this chapter, we proposed a novel procedure to automatically select image enhancement procedures by using singular value decomposition to extract features of an image. Forty-five industrial product images from literature and local companies were used in the experiment. The results showed that the contrast values had no significant differences with the literature. The study results implied that the system could automatically applied and effectively improve the image quality.

Keywords Visual inspection · Singular value decomposition · Feature database

1 Introduction

In practical applications, to increase the efficiency of the machine vision inspections, the inspection must use image enhancement to increase the contrast and decrease noise before detecting the objects. Most image enhancement methods are

---

C.-C. Chu
Department of Automation, Jabil Green Point, Taiwan, China
e-mail: chiencheng.chu@gmail.com

C.-C. Wang (✉)
Department of Industrial Engineering and Management, Ming Chi University of Technology, New Taipei city, Taiwan, Republic of China
e-mail: ieccwang@mail.mcut.edu.tw

B. C. Jiang
Department of Industrial Engineering and Management, Yuan Ze University, Jungli, Taiwan, Republic of China
e-mail: iebjiang@saturn.yzu.edu.tw

Y.-K. Lin et al. (eds.), *Proceedings of the Institute of Industrial Engineers Asian Conference 2013*, DOI: 10.1007/978-981-4451-98-7_9,
© Springer Science+Business Media Singapore 2013

independent and based on supervised techniques. For example, Wang et al. (2011) proposed multivariate analysis to automatically build an image enhancement model. However, the method must know the corresponding enhancement method of the training images. Therefore, a non-supervised image enhancement mechanism is needed in practice.

Image enhancement technology usually relies on differences in the grey levels and their structures to describe the image features. Yeh et al. (2003) proposed a histogram equalization scheme to enhance multi-layer ceramic chip capacitors (MLCC) surface defect features and a median filter was then used to denoise and maintain image sharpness. Ko (2002) used a fast Fourier transform and selected different kinds of masks to remove high-frequency data and thereby enhance fingerprint images. Kang and Hong (2000) utilized a wavelet transform to enhance medical ultrasound images and remove noise. Tsai and Lin (2002) used a Gaussian algorithm to smooth the images and make traditional coefficient correlation more effective. Szydłowski and Powałka (2012) used image normalization to overcome chatter images taken under different illumination conditions. Withayachumnankul et al. (2013) revealed the green channel of the original image. To assist the interpretation, the intensity of the image is inverted and then normalized to 0–255 in order to increase the contrast. The experiments show that the algorithm reveals cracks with high accuracy and high sensitivity.

From the above discussion, if fast image enhancement techniques can be proposed and put into practical application, these can shorten the time needed for detection and fill the demand of manufacturers for faster inspection. In this chapter, the extraction of image features through *singular value decomposition* (SVD) was proposed to use of these features as a basis for the construction of the database. The proposed procedure can quickly to automatically select the appropriate image enhancement methods.

2 Methodology

Assume that an image matrix size is $m \times n$. The matrix can find an SVD to make $A = U\Sigma V^T$, where U is an $m \times m$ unitary matrix, Σ is an $m \times n$ rectangular diagonal matrix with non-negative real numbers on the diagonal and V is an $n \times n$ unitary matrix. The diagonal entries $\Sigma_{n,n}$ of Σ are known as the singular values of A. Let $\Sigma_{n,n}$ return the value of location (n, n) in the matrix. The size of the matrix is $m \times m$ and $m > n$. In this study, $U\Sigma_{1,1}V^T$ was used to determine the basic structure for the image's features required for classification. Each image obtained by $U\Sigma_{1,1}V^T$ can obtain one feature matrix that most closely represents the original image.

After the SVD extraction, the inspected image may be divided according to two conditions. If the database already has a similar feature set, the classified image should be set to the feature's group. If the database system cannot find similar images using the SVD feature, the process sets the image as a new group. Finally, we obtain the highest matching process, and save it in the database for future use. The SSIM index calculation formula

$$SSIM(x, y) = \frac{\left(2\mu_x\mu_y + C_1\right)\left(2\sigma_{xy} + C_2\right)}{\left(\mu_x^2 + \mu_y^2 + C_1\right)\left(\sigma_x^2 + \sigma_y^2 + C_2\right)} \tag{1}$$

where μ_x is the mean of x, μ_y is the mean of y, σ_x is the standard deviation of x, σ_y is the standard deviation of y. C_1 used in order to avoid obtaining $\mu_x^2 + \mu_y^2$ close to or equal to zero. Let $C_1 = (K_1 L)^2$ and L be a dynamic range that is assumed by the image's pixel range. For example, for an 8-bit image, the dynamic range is from 0 to 255. At the same time, $K_1 \ll 1$ is a very small value, and this definition also applies to the contrast comparison function and structure function. Let $C_2 = (K_2 L)^2$, $K_2 \ll 1$, this design is also used in order to prevent the denominator from approaching or becoming equal to zero.

The contrast value and entropy were used to estimate the results and determine the proposed method compared to the method in literature. The contrast value was a method that uses frequency to evaluate the quality of an image. The definition of the contrast value is as follows

$$\sum_{f=0}^{Max} \frac{\left(f - \mu_f\right)^2 n_f}{M \times N} \tag{2}$$

where, $M \times N$ is the image size, μ_f is the mean of grey levels equal to f, n_f is the sum of grey levels equal to f.

Next, the enhancement process stresses the foreground and makes the background inconspicuous. In other words, it aims to filter noise and cause the useless information to disappear, thus improving the thresholding. Entropy can measure the degree of data, and higher entropy means that more information in a group of data may exist in the foreground. Thus, for the enhancement process, higher entropy is better. The definition of entropy is as follows

$$\sum_{f=0}^{L} \left(p(f) \times \ln p(f)\right) \tag{3}$$

where, $p(f) = n_f / (M \times N)$

3 Experiment Analysis

The 45 test images (Fig. 1) obtained from literature (Chen 2003; Tsai et al. 2003; Yeh et al. 2003; Chen 2005; Lin and Ho 2005) were used to set up the database and to create an automatic enhancement method. These are images that are mainly

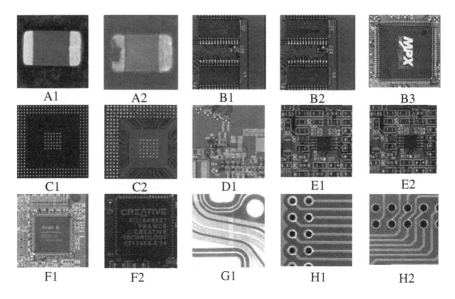

Fig. 1 The part of experiment images

obtained from electronic products, including capacitors, integrated circuit (IC), ball grid array (BGA), printed circuit board (PCB) and flexible printed circuit (FPC). Also, the research used common image enhancement methods to establish the automatic selection model for image enhancement. They included the uniform filter, histogram equalization, Gaussian filter, median filter and the sobel operator.

Tables 1 and 2 shows the analysis results of all images in the Fig. 1 and their calculated contrast values and entropy. From Tables 1 and 2, we can deduce that all of the entropy index results for the proposed methods are better than those in literature. The average entropy of the image relative to previous research increased

Table 1 Average contrast and entropy results using the proposed method

Image type	Proposed method					
	Contrast			Entropy		
	Before	After	Difference (%)	Before	After	Difference (%)
A	44.8453	45.0917	0.54	6.4288	6.4531	0.37
B	50.7090	45.1997	−10.86	6.7541	6.7390	−0.22
C	46.0046	42.5420	−7.53	5.9323	5.9951	1.06
D	36.6257	32.7884	−10.48	6.5143	6.3502	−2.52
E	58.3931	53.8714	−7.74	7.2699	7.3040	0.47
F	35.2855	30.8388	−12.60	6.4283	6.2788	−2.33
G	40.7573	37.9707	−0.683	5.4003	5.5108	2.05
H	52.9750	51.4671	−2.84	6.8421	6.9394	1.42

An Automatic Image Enhancement Framework for Industrial Products Inspection   77

Table 2 Average contrast and entropy results using the literature method

| Image type | Literature's method | | | | | |
| | Contrast | | | Entropy | | |
	Before	After	Difference (%)	Before	After	Difference (%)
A	44.8453	74.8553	66.92	6.4289	5.6038	−12.83
B	50.7090	45.1997	−10.86	6.7541	6.7390	−0.22
C	46.0046	42.5420	−7.53	5.9323	5.9951	1.06
D	36.4902	0.2274	−99.38	6.5143	0.3059	−95.30
E	58.3931	53.8714	−7.74	7.2699	7.3040	0.47
F	35.2855	0.2234	−99.37	6.4283	0.2979	−95.37
G	40.7573	71.5022	75.43	5.4003	4.4139	−18.27
H	52.9750	48.5487	−8.35	6.8421	6.8770	0.51

to 17.54 %. Using a paired t test, we compared the different significances for the proposed method and that in literature.

For contrast value, the 95 % CI for the mean difference was $(-19.67, 20.37)$ and a p value $= 0.968$. This result shows that there are no significant differences in the contrast value between the proposed method and that in literature. For entropy index, the proposed method of increase is greater than the literature methods.

4 Conclusions

This research proposed an SVD-based framework to choose suitable enhancement methods that can quickly and effectively solve practical inspection problems. When compared with the methods in literature, this method provides a better unsupervised learning mechanism. As opposed to supervised methods, this method does not require pre-learning or specific characteristics. Therefore, this application is more flexible, but requires more technical knowledge to realize further improvements.

Acknowledgments The National Science Council of Taiwan, Project Number NSC 100-2628-E-131-001, supported this research.

References

Chen CC (2003) The application of wavelet theory to the defect detection in printed circuit board. Thesis (Master's degree). National Cheng Kung University, Tainan, Taiwan

Chen CS (2005) Applications of near infrared image process technique in the alignment of GaAs wafer. Thesis (Master's degree). Yuan Ze University, Taiwan

Kang SC, Hong SH (2000) Design and implementation of denoising filter for echocardiographic images based on wavelet method. The conference on microtechnologies in medicine and biology, Lyon, pp 80–83

Ko T (2002) Fingerprint enhancement by spectral analysis techniques. The proceedings of the 31st applied imagery pattern recognition workshop, pp 133–139

Lin HD, Ho DC (2005) A new detection method based on discrete cosine transform for pinhole defects applied to computer vision systems. The 5th international conference on intelligent processing and manufacturing of materials, Monterey, California, USA

Szydłowski M, Powałka B (2012) Chatter detection algorithm based on machine vision. Int J Adv Manuf Tech 62:517–528

Tsai DM, Lin BT (2002) Defect detection of gold-plated surfaces on PCBs using entropy measures. Int J Adv Manuf Tech 20:420–428

Tsai DM, Lin CT, Chen JF (2003) The evaluation of normalized cross correlations for defect detection. Pattern Recogn Lett 24:2525–2535

Wang CC, Jiang BC, Chou YS, Chu CC (2011) Multivariate analysis-based image enhancement model for machine vision inspection. Int J Prod Res 49:2999–3021

Withayachumnankul W, Kunakornvong P, Asavathongkul C, Sooraksa P (2013) Rapid detection of hairline cracks on the surface of piezoelectric ceramics. Int J Adv Manuf Tech 64:1275–1283

Yeh CH, Shen TC, Wu FC (2003) A case study: passive component inspection using a 1D wavelet transform. Int J Adv Manuf Tech 22:899–910

Ant Colony Optimization Algorithms for Unrelated Parallel Machine Scheduling with Controllable Processing Times and Eligibility Constraints

Chinyao Low, Rong-Kwei Li and Guan-He Wu

Abstract In this paper, we consider the problem of scheduling jobs on unrelated parallel machines with eligibility constraints, where job-processing times are controllable through the allocation of a nonrenewable common resource, and can be modeled by a linear resource consumption function. The objective is to assign the jobs to the machines and to allocate the resource so that the makespan is minimized. We provide an exact formulation of the addressed problem as an integer programming model. As the problem has been proven to be NP-hard even for the fixed job-processing times, two ant colony optimization (ACO) algorithms based on distinct procedures, respectively, are also presented and analyzed. Numerical results show that both the proposed algorithms are capable of solving large-sized problems within reasonable computational time and accuracy.

Keywords Scheduling · Unrelated parallel machines · Eligibility constraints · Resource allocation · Ant colony optimization

1 Introduction

We consider the problem of scheduling jobs on unrelated parallel machines with eligibility constraints to minimize the makespan (the maximum completion time). The processing time of a job is dependent on both the machine assigned and the

---

C. Low (✉)
Institute of Industrial Engineering and Management, National Yunlin University of Science and Technology, Yunlin, Taiwan, Republic of China
e-mail: lowcy@yuntech.edu.tw

R.-K. Li · G.-H. Wu
Department of Industrial Engineering and Management, National Chiao Tung University, Hsinchu, Taiwan, Republic of China
e-mail: ghwu.iem99g@nctu.edu.tw

Y.-K. Lin et al. (eds.), *Proceedings of the Institute of Industrial Engineers Asian Conference 2013*, DOI: 10.1007/978-981-4451-98-7_10,
© Springer Science+Business Media Singapore 2013

amount of resource allocated. If job-processing times are fixed, the problem has been shown to be NP-Hard in the strong sense. Furthermore, Lenstra et al. (1990) showed that no approximation algorithm can achieve a worst-case ratio smaller than 3/2, unless $P = NP$.

In most classical machine scheduling models, job-processing times are generally treated as constant values and known in advance. However, in various realistic situations, jobs may also require, apart from machines, certain additional limited resources (e.g., manpower, electricity, catalyzer) for their performing, and the processing times can be considerably affected by consumption of such resources. In these situations, both the job scheduling and the distribution of limited resources to individual operations should be taken into account and coordinated carefully so as to optimize system performance.

The majority of works on scheduling models with controllable job-processing times have been presented for various single machine problems, such as those described in the review provided by Shabtay and Steiner (2007). Only a few studies have dealt with multi-processor systems. Jozefowska et al. (2002) presented a tabu search algorithm to deal with the identical parallel machine makespan problem under given resource constraints. Jansen and Mastrolilli (2004) proposed approximation algorithms for the problem of scheduling jobs on identical parallel machines, in which the processing times of the jobs were allowed to be compressed in return for compression cost. Shabtay and Kaspi (2006) examined the identical parallel machine scheduling problems with controllable processing times to minimize the makespan and total flow time. Mokhtari et al. (2010) suggested a hybrid discrete differential evolution algorithm combined with variable neighborhood search for a permutation flow shop scheduling problem, in which the processing times can be controlled by consumption of several types of resources. The objective was to minimize both makespan and total cost of resources.

As can be seen, most of the aforementioned studies assumed that each job can be processed on any machine. Nonetheless, this assumption usually deviates from the real-world situation. The presence of eligibility constraints is consistent with many manufacturing environments, for instance, a work center may be composed of different sets of machines with different capability for their performing, and so the jobs must be processed on the machines that can satisfy its process specification. Therefore, we concern on the general problem, where eligibility constraints may exist.

The remainder of this paper is organized as follows. The problem definition and formulation are presented in Sect. 2. In Sect. 3, we give the detailed steps of the proposed ACO algorithms. Computational results are shown in Sect. 4. We demonstrate that our applications of these ACO algorithms are efficient in obtaining near-optimal solutions for industrial sized problems. Finally, some concluding remarks are given in Sect. 5.

2 Problem Description and Formulation

There are n independent non-preemptive jobs have to be processed on m unrelated parallel machines. Each job j is considered available for processing at time zero, requires a single operation and can be processed only on a specific subset M_j of machines. The processing time of job j on machine i ($i \in M_j$, $j = 1, 2, \ldots, n$) is controllable, and can be modeled as a linear decreasing function of the amount of a nonrenewable resource, r_{ij}, used for its processing: $p_{ij}(r_{ij}) = b_{ij} - a_{ij} \cdot r_{ij}$, where b_{ij} and a_{ij} are the normal processing time and resource compression rate, respectively. It is assumed that $r_{ij} \in [\alpha_{ij}, \beta_{ij}]$, where α_{ij} and β_{ij} are given bounds, and the total resource consumption cannot exceed the limited value R, $R \geq \sum_{j=1}^{n} min\{\alpha_{ij} | i \in M_j\}$. Without loss of generality, we assumed that all b_{ij}, a_{ij}, α_{ij}, β_{ij} and R are positive integers. The objective is to determine the optimal assignment of the jobs to the machines and resource allocation, such that the schedule makespan C_{max} is minimized.

Let x_{ij} denote binary variables equal to 1 if job j is processed on machine i and 0 otherwise. Let y_j be the amount of resource used for processing job j. With the above notation the problem under consideration can be formulated as follows:

$$Min \ C_{max} \tag{1}$$

subject to

$$\sum_{i \in M_j} x_{ij} = 1 + m \cdot \sum_{i \notin M_j} x_{ij}, \quad \forall j = 1, 2, \ldots, n, \tag{2}$$

$$\sum_{i=1}^{m} (x_{ij} \cdot \alpha_{ij}) \leq y_j \leq \sum_{i=1}^{m} (x_{ij} \cdot \beta_{ij}), \quad \forall j = 1, 2, \ldots, n, \tag{3}$$

$$\sum_{j=1}^{n} y_j \leq R, \tag{4}$$

$$\sum_{j=1}^{n} x_{ij} \cdot (b_{ij} - a_{ij} \cdot y_j) \leq C_{max}, \quad \forall i = 1, 2, \ldots, m, \tag{5}$$

$$x_{ij} \in \{0, 1\}, \quad \forall i = 1, 2, \ldots, m, \quad \forall j = 1, 2, \ldots, n, \tag{6}$$

$$y_j \in \{0, 1, \ldots, R\}, \quad \forall j = 1, 2, \ldots, n. \tag{7}$$

Constraint (2) ensures that each job is assigned to exactly one of its eligible machines. Constraints (3) and (4) guarantee that the amount of resource allocated to jobs are within the resource limits. Constraint (5) states that the total processing on each machine is a lower bound on the makespan. Constraints (6) and (7) define the value ranges of the variables. This model includes $m + 2n + 1$ constraints, $m \cdot n$ binary variables and n standard variables.

3 Solution Procedures

The ACO algorithm (Dorigo and Gambardella 1997) is essentially a population-based metaheuristic that imitates the cooperative behavior of real ants to tackle combinatorial optimization problems. The basic component of ACO is the stochastic solution construction mechanism. At each iteration, the artificial ants in ACO choose opportunely the next solution component to be appended in the partially constructed ones based on the favorability (pheromone values) and the cost (heuristic information) of adding the component, until a complete solution to the problem is reached. Once all ants have constructed their solutions, some of them are used for performing an update of the pheromone values. This will allow the succeeding ants to generate high quality solutions over time.

In this section, we propose two ACO algorithms based on distinct solution construction mechanisms for solving the addressed problem. The main idea in the first algorithm (ACO-I) is to simultaneously determine the assignment of the jobs to the machines and the amount of resource allotted for their performing during the solution construction process; while the solution procedure of the second algorithm (ACO-II) is composed of two phases, in which an initial schedule is generated with the constant job-processing times, and then the total processing on each machine is compressed in terms of the makespan using a resource allocation mechanism.

3.1 Development of the First Proposed ACO algorithm (ACO-I)

3.1.1 Solution Construction

In ACO-I, the pheromone information is represented by the matrix $[\tau(i,j_r)]$ which describes the favorability of allocating r units of resource to job j on machine i. Let $\mathcal{U}$ be the sets of unscheduled jobs, and $\hat{M}_j$ be the set of the machines that are considered available for processing job j with respect to the eligibility and resource constraints. Given the amount R' of available resource, the upper bound on the consumed resource amount corresponding to machine-job pair (i,j) will be

$$\beta_{ij}^* = min\left\{ \left(R' - \sum_{j' \in \mathcal{U}, j' \neq j} min\{\alpha_{ij'} | i \in \hat{M}_{j'} \} \right), \beta_{ij} \right\} \tag{8}$$

As a result, the probability of allocating r units of resource to job j on machine $i \in \hat{M}_j$ can be computed as follows:

Ant Colony Optimization Algorithms

$$j_r = \begin{cases} \arg \max_{j \in U, r \in [\alpha_{ij}, \beta_{ij}^*]} \left\{ [\tau(i,j_r)]^\alpha \cdot [\eta(i,j_r)]^\beta \cdot [\delta(i,j_r)]^\gamma \right\} & \text{if } q \leq q_0 \\ \hat{J}_r & \text{otherwise} \end{cases}. \quad (9)$$

where the random variable $\hat{J}_r$ is selected according to the probability distribution given by

$$p(i,j_r) = \begin{cases} \dfrac{[\tau(i,j_r)]^\alpha \cdot [\eta(i,j_r)]^\beta \cdot [\delta(i,j_r)]^\gamma}{\sum_{k \in U, s \in [\alpha_{ik}, \beta_{ik}^*]} [\tau(i,k_s)]^\alpha \cdot [\eta(i,k_s)]^\beta \cdot [\delta(i,k_s)]^\gamma} & \text{if } j \in U, r \in [\alpha_{ij}, \beta_{ij}^*] \\ 0 & \text{otherwise} \end{cases}. \quad (10)$$

Of which, the heuristic information $\eta(i,j_r)$ and $\delta(i,j_r)$ are defined as

$$\eta(i,j_r) = \frac{1}{b_{ij} - a_{ij} \cdot r} \quad (11)$$

and

$$\delta(i,j_r) = \frac{R' - r}{\sum_{k \in U, k \neq j} \min \left\{ \alpha_{ik} \mid i \in \hat{M}_k \right\}} \quad (12)$$

Besides, α, β, and γ are, respectively, the parameters representing the relative influence of pheromone and heuristic information, q is a random number uniformly distributed in $[0, 1]$ and q_0 is a parameter that determines the exploitation and exploration properties of the algorithm.

3.1.2 Local Pheromone Update

After an ant has constructed a complete solution, the local pheromone update procedure is implemented to evaporate the pheromone values in each link (i,j_r) selected by the ant in order to avoid premature convergence. The update rule is given by

$$\tau(i,j_r) \leftarrow (1 - \theta) \cdot \tau(i,j_r) + \theta \cdot \tau_0 \quad (13)$$

where $0 < \theta < 1$ is the evaporation rate and τ_0 is the initial pheromone value.

3.1.3 Global Pheromone Update

Once all of the k ants have constructed their solutions (i.e., an iteration), the iteration-best solution s^{ib} and the best-so-far solution s^{bs} will update the pheromone values to enforce the exploitation of search. This update rule is defined as

$$\tau(i,j_r) \leftarrow \tau(i,j_r) + \rho \cdot \left\{ w^{ib} \cdot C^{ib}(i,j_r) + w^{bs} \cdot C^{bs}(i,j_r) - \tau(i,j_r) \right\} \quad (14)$$

Of which, the parameter ρ is evaporation rate $(0 < \rho \leq 1)$, $C^*(i,j_r)$ is set to 1 if link $(i,j_r) \in s^*$, and 0 otherwise. w^{ib} $(0 \leq w^{ib} \leq 1)$ and w^{bs} are the parameters $(w^{ib} + w^{bs} = 1)$ representing the relative influence of the iteration-best solution and the best-so-far solution. Note that the update rule presented here is based on the Hypercube framework (Blum and Dorigo 2004), which automatically rescales the pheromone values and thereby bound them to the interval [0, 1].

3.2 Description of the Second Proposed ACO Algorithm (ACO-II)

In contrast to the ACO-I, the solution procedure of the ACO-II consists of two stages: *assignment* then *allocation*. In the assignment stage, the job-processing times are regarded as constant (i.e., $\hat{p}_{ij} = b_{ij} - a_{ij} \cdot \alpha_{ij} \ \forall i \in M_j, j = 1, 2, \ldots, n$), and the pheromone information is represented as an $m \times n$ matrix where each element $\tau(i,j)$ of the matrix describes the favorability of assigning job j to machine i. Accordingly, the probability of choosing job j to be appended on the machine $i \in \hat{M}_j$ can be computed as follows:

$$j = \begin{cases} argmax_{j \in U}\left\{ [\tau(i,j)]^{\alpha} \cdot [\eta(i,j)]^{\beta} \right\} & \text{if } q \leq q_0 \\ \hat{J} & \text{otherwise} \end{cases} \tag{15}$$

where $\hat{J}$ is a random variable selected according to the probability distribution given by

$$p(i,j) = \begin{cases} \dfrac{[\tau(i,j)]^{\alpha} \cdot [\eta(i,j)]^{\beta}}{\sum_{k \in U}[\tau(i,k)]^{\alpha} \cdot [\eta(i,k)]^{\beta}} & \text{if } j \in U \\ 0 & \text{otherwise} \end{cases} \tag{16}$$

Of which, the heuristic information $\eta(i,j)$ is defined as

$$\eta(i,j) = \frac{1}{\hat{p}_{ij} = b_{ij} - a_{ij} \cdot \alpha_{ij}} \tag{17}$$

After the jobs on each machine have been specified, the update of the pheromone values is carried out according to

$$\tau(i,j) \leftarrow (1 - \theta) \cdot \tau(i,j) + \theta \cdot \tau_0 \tag{18}$$

Then, in the second stage we apply the following algorithm to determine the optimal resource allocation for the machine-job assignment generated. This resource allocation procedure was originally proposed by Su and Lien (2009) for identical parallel machine problem. Here, we generalize with slide modification the procedure to the case of unrelated machines. Define M_a as the set of machines with maximal completion time. Let $\{i,j\}$ be the machine-job pair determined in

Ant Colony Optimization Algorithms

the assignment stage and $\hat{R}$ be the amount of remaining available resource. A detail description of the algorithm is presented below.

Step 0 Set $r_{\{i,j\}} = \alpha_{\{i,j\}}$, $i \in M_j$, $j = 1, 2, \ldots, n$

Step 1 If $\hat{R} \geq |M_a|$, then go to Step 2; Otherwise, go to Step 5

Step 2 Determine the job j such that $\beta_{\{i,j\}} = 0$, then set $a_{\{i,j\}} = 0$. Identify job t on machine $k \in M_a$ with the largest $a_{\{k,t\}}$. If $a_{\{k,t\}} \neq 0$, then go to Step 3; Otherwise, go to Step 5

Step 3 Update $r_{\{k,t\}} = r_{\{k,t\}} + 1$, $\beta_{\{k,t\}} = \beta_{\{k,t\}} - 1$, $\hat{R} = \hat{R} - 1$ and $M_a = \{M_a\}\backslash\{k\}$, then go to Step 4

Step 4 If $\hat{R} \neq 0$ then return to Step 1, else go to Step 5

Step 5 Calculate the resource allocation $r_{\{i,j\}}$ and terminate the procedure.

Finally, the resulted iteration-best solution and the best-so-far solution are used for performing an update of the pheromone values. This update rule is expressed by

$$\tau(i,j) \leftarrow \tau(i,j) + \rho \cdot \left\{ w^{ib} \cdot C^{ib}(i,j) + w^{bs} \cdot C^{bs}(i,j) - \tau(i,j) \right\} \qquad (19)$$

4 Computational Results

The ACO algorithms shown in the previous section were coded in C++ and implemented on a 2.80 GHz AMD Athlon II CPU personal computer. The test problems we considered include the problems with 2, 3, 5, 10 machines and 3, 5, 10, 20, 30, 50 jobs. Normal processing times of jobs were generated from a discrete uniform distribution in the interval [1, 100]. The amount of available resource, the resource compression rate, and the bounds of consumed resource amount associated with each machine-job pair followed the discrete uniform distributions defined by the interval that satisfy the addressed resource constraints. For each problem size ($m \times n$), 3 test problems were generated and the algorithms were executed 10 times for each problem. Let $C_{max}(H)$ be the value of the objective function obtained by algorithm H, and thus the optimality gap can be defined as $G_{opt}(H) = C_{max}(H) - \text{optimum}/\text{optimum}$. Note that the optimal solution was obtained by solving the presented integer programming model with LINGO 11.0. For cases when an optimal solution cannot be obtained within a time limit 10,800 s (3 h), the optimality gap was calculated as $G_{opt}(H) = C_{max}(H) - C_{max}(ACO - I)/C_{max}(ACO - I)$. The computational results for all test problems are given in Table 1.

As the results show, both the algorithms performed well based on the averages of optimality gap and solution time for small sized problems. The algorithms attained optimal solutions in 9 problems out of 27 tested problems. Note that for all the small-sized problems evaluated, the solutions are the same in 10 runs of each

Table 1 Evaluation results of the proposed ACO algorithms

$m \times n$	Lingo		ACO-I		ACO-II	
	Avg. sol. time (s)	Avg. gap (%)	Avg. sol. time (s)	Avg. gap (%)	Avg. sol. time (s)	Avg. gap (%)
2×3	11.33	0.00	5.26	0.00	11.60	0.00
3×3	18.33	0.00	5.33	0.00	10.40	0.00
2×5	962.00	0.00	7.76	0.00	11.53	0.00
3×5	2,750.67	0.00	6.56	0.39	11.26	0.17
2×10	e. t. l.	–	18.23	–	26.70	2.89^{ψ}
5×20	e. t. l.	–	44.53	–	49.58	5.41^{ψ}
5×30	e. t. l.	–	117.20	–	93.13	4.36^{ψ}
5×50	e. t. l.	–	232.36	–	141.16	3.41^{ψ}
10×50	e. t. l.	–	228.40	–	121.13	5.15^{ψ}

e. t. l. = Exceed time limit; $^{\psi}$ gap calculated based on $C_{max}(ACO - I)$

test problem. This signified the robustness of the suggested algorithms. The results also evinced that the ACO-I generated similar, actually better on average, solutions in a short computation time as compared to the ACO-II. Additionally, the superiority of the ACO-I got more significant with the increase of the problem size. In summary, the suggested ACO algorithms can effectively solve the addressed problem to a certain scale in terms of the tradeoff between solution quality and computation time.

5 Conclusions

In this work, we introduce the problem of minimizing the makespan on a set of unrelated parallel machines with controllable processing times and eligibility constraints. This problem has been formulated as an integer programming model. Due to its computational complexity, two ACO-based heuristics are also presented and analyzed. Numerical results for problems with up to 50 jobs demonstrate that both the ACO algorithms can obtain the optimal or near optimal solutions for small and medium sized problems in a very short computation time, and the first proposed algorithm (ACO-I) outperforms the second proposed algorithm (ACO-II) in the case of relatively large-sized problems. Further research might extend our study to multi-stage scheduling problems, and consider other performance measures such as number of tardy jobs, maximum tardiness, and total weighted completion time.

Acknowledgments This work is partially supported by the National Science Council under Grant NSC-101-2221-E-224-070. The authors also gratefully acknowledge the helpful comments and suggestions of the reviewers, which have improved the presentation.

References

Blum C, Dorigo M (2004) The hyper-cube framework for ant colony optimization. IEEE Trans Syst Man Cybern 34(2):1161–1172

Dorigo M, Gambardella LM (1997) Ant colony system: a cooperative learning approach to the traveling salesman problem. IEEE Trans Evol Comput 1(1):53–66

Jansen K, Mastrolilli M (2004) Approximation schemes for parallel machine scheduling problems with controllable processing times. Comput Oper Res 31:1565–1581

Jozefowska J, Mika M, Rozycki R (2002) A heuristic approach to allocating the continuous resource in discrete continuous scheduling problems to minimize the makespan. J Sched 5:487–499

Lenstra JK, Shmoys DB, Tardos E (1990) Approximation algorithms for scheduling unrelated parallel machines. Math Program 46:259–271

Mokhtari H, Abadi INK, Cheraghalikhani A (2010) A multi-objective flow shop scheduling with resource-dependent processing times: trade-off between makespan and cost of resources. Int J Prod Res 49(19):5851–5875

Shabtay D, Kaspi (2006) Parallel machine scheduling with a convex resource consumption function. Eur J Oper Res 173:92–107

Shabtay D, Steiner G (2007) A survey of scheduling with controllable processing times. Discrete Appl Math 155:1643–1666

Su LH, Lien CY (2009) Scheduling parallel machines with resource-dependent processing times. Int J Prod Econ 117(2):256–266

General Model for Cross-Docking Distribution Planning Problem with Time Window Constraints

Parida Jewpanya and Voratas Kachitvichyanukul

Abstract The research studies a cross-docking distribution planning problem that consists of manufacturers, cross-docking centers and customers. It is focused on how to distribute and receive products within time interval restrictions of each node. This means that the manufacturer has specific time intervals for releasing products to be shipped to destinations, the cross-docking centers have time intervals to receive products from manufacturers and to release them to customers, and the customers also have their time intervals for receiving the products. A mixed integer programming model is formulated to deal with this time interval restrictions by including time window constraints at each level in the network. Also, the multiple types of products and consolidation of customer orders are considered. The objective function is to minimize the total cost which combines the transportation cost and inventory cost. A LINGO program was improved from Jewpanya and Kachitvichyanukul (General model of Cross-docking distribution planning problem. In: Proceedings of the 7th international congress on logistics and SCM systems, Seoul, 2012) to efficiently handle the problem with time window constraints. Some example problems are solved to demonstrate the optimal distribution plan of the cross-docking distribution planning problem under the limitation of each time window.

Keywords Cross-docking distribution planning problem · Cross-docking · Cross-docking center · Time window

---

P. Jewpanya (✉)
Department of Industrial Engineering, Rajamangala university of Technology Lanna,
Tak 63000, Thailand
e-mail: parida.jewpanya@gmail.com

V. Kachitvichyanukul
Industrial and Manufacturing Engineering, Asian Institute of Technology (AIT),
Pathumtani 12120, Thailand
e-mail: voratas@ait.ac.th

Y.-K. Lin et al. (eds.), *Proceedings of the Institute of Industrial Engineers Asian Conference 2013*, DOI: 10.1007/978-981-4451-98-7_11,
© Springer Science+Business Media Singapore 2013

1 Introduction

Cross-docking is one of the more important distribution techniques in the supply chain network. It is used to eliminate the storage cost by reducing the amount of products stored in the cross-docking station or cross-docking center by delivering to destinations as soon as possible (Santos et al. 2013). The cross-docking concept may also be used to improve the customer satisfaction level (Dondo and Cerdá 2013). Some example indicators for customer satisfaction level are on-time delivery and percentage of demand served. These mean that the quantity of goods must meet requirements within the customer expected time. Therefore, in a cross-docking distribution system, products are transported by manufacturers or suppliers to the cross-docking center where multiple products will be consolidated by customers order. After that, the delivery process will begin by sending those products to each customer. Therefore, typical operations at the cross-docking center include: (1) products arrived at the cross-docking center are unloaded at the receiving dock; (2) products are split into smaller lots and consolidated with other products according to customer orders and destinations; and (3) Products are moved to suitable location on the shipping dock and are loaded into truck for transporting to customer (Arabani et al. 2009).

As the function of cross-docking network is to reduce the inventory and to satisfy the customer need, not only that the quantities of products must meet the requirements but they must also be delivered within the time constraints of the network. This is because in real operations, suppliers, distribution center and customers may have specific time periods to ship and receive products. This is normally referred to as time window constraint. When time windows are fixed, the uncertainty of daily operations can be reduced and the service level can be improved. Ma et al. (2011) and Jai-oon (2011) studied the cross-docking distribution problem with time window constraint in the network. The model is for single product and has discrete time window constraint of manufacturers for shipping products to the destinations. Jewpanya and Kachitvichyanukul (2012) extended the model to handle multiple products and continuous time window constraints.

A good cross-docking distribution plan can help the network to reduce the transportation delay, to minimize the freight transfer time from supplier to customer, and to lower relevant costs such as transportation cost and inventory cost. This paper focuses on the problem of how to deliver products from manufacturers to customers to reduce such costs as transportation cost and inventory cost that also takes time constraints into consideration. It extends the cross-docking distribution planning model in Jewpanya and Kachitvichyanukul (2012) to include time windows for all parties: manufacturer time windows, cross-docking time windows and customer time windows as shown schematically in Fig. 1. Moreover, multiple products and consolidation of customer orders are included. A general mathematical model for cross-dock distribution planning is given in the next section.

General Model for Cross-Docking Distribution Planning Problem

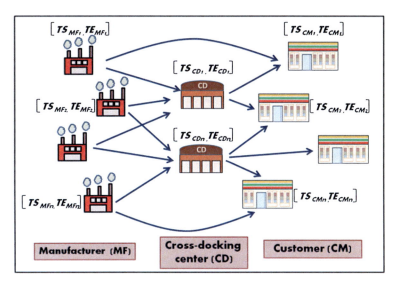

Fig. 1 Cross-docking distribution network with time windows

2 The Mathematical Model

The mathematical model for the cross-docking distribution planning problem with time window is based on the model from Jewpanya and Kachitvichyanukul (2012). This study extends the model to consider the restricted time interval of Manufacturers (i), Cross-docking center (k) and Customers (j) in the cross-docking network and the consolidation of products by customer orders. The constraints are added to the model to cover the general situations often occurred in the distribution network planning.

Indices:

i	Origin node	$i = 1, 2,..., I$
j	Destination node	$j = 1, 2,..., J$
k	Cross-docking center	$k = 1, 2,..., K$
r	Product type	$r = 1, 2,..., R$
t	Time	$T_{min} \leq t \leq T_{max}$

Parameters:

Q	The truck capacity
C'_{ij}	The set up cost of truck from location i to j
C''_{ik}	The set up cost of truck from location i to k
C'''_{kj}	The set up cost of truck from location k to j

(continued)

(continued)

D'_{ij}	The distance from manufacturer i to customer j
D''_{ik}	The distance from manufacturer i to cross-docking center k
D'''_{kj}	The distance from cross-docking center k to customer j
I	The cost of a unit distance
H	The cost of handling a unit product r for a unit time at cross-docking center
G'_{ij}	Total shipping time on route (i, j)
G''_{ik}	Total shipping time on route (i, k)
G'''_{kj}	Total shipping time on route (k, j)
τ_i^M	The starting time points of manufacturer i
φ_i^M	The ending time points of manufacturer i
τ_k^{CD}	The starting time to receive and release products at cross-docking center k
φ_k^{CD}	The ending time to receive and release products at cross-docking center k
τ_j^C	The starting time to receive products for customer j
φ_j^C	The ending time to receive products for customer j
T_{min}	The minimum times
T_{max}	The maximum times
$M_{j,r}$	The demand of product type r for customer j
$S_{i,r}$	The supply capacity of manufacturer i of product type r

Decision variables:

V'_{ijt}	The number of truck used on (i, j) at time t
V''_{ijt}	The number of truck used on (i, k) at time t
V'''_{kjt}	The number of truck used on (k, j) at time t
Z_{rkt}	The number of product r in cross-docking center k at time t
P'_{ijrt}	The quantity of product r delivery from manufacturer i to customer j at time t
P''_{ikrt}	The quantity of product r delivery from manufacturer i to cross-docking center k at time t
P'''_{kjrt}	The quantity of product r delivery from cross-docking center k to customer j at time t
τ'_{ij}	The feasible starting time points of manufacturer i that can deliver the product to customer j
φ'_{ij}	The feasible ending time points of manufacturer i that can deliver the product to customer j
τ''_{ik}	The feasible starting time points of manufacturer i that can send products to cross-docking center k
φ''_{ik}	The feasible ending time points of manufacturer i that can send products to cross-docking center k
τ'''_{kj}	The feasible starting time points of cross-docking center k that can deliver the product to customer j
φ'''_{kj}	The feasible ending time points of cross-docking center k that can deliver the product to customer j

General Model for Cross-Docking Distribution Planning Problem 93

Objective function:

$$\text{Minimize } f = (COST_{\text{transportation}} + COST_{\text{inventory}})$$

$$COST_{\text{Transportation}} = \sum_{i=1}^{I} \sum_{j=1}^{J} \sum_{t=T_{min}}^{T_{max}} [V'_{ijt} \cdot (C'_{ij} + I \cdot D'_{ij})] + \sum_{i=1}^{I} \sum_{k=1}^{K} \sum_{t=T_{min}}^{T_{max}} [V''_{ikt} \cdot (C''_{ik} + I \cdot D''_{ik})]$$

$$+ \sum_{k=1}^{K} \sum_{j=1}^{J} \sum_{t=T_{min}}^{T_{max}} [V'''_{kjt} \cdot (C'''_{kj} + I \cdot D'''_{kj})]$$

$$COST_{\text{inventory}} = \sum_{k=1}^{K} H \sum_{r=1}^{R} \sum_{t=T_{min}}^{T_{max}} Z_{r,k,t}$$

The two main costs included in the decision are transportation cost and inventory cost. The transportation cost consists of three types of links in the network. The first type of links are the links from manufacturers to cross-docking centers (i, k). The second type of links are from manufacturers directly to customers (i, j). And finally, links from cross-docking centers to customers. In each link, the costs that concern about the number of truck are considered; the setup cost of truck and the unit distance cost. The inventory cost is calculated from the change of inventory level in the cross-docking centers at time t, and the cost of handling a unit product for a unit time at cross-docking center.

The objective function is to be minimized subjected to the constraints as described in the next section.

2.1 Time Window Constraint

In real operations, suppliers, distribution centers and customers may have specific time periods to ship and receive products. This is the time window constraint. If it can be managed in proper way, it may lead to high distribution efficiency in the cross-docking network. Therefore, the model considers the time windows for all links between parties in the cross-docking network, customer $[\tau'_{ij}, \varphi'_{ij}]$, manufacturer to cross-docking center $[\tau''_{ik}, \varphi''_{ik}]$, and cross-docking center to customer $[\tau'''_{kj}, \varphi'''_{kj}]$. The important time window constraints are formulated below:

$$\tau'_{ij} = Min\, \{x | x = [(\tau^C_j - G'_{ij}), (\varphi^C_j - G'_{ij})]\, and\, x \in [\tau^M_i, \varphi^M_i]\} \quad \text{for all } i \text{ and } j \quad (1)$$

$$\varphi'_{ij} = Max\, \{x | x = [(\tau^C_j - G'_{ij}), (\varphi^C_j - G'_{ij})]\, and\, x \in [\tau^M_i, \varphi^M_i]\} \quad \text{for all } i \text{ and } j$$
$$(2)$$

$$\tau''_{ik} = Min\, \{x | x = [(\tau^{CD}_k - G''_{ik}), (\varphi^{CD}_k - G''_{ik})]\, and\, x \in [\tau^M_i, \varphi^M_i]\} \quad \text{for all } i \text{ and } k$$
$$(3)$$

$$\varphi_{ik}'' = Max\{x|x = [(\tau_k^{CD} - G_{ik}''), (\varphi_k^{CD} - G_{ik}'')] \text{ and } x \in [\tau_i^M, \varphi_i^M]\} \quad \text{for all } i \text{ and } k \tag{4}$$

$$\tau_{kj}''' = Min\{x|x = [(\tau_j^C - G_{kj}'''), (\varphi_j^C - G_{kj}''')] \text{ and } x \in [\tau_k^{CD}, \varphi_k^{CD}]\} \quad \text{for all } k \text{ and } j \tag{5}$$

$$\varphi_{kj}''' = Max\{x|x = [(\tau_j^C - G_{kj}'''), (\varphi_j^C - G_{kj}''')] \text{ and } x \in [\tau_k^{CD}, \varphi_k^{CD}]\} \quad \text{for all } k \text{ and } j \tag{6}$$

Time window constraints are separated into three groups. The first group include Constraints (1) and (2) describe the feasible starting time and ending time for manufacturers i to deliver the products directly to customers j within the receiving feasible starting time and ending time of customers $j[\tau_j^C, \varphi_j^C]$. The second group, constraints (3) and (4) are the feasible starting time and ending time for manufacturer to release products to cross-docking center. Another group includes Constraints (5) and (6) describe the feasible starting time and ending time for cross-docking center k to deliver the products to customer j within the receiving feasible starting time and ending time point of customers.

The first group of time window constraint, manufacturers can directly send products to customer within the customer expected time. For example, if the release time windows of manufacturers are: $[\tau_i^M, \varphi_i^M] = [\tau_1^M, \varphi_1^M] = [8.00, 8.50]$, $[\tau_2^M, \varphi_2^M] = [8.50, 10.50]$ and the customers has accepted time: $[\tau_j^C, \varphi_j^C] = [\tau_1^C, \varphi_1^C] = [14.00, 17.00]$ and suppose the shipping time from manufacturer to this customer is 7.50. Therefore, the release times of manufacturers that can deliver products to customers within expected time of customers $[\tau_{ij}', \varphi_{ij}']$ are from constraints (1) and (2) and the results are:

$$\tau_{11}' = Min\{x|x = [(14.00 - 7.50), (17.00 - 7.50)] = [6.50, 9.50] \text{ and}$$
$$[6.50, 9.50] \in [8.00, 8.50]\} = 8.00$$

$$\varphi_{11}' = Max\{x|x = [(14.00 - 7.50), (17.00 - 7.50)] = [6.50, 9.50] \text{ and}$$
$$[6.50, 9.50] \in [8.00, 8.50]\} = 8.50$$

Therefore, the release time of manufacturer1 that can send products to customer1 is [8.00, 8.50].

In the same way with the first group, the time window that manufacturer can release products to customer via the cross-docking center in constraints (3) and (4). For instance, the same manufacturer above send products to the cross-docking center that opens at 10.30 and closes at 11.45. The shipping time from manufacturer to this cross-docking center is 1.40. Therefore, the release time for manufacturer to send product to cross-docking center is [8.90, 10.50].

The last group, the time windows for the cross-docking centers, they are the time window that cross-docking center can send products to customer so that it can

General Model for Cross-Docking Distribution Planning Problem 95

reach the customers within expected receiving time window ($[\tau_{kj}''' \varphi_{kj}''']$). For example, the expected time of customer $[\tau_1^c, \varphi_1^c] = [14.00, 17.00]$. And the time the cross-docking can operate is from 10.30 to 11.45. The transportation takes time for 2.90. The release time window of cross-docking center to customer can be calculated following Eqs. (5) and (6).

$$\tau_{11}''' = Min\{x|x = [(14.00 - 2.90), (17.00 - 2.90)] = [11.10, 14.10] \ and$$
$$[11.10, \ 14.10] \in [10.30, 11.45]\} = 11.10$$

$$\varphi_{11}''' = Max\{x|x = [(14.00 - 2.90), (17.00 - 2.90)] = [11.10, 14.10] \ and$$
$$[11.10, \ 14.10] \in [10.30, 11.45]\} = 11.45$$

Thus, the release time window at cross-docking center that can send products to customer is [11.10, 11.45]. This time window model is more general when compare with Jewpanya and Kachitvichyanukul (2012). It can solve the realistic time window problem. This is because the model can deal with every party within the distribution network, i.e., manufacturers, cross-docking centers and customers.

2.2 Supply and Demand Constraint

To satisfy the customer needs, all of requirements must be carefully considered. The quantities of products that are sent to customers must meet the customer demand and they must not exceed the supply of manufacturer.

$$\sum_{i=1}^{I} \sum_{t=\tau_{ij}'}^{\varphi_{ij}'} P_{ijrt}' + \sum_{k=1}^{K} \sum_{t=\tau_{kj}''}^{\varphi_{kj}'''} P_{kjrt}''' = M_{j,r} \quad \text{for all } j \text{ and } r \tag{7}$$

$$\sum_{j=1}^{J} \sum_{t=\tau_{ij}'}^{\varphi_{ij}'} P_{ijrt}' + \sum_{k=1}^{K} \sum_{t=\tau_{ik}''}^{\varphi_{ik}''} P_{ikrt}'' \leq S_{i,r} \quad \text{for all } i \text{ and } r \tag{8}$$

Constraint (7) ensures that the number of products delivered to customer must meet the customer demand while constraint (8) confirms the total quantity of products shipped from manufacturers do not exceed the available supply.

2.3 Constraints of Inventory in Cross-Docking Center and Volume Constraints with Customer Order Consolidation

The cross-docking center can be seen as a warehouse where a reduced quantity of products is stored in a short term stock. Constraint (9) ensures that the flow of product at each cross-docking center at each time is non-negative. Constraint (10) indicates that there are no products in each cross-docking center at the starting time point. Constraint (11) states the number of product in each cross-docking center at time t.

$$Z_{r,k,t} \geq 0 \quad \text{for all } r \text{ and } k, \ \tau_k^{CD} \leq t \leq \varphi_k^{CD} \tag{9}$$

$$Z_{r,k,T_{min-1}} = 0 \quad \text{for all } r \text{ and } k \tag{10}$$

$$Z_{r,k,t} = Z_{r,k,t-1} + \sum_{i=1}^{I} P''_{ikrt} - \sum_{j=1}^{J} P'''_{kjrt} \quad \text{for all } k \text{ and } r, \ \tau_k^{CD} \leq t \leq \varphi_k^{CD} \tag{11}$$

$$\sum_{t=\tau'_{ij}}^{\varphi'_{ij}} \sum_{r=1}^{R} P'_{ijrt} \leq V'_{ijt} \cdot Q \quad \text{for all } i \text{ and } j \tag{12}$$

$$\sum_{t=\tau''_{ik}}^{\varphi''_{ik}} \sum_{r=1}^{R} P''_{ikrt} \leq V''_{ikt} \cdot Q \quad \text{for all } i \text{ and } k \tag{13}$$

$$\sum_{t=\tau'''_{kj}}^{\varphi'''_{kj}} \sum_{r=1}^{R} P'''_{kjrt} \leq V'''_{kjt} \cdot Q \quad \text{for all } k \text{ and } j \tag{14}$$

It is common to consolidate products from different manufacturers by customer order to improve transportation efficiency. Constraints (12), (13) and (14) combine products into a larger volume for each customer before it is allocated to truck which has a fixed capacity and this is done on a product basis with order consolidation.

3 Time Window Constraint for LINGO Model

Time window in cross-docking distribution planning model should consider the specific releasing time of manufacturers, the expected receiving time of customers, also the receiving and shipping time windows of cross-docking center. Time window constraint should be carefully handled because it can help to reduce the

General Model for Cross-Docking Distribution Planning Problem 97

inventories and waiting time of products in the storage site. Furthermore, this may lead to customer satisfaction.

In this study, the time window model in LINGO is extended from Jewpanya and Kachitvichyanukul (2012). The previous time window model deals only with manufacturer and customer time windows. That is not proper for the realistic operation. Therefore, this paper, the time window model of LINGO was designed to cover time windows for all parties follow the constraint in Sect. 2.1. Moreover, it is modified to reduce the running time by improving a part of time window constraint in the LINGO model.

For LINGO model, the time window was considered as an index in the program that was run from a beginning time (T_{min}) to an ending time (T_{max}) in very routes of network consist of, route from i to j and route i to k to j (for all i, j, k). Jewpanya and Kachitvichyanukul (2012) specified the beginning time is the time that first manufacturer can send products to the destinations. And, the ending is the time that the last customer can receive the products. For example, in Table 2, T_{min} is 8.00 and T_{max} is 17.00. From 8.00 to 17.00, there are 900 index combination that must be run in every route (i to j and i to k to j) in order to find the best answer. This make the running time quite long.

Therefore, the development in this study is to reduce the combination of time index in the LINGO model. The beginning time (T_{min}) and ending time (T_{max}) are separated into two groups. The first group is T_{min} and T_{max} of route manufacturers to customers (i to j). The second is for route manufacturers to cross-docking center to customers (i to k to j). In the first group, T_{min} is considered from the beginning that the first manufacturer can send products to the customers. For example, the same problem above (Table 2), the starting time considered (T_{min}) is 8.00. But for T_{max} will consider the time that last manufacturer can release the product to customers within the customer expected time. This can be calculated follows this:

$$T_{max} = \max(\varphi_j^C - G'_{ij}), \quad \text{for all } i, j$$

If the shipping time from manufacturer to customer (G'_{ij}) is showed at Fig. 2, T_{max} is 11.4. By considering this, the LINGO model needs to generate index

Fig. 2 Shipping time for example in Sect. 3

			CM1	CM2	CM3
Distance_(i,j)		MF1	450	408	390
		MF2	408	336	360
Shipping T_(i,j)		MF1	7.50	6.80	6.50
		MF2	6.80	5.60	6.00
		CD1			
Distance_(i,k)		MF1	120		
		MF2	108		
Shipping T_(i,k)		MF1	2.00		
		MF2	1.80		
		CM1	CM2	CM3
Distance_(k,j)	CD1	114	180	150
Shipping (k,j)	CD1	1.90	3.00	2.50

combination from 8.00 to 11.40 that is only 300 combinations in every possible route for link manufacturer to customer. It can be seen that the combination is reduced from 900 to 300 in each i to j.

The second group is for the link manufacturer to cross-docking center to customer (i to k to j). T_{min} is still same with the first group. For the ending time, T_{max} is indicated that last cross-docking center can release the product to customers within the customer expected time are investigated. And the calculation is this:

$$T_{max} = \max(\varphi_j^C - G'_{kj}), \quad \text{for all } k, j$$

Therefore, in the second group, T_{max} is 15.1. The index generated will be 710 combinations. It means that the time in every route in the second group (i to k to j) will run only 710 combinations. That also reduced from 900.

With the improvement, the solution can be seen in the next section, the illustrative examples to explain the operation of cross-docking distribution with time window constraint in Sect. 4.1 and Comparison of the computational time in Sect. 4.2.

4 Computation Results

4.1 The Distribution Planning Results

The cross-docking distribution planning problems are solved using the model in this paper in order to find the optimal distribution plan under the limitation of time window. There are three problems instances used as illustrative examples are

Table 1 Test problem instance

Instance	Test set name	Manufacturers	Cross-docking centers	Customers	No. of products type
1	2MF 1CD 3CM 2P	2	1	3	2
2	2MF 2CD 3CM 3P	2	2	3	3
3	3MF 2CD 4CM 3P	3	2	4	3

Table 2 Specific data for problem instance 1

2MF_1CD_3CM_2P		Manufacturer		Cross-docking center	Customer		
		MF1	MF2	CD1	CM1	CM2	CM3
Time	Starting time	8.00	8.50	10.00	14.00	14.45	12.00
	Ending time	8.50	10.50	13.00	17.00	17.00	16.50
Supply/demand	P1	2,000	200		250	300	500
	P2	600	2,500		300	600	200

General Model for Cross-Docking Distribution Planning Problem

Table 3 Specific data for problem instance 2

2MF_2CD_3CM_3P		Manufacturer		Cross-docking center		Customer		
		MF1	MF2	CD1	CD2	CM1	CM2	CM3
Time	Starting time	7.30	8.50	10.00	10.00	13.50	14.45	13.50
	Ending time	12.00	12.00	13.00	13.50	17.00	17.00	16.50
Supply/demand	P1	2,000	200			250	300	500
	P2	0	2,500			300	600	200
	P3	2,200	0			100	0	250

Table 4 Specific data for problem instance 3

3MF_2CD_4CM_3P		Manufacturer			Cross-docking center		Customer			
		MF1	MF2	MF3	CD1	CD2	CM1	CM2	CM3	CM4
Time	Starting time	7.30	8.50	8.50	7.00	8.00	13.50	14.45	13.50	15.00
	Ending time	8.50	10.50	10.00	15.50	14.20	17.00	17.00	17.00	17.00
Supply/demand	P1	2,000	200	0			250	300	500	0
	P2	0	2,500	600			300	600	200	0
	P3	2,200	1,000	400			100	0	250	650

Table 5 Distribution plan for problem instance 1

From	Depart time	To	Arrival time	Inventory	Inventory time	Amount of	
						P1	P2
MF1	8.30	CD1	10.30	–	–	500	–
MF1	8.50	CD1	10.50	–	–	350	–
MF1	9.39	CD1	11.19	–	–	–	200
MF2	9.65	CD1	11.45	–	–	–	600
MF2	10.30	CD1	12.10	–	–	200	300
CD1	10.30	CM3	12.80	–	–	500	–
–	–	–	–	CD1	10.50–11.44	350	–
CD1	11.19	CM3	13.69	–	–	–	200
CD1	11.45	CM2	14.45	–	–	300	600
–	–	–	–	CD1	11.45–12.09	50	–
CD1	12.11	CM1	14.01	–	–	250	300
Total costs			107,251.30				
Inventory cost			255.50				
Transportation cost			106,995.80				

given in Table 1. The key elements in the examples include number of manufacturers, cross-docking centers, customers and products.

Detailed information of each problem is showed in Tables 2, 3 and 4. Those problems are the cross-docking distribution planning problem that has the limitation of time in each node, manufacturer node, cross-docking center node and customer node. Moreover, it has the different supply and demand as indicated in the Table.

After solving the model using LINGO software, the distribution plan of the problem instance 1 is shown in the Table 5. The plan of distribution start at time 8.30 manufacturer 1 sends 500 units of product P1 to cross-docking center 1 and it arrives at time 10.30. Then, time 8.50, manufacturer 1 release products P1 350 units to the same cross-docking center arrive at 10.50. After that at time 9.39 and 9.65, manufacturer 2 ships 200 and 600 units of product P2 to the cross-docking center 1. They arrive at 11.19 and 11.45 respectively. The other details are shown in Table 5.

From the planning result in Table 5, the arrival time of every destination must be within their expected time window. For example, MF1 release P1 to CD1, the depart time from MF1 is 8.30 and it arrived CD1 at 10.30. In this case, the time window of CD1 is [10.00, 13.00]. It means that CD1 can receive the products because it arrives within expected time of CD1.

For instances 2 and 3, the distribution plan of products are indicated in Tables 6 and 7.

These results can provide the distribution plan of the cross-docking distribution planning problem that has the minimum total cost. Moreover, this plan satisfied the time window constraints of manufacturers, cross-docking centers and customers in the cross-docking network.

Table 6 Distribution plan for problem instance 2

From	Depart time	To	Arrival time	Inventory	Inventory time	Amount of		
						P1	P2	P3
MF1	8.50	CD1	10.50	–	–	850	–	350
MF2	10.01	CD2	11.59	–	–	200	–	–
MF2	10.50	CD1	12.30	–	–	–	300	–
MF2	10.50	CD2	12.08	–	–	–	800	–
–	–	–	–	CD1	10.50–10.98	850	–	–
–	–	–	–	CD1	10.50–10.99	–	–	350
–	–	–	–	CD1	10.99–11.44	550	–	–
–	–	–	–	CD1	11.00–11.59	–	–	100
CD1	11.00	CM3	13.50	–		300	–	250
–	–	–	–	CD1	11.45–11.59	250	–	–
CD1	11.45	CM2	14.45	–	–	300	–	–
CD2	11.59	CM3	13.51	–	–	200	–	–
CD1	11.60	CM1	13.50	–	–	250	–	100
CD2	12.08	CM2	14.58	–	–	–	600	–
CD2	12.08	CM3	14.00	–	–	–	200	–
CD1	12.30	CM1	14.20	–	–	–	300	–
Total costs		117,203.20						
Inventory cost		661.50						
Transportation cost		116,541.70						

General Model for Cross-Docking Distribution Planning Problem

Table 7 Distribution plan for problem instance 3

From	Depart time	To	Arrival time	Inventory	Inventory time	Amount of		
						P1	P2	P3
MF1	8.50	CD1	10.50	–	–	850	–	–
MF2	9.20	CD1	11.00	–	–	–	–	250
MF2	9.74	CD1	11.54	–	–	–	200	–
MF2	10.05	CD2	11.65	–	–	200	900	–
MF2	10.09	CD2	11.69	–	–	–	–	750
–	–	–	–	CD1	10.50–10.99	600	–	–
CD1	10.50	CM1	13.50	–	–	250	–	–
–	–	–	–	CD1	11.00–11.44	100	–	–
CD1	11.00	CM3	13.50	–	–	500	–	250
CD1	11.45	CM2	14.45	–	–	100	–	–
CD1	11.54	CM3	14.04	–	–	–	200	–
CD2	11.65	CM1	14.75	–	–	–	300	–
CD2	11.65	CM2	14.45	–	–	200	–	–
CD2	11.69	CM1	14.79	–	–	–	–	100
CD2	11.69	CM2	14.49	–	–	–	600	–
CD2	11.69	CM4	15.01	–	–	–	–	650
Total costs		162,612.30						
Inventory cost		241.50						
Transportation cost		162,853.80						

Table 8 Computational time

Instance	Test set name	Computational time (s)		Improve time (%)
		Previous solution	Current solution	
1	2MF 1CD 2CM 2P	60.50	0.70	98.84
2	2MF 1CD 2CM 3P	103.55	1.26	98.78
3	2MF 1CD 3CM 2P	143.32	23.85	83.35
4	2MF 2CD 3CM 3P	154.34	45.09	70.78
5	2MF 2CD 4CM 4P	204.40	55.46	72.86
6	3MF 1CD 2CM 2P	561.11	102.20	81.78
7	3MF 1CD 4CM 2P	879.70	239.98	72.72
8	3MF 2CD 4CM 3P	1249.94	522.78	58.17
9	3MF 2CD 5CM 4P	2390.03	908.57	61.98
10	4MF 2CD 5CM 5P	3398.22	1269.90	62.63

4.2 Comparison of the Computational Time

The LINGO model from Jewpanya and Kachitvichyanukul (2012) is improved to deal with all time window constraints and to reduce the running time of program that have explained in Sect. 3. Ten problem instances were run with the original model and with the revised model in this paper. The result shows in Table 8. It demonstrates that the computational time is improved from the previous work average 76.19 %.

5 Conclusion

This research presented a formulation of the cross-docking distribution planning model with multiple products, order consolidation and time windows that was extended from the previous work to consider all of time window in the cross-docking network, manufacturers, cross-docking center and customers. This model is linear and can be solved by LINGO for small problem sizes. Three small distribution planning problems are used to illustrate the model. The results indicate that this model can obtain the distribution plan with minimum cost and under the limitation of time windows. Moreover, ten instances were run with two models those are current model and previous model obtained from Jewpanya and Kachitvichyanukul (2012) by using LINGO to compare the computational time. The results show that the improved LINGO model can find solutions in much shorter time with average improvement of about 76.19 %.

References

Arabani ARB, Ghomi SMTF, Zandieh M (2009) A multi-criteria cross-docking scheduling with just-in-time approach. Int J Adv Manuf Technol

Dondo R, Cerdá J (2013) A sweep-heuristic based formulation for the vehicle routing problem with cross-docking. Comput Chem Eng 48:293–311

Jai-oon A (2011) A particle swarm optimization algorithm for cross-docking distribution networks. Master Thesis, Asian Institute of Technology, Thailand

Jewpanya P, Kachitvichyanukul V (2012) General model of Cross-docking distribution planning problem. In: Proceedings of the 7th international congress on logistics and SCM systems, Seoul, 7–9 June 2012

Ma H, Miao Z, Lim A, Rodrigues B (2011) Crossdocking distribution networks with setup cost and time window constraint. Omega 39:64–72

Santos FA, Mateus GR, Cunha ASD (2013) The pickup and delivery problem with cross-docking. Comput Oper Res 40:1085–1093

A New Solution Representation for Solving Location Routing Problem via Particle Swarm Optimization

Jie Liu and Voratas Kachitvichyanukul

Abstract This paper presents an algorithm based on the particle swarm optimization algorithm with multiple social learning terms (GLNPSO) to solve the capacitated location routing problem (CLRP). The decoding method determines customers clustering followed by depot location and ends with route construction. The performance of the decoding method is compared with previous work using a set of benchmark instances. The experimental results reveal that proposed decoding method found more stable solutions that are clustered around the best solutions with less variation.

Keywords Location routing problem · Particle swarm optimization · Solution representation · Decoding

1 Introduction

Location routing problem (LRP) is basically an integration of two sub-problems: a strategic location-allocation problem (LAP) and an operational vehicle routing problem (VRP) (Bruns 1998). LRP shares the common properties of LAP and VRP both of which are strong NP-hard problems. The comprehensive review on Location routing problem can be found in (Nagy and Salhi 2007). The synthesis of LRP study with a hierarchical taxonomy and classification scheme as well as reviews of different solution methodologies are given in (Min et al. 1998).

J. Liu (✉) · V. Kachitvichyanukul
Industrial System Engineering, School of Engineering and Technology, Asian Institute of Technology, P.O. Box 4, Klong Luang, Pathumtani 12120, Thailand
e-mail: st111504@ait.ac.th

V. Kachitvichyanukul
e-mail: voratas@ait.ac.th

Y.-K. Lin et al. (eds.), *Proceedings of the Institute of Industrial Engineers Asian Conference 2013*, DOI: 10.1007/978-981-4451-98-7_12,
© Springer Science+Business Media Singapore 2013

Liu and Kachitvichyanukul (2012) proposed a decoding method for solving general LRP based on GLNPSO framework (Pongchairerks and Kachitvichyanukul 2009). This paper improved the decoding method in Liu and Kachitvichyanukul (2012) and tested the method using the benchmark instances from the website of http://prodhonc.free.fr/Instances/instances_us.htm.

The remainder of this paper is organized as follows. Section 2 describes the problem and introduces the literature reviews. Section 3 describes the solution representation and the decoding method. The results of the computational experiments are provided in Sect. 4. Conclusions are given in Sect. 5.

2 Problem Definition and Related Works

This paper considers the discrete location-routing problem (CLRP) with the capacity constraints on both depot and route and with the homogeneous fleet type and unlimited number of vehicles. Early exact algorithms for the problem included: Laporte and Nobert (1981), Averbakh and Berman (1994), Laporte et al. (1986), and Ghosh et al. (1981).

Some of the more recent heuristic algorithms for LRP are cited here. Wu et al. (2002) decomposed LRP into two sub-problems of LAP and VRP and solved the two sub-problems in an iterative manner by simulated annealing based with a tabu list to avoid cycling. Prins et al. (2006a) combined greedy randomized adaptive search procedure (GRASP) with a path relinking mechanism to solve the LRP with capacitated depots and routes with homogeneous fleet and unlimited vehicle. Prins et al. (2006b) and Duhamel et al. (2008) proposed different memetic algorithms by hybridizing genetic algorithm with a local search procedure (GAHLS) and different chromosomes. Liu and Kachitvichyanukul (2012) proposed an algorithm based on GLNPSO for solving CLRP.

This paper proposed a new solution representation that extends the work by Liu and Kachitvichyanukul (2012). The solution representation and decoding method is given in the next section.

3 Solution Representation and Decoding Method

An indirect representation is used in this paper. A particle position $[\theta l_1, \theta l_2, ..., \theta l_H]$, is transformed into selected locations, customer assignment and service routes via the decoding process. For LRP problem with n depots and m customers, the dimension of the particle is n + m. Each dimension of the particle is initialized with random number between 0 and 1. Figure 1 illustrates a sample particle for an LRP problem with 3 depots and 10 customers.

A New Solution Representation for Solving Location

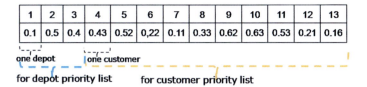

Fig. 1 Example of one particle for LRP with 3 depots and 10 customers

Fig. 2 Depots priority

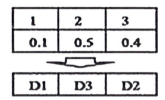

The decoding of depot priority and customer priority are shown in Figs. 2, and 3. Priority sequence is sorted in ascending order of the value of each dimension.

Given that the coordinate of a location i is denoted as (x_i, y_i), the Euclidian distance between any two locations i, j is given in Eq. (1).

$$d_{ij} = \sqrt{(x_i - x_j)^2 + (y_i - y_j)^2} \qquad (1)$$

Two proximity matrices are created for clustering of customers and depots. The customers proximity to the depots shown in Table 1 for the sample particle given in Fig. 1. Similarly, the customers proximity to other customers can be formed as shown in Table 2.

The clustering for the proposed decoding method is based on the customers selected from the priority list given in Fig. 3 so customer 4 is used as the center of the first group. From Table 2, the customers that are closed to customer 4 are customers 5, 3, and so on. The clustering of the customers based on Fig. 3 and Table 2 are shown in Fig. 4. The next unassigned customer on the priority list is customer 10, and it is used to form the second group, etc. The geometrical center of each group is calculated and the distance between the depots and the

1	2	3	4	5	6	7	8	9	10
0.43	0.52	0.22	0.11	0.33	0.62	0.67	0.53	0.21	0.16

| 4 | 10 | 9 | 3 | 5 | 1 | 2 | 8 | 6 | 7 |

Fig. 3 Decoding customers priority

Table 1 Proximity matrix of relative position between customers to depot

	Customers
D1	c3, c1, c5, c2, c4, c6, c8, c9, c10, c7
D2	c2, c9, c1, c5, c4, c7, c3, c8, c6, c10
D3	c6, c10, c7, c1, c9, c2, c5, c3, c8, c4

Table 2 Proximity matrix of relative position between pair customers

	Sequence of distance in ascending order
c1	9, 2, 3, 5, 7, 4, 8, 6, 10
c2	1, 9, 5, 3, 4, 7, 8, 6, 10
c3	5, 8, 4, 1, 2, 9, 6, 7, 10
c4	5, 3, 8, 2, 1, 9, 6, 7, 10
c5	4, 3, 2, 1, 8, 9, 7, 6, 10
c6	10, 1, 7, 8, 3, 9, 5, 2, 4
c7	9, 1, 2, 6, 5, 3, 10, 4, 8
c8	3, 4, 5, 1, 2, 6, 10, 9, 7
c9	7, 1, 2, 5, 3, 4, 6, 8, 10
c10	6, 8, 1, 3, 7, 9, 5, 2, 4

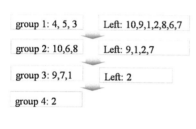
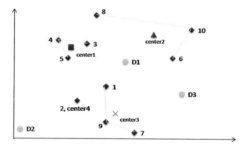

Fig. 4 Grouping process for all customers

geometrical centers are then used to assign customer group preference to depot as shown in Table 3. Finally, the routes are constructed by lowest cost insertion heuristic and the routes are given in Fig. 5.

4 Computational Experiments

Benchmark test datasets for general capacitated LRP are obtained from the website with URL, (http://prodhonc.free.fr/Instances/instances_us.htm). The objective is the total cost which is defined as the sum of the opening cost for depots, the fixed cost of routes, and the total distance times 100.

The decoding methods are implemented in C# programming language under Microsoft Visual studio. NET4.0. The GLNPSO algorithm from the ET-Lib object

Table 3 Proximity matrix of relative position between groups and each depot

	Group 1–4			
D1	g2	g3	g1	g4
D3	g2	g3	**g4**	**g1**
D2	g4	g3	g1	g2

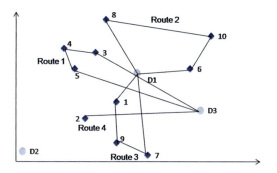 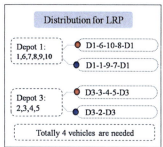

Fig. 5 The distribution for LRP example using the proposed decoding

Table 4 Parameters setting of GLNPSO for CLRP

Parameters	Value
Number of particle	20, 40, 50
Number of iteration	100 ~ 1,500
Number of neighborhood (NB)	5
Inertial weight (w)	Linearly decreasing from 0.9 to 0.4
Acceleration constant, cp, cg, cl, cn	1, 1, 1, 1

library is used (Nguyen et al. 2010). The experiments are carried out on a desktop computer with Intel (R) Core(TM)2 Quad CPU, Q9550 @ 2.83 GHz, 2.83 GHz, 3.00 GB of RAM.

The parameters used in the experiments are based on previous study by (Ai and Kachitvichyanukul 2008), as shown in Table 4, the number of particles and number of iteration are fixed with respective to different size of instance.

The two decoding methods are tested with 30 problem instances. The solutions and the comparisons with best known solution (BKS) and solutions by other heuristics are listed. The two main measurements are the percent deviation from the best known solution (DEV) and the computational time (CPU). The results are shown in Table 5.

As shown in Table 5, the proposed decoding yields 1 new BKS out of 30 problem instances. However, the variation of solution is much smaller with only 4 instances have DEV over 5 %, and only 2 instances DEV over 10 % while most of the instances are below 5 %. For type "a" instances, computer times by these two decoding methods are closed, but for type "b" instance, time required by the proposed decoding is much shorter.

Table 5 The solution quality and solution times

Instance	BKS	Liu and Kachitvichyanukul (2012)			Proposed decoding		
		Cost	DEV %	CPU	Cost	DEV %	CPU
20-5-1a	54,793	54,149	−1.18	1	55,034	0.44	1
20-5-1b	39,104	44,297	13.28	5	41,929	7.22	2
20-5-2a	48,908	50,369	2.99	1	48,895	−0.02	1
20-5-2b	37,542	41,511	10.57	5	38,966	3.79	2
50-5-1a	90,111	92,685	2.86	20	93,290	3.53	18
50-5-1b	63,242	63,485	0.38	58	64,418	1.86	28
50-5-2a	88,298	97,419	10.33	17	91,797	3.96	17
50-5-2b	67,340	70,819	5.17	50	68,949	1.94	26
50-5-2BIS	84,055	87,085	3.13	18	85,367	1.56	18
50-5-2bBIS	51,822	58,674	13.22	57	54,973	6.08	25
50-5-3a	86,203	90,761	5.29	20	87,962	2.04	18
50-5-3b	61,830	66,361	7.33	55	63,998	3.51	27
100-5-1a	275,993	295,690	7.14	152	288,328	4.47	120
100-5-1b	214,392	229,490	7.00	372	220,752	2.97	231
100-5-2a	194,598	193,946	−0.34	145	198,223	1.86	183
100-5-2b	157,173	156,246	−0.59	361	159,446	1.45	251
100-5-3a	200,246	206,087	2.91	148	206,710	3.23	184
100-5-3b	152,586	154,587	1.31	433	154,741	1.41	248
100-10-1a	290,429	285,164	−1.81	153	322,954	11.20	186
100-10-1b	234641	231,802	−1.21	370	272,556	16.12	247
100-10-2a	244,265	242,890	−0.56	163	248,806	1.86	184
100-10-2b	203,988	203,790	−0.10	426	207,502	1.72	249
100-10-3a	253,344	247,379	−2.35	160	262,405	3.58	185
100-10-3b	204,597	200,579	−1.96	530	209,540	2.42	249
200-10-1a	479,425	496,553	3.57	1,358	48,989	2.18	1,962
200-10-1b	378,773	384,811	1.59	3,160	385,785	1.85	2,383
200-10-2a	450,468	472,770	4.91	1,346	457,168	1.48	1,963
200-10-2b	374,435	386,670	3.28	3,682	381,297	1.83	2,366
200-10-3a	472,898	478,828	1.25	1,324	480,348	1.58	1,935
200-10-3b	364,178	368,360	1.15	2,993	370,746	1.80	2320

As seen in Table 5, decoding method from Liu and Kachitvichyanukul (2012) can reach more best known solutions than the proposed decoding method. However, the proposed decoding method is more consistent in that the average gap between the solutions and the BKS are much smaller. For decoding method from Liu and Kachitvichyanukul (2012), the preclustering of customers around the depot restricted the route construction that frequently leads to suboptimal routes. For medium and larger problem instances, more grouping combinations can be formed and the drawbacks had less effect on the search results. It can be seen from Table 5 that the proposed decoding method is more consistent and the average gap is less than 5 %. This is expected since the design of the decoding method is aimed at increasing the diversity of routes formed in the route construction step. For problem instances 2, 19, and 20, further investigation should be carried out to see if there is any problem specific situations that prevent the route construction to reach better routes.

5 Conclusion

This paper proposed a solution representation and decoding method for solving CLRP problem using GLNPSO. The proposed solution representation is evaluated on a set of benchmark problem instances. Based on the experimental results, the following conclusions are drawn:

- The proposed decoding method is more consistent than that by Liu and Kachitvichyanukul (2012), the average gap is below 5 % with only a few problem instances with gap higher than 5 %.
- Liu and Kachitvichyanukul (2012) can reach better solutions more often but the average performance is poorer.
- The comparison of computational time is inconclusive. More experiments with larger samples are required to make a definite conclusion.

References

Ai TJ, Kachitvichyanukul V (2008) A study on adaptive particle swarm optimization for solving vehicle routing problems. Paper in the APIEMS 2008 conference. Bali, Dec 2008

Averbakh I, Berman O (1994) Routing and location-routing p-delivery men problems on a path. Transp Sci 28:184–197

Bruns AD (1998) Zweistufige Standortplanung unter Berü cksichtigung von Tourenplanungsaspekten—Primale Heuristiken und Lokale Suchverfahren. Ph.D dissertation, Sankt Gallen University

Duhamel C, Lacomme P, Prins C, Prodhon C (2008) A memetic approach for the capacitated location routing problem. Paper presented in the EU/meeting 2008 workshop on metaheuristics for logistics and vehicle routing. France, 23–24 Oct 2008

Ghosh JK, Sinha SB, Acharya D (1981) A generalized reduced gradient based approach to round-trip location problem. Sci Manage Transp Systems 209–213

Laporte G, Nobert Y (1981) An exact algorithm for minimizing routing and operating costs in depot location. Eur J Oper Res 6:224–226

Laporte G, Nobert Y, Arpin D (1986) An exact algorithm for solving a capacitated location-routing problem. Ann Oper Res 6:291–310

Liu J, Kachitvichyanukul V (2012) Particle swarm optimization for solving location routing problem. In: Proceedings of the 13th Asia pacific industrial engineering and management systems conference. Phuket, Thailand, pp 1891–1899, Dec 2012

Min H, Jayaraman V, Srivastava R (1998) Combined location-routing problems: a synthesis and future research directions. Eur J Oper Res 108:1–15

Nagy G, Salhi S (2007) Location-routing: issues, models and methods. Eur J Oper Res 177:649–672

Nguyen S, Ai TJ, Kachitvichyanukul V (2010) Object library for evolutionary techniques ETLib: user's guide. High Performance Computing Group, Asian Institute of Technology, Thailand

Pongchairerks P, Kachitvichyanukul V (2009) Particle swarm optimization algorithm with multiple social learning structures. Int J Oper Res 6(2):176–194

Prins C, Prodhon C, Calvo RW (2006a) Solving the capacitated location-routing problem by a GRASP complemented by a learning process and a path relinking. Q J Oper Res 4(3):221–238

Prins C, Prodhon C, Calvo RW (2006b) A memetic algorithm with population management (MA|PM) for the capacitated location-routing problem. EvoCOP 2006:183–194

Wu TH, Low C, Bai JW (2002) Heuristic solutions to multi-depot location-routing problems. Comput Oper Res 29:1393–1415

An Efficient Multiple Object Tracking Method with Mobile RFID Readers

Chieh-Yuan Tsai and Chen-Yi Huang

Abstract RFID (Radio Frequency Identification) technology originally is designed for object identification. Due to its relatively low cost and easy deployment, RFID technology becomes a popular approach for tracking the positions of objects. Many researchers proposed variant object tracking algorithms to quickly locate objects. Although these algorithms are efficient for certain applications, their tracking methods are limited on multiple objects with fixed RFID readers or single object with mobile RFID readers. None of them focus on tracking multiple objects with mobile RFID readers. To bridge this gap, this study develops an efficient multiple object tracking method with mobile RFID readers. First, the omni-directional antenna in a mobile RFID reader is used to judge the annular regions at which objects locate by adjusting the reader's reading range. Second, the directional antenna in the mobile RFID reader is used to judge the circular sectors of objects according to the received signal strengths in each direction. Third, four picking strategies are proposed to decide whether an object is picked in current step or later step. Finally, the simulated annealing algorithm is performed for all objects in the list of current step to generate the picking sequence with shortest distance. The experiments show that the proposed tracking method can help users find multiple objects in shortest distance.

Keywords RFID · Multiple object tracking · Picking sequence · Simulated annealing (SA) algorithm

---

C.-Y. Tsai (✉)
Department of Industrial Engineering and Management, Yuan-Ze University, No. 135,
Yungtung Road, Chungli City, Taoyuan, Taiwan, Republic of China
e-mail: cytsai@saturn.yzu.edu.tw

C.-Y. Huang
Department of Production, Powertech Technology Inc., No. 26, Datong Rd,
Hsinchu Industrial Park, Hukou, Hsinchu, Taiwan, Republic of China
e-mail: mattcyhuang@pti.com.tw

Y.-K. Lin et al. (eds.), *Proceedings of the Institute of Industrial Engineers Asian Conference 2013*, DOI: 10.1007/978-981-4451-98-7_13,
© Springer Science+Business Media Singapore 2013

1 Introduction

In recent years, many object tracking technologies has been developed such as GPS, infrared technology, radar, ultrasonic, and Radio Frequency Identification (RFID). Each technology has its own advantage and limitation. For example, GPS localization technology is popular in navigation field but not suitable for indoor environment. The equipment cost for infrared technology is low but line-of-sight and short-range signal transmission is its major limitations. RFID is a communication technology which receives many attentions in the fields of entrance management, logistics and warehouse management, medical care, and others.

In recent years, RFID technology has been successfully applied to object localization. However, a lot of RFID readers might be required to precisely estimate the accurate object position. Lionel et al. (2004) proposed a LANDMARC method that uses reference tags to reduce the cost and to improve the localization accuracy. Shih et al. (2006) and Zhao et al. (2007) proposed different approaches to improve LANDMARC method. Another important research area is to search objects using mobile RFID reader. That is, a user holds a mobile RFID reader and moves in the searching area to find objects. When the user receives the signal from an object, he/she will know the object is within the range. However, a user needs to move constantly with try and error approach to determine object position. Song et al. (2007) aimed at the mobile localization and proposed a prototype system. Bekkali et al. (2007) used two mobile readers to locate objects and used a filter to reduce the localization variation.

To improve the performance of object searching using mobile RFID readers, this study develops an efficient multiple object tracking method with mobile RFID readers. First, the omni-directional antenna in a mobile RFID reader is used to judge the annular regions at which objects locate by adjusting the reader's reading range. Second, the directional antenna in the mobile RFID reader is used to judge the circular sectors of objects according to the received signal strengths in each direction. Third, four picking strategies are proposed to decide whether an object is picked in current step or later step. Finally, the simulated annealing algorithm is performed for all objects in the list of current step to generate the picking sequence with shortest distance.

2 Methodology

2.1 Objects Tracking in Two Stages

In the first stage, a user takes a mobile RFID reader to scan objects when he/she moves to a desired position (called detecting point). The received signal strength (RSS) in the reader is used to determine the distance between the mobile reader and the object. That is, the annual region at which an object locates can be

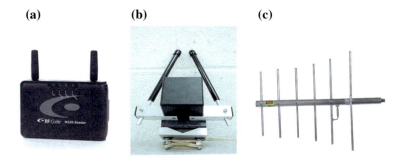

Fig. 1 a The mobile RFID reader. **b** Omni antenna. **c** Yagi antenna

determined. In the second stage, the direction (circular sector) of the object can be determined if the signal is received from the direction. By matching both annual region and circular sector, object position can be estimated. To achieve the above goal, this research adopts the mobile reader having two channels. Each channel connects to different antennas having different reading ranges and functions. Omni antenna is used in the first stage to scan objects, while Yagi Antenna is used to point out object direction in the second stage. Figure 1 visually shows the mobile RFID reader, Omni antenna, and Yagi antenna used in this study.

2.2 Object Picking

After finishing the first and second stage, a user will know which sub-area that an object is in. However, the precise position of the target object in the sub-area is not clear unless the user moves to the sub-area and takes a look. Therefore, this research suggests the user move to the center of the sub-area first, then moves along the arc of the sub-area to see where the target object is. If the object is found, then he/she will pick the object. Figure 2 shows a user moves toward left first (route 1). If the user could find the target object, he/she will pick the object and go to the next target object sub-region; otherwise, the user moves back the central point and search objects in the reversed direction (route 2).

2.3 Searching Route

As mentioned in Sect. 2.1, a user moves a fixed distance to detecting point then scan whether there is object in the current reading range. Since the exact location of an object is unknown, this research adopts the creeping line search style with S route to find objects as shown in Fig. 3. Starting from the initial position, a user

Fig. 2 Object picking in the sub-area

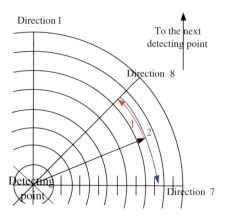

Fig. 3 Searching objects with S route

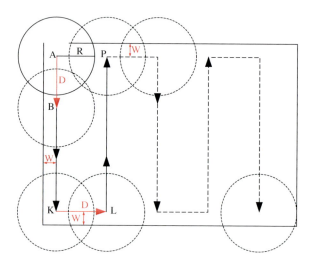

moves a fixed distance D each time and operates the mobile reader for object detection in the detecting position. It is clear that there is overlapped scanning area in two neighboring search points so that no area will be missed.

2.4 Path Planning

After generating the pick list, the system will arrange the picking order. Starting from detecting point A, a user picks all objects in the list, and then move to detecting point B. However, in each detecting point, the system will generate a shortest-path. Some assumptions are made to have the shortest path planning: (1) Moving path between target objects is without directionality, in other words, is a symmetric TSP problem. (2) All objects could be picked directly (Complete Graph). (3) The sub-area center is known in advance. (4) Each object can be picked only once. (5) N objects been picked from A to B is known. To solve the above problem, this research uses the simulated annealing (SA) algorithm to obtain the shortest path.

3 Case Study

The multiple objects searching method is programed using Microsoft Visual Studio 2005. The simulation program will derive the quantity and coordination of detecting points if the size of searching area is provided. In each detecting point, the program will determine how many objects should be picked by four strategies. Under the same strategy, simulated annealing algorithm will calculate the optimal picking path. Figure 4 shows the interface of the developed multiple objects searching program.

For example, the largest read range for a reader is 80 m in which each power level of the reader is 10 m. The searching area is 540*540 m^2. A user starts to search objects at (50, 0). Figure 5 shows the user moving path. The red block expresses the detecting point, and the black route means the path the user will follow. The start point of this route is from (50, 50), pass through (50, 160), and to (50, 270).

Fig. 4 User interface of the system

Fig. 5 The picking path example

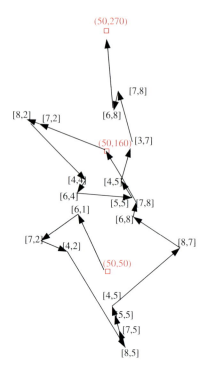

4 Conclusions

To reduce the searching space and time for finding objects, this research proposes a multiple objects search method. Each object is embedded with an active RFID tag. A user holds the mobile RFID reader installed with Omni antenna and Yagi antenna to detect objects. According to two-stage approach, a user can know which sub-area an object is. The final pick route will be determined by the simulated annealing method. Currently, only squared area is considered. It will be worthwhile to develop a searching method for any kind of shape. In addition, no obstacle is considered in current method. However, obstacles might prevent user's movement. It is suggested that future works can include obstacle consideration.

Acknowledgments This work was partially supported by the National Science Council of Taiwan No. NSC98-2221-E-155-16.

References

Bekkali A, Sanson H, Matsumoto M (2007) RFID indoor positioning based on probabilistic RFID map and Kalman filtering. Paper presented at the 3rd IEEE international conference on wireless and mobile computing, networking and communications, pp 21–21

Lionel MN, Liu Y, Lau YC, Patil AP (2004) LANDMARC: indoor location sensing using active RFID. Wirel Netw 10:701–710

Shih ST, Hsieh K, Chen PY (2006) An improvement approach of indoor location sensing using active RFID. Paper presented at the first international conference on innovative computing, information and control, pp 453–456

Song J, Haas CT, Caldas CH (2007) A proximity-based method for locating RFID tagged objects. Adv Eng Inform 21:367–376

Zhao Y, Liu Y, Ni LM (2007) VIRE: Active RFID-based localization using virtual reference elimination. Paper presented at the international conference on parallel processing, p 56

A New Bounded Intensity Function for Repairable Systems

Fu-Kwun Wang and Yi-Chen Lu

Abstract Lifetime failure data might have a bathtub-shaped failure rate. In this study, we propose a new model based on a mixture of bounded Burr XII distribution and bounded intensity process, to describe a failure process including a decreasing intensity phase, an increasing phase, and an accommodation phase for repairable systems. The estimates of the model parameters are easily obtained using the maximum likelihood estimation method. Through numerical example, the results show that our proposed model outperforms other existing models, such as superposed power law process, Log-linear process-power law process, and bounded bathtub intensity process with regard to mean square errors.

Keywords Bathtub-shaped failure rate · Bounded intensity function · Maximum likelihood estimation · Repairable system

1 Introduction

In some situations, lifetime failure data might have a bathtub-shaped failure rate. These models can be categorized into three types to describe the bathtub-shaped intensity function: a single model (Crevecoeur 1993; Xie et al. 2002; Lai et al. 2003; Liu and Wang 2013), mixture models (Lee 1980; Pulcini 2001a, b; Guida and Pulcini 2009), and a mixture of two distributions (Jiang and Murthy 1998; Block et al. 2008, 2010).

F.-K. Wang (✉) · Y.-C. Lu
Department of Industrial Management, National Taiwan University of Science and Technology, Taipei 106, Taiwan, Republic of China
e-mail: fukwun@mail.ntust.edu.tw

Y.-C. Lu
e-mail: D10101009@mail.ntust.edu.tw

Y.-K. Lin et al. (eds.), *Proceedings of the Institute of Industrial Engineers Asian Conference 2013*, DOI: 10.1007/978-981-4451-98-7_14,
© Springer Science+Business Media Singapore 2013

In our literature review, some mixture models propose to describe the bathtub-shaped intensity such as the superposed-power law process (S-PLP) model by Pulcini (2001a, b) and the bounded bathtub intensity process (BBIP) model by Guida and Pulcini (2009). The failure intensity functions of S-PLP model and BBIP model are defined as

$$\lambda(t)_{S-PLP} = \lambda_1(t) + \lambda_2(t) = \frac{\beta_1}{\alpha_1}(\frac{t}{\alpha_1})^{\beta_1-1} + \frac{\beta_2}{\alpha_2}(\frac{t}{\alpha_2})^{\beta_2-1}, \quad t \geq 0; \; \alpha_1, \beta_1, \alpha_2, \beta_2 > 0,$$

(1)

and

$$\lambda(t)_{BBIP} = \lambda_1(t) + \lambda_2(t) = ae^{(\frac{-t}{b})} + \alpha[1 - e^{(\frac{-t}{\beta})}], \quad a, b, \alpha, \beta > 0, \; t \geq 0.$$

(2)

However, for the increasing intensity phase when the system age grows, the failure intensity should tend to become constant in wear-out conditions for large system ages, if the repairs were exactly minimal. This result is often known as Drenick's theorem. It can be noted that failure intensity should be bounded and tend to become constant in the failure pattern of deteriorating systems. The S-PLP model cannot be applied to Drenick's theorem and has two drawbacks: $\lambda(t)_{S-PLP} \xrightarrow{t \to 0} \infty$ and $\lambda(t)_{S-PLP} \xrightarrow{t \to \infty} \infty$. Therefore, Guida and Pulcini (2009) proposed the BBIP model to satisfy Drenick's theorem and overcome these two drawbacks. This BBIP model has overcome the drawbacks of $\lambda(t)_{BBIP} \xrightarrow{t \to 0} a$ and $\lambda(t)_{BBIP} \xrightarrow{t \to \infty} \alpha$.

However, Krivtsov (2007) showed that the NHPP's rate of occurrence of failures formally coincides with the intensity function of the underlying lifetime distribution. Therefore, other lifetime distributions (lognormal, normal, Gumbel, a mixture of underlying distributions, etc.) could be chosen for the ROCOF of the respective NHPPs. This finding motivates us to formulate a new mixture model based on the Burr XII distribution in order to describe the bathtub-shaped failure rate.

2 New Model

The proposed bounded model is obtained by combining two independent NHPPs, the first being a bounded intensity process with decreasing intensity function

$$\lambda_1(t) = a + \frac{kct^{c-1}}{1 + t^c}, \quad a, c, k > 0, \; t \geq 0,$$

(3)

where a is a positive constant and k and c are the shape parameters; the second being a bounded intensity process (Pulcini 2001a, b) with increasing bounded intensity function

A New Bounded Intensity Function for Repairable Systems 121

$$\lambda_2(t) = \alpha[1 - e^{(\frac{-t}{\beta})}], \quad \alpha, \beta > 0, \ t \geq 0. \tag{4}$$

Thus, the intensity function of the proposed model is established as

$$\lambda(t) = \lambda_1(t) + \lambda_2(t) = a + \frac{kct^{c-1}}{1 + t^c} + \alpha[1 - e^{(\frac{-t}{\beta})}], \quad a, c, k, \alpha, \beta > 0, \ t \geq 0, \tag{5}$$

where the parameters a and $a + \alpha$ are the values of the failure intensity at t = 0 and $t \to \infty$, respectively. The first derivative is equal to α/β at t = 0. The mean cumulative number of failures for the proposed model is given by

$$m(t) = \int_0^t \left(a + \frac{kcu^{c-1}}{1 + u^c} + \alpha[1 - e^{(\frac{-u}{\beta})}] \right) du = at + k \ln(1 + t^c) + \alpha\beta \left(\frac{t}{\beta} - 1 + e^{\frac{-t}{\beta}} \right). \tag{6}$$

3 Parameters Estimation

If the data are time-truncated (say T), then the likelihood function is given as

$$L(a, c, k, \alpha, \beta | t_1, t_2, \ldots, t_n) = \prod_{i=1}^{n} \lambda(t_i) \times e^{-m(T)}. \tag{7}$$

Then, the log-likelihood function is given by

$$\ln L = \sum_{i=1}^{n} \ln \left[a + \frac{kct_i^{c-1}}{1 + t_i^c} + \alpha(1 - e^{\frac{-t_i}{\beta}}) \right] - [aT + k \ln(1 + T^c)] - \alpha\beta \left[\frac{T}{\beta} - 1 + e^{\frac{-T}{\beta}} \right]. \tag{8}$$

The parameter estimates in (8) can be determined by general-purpose optimization function from R Development Core Team (2013).

In testing for a trend with operating time for time-truncated data, the Laplace statistic and the MIL-HDBK-189 statistic are defined as

$$LA = \frac{\sum_{i=1}^{n} t_i - \frac{nT}{2}}{T\sqrt{\frac{n}{12}}} \sim N(0, 1), \tag{9}$$

and

$$Z = 2 \sum_{i=1}^{n} \ln(\frac{T}{t_i}) \sim \chi^2(2n). \tag{10}$$

Firstly, we assess the presence or absence of a trend in these data. In this study, we evaluate the Laplace statistic and MIL-HDBK-189 statistic, which are able to provide evidence of a monotonic trend when the null hypothesis is an HPP (Vaurio 1999). From these studies, we can conclude that LA and Z are effective in detecting monotonic trends with efficiency better or roughly equal to many other tests that are mathematically more complex (Kvaloy and Lindqvist 1998).

Glaser (1980) has obtained sufficient conditions to ensure whether a lifetime model has a bathtub failure rate or not. In this study, we use a graphical method based on the total time on test (TTT) transform (Barlow and Campo 1975; Bergman and Klefsjo 1982). It has been shown that the failure rate is increasing (decreasing) if the scaled TTT-transform is concave (convex). In addition, for a distribution with bathtub (unimodal) failure rate, the TTT-transform is initially convex (concave) and then concave (convex) (Aarset 1987).

We evaluate the performance of the model using the mean of squared errors (MSE). The MSE is defined as

$$MSE = \frac{1}{n}\sum_{i=1}^{n}[\hat{m}(t_i) - y_i]^2, \quad (11)$$

where $\hat{m}(t_i)$ is the estimated cumulative number of failures at time t_i obtained from the model and y_i is the total number of failures observed at time t_i according to the actual data. For MSE, the smaller the metric value, the better the model fits.

4 Illustrative Examples

The data concerned the powertrain system of a bus (Guida and Pulcini 2009). We take the travel distance as the time (unit = 1,000 km) to compare the proposed model with the S-PLP, LLP-PLP, and BBIP models; the parameter estimates were

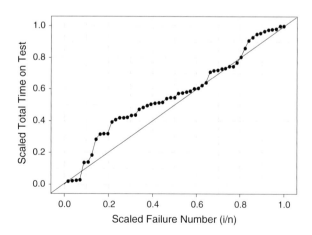

Fig. 1 TTT plot on the 55 observations in Bus 510

Table 1 Comparison of different models for example 1

Model	Estimated parameters	log-likelihood	MSE
S-PLP	$\alpha_1 = 7.32$ $\beta_1 = 0.6919$ $\alpha_2 = 48.62$ $\beta_2 = 1.754$	−161.639	10.81
LLP-PLP	$a = 0.3443$ $b = 10.0602$ $\alpha = 33.3399$ $\beta = 1.5962$	−160.268	12.93
BBIP	$a = 0.3693$ $b = 10.1103$ $\alpha = 0.2016$ $\beta = 150.82$	−159.6792	9.10
Proposed model	$a = 1.54 \times 10^{-17}$ $c = 53.57$ $k = 0.0174$ $\alpha = 0.2045$ $\beta = 168.71$	−160.9126	9.01

Note $a = 1.54 \times 10^{-17} \approx 0$

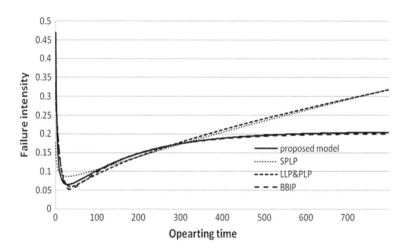

Fig. 2 Plots of $\lambda(t)$ against time t for example 1

obtained using general-purpose optimization function from R Development Core Team (2013).

Example 1: We considered the data that is time-truncated at T = 394.064. The Laplace statistic and the MIL-HDBK-189 statistic for the above data are LA = 1.61 and Z = 90.31, respectively, which provide a result in the trend test. The p-values of Laplace statistic and MIL-HDBK-189 statistic are 0.054 and 0.086, respectively; therefore, we can know the data should be non-monotone. The

TTT plot is shown in Fig. 1. The scaled TTT-transform is initially convex and then concave, which indicates that the data have a bathtub-shaped failure rate characteristic.

The comparative results of our proposed model and the other models (S-PLP, LLP-PLP and BBIP) are shown in Table 1. It appears that our proposed model fits these data better than the other models in MSE. Figure 2 depicts the plots of $\lambda(t)$ against time t for the different models and shows a bathtub-shaped curve during test time. The use of the S-PLP and LLP-PLP models could lead to wrong conclusions in such situations because their failure intensity tends to infinity as t tends to infinity. Thus, it appears that the bounded type model provides a better representation than the no bounded model or partially bounded model for situations of the large system ages.

5 Conclusions

The S-PLP model is one of the mixture models used for the bathtub-shaped failure intensity during the development test phase for a system. However, the intensity function of the S-PLP model does not satisfy Drenick's theorem and has two unrealistic situations for the bathtub-shaped failure intensity case. Recently, Guida and Pulcini (2009) proposed the BBIP model to overcome these two drawbacks. In this study, we have proposed a new mixture model for bathtub-shaped failure intensity. Results obtained from numerical examples show that our proposed model outperforms the other models in terms of MSE for Bus510. We conclude that our proposed model can compete with BBIP in some bathtub-shaped failure intensities for a repairable system.

Acknowledgments The authors are grateful for financial support from the National Science Council in Taiwan under the Grant NSC-101-2221-E011-050-MY3.

References

Aarset MV (1987) How to identity a bathtub hazard rate. IEEE Trans Reliab 36:106–108

Barlow RE, Campo R (1975) Total time on test processes and applications to failure data analysis. Siam, Philadelphia

Bergman B, Klefsjo B (1982) A graphical method applicable to age replacement problems. IEEE Trans Reliab 31:478–481

Block HW, Li Y, Savits TH, Wang J (2008) Continuous mixtures with bathtub-shaped failure rates. J Appl Probab 45:260–270

Block HW, Langberg NA, Savits TH (2010) Continuous mixtures of exponentials and IFR gammas having bathtub-shaped failure rates. J Appl Probab 47:899–907

Crevecoeur GU (1993) A model for the integrity assessment of ageing repairable systems. IEEE Trans Reliab 42:148–155

Glaser RE (1980) Bathtub and related failure rate characterizations. J Am Stat Assoc 75:667–672

Guida M, Pulcini G (2009) Reliability analysis of mechanical systems with bounded and bathtub shaped intensity function. IEEE Trans Reliab 58:432–443

Jiang R, Murthy DNP (1998) Mixture of Weibull distribution—parametric characterization of failure rate function. Appl Stoch Models Data Anal 14:47–65

Krivtsov VV (2007) Practical extensions to NHPP application in repairable system reliability. Reliab Eng Syst Saf 92:560–562

Kvaloy JT, Lindqvist B (1998) TTT-based tests for trend in repairable system data. Reliab Eng Syst Saf 60:13–28

Lai CD, Xie M, Murthy DNP (2003) A modified Weibull distribution. IEEE Trans Reliab 52:33–37

Lee L (1980) Testing adequacy of the Weibull and log linear rate models for a poisson process. Technometrics 22:195–199

Liu J, Wang Y (2013) On Crevecoeur's bathtub-shaped failure rate model. Comput Stat Data Anal 57:645–660

Pulcini G (2001a) Modeling the failure data of repairable equipment with bathtub failure intensity. Reliab Eng Syst Saf 71:209–218

Pulcini G (2001b) A bounded intensity process for the reliability of repairable equipment. J Qual Technol 33:480–492

R Development Core Team (2013) R: a language and environment for statistical computing. R Foundation for Statistical Computing, Vienna

Vaurio JK (1999) Identification of process and distribution characteristics by testing monotonic and non-monotonic trends in failure intensities and hazard rates. Reliab Eng Syst Saf 64:345–357

Xie M, Tang Y, Goh TN (2002) A modified Weibull extension with bathtub-shaped failure rate function. Reliab Eng Syst Saf 76:279–285

Value Creation Through 3PL for Automotive Logistical Excellence

Chin Lin Wen, Schnell Jeng, Danang Kisworo, Paul K. P. Wee and H. M. Wee

Abstract Change is the only constant in today's business environment. Flexibility and adaptability have become key factors of organizational success. In today's global business environment, organizations not only faced with issues of where to source their parts but also how to ship and store these parts effectively and efficiently. This issue is magnified as the complication of the products produced increases. Automotive industry has highly complicated parts which are sourced from around the globe. Through 3PL strategies, an automotive manufacturer can minimize downtime risk, reduce delivery lead time and ultimately improve its ability to adjust to the changing market demand. In this paper, we investigate the logistic factors affecting the automotive industry and discuss the challenges. An actual case study from DB Schenker, a German 3PL company, is used to illustrate the benefits of an effective VMI, information technology and inventory tracking in the supply chain. It can synchronize information and physical flow of goods across the supply chain. DB Schenker uses state-of-the-art storage and order-picking technologies in order to meet the very precise production time schedule, while keeping warehouse costs to a minimum.

Keywords VMI · Logistics · Excellence · Supply chain · Inventory · 3PL

C. L. Wen (✉) · S. Jeng · D. Kisworo · P. K. P. Wee · H. M. Wee
Department of Industrial and Systems Engineering, Chung Yuan Christian University,
Chung Li, Taiwan, Republic of China
e-mail: chinlin@cycu.edu.tw

H. M. Wee
e-mail: weehm@cycu.edu.tw

1 Introduction

At the start of the second automotive century, the global car industry finds itself in a major phase of transition and arguably, in one of its most interesting times. Beyond the days of Henry Ford, high-variety mass production, and the early adoption of lean production concepts, the global car industry currently faces significant uncertainty of how to regain profit-ability, which averages at less than 4 % EBIT (Earning Before Interest and Taxes) at present. The industry is threatened by global production over-capacity and rising stock levels of unsold vehicles. Vehicle manufacturers have attempted to meet future challenges through a series of (often doubtful) global mergers and acquisition, hoping for better economies of scale through platform and component sharing. At the same time, previous core competencies, such as component or module assembly, are being outsourced to large first tier suppliers some of which have already over-taken their vehicle manufacturer customers in terms of turnover and size. Further uncertainty stems from the European End-of-Life-of-Vehicle legislation, requiring manufacturers to recycle 95 % of the total vehicle by 2015. If implemented in its current form, this will make this reverse flow of parts and materials an integral part of the supply chain.

2 Related Literature

2.1 Third Party Logistic

A third-party logistics provider (abbreviated 3PL, or sometimes TPL) is a firm that provides service to its customers of outsourced (or "third party") logistics services for part, or all of their supply chain management functions. Third party logistics providers typically specialize in integrated operation, warehousing and transportation services that can be scaled and customized to customers' needs based on market conditions and the demands and delivery service requirements for their products and materials. Often, these services go beyond logistics and included value-added services related to the production or procurement of goods, i.e., services that integrate parts of the supply chain. Then the provider is called third-party supply chain management provider (3PSCM) or supply chain management service provider (SCMSP). Third Party Logistics System is a process which targets a particular Function in the management. It may be like warehousing, transportation, raw material provider, etc.

3PL logistics service advantages e.g., provide quality delivery on services for the local retail market. Has multiple routes, parcels cover districts. All employees must receive on-the-job training to ensure the quality of logistics services. Has a customer hotline, with the delivery of purchase to meet customers needs for logistics.

LSPs provide for their customers traditional logistics services such as transportation and warehousing, and supplementary services such as order administration and track-and-trace services suppliers and customers in supply chains (Hertz and Alfredsson 2003). They handle larger shares of their customers' activities and are therefore significant actors in supply chains.

The pursuit of improved efficiency performance in logistics operations is a constant business challenge (Tezuka 2011). One initiative that is proving productive and allows businesses to concentrate on their core competencies is the outsourcing of the logistics function to partners, known as third-party logistics (3PL) providers. 3PL providers provide an opportunity for businesses to improve customer service, respond to competition and eliminate assets (Handfield and Nichols 1999). Many 3PL providers have broadened their activities to provide a range of services that include warehousing, distribution, freight forwarding and manufacturing.

2.2 Automotive Industry

In the recent years, the automotive industry has shown an increased interest in lead time reduction. Shorter lead times (Erdem et al. 2005) Increase responsiveness to market changes and Reduce pipeline inventory, and Improve customer satisfaction.

The total lead time depends on: the time between receiving an order from the customers or dealers and launching the production of this order, the manufacturing lead time, and the time to ship the final product to customers. With the JIT systems' emphasis on balancing the mixed model assembly lines, especially on reducing the variation of the consumption rate of the parts (Kubiak 1993).

3 Case Study

The case study is developed base on the following questions.

1. How company makes sure that the several tier on suppliers have capable information system to automated messaging when they have shortages of material?
2. How to make all the dealers confidence with 3PL to work together in outbound logistic?

4 Discussion

4.1 Vendor Managed Inventory

The supplier decides on the appropriate inventory levels of each of the products and the appropriate inventory policies to maintain these levels. In most cases, the customers tend to avail themselves of a Third Party Logistics provider to manage their vendor-managed-inventory hub.

Vendor managed inventory (VMI) is a business models in which the buyer of a product (business) provides certain information to a vendor (supply chain) supplier of that product and the supplier takes full responsibility for maintaining an agreed inventory of the material, usually at the buyer's consumption location. A third-party logistics provider can also be involved to make sure that the buyer has the required level of inventory by adjusting the demand and supply gaps (Southard and Swenseth 2008).

As a symbiotic relationship, VMI makes it less likely that a business will unintentionally become out of stock of a good (Cetinkaya 2000) and reduces inventory in the supply chain (Yao et al. 2007). Furthermore, vendor (supplier) representatives in a store benefit the vendor by ensuring the product is properly displayed and store staffs are familiar with the features of the product line, all the while helping to clean and organize their product lines for the store.

One of the keys to making VMI work is shared risk. In some cases, if the inventory does not sell, the vendor (supplier) will repurchase the product from the buyer (retailer). In other cases, the product may be in the possession of the retailer but is not owned by the retailer until the sale takes place, meaning that the retailer simply houses (and assists with the sale of) the product in exchange for a prede-termined commission or profit (sometimes referred to as consignment stock). A special form of this commission business is scan-based trading whereas VMI is usually applied but not mandatory to be used.

DB Schenker was one of the first 3PLs to develop VMI and supplier park solutions for the Electronics and Automotive industries. DB Schenker apply the latest techniques and systems to provide services ranging from material consoli-dation at origin, order management, inbound transportation, line-side feeding, order sequencing, quality assurance, kitting and pre-production preparation or un-packaging/packaging services in addition to a range of tailored solutions to support your production.

4.2 Information Technology

Strong information technology systems are increasingly a core competence to ensure competitiveness. Bar Code and Pick by Voice are valuable technologies for tracking inventory in the supply chain. It can synchronize information and physical flow of goods across the supply chain.

A barcode is an optical machine-readable representation of data relating to the object that contains information (as identification) about the object it labels. Originally, barcodes systematically represented data by varying the widths and spacing of parallel lines and sometimes numerals that is designed to be scanned and read into computer memory. Voice-directed warehousing (VDW) refers to the use of the speech recognition and speech synthesis software to allow workers to communicate with the Warehouse Management System in warehouses and distribution centers.

One of the great productivity benefits of voice-based systems is that they allow operators to do two things at once whereas other media used in warehouses such as paper or radio frequency terminals tend to require that you surrender the use of at least one hand or you have to stop and read something before proceeding. This productivity benefit tends to be inversely related to the fraction of an operator's day spent walking. Information technology is a vital element for providing professional and reliable forwarding and logistics services around the world. DB Schenker is capable of efficient electronic interchange of all relevant shipping information, lead times, performance indicators, etc. Unified worldwide information and communication systems are the backbone to handle shipments not only physically, but also informational to the satisfaction of the clientele. Sophisticated methods such as SWORD, EDI, tracking incl. electronic proofs of delivery (ePOD), barcoding /scanning and mobile communications ensure efficient processes and smooth flow of information between DB Schenker, the local subsidiaries and branches and not least the customers.

5 Conclusion

The thrust of any 3PL-generated VMI program is the reduction of inventory while maintaining customer service and production leveled. The economy continues to put pressure on many industries and the automotive industry is no exception. One of the key supply chain issues facing automakers today involves long order-to-delivery lead times and unreliable production schedules that lead to excess inventory throughout the value chain. Providing a sophisticated production supply to the plant, 3PL has had a significant role to play in the automotive manufacturer's new record. Main recipes for success are the Just-in-Time and Just-in-Sequence strategies, KANBAN, VMI hub, warehouse management system and Value Added Services. JIT has served as a very effective method in the automotive industry for managing functions such as production, purchasing, inventory control, warehousing and transportation. It's potential benefits are limitless. Sustainability is becoming an increasingly important criterion for automotive logistics. 3PLs have to be adaptable and be able to offer solutions at early stages. JIT II is an extension of the vendor managed inventory concept. This could be further discussed in the future on how a 3PL could contribute its solution to this concept.

References

Cetinkaya S, Lee C-Y (2000) Stock replenishment and shipment scheduling for vendor-managed inventory systems. Manage Sci 46:217–232

Erdem E, Reha U, Paul VP, George B, Subramanian K, Jeffrey DT (2005) Outbound supply chain network design with mode selection, lead times and capacitated vehicle distribution centers. Eur J Oper Res 165:182–206

Handfield RB, Nichols EL Jr (1999) Introduction to supply chain management. Prentice Hall, Upper Saddle River

Hertz S, Alfredsson M (2003) Strategic development of third party logistics providers. Ind Mark Manage 32:139–149

Kubiak W (1993) Minimizing variation of production rates in just-in-time system: a survey. Eur J Oper Res 66:259–271

Southard PB, Swenseth SR (2008) Evaluating vendor-managed inventory (VMI) in non-traditional environments using simulations. Int J Prod Econ 116:275–287

Tezuka K (2011) Rationale for utilizing 3PL in supply chain management: a shippers' economic perspective. IATSS Res 35(1):24–29

Yao Y, Evers PT, Dresner ME (2007) Supply chain integration in vendor-managed inventory. Decis Support Syst 43:663–674

Dynamics of Food and Beverage Subsector Industry in East Java Province: The Effect of Investment on Total Labor Absorption

Putri Amelia, Budisantoso Wirjodirdjo and Niniet Indah Arvitrida

Abstract Gross Regional Domestic Product (GRDP) gives an overview of economic development performance in over time. The GRDP value in East Java province is increasing in every year, but this value cannot reduce unemployment rate. The existing of economic development condition is not able to absorb the large number of unemployment where the number of work force is also increasing in the period of time. These phenomena will influence social and economic problems in a region. Investment is one of ways that could be used in order to increase the field of business and reduce the unemployment rate. In this paper, describes the industrial development in food and beverage industries subsector in East Java. By using system dynamic approach, it will construct the change of business investment related to labor absorption in East Java which is affected by economic and climate change of industrial business factors. This research will consider several policy scenarios that have been taken by government such as changes in the proportion of infrastructure funds, the proportion of aid investment by government and licensing index. This scenario could reach the main objective of the system.

Keywords Food and beverage subsector industry · Investment · Labor force · System dynamic

---

P. Amelia (✉) · B. Wirjodirdjo · N. I. Arvitrida
Department of Industrial Engineering, Institut Teknologi Sepuluh Nopember Surabaya, Surabaya, Indonesia
e-mail: putri.amelia.ie@gmail.com

B. Wirjodirdjo
e-mail: wirjodirdjo@gmail.com

N. I. Arvitrida
e-mail: arvietrida@gmail.com

Y.-K. Lin et al. (eds.), *Proceedings of the Institute of Industrial Engineers Asian Conference 2013*, DOI: 10.1007/978-981-4451-98-7_16,
© Springer Science+Business Media Singapore 2013

1 Introduction

Economic growth is one of indicators to see development results that has been done. It defines as the increasing value of goods and services that produced by every sector of the economy in a region. Many ways have been done by the local government in order to develop economy in a region. One of the ways that taken by government is making local government law number 22 of 1999 about regional autonomy and law number 32 and 33 of 2004 about fiscal balancing between central and local government. Then, economic growth in each region especially in each province in Indonesia will be more optimal.

The GRDP value in each region will give an overview of economic development performance in every year. Base on Central Bureau of Statistics in Indonesia shows five major islands in Indonesia which has the highest GRDP value is Java Island. On this island, the highest GRDP value is DKI Jakarta province which has Rp 862,158,910,750,000, and then the next highest province is an East Java province Rp 778,455,772,460,000.

The GRDP value in each region will give the overview of economic development performance in every year which consist agriculture, mining and quarrying, manufacture industry, electricity gas and water supply, construction, trade hotel and restaurant transportation and communication, banking and other financial intermediaries, and services sector. From nine economic sectors, manufacturing industry sector is an important sector from other sector development. The increase output of manufacturing industry will affect the increase number of products in other sectors. Therefore, the manufacturing sector is often regarded as the leading sector from another sector in economic sector (Arsyad 2000).

In the GRDP comparison at current price between DKI Jakarta and East Java province, it is known that manufacturing industry sector is the third-largest contributing sector in economic sector. Moreover, the amount of manufacturing industry GRDP sector in DKI Jakarta province is least than in East Java province. East Java province has a larger industrial area than DKI Jakarta province which is 7.40380 ha. Therefore, East java province has a bigger opportunity to open the industrial business than in the DKI Jakarta province.

The amount of GRDP contribution in current market still cannot absorb the large number of unemployment in East Java province. In fact, economic development has not been able to create the increase of job opportunities faster than the labor force. Based on Table 1 it showed that unemployed people in East Java have not absorbed fully in every sector in the economy sector.

Table 1 East java province labor force in 2005–2009

Period	Labor force	Employed	Unemployed
2009	20338568	19305056	1033512
2008	20178590	18882277	1296313
2007	20177924	18811421	1366503
2006	19244959	17669660	1575299

Food and beverage industry subsector is part of the manufacturing industry sector which gives the significant contribution in the economic development. In addition, this sector also contributed a great contribution in employment. The manufacturing industry contribution to the economic development in four consecutive years from 2005 to 2008 is 12.4, 14, 14.5, and 15.4 %. The amount of the contribution sector was offset by the labor contribution in East Java province which is 20, 21, 21, and 21 % (BPS 2010a, b). This value gives a larger contribution than economic development contribution. Moreover, food and beverage industry is a kind of labor-intensive industries. Labor-intensive industries are one of the type industries which oriented to the largest labor rather than in the capital industry for purchasing a new technology. If this subsector is continued to be developed, it will get the large labor absorption.

Investment is an attempt that can be done to support the increase of industries tin a region. The increase of investment will influence production capacity in industrial sector. Then, it will also impact the increase of output value in the regional development sectors, economic growth, and the personal real income (Samuelson 1995). Due to the increase of personal real income, it will affect the purchasing of goods and services produced in a region. Moreover, it will influence the living standard and community prosperity in a region (Yasin et al. 2002). Eventually, the huge number of demand can encourage the number of investment projects in an area.

Investment activities, investors always have many considerations in investment decisions. One of the factors is investment climate. The Circumstances of investment climate depends on the potential and the economic crisis (Sarwedi 2005). Based on the existing data on the Regional Autonomy Committee (2008), East Java province has a rating of 6 from 33 provinces in Indonesia in case of investment climate index. Based on these data it can be seen that the province of East Java province is still not capable of being a top choice in making investments. Although, there are various regulations and policies that are made by central and local governments in order to increase investment attractiveness such as infrastructure policies, regulations regarding business licensing, tax and others.

In short, in order to increase the investment, it is needed the favorable investment climate. The right of investment climate will trigger the increment of investment industry. Then it will also affect the number of labor absorption and reduce the unemployed. In fact the investment climate in the region is always changing according to the state economy, and other factors that influence it. As a result, it will impact the uncertainty and dynamics of the industrial investment and also change the unemployed people in a region. Therefore, in this research will be discussed about the influence food and beverage industry sub-sectors investment to the labor absorption in East Java by using systems dynamic approach. This method could describe climate change investment due to changes in business and economic factors in East Java. Simulation model will also be considered the regulations and current policies that have been made by central government or local government. Then, in order to know the hole of the system, it can be known the main variable main variables that can affect the amount of the food and beverage industry investment. Finally, it can be known the right policy that can be conduct to the system through the alternative policy scenario.

2 Conceptual Model of System Dynamic

Conceptualization model begins with identifying variables that interact and influence in the system. Identification variables are given by knowing the characteristics and behavior of the investment system in food and beverage industry. Variables that exist in the study were obtained from the literature, brainstorming, and government ministry such as the Department of Trade and Industry, Investment Coordinating Board and Central Bureau of Statistics.

Moreover, the indentified variable has been done in order to analyze the cause and effect relationship between variable in system. In Fig. 1, it will provide a general overview of the system. There are 2 kind of loop which is negative and positive loop. Positive loop was illustrated by the red line flow, and the negative loop was illustrated by the blue line flow. One example of a positive causal loop consists "investment-production-infrastructure availability-energy cost-input-value added" variable (Fig. 1). Then, One example of a negative causal loop consists "investment-other cost-input-value added" variable (Fig. 1).

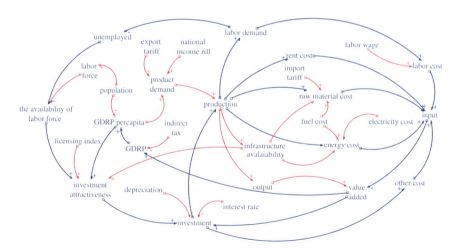

Fig. 1 Causal loop diagram

3 System Dynamic Simulation

In the food and beverage industry system, there are 6 main modules which consists investment module, industry module, labor module, GRDP module, transportation infrastructure module, and investment attractiveness module (Fig. 2). Moreover, 6 main modules will interact with each other. In each main module contains many variables which are interacting with other variables. In this system, the period of simulation will run from 2006 to 2025. The long period of the system is based on the long term development plan targets by East Java government (Fig. 3).

4 Verification and Validation

Verification model aims to examine the error and ensure the model that is built according to logic research problem. Verification is done by checking formulations (equations) and checking unit variables. Otherwise, validation model aims to examine that the model could represent the real system. The validation test consists of structure model test, model parameters test, extreme conditions test, and behavior model test using Barlas method (1996).

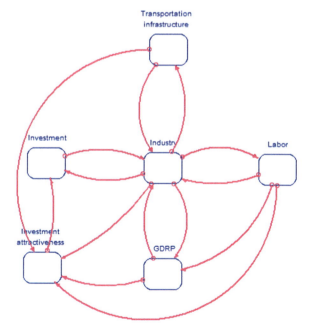

Fig. 2 Main module

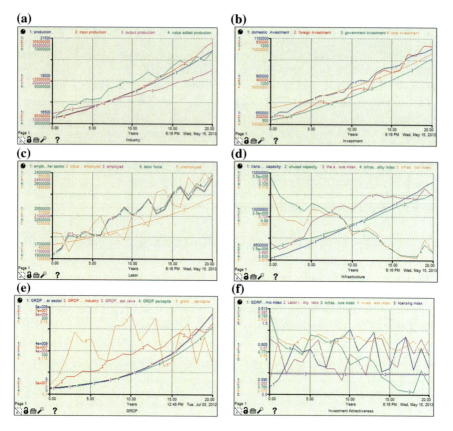

Fig. 3 Simulation model. **a** Industry sub module, **b** Investment sub module, **c** Labor sub module, **d** Transportation infrastructure sub module, **e** GRDP sub module, **f** Investment attractiveness sub module

5 Policies Scenario

Improvements scenario are done to increase food and beverage industrial investment and reduce the unemployment. In each scenario, it considers 2 scenarios which increases and decreases the value of decision variables.

5.1 Scenario 1: Changes in Infrastructure Funds

The value of infrastructure fund variable will give a major impact on the infrastructure availability and value added that has been produced by all sector. In this scenario will conduct two scenarios which increase the value to be 0.15 in the first scenario, and decrease the value to be 0.05 in the second scenario (Table 2).

Dynamics of Food and Beverage Subsector Industry 139

Table 2 Scenario 1: changes in infrastructure funds

Control variable	Lower limit	Upper limit	Existing	Scenario 1	Scenario 2
Infrastructure	0	1	0.1	0.15	0.05

Table 3 Scenario 2: changes in investment assistance fund

Control variable	Lower limit	Upper limit	Existing	Scenario 1	Scenario 2
Infrastructure proportion	0	1	0.000017	0.000034	0.000005

Table 4 Scenario 3: changes in regional licensing index

Control variable	Lower limit	Upper limit	Existing	Scenario 1	Scenario 2
Licensing index	0	1	0.73	0.83	0.5

Based on the simulation results is known that the growth rate of food and beverage industrial investment scenario 1 is increasing 0.08 %. Otherwise, the growth rate of food and beverage industrial investment scenario 1 is decreasing 0.02 %. In the terms of workforce, it is known that the unemployment rate on average each year in the first scenario has decreased 0.01 % and it will increase 0.01 in the second scenario.

5.2 Scenario 2: Changes in Investment Assistance Fund

In every year, the government has provided investment funds especially on small business loans. On this scenario will be changes in the proportion of investment funds that provided by the local government. Based on the simulation results, it is known that the growth rate of food and beverage industrial investment in scenario 1 is increasing 0.001 %. Then the growth rate of food and beverage industrial investment in scenario 2 is decreasing 0.0009 %. In the terms of workforce, it is known that the unemployment rate in the first scenario decreased 0.0004 %, and then the value in the second scenario increased 0.0002 % (Table 3).

5.3 Scenario 3: Changes in Regional Licensing Index

In this system, licensing has a role to increase the investment attractiveness in a region. Based on the simulation results, it known that growth rate investment in scenario 1 increased 0.02 %, and in scenario 2 decreased 1.4 %. In the terms of workforce, it is known that the unemployment rate in the first scenario decreased 0.004 %, then an increase in the second scenario of 0.01 % (Table 4).

6 Conclusions and Research Recommendations

The development food and beverage industry depends on the amount of investment in this subsector. The amount of investment depends on opportunities for foreign and domestic investment. In order to know the right policy that related to investment. In this research, there are three scenarios which consists changes in the proportion of infrastructure funds, changes in investment assistance fund and changes in the investment index. Based on a scenario, it showed that the most influential to increase investment and labor absorption is first scenario which is increasing proportion of infrastructure funds.

Recommendation for future research can focus in the special industrial scale such as large and medium industries or micro and small industries. In addition, the future research can conduct about other sectors that can give the most effect to food and beverage industry sectors such as trade and transport sectors.

Acknowledgments This research is supported by the Department of Trade and Industry in East Java province and East Java Investment Coordinating Board to Mr. Arya as Department Industry in East Java province and Mrs. Desi as staff East java Investment Coordinating Board for their supports of domain knowledge and data collection.

References

Arsyad L (2000) Ekonomi Pembangunan. Sekolah Tinggi Ilmu Ekonomi Yayasan keluarga Pahlawan Negara. S. YKPN. Yogyakarta

Barlas Y (1996) Formal aspects of model validity and validation in system dynamics. System Dynamics Review 12:3. Wiley

BPS (2010) Keadaan Angkatan Kerja di Provinsi Jawa Timur 2010. Surabaya, Badan Pusat Statistik Jawa Timur

BPS (2010) Pertumbuhan Ekonomi Jawa Timur Tahun 2010. Surabaya, Badan Pusat Statistik Jawa Timur

Samuelson (1995) Intellectual property rights and the global information economy. J Commun ACM 39:7

Sarwedi (2005) Investasi Asing Langsung di Indonesia dan Faktor yang mempengaruhinya. Jurnal Akuntasi dan Keuaangan 4:17–35

Yasin, Baihagi et al (2002) Pemograman visual 2002. Surabaya

Solving Two-Sided Assembly Line Balancing Problems Using an Integrated Evolution and Swarm Intelligence

Hindriyanto Dwi Purnomo, Hui-Ming Wee and Yugowati Praharsi

Abstract Assembly line balancing problem (ALBP) is an important problem in manufacturing due to its high investment cost. The objective of the assembly line balancing problem is to assign tasks to workstations in order to minimize the assembly cost, fulfill the demand and satisfy the constraints of the assembly process. In this study, a novel optimization method which integrates the evolution and swarm intelligence algorithms is proposed to solve the two-sided assembly line balancing problems. The proposed method mimics the basic soccer player movement where there are two main movements, the *move off* and the *move forward*. In this paper, the *move off* and the *move forward* are designed based on the specific features of two-sided assembly line balancing problems. Prioritize tasks and critical tasks are implemented in the *move off* and *move forward* respectively. The performance of the proposed method is compared to the heuristic and ant colony based method mentioned in the literature.

Keywords Move off · Move forward · Two-sided assembly lines

H. D. Purnomo (✉) · Y. Praharsi
Department of Information Technology, Satya Wacana Christian University,
Salatiga 50711, Indonesia
e-mail: hindriyanto.purnomo@staff.uksw.edu

Y. Praharsi
e-mail: yugowati.praharsi@staff.uksw.edu

H.-M. Wee
Department of Industrial and System Engineering, Chung Yuan Christian University,
32023 Chungli, Taiwan, Republic of China
e-mail: weehm@cycu.edu.tw

Y.-K. Lin et al. (eds.), *Proceedings of the Institute of Industrial Engineers Asian Conference 2013*, DOI: 10.1007/978-981-4451-98-7_17,
© Springer Science+Business Media Singapore 2013

1 Introduction

Assembly line balancing problem (ALBP) is an important problem in manufacturing due to the higher investment cost; it comprises all the decisions and tasks including the system capacity (Boysen et al. 2007). An assembly line is a sequential workstation that is connected by material handling system (Askin and Standridge 1993). Based on the task orientation, ALBP is commonly classified into one-sided assembly line and two-sided assembly line. One-sided assembly line balancing problems (OALBP) is the most widely studied assembly line balancing problem (Lee et al. 2001) while little attention has been given for the two-sided assembly line balancing problems (TALBP). However, TALBPs are very important for large-sized products such as buses and truck (Kim et al. 2000, 2009).

The objective of the assembly line balancing problems is to assign tasks to workstations in order to minimize the assembly cost, fulfill the demand and satisfy the constraints of the assembly process. The ALBP is an NP-hard combinatorial problem (Gutjahr and Nemhauser 1964). Therefore, heuristic and meta-heuristic methods are commonly used to solve these problems; they are: genetic algorithm (Kim et al. 2000, 2009), ant colony optimization (Simaria and Vilarinho 2009), simulated annealing (Özcan 2010), particle swarm optimization (Nearchou 2011), and bee colony optimization (Özbakır and Tapkan 2011).

In this paper, we proposed a new method that combines the basic concept of evolution algorithm and swarm intelligence algorithm to solve two-sided assembly line balancing problem. The method mimics the soccer player movements in deciding their position based on the ball and other player's position. The rest of the paper is organized as follows. Section 2 describes the TALBP model. Section 3 explains the proposed method. Section 4 explains the implementation of the proposed method in TALBP. Section 5 presents the experiment and discusses the results. Conclusion and future directions are presented in Sect. 6.

2 Two-sided Assembly Line Balancing Problem

The two-sided assembly lines consist of workstations that are located in each side of the line, the left side and the right side. A mated workstation performs different task simultaneously on the same individual product. The tasks directions are classified into: left (L), right (R) and either (E). The consideration of the assembly direction may result in idle time (Özcan 2010) or interference (Simaria and Vilarinho 2009).

The primary objective of the TALBP is minimizing the number of workstations for a given cycle time C which is equivalent to reducing the idle time d as much as possible. Assembly lines with a small amount of idle time offer high efficiency. The assembly line efficiency is defined as:

$$WE = \frac{\sum_{i=1}^{n} t_i}{wC}; \quad i \in I \tag{1}$$

Besides decreasing the idle time, the tasks should be distributed as equally as possible to balance the workload of the workstations. The average idle time in a workstation is given by:

$$\bar{d} = \frac{wC - \sum_{i=1}^{n} t_i}{w} \tag{2}$$

Then, the workload balance between workstations is formulated as:

$$WB = \frac{\sqrt{\sum_{k=1}^{w} \left(C - \sum_{i=1}^{n} x_{ik} t_i - \bar{d}\right)^2}}{wC} \tag{3}$$

Equation 1 is a maximization function while Eq. 3 is a minimization function. The objective of the proposed TALBP model is defined as:

$$Maximize\ F = \alpha WE - \beta WB \tag{4}$$

where α and β are the weight for work efficiency and workload balance respectively.

3 The Proposed Method

$P(X) = P_0; S(X) = S_0;$ */* Initialize players and substitute players*
$B = best\ (P(X))$ */* Initialize ball dribbler*
$P_b(X) = P\ (X)$ */* initialize player's best position*
While *termination criteria not met*

    For $i = 1: p$

        If *random* $\leq m$

            $X_t^i = move\_off(\ X_{t-1}^i)$
            If *random* $\leq l$
                $X_t^i = move\_forward\ (\ X_t^i, X_b^i, B)$
        Endif
        Else

            $X_t^i = move\_forward(\ X_{t-1}^i, X_b^i, B)$
        Endif
    Endfor
    Update_ball_dribbler(P, B)
    If *random* $< k$

Subs (*P, S*)
Endif
Update_ player_best_position(*P, P_b*)
Update_subs_player_knowledge(*P, S*)
Endwhile

4 Implementation of the Proposed Method for TALBP

4.1 Initialization

In this paper, we adopt the COMSOAL heuristic method proposed by Arcus (1966) to encode the problems into the proposed method. In assigning tasks into a workstation, a modification of '*assignable task*' by Gamberini et al. (2006) is implemented as '*prioritize tasks*'. In a prioritize task, its predecessor tasks have been assigned and its processing time is shorter than the remaining workstation available time. Prioritize tasks can minimize the chance of creating long idle time due to opening new workstation.

The procedure for the initialization is as follow:

Step 1 Open the first mated-workstation. The left side workstation $w_l = 1$ while the right workstation $w_r = 2$. List all unassigned tasks UT

Step 2 From UT, list all prioritize tasks (*PT*). If $PT = \emptyset$, then set PT as list of tasks in which their predecessor have been assigned

Step 3 Select a task from PT based on the selection rules

Step 4 If the selected task has specific operations direction; put the task in the appropriate side. If the finish time of the task is less than the cycle time C, put the task in the current workstation, otherwise open new workstation based on the task specific operations direction. If the selected task can be placed in either side, select the side in which it completes early. If both sides have the same finish time, then select the side randomly

Step 5 Remove the selected task for UT

Step 6 If all tasks have been assigned, then stop, otherwise, go to step 2

In step 3, five selection rules are used by the same probability. (1) MAX-RPW that selects a task having the maximum Ranked Positional Weight (RPW). (2) MAX-TFOL that select a task having the maximum total successors. (3) MAX-FOL that selects a task having the maximum number of direct successors. (4) MAX-DUR that selects a task having the maximum operation time and (5) random selection.

Fig. 1 *Move off* for TALBP

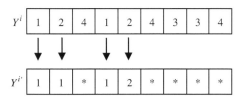

4.2 The Move Off

In this paper, the *move off* is performed by reassigning tasks start from the randomly selected mated-workstations. Assume the selected mated-workstation $r = 1$ (the mated-workstation consist of workstation 1 and 2), then the reassignment process is started from workstation 3. *The move off* procedure is illustrated in Fig. 1.

4.3 The Move Forward

In the *move forward,* the ball position B, the current player position Y^i and the player best position Y^i_b are considered to decide the new player position. The *move forward* will transmit the tasks that are assigned in the same workstation in all the three position mentioned above (B, Y^i and Y^i_b) while the remaining tasks will be reassigned. In reassigning the tasks, '*critical task*' by Kim et al. (2009), is considered. Critical task is such a task in which at least one of its predecessors and one of its successors are assigned in the same pair of workstation. This implies than the critical task must be assigned in the same workstation with its predecessor and successor; therefore, it should be sequenced early.

5 Experimental Results and Discussion

Five benchmark problems, P12, P16, P24, P65, and P148 are used to measure the performance of the proposed method. Problems P12 and P24 can be found in Kim et al. (2000), while P16 is described in Kim et al. (2009). P 65 is formulated by Lee et al. (2001), while P148 is created by Bartholdi (1983). The modification of P148 as in Lee et al. (2001) is used where the processing time of tasks 79 and 108 were changed from 281 to 111 and from 383 to 43. The parameter of the proposed method was set as the following: probability of *move off* $m = 0.5$, probability of *move forward* after the *move off* $l = 0.3$ and the probability of substitution $k = 0.1$.

Table 1 Result for small size problems

Problem	C	LB	MIP	The proposed method		
				Mean	Min	Max
P12	5	5	**6**	**6.0**	6	6
	7	4	**4**	**4.0**	4	4
P16	16	6	**6**	**6.0**	6	6
	22	4	**4 .**	**4.0**	4	4
P24	18	8	**8**	**8.0**	8	8
	25	6	**6**	**6.0**	6	6

The experiments were run 10 times for each problem. The number of maximum objective function evaluation for P12 was 500, P16 and P24 were 5000, P65 was 10,000, and P148 was 20,000. The number objective function evaluation means the number of objective function being evaluated during the search process (Table 1).

For the small size problems, the proposed method is compared to the Mixed Integer Programming, which is an exact solution method. Columns C and LB represent the given cycle time and the lower bound respectively. The results of the experiments show that the proposed method achieved the optimal solution for all given problems (Table 2).

For the large size problems, the proposed method is compared to heuristic rules H and group assignment G proposed by Lee et al. (2001) and the 2-ANTBAL proposed by Simaria and Vilarinho (2009). The experiment results show that the proposed method clearly outperforms the heuristic rules and group assignment procedure as it produces better solution or at least the same, for all problems. When compared to the 2-ANTBAL, the proposed method perform better in

Table 2 Result for large size problems

Problem	C	LB	2-ANTBL (Simaria and Vilarinho 2009)			Lee et al. (2001)		The proposed method		
			Mean	Min	Max	H	G	Mean	Min	Max
P65	326	16	**17.0**	17	17	17.7	17.4	**17.0**	17	17
	381	14	14.8	14	15	15.7	15.0	**14.7**	14	15
	435	12	**13.0**	13	13	14.0	13.4	**13.0**	13	13
	490	11	12.0	12	12	12.1	12.0	**11.2**	11	12
	544	10	10.8	10	11	11.5	10.6	**10.2**	10	11
P148	204	26	**26.0**	26	26	27.8	27.0	**26.0**	26	27
	255	21	**21.0**	21	21	22.0	21.0	**21.0**	21	21
	306	17	18.0	18	18	19.3	18.0	**17.0**	17	18
	357	15	15.4	15	16	16.0	**15.0**	**15.0**	15	15
	408	13	14.0	14	14	14.0	14.0	**13.0**	13	13
	459	12	**12.0**	12	12	12.1	13.0	**12.0**	12	12
	510	11	**11.0**	11	11	12.0	**11.0**	**11.0**	11	11

6 problems, perform the same in 6 problems (mainly because the solutions are already the optimal solutions).

Generally, we can conclude that in term of the number of workstations, the proposed method outperforms the 2-ANTBAL proposed by Simaria and Vilarinho (2009) and both approached methods proposed by Lee et al. (2001).

6 Conclusions

In this study, we develop a new meta-heuristic method based on the integration of evolution algorithm and swarm intelligence algorithm to solve two-sided assembly line balancing problems. In order to obtain a good solution, we proposed a method that mimics the movement behavior of soccer player. It incorporates the information sharing mechanism as well as the evolution operators in its two basic movements, the *move off* and the *move forward*. The movements are used to balance the intensification and diversification.

The basic principle COMSOAL heuristic is adopted in the *move off* and *moves forward*. The prioritize tasks and critical task are implemented in the move off and move forward respectively. The superior performance of the proposed method was illustrated in the experiments results. The proposed method is restricted to simplified model of soccer player movements; therefore, more research is necessary to elaborate and improve the method. For further research, the proposed method can be implemented on mixed model and stochastic assembly lines.

References

Arcus A (1966) COMSOAL: a computer method of sequencing operations for assembly lines. Int J Prod Res 4:259–277

Askin RG, Standridge CR (1993) Modeling and analysis of manufacturing systems. Wiley, Florida

Bartholdi JJ (1983) Balancing two-sided assembly lines: a case study. Int J Prod Res 31:2447–2461

Boysen N, Fliedner M, Scholl A (2007) A classification of assembly line balancing problems. Eur J Oper Res 183(2):674–693

Gamberini R, Grassi A, Rimini B (2006) A new multi-objective heuristic algorithm for solving the stochastic assembly line re-balancing problem. Int J Prod Econ 102:226–243

Gutjahr AL, Nemhauser GL (1964) An algorithm for the line balancing problem. Manage Sci 11(2):308–315

Kim YK, Kim Y, Kim YJ (2000) Two-sided assembly line balancing: a genetic algorithm approach. Prod Plann Control 11:44–53

Kim YK, Song WS, Kim JH (2009) A mathematical model and a genetic algorithm for two-sided assembly line balancing. Comput Oper Res 36(3):853–865

Lee TO, Kim Y, Kim YK (2001) Two-sided assembly line balancing to maximize work relatedness and slackness. Comput Ind Eng 40(3):273–292

Nearchou AC (2011) Maximizing production rate and workload smoothing in assembly lines using particle swarm optimization. Int J Prod Econ 129(2):242–250

Özbakır L, Tapkan P (2011) Bees colony intelligence in zone constrained two-sided assembly line balancing problem. Expert Syst Appl 38(9):11947–11957

Özcan U (2010) (2010) Balancing stochastic two-sided assembly lines: a chance-constrained, piecewise-linear, mixed integer program and a simulated annealing algorithm. Eur J Oper Res 205(1):81–97

Purnomo HD, Wee HM (2012) Soccer game optimization: an innovative integration of evolutionary algorithm and swarm intelligence algorithm. In: Vasant P (eds) Meta-Heuristics optimization algorithms in engineering, business, economics, and finance. IGI Global, Pennsylvania

Simaria AS, Vilarinho PM (2009) 2-ANTBAL: an ant colony optimization algorithm for balancing two-sided assembly lines. Comput Ind Eng 56(2):489–506

Genetic Algorithm Approach for Multi-Objective Optimization of Closed-Loop Supply Chain Network

Li-Chih Wang, Tzu-Li Chen, Yin-Yann Chen, Hsin-Yuan Miao, Sheng-Chieh Lin and Shuo-Tsung Chen

Abstract This paper applies multi-objective genetic algorithm (MOGA) to solve a closed-loop supply chain network design problem with multi-objective sustainable concerns. First of all, a multi-objective mixed integer programming model capturing the tradeoffs between the total cost and the carbon dioxide (CO_2) emission is developed to tackle the multi-stage closed-loop supply chain design problem from both economic and environmental perspectives. The multi-objective optimization problem raised by the model is then solved using MOGA. Finally, some experiments are made to measure the performance.

Keywords Multi-objective · Genetic algorithm · Closed-loop supply chain

---

L.-C. Wang (✉) · S.-C. Lin
Department of Industrial Engineering and Enterprise Information, Tunghai University, Taichung 40704, Taiwan, Republic of China
e-mail: wanglc@thu.edu.tw

T.-L. Chen
Department of Information Management, Fu Jen Catholic University, New Taipei 24205, Taiwan, Republic of China
e-mail: chentzuli@gmail.com

Y.-Y. Chen
Department of Industrial Management, National Formosa University, Yunlin County 632, Taiwan, Republic of China
e-mail: yyc@nfu.edu.tw

H.-Y. Miao
Department of Electrical Engineering, Tunghai University, Taichung 40704, Taiwan, Republic of China

L.-C. Wang · H.-Y. Miao · S.-T. Chen
Tunghai Green Energy Development and Management Institute (TGEI), Tunghai University, Taichung 40704, Taiwan, Republic of China

Y.-K. Lin et al. (eds.), *Proceedings of the Institute of Industrial Engineers Asian Conference 2013*, DOI: 10.1007/978-981-4451-98-7_18,
© Springer Science+Business Media Singapore 2013

1 Introduction

Recent years, carbon asset became a critical subject to global enterprises. Global enterprises need to provide effective energy-saving and carbon-reduction means to meet the policies of Carbon Right and Carbon Trade (Subramanian et al. 2010). In this scenario, forward and reverse logistics have to be considered simultaneously in the network design of entire supply chain. Moreover, the environmental and economic impacts also need to be adopted and optimized in supply chain design (Gunasekaran et al. 2004; Srivastava 2007). Chaabane et al. (2012) introduce a mixed-integer linear programming based framework for sustainable supply chain design that considers life cycle assessment (LCA) principles in addition to the traditional material balance constraints at each node in the supply chain. The framework is used to evaluate the tradeoffs between economic and environmental objectives under various cost and operating strategies in the aluminum industry.

This paper applies multi-objective genetic algorithm (MOGA) to solve a closed-loop supply chain network design problem with multi-objective sustainable concerns. First of all, a multi-objective mixed integer programming model capturing the tradeoffs between the total cost and the carbon dioxide emission is developed to tackle the multi-stage closed-loop supply chain design problem from both economic and environmental perspectives. Then, the proposed MOGA approach is used to deal with multi-objective and enable the decision maker to evaluate a greater number of alternative solutions. Based on the MOGA approach, the multi-objective optimization problem raised by the model is finally solved.

The remainder of this paper is organized as follows. Section 2 introduces the multi-objectives closed-loop supply chain model and the proposed MOGA. Experiments are conducted to test the performance of our proposed method in Sect. 3. Finally, conclusions are summarized in Sect. 4.

2 Multi-Objective Closed-Loop Supply Chain Design

2.1 Problem Statement and Model Formulation

This section will firstly propose a multi-objectives closed-loop supply chain (MOCLSCD) model to discuss the relationship of forward and reverse logistics, the plant locations of forward and reverse logistics, the capacity of closed-loop logistics, and carbon emission issues. Next, decision makers have to determine the potential location and quantity of production and recycling units in forward and reverse logistics, furthermore, design the capacity and production technology level.

This investigation assumes manufacturers will recycle, reuse, and refurbish the EOL products though the RC process. The EOL product collection includes that product recycling from customers or used product markets. Any EOL products which cannot be used or recycled will leave the supply chain via disposal process.

Thus, a multi-stage and multi-factory closed-loop supply chain structure will be formed as Fig. 1.

Decision makers have to determine the potential location and quantity of recycling units, furthermore, design the capacity based on recycle quantity from customers as shown in Fig. 2a. The production unit capacities may expand or shrink due to uncertain material supply, customer demand, and recycling in Fig. 2b. Reverse logistics and capacity increasing result in more carbon emission in closed-loop supply chain than open-loop supply chain. Effective means such as eco-technologies or lean management are necessary for manufacturers to reduce carbon emission at a lower cost.

Based on the above assumptions and challenges, this paper will discuss the relationship of forward and reverse logistics, the plant locations of reverse logistics, the capacity of closed-loop logistics, and carbon emission issues. The proposed new closed-loop supply chain model is expected to reach the environmental and economical benefit by considering various levels of carbon emission manufacturing process and invested cost. The assumptions used in this model are summarized as follows:

1. The number of customers and suppliers and their demand are known.
2. Second market is unique.
3. The demand of each customer must be satisfied.
4. The flow is only allowed to be transferred between two consecutive stages.
5. The number of facilities that can be opened and their capacities are both limited.
6. The recovery and disposal percentages are given.

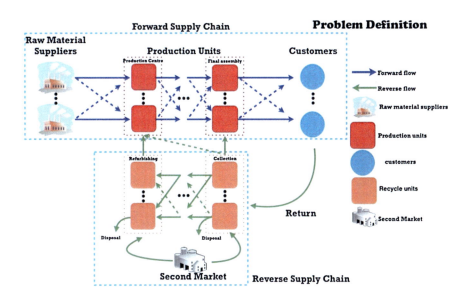

Fig. 1 The structure of sustainable closed-loop supply chain

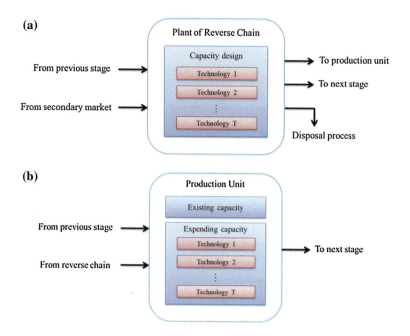

Fig. 2 The characteristics of closed-loop supply chain. **a** Recycling unit in reverse supply chain. **b** Production unit in forward supply chain

Multi-Objective Closed-Loop Supply Chain Design (MOCLSCD) Problem:
The purpose of the MOCLSCD model aims to identify the trade-off solutions between the economic and environmental performances under several logistic constraints. The economic objective, $F_1 = PC + BC + MC + CEC + TC + DC$, is measured by the total closed-loop supply chain cost. The environmental objective, $F_2 = PCOE + BCOE + TCOE$, is measured by the total carbon (CO_2) emission in all the closed-loop supply chain.

Economic objective (F_1):

1. Total material purchasing cost (PC)
2. Total installation cost (BC)
3. Total production cost (MC)
4. Total capacity expansion cost (CEC)
5. Total transportation cost (TC)
6. Total disposal cost (DC).

Environmental objective (F_2):

1. Total production carbon emission (PCOE)
2. Total installation carbon emission (BCOE)
3. Total transportation carbon emission (TCOE).

Genetic Algorithm Approach 153

Constraints:

1. Material supply constraints
2. Flow conservation constraints
3. Capacity expansion and limitation constraints
4. Transportation constraints
5. Domain constraints.

2.2 Multi-Objective Genetic Algorithm

In this section, a novel multi-objective genetic algorithm (MOGA) with ideal-point non-dominated sorting is designed to find the optimal solution of the proposed MOCLSCD model. First of all, the proposed ideal-point non-dominated sorting for non-dominated set is as follows.

a. Calculate $\left(F_1', F_2'\right)_{x_i} = \left(\frac{F_1 - F_1^{\min}}{F_1^{\max} - F_1^{\min}}, \frac{F_2 - F_2^{\min}}{F_2^{\max} - F_2^{\min}}\right)$ and the distance $dist_{x_i} = \sqrt{F_1'^2 + F_2'^2}$ between $\left(F_1', F_2'\right)_{x_i}$ and zero, where $(F_1, F_2)_{x_i}$ is an element in non-dominated set; $F_1^{\max}$ and $F_2^{\max}$ are the highest values of the first and second objectives among experiments; $F_1^{\min}$ are $F_1^{\min}$ the lowest values of the first and second objectives among experiments.
b. Sort the distances $dist_{x_i}, \forall i$

which leads to the proposed MOGA in the following.

1. Chromosome representation for initial population: The chromosome in our MOGA implementation is divided into three segments namely **Popen, Pcapacity**, and **Ptechnology**, according to the structure of the proposed MOC-LSCD model. Each segment is generated randomly.
2. Flow assign and fitness evaluation: The flow assign between two stages depends on the values of the two objectives F_1, F_2. Respectively, the flow with minimum values of F_1, F_2 simultaneously will be assigned firstly. On the point of this view, the proposed ideal-point non-dominated sorting is performed again on the priority of flow assign. The fitness values of F_1 and F_2 are then calculated.
3. Crossover operators: Two-Point Crossover is used to create new offsprings of the three segments **Popen, Pcapacity**, and **Ptechnology**.
4. Mutation operators: The mutation operator helps to maintain the diversity in the population to prevent premature convergence.
5. Selection/Replacement strategy: The selection method adopts the $(\mu+\lambda)$ method suggested by Horn et al. (1994) and Deb (2001).
6. Stopping criteria: There are two stopping criteria proposed. Due to the time constraints in the real industry, the first stopping criterion of the proposed MOGA approach is the specification of a maximum number of generations. The algorithm will terminate and obtain the near-optimal solutions once the

iteration number reaches the maximum number of generations. In order to search better solutions without time-constraint consideration, the second stopping criterion is the convergence degree of the best solution. While the same best solution has not been improved in a fixed number of generations, the best solution may be convergent and thus the GA algorithm is automatically terminated (Fig. 3).

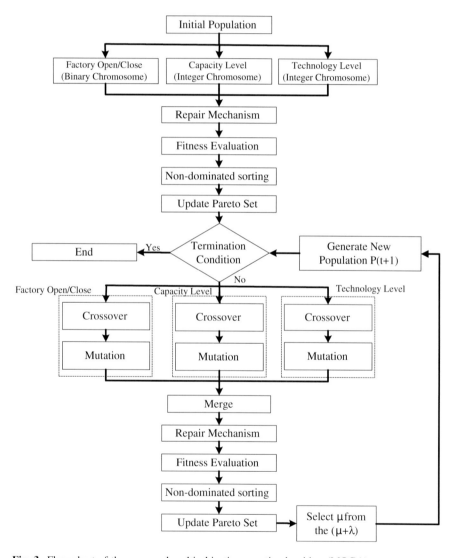

Fig. 3 Flow chart of the proposed multi-objective genetic algorithm (MOGA)

Fig. 4 Non-dominated set for CPLEX method, PSO method, and the propose MOGA

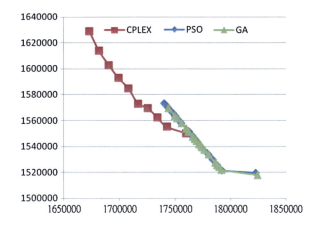

3 Experimental results

This section gives the experimental results. As an example, we implement the MOGA approach on the MOCLSCD model with 2 stages, 2 facilities in each stage, 4 capacity levels, and 4 technology levels. Figure 4 shows the non-dominated set for CPLEX method, PSO method, and the propose MOGA. Figure 5 shows the optimal 2 stages-2 facilities supply chain design results.

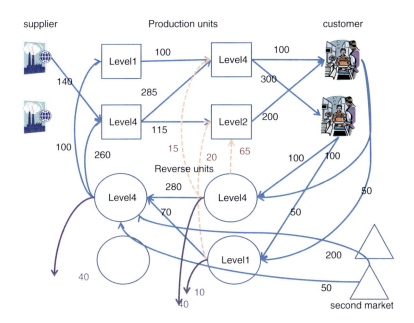

Fig. 5 Optimal 2 stages-2 facilities supply chain design

4 Conclusions

In this work, we firstly proposed a multi-objectives closed-loop supply chain (MOCLSCD) model. Next, a novel multi-objective genetic algorithm (MOGA) with ideal-point non-dominated sorting is used to solve the model. The experimental results show the efficiency of the proposed MOGA when comparing with CPLEX method and PSO method.

References

Chaabane A, Ramudhin A, Paquet M (2012) Design of sustainable supply chains under the emission trading scheme. Int J Prod Econ 135(1):37–49

Deb K (2001) A Fast and Elitist Multiobjective Genetic Algorithm: NSGA-II. IEEE Trans Evol Comput 6(2):182–197

Gunasekaran A, Patel C, McGaughey RE (2004) A framework for supply chain performance measurement. Int J Prod Econ 87(3):333–347

Horn J, Nafpliotis N, Goldberg DE (1994) A niched pareto genetic algorithm for multiobjective optimization. Paper presented at the IEEE world congress on computational intelligence

Srivastava SK (2007) Green supply-chain management: a state-of-the-art literature review. Int J Manage Rev 9(1):53–80

Subramanian R, Talbot B, Gupta S (2010) An approach to integrating environmental considerations within managerial decision-marking. J Ind Ecol 14(3):378–398

Replacement Policies with a Random Threshold Number of Faults

Xufeng Zhao, Mingchih Chen, Kazunori Iwata, Syouji Nakamura and Toshio Nakagawa

Abstract Most systems fail when a certain amount of reliability quantities have exceeded their threshold levels. The typical example is cumulative damage model in which a system is subjected to shocks and suffers some damage due to shocks, and fails when the total damage has exceeded a failure level K. This paper proposes the following reliability model: Faults occur at a nonhomogeneous Poisson process and the system fails when N faults have occurred, which could be applied to optimization problems in computer systems with fault tolerance, and we suppose that the system is replaced before failure at a planned time T. Two cases where the threshold fault number N is constantly given and is a random variable are considered, we obtain the expected cost rates and discuss their optimal policies.

Keywords Replacement · Constant threshold · Random threshold · Faults

---

X. Zhao (✉) · T. Nakagawa
Aichi Institute of Technology, Toyota, Japan
e-mail: g09184gg@aitech.ac.jp

T. Nakagawa
e-mail: toshi-nakagawa@aitech.ac.jp

M. Chen
Fu Jen Catholic University, New Taipei, Taiwan
e-mail: 081438@mail.fju.edu.tw

K. Iwata
Aichi University, Nagoya, Japan
e-mail: kazunori@vega.aichi-u.ac.jp

S. Nakamura
Kinjo Gakuin University, Nagoya, Japan
e-mail: snakam@kinjo-u.ac.jp

Y.-K. Lin et al. (eds.), *Proceedings of the Institute of Industrial Engineers Asian Conference 2013*, DOI: 10.1007/978-981-4451-98-7_19,
© Springer Science+Business Media Singapore 2013

1 Introduction

Most systems often fail when a certain amount of quantities due to various causes have exceeded a threshold level. The typical example is the cumulative damage model in which systems are subjected to shocks and suffer some damage due to shocks, and fail when the total damage has exceeded a failure level (Nakagawa 2007). This is called the cumulative process in stochastic processes (Cox 1962; Nakagawa 2011). There are continuous wear models in which the total damage increases with time t and the system fails when it has exceeded a failure level (Reynols and Savage 1971; Lehmann 2010). In some actual cases, an amount of damage due to each shock might not be estimated exactly, and so that, the total damage could not be estimated statistically. In these cases, we could estimate the damage on system only by counting the number of shocks up to present.

However, when shock occurrence becomes not so clear for counting and varies according to some stochastic process, we need to estimate its approximate cumulative distribution function. So that it will become much difficult to perform maintenance for this kind of failure mechanism. Take fault tolerance in computer systems (Adb-El-Barr 2007) as an example, the system will fail or be a low efficiency state when total number of faults exceeds some level, but this level is actually a non-transparent variable for system maintainers. There has been summarized many maintenance models to improve computer reliability in (Nakamura and Nakagawa 2010); However, almost all discussions have based on constant failure threshold level.

We propose that almost all failure thresholds are random for any failure mechanism, and this paper considers the following models: Faults occur at a nonhomogeneous Poisson process (Nakagawa 2011), and the system fails when n faults have occurred. We suppose that the system is replaced before failure at a planned time T and obtain analytically expected replacement cost rate and its optimal policy. A threshold number n in the above model might be actually uncertain, but could be estimated statistically by using the past data. When n is a random variable N, we obtain expected cost rate and make similar discussion of deriving optimal policy.

2 Model Formulation

We consider the following operating system: Faults of the system occur at a non-homogeneous Poisson process with a mean-value function $H(t) \equiv \int_0^t h(u)du$, and it fails when n $(n = 1, 2, \cdots)$ faults have occurred and undergoes only minimal repair before its failure. Letting $M(t)$ be the number of faults in $[0, t]$ (Nakagawa 2007),

$$H_j(t) \equiv \Pr\{M(t) = j\} = \frac{[H(t)]^j}{j!} e^{-H(t)} \quad (j = 0, 1, 2, \cdots), \tag{1}$$

and $E\{M(t)\} = H(t)$.

2.1 Constant n

Suppose that the system is replaced before failure at a planned time T $(0 < T \leq \infty)$, then the probability that the system is replaced at time T before fault $n (n = 1, 2, \cdots)$ is

$$\Pr\{M(T) \leq n - 1\} = \sum_{j=0}^{n-1} H_j(T),$$

and the probability that it is replaced at fault n before time T is

$$\Pr\{M(T) \geq n\} = \sum_{j=n}^{\infty} H_j(T).$$

Thus, the mean time to replacement is

$$T \Pr\{M(T) \leq n - 1\} + \int_0^T t \, d\Pr\{M(t) \geq n\} = \sum_{j=0}^{n-1} \int_0^T H_j(t) dt. \tag{2}$$

Therefore, the expected cost rate is

$$C_1(T; n) = \frac{c_N - (c_N - c_T) \sum_{j=0}^{n-1} H_j(T)}{\sum_{j=0}^{n-1} \int_0^T H_j(t) dt}, \tag{3}$$

where c_T = replacement cost at time T and c_N = replacement cost at fault n with $c_N > c_T$. Clearly,

$$C_1(T; \infty) \equiv \lim_{n \to \infty} C_1(T; n) = \frac{c_T}{T}, Z$$

which decreases with T from ∞ to 0.

When $n = 1$, $C_1(T; 1)$ agrees with the expected cost rates of an age replacement (Nakagawa 2005).

We find an optimal T_n^* $(0 < T_n^* \leq \infty)$ which minimizes $C_1(T; n)$ for a specified n $(1 \leq n < \infty)$. It can be easily seen that

$$C_1(0; \infty) \equiv \lim_{T \to 0} C_1(T; n) = \infty,$$

$$C_1(\infty; \infty) \equiv \lim_{T \to \infty} C_1(T; n) = \frac{c_N}{\sum_{j=0}^{n-1} \int_0^{\infty} H_j(t) dt}. \tag{4}$$

Differentiating $C_1(T; n)$ with respect to T and setting it equal to zero,

$$Q(T;n) \sum_{j=0}^{n-1} \int_0^T H_j(t)dt - \sum_{j=n}^{\infty} H_j(T) = \frac{c_T}{c_N - c_T}, \tag{5}$$

where

$$Q(T;n) \equiv \frac{h(T)H_{n-1}(T)}{\sum_{j=0}^{n-1} H_j(T)},$$

which decreases strictly with n from $h(T)$ to 0. So that, the left-hand side of (5) goes to 0 as $n \to \infty$, i.e., T_n^* becomes large for large n.

Letting $L_1(T;n)$ be the left-hand side of (5),

$$L_1(0;n) \equiv \lim_{T \to 0} L_1(T;n) = 0,$$

$$L_1(\infty;n) \equiv \lim_{T \to \infty} L_1(T;n) = Q(\infty,n) \sum_{j=0}^{n-1} \int_0^{\infty} H_j(t)dt - 1,$$

$$L_1'(T;n) = Q'(T,n) \sum_{j=0}^{n-1} \int_0^T H_j(t)dt.$$

Therefore, if $Q(T;n)$ increases strictly with T and

$$Q(\infty,n) \sum_{j=0}^{n-1} \int_0^{\infty} H_j(t)dt > \frac{c_T}{c_N - c_T},$$

then there exists a finite and unique T_n^* $(0 < T_n^* \leq \infty)$ which satisfies (5), and the resulting cost rate is

$$C_1(T_n^*; n) = (c_N - c_T)Q(T_n^*; n). \tag{6}$$

Furthermore, it is assumed that $h(T)$ increases strictly for $n = 1$ and increases for $n \geq 2$. Then, prove that $Q(T; n)$ increases strictly with T. It is trivial that $Q(T; 1)$ increases strictly. For $n \geq 2$,

$$\left(\frac{[H(T)]^{n-1}/(n-1)!}{\sum_{j=0}^{n-1} \{[H(T)]^j/j!\}} \right)'$$

$$= \frac{h(T)\{[H(T)]^{n-2}/(n-1)!\} \sum_{j=0}^{n-1} (n-1-j)\{[H(T)]^j/j!\}}{\left(\sum_{j=0}^{n-1} \{[H(T)]^j/j!\} \right)^2} > 0,$$

which implies that $Q(T; n)$ increases strictly.

2.2 Random N

Next, a threshold number n is not constant and is a random variable N, i.e., $\Pr\{N = n\} \equiv p_n (n = 1, 2, \cdots)$ and $P_n \equiv \sum_{j=1}^{n} p_j$, where $P_\infty = \sum_{j=1}^{\infty} p_j \equiv 1$. Then, the expected cost rate in (3) is easily rewritten as

$$C_1(T; N) = \frac{c_N - (c_N - c_T) \sum_{n=1}^{\infty} p_n \sum_{j=0}^{n-1} H_j(T)}{\sum_{n=1}^{\infty} p_n \sum_{j=0}^{n-1} \int_0^T H_j(t) dt}, \tag{7}$$

and (5) is

$$Q(T; N) \sum_{n=1}^{\infty} p_n \sum_{j=0}^{n-1} \int_0^T H_j(t) dt - \sum_{n=1}^{\infty} p_n \sum_{j=0}^{n-1} H_j(T) = \frac{c_T}{c_N - c_T},$$

where

$$Q(T; N) \equiv \frac{h(T) \sum_{n=1}^{\infty} p_n H_{n-1}(T)}{\sum_{n=1}^{\infty} p_n \sum_{j=0}^{n-1} H_j(T)}.$$

If $Q(T; N)$ increases strictly with T and

$$Q(\infty; N) \sum_{n=1}^{\infty} p_n \sum_{j=0}^{n-1} \int_0^\infty H_j(t) dt > \frac{c_N}{c_N - c_T},$$

then there exists an optimal T_N^* $(0 < T_N^* < \infty)$ which satisfies above conditions, and the resulting cost rate is

$$C_1(T_N^*; N) = (c_N - c_T) Q(T_N^*; N). \tag{9}$$

It is assumed that $h_{n+1} \equiv p_{n+1}/(1 - P_n) (n = 0, 1, 2 \cdots)$, which is called the discrete failure rate (Nakagawa 2005), and increases strictly with n. This means that the probability of exceeding a threshold number N increases with the number of faults. Then, we prove (Nakagawa 2011) that

$$\frac{\sum_{n=1}^{\infty} p_n \{[H(T)]^{n-1}/(n-1)!\}}{\sum_{n=1}^{\infty} p_n \sum_{j=0}^{n-1} \{[H(T)]^j/j!\}} = \frac{\sum_{n=0}^{\infty} p_{n+1} \{[H(T)]^n/n!\}}{\sum_{n=0}^{\infty} (1 - P_n) \{[H(T)]^n/n!\}}$$

increases strictly with T.

Differentiating above equation with respect to T,

$$\frac{h(T)}{\left(\sum_{n=0}^{\infty} (1 - P_n) \{[H(T)]^n/n!\} \right)^2} \left\{ \sum_{n=1}^{\infty} p_{n+1} \frac{[H(T)]^{n-1}}{(n-1)!} \sum_{j=0}^{\infty} (1 - P_j) \frac{[H(T)]^j}{j!} \right. $$
$$\left. - \sum_{n=1}^{\infty} (1 - P_n) \frac{[H(T)]^{n-1}}{(n-1)!} \sum_{j=0}^{\infty} p_{j+1} \frac{[H(T)]^j}{j!} \right\}.$$

The expressions within the bracket of the numerator is

$$\sum_{n=1}^{\infty} \frac{[H(T)]^{n-1}}{(n-1)!} \sum_{j=0}^{\infty} \frac{[H(T)]^{j}}{j!} (1 - P_n)(1 - P_j)(h_{n+1} - h_{j+1})$$

$$= \sum_{n=1}^{\infty} \frac{[H(T)]^{n-1}}{(n-1)!} \sum_{j=0}^{n-1} \frac{[H(T)]^{j}}{j!} (1 - P_n)(1 - P_j)(h_{n+1} - h_{j+1})$$

$$+ \sum_{n=1}^{\infty} \frac{[H(T)]^{n-1}}{(n-1)!} \sum_{j=n}^{\infty} \frac{[H(T)]^{j}}{j!} (1 - P_n)(1 - P_j)(h_{n+1} - h_{j+1})$$

$$= \sum_{n=1}^{\infty} \frac{[H(T)]^{n-1}}{(n-1)!} \sum_{j=0}^{n-1} \frac{[H(T)]^{j}}{j!} (1 - P_n)(1 - P_j)(h_{n+1} - h_{j+1})(n - j) > 0,$$

which implies that when h_{n+1} increases strictly, $Q(T; N)$ also increases strictly with T.

We have neglected the cost for minimal repair due to each fault. The expected number of minimal repair until replacement is

$$\sum_{j=0}^{n-1} j H_j(T) + (n - 1) \sum_{j=n}^{\infty} H_j(T) = n - 1 - \sum_{j=0}^{n-1} (n - 1 - j) H_j(T).$$

Letting c_M be the cost for minimal repair, the expected cost rate in (3) is

$$\tilde{C}_1(T; n) = \frac{c_N - (c_N - c_T) \sum_{j=0}^{n-1} H_j(T) + c_M \left[n - 1 - \sum_{j=0}^{n-1} (n - 1 - j) H_j(T) \right]}{\sum_{j=0}^{n-1} \int_0^T H_j(t) dt},$$

$$(11)$$

which agrees with that of the periodic replacement when $c_N = c_T$ (Nakagawa 2005). Furthermore, the expected cost rate in (7) is

$$\tilde{C}_1(T; N) = \frac{c_N - (c_N - c_T) \sum_{n=1}^{\infty} p_n \sum_{j=0}^{n-1} H_j(T) + c_M[n - 1 - \sum_{j=0}^{n-1} (n - 1 - j) H_j(T)]}{\sum_{n=1}^{\infty} p_n \sum_{j=0}^{n-1} \int_0^T H_j(t) dt}.$$

$$(12)$$

As further studies, we could make similar discussions for deriving optimal policies which minimize the expected cost rates $\tilde{C}_1(T; n)$ and $\tilde{C}_1(T; N)$.

Table 1 Optimal T_n^* and T_N^* when $H(t) = t^2, p_n = pq^{n-1}$ and $n = 1/p$

n	$c_N/c_T = 2$		$c_N/c_T = 5$		$c_N/c_T = 10$	
	T_n^*	T_N^*	T_n^*	T_N^*	T_n^*	T_N^*
2	1.160	0.857	0.719	0.413	0.564	0.274
3	1.349	0.799	0.949	0.390	0.795	0.259
4	1.541	0.773	1.161	0.380	1.008	0.235
5	1.725	0.759	1.357	0.374	1.205	0.249
7	2.063	0.743	1.710	0.368	1.558	0.245
10	2.510	0.732	2.170	0.364	2.020	0.242
15	3.149	0.723	2.822	0.360	2.674	0.240
20	3.701	0.719	3.382	0.358	3.235	0.239
50	6.163	0.712	5.866	0.355	5.723	0.237

3 Numerical Example

Suppose that $p_n = pq^{n-1} (n = 1, 2, \cdots; 0 < p \leq 1)$ with mean $1/p$, and $h(t)$ increases strictly to $h(\infty) = \lim_{t \to \infty} h(t)$. Then, (8) is

$$h(T) \int_0^T e^{-pH(t)} dt - [1 - e^{-pH(T)}] = \frac{c_T}{c_N - c_T}, \tag{13}$$

whose left-hand side increases strictly with T from 0 to $h(T) \int_0^T e^{-pH(t)} dt - 1$. Therefore, if $h(\infty) > c_N/(c_N - c_T) \int_0^\infty e^{-pH(t)} dt$, then there exists a finite and unique T_N^* $(0 < T_N^* < \infty)$ which satisfies (13), and the resulting cost rate is

$$C_1(T_N^*; N) = (c_N - c_T)h(T_N^*). \tag{14}$$

It can be easily seen that T_N^* increases with p to the optimal time of an age replacement (Nakagawa 2005) for $p = 1$, because the left-hand side decreases with p.

Table 1 presents optimal T_n^* and T_N^* when $H(t) = t^2$, $p_n = pq^{n-1}$ and $n = 1/p$ for $c_N/c_T = 2, 5, 10$ and $n = 2, 3, 4, 5, 7, 10, 15, 20, 50$.

4 Conclusions

We have discussed replacement models with constant and random threshold number of faults. That is, faults occur at a nonhomogeneous Poisson process and the system fails when a constant number n faults and random number N faults have occurred. We have supposed that the system is replaced before failure at a planned time T and obtained the expected cost rates and discussed their optimal policies. From numerical analysis, we have found that both optimal times T_n^* and T_N^* have the same properties with cost rate c_N/c_T but different for $n = 1/p$. Replacement models with random threshold failure level would be a more practical extension of classical methods, which could be done further in the following studies.

Acknowledgments This work is partially supported by Grant-in-Aid for Scientific Research (C) of Japan Society for the Promotion of Science under Grant No. 22500897 and No. 24530371; National Science Council of Taiwan NSC 100-2628-E-0330-002.

References

Abd-El-Barr M (2007) Reliable and fault-tolerant. Imperial College Press, London
Cox DR (1962) Renewal theory. Methuen, London
Lehmann A (2010) Failure time models based on degradation process. In: Nikulin MS et al (eds) Advances in degradation modeling. Birkhauser, Boston, pp 209–233
Nakagawa T (2005) Maintenance theory of reliability. Springer, London
Nakagawa T (2007) Shock and damage models in reliability theory. Springer, London
Nakagawa T (2011) Stochastic processes with applications to reliability theory. Springer, London
Nakamura S, Nakagawa T (2010) Stochastic Reliability Modeling, Optimization and Applications. World Scientific, Singapore
Reynols DS, Savage IR (1971) Random wear models in reliability theory. Adv Appl Probab 3:229–248

A Multi-Agent Model of Consumer Behavior Considering Social Networks: Simulations for an Effective Movie Advertising Strategy

Yudai Arai, Tomoko Kajiyama and Noritomo Ouchi

Abstract It is essential for a firm to understand consumer behavior in order to advertise products efficiently on a limited budget. Nowadays, consumer behavior is highly complex because consumers can get a lot of information about products from not only firm's advertising but also social networking services. The purposes of this study are to construct consumer behavior model considering social networks and to demonstrate an effective weekly advertising budget allocation in order to increase the number of adopters of products. First, we developed a multi-agent model of consumer behavior taking the movie market as an example. In our model, each agent decides whether or not to watch a movie by comparing the weighted sum of "individual preference" and "effects of advertising and word-of-mouth (WOM)" with "individual threshold." The scale-free network is used to describe social networks. Next, we verified the accuracy of the model by comparing the simulation results with the actual sales figures of 13 movies. Finally, we showed an effective weekly advertising budget allocation corresponding to movie type by simulations. Furthermore, it was demonstrated that the weekly advertising budget allocation gives greater impact on the number of adopters of products as social networks grow.

Keywords Consumer behavior · Multi-agent simulation · Decision making · Advertising strategy · Social networks · Movie market

---

Y. Arai (✉)
Graduate School of Science and Engineering, Aoyama Gakuin University,
5-10-1 Fuchinobe, Chuo-ku, Sagamihara-shi, Kanagawa 252-5258, Japan
e-mail: c5612119@aoyama.jp

T. Kajiyama · N. Ouchi
Department of Industrial and Systems Engineering, Aoyama Gakuin University, 5-10-1
Fuchinobe, Chuo-ku, Sagamihara-shi, Kanagawa 252-5258, Japan
e-mail: tomo@ise.aoyama.ac.jp

N. Ouchi
e-mail: ouchi@ise.aoyama.ac.jp

Y.-K. Lin et al. (eds.), *Proceedings of the Institute of Industrial Engineers Asian Conference 2013*, DOI: 10.1007/978-981-4451-98-7_20,
© Springer Science+Business Media Singapore 2013

1 Introduction

Understanding consumer behavior is essential for firms in order to advertise their products efficiently on a limited budget. Nowadays, consumer behavior is highly complex because consumers can get a lot of information about products from not only firm's advertising but also social networking services such as Twitter and Facebook.

Lots of studies have been made on consumer behavior models including diffusion models (e.g. Mansfield 1961; Bass 1969; Mahajan et al. 1990; Libai et al. 2009). However, there are few works that tried to develop the model incorporating social networks' effects. Meanwhile, there are several studies linking consumer behavior models and firms' strategies (e.g. Kamimura et al. 2006; Tilman et al. 2007). For example, Kamimura et al. (2006) demonstrated the effective way of weekly advertising budget allocation in the movie market. However, it does not consider the existence of social networks. In addition, the characteristics of movies are not considered although the effective way of weekly advertising budget allocation might depend on the characteristic of products.

The purposes of this study are to construct consumer behavior model considering social networks and to demonstrate an effective weekly advertising budget allocation in order to increase the number of adopters of products. In this study, we pick up consumer behavior in the movie market because it is strongly influenced by the effects of advertising and word-of-mouth (WOM).

Section 2 presents model construction. In Sect. 3, we demonstrate simulation results. Finally, Sect. 4 summarizes new findings, implications for firm's strategy.

2 Model Construction

In this study, we developed a consumer behavior model considering social networks' effects by using multi-agent simulation taking the movie market as an example. Multi-agent simulation can simulate the actions and interactions of autonomous agents with a view to assessing their effects on the system as a whole.

In our model, 1,000 agents are created in a grid size of 50×50 cells. Agent i decides whether or not to watch a movie at time t by comparing the weighted sum of "individual preference $(P_{i,t})$" and "effects from advertising and WOM $(E_{i,t})$" with "individual threshold (T_i)." By using weighted parameter (a_i), the condition where agent i decides to watch a movie is defined as depicted in inequality (1).

$$a_i \times P_{i,t} + (1 - a_i) \times E_{i,t} > T_i \tag{1}$$

"Individual preference $(P_{i,t})$" represents agent i's degree of preference of the movie. $P_{i,t}$ is computed by Eq. (2).

$$P_{i,t} = (pb + adv + pre) * F^P(t) \times I_i^P \times D_i^P \tag{2}$$

where pb: the effect of production budget, adv: the effect of advertising budget, pre: the effect of the previous movie's sales figure, $F^P(t)$: variation of interest for a new movie over time, I_i^P: individuality of agent i, and D_i: dummy variable denoting whether or not agent i has an interest in the genre of the movie ($D_i = 1$ if agent i has an interest, $D_i = 0$ if agent i does not have an interest).

Only if agent i has an interest in the genre of the movie, individual preference is computed. $F^P(t)$ is a decreasing function of t. In our model, we set $F^P(t) - 0.94^t$ according to Kamimura et al. (2006).

"Effects of advertising and WOM $(E_{i,t})$" is the sum of "effects of advertising $(A_{i,t})$" and "effects of WOM $(W_{i,t})$" as depicted in Eq. (3).

$$E_{i,t} = A_{i,t} + W_{i,t} \tag{3}$$

$A_{i,t}$ is computed by Eq. (4).

$$A_{i,t} = adv \times R^A \times I_i^A \times G^A(t) \tag{4}$$

where R^A: reliability of information source (mass media), I_i^A: agent i's susceptibility of advertising, $G^A(t)$: variation of advertising effect over time.

We set $G^A(t) = B^A \times F^A(t)$. B^A is a fixed value. $F^A(t) = 0.7\,(t<0), F^A(t) = 0.94^t\,(t \geq 0)$, following to Kamimura et al. (2006). In Sect. 3.2, we change $F^A(t)$ corresponding to advertising budget allocation patterns.

Regarding $W_{i,t}$, in this study, we consider three resources of WOM; "neighbors," "Internet sites" and "social networks." Thus, $W_{i,t}$ is the sum of the effects of neighbors WOM $\left(\sum_{j=1}^{N_N} w_{i,j,t}\right)$, the effects of Internet WOM $\left(\sum_{k=1}^{N_I} w_{i,k,t}\right)$ and the effects of social networks WOM $\left(\sum_{l=1}^{N_S} w_{i,l,t}\right)$ as depicted in Eq. (5).

$$W_{i,t} = \sum_{j=1}^{N_N} w_{i,j,t} + \sum_{k=1}^{N_I} w_{i,k,t} + \sum_{l=1}^{N_S} w_{i,l,t} \tag{5}$$

where $w_{i,j,t}$, $w_{i,k,t}$, and $w_{i,l,t}$: the effect of WOM agent i received from agent j, k and l, respectively, N_N, N_I and N_S: the number of agents who can affect agent i's effects of neighbors WOM, Internet WOM and social networks WOM, respectively.

In this model, N_N and N_I are the number of agents who are within 3×3 cells, 21×21 cells around agent i, respectively. N_S is the number of agents who have a link with agent i and are within 21×21 cells around agent i.

The scale-free network (Barabashi and Albert 1999) can be used to describe social networks. We create the links between agents, which denote the social networks, by using the scale-free network.

$w_{i,j,t}$ is computed as depicted in Eq. (6). $w_{i,k,t}$ and $w_{i,l,t}$ are computed in a similar way.

Table 1 The differences of three WOM resources

	B^W	R^W
Neighbor	High (0.32)	High (2)
Internet site	Low (0.15)	Low (1)
Social networks	Medium (0.24)	Medium (1.5)

(): values used in the model

Table 2 Parameters for individual threshold

Category	Proportion (%)	α	β
Innovator	2.5	4	1
Early adopter	13.5	6	3
Early majority	34	9	6
Late majority	34	11	6
Laggards	16	13	3

$$w_{i,j,t} = \begin{cases} (pb + adv + pre) \times R^W \times O_j \times I_i^w \times G^W(t)(before\ release) \\ RatingScore \times R^W \times O_j \times I_i^w \times G^W(t)(after\ release) \end{cases} \quad (6)$$

where R^W: reliability of information source, I_i^W: agent i's susceptibility to WOM, O_j: agent j's ability to impart information, $G^W(t)$: variation of frequency of chats about the movie over time.

We set $G^W(t) = B^W \times F^W(t)$. B^W is a fixed value. $F^W(t) = e^t(t < 0), F^W(t) = 0.94^t(t \geq 0)$, following to Kamimura et al. (2006). Because it can be considered that the values of B^W and R^W are different depending on the information sources, we set these parameters as shown in Table 1.

Individual threshold is computed by Eq. (7).

$$T_i = \alpha + \beta \times r \quad (7)$$

where α, β: coefficient determining individual threshold, r: an uniform random number between 0 and 1.

Agents are characterized as innovators, early adopters, early majority, late majority and laggards according to Rogers' diffusion theory (Rogers 1962). The values of α and β differ depending on these characteristics as shown in Table 2.

3 Simulations

3.1 Model Validation

We validated the proposed model by comparing the simulation results (S) with actual data (A) in terms of the cumulative number of movie viewers every week. As sample data, we selected 13 movies as listed in Table 3. They include the top five box-office movies in 2010 and "Kings Speech", the academy award for best

Table 3 List of 13 movies

Title	Sales figure	Production budget	Rating score
Toy story 3	$415,004,880	$200,000,000	8.84
Alice in wonderland	$334,191,110	$200,000,000	7.18
Iron man 2	$312,433,331	$170,000,000	7.27
The twilight saga: eclipse	$300,531,751	$68,000,000	6.86
Inception	$292,568,851	$160,000,000	8.25
Harry potter and the deathly hallows: part I	$294,980,434	$125,000,000	7.92
Shrek forever after	$238,319,043	$165,000,000	5.75
The book of eli	$94,822,707	$80,000,000	6.33
The social network	$96,619,124	$40,000,000	8.23
The town	$92,173,235	$37,000,000	8.54
Red	$90,356,857	$60,000,000	7.8
Percy jackson and the olympians: the lightning thief	$88,761,720	$95,000,000	7.03
The king's speech	$138,797,449	$15,000,000	8.25

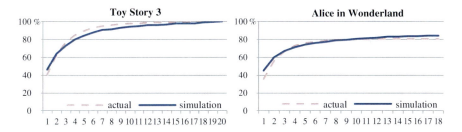

Fig. 1 Trends in actual data and simulation results

motion picture of 2010. The data in Table 3 are available at "The Numbers (http://www.the-numbers.com)". In addition, in our study "Rotten Tomatoes (http://www.rottentomatoes.com/)" was also referred.

We conducted simulations 30 times in each movie. Simulations period is from four weeks before the release date ($t = -4$).

Figure 1 shows examples of trends in cumulative adoption rate of actual data and simulation results. Cumulative adoption rate is calculated by dividing the number of cumulative movie viewers in each week by the total movie viewers of Toy Story 3, which number of total movie viewers is the largest.

Table 4 summarizes the mean error rate. The error rate (E) is computed by Eq. (8).

$$E^k = \frac{|S^k - A^k|}{A^k} \qquad (8)$$

As shown in Table 4, mean error rates can be considered to be small. These results demonstrate that our proposed model can describe the actual consumer behavior.

Table 4 Mean error rates

Movie	Mean error rate (%)
Toy story 3	3.73
Alice in wonderland	3.99
Iron man 2	2.37
The twilight saga: eclipse	9.51
Inception	4.01
Harry potter and the deathly hallows: part I	3.60
Shrek forever after	8.56
The book of eli	6.60
The social network	9.43
The town	8.67
Red	7.07
Percy jackson and the olympians: the lightning Thief	6.24
The king's speech	11.40

3.2 Application to Advertising Strategy

To clarify the effective way of weekly advertising budget allocation, simulations were conducted by using proposed model. We set four patterns of advertising budget allocation, (1) focusing on around a release date (standard), (2) focusing on before a release date (before), (3) focusing on after a release date (after) and (4) same weight during period (constant), as demonstrated in Fig. 2.

Additionally, we divide movies into four types by consumers' expectation for a movie and rating score as shown in Table 5. Expectation is estimated by the effect of production budget, advertising budget and previous movie's sales figure. We analyze the impact of characteristic of movies on the number of movie viewers.

Furthermore, in order to clarify the impacts of social networks on the number of movie viewers, we set the following two conditions; (a) social networks WOM is not considered, (b) social networks WOM is considered.

Figure 3 compares the total number of movie viewers in each situation with the one in the case of (Type 1, Standard, (a) not considered).

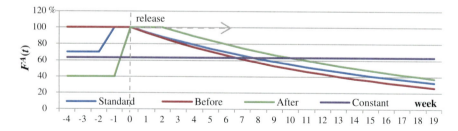

Fig. 2 Four patterns of advertising budget allocation

Table 5 Four movie types

Type	Expectation	Rating score
1	High	High
2	Low	High
3	High	Low
4	Low	Low

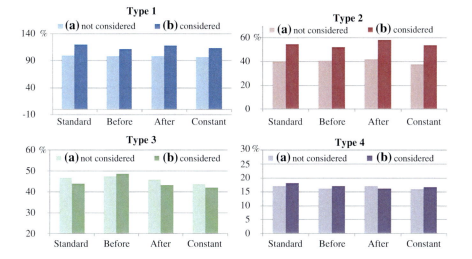

Fig. 3 The comparison of simulation results

In type 2, allocating the budget mainly after release is the best way, while standard allocation is the best way in type 1 and type 4. This might be because a positive WOM is stimulated by effects of advertising after release, and it leads more people to watch a movie. On the other hand, in type 3, allocating the budget mainly before release becomes the best allocation. This result suggests that it is effective to acquire lots of movie viewers by effects of advertising before negative WOM spreads.

In addition, the difference of number of movie viewers between allocation patterns becomes larger when social networks WOM is considered.

4 Conclusions

We constructed a consumer behavior model considering social networks and verified that our proposed model can describe the actual consumer behavior. Then we applied the model to an advertising strategy by conducting the simulation of effective weekly advertising budget allocation. As a result, we clarified that there are effective weekly advertising budget allocations corresponding to the characteristic of movie products. In addition, it was demonstrated that the difference of weekly advertising budget allocation give a lager impact on the number of adopters as social networks grow.

As future work, applying the model to other market should be required.

References

Barabasi AL, Albert R (1999) Emergence of scaling in random networks. Science 286:509–512
Bass FM (1969) A new product growth model for consumer durables. Manage Sci 15:215–227
Kamimura R, Masuda H, Arai T (2006) An agent-based model of consumer behaviour: application to the movie market. Jpn Ind Manage Assoc 57:450–469 (in Japanse)
Libai B, Muller E, Peres R (2009) The diffusion of services. J Mark Res 46(April):163–175
Mahajan V, Muller E, Bass FM (1990) New-product diffusion models in marketing: a review and direction for research. J Mark 54(January):1–26
Mansfield E (1961) Technical change and the rate of imitation. Econometric Soc 29(4):741–766
Rogers EM (1962) Diffusion of innovations. The Free Press, New York
Tilman AS, Gunter L, Jurgen R (2007) Agent-based simulation of consumer behavior in grocery shopping on a regional level. J Bus Res 60:894–903

Government Subsidy Impacts on a Decentralized Reverse Supply Chain Using a Multitiered Network Equilibrium Model

Pin-Chun Chen and I-Hsuan Hong

Abstract Government subsidies to reverse supply chains can play important roles in driving or curtailing the flows of recycled items. We examine the impacts of exogenous subsidies on recycled material flows in a decentralized reverse supply chains where each participant acts according to its own interests. We outline a multitiered model of the supply network from sources of scrap electronics, collectors, processors and demand markets. The individual behavior of each player is governed by participants' optimality conditions, which are mathematically transformed into a variational inequality formulation. The modified projection method is utilized for solving the equilibrium quantities and prices of each participant. We investigate the impact of alternate schemes of government subsidies on decisions of the equilibrium quantities, prices and the total amount collected. For the case studied in this paper, the best tier selection between collectors and processors for government subsidies in terms of the total collected amount is located in collectors in laptop reverse supply chains.

Keywords Reverse supply chain · Government subsidy · Variational inequality · Modified projection method

---

P.-C. Chen (✉) · I.-H. Hong
Institute of Industrial Engineering, National Taiwan University, No. 1, Sec. 4,
Roosevelt Road, Taipei 10617, Taiwan
e-mail: r00546006@ntu.edu.tw

I.-H. Hong
e-mail: ihong@ntu.edu.tw

Y.-K. Lin et al. (eds.), *Proceedings of the Institute of Industrial Engineers Asian Conference 2013*, DOI: 10.1007/978-981-4451-98-7_21,
© Springer Science+Business Media Singapore 2013

1 The Decentralized Reverse Supply Chain Network

Government provides subsidies for recycling in reverse supply chains for environmental consciousness and further usages. Collectors and processors can obtain subsidies from government as incentives which effectively assist in recycling and recovery of e-scrap products. In Taiwan, a semi-official organization supervised by the Environmental Protection Administration (EPA) collects processing fee and subsidizes the associated collectors and processors, yet the discussion on the appropriate subsidy target is neglected: Some research related to the government subsidized e-scrap reverse supply chain regarded the organization between the e-scrap source and the refurbished product demand market as only one party and some classed the recycling system into two or more parties, yet so far the allocation of government subsidies has been overlooked.

The purpose of this study is to analyze how to appropriately allocate government subsidies to collectors and processors in a reverse supply chain consisting of four tiers of decision makers: sources of e-scrap, collectors, processors, and demand markets. In Sect. 2, we present a four-tiered network equilibrium model for reverse supply chain management. In Sect. 3, the case study based on a laptop computer reverse production system in Georgia, United States is demonstrated and the impact of the government subsidy policies on recycled e-scrap products is discussed. The summary is drawn in Sect. 4.

2 The Multitiered Network Equilibrium Model for Decentralized Reverse Supply Chain Management

A multitiered network model is developed for investigating the impact of government subsidies on the behaviors of individual entities in a reverse supply chain. The network consists of four tiers of entities, including r sources of e-scrap products, m collectors, n processors and o demand markets. A source of e-scrap products is denoted by h. A collector is denoted by i, who collects e-scrap products and may further disassemble e-scrap products to separate components and materials from e-scrap products. A processor is denoted by j, who converts e-scrap products into more fungible commodities that are sold to customers in demand markets. Let S_{ij} denote the government subsidy that collector i can receive by selling e-scrap products collected to processor j, while S_{jk} denotes the government subsidy that processor j can receive by dissembling or refurbishing processes and then selling e-scrap products to demand market k.

The objective of the sources of e-scrap, collectors and processors is to maximize their own profits, while the customers in the demand markets have to satisfy the market equilibrium conditions. From definitions of variational inequality problem (Ben-Arieh 1999) and supply chain network Cournot-Nash equilibrium

Government Subsidy Impacts   175

(Nagurney and Yu 2012; Nagurney and Toyasaki 2005), we propose the variational inequality formulation to specifically discuss the government subsidy impacts on the total collected amount in a e-scrap reverse supply chain.

2.1 The Behavior of the Sources of E-scrap Products

The sources make profit from selling e-scrap products to collectors. Let q_{hj} denote the nonnegative amount of e-scrap products that is allocated from source h to collector i, where $h = 1,\ldots,\ r;\ i = 1,\ldots,\ m$. We denote the transaction cost between source h and collector i by c_{hj} and assume that

$$c_{hi} = c_{hi}(q_{hi}),\ h = 1,\ldots,r;\ i = 1,\ldots,m. \tag{1}$$

Each source receives revenue from selling e-scrap products to collectors. We let p^*_{1hi} denote the price that source h charges collector i for a unit of e-scrap products. Each source h seeks to maximize the total profit with the optimization problem by

$$\underset{p_{1hi},q_{hi}}{\text{Max}} \sum_{i=1}^{m} p^*_{1hi} q_{hi} - \sum_{i=1}^{m} c_{hi}(q_{hi}) \tag{2}$$

subject to

$$q_{hi} \geq 0,\ i = 1,\ldots,m. \tag{3}$$

Equation (2) states that a source's profit is equal to sales revenues minus the costs associated with transactions between sources and collectors.

2.2 The Behavior of the Collectors

Collectors purchase e-scrap products from the sources, and sell them to processors after sorting or dismantling e-scrap products. Collectors transact both with the sources and processors. Let q_{ij} denote the nonnegative amount of materials from collector i to processor j, where $i = 1,\ldots,m;\ j = 1,\ldots,n$. Each collector receives revenue from selling e-scrap products to processors after initial sorting or disassembling processes. We let p^*_{2ij} denote the price that collector i charges processor j for a unit of e-scrap products. In addition, we recall that the price p^*_{1hi} denotes the price that collector i pays source j for a unit of e-scrap products. We denote the transaction cost between collector i and processor j by c_{ij} and assume that

$$c_{ij} = c_{ij}(q_{ij}),\ i = 1,\ldots,m;\ j = 1,\ldots,n. \tag{4}$$

A transportation cost paid by collector i between each collector i and source h, denoted by $\hat{c}_{hi}$, and assume that

$$\hat{c}_{hi} = \hat{c}_{hi}(q_{hi}), \ h = 1,\ldots,r; \ i = 1,\ldots,m. \tag{5}$$

The cost to collect e-scrap products at collector i is denoted by c_i and is assumed to be

$$c_i = c_i(Q^2), \ i = 1,\ldots,m. \tag{6}$$

A collector can receive a government subsidy. We let S_{ij} denote the subsidy that collector i can receive by selling e-scrap products to processor j and assume that

$$S_{ij} = S_{ij}(q_{ij}), \ i = 1,\ldots,m; \ j = 1,\ldots,n. \tag{7}$$

A collector seeks to maximize the total profits and collector i's optimization problem can be formulated as

$$\underset{q_{hi},q_{ij}}{\text{Max}} \ \sum_{j=1}^{n} p_{2ij}^* q_{ij} - \sum_{h=1}^{r} p_{1hi}^* q_{hi} - \sum_{j=1}^{n} c_{ij}(q_{ij}) - \sum_{h=1}^{r} \hat{c}_{hi}(q_{hi}) - c_i(Q^2) + \sum_{j=1}^{n} S_{ij}(q_{ij}) \tag{8}$$

subject to

$$\sum_{j=1}^{n} q_{ij} \le \sum_{h=1}^{r} q_{hi}, \tag{9}$$

$$q_{hi} \ge 0, \ h = 1,\ldots,r; \ q_{ij} \ge 0, \ j = 1,\ldots,n. \tag{10}$$

Equation (8) states that a collector's profit is equal to the sales revenue plus the subsidies minus the costs.

2.3 The Behavior of the Processors

Processors purchase e-scrap products from collectors, and sell them to the demand markets after processing e-scrap products. The processes include smelting or refining materials into pure metal streams (Sodhi and Reimer 2001) and refurbishing or recovering the e-scrap products. Let q_{jk} denote the nonnegative amount of materials that is allocated from processor j to demand market k where $j = 1,\ldots,n; \ k = 1,\ldots,o$.

Each processor receives revenue from selling processed e-scrap products to different demand markets. We let p_{3jk}^* denote the price that processor j charges consumers at demand market k for a unit of e-scrap products. The transaction cost between processor j and demand market k is denoted by c_{jk} and is assumed to be

$$c_{jk} = c_{jk}(q_{jk}), \ j = 1,\ldots,n; \ k = 1,\ldots,o. \tag{12}$$

We denote a transportation cost paid by processor j between processor j and collector i, denoted by $\hat{c}_{ij}$, and assume that

Government Subsidy Impacts

$$\hat{c}_{ij} = \hat{c}_{ij}(q_{ij}), \ i = 1, \ldots, m; \ j = 1, \ldots, n. \tag{13}$$

A processor has a processing cost to smelt, refine, refurbish or recover e-scrap products at processor j is denote by c_j and is assumed to be

$$c_j = c_j(Q^3), \ j = 1, \ldots, n. \tag{14}$$

A processor can receive a government subsidy. We let S_{jk} denote the subsidy that processor j can receive by selling one unit of processed e-scrap products to demand market k and assume that

$$S_{jk} = S_{jk}(q_{jk}), \ j = 1, \ldots, n; \ k = 1, \ldots, o. \tag{15}$$

A processor seeks to maximize its total profits and processor j's optimization problem can be formulated as

$$\underset{q_{ij}, q_{jk}}{\text{Max}} \sum_{k=1}^{o} p_{3jk}^* q_{jk} - \sum_{i=1}^{m} p_{2ij}^* q_{ij} - \sum_{k=1}^{o} c_{jk}(q_{jk}) - \sum_{i=1}^{m} \hat{c}_{ij}(q_{ij}) - c_j(Q^3) + \sum_{k=1}^{o} S_{jk}(q_{jk}) \tag{16}$$

subject to

$$\sum_{k=1}^{o} q_{jk} \leq \sum_{i=1}^{m} q_{ij}, \tag{17}$$

$$q_{ij} \geq 0, \ i = 1, \ldots, m; \ q_{jk} \geq 0, \ k = 1, \ldots, o. \tag{18}$$

Equation (16) states that a processor's profit is equal to the sales revenue plus the subsidies minus the costs.

2.4 The Demand Markets and the Equilibrium Conditions

Consumers at demand markets transact with processors. We denote the transportation cost between demand market k and processor j by $\hat{c}_{jk}$, and assume that

$$\hat{c}_{jk} = \hat{c}_{jk}(q_{jk}), \ j = 1, \ldots, n; \ k = 1, \ldots, o, \tag{19}$$

We assume that the demand function of demand market k is

$$d_k = d_k(p_4), \ k = 1, \ldots, o, \tag{20}$$

where p_4 is the vector of the collective price of demand markets, and we denote the price of demand market k by p_4.

The equilibrium conditions for consumers at demand market k are identical to the well-known spatial equilibrium conditions (Nagurney et al. 2002) and are dictated by the following conditions for all processors.

$$p_{3jk}^* + \hat{c}_{jk}(q_{jk}) \begin{cases} = p_{4k}^*, & \text{if } q_{jk}^* > 0, \\ \geq p_{4k}^*, & \text{if } q_{jk}^* = 0, \end{cases} \tag{21}$$

$$d_k(p_4^*) \begin{cases} = \sum_{j=1}^{n} q_{jk}^*, & \text{if } p_{4k}^* > 0, \\ \leq \sum_{j=1}^{n} q_{jk}^*, & \text{if } p_{4k}^* = 0. \end{cases} \tag{22}$$

Conditions (21) guarantee that the transaction isn't unprofitable; Conditions (22) guarantee that demand is feasible. The modified projection method (see Ben-Arieh 1999; Ceng et al. 2010; Nagurney and Zhang 1997; Zeng and Yao 2006) is utilized to solve the proposed variational inequality formulation.

3 Case Study of the Laptop Computer Reverse Supply Chain

3.1 Case Study Overview and Data Input

Based on Hong et al. (2012), we model the laptop computer reverse supply chain in Georgia as a four-tier network, which consists of twelve sources of e-scrap, twelve collectors, three processors and three demand markets. Table 1 gives the detailed geographical information about the four tiers in Georgia's case.

Given the ground and ocean transportation costs from Hong et al. (2012), we estimate the unit transportation costs between tiers. We use the information given

Table 1 The geographical information of the laptop computer reverse supply chain

Node type	Node	State/country	County/city
Sources of e-scrap (tier 1) and collectors (tier 2)	$S_1 = C_1$	Georgia	Gordon
	$S_2 = C_2$	Georgia	White
	$S_3 = C_3$	Georgia	DeKalb
	$S_4 = C_4$	Georgia	Meriwether
	$S_5 = C_5$	Georgia	Oconee
	$S_6 = C_6$	Georgia	Bibb
	$S_7 = C_7$	Georgia	Richmond
	$S_8 = C_8$	Georgia	Chattahoochee
	$S_9 = C_9$	Georgia	Toombs
	$S_{10} = C_{10}$	Georgia	Dougherty
	$S_{11} = C_{11}$	Georgia	Ware
	$S_{12} = C_{12}$	Georgia	Chatham
Processors (tier 3) and demand markets (tier 4)	$P_1 = D_1$	Georgia	Marietta
	$P_2 = D_2$	Tennessee	Nashville
	$P_3 = D_3$	Nigeria	Lagos

Government Subsidy Impacts

Table 2 The impact of subsidy on total material flows

Total material flows (laptop unit)	Scenario 1 $a = 0$ $b = 2.88$	Scenario 2 $a = 2.88$ $b = 0$	Scenario 3 $a = 1.44$ $b = 1.44$	Scenario 4 $a = 2.88$ $b = 2.88$
1. $b_{4k} = -1$; $b_{-4k} = 0, 0$	5332.06	6367.92	5724.42	6403.08
2. $b_{4k} = -1$; $b_{-4k} = -0.5, -0.5$	2774.51	3280.59	2980.51	3349.53
3. $b_{4k} = -1$; $b_{-4k} = -0.8, -0.5$	2346.70	2770.19	2506.48	2843.15
4. $b_{4k} = -1$; $b_{-4k} = -0.8, -0.8$	2006.60	2379.20	2161.50	2446.50
5. $b_{4k} = -1$; $b_{-4k} = -1, -1$	1625.89	1972.08	1773.92	2037.92

in Khetriwal et al. (2009) to estimate the unit collection cost, unit processing cost and transaction cost, and data in Yoshida et al. (2009) to estimate market demand.

We assume that demand functions are linear functions with a slope vector $b = [b_{4k} \, b_{-4k}]$ and a constant intercept a, where b_{4k} denotes the coefficient of price in demand market $k = 1, 2, 3$ and b_{-4k} denotes the coefficients of prices exporting to others. We vary b_{-4k} as five different combinations to study the impact of other demand markets on the total material flow. The demand functions are $d_k(p_4) = -0.973 \, p_{4k} + b_{-4k} \, p_{-4k} + 580$, $k = 1, 2, 3$. We define the government subsidy in a linear form and assume that the subsidy functions for collector i and processor j are $S_{ij} = aq_{ij}$, $i = 1, 2,...,12$; $j = 1, 2, 3$, $S_{jk} = bq_{jk}$, $j = 1, 2, 3$; $k = 1, 2, 3$, where a and b are indicated in Table 2.

3.2 Case Study Result and the Impact of Subsidies on Flows

We investigate the impact of government subsidies on the total collected amount, where the government only puts the subsidy at collectors (Scenario 1), processors (Scenario 2), both equally at collectors and processors (Scenario 3), and at collectors and processors with a double subsidy (Scenario 4). The corresponding total material flows under each scenario with different combinations on demand functions are shown in Table 2.

4 Summary

We investigate the individual behaviors of a decentralized reverse supply chain system, and formulate the general network as a variational inequality problem to obtain the equilibrium prices and material flows between tiers. From a large-scale case study, we construct a multitiered reverse supply chain network of laptop computers in Georgia, USA under four different subsidy scenarios. Several insights are drawn: (1) government subsidies efficiently encourage collectors and processors to recycle and remanufacture scrap laptop computers; (2) a higher level of subsides results in a higher total collection flow; (3) the scenario allocating all subsidies to collectors (upstream) outperforms the scenario allocating subsidies to processors (downstream) in terms of the total flow.

Acknowledgments This work was supported in part by the National Science Council, Taiwan, under Grant NSC99-2221-E-002-151-MY3.

References

Ben-Arieh D (1999) Network economics: a variational inequality approach, 2nd edn. Kluwer, Dordrecht

Ceng LC, Hadjisavvas N, Wong NC (2010) Strong convergence theorem by a hybrid extra gradient-like approximation method for variational inequalities and fixed point problems. J Glob Optim 46(4):635–646

Hong IH, Ammons JC, Realff MJ (2012) Examining subsidy impacts on recycled WEEE material flows. In: Goodship V, Stevels A (ed) Waste Electrical and Electronic Equipment (WEEE) Handbook, 1st edn. Woodhead Publishing, Cambridge, UK

Khetriwal DS, Kraeuchi P, Widmer R (2009) Producer responsibility for e-waste management: key issues for consideration—Learning from the Swiss experience. J Environ Manage 90(1):153–165

Nagurney A, Toyasaki F (2005) Reverse supply chain management and electronic waste recycling: a multitiered network equilibrium framework for e-cycling. Transp Res Part E 41(1):1–28

Nagurney A, Yu M (2012) Sustainable fashion supply chain management under oligopolistic competition and brand differentiation. Int J Prod Econ 135(2):532–540

Nagurney A, Zhang D (1997) Projected dynamical systems in the formulation, stability analysis, and computation of fixed-demand traffic network equilibria. Transp Sci 31(2):147–158

Nagurney A, Dong J, Zhang D (2002) A supply chain network equilibrium model. Transp Res Part E 38(5):281–303

Sodhi MS, Reimer B (2001) Models for recycling electronics end-of-life products. OR Spectrum 23(1):97–115

Yoshida A, Tasaki T, Terazono A (2009) Material flow analysis of used personal computers in Japan. Waste Manage (Oxford) 29(5):1602–1614

Zeng LC, Yao JC (2006) Strong convergence theorem by an extragradient method for fixed point problems and variational inequality problems. Taiwanese J Math 10(5):1293–1303

A Capacity Planning Method for the Demand-to-Supply Management in the Pharmaceutical Industry

Nobuaki Ishii and Tsunehiro Togashi

Abstract In the pharmaceutical industry, in order to secure a reliable supply of drags, the manufacturer tends to possess a production capacity much higher than the market demand. However, because of the severe competition and increasing product variety, the demand-to-supply management, which strategically supplies products to the market based on a collaborative strategy with the sales function and the production function in order to maximize profit, becomes a critical issue for any manufacturer in regards to improving his competitiveness and sustainability. In this paper, we propose a capacity planning method and a tool, developed with an engineering consulting company that can be used to support the demand-to-supply management in the pharmaceutical industry. The method synthesizes an initial manufacturing process structure and the capacity of each process unit based on the demand forecast and candidate equipment specifications at the first step. Then it improves the process structure in a step-by-step fashion at the second step. The developed tool supports the development and evaluation of the process structure from the perspective of the utilization of each process unit, investment cost, operation cost, and so on. We show the effectiveness of the developed method and the tool through a case study.

Keywords Demand-to-supply management · Engineering economy · Facilities planning

---

N. Ishii (✉)
Faculty of Information and Communications, Bunkyo University, 1100 Namegaya,
Chigasaki 2538500 Saitama, Japan
e-mail: ishii@shonan.bunkyo.ac.jp

T. Togashi
CM Plus Corporation, 6-66 Sumiyoshi, Naka-Ku, Yokohama 2310013, Japan
e-mail: togashi@cm-plus.co.jp

Y.-K. Lin et al. (eds.), *Proceedings of the Institute of Industrial Engineers Asian Conference 2013*, DOI: 10.1007/978-981-4451-98-7_22,
© Springer Science+Business Media Singapore 2013

1 Introduction

Over the last several decades, manufacturing industries have faced big changes in the business environment such as increasing global competition, increasing demand uncertainties, and so on. To survive and succeed in this severe business environment, manufacturing industries have been keen to improve their competitiveness with the introduction of advanced manufacturing technologies. For instance, SCM (Supply Chain Management) (Simchi-Levi et al. 2008) is expected to shorten lead times, reduce inventory levels, increase sales, and improve their capability to quickly respond to the market.

However, the shorted lead times, reduced inventory levels, and increased sales do not always attain the profit maximization. As Matsui (2009) pointed out by using the pair-strategic chart, the collaborative decision making among different functions, especially the sales function and the production function, is necessary to attain the profit maximization. Thus, most industries have recognized that the demand-to-supply management, which strategically supplies products to the market based on a collaborative strategy with the sales function and the production function in order to maximize profit, is critical for implementing successful SCM and for establishing a sustainable company in today's business environment.

It is also the same in the pharmaceutical industry. In the past, in order to secure a reliable supply of drags, the pharmaceutical manufacturer possessed a production capacity much higher than the market demand. Strong demand and long product life cycles also justify such excess capacity. However, the increase of cost and term length in the research and development of new drugs in the market carries substantial risk in the pharmaceutical industry. In addition, because of the popularity of generic drugs, most pharmaceutical manufacturers are under increasing pressure to change from mass production to small-lot production, and also to reduce manufacturing cost. For these reasons, successful implementation of the demand-to-supply management is a critical issue in the pharmaceutical industry. In particular, the advanced capacity planning for flexible production in response to the demand variation is required.

In this paper, we propose a capacity planning method and a tool, jointly developed by an engineering consulting company, CM Plus Corporation in Yokohama, Japan, intended to support the demand-to-supply management in the pharmaceutical industry. The method synthesizes an initial manufacturing process structure and the capacity of each process unit based on the demand forecast and candidate equipment specifications at the first step. Then it improves the process structure in a step-by-step fashion at the second step. The developed tool supports the development and evaluation of the process structure from the perspective of the utilization of each unit, investment cost, operation cost, and so on. We show the effectiveness of the developed method and the tool through a case study.

2 Overview of the Demand-to-Supply Management and Capacity Planning

One of the goals of demand-to-supply management is to improve corporate sustainability under today's uncertain and unforeseeable market conditions. The overall framework of demand-to-supply management consists of the business lifecycle management, the demand-to-supply strategic management, and demand-to-supply management (Hayashi et al. 2009) as depicted in Fig. 1.

The business lifecycle management creates products and markets, and designs a business plan from a long-term strategic perspective. Any company must decide the right withdrawal time from the market of matured products. The system for the long-term strategic perspective supports decisions for product release and revision as well as decisions for product portfolio management to improve corporate sustainability.

The demand-to-supply strategic management creates the demand development plan and the supply system plan, which are made based on the collaborative work by the sales function and the production function for maximizing profit from a mid-term perspective. The demand development plan includes the marketing strategy, pricing strategy, etc. The supply system plan includes the structure of production system, capacity expansion, outsourcing, etc.

From the short-term perspective, the demand-to-supply management creates the collaborative demand-to-supply plan, based on the strategic map developed using the concept of a pair-matrix table (Matsui 2002). The collaborative demand-to-supply plan is used to control the production rate as well as demand rate to maximize the profit in the uncertain business environment. An application software tool for the collaborative demand-to-supply planning, called the *planner*, is

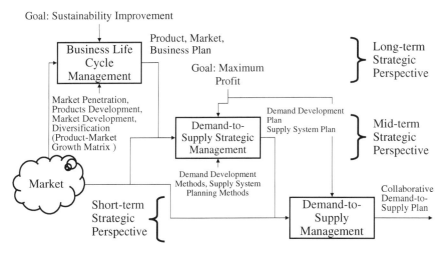

Fig. 1 An overall framework for demand-to-supply management

designed and demonstrated (Matsui et al. 2006; Hayashi et al. 2009). In most companies, Enterprise Resource Planning (Monk and Wagner 2008) has been widely implemented for corporate core information systems since the middle of the 1990s. ERP is a type of software application that fully integrates core business functions, including transaction processing, management information, etc. However, most ERP packages do not include the concept of demand-to-supply management. The *planner* is expected to complement the demand-to-supply management functions with EPR.

As a supply system planning method in the demand-to-supply strategic management from a mid-term strategic perspective, Ishii et al. (1997) developed a process synthesis method in consideration of the product lifecycle scenario and equipment reliability. Ishii (2004) developed a heuristic procedure for planning flexible production systems. Matsui et al. (2007) applied the procedure to making a mid-term supply system plan. In addition, Ohba et al. (2013) analyzed a rate of utilization on process unit for maximizing profit from the demand-to-supply strategic management perspective. However, those methods show no procedure and tool to design and assess the process structure and capacity in detail.

The capacity planning method and the tool explained in this paper are developed to design and evaluate the process structure and capacity of each process unit for the pharmaceutical manufacturing process. It works within the supply system planning method in the demand-to-supply strategic management from a mid-term strategic perspective. The tool supports the design and evaluation of the process structure from the perspective of utilization of each unit, investment cost, operation cost, and so on.

3 A Capacity Planning Method

The capacity planning method consists of the steps as shown in Fig. 2.

Product mix analysis evaluates the middle-term demand forecast on products. Material balancing calculates the quantities of raw materials required and products produced in consideration of the standard yield ratio in each process unit. Material balances provide the basic data for sizing equipment. Lot sizing determines production lot size of each product, and equipment allocation selects equipment in each process unit. Since lot sizing affects the number of changeovers and required capacity of each process unit, lot sizing and equipment allocation must be decided concurrently. Operation analysis evaluates the utilization, production time, and so on based on the lot sizing and equipment allocation of each lot in each process unit.

The method determines an initial manufacturing process structure and capacity of each process based on the demand forecast from product mix analysis and specifications in the equipment list at the first step. Then it improves the process structure in a step-by-step fashion at the second step by changing lot sizing and equipment allocation repeatedly until the utilization, production time, and lot size

A Capacity Planning Method

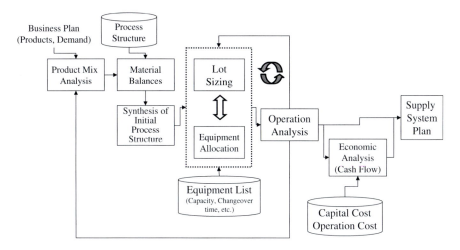

Fig. 2 An overall framework of capacity planning method

are validated in the operation analysis. When the utilization, production time, and lot size are not validated, the product mix analysis is reconsidered so as to match feasible supply capacity.

Figure 3 shows the structure of the capacity planning tool, which is jointly developed by CM Plus Corporation in Yokohama, Japan and Bunkyo University, based on the pharmaceutical manufacturing model.

The tool, where the pharmaceutical manufacturing process is embedded, has the capability to synthesize an initial manufacturing process structure and the capacity of each process unit, and to simulate the utilization, production time, and lot size. It also provides supports for validating lot sizing and equipment allocation in each process unit, and for improving the process structure.

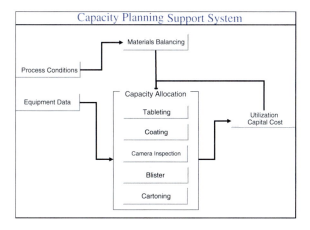

Fig. 3 Structure of the capacity planning tool

4 An Example Problem

In this paper, we use the drag manufacturing process for solid dosage forms, shown in Fig. 4, as an example problem for demonstrating the effectiveness of the developed capacity planning method and the tool.

In this example, we assume 16 products are manufactured from four kinds of tablets. Namely, final products are diversified by the packaging variation as shown in Fig. 5. The amount of total annual production assumed 2,100 million tablets (MM Tablet). In addition, the amount of annual production of each product is shown in Table 1. The equipment specifications to be selected by the capacity planning method are assumed as shown in Table 2.

Table 3 shows the capacity plan, i.e., selected equipment and the number products for manufacturing in each process unit, allocated by the developed capacity planning method and the tool. In addition, utilization of equipment is shown in Table 3. In this case, the process structure is made so as to attain around 85 % utilization by the minimum investment cost.

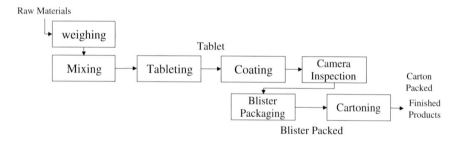

Fig. 4 A drug manufacturing process for solid dosage forms

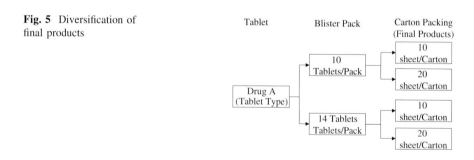

Fig. 5 Diversification of final products

A Capacity Planning Method

Table 1 Annual production of each product

Product			Annual production (MM Tablet/Year)	Conversion (Tablet/Carton)
Tablet	Blister type (Tablet/Pack)	Carton type (Sheet/Carton)		
X	10	A: 10; B: 20	A: 300; B: 200	A: 100; B: 200
	14	A: 10; B: 20	A: 300; B: 200	A: 140; B: 280
Y	10	A: 10; B: 20	A: 200; B: 100	A: 100; B: 200
	14	A: 10; B: 20	A: 200; B: 100	A: 140; B: 280
Z	10	A: 10; B: 20	A: 100; B: 100	A: 100; B: 200
	14	A: 10; B: 20	A: 50; B: 50	A: 140; B: 280
W	10	A: 10; B: 20	A: 50; B: 50	A: 100; B: 200
	14	A: 10; B: 20	A: 50; B: 50	A: 140; B: 280

Table 2 Equipment list

Process	Equipment	Capacity	Unit	Changeover (Hr/Co)	Investment cost (MMJPY)
Tableting	#1 Tablet press	1,667	Tablet/Min.	0.5	31
	#2 Tablet press	3,333		1.0	50
	#3 Tablet press	6,667		1.5	80
Coating	#1 Coater	660,000	Tablet/Cycle	6.0	94
	#2 Coater	937,500		6.0	120
	#3 Coater	1,875,000		6.0	195
Camera inspection	#1 Camera inspection	2,000	Tablet/Min.	0.5	37
	#2 Camera inspection	4,000		1.0	60
	#3 Camera inspection	6,000		1.5	80
Blister packaging	#1 Blister packer	3,000	Tablet/Min.	1.0	62
	#2 Blister packer	6,000		2.0	100
Cartoning	#1 Cartoner	100	Carton/Min.	0.5	32
	#2 Cartoner	200		1.0	53
	#3 Cartoner	300		1.5	70

Table 3 A Capacity Plan (T:Tablet C:Carton Y:Year)

Process	Equipment	No. of equipment	No. of allocated Products	Production (MM T/Y)	Utilization (%)	
					Production	Changeover
Tableting	#3 Tablet press	6	4	2,645	68.9	14.1
Coating	#3 Coater	4	4	2,381	39.7	39.7
Camera inspection	#2 Camera inspection	2	2	510	66.4	6.25
	#3 Camera inspection	4	2	1,632	70.9	14.1
Blister pack	#1 Blister packer	1	2	202	70.2	10.0
	#2 Blister packer	5	6	1,919	66.6	19.0
Carton	#3 Cartoner	1	16	14 (MM C/Y)	50.2	28.1

5 Conclusion

In this paper, we proposed a capacity planning method and a tool, jointly developed by an engineering consulting company, CM Plus Corporation in Yokohama, Japan, intended to support the demand-to-supply management in the pharmaceutical industry. The method synthesizes an initial manufacturing process structure and the capacity of each process unit based on the demand forecast and candidate equipment specifications of each process unit at the first step. Then it improves the process structure in a step-by-step fashion at the second step. We showed the effectiveness of the developed method and the tool through a case study.

There are several issues that require future development. For example, an algorithm to find optimum lot sizing and equipment allocation simultaneously should be developed. In addition, the function that evaluates the process structure using production scheduler should be developed to improve the accuracy of the capacity plan.

References

Hayashi A, Ishii N, Matusi M (2009) A theory and tools for collaborative demand-to-supply management in the SCM Age. Oper Suppl Chain Manage 2:111–124

Ishii N, Fuchino T, Muraki M (1997) Life cycle oriented process synthesis at conceptual planning phase. Comput Chem Eng 21:S953–S958

Ishii N (2004) A heuristic procedure for planning flexible production systems for competitive supply chains. Paper presented at the 1st international congress on logistics and SCM systems, Waseda University, Tokyo, 2004

Matsui M (2002) A management game theory: economic traffic, leadtime and pricing setting. J Jpn Ind Manage Assoc 53:1–9

Matsui M, Takahashi Y, Wang Z (2006) A construction and theory of the manager-aided planner for collaborative demand and supply. J Jpn Ind Manage Assoc 57:120–131

Matsui M, Nakamura T, Ishii N (2007) A demand-to-supply development system for sustainable business strategies and a case study. J Jpn Ind Manage Assoc 58:307–316

Matsui M (2009) Manufacturing and service enterprise with risks. Springer, New York

Monk EF, Wagner BJ (2008) Concepts in enterprise resource planning, 3rd edn. Course Technology, Boston

Ohba M, Matsui M, Ishii N, Yamada T (2013) Case study of generalized material flow cost accounting in consideration of a rate of utilization. Submitted to UTCC International journal of business and economics

Simchi-Levi D, Kaminsky P, Simchi-Levi E (2008) Designing and managing the supply chain: concepts, strategies and case studies. Irwin McGraw-Hill, New York

Storage Assignment Methods Based on Dependence of Items

Po-Hsun Kuo and Che-Wei Kuo

Abstract According to historical customer orders, some items tend to be ordered at the same time, i.e., in the same orders. The correlation of items can be obtained by the frequency of these items present at the same orders. When the dependent items are assigned to adjacent storage locations, order pickers will spend less time to complete customer orders, compared to the other storage assignment which treats items independently. This research provides optimization models to make sure that the storage locations of highly dependent items are nearby. Although the provided nonlinear integer programming can be transformed to a linear integer model, it is still too complex to deal with large problems. For a large number of items, two heuristic algorithms are proposed according to properties of optimal solutions in small problems. Numerical experiments are conducted to show the results of the proposed algorithms with the comparison of the random and class-based storage assignment.

Keywords Storage assignment · Item dependence · Order picking · Optimization models · Heuristic algorithms

1 Introduction

Order picking is one of the most popular research topics regarding warehouse operations since time and labors used are much more intense compared to the other operations in the warehouse (Heragu 1997). Therefore, significantly reducing cost

---

P.-H. Kuo (✉) · C.-W. Kuo
Industrial Management, National Taiwan University of Science and Technology,
#No. 43, Sec. 4, Keelung Road., Da'an District, Taipei City 106, Taipei, Taiwan
e-mail: phkuo@mail.ntust.edu.tw

C.-W. Kuo
e-mail: grandtheftauto.gta@hotmail.com

Y.-K. Lin et al. (eds.), *Proceedings of the Institute of Industrial Engineers Asian Conference 2013*, DOI: 10.1007/978-981-4451-98-7_23,
© Springer Science+Business Media Singapore 2013

in order picking processes will result in significant improvement in the supply chain. One way to make this improvement is to develop a more efficient storage assignment method. Traditional methods existed for storage assignment policies in warehouses are random storage, class-based storage, and dedicated storage. The class-based storage and dedicated storage locate fast-moving items close to I/O point so that the distance involved in picking routes can be reduced. In a unit load order picking perspective, class-based and dedicated storage are indeed effective because items are independent. However, when all items in an order should be retrieved together, the distances between these items become critical, but class-based and dedicated storage which only consider independent item demands are less effective.

This research focuses on the importance of item correlations as a way of reducing the total travel distance in one picking route. The proposed optimization model ensures that items which tend to be ordered together are located nearby. However, due to the complexity of the problem, it is practically not easy to obtain optimal solutions for large problems. Therefore, following some properties of the optimal solutions in small problems, two heuristic algorithms are proposed to deal with large problems. Finally, numerical experiments are conducted to compare the performances between proposed heuristic algorithms and traditional storage assignment policies such as random and class-based storage.

2 Literature Review

The storage part of a warehouse consists of SKU-department assignment, zoning, and storage location assignment. The decisions regarding to SKU-department assignment are assigning items to different warehouse departments and space allocation. In a warehouse with zones, the assignment of SKUs to zones and pickers to zones are also important issues. The storage location assignment can have various decisions to make according to policies used in each warehouse, such as random, class-based and dedicated storage.

A number of methods of order picking can be employed in a warehouse, such as single-order picking, batching and sort-while-pick, batching and sort-after-pick, single-order picking with zoning, and batching with zoning (Yoon and Sharp 1996). Each order picking method consists of some or all of the following basic steps: batching, routing and sequencing, and sorting Gu et al. (2007).

According to Tompkins et al. (2003), order picking is the process of picking products from the storage locations to fill customer orders, and is known as the most important activity in warehouses. Surveys have shown that order picking consumes more than 50 % of the activities in a warehouse. It is therefore not surprising that order picking is the single largest expense in warehouse operations (Heragu 1997).

The most common objective of order-picking systems is to maximize the service level subject to resource constraints such as labor, machines, and capital

(Goetschalckx and Ashayeri 1989). Also, short order retrieval times imply high flexibility in handling late changes in orders. Minimizing the order retrieval time (or picking time) is, therefore, a need for any order-picking system.

There are numerous ways to assign products to storage locations within the storage areas. Three most frequently used storage assignment are: random storage, dedicated storage, and class-based storage. For the importance of reducing order picking time, however, traditional literatures miss the issue of dependence in customer demand. It has come to our attention that if items in the same order are located as close to one another as possible, the travel distance could be significantly reduced. Rosenwein (1994) describes an optimization model that identifies clusters of warehouse items that tend to be ordered together. Jane and Laih (2005) construct a model using a similarity measurement which measures the co-appearance of any two items in customer orders, and then develop a heuristic algorithm to balance the workload among all pickers in different zones. Unlike any other clustering or family grouping literatures where a storage area is separated into several clusters and random assignment is applied within the clusters, this paper focuses on assigning related items as close as possible to achieve minimum travel distance during one single pick.

3 Model Establishment

Travel distance will decrease if these correlated items are located around each other. The proposed method is to assign these items close to each other so the order picker can travel a shorter distance while retrieving multiple items during one picking route. This paper focuses on single block, single-level warehouses with manual order picking operations in which the order picker walks or drives in the pick area to retrieve items. Picked items are placed on a pick device and the order picker takes it with him/her and proceeds to the next item. Every storage location has the same size, and the length and width of a storage location are the same. The I/O point is located at the left lower corner of the warehouse.

Notation used in the optimization model and heuristic algorithms is listed as follows:

N Number of items = Number of locations in the warehouse

a, b Index of items, $a,b \in \{1, 2,3,...,N\}$

i, j Index of locations, $i, j \in \{1,2,3,...,N\}$

$$X_{a,i} = \begin{cases} 0, & \textit{If item a is not assigned to location i} \\ 1, & \textit{If item a is assigned to location i} \end{cases}$$

$$X_{b,j} = \begin{cases} 0, & \textit{If item b is not assigned to location j} \\ 1, & \textit{If item b is assigned to location j} \end{cases}$$

$D_{i,j}$ Distance from location i to location j

$P_{a,b}$ Probability of both item a and b being requested in the same order, which can be obtained from the co-appearance frequency of a and b in the warehouse's historical data.

The major existed storage policy which also considers item popularity is class-based storage. The reason why this model is developed is that an order may contain more than one item. When there are multiple items in an order, it is not suitable for class-based storage because the probability of all items being in the same class is low. The items in an order may be located across the warehouse, in order to reduce the travel distance involved in an order, the correlations of any two items become important information. In this model, the correlation information of two items a and b, $P_{a,b}$, is considered. This information can be obtained from the co-appearance frequency of any two items in same orders from the warehouse's historical statistic data. The reason of using exactly two items' correlation and not three or more items' correlation is that the problem will become too complex and virtually impractical to solve. If only one item frequency is considered, it will be the same as class-based storage.

$$\text{Minimize} \sum_{a<b} \sum_{i \neq j} P_{a,b} \cdot D_{i,j} \cdot Y_{a,b,i,j} \tag{1}$$

$$\text{Subject to} \sum_{i} X_{a,i} = 1, \quad a = 1, \ 2, \dots, N \tag{2}$$

$$\sum_{a} X_{a,i} = 1, \quad i = 1, \ 2, \dots, N \tag{3}$$

$$Y_{a,b,i,j} \leq X_{a,i}, \ a<b \in (1, \dots, \ N), \ i \neq j \in (1, \ \dots, \ N) \tag{4}$$

$$Y_{a,b,i,j} \leq X_{b,j}, \ a<b \in (1, \dots, N), \ i \neq j \in (1, \dots, N) \tag{5}$$

$$\sum_{a} \sum_{b} \sum_{i} \sum_{j} Y_{a,b,i,j} = N(N-1)/2 \tag{6}$$

$$Y_{a,b,i,j} \geq 0, \ a<b \in (1, \dots, N), \ i \neq j \in (1, \dots, N) \tag{7}$$

The objective of the model, (1), is to assign items with higher correlations as close as possible. Equations (2) and (3) indicate that item a will be located in one of the storage locations, and location i is occupied by one of N items. Equations (4) and (5) are required because $Y_{a,b,i,j}$ cannot exceed either $X_{a,i}$ or $X_{b,j}$ due to the assumption that the variables are binary. Given that the constant N is the total number of items or locations in the warehouse, as Eq. (6) shows, the sum of all $Y_{a,b,i,j}$ should be exactly the number of all possible combinations of any two items or locations, that is C_2^N. Otherwise, all $Y_{a,b,i,j}$'s will be zero due to the minimization objective.

Storage Assignment Methods Based on Dependence of Items

The problem can relate to a knapsack problem because the decision of the problem is whether an item needs to be put in a location or not. A knapsack problem is an example of a NP-complete problem and a knapsack problem can be solved in pseudo-polynomial time by dynamic programming. However, any single item has limited optional locations (based on the properties provided) because the decision for the assigned locations will affect the objective function value. In other words, multiple locations lead to a large number of possible solving combinations and increase the complexity of the problem. As a result, the problem is NP-hard. It is clear that finding optimal solutions to large problems will take a lot of time. Therefore, two algorithms are developed to find heuristic solutions of large problems.

In Algorithm 1, the warehouse is transformed into one-dimensional scale before the assigning process begins. The reason is to reduce the number of situations of any two locations with equal distances in the two-dimensional warehouse. After the assignment is finished, the one-dimensional assignment result is transformed back to its original two-dimensional warehouse locations. As a result, if the warehouse shape is closer to one-dimension scale, for instance, flat and wide, algorithm 1 will be more effective.

Algorithm 1
1. Number the location closest to I/O as 1.
2. Number the other storage locations based on the sequence of $D_{1,j}$, $j \neq 1$. After the numbering scheme, the two-dimensional warehouse becomes one-dimensional scale.
3. Assign items A and B with least popularities to location 1 and location N. The set of assigned items is $S = \{A, B\}$.
4. Assign item C with the largest popularity to the central location of the sequence. $S = S + C$.
5. Search for the first gap from location 1. The gap indicates that locations between the assigned items X and Y are not occupied.
6. Assign item i with the maximal value of $P_{X,i} + P_{Y,i}$, $i \neq S$ to the center of the gap. $S = S + i$ (the new assigned item).
7. If all items are assigned, go to step 8. Otherwise, go to step 5.
8. Transform the one-dimensional scale to the original warehouse configuration.

In algorithm 1, the distance factor of the objective function is simplified by transforming the warehouse into one-dimensional sequence. However, this procedure does not accurately interpret the meaning of the objective function that both correlation probability (P) and distance (D) should be considered at the same time. Therefore, another algorithm, algorithm 2, is developed to check the effect of considering both factors simultaneously.

Algorithm 2
1. Assign items A and B with least popularities to location 1 and N, where location 1 is closest to I/O and location N is furthest from I/O. The set of assigned items is $S = \{A, B\}$.

2. Assign item C with the largest popularity to the central location. $S = S + C$.
3. Assign item i to location l with $\underset{i \notin S, l \notin l_S}{Min} \sum_X \sum_{l_X} P_{X,i} \cdot D_{l_X,l}$ for all unassigned items

i and locations l. $X \in S$ (assigned items), and the location of assigned item X is $l_X \in l_S$ (occupied locations), $S = S + i$ (the new assigned item).
4. Repeat step 3 until all items are assigned.

4 Numerical Examples

In this section, a set of small examples with 8 items are presented. Three $P_{a,b}$ matrixes are randomly generated from 0 to 0.5. The shapes of warehouses used in the example are 2×4 and 4×2 (columns $\times$ rows), and the configuration of parameters w_a (the width of each pick aisle), w_c (the width of each cross aisle), and w_s (the length and width of one storage location) are listed in Table 1. In the example, the optimal solutions obtained from AMPL optimization software are presented to compare with both algorithms and random and class-based storage.

In the above examples, algorithm 2 performs better than algorithm 1 in the 2×4 cases, and algorithm 1 does better than algorithm 2 in the 4×2 case. The reason might be the shape property mentioned in Sect. 3, algorithm 1 will be more effective if the warehouse is wide and flat.

Compared to the optimal solutions, the algorithms appear to have very small differences, and the other storage policies also have small differences. The reason for such small differences might be the objective function containing too much information so that the improvement in the algorithms is hard to see.

Table 1 Examples of a warehouse with 8 items

Config.	Results					
		Random	Class-based	Alg. 1	Alg. 2	Optimum
2×4	Pab 1	1.92958	2.11296	1.93996	1.86759	1.7025
$w_s = w_c = 1$	Pab 2	2.03775	1.78435	1.84521	2.0022	1.64872
$w_a = 2$	Pab 3	1.95539	2.03687	1.85773	1.84422	1.63214
4×2	Pab 1	4.26611	5.0153	3.81442	3.84555	3.66269
$w_s = w_c = 1$	Pab 2	4.27593	3.84111	3.76357	3.86013	3.57586
$w_a = 2$	Pab 3	4.26611	3.94976	3.59457	3.96918	3.35801
2×4	Pab 1	3.29724	3.5391	3.29698	2.96615	2.91804
$w_s = w_c = 2$	Pab 2	3.52026	3.08925	3.05077	2.96401	2.81039
$w_a = 3$	Pab 3	3.35083	3.56994	3.09147	3.25288	2.73515

5 Conclusions

This research wants to improve the efficiency of order picking processes. Order picking is one of the most expensive operations in the warehouse. Therefore, it is also a critical part of the warehouse, even in the supply chain. One way to improve the order picking processes is to enhance the efficiency of storage assignment policies. In this research, a linear integer programming model is developed to ensure the highly related items are located nearby. As mentioned in Sect. 3, this model is NP-hard, so it is very difficult to solve large problems. Therefore, two heuristic algorithms are proposed to deal with large problems. After the algorithms are proposed, the data results are compared with random and class-based storage policies.

In this research, the distance function in Sect. 3 provides a way to obtain the shortest distance between any two locations in the warehouse, which can be very useful in developing storage assignment policies. Furthermore, the main difference from the other literatures is that this research considers items' correlations. Unlike the clustering assignment methods, there is no cluster in the developed assignment methods. Every item is assigned to the location according to certain calculations in order to shorten the distances between correlated items. The information of interrelated items is an important factor to optimize storage assignment methods, and applying this information to improve order picking process is the main contribution of this research.

References

Goetschalckx M, Ashayeri J (1989) Classification and design of order picking systems. Logistics World June, pp 99–106

Gu J, Goetschalckx M, McGinnis LF (2007) Research on warehouse operation: a comprehensive review. Eur J Oper Res 177:1–21

Heragu SS (1997) Facilities design. PWS Publishing Company, Boston

Jane CC, Laih YW (2005) A clustering algorithm for item assignment in a synchronized zone order picking system. Eur J Oper Res 166:489–496

Rouwenhorst B, Reuter B, Stockrahm V, van Houtum G, Mantel R, Zijm W (2000) Warehouse design and control: framework and literature review. Eur J Oper Res 122(3):515–533

Tompkins JA, White JA, Bozer YA, Frazelle EH, Tanchoco JMA (2003) Facilities planning. Wiley, Hoboken

Yoon CS, Sharp GP (1996) A structured procedure for analysis and design of order pick systems. IIE Trans 28:379–389

Selection of Approximation Model on Total Perceived Discomfort Function for the Upper Limb Based on Joint Moment

Takanori Chihara, Taiki Izumi and Akihiko Seo

Abstract The aim of this study is to formulate the relationship between the total perceived discomfort of the upper limb and perceived discomforts of each degree of freedom (DOF). The perceived discomforts of each DOF were formulated as functions of the joint moment ratio based on the results of previous study, and then the function approximation model for the total perceived discomfort was investigated. The summary score of the rapid upper limb assessment (RULA), which is assumed as the total perceived discomfort, and the perceived discomforts of each DOF were taken as the objective and explanatory variables respectively. Three approximation models (i.e., the average, maximum, and radial basis function (RBF) network) were compared in terms of the accuracy of predicting the total perceived discomfort, and the RBF network was selected because its average and maximum error were lowest.

Keywords Perceived discomfort · Function approximation · RULA · Radial basis function network · Psychophysics · Biomechanics

1 Introduction

The physical workload should be evaluated quantitatively and objectively so as to design a work environment that reduces the workload and prevents musculo-skeletal disorders. In addition, the time that can be allocated to improve work

---

T. Chihara (✉) · A. Seo
Tokyo Metropolitan University, 6-6, Asahigaoka, Hino, Tokyo, Japan
e-mail: chihara@sd.tmu.ac.jp

A. Seo
e-mail: aseo@sd.tmu.ac.jp

T. Izumi
Graduate School of Tokyo Metropolitan University, 6–6, Asahigaoka, Hino, Tokyo, Japan
e-mail: izumi-taiki@sd.tmu.ac.jp

Y.-K. Lin et al. (eds.), *Proceedings of the Institute of Industrial Engineers Asian Conference 2013*, DOI: 10.1007/978-981-4451-98-7_24,
© Springer Science+Business Media Singapore 2013

environments are decreasing with each passing year, in conjunction with the shortening of the development period. That is, an ergonomic physical load evaluation should be performed effectively in a short time. Biomechanical analysis evaluates the physical load based on the equilibrium of force with a rigid link model of the human body (Chaffin et al. 2006). The reactive moment on each joint against external forces (hereafter referred to as "the joint moment") is regarded as the indicator of physical load. The joint moment can be calculated by computer simulation such as commercial digital human software (LaFiandra 2009); therefore, the evaluation with biomechanical analysis can be applied to the efficient ergonomic design of work environment (Chaffin 2007).

In our previous study, we formulated the relationship between the perceived discomfort and joint moment for twelve joint motion directions of the upper limb (Chihara et al. 2013). However, the study has not investigated the evaluation method for total perceived discomfort of multiple joint moments. The total perceived discomfort function should be formulated so as to determine the order of multiple design solutions of work environment. Several observational methods such as RULA and OWAS are used to assess the total physical workload (McAtamney and Corlett 1993; Karhu et al. 1977). These methods are easy to use, but they cannot perform detailed evaluation of total workload, because their worksheets roughly classify the postures of workers and lifting weights. In addition, the observation methods consider only the weight of load handled, but they do not consider the direction of force.

The objective of the present study was to formulate the relationship between the total perceived discomfort of the upper limb and perceived discomforts of each joint motion. The biomechanical analysis was performed based on the classification of postures and load in the RULA. The summary score of RULA was set as the objective variable, and the perceived discomforts of each degree of freedom (DOF) were set as the explanatory variables. The response surfaces of total perceived discomfort were approximated by three different approximation models: the average, maximum, and radial basis function network (RBF) (Orr 1996). The accuracy of response surfaces was compared, and the proper approximation model was investigated.

2 Method

2.1 Selection of Calculating Condition from RULA

The calculating conditions of biomechanical analysis (i.e., the posture of the upper limb and the weight of load handled) were determined based on the posture and load classification of the RULA. In the RULA method, the sub-summary scores are calculated for the two groups: the arm and wrist (Group A), and the neck, trunk, and leg (Group B). Then the summary score is calculated by sum of the two

Selection of Approximation Model

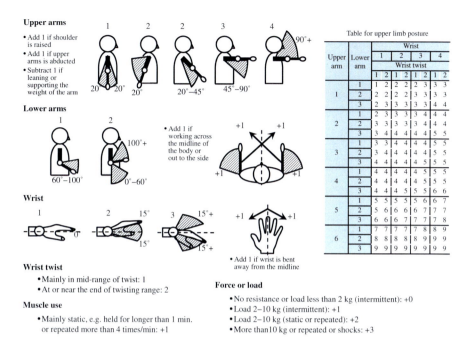

Fig. 1 RULA sheet for the arm and wrist (Group A) (*Source* McAtamney and Corlett (1993) RULA: a survey method for the investigation of work-related upper limb disorders, Appl Ergon, 24(2), 91–99)

sub-summary scores. In this study, we focused on the total perceived discomfort of the upper limb; hence, the criteria of Group A in the RULA were used to select the calculating conditions. The calculation of sub-summary score for Group A consists of four parts on the upper limb posture (i.e., the upper arm position, lower arm position, wrist position, and wrist twist), one part on duration of muscle use, and one part on the amplitude of load handled. Each part has several ranges which are assigned a score as shown in Fig. 1. First, the score for the posture part is determined based on the worksheet for the posture score shown in Fig. 1. The scores for the duration of muscle use and the amplitude of load are added to that of the posture part, and then the sub-summary score for the upper limb is obtained.

In this study, the levels of each part for the biomechanical analysis were determined as shown in Table 1. Among the six parts of Group A, the wrist twist and the duration of muscle use were ignored, because these conditions have not been considered in the previous research (Chihara et al. 2013). With respect to the remaining four categories, we selected the almost middle value of the ranges as the levels for calculating conditions. In addition, to cover all scores of RULA efficiently, the one condition was selected when the multiple conditions had the same score. Then, the five, three, four and three levels were determined for the upper arm position, lower arm position, wrist position, and amplitude of load handled respectively; thus, the number of calculating conditions was 180. It should be note

Table 1 Calculating conditions of biomechanical analysis

Category	Items of each category	Level 1	Level 2	Level 3	Level 4	Level 5
Upper arm	Flexion angle of shoulder joint (deg)	0	0	65	135	135
	Abduction angle of shoulder joint [deg]	0	45	0	0	45
Lower arm	Flexion angle of elbow joint (deg)	80	120	120	NA	NA
	Internal rotation angle of shoulder joint (deg)	0	0	40	NA	NA
Wrist	Flexion angle of wrist joint (deg)	0	10	50	50	NA
	Ulnar deviation angle of wrist joint (deg)	0	0	0	30	NA
Force	Load (N)	0	59	137	NA	NA

that the sub-summary score of calculating conditions depicted in Table 1 range from 1 to 10. In addition, the sub-summary score is normalized to [0, 1], and the normalized score is used as the objective variable for the function approximation.

2.2 Biomechanical Analysis for Selected Conditions

The biomechanical analyses were conducted for the selected 180 conditions so as to calculate the joint moments. The digital human model for the analysis was constructed based on the 50 percentile of Japanese male (Kouchi and Mochimaru 2005). That is, 171 cm in body height and 63.7 kg in body weight. The joint moments for each DOF were calculated by the biomechanical analysis. Then the calculated joint moments were divided by the maximum joint moment of each joint motion direction (hereafter referred to as "the joint moment ratio r ($r = [0, 1]$)"). Here, the maximum joint moments that human can exert were quoted from Chaffin et al. (2006). The perceived discomfort of each joint motion direction was calculated by the following equations (Chihara et al. 2013):

$$f_1(r) = \frac{0.986}{1 + \exp\{-7.56(r - 0.354)\}} \tag{1}$$

$$f_2(r) = \frac{0.978}{1 + \exp\{-9.93(r - 0.234)\}} \tag{2}$$

where f_1 and f_2 denote the perceived discomfort scores for except the elbow flexion and the elbow flexion respectively. In addition, it should be note that f_1 and f_2 range from 0 to 1; and the higher score indicates the higher physical load. In this study, the upper limb has 6 DOF: the three DOF of the shoulder joint (i.e., extension–flexion, adduction–abduction, and internal rotation–external rotation), one DOF of the elbow joint (i.e., extension–flexion), and two DOF of the wrist joint (i.e., extension–flexion and ulnar deviation–radial deviation). Therefore, the perceived discomforts of six DOF were calculated.

2.3 Approximation of Total Discomfort Function and Evaluation of Appropriate Model

We assume that the total discomfort is dominated by the average of discomforts of each DOF in the range that the discomforts of each DOF are relatively low. However, the total discomfort may be dominated by the maximum value of discomforts of each DOF in the range that one or more discomforts of each DOF is relatively high. In addition, it is possible that the relationship between the total discomfort and the discomforts of each DOF is a weakly nonlinear function. Therefore, in this study, the three function approximation models were used: the average model, maximum model, and RBF. The RBF performs well in terms of accuracy and robustness, irrespective of the degree of nonlinearity (Jin et al. 2001). Among the three models, the average and maximum models are set as follows:

$$T = \frac{a \cdot \sum_{i=1}^{6} w_i}{6} \tag{3}$$

$$T = a \cdot \max_i w_i \tag{4}$$

where T and w_i denote the total perceived discomfort (i.e., the objective variable) and the perceived discomfort of ith DOF (i.e., the explanatory variable), respectively, and a is regression coefficient. The normalized sub-summary score of RULA is set as the objective variable, and the perceived discomforts of the upper limb are set as the explanatory variables. The regression coefficient is obtained by the least-square method. The parameters proposed by Kitayama and Yamazaki (2011) are adopted for the prediction by the RBF. Please refer to Orr (1996) about the details of RBF. In addition to all calculating conditions, the response surfaces were approximated for two groups of calculating conditions that were divided based on the amplitude of discomforts of each DOF. In this study, the calculating conditions was divided into the low discomfort group that the all discomfort scores of each DOF below 0.5, and the high discomfort group that one or more discomfort scores exceeds 0.5.

The accuracy of the three response surfaces were compared by the average absolute error (AAE). The AAE for ith function model was calculated as follows:

$$AAE_i = \frac{\sum_{j=1}^{n} |T_j - \hat{T}_{ij}|}{n} \tag{5}$$

where, T_j and $\hat{T}_{ij}$ denote the normalized sub-summary score and the approximated total perceived discomfort for j-th calculating condition of the biomechanical analysis n is the number of the calculating conditions that are used for constructing the response surfaces. The AAEs of response surfaces are compared between the three approximation models. One-way ANOVA was conducted at the 5 % significance level, and post hoc tests were carried out to compare the three models.

Table 2 Regression coefficient of the average and maximum models

Approximation models	Low group	High group	All
Average	3.53	1.70	1.79
Maximum	1.86	0.745	0.783

3 Result

The total perceived discomfort was predicted by using the calculating data of the low discomfort group, high discomfort group, and all data. Here, low group and high group had 77 and 103 data respectively. Table 2 shows the regression coefficients of the average and maximum models for each data set group. The result of ANOVA shows that there is the main effect of approximation model irrespective of the data set groups. Figure 2 shows the AAE of the three approximation models. Among the three models, the AAE of RBF is significantly lower than that of the average and maximum models irrespective of the data set group. Compared with the average and maximum models, in the case of low group, there is no significant difference, but the AAE of the average model is lower than that of the maximum model. In contrast, in the case of high group, the AAE of the maximum group is lower than that of the average model with 1 % significant level. Moreover, in the case of all calculating data, the AAEs of the average and maximum models are almost the same.

4 Discussion

According to Fig. 2, the accuracy of the average model is greater than that of the maximum model in the case of low discomfort group. Therefore, in the range that the discomforts of DOF are relatively low, the total discomfort is more affected by the average of discomforts of each DOF than by the maximum value of them. In

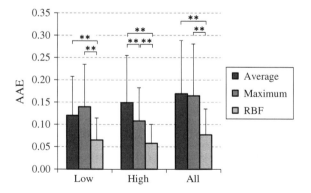

Fig. 2 Comparison of average absolute error between the three approximation model

addition, the accuracy of the maximum model is greater than that of the average model in the case of high discomfort group; hence, the total discomfort is affected by the maximum value of the discomforts of each DOF in the range that the discomforts of DOF are relatively high. This is consistent with the premise that the relationship between the total discomfort and discomforts of each DOF varies by the amplitude of discomforts of each DOF.

The AAEs of RBF are lower than that of the average and maximum models irrespective of the amplitude of perceived discomfort. The RBF can predict nonlinear functions with good accuracy; therefore, the relationship between the total perceived discomfort and discomforts of each DOF may be nonlinear. In addition, the RBF shows the best accuracy in the case of all calculating condition. The AAE of the RBF for the all condition is approximately 8 %, so that the response surface predicted by the RBF perhaps has the sufficient accuracy. Thus, the RBF is preferable approximation method as the total perceived discomfort function among the three approximation models. However, lots of data sets are required to construct the response surface with sufficient accuracy by using the RBF. Therefore, when the sufficient number of data sets can be obtained, the RBF is recommended. However, when a number of data sets are hardly obtained, the average and maximum models may be used depend on the amplitude of discomforts of each DOF. That is, when the discomforts of each DOF are relatively low, the average model is recommended; conversely, when they are relatively high, the maximum model is recommended.

5 Conclusions

In this study, function approximation model for the total perceived discomfort of the upper limb is investigated. The major findings are as follows:

1. In the range that the discomforts are relatively low, the average model provides better fit as the total perceived discomfort function than the maximum model. Conversely, in the range that the discomforts are relatively high, the maximum model provides better fit than the average model.
2. The AAE of the RBF is the lowest among the three function approximation models irrespective of the data set group. Therefore, the RBF is the preferable approximation model among the three models.

Acknowledgments This work was supported by JSPS KAKENHI Grant Number 24760123.

References

Chaffin DB (2007) Human motion simulation for vehicle and workplace design. Hum Factors Ergon Manuf 17(5):475–484

Chaffin DB, Andersson GBJ, Martin BJ (2006) Occupational biomechanics, 4th edn. Wiley, New York

Chihara T, Izumi T, Seo A (2013) Perceived discomfort functions based on joint moment for various joint motion directions of the upper limb. Appl Ergon. doi:10.1016/j.apergo.2013.04.016

Jin R, Chen W, Simpson TW (2001) Comparative studies of metamodeling techniques under multiple modeling criteria. Struct Multidisc Optim 23(1):1–13

Karhu O, Kansi P, Kuorinka I (1977) Correcting work postures in industry: a practical method for analysis. Appl Ergon 8(4):199–201

Kitayama S, Yamazaki K (2011) Simple estimate of the width in Gaussian kernel with adaptive scaling technique. Appl Soft Comput 11(8):4726–4737

Kouchi M, Mochimaru M (2005) AIST Anthropometric database, national institute of advanced industrial science and technology. H16PRO 287. http://riodb.ibase.aist.go.jp/dhbodydb/91-92/. Accessed 15 May 2013 (in Japanese)

LaFiandra M (2009) Methods, models, and technology for lifting biomechanics. In: Duffy VG (ed) Handbook of digital human modeling. CRC Press, New York, pp 8-1–8-27

McAtamney L, Corlett EN (1993) RULA: a survey method for the investigation of work-related upper limb disorders. Appl Ergon 24(2):91–99

Orr MJL (1996) Introduction to radial basis function networks. http://anc.ed.ac.uk/rbf/papers/intro.ps.gz. Accessed 15 May 2013

Waiting as a Signal of Quality When Multiple Extrinsic Cues are Presented

Shi-Woei Lin and Hao-Yuan Chan

Abstract While quality of a product or a service is considered one of the most important factors that influence consumer satisfaction, evaluating and determining product or service quality can be difficult for many consumers. People thus usually rely on extrinsic cues or surrogate signals of quality to tackle the information asymmetry problems associated with product/service quality. Unfortunately, research which has empirically documented the link between quality signals and perceived quality focus mainly on the situation where there exists only a single extrinsic cue. This study aims to investigate the interaction effect between multiple cues or signals on perceived quality. In particular, "waiting" or "queuing" in this study is no longer treated as a phenomenon that solicits disutility or negative emotions, but considered a signal of quality that has positive effect on consumer evaluation or satisfaction. Furthermore, this study hypothesized that the "waiting" can only be a positive signal under some specific situations especially when other quality signals (i.e., price and guidance) co-exist, and used experiments to rigorously test the hypotheses. By considering multiple cues simultaneously, this study lead to a better understanding of when and to what extent waiting can be use as a quality signal, and thus extend the original theory proposed by other researchers.

Keywords Perceived quality · Satisfaction · Signaling · Wait · Price

---

S.-W. Lin (✉)
Department of Industrial Management, National Taiwan University of Science and Technology, Taiwan, Japan
e-mail: shiwoei@mail.ntust.edu.tw

H.-Y. Chan
College of Management, Yuan Ze University, Taiwan, Japan
e-mail: s1007126@mail.yzu.edu.tw

Y.-K. Lin et al. (eds.), *Proceedings of the Institute of Industrial Engineers Asian Conference 2013*, DOI: 10.1007/978-981-4451-98-7_25,
© Springer Science+Business Media Singapore 2013

1 Introduction

Although many researchers have empirically verified the causal effect between product quality and customer satisfaction and/or willingness to buy (Baker et al. 2002; Cronin and Taylor 1992), pointed out that the quality of a product or service is sometimes very difficult to observe or evaluate objectively and customers usually relies on some extrinsic cues or signals to determine the quality (Boulding et al. 1993). In the literature related to the signals or indicators of quality, one of the most common themes focuses on the positive relationship between price and perceived quality, as well as the contextual or situational factors that may moderate this relationship. Near 100 relevant studies in the past 30 years have been reviewed and summarized by Brucks et al. (2000).

Other than the effects of price or brand name, Giebelhausen et al. (2011) proposed and experimentally showed that waits can also function as a signal of quality. After all, it is easy to find circumstances in which consumers are willing to wait. However, the relationship between waiting and quality perception of a product or service is relatively unclear, compared to the relationship between price or brand name and perceived quality. In traditional management point of view and in most academic literature, waiting is usually described as a phenomenon that cause negative emotions of consumers and have negative impacts on consumers' evaluation of products or services (Berry et al. 2002; Hui and Tse 1996; Baker and Cameron 1996). However, in recent years, business practitioners and marketing managers may deliberately create "must wait" situation or even deliberately increase the waiting time. For example, store may provide only limited space for waiting so that the customers must line up on the sidewalk. Giebelhausen et al. (2011) provided an explanation for this phenomenon. They believed that letting customer wait can increase the perceived quality (of a product or service), satisfaction, and intention to buy. In other words, while waiting may have negative emotional impact, it can also be treated as a positive signal of quality.

Although many theoretical and empirical research have been conducted to explore or investigate variables that can be used as signals of quality, most of the investigations focus on investigating one specific factor. However, several signals of quality usually exist simultaneously in the real world business practices. The one-factor-at-a-time approach used in previous studies may overlook important interactions between different signals. For example Monroe and Krishnan (1985) found that the effect of price on quality perception is moderated by the variable brand name. Without a factorial experiment design, this interaction cannot be easily identified.

Thus, in this study, we use a similar paradoxical view of waiting proposed by Giebelhausen et al. (2011) to study the positive effect of waiting on quality perception and purchase intention. In particular, we intend to investigate the interactions between wait and price, and try to determine that whether effect of wait as a indicator of quality will be suppressed or be strengthened when the signal of price is considered concurrently.

By using rigorous experiments to verify the main and interactions effects of wait and price, the main contribution of the present research include: (1) confirm that waits can have a positive impact on quality perception and at the same time identify the underlying mechanism through which the effect operates to make sure the effect is not caused by other confounding factors or lurking variables, (2) experimentally evaluate the signal effect of wait under different prices to test whether the "economic value of wait" is moderated by price, which is a commonly used signal and is sometimes called a "surrogate for quality" in the absence of other information. This study thus lead to a better understanding of when and to what extent waiting can be use as a quality signal, and thus extend the original theory proposed by other researchers.

2 Methods

2.1 Research Model

Based on the literature review, price and waiting may function as signals for quality when there is information asymmetry between buyers and sellers (Spence 2002; Volckner and Hofmann 2007; Giebelhausen et al. 2011). Furthermore, Kirmani and Rao (2000) pointed out that a customer usually consider several intrinsic and extrinsic cues or signals (e.g., price, warranty, country of origin, brand name) simultaneous to form his/her quality perception when evaluating the quality of a product or a service. However, there may be some interaction effects between various cues or signals of quality. Miyazaki et al. (2005) suggested that when the information presented in two or more signals is consistent, these signaling factors may complement each other, further increasing a customer's quality perception. On the other hand, if the information from multiple signals is inconsistent, Miyazaki et al. (2005) suggested that the signal containing negative information may become more dominant.

Therefore, this study aims to examine whether there is an interaction effect between extrinsic quality signals price and wait. In particular, we are also interested in investigating the effect of waits on quality perception and satisfaction when it is co-existed with different signals or under different contexts. Based on the research framework, we summarized the four hypotheses proposed as follows.

Hypothesis 1 There is a positive main effect of a wait such that quality perception will be greater when a wait is present than when a wait is absent.

Hypothesis 2 There is a positive main effect of price such that quality perception will be greater for the high-priced product (or service) than the low-priced product (or service).

Hypothesis 3 There is an interaction between price and the presence of a wait. In particular, for a low price service, the presence of a wait increases perceived quality. However, for a high price service, the presence of wait has smaller effect on perceived quality.

Hypothesis 4 Perceived quality function as a mechanism by which price and the presence of a wait influences purchase intentions.

2.2 Design and Participants

According to the literature, consumers usually rely on the extrinsic cues or alternative signals to determine the quality of a product or a service when there is no sufficient information (about the product or service) available or when the product or service quality is ambiguous. This study thus focuses on the cases where the consumers are making their first-time consumption and are not very familiar with the product or service.

In particular, the experiment design of this study allow participants (i.e., consumers) received two signals (price and wait) simultaneously. Furthermore, different contexts (the scenario related to the product or service consumption) are constructed to mimic the realistic consumption environments. A survey are finally employed to collect the measures of quality perception and satisfaction of the participants, and statistical tests were conducted to determine whether proposed hypotheses can be supported by the data.

This study use a two-factor within-subject experimental design. To avoid the carry-over effect and to make participants less easy to see the whole picture of the experiment (and avoid the possible biases that may have caused), two different decision contexts are considered in this study. Both decision contexts is related to restaurant service settings, but one is serving the western style food and the other is serving the Japanese noodle soup. The study utilized a 2 (price: high, low) × 2 (wait: absent, present) design. The price levels or the levels of stimuli are based on the market research of the general price range of the same products or services in Taiwan, and different menus each containing a sequence of dishes are designed and presented to the participants. The low-price and high-price settings for the menu of steak house (western style food) are around NT\$200 and NT\$1200, respectively. The low-price and high-price settings for the menu of Japanese noodle house are around NT\$100 and NT\$300, respectively. For manipulating the factor of wait in the experiment, in the "wait absent" condition, the scenario indicated that the waiting area was empty and the customer can be seated right away. However, in the "wait present" condition, the customer is notified that no more reservation can be taken and he/she need to wait 25–30 min to be seated.

While the within-subject design employed in this study has the strength of making experiment more efficient, there are also threats to the internal validity of this design such as the carry over or order effects. It is possible that effects from previous treatments (scenario) may still be present when testing new treatment (scenario), thus affecting or biasing the outcome. One solution to the problem of carryover effects is to counterbalance the order of treatment levels. Thus, different subjects are randomly assigned to different scenarios in different orders.

3 Preliminary Results

SPSS statistical package was used to perform the analysis. Quality perception measures of participants were analyzed by means of the two-way within-subject ANOVA with two levels of price (high, low) and two levels of wait (absent, present). All main effects were found to be statistically significant. The main effect of price showed that high price led to higher quality perception than did low price, $F(1, 171) = 25.930$, $p < 0.01$. The main effect of wait also showed that the presence of wait led to higher quality perception than did the absence of wait, $F(1, 171) = 28.622$, $p < 0.01$. Our hypotheses 1 and 2 are supported by the preliminary results of the data analysis.

On the other hand, Although we hypothesize that here is an interaction between price and the presence of wait (partially based on Miyazak (Miyazaki et al. 2005)), the interaction effect between price and wait on quality perception was not significant ($F(1, 171) = 0.244$, $p = 0.622$). In particular, while we suggested for a low price service, the presence of a wait increases perceived quality, and for a high price service, the presence of wait has smaller effect on perceived quality, no significant trend were identified. The seemly parallel interaction plots also confirmed this finding.

4 Conclusion

Multiple signals or cues of quality usually exist simultaneously in the real world business practices, but most of the study conducted in this field use the one-factor-at-a-time approach by focusing on one specific factor. In this study, We intend to investigate the important interactions between different signals which were usually over-looked. In particular, we aim to investigate interactions between wait and price, and try to determine that whether effect of wait as a indicator of quality will be suppressed or be strengthened when the signal of price is considered concurrently.

Although significant price and wait effects were found in our analysis, showing that both price and wait can function as signals of quality. The interaction effect between price and wait on quality perception was not significant. Other variables or moderators, such as the consumer's motivation, might need to be taken into consideration to further clarify this issue.

Acknowledgments This work is partially supported by the National Science Council of Taiwan under grant number NSC101-2410-H-011-034-MY2. Any opinions, findings, and conclusions or recommendations expressed in this material are those of the authors and do not necessarily reflect the views of the sponsors.

References

Baker J, Cameron M (1996) The effects of the service environment on affect and consumer perception of waiting time: an integrative review and research propositions. J Acad Market Sci 24:338–349

Baker J, Parasuraman A, Grewal D, Voss GB (2002) The influence of multiple store environment cues on perceived merchandise value and patronage. J Marketing 66:120–141

Berry LL, Seiders K, Grewal D (2002) Understanding service convenience. J Marketing 66:1–17

Boulding W, Kalra A, Staelin R, Zeithaml VA (1993) A dynamic process model of service quality: from expectations to behavioral intentions. J Marketing Res 30:7–27

Brucks M, Zeithaml VA, Naylor G (2000) Price and brand name as indicators of quality dimensions for consumer durables. J Acad Market Sci 28:359–374

Cronin JJ, Taylor SA (1992) Measuring service quality: a reexamination and extension. J Marketing 56:55–68

Giebelhausen M, Robinson S, Cronin J (2011) Worth waiting for: increasing satisfaction by making consumers wait. J Acad Market Sci 39:889–905

Kirmani A, Rao AR (2000) No pain, no gain: A critical review of the literature on signaling unobservable product quality. J Marketing 64:66–79

Hui MK, Tse DK (1996) What to tell consumers in waits of different lengths: an integrative model of service evaluation. J Marketing 60:81–90

Miyazaki AD, Grewal D, Goodstein RC (2005) The effect of multiple extrinsic cues on quality perceptions: a matter of consistency. J Consum Res 32:146–153

Monroe KB, Krishnan R (1985) The effect of price on subjective product evaluations. Perceived quality: how consumers view stores and merchandise. J Marketing Res 28:209–232

Spence M (2002) Signaling in Retrospect and the Informational Structure of Markets. Am Econ Rev 92:434–459

Völckner F, Hofmann J (2007) The price-perceived quality relationship: A meta-analytic review and assessment of its determinants. Market Lett 18:181–196

Effect of Relationship Types on the Behaviors of Health Care Professionals

Shi-Woei Lin and Yi-Tseng Lin

Abstract Human's behavior and attitude can be highly influenced by two types of relationship, communal relationship and exchange relationship, and the moral-oriented social norms and the money-oriented market norms applied mechanically in these two relationships, respectively. While there is a great deal of general literature discussing the effect of relationship types on interpersonal interaction, there are limited number of studies focusing on the relationship types between organizations and their members and whether the introduction of monetary incentives affect the relationship types. Taking healthcare industry as an example, this study aims to explore how the types of relationship (communal vs. exchange relationship) between hospitals and medical staffs influence their attitude. Furthermore, we also want to investigate whether different types of reward (monetary vs. nonmonetary incentives) provided by hospitals affect or alter the types of relationships a medical staff originally had. We expect the results of this study can provide some suggestions for designing compensation plan in healthcare industry and important general managerial implications to managers in other industries.

Keywords Communal relationship · Exchange relationship · Social norm · Incentive · Healthcare industry

---

S.-W. Lin (✉)
Department of Industrial Management, National Taiwan University of Science and Technology, Taipei, Taiwan
e-mail: shiwoei@mail.ntust.edu.tw

Y.-T. Lin
College of Management, Yuan Ze University, Taipei, Taiwan
e-mail: s1007116@mail.yzu.edu.tw

Y.-K. Lin et al. (eds.), *Proceedings of the Institute of Industrial Engineers Asian Conference 2013*, DOI: 10.1007/978-981-4451-98-7_26,
© Springer Science+Business Media Singapore 2013

1 Introduction

In human society, whether it is about the interaction of the interpersonal relationship in our daily life or about the consumption, competition and cooperation in business practices, our behavior and attitude can be highly influenced by the social factors such as the two main types of relationship, communal relationship and exchange relationship. Because these two relationship types can be differentiated based on the underlying norms or motivation for giving benefits to the partner, the moral-oriented social norms and the money-oriented market norms can thus be applied mechanically in the communal relationship and exchange relationship, respectively.

Clark and Mills (1979) is considered the first article to propose the qualitative distinction between communal and exchange relationships. Since then, researchers in psychology, sociology, and business have been conducted many studies to experimentally or empirically investigate how the communal relationship and exchange relationship influence the patterns of human behavior. Clark and Mills (1993) experimentally demonstrate the validity of the distinction between the communal of relationship and exchange relationship relations. Furthermore, while Clark and Mills (1979, 1993) suggested that the two major relationships are mutually exclusive. Johnson and Grimm (2010) put forward a different viewpoint. They think communal relationship and exchange relationship are not diametrically opposed to each other and are not located at the different ends of the spectrum of the relationship. In other words, Johnson and Grimm (2010) believe that in the pursuit of different personal goals, an individual person is indeed possible to possess both relationships.

In addition to the discussion and investigation of basic concepts and characteristics of the two types of relationships, many researchers have also examined the impact of the introduction of the monetary reward on the communal and exchange relationships. For example, Heyman and Ariely (2004) found that money (or even when people were subliminally primed to think about money) will make people feel more self-sufficient, and thus become more isolated and become (more selfish) and less willing to help others. Heyman and Ariely (2004) also found that as long as the emphasis is on monetary rewards or exchanges, the market norms will be triggered automatically. They further pointed out that social norm is usually the motivations that really can drive the members in an organization to maximize their efforts.

Early research related to communal and exchange relationships focuses more on investigating the relationships between people (e.g., in the study of friendship and love). In recent years, researchers have extend the theoretical framework and these two types of relationships to explore and explain the relationship formed between people and brands (see, for example, Aggarwal 2004). Heyman and Ariely (2004) discussed another possible extension of this framework to the relationship between an organization and people within the organization. It should be noted, however, that there have been few attempts to test the effects of (organization) relationship norms on employee's attitude and behavior.

In the healthcare industry, in light of recent major changes in the medical marketplace make the health care administrators focus more on the efficiency of the operations of a health care institutes. The efficiency may be achieved by more stringent control and more immediate monetary incentives for performance of the health care professionals. The relationship between health care professionals and the organizations (or administrations) thus gradually shift from communal relationship to exchange relationship. This study thus aim to investigate how the financial incentives and non-financial incentives affects work attitude or behavior of health care workers (e.g., physicians, nurses) with different relationship norms.

In particular, this study use an experiment to explore how the types of relationship (communal vs. exchange relationship) between hospitals and medical staffs influence their attitude. Furthermore, we also investigate whether different types of reward (monetary vs. nonmonetary incentives) provided by hospitals affect or alter the types of relationships a medical staff originally had. This study adopt the psychology priming skills to trigger the relationship norms between the medical staffs and organizations, and use a questionnaire as a measure scale to evaluate the participants responses or evaluation of the organization's action (reward).

By experimentally testing the motivational crowding effect in health care professionals, this study intends to check whether the extrinsic motivators such as monetary incentives can undermine intrinsic motivation. After all, if the relationship norm and incentives are inconsistent, the crowding out effect may actually lower the performance when extra incentives are provided. We expect the results of this study can provide some suggestions for designing compensation plan in healthcare industry and important general managerial implications to managers in other industries.

2 Conceptual Framework and Experiment

2.1 Conceptual Framework and Hypothesis

The conceptual model being proposed in this study is that healthcare professionals (i.e., employees of a healthcare institute) evaluate the healthcare institute and its actions depending upon whether the actions violate or conform to the norm of their relationship (with the organization). However, it is important to note that employee-organization relationships are different from interpersonal relationships in several respects. For example, healthcare professionals' relationships with institutes almost always involve some degree of monetary exchange.

The study thus examine participants' (i.e., doctors' and nurses') reactions to being provided different types of incentives (monetary or financial incentives vs. non-monetary incentives) offered by the organization. The participants were first exposed to a description of a prior relationship between a doctor or a nurse and a hypothetical hospital. These descriptions were used to trigger either the communal or exchange relationships and the norms associated with these relationships. Next, the scenario described the hospital's incentive scheme (i.e., rewards) for the extra effort made by the medical professionals. The very timely and direct pecuniary reward in response to the extra effort violates the norms of communal relationship, but conform to the norms of exchange relationship. Based on the conceptual framework discussed above, several hypothesis are formulated as follows.

Effect of relationship type on the attitude of health care professionals toward the hospital

Based on the review of the literature, this study suggests that the relationship type between a health care professionals and the hospital will affect the attitude and the evaluation of the medical staffs on the organization. Hence, the following hypotheses are proposed:

Hypotheses 1 Relationship types have a significant impact on the health care professionals' attitude and evaluation of the hospital.

Hypotheses 1a Relative to health care professionals in an exchange relationship, those in a communal relationship with the hospital will evaluate the hospital more positively.

Effect of relationship type on the attitude of health care professionals toward different incentive or rewarding scheme

This study also suggests that the conformity and violation between relationship norms and the organization's actions (or strategies) will have significant impacts on the health professionals evaluation of the actions (or strategies) of the organization. In other words, when monetary rewards is provided, it is predicted that the communal medical staffs evaluate this rewarding policy negatively relative to exchange type medical staffs. Hypotheses 2a, 2b are as thus proposed as follows:

Hypothesis 2a Relative to subjects (i.e., healthcare professionals) in an exchange relationship, those in a communal relationship with an organization (i.e., hospital) will evaluate the monetary reward more negatively than when a non-monetary reward scheme is imposed.

Hypothesis 2b Relative to subjects (i.e., healthcare professionals) in an exchange relationship, those in a communal relationship with an organization (i.e., hospital) will evaluate the organization more negatively than when a non-monetary reward scheme is imposed.

2.2 Designs and Experiments

The experiment was a 2×2 between-subject design with Relationship Type (communal, exchange) and Reward Type (monetary incentive, non-monetary incentive) as the between-subject conditions. Although doctors and nurses are both

Effect of Relationship Types on the Behaviors

work for the benefits of patients, their duties in a hospital can be quite different. The underlying mechanism of relationship building and the effects of relationship norms on their attitude or behavior might also be different. Thus, we conducted separate experiments to test the effects on doctors and nurses. They are, in total, 80 doctors and 80 nurses from hospitals in Taiwan were recruited through notices posted on the web site. The experiment (read the scenario and answer the questions) usually took about 10 min.

The questionnaire used to measure subjects' or participants' attitude or behaviors is divided into three parts. The first part exposed a description of the (prior) relationship between the healthcare professional and a hypothetical hospital. When reading this description, the participant was asked to projected himself or herself into the condition of the scenario. In other words, the description was served as a stimulus of priming and was aimed at triggering the relationship norms. After the description, several questions were used to check the success or failure of priming. The second part is the description of the incentive scenarios or incentive schemes proposed by the hospitals (which may conform or violate the relationship norms of specific groups of participants). At the end of second part, participants responded to questions about the evaluation of the hospital's incentive scheme and the evaluation measure of the hospital, which are aimed at testing hypotheses discussed above. All items in part 2 were measured on a seven-point scale. Finally, the third part include demographic variables used for measure the characteristics of the participants as well as the characteristics of the healthcare institute where the participant works.

3 Preliminary Results

3.1 Manipulation Check

After the participants were exposed to the relationship manipulations, they were administered a questionnaire designed to assess the effectiveness or success of priming (i.e., the relationship manipulation). We found that after priming, communal participants felt that the organization gives them warm feelings, the organization cares about the needs of their employees, and the employees in the organization are happy to help others without asking for an instant and comparable benefits for return. To further assess whether the relationship manipulations actually occurred, the participants were asked to imagine the hospital coming alive and becoming a person. Most of the communal participants considered their relationship with the organization as "friend".

Similarly, we found that after priming, exchange participants felt that they pay more attention to their own interests than the overall interests of the hospital, and they carefully calculate their gain and loss to make sure they get fair salaries. When these exchange participants were asked to imagine the hospital coming alive and becoming a person, most of them considered their relationship with the organization as "businessman". Thus, the manipulation check showed that the priming for both the communal relationship and exchange relationship (and the corresponding norms) was successful.

3.2 Reaction to Incentive Scheme and Overall Evaluation of the Hospital

There are, in total, six questions used to evaluate participants' evaluation of the incentive scheme proposed by the hospital and the hospital as a whole. Both measures (each created by combining three questions) achieved reasonable reliability and validity. The results showed that when there is a conflict or violation between the relationship norms and incentives (e.g., communal relationship norm and monetary incentive scheme or exchange relationship norm and non-monetary incentive scheme), participants tend to evaluate the scheme and the hospital as a whole relatively negative. On the contrary, the conformity of relationship norms and incentive schemes lead to relatively positive evaluations. These findings are consistent with what are predicted by the conceptual model.

4 Conclusions

Taiwan's brain drain is considered one of the most alarming threats to its sustainable development. In healthcare industry, the brain drain issue is even more serious. The administrators or managers of healthcare organization needs to shoulder its share of responsibility by creating a more attractive environment for doctors and nurses. According to the results of this study, incentive schemes can be effective mainly when they are conform with the underlying relationship norms of the doctors and nurses. A poor incentive system without taking relationship norms into consideration thus may decrease the healthcare professionals' evaluation of the hospital, their job satisfaction and their organizational commitment in the long run.

The results of this study thus can provide some guidelines for designing incentive schemes in healthcare industry and offer important general managerial implications to managers in other industries.

Acknowledgments This work is partially supported by the National Science Council of Taiwan under grant number NSC101-2410-H-011-034-MY2. Any opinions, findings, and conclusions or recommendations expressed in this material are those of the authors and do not necessarily reflect the views of the sponsors.

References

Aggarwal P (2004) The effects of brand relationship norms on consumer attitudes and behavior. J Consum Res 31:87–101

Clark MS, Mills JR (1979) Interpersonal attraction in exchange and communal relationships. J Pers Soc Psychol 37:12–24

Clark MS, Mills JR (1993) The difference between communal and exchange relationships—what it is and is not. Pers Soc Psychol Bull 19:684–691

Heyman J, Ariely D (2004) Effort for Payment—A tale of two markets. Psychol Sci 15:787–793

Johnson JW, Grimm PE (2010) Communal and exchange relationship perceptions as separate constructs and their role in motivations to donate. J Consum Psychol 20:282–294

A Simulation with PSO Approach for Semiconductor Back-End Assembly

James T. Lin, Chien-Ming Chen and Chun-Chih Chiu

Abstract This paper studies a dynamic parallel machine scheduling problem in a hybrid flow shop for semiconductor back-end assembly. The facility is a multi-line, multi-stage with multi-type parallel machine group, and orders scheduled with different start time. As a typical make-to-order and contract manufacturing business model, to obtain minimal manufacturing lead time as main objective and find an optimal assignment of production line and machine type by stage for each order as main decisions. Nevertheless, some production behavior and conditions increase the complexity, and including order split as jobs for parallel processing and merged completion for shorten lead time. Complying quality and traceability requirement so each order only can be produced from one of qualified line(s) and machine type(s) and all jobs with the same order can only be produced in same assigned line and machine type with stochastic processing time. Lead time is counted from order start time to completion, including sequence dependent setup times. As a NP-hard problem, we proposed a simulation optimization approach, including an algorithm, particle swarm optimization (PSO) to search optimal assignment which achieving expected objective, a simulation model to evaluate performance, and combined with optimal computing budget allocation (OCBA) to reduce replications. It provides a novel applications using simulation optimization for semiconductor back-end assembly as a complex production system.

J. T. Lin (✉) · C.-M. Chen · C.-C. Chiu
Department of Industrial Engineering and Engineering Management,
National Tsing Hua University, Hsinchu, Taiwan, ROC
e-mail: jtlin@ie.nthu.edu.tw

C.-M. Chen
e-mail: d937811@oz.nthu.edu.tw

C.-C. Chiu
e-mail: vhasb3210@gmail.com

Y.-K. Lin et al. (eds.), *Proceedings of the Institute of Industrial Engineers Asian Conference 2013*, DOI: 10.1007/978-981-4451-98-7_27,
© Springer Science+Business Media Singapore 2013

Keywords Dynamic parallel machine scheduling · Simulation optimization · Particle swarm optimization · Optimal computing budget allocation · Semiconductor back-end assembly

1 Introduction

The semiconductor industry was originally highly integrated from IC design to manufacturing by integrated device manufacturers (IDMs). However, this industry has been decentralized approximately for thirty years because of technological differences, core business focus, and cost scale. The industry was split into the IC design house (i.e. fabless), which focuses on the value creation of IC functions for devices or applications, manufacturing also separate wafer foundry and probe (or wafer sort) as front-end, and assembly and final test as back-end. Back-end assembly is the process to pick up die from wafer and package integrated circuits (ICs), which are key components of most electronic devices. The back-end assembly may be an offshore of IDM or an individual company serves as a virtual factory for fabless and IDM. The lead time of front-end is relative longer than back-end and customers will request back-end assembly to provide a short but robust lead time service to absorb fluctuation instead of physical ICs inventory. It becomes the key as an order winner in this kind of typical make-to-order (MTO) and contract-manufacturing model.

Some criteria of order winner are delivery reliability and speed. In general, back-end assembly adopt bottleneck scheduling to schedule the start time and estimate ship of date but more challenges from practice is that bottleneck shifting. There are two reasons caused. The first one is parallel processing of jobs. There are too many jobs split by order for shorten lead time. The second is dynamic routing with conditions for these jobs. While lots of jobs in the production system, each job has alternatives in production line and machine type, and thus the possibility of bottleneck shifting is raised. In this research we proposed a methodology using simulation optimization by particle swarm optimization. We aim to obtain minimal manufacturing lead time through optimal assignment of production line and machine type for each order. Before proposed it, the production system of back-send assembly is stated below and problem statement in later section.

1.1 System Description of Back-End Assembly

The production system of back-end assembly is a classical product layout. Through these processes, the die in the wafer will be picked up and packaged as IC. The typical main assembly operations include multiple stages as Fig. 1. "Tape", "back grind", "de-tape/mount" and "saw" are the wafer base operations.

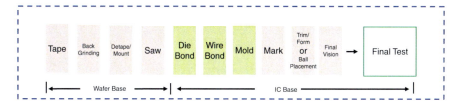

Fig. 1 Main operations of back-end assembly

After these operations, the die on the wafer are able to pick up at "die bond" stage. "Die bond" attaches a die to a substrate or lead frame by using adhesives. In the "wire bond" stage, the lead is bound, the die is held, and the substrate or lead frame is linked by using gold or copper wires, thus connecting the die to the outer circuit. In the "mold" stage the circuits of components are molded to protect them from outside forces and strengthen their physical characteristics. "Mark," "trim/ form" or "ball placement," and "final vision" are straightforward operations that complete the assembly process (Zhang et al. 2007).

The input of the production system is wafer. Considering quality traceability, the wafer with a specific lot number and each lot number has around twenty five wafers based foundry process. Customers will release the order to back-end assembly with identified demand quantity and wafers lot numbers as input. Back-end assembly will schedule them the start time using bottleneck (normally is wire bonder) scheduling daily. Considering management span, a flow shop will set five hundreds to one thousand wire bonders, i.e. if demand is over than the line size, there will be another production line as product layout base, and thus there will be a multi-line environment. As previous description, a lot of stages are in production line. Each stage has different machine types with different numbers cause of precision level of product specification or equipment technology evolution. During the process, the order split size is based on movement quantity of a job. In general, the movement quantity is determined by how many die in a magazine which contains some pieces of lead frame or substrate which contains some pieces of die. Thus, a lot of jobs spilt from orders are in the system. The system output is IC. From system management, manufacturing lead time is one of system performance indicator. The complexity comes from these jobs with alternatives in line and machine type selection as dynamic routing.

2 Problem Statement

From system description, some characteristics are summarized by three perspectives. The first is physical layout, including numbers of production line and numbers at each stage by machine type and line. It is also a system constraint by capacity view. The second is product attributes, including capability, split size rules and processing times. Capability means that each product can be produced in specific line and machine time from qualification concept. Split size rule is to determine movement quantity of a job so that the jobs of an order can be parallel processed for lead time shorten. Processing time is defined by product and machine type at each stage. It is a stochastic times with a distribution. The third is related with the physical entity flow in the system. An order introduced some information demand quantity, start time and product type. Once an order enters into production line, the whole lots are wafer base before die bond stage. After die bond, the order is split as a job per predefined split rule and identified by a serial number based on original order number. As variety order size and predefined split rule, each order generated different numbers of job and parallel processed in the production system. For traceability and quality concerns, each job not only constrained into predefined (qualified) line and machine type for production but also has to keep same line and machine type for the same order. Only all jobs completed, and the order completed to count manufacturing lead time. During process, if jobs between in and out are different product type, and there will be a sequence dependant setup times.

Even there are lots of stages in production system, the most critical stages with regard to manufacturing lead time are die bond, wire bond and molding. We will include these three stages as our study assumption. A lot of orders and jobs in the production system and constrained by capacity, capability and physical entity flow rules. The decisions are order assignment to which production line and to which machine type for their counter jobs and obtain minimal average manufacturing lead time.

3 Literature Review

More attentions on dynamic scheduling since the production system with some events, like processing time is non-deterministic, loading limit and dynamic routing (Suresh et al. 1993). This problem is an extension of hybrid flow shop scheduling problem (HFSSP). Ribas et al. (2010) recently reviewed published papers about this topic and classified the HFSSP from the production system and solution procedure perspectives. As a NP-hard problem, simulation optimization is a general approach. Carson and Maria (1997) defined simulation optimization as the process of finding the best input variable values among the possibilities without explicitly evaluating each value. The use of heuristic methods such as GA, evolutionary strategy, simulated annealing, tabu search, or simplex search is common. Chaari et al. (2011) adopted the genetic algorithm (GA) for the demand uncertainty of robust HFSSP. Chaudhry and Drake (2009) used the GA for machine scheduling and worker assignment to minimize total tardiness. Chen and Lee (2011) presented a GA-based job-shop scheduler for a flexible multi-product, parallel machine sheet metal job shop. Anyway, simulation approach consumed times since the system is stochastic. To reduce simulation times, Chen et al. (2000), Chen and Lee (2011) proposed optimal computing budget allocation (OCBA) as the ultimate simulation resource and applied in product design (Chen et al. 2003).

The PSO algorithm shares many similarities with evolutionary computation techniques such as GAs. It was proposed by Kennedy and Eberhart. The system is initialized with a population of random solutions and searchers for optima by updating generations. However, unlike GA, the PSO algorithm has no evolutionary operators, such as crossover and mutation. In the PSO algorithm, the potential solutions, called particles, move through the problem space by following the current optimum particles. Kuo and Yang (2011) compared GA and PSO based algorithm and found the PSO is better than GA base.

4 Proposed Methodology and Illustration

4.1 Simulation Optimization Methodology Based PSO Combined with OCBA

As Fig. 2, we proposed a simulation optimization methodology based on PSO combined with OCBA to find the optimal assignment achieving our objective. We apply PSO as our searching engine for optimization and apply a simulation tool to evaluate performance or fitness. We also apply OCBA for simulation replications reduction. The basic steps are stated as below:

Step 1: *Parameters setting for PSO and OCBA*

The parameters of PSO include population size, weight, learning factor $c1$ and $c2$. For OCBA, parameters are initial replications of simulation (n_0), total simulation budget (T), incremental replications (Δ) and P{CS}*.

Step 2: *Initial population*

According the parameter of population size, it generated multiple alternatives for further evaluation.

Step 3: *Evaluation of performance or fitness and combined with OCBA*

There are three sub-steps. The first is to evaluate all alternatives of initial population using n_0 times of replications for each one. The second is to calculate P{CS} and check if reach the P{CS}*. If achieved, go to step 4 or do more replications based on Δ and continue to the second sub-step.

Step 4: *Update the Pbest and Gbest*

For each particle, it will generate a new fitness and compared with before, and the better one is the Pbest; for particle, it will find the best Pbest as Gbest.

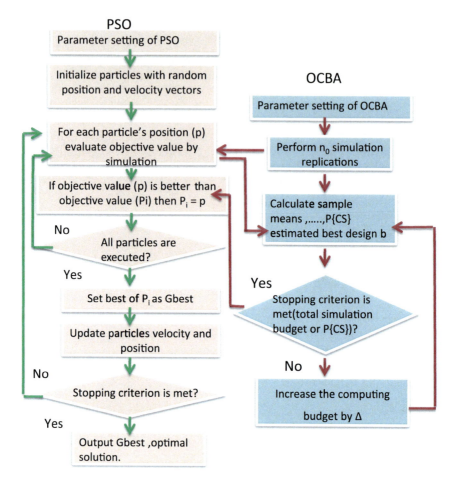

Fig. 2 Simulation optimization methodology based on PSO combined with OCBA

Step 5: *Update velocity and position according to* Eq. (1) *and* (2)

Each particle was updated for velocity and position.

$$v_{i,new} = w \cdot v_{i,old} + c_1\, rand() \cdot (P_i - X_{i,old}) + c_2\, rand() \cdot (G - X_{i,old}) \quad (1)$$

$$X_{i,new} = X_{i,old} + v_{i,new} \quad (2)$$

Step 6: *Check if met condition of termination*

As the loop, while if it met condition of termination. In general, the generation of PSO or T reached. If it is not optimal, go to step.

4.2 Encoding and Decoding Design

The PSO is normally adopted in a continuous space. As our study, it is discrete type so we have to encode using a special approach for encoding and decoding. We use a matrix table to define all alternatives in line assignment and machine type assignment for each order in Tables 1, 2 and 3. In the Fig. 3, we separate two segments for line and machine type assignment. First segment means line assignment for each order; and second one means machine assignment by die bonder, wire bonder and mold system for each order. Using order 2 as an example, original line assign number is 1.2, refers table x, it was decoded as line 2. For machine type, the original number for wire bonder is 1.3, round up to 2. Since it is assigned in line 2, so refers table z, it was decoded as machine type, DB4.

4.3 Illustration by Simple Case

The simple case includes three orders and related parameters were set: $w = 0.8$; $c1 = 1$; $c2 = 2$; $n_0 = 10$; $\Delta = 100$; $T = 300$; $P\{CS\} = 0.9$. Available line(s) and machine type(s) list as Tables 1, 2 and 3. We adopt three approaches and compared their performance of manufacturing lead time and replications in Table 4.

Table 1 Available line assignment for orders

	Order 1	Order 2	Order 3
Numbers of available line	1	2	1
Position 1	Line 2	Line 1	Line 1
Position 2	–	Line 2	–

Table 2 Available machine type assignment for orders in line 1

	Order 1			Order 2			Order 3		
Stage	DB	WB	MD	DB	WB	MD	DB	WB	MD
Numbers	3	2	2	2	2	3	1	3	2
Position 1	DB1	WB1	MD1	DB1	WB2	MD2	DB3	WB1	MD2
Position 2	DB2	WB3	MD4	DB2	WB3	MD3	–	WB3	MD4
Position 3	DB3	–	–			MD4	–	WB4	–

Table 3 Available machine type assignment for orders in line 2

	Order 1			Order 2			Order 3		
Stage	DB	WB	MD	DB	WB	MD	DB	WB	MD
Numbers	1	3	2	2	2	1	2	2	3
Position 1	DB2	WB2	MD1	DB1	WB1	MD3	DB2	WB2	MD1
Position 2	–	WB3	MD2	DB4	WB3	–	DB4	WB3	MD2
Position 3	–	WB4	–	–	–	–	–	–	MD4

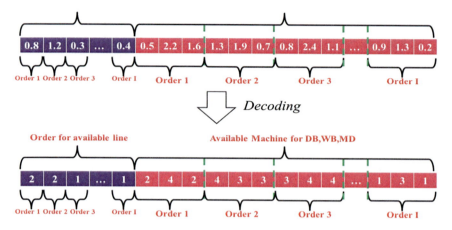

Fig. 3 Encoding and decoding illustration

Table 4 Comparison among different approaches

Approaches	Assignment	Lead time (s)	Replications	p-value
Exhaustive method	(1,1,2,2,3,1,1,4,1,3,3,6)	123563.2	306900	–
PSO	(2,1,2,4,3,6,1,4,5,3,3,6)	124379.7	54900	0.067355879
PSO + OCBA	(2,1,1,4,3,6,2,4,1,3,3,1)	124477.4	34740	0.076163638

Compared with exhaustive method

PSO can find an optimal solution but less replication. If it is combined with OCBA, it reduced more replications. Both PSO base approaches are statistically significant differences compared with exhaustive method by t-test. The p-value is greater than 0.05.

5 Conclusions

This is a NP-hard problem since the system is complex with many characteristics of back-end assembly, especially faced a dynamic routings with sequence dependent setup times. Simulation is visible so practitioners preferred but lack of optimization, and it also has concerns on long replications cause of stochastic properties. As our study, it conquers the weakness and obtains a good result without statistically significant differences with exhaustive method. This study inspired us to apply this methodology in system performance improvement if the system behavior, constraints, objectives and decision variables and optimization approach can be well defined. As a simulation base approaches, it also can compare more scenarios and explore the influence on other performance like utilization, delivery robustness and predict coming bottleneck shifting issue heads up some actions. It is helpful for practitioners using academic approach to improve system performance. It also provides opportunity to find more insights in relationship between system performance and system characteristics.

References

Carson Y, Maria A (1997) Simulation optimization: methods and Applications In: Proceedings of the 1997 winter simulation conference, pp 118–126

Chaari T, Chaabane S, Loukil T, Trentesaux D (2011) A genetic algorithm for robust hybrid flow shop scheduling. Int J Comput Integr Manuf 24(9):821–833

Chaudhry IA, Drake PR (2009) Minimizing total tardiness for the machine scheduling and worker assignment problems in identical parallel machines using genetic algorithms. Int J Adv Manuf Technol 42:581–594

Chen CH, Donohue K, Yucesan E, Lin J (2003) Optimal computing budget allocation for Monte Carlo simulation with application to product design. Simul Model Pract Theory 11:57–74

Chen CH, Lee LH (2011) Stochastic simulation optimization—an optimal computing budget allocation. World Scientific Publishing Co. Ptd. Ltd, Singapore

Chen CH, Lin J, Yucesan E, Chick SE (2000) Simulation budget allocation for further enhancing the efficiency of ordinal optimization. Discrete Event Dyn Syst 10:251–270

Kuo RJ, Yang CY (2011) Simulation optimization using particle swarm opti algorithm with application to assembly line design. Appl Soft Comput 11:605–613

Ribas I, Leisten R, Framinan JM (2010) Review and classification of hybrid flow shop scheduling problems from production system and a solutions procedure perspective. Comput Oper Res 37:1439–1454

Suresh et al (1993) Dynamic scheduling—A survey of research. Int J Prod Econ 32:53–63

Zhang MT, Niu S, Deng S, Zhang Z, Li Q, Zheng L (2007) Hierarchical capacity planning with reconfigurable kits in global semiconductor assembly and test manufacturing. IEEE Trans Autom Sci Eng 4(4):543–552

Effect of Grasp Conditions on Upper Limb Load During Visual Inspection of Objects in One Hand

Takuya Hida and Akihiko Seo

Abstract Automated visual inspection systems based on image processing technology have been introduced to visual inspection processes. However, there are still technical and cost issues, and human visual inspection still plays a major role in industrial inspection processes. When a worker inspects small parts or products (e.g., lens for digital cameras, printed circuit boards for cell phones), they suffer from an upper limb load caused by handling objects in one hand and maintaining this awkward posture. Such workload causes damage to the hands, arms, and shoulders. So far, few studies have elucidated the effect of upper limb loads. Therefore, we conducted an experiment where the subjects were assumed to be visually inspecting small objects while handling them with one hand, and investigated the effect of grasp conditions on the upper limb load during tasks. We used electromyography, the joint angle, and subjective evaluation as evaluation indices. The results showed that the upper limb load due to the grasp condition differed depending on the upper limb site. Therefore, it is necessary to consider not only the muscle load but also the awkward posture, the duration of postural maintenance, and subjective evaluation when evaluating the upper limb load during such tasks.

Keywords Visual Inspection · Muscle Load · Joint Angle · Awkward Posture

---

T. Hida (✉) · A. Seo
Graduate School of System Design, Tokyo Metropolitan University, 6-6 Asahigaoka, Hino-shi, Tokyo 191-0065, Japan
e-mail: hida-takuya1@sd.tmu.ac.jp

A. Seo
e-mail: aseo@sd.tmu.ac.jp

Y.-K. Lin et al. (eds.), *Proceedings of the Institute of Industrial Engineers Asian Conference 2013*, DOI: 10.1007/978-981-4451-98-7_28,
© Springer Science+Business Media Singapore 2013

1 Introduction

Visual inspection is a method to certify the quality of products in industries (Cho et al. 2005). By visual inspection, an inspector evaluates the surface characteristics of objects. These characteristics are divided into two types: one that can be defined and detected as a physical quantity, and the other that cannot. The latter type can only be evaluated by human, and there remain major needs for human visual inspection.

Ergonomics research has been trying to alleviate the physical and mental workloads that inspectors suffer from. Those loads come from the long time of fixation of sight on objects and maintenance of working posture (Liao and Drury 2000). Inspectors stay sitting and use a microscope or a magnifier (Yeow and Sen 2004). These are mainly mental work which results in mental workload (Lee and Chan 2009) and visual fatigue (Jebaraj et al. 1999). Additionally, in recent years, there is another style of inspection which involves grasping and handling of parts or products by the inspector. Repetitive motions, maintenance of the elevation of upper limb, and awkward postures are observed in this new style, which are suspected to cause more upper limb disorders than in the traditional style of inspection.

The authors have so far conducted researches on the upper limb load during inspections with handling of objects (Hida and Seo 2012; Hida et al. 2013a, b). This research targets the visual inspection of rather small objects (e.g., camera lenses and printed circuit boards of cellular phone). This sort of work does not overload the musculoskeletal system of the inspector thanks to the small weight of the objects. However, there are still considered to cause large workload on the upper limb as it involves handling of objects by one hand and maintenance of the elevation of upper limb. In order to clarify this workload, we conducted an experiment where the subjects were assumed to be visually inspecting small objects while handling them with one hand, and investigated the effect of grasp conditions on the upper limb load during tasks. We used electromyography, joint angle, and subjective evaluation as evaluation indices.

2 Method

2.1 Subject

The subjects were 6 healthy male and 6 healthy female university students with no pain in their upper limbs. The means and standard deviations of age, height, and body mass were 22.5 ± 0.8 years, 166.6 ± 10.7 cm, and 55.8 ± 12.4 kg, respectively. All the subjects were right-handed. The experiment was conducted on the right upper limb.

Fig. 1 Work posture

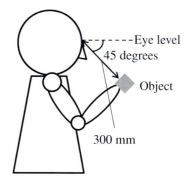

Fig. 2 Visible range of symbol

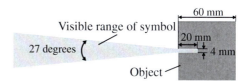

2.2 Experimental Procedure

The subjects grasped an inspection object at 45 degrees downward from the eye level with right hand in a sitting posture (see Fig. 1). The shape of the object was a cube with a side length of 60 mm and a weight of 8 g. The object had a hole which was 4 mm in diameter and 20 mm in depth. One symbol out of five was placed at the bottom of the hole. As it was at the bottom, visible range of symbol was limited (see Fig. 2). The subjects were asked to inspect the object visually and read out the shape of the symbol. Experimental conditions were divided into six grasp types (see Fig. 3). These were randomized to eliminate the effect of the order.

2.3 Measurement Data and Analysis Method

In this research, surface electromyography (called EMG) was used to evaluate the muscle load of the upper limb by the upper limb motion. The percentage of maximum voluntary contraction (%MVC) was calculated using the ratio of muscle

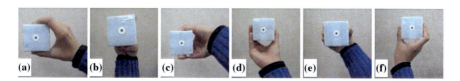

Fig. 3 Grasp types

activity for a task to the MVC. Moreover, joint angle was measured to evaluate the joint load of the upper limb by postural maintenance. In addition to these quantitative evaluations, subjective evaluation was also collected. We applied an analysis of variance (ANOVA) for each measurement value. The factors were subject and grasp type. In addition, Tukey's test was employed as a follow-up test of the main effect. The significance level was set at 5 % for both tests.

(1) EMG

EMG was measured by using an active electrode (SX230-1000; Biometrics Ltd.). The muscle measurement sites of EMG were as follows: upper part of the trapezius muscle, clavicular part of the pectoralis major muscle, middle part of the deltoid muscle, the biceps brachii muscle, the round pronator muscle, the flexor carpi ulnaris muscle, the extensor carpi radialis longus muscle, and thenar muscle. Upper part of the deltoid muscle performs elevation of scapula; clavicular part of the pectoralis major muscle performs flexion and adduction of shoulder joint; middle part of the deltoid muscle performs abduction of the shoulder joint; the biceps brachii muscle performs flexion of elbow joint; the round pronator muscle performs pronation of forearm; the flexor carpi ulnaris muscle performs flexion and adduction of wrist joint; the extensor carpi radialis longus muscle performs extension and abduction of wrist joint; and the thenar muscle performs abduction, adduction, flexion, and opposition of thumb (Criswell 2010). Using all of the above EMG sites, we were able to evaluate the upper limb load during the task.

The measured EMG was converted to %MVC for each muscle. The muscle load was then evaluated by means of the average of %MVC for each subject.

(2) Joint angle

To measure the joint angle, we used 3D sensor (9 Degrees of Freedom Razor IMU; SparkFun Electronics). This sensor can measure triaxial terrestrial magnetism, triaxial acceleration, and triaxial angular velocity. The measurement position were upper arm, forearm and back of the hand. It is possible to calculate each joint angle from sensor orientation and inclination angle.

(3) Subjective evaluation

Subjective fatigue was evaluated in four body areas: neck, shoulder, elbow, and wrist. In addition, overall subjective fatigue of upper limb and difficulty of taking a posture were also evaluated. The range of rating scale is 1 (feel no fatigue at all or feel no difficulty at all) to 5 (feel highly fatigued or feel so difficult).

3 Result and Discussion

3.1 EMG

Figures 4, 5, and 6 show %MVC of clavicular part of the pectoralis major muscle, the flexor carpi ulnaris muscle, and the extensor carpi radialis longus muscle respectively. Every measurement of %MVC showed a significant main effect with respect to grasp types.

Effect of Grasp Conditions

Fig. 4 %MVC of clavicular part of the pectoralis major m

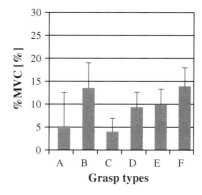

Fig. 5 %MVC of the flexor carpi ulnaris m

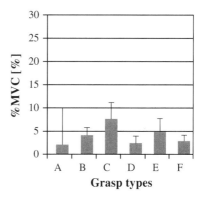

Fig. 6 %MVC of the extensor carpi radialis longus m

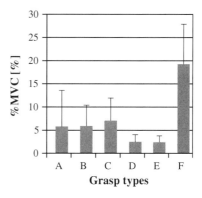

The %MVC of the clavicular part of the pectoralis major muscle was greater with the grasp types B and F. These grasp types require the subject to place his/her forearm in parallel to his/her midline. We consider this to be the cause of the observed higher muscle load, as the clavicular part of the pectoralis major muscle performs the horizontal flexion of shoulder joint. The %MVC of the flexor carpi

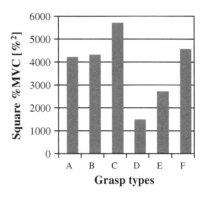

Fig. 7 Square sum of %MVC

ulnaris muscle was the greatest with the grasp type C. The grasp type C requires the subject to keep his/her line of sight perpendicular to the surface to be inspected by means of the ulnar and palmar flexion of wrist joint. These movements are performed by the flexor carpi ulnaris muscle, hence the observed highest load. The %MVC of the extensor carpi radialis longus muscle was the greatest with the grasp type F. The grasp type F requires the subject to keep his/her line of sight perpendicular to the surface to be inspected by means of the radial and dorsal flexion of wrist joint. These movements are performed by the extensor carpi radialis longus muscle, hence the observed highest load.

As above, the body site where muscle load is greater is different by means of grasp type. Therefore we have to evaluate these values synthetically, not individually. To achieve it, we calculated the square sum of all the %MVCs for each level, and took the sum as the evaluation index of the integral workload of the upper limb muscles (see Fig. 7). As the result of this calculation, the grasp type C had the greatest value for the square sum, and the grasp type D had the least. Therefore, with respect to EMG, we can see that the grasp type D has the least workload on muscles, and we recommend it.

3.2 Joint Angle

In order to evaluate the joint angle, we do not deal with each measurement at an individual joint, but define a value of cumulative displacement of joint angles and use it. The definition goes as follows. First we take, for each joint, the difference between the joint angle of a neutral position and the joint angle of working posture. Note that the joint may have more than one degrees of freedom (DOFs); in that case, each DOF has its own neutral position and we take the difference for each neutral position. Second, we take the absolute values of the differences. The cumulative displacement is then defined to be the sum of these absolute values. Details of the DOFs for joints are as follows. Shoulder joint has three DOFs (abduction–adduction, flexion–extension, and internal-external rotation), elbow

Fig. 8 Cumulative displacement of joint angle

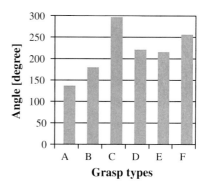

Fig. 9 Pronation-supination angle

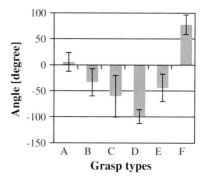

joint has one DOF (flexion–extension), forearm has one DOF (pronation-supination), and wrist joint has two DOFs (radial-ulnar flexion and palmar- dorsal flexion). We give the values of cumulative displacement in Fig. 8. Figure 9 gives the pronation-supination angles of forearm.

As a result, the cumulative displacement was the greatest for the grasp type C and the least for the grasp type A. Looking into each joint, the grasp type C had the greatest displacement along all the DOFs of shoulder joint and along one DOF of wrist joint.

3.3 Subjective Evaluation

Figure 10 shows the subjective fatigue of the overall upper limb. Figure 11 shows the difficulty of taking a posture. Both subjective evaluations indicated the highest score with the grasp type C, and the lowest with the grasp type A. With the grasp type C, the trend of the subjective evaluation was consistent with the trend of EMG or joint angle.

Fig. 10 Subjective fatigue of the overall upper limb

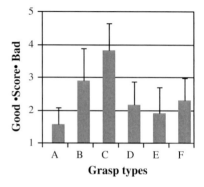

Fig. 11 Difficulty of taking a posture

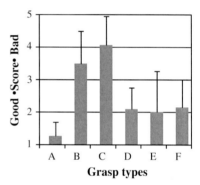

3.4 Comprehensive Evaluation

Three evaluation indices above coincide with others about the source of the greatest workload on the upper limb, which is the grasp type C. It is clear from this fact that the grasp type C should be avoided as it is the worst in terms of muscle load, working posture, and subjective evaluation. On the other hand, these indices give different answers about the best grasp type which poses the least upper limb load. Namely, the grasp type D was the best in terms of muscle load, and the grasp type A was the best in terms of working posture and subjective evaluation.

This difference can be reasoned by the pronation-supination angle of forearm (see Fig. 9). With the grasp type D, supination angle is almost at the maximum range of motion although muscle load is small. This large movement of forearm should have affected negatively on working posture and subjective evaluation. In other words, the grasp type D is the best if muscle load alone is evaluated, while the grasp type A is recommended after taking into account the loads on joints by supination, subjective load, and the difficulty of taking working posture.

4 Conclusions

We conducted an experiment on the upper limb load during a sort of human visual inspection which involves the grasping of the object with one hand. We evaluated six different types of grasp. The evaluation indices were electromyography, joint angle, and subjective evaluation. As a result, the grasp type C was evaluated to be the worst and we recommend to avoid it. The best and recommended grasp type differed between evaluation indices. It was grasp type D in terms of muscle load, and grasp type A in terms of working posture and inspector's subjective evaluation. Therefore, in order to evaluate the workload of inspection and similar prolonged light load works, there would always be a need for the subjective evaluation of posture taking and keeping.

References

Cho CS, Chung BM, Park MJ (2005) Development of real-time vision-based fabric inspection system. IEEE Trans Ind Electron 52(4):1073–1079

Criswell E (ed) (2010) Cram's introduction to surface electromyography, 2nd edn. Jones and Bartlett Publishers, Massachusetts

Hida T, Seo A (2012) Development of evaluation function for estimating upper-limb load during visual inspection. Proceedings of the 13th Asia pacific industrial engineering and management systems conference, Phuket, Thailand, 2–4 December 2012

Hida T, Chihara T, Seo A (2013a) Effect of object size and grasp position on upper limb load during visual inspection (in Japanese). J Japan Ind Manag Assoc 63(4):217–244

Hida T, Chihara T, Seo A (2013b) Effect of speed and scanning direction on upper limb load during visual inspection (in Japanese). J Japan Ind Manag Assoc 64(2):1–8

Jebaraj D, Tyrrell RA, Gramopadhye AK (1999) Industrial inspection performance depends on both viewing distance and oculomotor characteristics. Appl Ergon 30(3):223–228

Lee FC, Chan AH (2009) Effects of magnification methods and magnifier shapes on visual inspection. Appl Ergon 40(3):410–418

Liao MH, Drury CG (2000) Posture, discomfort and performance in a VDT task. Ergonomics 43(3):345–359

Yeow PH, Sen RN (2004) Ergonomics improvements of the visual inspection process in a printed circuit assembly factory. Int J Occup Saf Ergon 10(4):369–385

A Process-Oriented Mechanism Combining Fuzzy Decision Analysis for Supplier Selection in New Product Development

Jiun-Shiung Lin, Jen-Huei Chang and Min-Che Kao

Abstract The selection of well-performed supplier involved in new product development (NPD) is one of the most important decision issues in the contemporary industrial field, in which the collaborative design is common. This paper proposes a systematical process-oriented mechanism combining fuzzy arithmetic operations for solving supplier selection problem in the NPD stage. In the proposed mechanism, the Design Chain Operations Reference model (DCOR) developed by Supply Chain Council (SCC) is adopted to describe NPD processes between the business and the candidate suppliers. These processes are deployed by four-level framework, including the top level, the configuration level, the process element level, and the implementation level. Then, the design structure matrix (DSM) is used to analyze the process relationship based on the results from the implementation level. The original DSM is partitioned by the Steward's method to get reordered DSMs with the interactive process information with respect to the metrics provided by the DCOR. The fuzzy decision analysis is executed to obtain the weighted aggregated scores with respect to the DCOR metrics and to select the best suppliers for different components. Finally, a practical case in Taiwan is demonstrated to show the real-life usefulness of the proposed mechanism.

Keywords New product development (NPD) · Supplier selection · DCOR · Design structure matrix (DSM) · Fuzzy decision analysis

---

J.-S. Lin (✉) · M.-C. Kao
Department of Industrial Engineering and Management, Ming Chi University
of Technology, New Taipei, Taiwan, Republic of China
e-mail: jslin@mail.mcut.edu.tw

M.-C. Kao
e-mail: kao_mio@yahoo.com.tw

J.-H. Chang
Department of Logistics Management, Tungnan University, New Taipei, Taiwan,
Republic of China
e-mail: jhchang@mail.tnu.edu.tw

Y.-K. Lin et al. (eds.), *Proceedings of the Institute of Industrial
Engineers Asian Conference 2013*, DOI: 10.1007/978-981-4451-98-7_29,
© Springer Science+Business Media Singapore 2013

1 Introduction

Dramatically global competition and consumers' needs diversity have resulted in shortening product life cycle and increasing complexity in new product development (NPD). To maintain competitive advantages, many companies have placed an emphasis on cultivating the core competences of product development. Selecting the best suppliers involved early in the NPD stage, which is also called the early supplier involvement (ESI), plays a strategic role in the degree to which an organization is able to achieve its goal of developing a new product. Apparently, it is a major factor in customer satisfaction, product quality, cost reduction, risk sharing (Johnsen 2009). Therefore, developing a mechanism to select well-performed suppliers involved early in the NPD stage is one of the most important decision issues in the contemporary industrial field.

In the past several decades, the supplier selection problem has received considerable attention from both practitioners and researchers (De Boer et al. 2001; Ho et al. 2010). From the viewpoint of contemporary supply chain management, many companies regard their suppliers as partners, not adversaries. Hence, an interesting topic relating to the ESI has gradually received more attention in the past two decades. Obviously, the supplier selection involved in the NPD stage is essentially the multi-criteria decision-making (MCDM) process. In many real-world situations, the MCDM process usually involves uncertainty. Zadeh (1965) was first to present fuzzy set theory. This theory provided a good methodology for handling imprecise and vague information. Kumar et al. (2004) provided a fuzzy goal programming approach for solving a vender selection problem with three objectives. Tang et al. (2005) applied the fuzzy synthesis evaluation method to assess the design scheme in part deployment process. They determined eight influencing factors for facilitating the selection of suppliers involved in NPD projects. Carrera and Mayorga (2008) considered three indicators: supplier characteristics, supplier performance, and project characteristics. They applied modular fuzzy inference system to the supply chain for the selection of suitable suppliers engaged in the NPD. Oh et al. (2012) proposed a decision-making framework using a fuzzy expert system in portfolio management for dealing with the uncertainty of the fuzzy front-end of product development.

Most previous researches on supplier selection involved in the NPD did not consider interactive process information among the design chain members. To integrate this information into the decision of selecting the best suppliers involved in the NPD, this paper proposes a process-oriented mechanism combining fuzzy arithmetic operations for solving supplier selection problem in the NPD stage, in which simultaneously takes the interactive process information and linguistic information into consideration in the decision-making process. The proposed mechanism can provide practitioners with selecting the best suppliers involved early in the NPD stage.

2 Description of the NPD Process Information

Implementing a NPD project requires the team consisting of members from a variety of companies at different locations. They work together to solve specific product issues and exchange information each other. Hence, a well-defined information exchange and communication platform is necessary for the members involved in NPD collaboration so as to achieve the synergy. In 2006, Design Chain Operation Reference Model (DCOR) was first introduced by Supply Chain Council (SCC). The DCOR can be adequately used as a platform for the design chain partners.

The DCOR is a process-oriented four-level deployment architecture, as illustrated in Fig. 1. The Level 1, called the top level/process types, consists of five management processes (i.e., plan (P), research (R), design (D), integrate (I), and amend (A)). The Level 2, called the configuration level/process Categories, includes product refresh, new product, and new technology. For example, based on the research (R) process of the Level 1, the processes of the level 2 are divided into three categories: research product refresh (R1), research new product (R2), and research new technology (R3). The Level 3, called the process element level/decompose processes, deployed from the Level 2, includes input, output, metrics, and best practices. According to the company's requirements, these processes are deployed to the Level 4, called the implementation level/decompose process elements.

The DCOR also defines five process performance metrics that can be used as indices for selecting the best suppliers in the NPD stage. The five metrics are (1) reliability, (2) responsiveness, (3) agility, (4) cost, and (5) asset. Each metric includes several sub-metrics (Nyere 2009). The DCOR provides practitioners with a common reference model that allows the effective communication between companies and their design chain partners. It is important that benefits and competition of a company can be improved by applying collaborative new product development and design.

Although the DCOR provides a process reference model for the practitioners, it does not systematically present the interactive process information among the design chain members. The Design Structure Matrix (DSM) is suited for converting the process information of the Level 4 in the DCOR to the interactive process information we want. By using the partition rule presented by Steward (1981), the process information is divided into three types: parallel, sequential, and coupled/interdependent processes. The DSM is able to capture the interactive processes (i.e., coupled/interdependent processes). In this paper, these coupled/interdependent processes are applied to express the interactive process information from the Level 4 of the DCOR in the NPD stage.

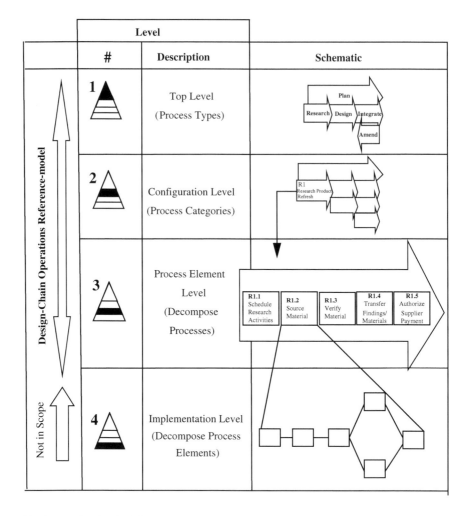

Fig. 1 Four levels of DCOR

3 The Proposed Process-Oriented Mechanism

In this section, we provide a process-oriented mechanism combining fuzzy decision analysis for solving supplier selection problem in NPD. Twelve steps of the proposed mechanism are presented as follows:

Step 1. Define the scope of supplier selection in the NPD stage for an enterprise.
Step 2. According to the requirements of the enterprise, establish D-L1-M (DCOR Level 1 Model) based on the top level of the DCOR.

A Process-Oriented Mechanism Combining Fuzzy Decision Analysis 243

Step 3. Generate D-L2-M (DCOR Level 2 Model) based on the configuration level of the DCOR.

Step 4. Form D-L3-M (DCOR Level 3 Model) based on the process element level of the DCOR.

Step 5. Build D-L4-M (DCOR Level 4 Model) based on the implementation level of the DCOR.

Step 6. Transform D-L4-M into the DSM, called the original DSM.

Step 7. Get the reordered (partitioned) DSM by applying the Steward's method (Steward 1981) to partition the original DSM.

Step 8. Construct fuzzy evaluation representation based on the DCOR metrics. The linguistic weighting variables are used to assess the importance of the metrics and the linguistic rating variables are utilized to evaluate the performance for candidate suppliers with respect to each metric. The linguistic weighting variables and the linguistic rating variables are measured by nine linguistic terms, respectively. For simplicity, but without loss of generality, these linguistic terms are assumed in this paper to be represented as triangular fuzzy number (TFN) expressed by a triple (a, b, c).

Step 9. Determine fuzzy weights of the metrics and fuzzy scores for candidate suppliers with ferred to TFNs. The fuzzy weight of metric j is denoted as $\tilde{w}_j = (w_{j1}, w_{j2}, w_{j3})$. Hence, the fuzzy weight matrix is $\tilde{W} = [\tilde{w}_j]$ of size $1 \times n$. The fuzzy score for candidate supplier i with respect to metric j is denoted as $\tilde{x}'_{ij} = (x'_{ij1}, x'_{ij2}, x'_{ij3})$.

Step 10. Form fuzzy decision matrix by normalizing fuzzy scores for each candidate supplier. The fuzzy decision matrix is $\tilde{X} = [\tilde{x}_{ij}]$ of size $m \times n$, where $\tilde{x}_{ij}$ denotes normalized fuzzy score of candidate supplier i with respect to metric j.

Step 11. Get fuzzy weighted composite scores for each candidate supplier ($\tilde{c}_i$) and construct the weighted composite fuzzy score matrix by $\tilde{C} = [\tilde{x}_{ij}] \cdot [\tilde{w}_j]^t$, $i = 1, 2, , m$, $j = 1, 2, , n$.

Step 12. Defuzzify the fuzzy weighted composite scores into the crisp real values. The defuzzified value corresponding $\tilde{c}_i$ is calculated by $c_i = (c_{i1} + c_{i2} + c_{i2} + c_{i3})/4$. Select the best supplier with the highest defuzzified value.

4 A Practical Application

In this section, a practical application is used to demonstrate the real-world usefulness of the proposed mechanism. The company Y, founded in 1984, is a world-class computer manufacturing company in Taiwan. To meet rapidly growing demand of the 4G LTE (Long Term Evolution) tablet personal computer (LTE-T PC) in the near future, the company Y is planning to select the best suppliers involved early in the NPD of the LTE-TPC so that it can be launched earlier than other competitors. The five main modules of the LTE-TPC are the the mainboard module (candidate suppliers are AS and GA), the LCD module (candidate suppliers are CM, HA, and AU), the Wi-LAN and Wi-WAN module (candidate suppliers are BR and IN), the Touch sensor module (candidate suppliers are TP, YF, JT, and HT), and the NFC module (candidate suppliers are NFC, WT, and BR). It is critical to select the abovementioned suppliers involved early in the NPD (Step 1). We recommend that the decision makers of the company Y adopt the proposed mechanism since it is suitable for the NPD of the LTE-TPC. The product development meeting including the project manager, key parts' engineers, procurement staffs, and related suppliers is convened, and then the D-L1-M (Step 2) is established based on the LTE-TPC schedule for time to market. The D-L2-M (Step 3), the D-L3-M (Step 4), and the D-L4-M (Step 5) are deployed in order. Due to the limit of the paper length, only the D-L1-M associated with the LCD module is presented here, as displayed in Fig. 2.

After building the D-L4-M, all the process information between the company Y and its suppliers are completely presented. To further obtain the interactive processes, the D-L4-M is transformed into the DSM (Step 6). Steward's method is adopted to partition the DSM to get the reordered DSM, as shown in Fig. 3. Then, the interdependent DSM is extracted from the reordered DSM (Step 7).

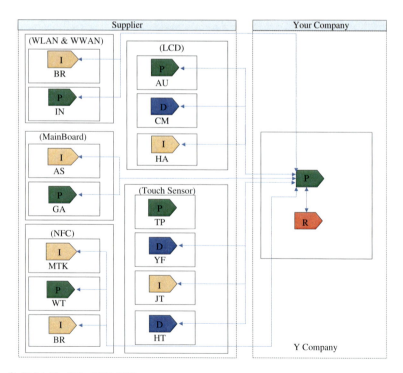

Fig. 2 D-L1-M of the LTE-TPC

A Process-Oriented Mechanism Combining Fuzzy Decision Analysis

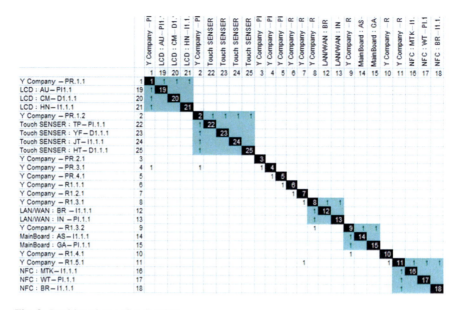

Fig. 3 Partitioned (reordered) DSM

Table 1 Linguistic variables for the weighting and rating

Linguistic weighting	Linguistic rating	TFN (a, b, c)
Definitely low (DL)	Definitely poor (DP)	(0, 0, 0.125)
Very low (VL)	Very poor (VP)	(0, 0.125, 0.25)
Low (L)	Poor (L)	(0.125, 0.25, 0.375)
Medium low (ML)	Medium poor (MP)	(0.25, 0.375, 0.5)
Medium (M)	Fair (F)	(0.375, 0.5, 0.625).
Medium high (MH)	Medium good (MG)	(0.5, 0.625, 0.75)
High (H)	Good (G)	(0.625, 0.75, 0.875)
Very high (VH)	Very good (VG)	(0.75, 0.875, 1)
Definitely high (DH)	Definitely good (DG)	(0.875, 1, 1)

Based on the DCOR, the structure of the metrics, including five main metrics and their corresponding submetrics is proposed to evaluate the best NPD module suppliers. The linguistic weighting and rating variables are measured by nine linguistic terms, as shown in Table 1 (Step 8). The fuzzy weights of the metrics and scores of candidate suppliers for the LCD module are determined by the members participated in the LTE-TPC project team. The results are shown in Tables 2 and 3 (Step 9). From Tables 2 and 3, we can get fuzzy weight matrix and fuzzy decision matrix (Step 10).

According to Step 11, the weighted composite fuzzy score matrix can be obtained. For example, the fuzzy weighted composite fuzzy score of candidate supplier CM for the LCD module is

246   J.-S. Lin et al.

Table 2 The fuzzy weight of the metrics

Weighting	Expert			
	e_1	e_2	Ee_3	Mean
Reliability	H	VH	H	(0.67, 0.792, 0.917)
Responsiveness	H	MH	MH	(0.54, 0.667, 0.792)
Agility	M	MH	M	(0.42, 0.542, 0.667)
Cost	VH	VH	DH	(0.79, 0.917, 1)
Assets	VL	L	VL	(0.04, 0.167, 0.292)

Table 3 The fuzzy ratings of three candidate suppliers for the LCD module

Rating	Item	Supplier		
		CM	HA	AU
Reliability	% On-time to commit	MG	G	G
	# of design errors	VG	G	G
	Document complete	G	MG	MG
	Perfect Integration	G	MG	VG
	Mean	(0.625, 0.75, 0875)	(0.563, 0.688, 0.813)	(0.625, 0.75, 0. 875)
Responsiveness	Research cycle time	G	F	G
	Design cycle time	G	F	G
	Integrate cycle time	G	MG	G
	Mean	(0.625, 0.75, 0.875).	(0.417, 0.542, 0.667)	(0.625, 0.75, 0.875)
Agility	Pilot build time	G	MG	MG
	Amend cycle time	MG	G	F
	Re-plan frequency	MG-	F	MG
	Mean	(0.542, 0.667, 0.792)	(0.5, 0.625, 0.75)	(0.458, 0.583, 0.708)
Cost.	Plan Cost	G	G	VG
	Research Cost	G	VG	G
	Design Cost	G	G	VG
	Integrate cost	G	G	G
	Amend cost	MG	MG	MG
	Mean	(0.6, 0.725, 0.85)	(0.625, 0.75, 0.875)	(0.65, 0.775, 0.9)

$$\tilde{C}_{CM} = [(0.625, 0.75, 0.875)(0.625, 0.75, 0.875)(0.542, 0.667, 0.792)(0.6, 0.725, 0.85)(0.475, 0.6, 0.725)] \cdot$$
$$[(0.67, 0.792, 0.917)(0.54, 0.66, 0.792)(0.42, 0.542, 0.667)(0.79, 0.917, 1.00)(0.04, 0.167, 0.292)]^t$$
$$= (1.476, 2.219, 3.08).$$

Similarly, $\tilde{C}_{HA} = (1.326, 2.04, 2.87)$, $\tilde{C}_{AU} = (1.482, 2.24, 3.09)$. Then, the defuzzified value is calculated by using Step 12. These resulting values are $C_{CM} = 2.249$, $C_{HA} = 2.069$, and $C_{AU} = 2.263$. As a result, the best supplier for the LCD module is AU since it has the highest crisp real value.

5 Conclusions

The supplier selection involved in the NPD stage is a complicated MCDM problem, in which ambiguous and linguistic decision information is usually encountered. Based on the DCOR structure, this paper have developed a process-oriented mechanism combining fuzzy arithmetic operations for supporting a company to select the best suppliers involved early in the NPD stage. The four-level deployment framework of the DCOR can clearly and completely express all NPD process information among design chain members. The DSM can adequately be used to transform the process information deployed from the DCOR into the interactive process information. Such interactive information is a critical communication media between the company and its suppliers across the design chain. The main contribution of this paper is to incorporate the NPD interactive process information into the supplier selection mechanism. Based on the practical illustration, it is concluded that the proposed mechanism can be a useful methodology for making supplier selection decision.

Acknowledgments The authors would like to thank the Ming Chi University of Technology for financially supporting this study under Contract No. 102-Academic Research Subvention-M-19.

References

Carrera DA, Mayorga RV (2008) Supply chain management: a modular fuzzy inference system approach in supplier selection for new product development. J Intell Manuf 19:1–12

De Boer L, Labro E, Morlacchi P (2001) A review of methods supporting supplier selection. Eur J Purch Supply Manag 7:75–89

Ho W, Xu X, Dey PK (2010) Multi-criteria decision making approaches for supplier evaluation and selection: a literature review. Eur J Purch Supply Manag 202:16–24

Johnsen TE (2009) Supplier involvement in new product development and innovation: taking stock and looking to the future. J Purch Supply Manag 15:187–197

Kumar M, Vrat P, Shankar R (2004) A fuzzy goal programming approach for vendor selection problem in a supply chain. Comput Ind Eng 46:69–85

Nyere J (2009) DCOR 2.0 framework update next release of the design chain operational reference-model. http://ebookbrowse.com/dcor-2009-05-13-vfinal-pdf-d287392716

Oh J, Yang J, Lee S (2012) Managing uncertainty to improve decision-making in NPD portfolio management with a fuzzy expert system. Expert Syst Appl 39:9868–9885

Steward DV (1981) The design structure system: a method for managing the design of complex systems. IEEE Trans Eng Manag 28:71–74

Tang J, Zang YE, Tu Y, Chen Y, Dong Y (2005) Synthesis, evaluation, and selection of parts design scheme in supplier involved product development. Concurr Eng 13:277–289

Zadeh LA (1965) Fuzzy sets. Inf. Control 8:338–353

Reliability-Based Performance Evaluation for a Stochastic Project Network Under Time and Budget Thresholds

Yi-Kuei Lin, Ping-Chen Chang and Shin-Ying Li

Abstract This study develops a performance indicator, named project reliability, to measure the probability that a stochastic project network (SPN) can be successfully completed under both time and budget thresholds. The SPN is represented in the form of AOA (activity-on-arc) diagram, in which each activity has several possible durations with the corresponding costs and probability distribution. From the perspective of minimal path, two algorithms are proposed to generate upper and lower limit vectors which satisfy both time and budget, respectively. Next, the project reliability is evaluated in terms of such upper and lower limit vectors. The procedure of reliability evaluation can be applied to the SPN with arbitrary probability distribution. Such an indicator is a beneficial factor of the trade-off between the time and budget in a decision scenario.

Keywords Project reliability · Stochastic project network (SPN) · Minimal path · Activity-on-arc (AOA)

1 Introduction

This paper discusses a novel performance index for a stochastic project network (SPN), called project reliability, and defines it as the probability that the project is completed within the given time and budget. A network technique is proposed to

---

Y.-K. Lin (✉) · P.-C. Chang · S.-Y. Li
Department of Industrial Management, National Taiwan University of Science and Technology, Taipei, Taiwan, Republic of China
e-mail: yklin@mail.ntust.edu.tw

P.-C. Chang
e-mail: D9901001@mail.ntust.edu.tw

S.-Y. Li
e-mail: M10101019@mail.ntust.edu.tw

Y.-K. Lin et al. (eds.), *Proceedings of the Institute of Industrial Engineers Asian Conference 2013*, DOI: 10.1007/978-981-4451-98-7_30,
© Springer Science+Business Media Singapore 2013

evaluate the project reliability. For the project manager, such an index is regarded as a decision factor of the trade-off between the given time and budget. Relevant literatures are reviewed as follows.

A project can be modeled as a project network (a graph with nodes and arcs) to portray the interrelationships among the activities of a project. A project network can be represented in either activity-on-arc (AOA) diagram. In AOA diagram, each activity is represented by an arc. A node is used to separate an activity from each of its immediate predecessors. Hiller and Liberman (2005) indicated AOA has fewer nodes than AON (activity-on-node) for the same project and thus most project networks are represented in AOA. A project is affected by uncertainties, and thus activity durations of the project should be stochastic (Pontrandolfo 2000). Such a project is usually modeled as an SPN, in which each activity has several durations with a probability distribution. The minimal path (MP) technique is a widely used to evaluate the reliability of a network, where an MP is a sequence of arcs from a source to a sink which contains no cycle (Zuo et al. 2007; Lin 2008). It implies that an MP is a path whose proper subsets are no longer paths (Lin et al. 1995). Since MP is based on the AOA diagram, it is a practical way to evaluate reliability for the SPN in terms of MP.

2 Stochastic Project Network Model

This study proposes an MP-based procedure that firstly finds all upper and lower limit vectors for the given time and budget in terms of MP, in which the activity durations and their corresponding costs and probability distributions are proposed by the activity contractors. Subsequently we evaluate the project reliability through such vectors for an SPN.

Assumptions:
1. The duration of each a_i is an integer value: $x_{i1} < x_{i2} < x_{i3} < \ldots < x_{iw_i}$ with given probability distribution and its corresponding cost c_i takes a value: $c_{i1} > c_{i2} > c_{i3} > \ldots > c_{iw_i}$.
2. The activity durations of different a_i are s-independent.

2.1 Project Model Construction and Reliability Definition

Let $G = (\mathbf{N}, \mathbf{A})$ denote an SPN with $\mathbf{N}$ representing the set of nodes and $\mathbf{A} = \{a_i | i = 1, 2, \ldots, n\}$ representing the set of arc (activities), where n is the number of activities. The duration vector $X = (x_1, x_2, \ldots, x_n)$ is defined as the current duration state of G where x_i represents the current duration of activity a_i.

The proposed performance index, project reliability, is defined as the probability that the project completion time and cost under X does not exceed T and B,

Reliability-Based Performance Evaluation   251

respectively. The project reliability is represented as $R_{T,B}(X) = \Pr\{X|$ $T(X) \leq T$ and $B(X) \leq B\} = \Pr\{X|X \in X_{T,B}\}$, in which $X \in X_{T,B}$ means that X satisfies (T, B). In order to calculate such a probability, we proposed a methodology for (T, B)-UL and (T, B)-LL to depict the boundaries of $X_{T,B}$. That is, the project reliability is equal to the probability of X between (T, B)-UL and (T, B)-LL.

2.2 Upper Limit Vector and Lower Limit Vector

Time constraint T is utilized to bound the maximal duration vectors among all $X \in X_{T,B}$; while the budget constraint is utilized to bound the minimal duration vectors. Following are definitions for (T, B)-UL and (T, B)-LL of $X_{T,B}$.

Definition 1 A duration vector X is a (T, B)-UL if $X \in X_{T,B}$ and $T(Y) > T$ for each duration vector Y such that $Y > X$.

Definition 2 A duration vector X is a (T, B)-LL if $X \in X_{T,B}$ and $B(Y) > B$ for each duration vector Y such that $Y < X$.

In order to generate all (T, B)-UL and (T, B)-LL, we define a pseudo duration vector $Z = (z_1, z_2, \ldots, z_n)$ and a pseudo cost vector $V = (v_1, v_2, \ldots, v_n)$, respectively. Each pseudo duration z_i satisfies $z_i \in \{x_{i1}, x_{i1} + 1, x_{i1} + 2, \ldots\}$, $i = 1, 2,$ $\ldots, n$. Similarly, each pseudo cost v_i satisfies $v_i \in \{c_{iwi}, c_{iwi} + 1, c_{iwi} + 2, \ldots\}$, $i = 1, 2, \ldots, n$. In order to obtain all (T, B)-UL, for each mp_j, we first generate the pseudo duration vector Z satisfying $\sum_{a_i \in mp_j} z_i = T$. Subsequently, for each Z, generate the largest X such that $B(X) \leq B$ and $X \leq Z$. Such X is regarded as a (T, B)-UL candidate. Hence we generate all pseudo cost vectors V such that $v_1 + v_2 + \ldots + v_n = B$. Subsequently, for each V, we find the largest C such that $C \leq V$. The corresponding X of such C is regarded as a (T, B)-LL candidate if it satisfies $T(X) \leq T$. Based on Lemmas 1 and 2, the following Algorithms I and II are proposed to generate all (T, B)-UL and (T, B)-LL, respectively.

Algorithm I: Generate all (T, B)-UL

Step 1. Find all pseudo duration vectors $Z = (z_1, z_2, \ldots, z_n)$ satisfying both constrains (1) and (2).

$$\sum_{a_i \in mp_j} z_i = T \quad j = 1, 2, \ldots, m, \text{ and} \tag{1}$$

$$z_i \geq x_{i1} \text{ for } i = 1, 2, \ldots, n, \text{ where } z_i \text{ is an integer value.} \tag{2}$$

Step 2. Utilize the following equation to find the largest duration vector X such that $X \leq Z$.

$$x_i = \begin{cases} x_{iw_i} & \text{if } x_{iw_i} \le z_i \\ x_{it} & \text{if } x_{it} \le z_i < x_{i(t+1)} \end{cases} \quad i = 1, 2, \ldots, n. \tag{3}$$

Step 3. Remove those X with $B(X) > B$. Then the remainder is the set of (T, B)-UL candidates.

Step 4. Find all (T, B)-UL from these candidates by using the following procedure, i.e. to remove the non-maximal ones from the candidate set.

Algorithm II: Generate all (T, B)-LL

Step 1. Find all pseudo cost vectors $V = (v_1, v_2, \ldots, v_n)$ satisfying both constrains (4) and (5).

$$v_1 + v_2 + \ldots + v_n = B, \tag{4}$$

$$v_i \ge c_{iw_i}, \quad i = 1, 2, \ldots, n, \tag{5}$$

where v_i is an integer value.

Step 2. Utilize the following equation to find the largest cost vector C for each V such that $C \le V$.

$$c_i = \begin{cases} c_{i1} & \text{if } v_i \ge c_{i1} \\ c_{it} & \text{if } c_{i(t-1)} > v_i \ge c_{it} \end{cases} \quad i = 1, 2, \ldots, n. \tag{6}$$

Step 3. Transform each C from Step 2 to the corresponding X and then remove those X with $T(X) > T$. Then the remainder is the set of (T, B)-LL candidates.

Step 4. Find all (T, B)-LL from these candidates by using the following procedure, i.e. to remove the non-minimal ones from the candidate set.

2.3 Evaluation of Reliability in Stochastic Project Network

We assume all (T, B)-UL from algorithm I are $X_1^u, X_2^u, \ldots, X_p^u$ and all (T, B)-LL from algorithm II are $X_1^l, X_2^l, \ldots, X_q^l$. Let $U_{T,B} = \cup_{i=1}^p \{X | X \le X_i^u\}$ and $L_{T,B} = \cup_{i=1}^q \{X | X \le X_i^l\}$. Since (T, B)-UL and (T, B)-LL are the upper limit and lower limit vectors of $X_{T,B}$, respectively, $X_{T,B}$ can be described as

$$X_{T,B} = \{U_{T,B} \backslash L_{T,B}\} \cup \{X_1^l, X_2^l, \ldots, X_q^l\}, \tag{7}$$

where the first term $\{U_{T,B} \backslash L_{T,B}\}$ implies that all (T, B)-LL are deleted by subtracting $L_{T,B}$ because $X_i^l \in L_{T,B}$ for $i = 1, 2, \ldots, q$. Note that, these (T, B)-LL

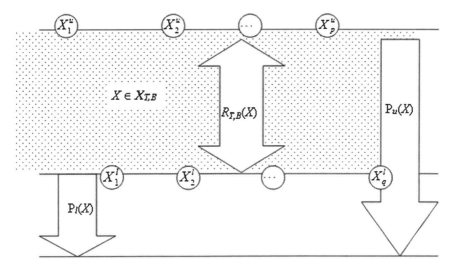

Fig. 1 Project reliability

belonging to $L_{T,B}$ also satisfy $X_{T,B}$, and we have to include them into $X_{T,B}$ by adding the second term $\{X_1^l X_2^l, \ldots, X_q^l\}$. Figure 1 illustrates the project reliability, in which $P_u(X) = \Pr\{U_{T,B}\} = \Pr\{\cup_{i=1}^{p}\{X|X \leq X_i^u\}\}$ and $P_l(X) = \Pr\{L_{T,B}\} = \Pr\{\cup_{i=1}^{q}\{X|X < X_i^l\}\}$. Therefore, the project reliability is

$$R_{T,B}(X) = \Pr\{X|X \in X_{T,B}\} = P_u(X) - P_l(X)$$
$$= \Pr\{\cup_{i=1}^{p}\{X|X \leq X_i^u\}\} - \Pr\{\cup_{i=1}^{q}\{X|X \leq X_i^l\}\} + \sum_{i=1}^{q}\Pr(X_i^l). \quad (8)$$

3 Illustrative Example

We utilize a simple project network composed of five activities and represented in the form of AOA diagram as shown in Fig. 2. This example demonstrates the procedure to generate the upper and lower limit vectors and to evaluate the project reliability. The activity cost and the activity duration of each activity are shown in

Fig. 2 Simple project network

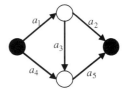

Table 1 Activity cost, activity duration and probability of each arc

a_i	Duration (weeks)	Cost (US\$ 100)	Probability	a_i	Duration (weeks)	Cost (US\$ 100)	Probability
1	2	8	0.3	4	2	6	0.3
	3	7	0.5		3	5	0.3
	4	6	0.2		4	4	0.4
2	3	8	0.25	5	3	8	0.35
	4	7	0.5		4	7	0.4
	5	6	0.25		5	6	0.25
3	2	7	0.2				
	3	6	0.5				
	4	5	0.3				

Table 1. The project network has three minimal paths, say $mp_1 = \{a_1, a_2\}$, $mp_2 = \{a_1, a_3, a_5\}$, and $mp_3 = \{a_4, a_5\}$. The case illustrates the method that generates the project reliability of the duration within 11 weeks and the cost within US\$ 3,400.

Algorithm I: Generate all (11, 3,400)-UL

Step 1. Find all pseudo duration vectors $Z = (z_1, z_2,..., z_n)$ satisfying

$$\begin{cases} z_1 + z_2 = 11 \\ z_1 + z_3 + z_5 = 11, \text{ and} \\ z_4 + z_5 = 11 \end{cases} \tag{9}$$

$$z_1 \geq 2, z_2 \geq 3, z_3 \geq 2, z_4 \geq 2, z_5 \geq 3. \tag{10}$$

Thus, we have 15 pseudo duration vectors shown in the first column of Table 2. Steps 2 to 4. We obtain six (11, 3,400)-UL: $X_3 = (4, 5, 2, 4, 5)$, $X_7 = (3, 5, 3, 4, 5)$, $X_8 = (4, 5, 3, 4, 4)$, $X_{10} = (4, 5, 4, 4, 3)$, $X_{11} = (3, 5, 4, 4, 4)$ and $X_{12} = (2, 5, 4, 4, 5)$ by applying Algorithm I. The calculation process is summarized in Table 2.

Algorithm II: Generate all (11, 3,400)-LL

Step 1. Find all pseudo cost vectors $V = (v_1, v_2,..., v_n)$ satisfying

$$v_1 + v_2 + v_3 + v_4 + v_5 = 34 \tag{11}$$

$$v_1 \geq 6, v_2 \geq 6, v_3 \geq 5, v_4 \geq 4, v_5 \geq 6 \tag{12}$$

Steps 2 to 4. We obtain 30 (11, 3,400)-LL which are summarized in Table 3. By adopting Algorithm II, Tables 2 and 3 shows the obtained (11, 3,400)-UL and (11, 3,400)-LL, respectively. The project reliability of duration within 11 weeks and cost within US\$ 3,400 is $P_u(X) - P_l(X) = 0.8985 - 0.065175 = 0.833325$.

Reliability-Based Performance Evaluation

Table 2 The (11, 3,400)-UL

Step 1	Step 2	Step 3	Step 4
$Z_1 = (2, 9, 2, 4, 7)$	$X_1 = (2, 5, 2, 4, 5)$	$B(X_1) = 3,100$	No, $X_3 > X_1$
$Z_2 = (2, 9, 3, 5, 6)$	$X_2 = (2, 5, 3, 4, 5)$	$B(X_2) = 3,000$	No, $X_3 > X_2$
$Z_3 = (2, 9, 4, 6, 5)$	$X_3 = (2, 5, 4, 4, 5)$	$B(X_3) = 2,900$	Yes, X_3 is a (11, 3,400)-UL.
$Z_4 = (2, 9, 5, 7, 4)$	$X_4 = (2, 5, 4, 4, 4)$	$B(X_4) = 3,000$	No, $X_3 > X_4$
$Z_5 = (2, 9, 6, 8, 3)$	$X_5 = (2, 5, 4, 4, 3)$	$B(X_5) = 3,100$	No, $X_3 > X_5$
$Z_6 = (3, 8, 2, 5, 6)$	$X_6 = (3, 5, 2, 4, 5)$	$B(X_6) = 3,000$	No, $X_7 > X_6$
$Z_7 = (3, 8, 3, 6, 5)$	$X_7 = (3, 5, 3, 4, 5)$	$B(X_7) = 2,900$	Yes, X_7 is a (11, 3,400)-UL.
$Z_8 = (3, 8, 4, 7, 4)$	$X_8 = (3, 5, 4, 4, 4)$	$B(X_8) = 2,900$	Yes, X_8 is a (11, 3,400)-UL.
$Z_9 = (3, 8, 5, 8, 3)$	$X_9 = (3, 5, 4, 4, 3)$	$B(X_9) = 3,000$	No, $X_8 > X_9$
$Z_{10} = (4, 7, 2, 6, 5)$	$X_{10} = (4, 5, 2, 4, 5)$	$B(X_{10}) = 2,900$	Yes, X_{10} is a (11, 3,400)-UL.
$Z_{11} = (4, 7, 3, 7, 4)$	$X_{11} = (4, 5, 3, 4, 4)$	$B(X_{11}) = 2,900$	Yes, X_{11} is a (11, 3,400)-UL.
$Z_{12} = (4, 7, 4, 8, 3)$	$X_{12} = (4, 5, 4, 4, 3)$	$B(X_{12}) = 2,900$	Yes, X_{12} is a (11, 3,400)-UL.
$Z_{13} = (5, 6, 2, 7, 4)$	$X_{13} = (4, 5, 2, 4, 4)$	$B(X_{13}) = 3,000$	No, $X_{12} > X_{13}$
$Z_{14} = (5, 6, 3, 8, 3)$	$X_{14} = (4, 5, 3, 4, 3)$	$B(X_{14}) = 3,000$	No, $X_{12} > X_{14}$
$Z_{15} = (6, 5, 2, 8, 3)$	$X_{15} = (4, 5, 2, 4, 3)$	$B(X_{15}) = 3,100$	No, $X_{12} > X_{15}$

Table 3 The (11, 3,400)-LL

Step 1	Step 2	Step 3		Step 4
$V_1 = (6, 6, 5, 4, 13)$	$C_1 = (6, 6, 5, 4, 8)$	$X_1 = (4, 5, 4, 4, 3)$	$T(X_1) = 11$	30 (11, 3,400)-LL: $X_{52} = (4, 4, 2, 2, 3)$,
$V_2 = (6, 6, 5, 5, 12)$	$C_2 = (6, 6, 5, 5, 8)$	$X_2 = (4, 5, 4, 3, 3)$	$T(X_2) = 11$	$X_{73} = (4, 3, 3, 2, 3)$, $X_{77} = (4, 3, 2, 3, 3)$,
$V_3 = (6, 6, 5, 6, 11)$	$C_3 = (6, 6, 5, 6, 8)$	$X_3 = (4, 5, 4, 2, 3)$	$T(X_3) = 11$	$X_{78} = (4, 3, 2, 2, 4)$, $X_{136} = (3, 5, 2, 2, 3)$,
$V_4 = (6, 6, 5, 7, 10)$	$C_4 = (6, 6, 5, 6, 8)$	$X_4 = (4, 5, 4, 2, 3)$	$T(X_4) = 11$	$X_{157} = (3, 4, 3, 2, 3)$, $X_{161} = (3, 4, 2, 3, 3)$,
$V_5 = (6, 6, 5, 8, 9)$	$C_5 = (6, 6, 5, 6, 8)$	$X_5 = (4, 5, 4, 2, 3)$	$T(X_5) = 11$	$X_{162} = (3, 4, 2, 2, 4)$, $X_{172} = (3, 3, 4, 2, 3)$,
$V_6 = (6, 6, 5, 9, 8)$	$C_6 = (6, 6, 5, 6, 8)$	$X_6 = (4, 5, 4, 2, 3)$	$T(X_6) = 11$	$X_{176} = (3, 3, 3, 3, 3)$, $X_{177} = (3, 3, 3, 2, 4)$,
$\vdots$	$\vdots$	$\vdots$	$\vdots$	$\vdots$
$V_{329} = (12, 7, 5, 4, 6)$	$C_{329} = (8, 7, 5, 4, 6)$	$X_{329} = (2, 4, 4, 4, 5)$	$T(X_{329}) = 11$	$X_{245} = (2, 3, 3, 4, 3)$, $X_{246} = (2, 3, 3, 3, 4)$,
				$X_{247} = (2, 3, 3, 2, 5)$,
$V_{330} = (13, 6, 5, 4, 6)$	$C_{330} = (8, 6, 5, 4, 6)$	$X_{330} = (2, 5, 4, 4, 5)$	$T(X_{330}) = 11$	$X_{248} = (2, 3, 2, 4, 4)$, and $X_{249} = (2, 3, 2, 3, 5)$.

4 Conclusions and Future Research

This study proposes the project reliability to be a performance index which is utilized to measure whether the project can be completed under the given time and budget or not. A procedure integrating two algorithms is developed to evaluate the project reliability and is illustrated through a simple project network. In particular, the proposed procedure can be applied to the SPN with arbitrary probability distribution. Activity duration and its corresponding cost are affected by the employed resources, such as the system engineers, system analysts, and consultants in the DR project. Based on this paper, future research may take the resources allocation into consideration for the project reliability maximization. Moreover, it is an important issue to consider the trade-off among the time, budget and project reliability, and it can be solved by using multi-objective optimization approaches.

References

Hiller FS, Liberman GJ (2005) Introduction to operations research, 8/e. McGraw-Hill, New York

Lin YK (2008) Project management for arbitrary random durations and cost attributes by applying network approaches. Comput Math Appl 56:2650–2655

Lin JS, Jane CC, Yuan J (1995) On reliability evaluation of a capacitated-flow network in terms of minimal pathsets. Networks 25:131–138

Pontrandolfo P (2000) Project duration in stochastic networks by the PERT-path technique. Int J Proj Manage 18:215–222

Zuo MJ, Tian Z, Huang HZ (2007) An efficient method for reliability evaluation of multistate networks given all minimal path vectors. IIE Trans 39:811–817

System Reliability and Decision Making for a Production System with Intersectional Lines

Yi-Kuei Lin, Ping-Chen Chang and Kai-Jen Hsueh

Abstract A three-phase procedure is proposed to measure the performance of a production system with intersectional lines by taking reworking actions into account. In particular, for a production system with intersectional lines, common station shares its capacity to all lines when processing. Hence, it is important to analyze the capacity of the common station while performance evaluation. This study addresses the system reliability as a key performance indicator to evaluate the probability of demand satisfaction. First, the production system is constructed as a production network (PN) by the graphical transformation and decomposition. Second, capacity analysis of all stations is implemented to determine the input flow of each station based on the constructed PN. Third, a simple algorithm is proposed to generate all minimal capacity vectors that stations should provide to satisfy the given demand. We evaluate the system reliability in terms of such minimal capacity vectors. A further decision making issue is discussed to decide a reliable production policy.

Keywords Production system · Intersectional lines · System reliability · Decision making · Production policy

---

Y.-K. Lin (✉) · P.-C. Chang · K.-J. Hsueh
Department of Industrial Management, National Taiwan University of Science
and Technology, Taipei, Taiwan, Republic of China
e-mail: yklin@mail.ntust.edu.tw

P.-C. Chang
e-mail: D9901001@mail.ntust.edu.tw

K.-J. Hsueh
e-mail: M10101018@mail.ntust.edu.tw

Y.-K. Lin et al. (eds.), *Proceedings of the Institute of Industrial
Engineers Asian Conference 2013*, DOI: 10.1007/978-981-4451-98-7_31,
© Springer Science+Business Media Singapore 2013

1 Introduction

This utilizes an AOA (activity-on-arrow) network diagram approach to construct the manufacturing system. In particular, each station in the production system has stochastic capacity levels (i.e., multi-state) due to machine failure, partial failure, and maintenance (Lin 2009; Lin and Chang 2011). Therefore, the manufacturing system is also multi-state and we can treat it as the so-called stochastic-flow network (Aven 1985; Hudson and Kapur 1985; Xue 1985; Zuo et al. 2007). Considering reworking actions, an effective methodology is proposed to evaluate the system reliability for such a production system. We propose a graphical methodology to model the production system as a stochastic-flow network, named production network (PN) herein. In particular, intersectional lines are considered. That is, different lines share the capacity of common station. Based on the PN, we derive the minimal capacity vectors that should be provided to satisfy the demand. In terms of such vectors, the production manager can calculate the system reliability and determine a reliable production strategy.

2 Methodology

2.1 Assumptions

1. The capacity of each station is a random variable according to a given probability distribution.
2. The capacities of different stations are statistically independent.
3. Each inspection point is perfectly reliable.

2.2 A Three-Phase Evaluation Procedure

Phase I: An AOA diagram is adopted to form a production system. Each arrow is regarded as a station of workers and each node denotes an inspection point following the station. We utilize a graphical transformation and decomposition to construct the production system as a PN.

Phase II: We determine the input amount of raw materials to satisfy the given demand d. A subsequent calculation for the maximum output of each line is assist to decide the demand assignment. Once all possible demand assignments are obtained, the input flow of each station is also derived.

Phase III: According to the results of capability analysis, a simple algorithm is developed to generate all minimal capacity vectors that satisfy demand d. In terms of such vectors, the system reliability can be derived by several existed methods easily. Moreover, the production manager can determine a reliable strategy to assign the output for each line in terms of every single minimal capacity vector.

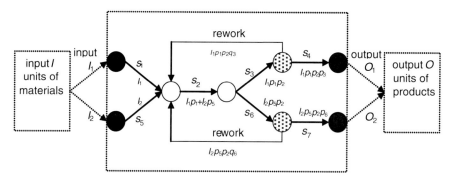

Fig. 1 Two lines in a manufacturing system

3 Network Model Construction (Phase I)

3.1 Transformation and Decomposition

Let (**N**, **S**) be a PN with two lines, where **N** represents the set of nodes and **S** = $\{s_i | i = 1, 2, \ldots, n_1 + n_2\}$ represents the set of arrows with n_1 (resp. n_2) is the number of stations in L_1 (resp. L_2). Each station s_i possesses a distinct success rate p_i (i.e., the failure rate $q_i = 1 - p_i$). Two lines, say L_1 (consisting of s_1, s_2, s_3, and s_4; $n_1 = 4$) and L_2 (consisting of s_5, s_6, and s_7; $n_2 = 3$), are layout in the production system. The common station s_2 has to process WIP from both s_1 and s_5 and thus $L_2 = \{s_5, s_2, s_6, s_7\}$ as shown in Fig. 1. The input amount of each station is shown under each arrow, where I_1 (resp. I_2) is the input amount of raw materials and O_1 (resp. O_2) is units of output product for L_1 (resp. L_2). To distinguish the input flows from the regular process or the reworking process, we transform the production system into a PN as Fig. 2, in which a dummy-station s_i' is set to denote the station s_i doing the reworking action. Note that since s_2 serves as a common station, it appears in both lines.

To analyze the PN in terms of paths, we decompose each line L_j into two paths, named the general manufacturing path $L_j^{(G)}$ and the reworking path $L_j^{(R|r,r-k)}$, where r and k are the indices to denote that defective WIP output from the rth sequenced station are reworked starting from previous k stations. Take Fig. 2 for instance, the set $L_1^{(G)} = \{s_1, s_2, s_3, s_4\}$ in line L_1 would be a general manufacturing path. On the other side, the path with reworking action is $L_1^{(R|3,3-1)} = \{s_1, s_2, s_3, s_2', s_3', s_4'\}$ where (R|3,3−1) denotes that defective WIP output from the third sequenced ($r = 3$) station s_3 in the line L_1 is reworked starting from previous one ($k = 1$) station (i.e., starting from s_2). For the special case that k is set to be zero, it implies that the defective WIP is reworked at the same station (rth sequenced station).

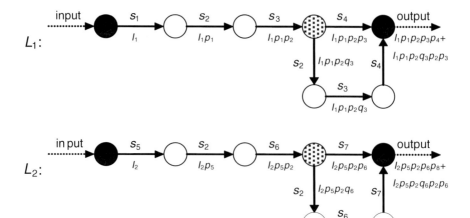

Fig. 2 The transformed manufacturing network for Fig. 2

4 Capability Analysis (Phase II)

For each line L_j, suppose that I_j units of raw materials are able to produce O_j units of product; we intend to obtain the relationship between I_j and O_j fulfilling $O_j \geq d_j$ and $\sum_{j=1}^{2} d_j = d$, where d_j is the assigned demand for line L_j. Two sets, say Ψ_i^{prior} and Ψ_i^{post}, are defined to record the stations prior to and posterior to the station s_i, respectively. The following equation guarantees the PN can produce exact sufficient output O_j that satisfies demand d_j.

$$I_j = d_j \Big/ \left\{ \left(\prod_{t: s_t \in L_j^{(G)}} p_t \right) + \delta_j \left(\prod_{t: s_t \in L_j^{(R|r,r-k)}} p_t \right) \right\}, \quad j = 1, 2. \quad (1)$$

The maximum output of each line is determined by the bottleneck station in that line. Thus, we assume the input WIP entering each station s_i exact equals the maximal capacity M_i. Then derive the potential output amount of s_i, say φ_i, in terms of its success rates p_i and success rates of its follow-up stations (i.e., $\left(\prod_{t: s_t \in \Psi_i^{\text{post}}} p_t \right)$). By comparing the potential output amount of all stations, the minimum one is the bottleneck of the line. The determination of maximum output is shown as following equation:

$$\varphi_i = M_i p_i \left\{ \left(\prod_{t: s_t \in \Psi_i^{\text{post}}} p_t \right) + \alpha_i \left(\prod_{t: s_t \in \Psi_i^{\text{post}} \cap \Psi_{\gamma_j}^{\text{prior}}} p_t \right) \right.$$
$$\left. (1 - p_{\gamma_j}) \left(\prod_{t: s_t \in L_j^{(R|r,r-k)}} p_t \right) \right\}, \quad (2)$$

where the index α_i equals 1 for s_i not on the reworking path; otherwise α_i equals 0.

System Reliability and Decision Making

To determine the input flow of each station, we utilize $f_{j,i}^{(G)}$ and $f_{j,i}^{(R|r,r-k)}$ to represent the input amount of the ith station in the jth line for $L_j^{(G)}$ and $L_j^{(R|r,r-k)}$, respectively. The input raw materials/WIP processed by the ith station s_i should satisfy the following constraint,

$$\sum_{j=1}^{2} \left(f_{j,i}^{(G)} + f_{j,i}^{(R|r,r-k)} \right) \le M_i, \ i \ = \ 1, 2, \ \ldots, \ n_1 \ + \ n_2. \tag{3}$$

Constraint (3) ensures that the total amount of input flow entering station s_i does not exceed the maximal capacity M_i. The term $\sum_{j=1}^{2} \left(f_{j,i}^{(G)} + f_{j,i}^{(R|r,r-k)} \right)$ is further defined as the loading of each station, say w_i.

Let x_i denote the capacity of each station s_i. The capacity x_i of each station s_i is a random variable and thus the PN is stochastic, where x_i takes possible values $0 = x_{i1} < x_{i2} < \ldots < x_{ic_i} = M_i$ for $i = 1, 2, \ldots, n_1 + n_2$ with c_i denoting number of possible capacities of s_i. Under the state $X = (x_1, x_2, \ldots, x_{n_1+n_2})$, it is necessary that $x_i \ge w_i$ to guarantee that s_i can process the input raw materials/WIP.

5 Performance Evaluation and Decision Making (Phase III)

Given the demand d, the system reliability R_d is the probability that the output product from the PN is not less than d. Thus, the system reliability is $\Pr\{X|V(X) \ge d\}$, where $V(X)$ is defined as the maximum output under X. Any minimal vector Y in the set $\{X|V(X) \ge d\}$ is claimed to be the minimal capacity vectors for d. That is, Y is a minimal capacity vector for d if and only if (1) $V(Y) \ge d$ and (2) $V(Y') < d$ for any capacity vectors Y' such that $Y' < Y$. Given Y_1, Y_2, ..., Y_h, the set of minimal capacity vectors satisfying demand d, the system reliability R_d is $R_d = \Pr\left\{ \bigcup_{v=1}^{h} B_v \right\}$, where $B_v = \{X|X \ge Y_v\}$.

Given two intersectional lines with a common station s_c installed in L_1, we have $L_1 = \{s_1, s_2, \ldots, s_c, \ldots, s_{n_1}\}$ and $L_2 = \{s_{n_1+1}s_{n_1+2}, \ldots, s_c, \ldots, s_{n_1+n_2}\}$. The minimal capacity vectors for d can be derived by the following steps.

Step 1 Find the maximum output for each path.

$$O_{1,\max} = \min\{\varphi_i|i : s_i \in L_1\} \text{ and } O_{2,\max} = \min\{\varphi_i|i : s_i \in L_2\}. \tag{4}$$

Step 2 Find the maximum output of the common station s_c.

$$O_{c,\max} = \max\{(\varphi_c|s_c \in L_1), (\varphi_c|s_c \in L_2)\}. \tag{5}$$

Step 3 Find the demand assignment (d_1, d_2) satisfying $d_1 + d_2 = d$ under constraints $d_1 \le O_{1,\max}$ and $d_2 \le O_{2,\max}$.

Step 4 For each demand pair (d_1, d_2), do the following steps.

4.1. Determine the amount of input materials for each line by

$$
\begin{aligned}
I_1 &= d_1 \Big/ \left\{ \left(\prod_{t: s_t \in L_1^{(G)}} p_t \right) + \delta_1 \left(\prod_{t: s_t \in L_1^{(R|r,r-k)}} p_t \right) \right\} \text{ and } I_2 \\
&= d_2 \Big/ \left\{ \left(\prod_{t: s_t \in L_2^{(G)}} p_t \right) + \delta_2 \left(\prod_{t: s_t \in L_2^{(R|r,r-k)}} p_t \right) \right\}.
\end{aligned}
\tag{6}
$$

4.2. Determine the input flows for each station s_i.

$$
f_{j,i}^{(G)} = I_j \prod_{t: s_t \in \Psi_i^{\text{prior}}} p_t \text{ for } i \text{ such that } s_i \in L_j^{(G)} \text{ and}
$$

$$
f_{j,i}^{(R|r,r-k)} = I_j \delta_j \prod_{t: s_t \in \Psi_i^{\text{prior}}} p_t \text{ for } i \text{ such that } s_i \in L_j^{(R|r,r-k)}.
\tag{7}
$$

4.3. Transform input flows into stations' loading vector $W = (w_1, w_2, \ldots, w_c, \ldots, w_{n_1+n_2})$ via

$$
w_i = \sum_{j=1}^{2} \left(f_{j,i}^{(G)} + f_{j,i}^{(R|r,r-k)} \right).
\tag{8}
$$

4.4. For each station, find the smallest possible capacity x_{ic} such that $x_{ic} \geq w_i > x_{i(c-1)}$. Then $Y = (y_1, y_2, \ldots, y_c, \ldots, y_{n_1+n_2})$ is a minimal capacity vector for d where $y_i = x_{ic}$ for all i.

Step 5 Apply RSDP algorithm to derive the system reliability (Zuo et al. 2007).

6 Example

Given a production system with a common station in the form of AOA diagram (see Fig. 3). Two intersectional lines, say $L_1 = \{s_1, s_2, s_3, s_4, s_5, s_6, s_7\}$ and $L_2 = \{s_8, s_2, s_9, s_{10}, s_{11}, s_{12}, s_{13}\}$, are producing products through seven stations. The success rate and capacity data of each station is given in Table 1. We consider the PN that has to satisfy the demand $d = 240$ products per hour, in which output products are packaged into a box in terms of 24 units. By the proposed algorithm, three minimal capacity vectors for $d = 240$ are derived. The derived minimal capacity vectors are shown in Table 2. In terms of these vectors, we obtain the system reliability $R_{240} = 0.91324$ by the RSDP algorithm.

The production manager could further evaluate the satisfaction probability for each demand pair in terms of its corresponding minimal capacity vector. This probability is denoted as R_{d_1,d_2} for each demand pair $D = (d_1, d_2)$. Table 2 shows

System Reliability and Decision Making

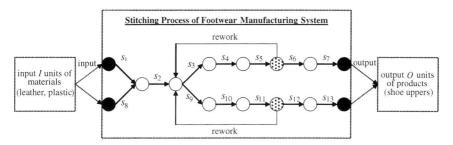

Fig. 3 A production system with common station

Table 1 Success rate and capacity probability distribution of stations

Station	Success rate	Capacity	Probability	Station	Success rate	Capacity	Probability
s_1	0.954	0	0.001	s_8	0.961	0	0.002
		60	0.002			60	0.003
		120	0.003			120	0.005
		180	0.006			180	0.013
		240	0.988			240	0.977
s_2^b	0.971	0	0.002				
		180	0.005				
		360	0.993				
s_3	0.983	0	0.001	s_9	0.978	0	0.001
		30	0.001			30	0.001
		60	0.001			60	0.001
		90	0.003			90	0.004
		120	0.003			120	0.005
		150	0.003			150	0.010
		180	0.988			180	0.978
s_4	0.965	0	0.002	s_{10}	0.972	0	0.003
		80	0.012			80	0.015
		160	0.986			160	0.982
s_5	0.975	0	0.003	s_{11}	0.975	0	0.001
		120	0.005			120	0.002
		240	0.992			240	0.997
s_6	0.986	0	0.003	s_{12}	0.989	0	0.002
		60	0.003			60	0.002
		120	0.005			120	0.008
		180	0.008			180	0.015
		240	0.981			240	0.973
s_7	0.991	0	0.001	s_{13}	0.993	0	0.002
		36	0.001			36	0.003
		72	0.003			72	0.005
		108	0.005			108	0.007
		144	0.010			144	0.010
		180	0.012			180	0.010
		216	0.968			216	0.963

Table 2 Demand pair and corresponding satisfaction probability

Demand pair (strategy)	Minimal capacity vector	Satisfaction probability
$D_1 = (96, 144)$	$Y_1 = (120, 360, 120, 160, 120, 120, 108, 180, 180, 160, 240, 180, 180)$	$R_{96,144} = 0.87189$
$D_2 = (120, 120)$	$Y_2 = (180, 360, 150, 160, 240, 180, 144, 180, 150, 160, 240, 180, 144)$	$R_{120,120} = 0.87124$
$D_3 = (144, 96)$	$Y_3 = (180, 360, 180, 160, 240, 180, 180, 120, 120, 160, 120, 120, 108)$	$R_{144,96} = 0.88358$

R_{d_1,d_2} for different demand pairs. The results indicate that $D_3 = (144, 96)$ with $R_{144,96} = 0.88358$ would be a better strategy to produce products since it has higher satisfaction probability. Thus, the production manger may decide to produce 144 pairs of shoe uppers per hour by L_1 and 96 pairs of shoe uppers per hour by L_2.

7 Conclusion

This paper presents a three-phase procedure to construct and evaluate the performance of production system. We evaluate the probability of demand satisfaction for the production system, where the probability is referred to as the system reliability. Such a quantitative performance indicator is scientific from the management perspective. In addition, different demand assignments may result different satisfaction probabilities. A reliable decision making for production strategy relies on the overall capability analysis.

References

Aven T (1985) Reliability evaluation of multistate systems with multistate components. IEEE Trans Reliab 34:473–479

Hudson JC, Kapur KC (1985) Reliability bounds for multistate systems with multistate components. Oper Res 33:153–160

Lin YK (2009) Two-commodity reliability evaluation of a stochastic-flow network with varying capacity weight in terms of minimal paths. Comput Oper Res 36:1050–1063

Lin YK, Chang PC (2011) Reliability evaluation of a manufacturing network with reworking action. Int J Reliab Qual Saf Eng 18:445–461

Xue J (1985) On Multistate System Analysis. IEEE Trans Reliab 34:329–337

Zuo MJ, Tian Z, Huang HZ (2007) An efficient method for reliability evaluation of multistate networks given all minimal path vectors. IIE Trans 39:811–817

Customer Perceptions of Bowing with Different Trunk Flexions

Yi-Lang Chen, Chiao-Ying Yu, Lan-Shin Huang, Ling-Wei Peng and Liang-Jie Shi

Abstract Bowing has traditionally been used to signify politeness and respect. A chain of restaurants in Taiwan has recently required servers to bow at a 90° angle to increase the customers' sense of being honored. Whether bowing at 90° is accepted by most consumers remains to be determined. This study analyzed 100 valid responses by questionnaire to determine consumer feelings regarding different degrees of trunk flexion when receiving bows. Results show that respondents typically believe that bowing at 30° was the most satisfactory, followed by 45°. Bowing at 45° or 60° causes customers to feel honored, and bowing at 90° induces the feelings of surprise and novelty but produces the lowest proportions of agreement to the at ease, necessary, and appropriate items. Previous studies had well validated that, when trunk is flexed to 90°, the posture is harmful because of the higher spinal loading. The finding of this study can be provided to the restaurant industry as a reference for service design from the perspective of server's health.

Keywords 90° bowing · Consumer perception · Trunk flexion

Y.-L. Chen (✉) · C.-Y. Yu · L.-S. Huang · L.-W. Peng · L.-J. Shi
Department of Industrial Engineering and Management, Ming Chi University
of Technology, 84 Gungjuan Rd, Taishan, New Taipei 24301, Taiwan
e-mail: ylchen@mail.mcut.edu.tw

C.-Y. Yu
e-mail: M01258007@mail.mcut.edu.tw

L.-S. Huang
e-mail: U98217017@mail.mcut.edu.tw

L.-W. Peng
e-mail: U98217029@mail.mcut.edu.tw

L.-J. Shi
e-mail: U98217013@mail.mcut.edu.tw

Y.-K. Lin et al. (eds.), *Proceedings of the Institute of Industrial Engineers Asian Conference 2013*, DOI: 10.1007/978-981-4451-98-7_32,
© Springer Science+Business Media Singapore 2013

1 Introduction

Traditional service industries place heavy emphasis on service encounters to raise customer satisfaction (Bitner et al. 1994). Because small service-based businesses (e.g., restaurants) have low entry requirements and engage in fierce competition, operators are constantly considering how to surprise and delight customers. In Eastern cultures, particularly in China and Japan, a greater angle of trunk flexion while bowing represents greater politeness and respect. A chain of restaurants in Taiwan has recently required servers to bow at a 90° angle when arriving at and leaving a table to increase the customers' sense of being honored, which has been established as a unique service feature. However, the servers in these restaurants bow at a 90° angle approximately 500 times a day, frequently using a cervical spine extension posture to maintain eye contact with customers. The bowing with trunk flexion 90°, undoubtedly, has been well-recognized as an awkward and harmful posture to the individual (DeLitto and Rose 1992; Holmes et al. 1992; Chen 1999).

Bowing has traditionally been used to signify politeness and respect. Lai (2009) indicated that bowing to a greater angle of trunk flexion produces better customer perceptions (of care and respect), and also indirectly improves service effects. However, whether bowing at 90° is accepted by most consumers remains to be determined.

This study therefore used a consumer perception survey to determine the perceptions and feelings of consumers regarding to provide opinions on the optimal angle of trunk flexion during bows, and also performed a bowing experiment to examine the differences in spinal curves and trunk muscle activities under five trunk positions (from upright to 90°).

2 Methods

2.1 The Consumer Perception Survey

This study distributed 122 questionnaires to determine consumer perceptions regarding different degrees of trunk flexion while bowing. Excluding unreturned and ineffective questionnaires, 100 valid responses were retrieved for an effective recovery rate of 82 %. The questionnaire comprised two sections. The first section surveyed the basic information of the respondents, including gender, age, occupation, salary level, and education level. The second section used different semantic feeling questions, asking respondents to select the bowing angle, with which they most agreed from among five angles (0, 30, 45, 60, and 90°). The content validity of the survey was confirmed using a pilot study and comments obtained from three marketing experts. In addition, to increase survey validity, photographs of five types of bows were provided along with the survey during implementation (Fig. 1).

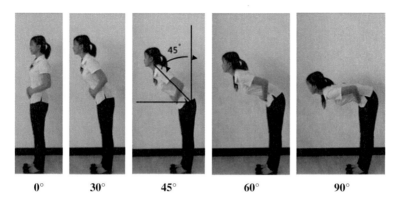

Fig. 1 Depictions of five bowing angles provided in the questionnaire

2.2 Survey Reliability

The second portion of this study consisted of an analysis of the reliability of the consumer questionnaire regarding bowing, as shown in Table 1. Typically, Cronbach α values less than 0.35 indicate low reliability and require rejection, whereas α values greater than 0.7 indicate high reliability. The reliability values of the questions in our survey were all greater than 0.8 (Table 1), indicating high validity, and showing that agreement among respondents on consumer feelings of different bowing angles exhibited high internal consistency.

2.3 Survey Criterion-Related Validity

The survey questionnaire used overall satisfaction as a criterion (Question 7) for performing validity analysis. The Chi-square test indicated that, aside from Question 6 (novelty), which showed no significant correlation with the criterion ($p > 0.05$), the other questions exhibited high correlation with the criterion ($p < 0.001$). These results show that, aside from Question 6, all of the questions could directly explain overall satisfaction.

Table 1 Internal consistency of survey questions in the questionnaire

Questions	Cronbach's α
1. Bowing at this angle makes me feel at ease	0.835
2. I feel that bowing at this angle is necessary	0.828
3. I feel that bowing at this angle is a pleasant surprise	0.837
4. Bowing at this angle makes me feel honored	0.848
5. I feel that bowing at this angle is appropriate	0.826
6. I feel that bowing at this angle is novel	0.862
7. Overall, I am satisfied when receiving a bow at this angle	0.816

3 Results and Discussion

3.1 Basic Sample Information of Survey

Of the 100 respondents in this study, men and women each constituted approximately half. 58 % of respondents were single. 27 % of the respondents ranged between 21 and 25 years of age, and 20 % ranged between 26 and 30 years of age. Half (50 %) of the respondents had an undergraduate degree. Approximately one-third were employed in manufacturing industry, and 27 % were students. 31 % of respondents had monthly incomes of ranged between NT$ 30,000 and 40,000.

3.2 Consumer Feelings Regarding Different Bowing Angles

Respondents were asked to select the angle with which they most agreed from five bowing angles to determine consumer feelings toward different bowing angles; the results are shown in Table 2. Most of the respondents considered that bowing at 90° gave them feelings of pleasant surprise and novelty (55 and 79 %). However, 45 and 60° achieved the effect of making consumers feel honored (60 %), which is twice the proportion of the 90° bows (30 %). In summary, although 90° bows are most capable causing people to feel pleasant surprise and novelty, 65 % of the respondents were most satisfied with bowing angles of 30 and 45°, significantly greater than the 20 % that most preferred 90°. In particular, respondents considered that bowing at 30° caused consumers to feel at ease (49 %), and were necessary (63 %) and appropriate (47 %). If the frequencies for bowing at 45° were added, then the proportions for the at ease, necessary, and appropriate items would subsequently be 67, 70, and 72 %, respectively, presenting an overall satisfaction of 75 %. In contrast, for the at ease, necessary, and appropriate items, bowing at 90° received the lowest proportion of agreement, with only 5, 1, and 3 %, respectively. The famous insurance saleswoman Shibata Kazuko, who was ranked

Table 2 Frequency distribution for consumers expressing agree the most for different bowing angles

Questions	Agree the most				
	0°	30°	45°	60°	90°
1. Bowing at this angle makes me feel at ease	4	49	18	24	5
2. I feel that bowing at this angle is necessary	0	63	17	19	1
3. I feel that bowing at this angle is a pleasant surprise	0	8	14	23	55
4. Bowing at this angle makes me feel honored	0	10	21	39	30
5. I feel that bowing at this angle is appropriate	0	47	25	25	3
6. I feel that bowing at this angle is novel	0	3	6	12	79
7. Overall, I am satisfied when receiving a bow at this angle	0	35	30	15	20

first in the Japanese insurance industry for 30 straight years, insisted on bowing no more than 45° even in Japan, a society renowned for expressing politeness through 90° bows to maintain a feeling of appropriateness between both parties (Lai and Lin 2012); this is consistent with the results of our study.

3.3 Cross-Analysis of Respondent Traits and Bowing Angles

To determine whether respondents' traits influenced their views of different bowing angles, we used the Chi-square test to analyze the agree the most questions. The results showed that gender did not influence respondents' feelings of different bowing angles ($p > 0.05$). Marital status and age had significant influences on the at ease, necessary, appropriate, and overall satisfied items (all $p < 0.05$). In general, married respondents were more satisfied with bowing angles of 60°, and unmarried respondents preferred 30°. Respondents over the age of 31 mostly preferred 60 and 90° bowing angles, and those under 30 mostly preferred 30°. The results for marital status may imply age factors. The level of education influenced at ease, appropriate, and overall satisfaction (all $p < 0.05$). Respondents with a junior high school education or lower were more satisfied with 60°, and those with a higher level of education preferred 30°. The pleasant surprise, honor, and novelty items were not influenced by respondent traits. Most respondents selected 60° and 90° for these items, demonstrating that restaurant operators who use standardized 90° bowing angles as a form of marketing can achieve the effects of attracting customers, and creating an atmosphere of surprise and amazement. However, whether or not this bowing design satisfies customers remains to be discussed. In summary, whether male or female, the well-educated unmarried younger group tended to be more accepting of smaller bowing angles (e.g., 30°); respondents in other groups tended to prefer greater bowing angles (e.g., 60 or 90°).

The posture bowing at 90° would cause the lumbar spine to bend to its limitation and this 'bow-out' posture of the back would increase the stretch on the posterior elements of the lumbar spine and thereby raise the stress on these structures (Lee and Chen 2000). That is, the lower back ES muscles become silent or inactive and the forward bending moment is counteracted by the passive tension of the muscles as well as the shorter lever arms between the discs and the posterior spinal ligaments (DeLitto and Rose 1992; Holmes et al. 1992). This posture undoubtedly would result in higher lower back loading.

4 Conclusion

If service industry operators endeavor to provide customers with a sense of honored, bowing at 45 or 60° angles yields optimal effects. In contrast, bowing at 90° has the greatest capacity to evoke pleasant surprise and novelty. Respondents were most satisfied with a bowing angle of 30°, followed by 45°. Results show that bowing at 90° received the smallest proportion of agreement for the at ease, necessary, and appropriate items. Moreover, this harmful and awkward posture would injure server's spine, especially when daily repetitive bowings. The results of this study can provide a trade-off consideration for service operators in designing service encounters.

References

Bitner MJ, Booms BH, Mohr LA (1994) Critical service encounters: the employee's viewpoint. J Mark 58:95–106

Chen YL (1999) Geometric measurements of the lumbar spine in Chinese men during trunk flexion. Spine 24:666–669

Delitto RS, Rose SJ (1992) An electromyography analysis of two techniques for squat lifting and lowering. Phys Ther 71:438–448

Lai YX (2009) A study of job standardization on service encounter in resort hotel, Master's thesis, Graduate Institute of Recreation, Tourism, and Hospitality Management. National Chiayi University

Lai CH, Lin CS (2012) Insurance queen, bowing no more than 90 degrees. Bus Wkly 1281:144

Lee YH, Chen YL (2000) Regressionally determined vertebral inclination angles of lumbar spine in static lifts. Clin Biomech 15:678–683

Holmes JA, Damaser MS, Lehman SL (1992) Erector spinae activation and movement dynamics about the lumbar spine in lordotic and kyphotic squat-lifting. Spine 17:327–334

A Pilot Study Determining Optimal Protruding Node Length of Bicycle Seats Using Subjective Ratings

Yi-Lang Chen, Yi-Nan Liu and Che-Feng Cheng

Abstract This study preliminarily investigated the subjective discomfort and riding stability by requiring ten participants to ride straight-handles bicycles equipped with five seat-protruding node lengths (PNLs, 0–12 cm, in increments of 3 cm) of seats for 20 min. Results indicated that seat PNL caused differences in the participants' subjective discomfort and stability scores. The various PNLs had significantly positive (r = 0.910, $p < 0.01$) and negative (r = -0.904, $p < 0.05$) correlations to the subjective discomfort rating for the perineum and ischial tuberosity, respectively. However, various PNLs did not affect riding stability during cycling. The findings of this study suggest that a 6 cm PNL is the optimal reference for bicycle seat designs.

Keywords Protruding node lengths (PNL) · Posture analysis · Subjective ratings

1 Introduction

Previous studies have primarily focused on riding efficiency, bicycle types, and frame size. In recent years, the focus of bicycle-related research has gradually shifted to seat (or saddle) design (Groenendijk et al. 1992; Bressel et al. 2009). During cycling, the contact between the seat and the buttocks is the critical cause

---

Y.-L. Chen (✉) · Y.-N. Liu · C.-F. Cheng
Department of Industrial Engineering and Management, Ming Chi University
of Technology, 84 Gungjuan Rd, Taishan, New Taipei 24301, Taiwan
e-mail: ylchen@mail.mcut.edu.tw

Y.-N. Liu
e-mail: i770314@hotmail.com

C.-F. Cheng
e-mail: m01258004@mail.mcut.edu.tw

Y.-K. Lin et al. (eds.), *Proceedings of the Institute of Industrial Engineers Asian Conference 2013*, DOI: 10.1007/978-981-4451-98-7_33,
© Springer Science+Business Media Singapore 2013

of discomfort and pain. Groenendijk et al. (1992) showed that the pressure distribution on the seat demands that pads support the pelvic bones. Richmond (1994) noted that handlebars that are set too low can also induce compression neuropathy and certain overuse symptoms. Nakamura et al. (1995) examined four long-period and long-distance bicycle-commuting Japanese male students and found that a nodule had developed near each of their coccygeal regions and the shape corresponded to the saddle of the bicycle.

Previous studies on bicycle seats mainly focused on analyzing traditional seats with a protruding node length (PNL; Bressel and Cronin 2005; Bressel et al. 2007) or analyzed seat pressure distribution for various commercial seats (Bressel et al. 2009). In summary, the most obvious difference in bicycle seats is whether they feature protruding nodes. Whether a bicycle seat has a protruding node significantly influences the rider's comfort when cycling. Traditional seats with a longer protruding node provide riding stability (Bressel et al. 2009). However, they increase the pressure on the perineal/groin regions. Non-protruding node seats can minimize pressure to the anterior perineum (Lowe et al. 2004), but they may also increase the risk of falling injuries if stability is compromised. Therefore, an appropriate PNL design is critical.

In this study, we hypothesized that an optimal PNL may exist between the traditional and non-PNL seats and can properly maintain a degree of body stability, reduce discomfort in the perineal region. We collected data on the riders' subjective discomfort ratings and riding stability after the participants had ridden a straight-handles bicycle for 20 min under five PNL conditions to identify the optimal PNL.

2 Methods

2.1 Participants

Ten male university students participated in this study. Their mean (SD) age was 22.8 (1.7) years, and the ages ranged from 21 to 26 years. Their mean (SD) height and weight were 171.6 (8.3) cm and 65.5 (6.3) kg, respectively. None of the participants had any history of musculoskeletal injury or pain and maintained good exercise habits, physical stamina, and bicycling habits. The participants were informed of the test procedures and were paid for their participation.

2.2 Experimental Bicycle Seats and Handles

Five of the bicycle seat types used in this study were made by a bicycle manufacturer (original type no: 6091010, Giant, Taichung, Taiwan). Except for PNL, all

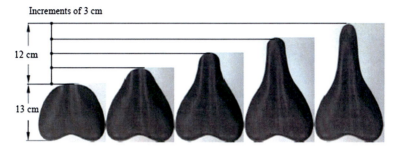

Fig. 1 Schematic diagram of the five types of PNL seat cushions used in this study

other seat design factors remained unchanged (e.g., shape, material, and structure). The width of the seats was the commercially adopted 16 cm (Bressel and Cronin 2005), and the length of the seat-protruding nose ranged from 0 to 12 cm, and each 3 cm interval was established as a level, as shown in Fig. 1. We referenced the study of Chen and Yu (2012) and selected the straight-handles as the handle design, which was the most frequently used in Chen and Yu's survey in Taiwan.

2.3 Subjective Discomfort and Stability Rating

This study modified the scale by Bressel et al. (2009) for the subjective discomfort of various body parts and cycling stability ratings. The subjective assessments were performed using a continuous visual analogue scale (Bressel et al. 2009). The scale was 10 cm in length and was modeled after comfort scales modified by this study form Mundermann et al. (2002). The left end of the scale was labeled no discomfort at all and the right end was labeled extreme discomfort. The levels of discomfort experienced in each cyclist's wrist, neck, lower back, perineum, and ischial tuberosity was rated. Regarding cycling stability, the scale correspondingly ranged from not unstable at all to most unstable seat imaginable. After the participants rode a bicycle for 20 min, they were immediately asked to rate the discomfort and stability scores.

2.4 Experimental Design and Procedure

The 10 participants simulated cycling for 20 min under 5 seat PNLs [0, 3, 6, 9, and 12 cm]). During the final minute, their cycling posture was randomly recorded. At the end of each ride, the subjective discomfort for various body parts and stability rating scores were immediately self-reported by the participants. The order of implementation for the various bicycle setup combinations was random.

The participants were required to wear cycling shirts. Prior to the experiment, the participants had to warm up for 5 min to acclimate to the 120-W impedance setting and the 15 ± 5 km/h pedaling speed. Each participant adjusted the seat to the most comfortable height and the seat height was set at 100 % of the trochanteric leg length. To prevent effects from fatigue, the participants had a maximum of two tests per day with a minimum of 2 h intervals between each test.

2.5 Statistical Analysis

This study used SPSS 17.0 statistical software for the statistical analyses. The statistical significance level was set at 0.05, and each participant was considered a block. One-way analysis of variance (ANOVA) was performed to clarify the effects of the PNL variable on subjective discomfort rating and stability scores, and Duncan's Multiple-range test (Duncan's MRT) was conducted for post hoc comparisons. In addition, a Pearson product-moment correlation was used to explore the PNL and subjective discomfort rating values.

3 Results

Results of one-way ANOVA show that the PNL significantly influenced only the subjective discomfort scores of the perineum ($p < 0.01$) and the ischial tuberosity ($p < 0.05$), and did not affect the other body parts, as shown in Table 1. The shorter PNL that was used indicated a lower score of discomfort for the perineum, but a higher score of discomfort for the ischial tuberosity. The results also showed that the PNL had significantly positive and negative correlations to the discomfort scores of the perineum ($r = 0.910$, $p < 0.01$) and ischial tuberosity ($r = 0.904$, $p < 0.05$), respectively. Regarding riding stability, the participants felt no difference existing among these 5 PNL conditions ($p > 0.05$, as shown in Table 1).

4 Discussion

Previous studies have not conducted systematic research on the PNLs of bicycle seats. This study collected the subjective rating values for various PNL conditions to determine the optimal PNL. Varied PNLs affected the trade-off discomfort between the perineum and ischial tuberosity regions. Because the riding stability was not significantly affected by PNL variable, we recommend that the 6 cm PNL be referenced as the optimal seat design.

The results indicated that various PNLs significantly influenced the discomfort level for the perineum and ischial tuberosity, but had no effect on the other body

A Pilot Study Determining Optimal Protruding Node Length

Table 1 ANOVA and Duncan MRT for the PNL's effect on subjective discomfort and stability

Variables	N	DF	F	p value	PNL (cm)[b]	Mean (SD)[a]	Duncan groups
Discomfort on							
Neck/shoulders	50	4	1.2	0.446	–	–	–
Wrist	50	4	1.0	0.430	–	–	–
Lower back	50	4	0.8	0.538	–	–	–
Perineum	50	4	3.1	$p < 0.05$	0	3.93 (2.02)	A
					3	4.04 (2.17)	A
					6	4.17 (1.97)	A
					9	4.70 (2.33)	B
					12	4.97 (2.18)	B
Ischial tuberosity	50	4	3.6	$p < 0.01$	0	5.12 (2.01)	A
					3	4.60 (1.86)	B
					6	4.33 (1.94)	B
					9	4.10 (1.90)	B
					12	3.69 (1.77)	C
Stability	50	4	3.6	0.284	0	3.50 (2.04)	A
					3	3.32 (1.96)	A
					6	3.18 (1.84)	A
					9	2.89 (1.90)	A
					12	2.93 (1.77)	A

[a] subjective scores
[b] protruding node length

parts (as shown in Table 1). Figure 2 shows that the discomfort scores for the perineum and ischial tuberosity had positive (r = 0.910, p < 0.01) and negative (r = −0.904, p < 0.05) correlations with the PNL, respectively. In other words, the traditional seat (PNL = 12 cm) and non-PNL seat may generate greater

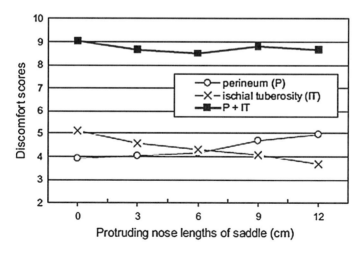

Fig. 2 Perineum and ischial tuberosity discomfort ratings when applying various PNLs

discomfort in the perineum and ischial tuberosity, respectively. When PNL is equal to 6 cm, the cyclists may subjectively perceive the less discomforts of both the perineum and ischial tuberosity regions. Figure 2 also shows that the sums of the discomfort scores for these two body regions remained nearly unchangeable. How discomfort can be appropriately distributed between the perineum and ischial tuberosity regions may depend on the PNL.

In the analysis, the participants felt no difference in riding stability. This may be the bicycling with straight-handles that used in this study. Chen and Yu (2012) found that the handles height would influence the trunk flexion during cycling. In other words, lower handles would lead more trunk forward flexion. Chen and Yu found a difference of $30°$ when cycling with straight-handles compared to that with drop-handles. Whether a more inclined trunk position would cause the cycling more unstably merits further clarification.

Considering both discomfort on the perineum/ischial tuberosity and perceived stability, this study suggests that 6 cm PNLs should be used as a reference for bicycle seat designs, especially for straight-handles bike. However, this PNL seat should be further validated in actual road cycling and evaluated by pressure distribution on the seat.

5 Conclusion

No previous study has systematically examined the effects of PNL on body discomfort and cycling stability. This study collected data on the subjective rating values at various PNL conditions. The results showed that, when PNL = 6 cm, the discomfort between the perineum and ischial tuberosity regions achieved a more favorable distribution, subsequently providing a sufficient degree of stability for the rider. The finding can be served as a reference for bicycle seat design.

References

Bressel E, Bliss S, Cronin J (2009) A field-based approach for examining bicycle seat design effects on seat pressure and perceived stability. Appl Ergon 40:472–476

Bressel E, Cronin J (2005) Bicycle seat interface pressure: reliability, validity, and influence of hand position and workload. J Biomech 38:1325–1331

Bressel E, Reeve T, Parker D, Cronin J (2007) Influence of bicycle seat pressure on compression of the perineum: a MRI analysis. J Biomech 40:198–202

Chen YL, Yu ML (2012) A preliminary field study of optimal trunk flexion by subjective discomfort in recreational cyclists. J Chinese Inst Ind Eng 29:526–533

Groenendijk MC, Christiaans HCM, Van Hulten CMJ (1992) Sitting comfort on bicycles. In: Megaw ED (ed) Contemporary ergonomics. Taylor and Francis, London, pp 551–557

Lowe BD, Schrader SM, Breitenstein MJ (2004) Effect of bicycle saddle designs on the pressure to the perineum of the bicyclist. Med Sci Sports Exerc 36:1055–1062

Mundermann A, Nigg BM, Stefanyshyn DJ, Humble RNR (2002) Development of a reliable method to assess footwear comfort during running. Gait Posture 16:38–45

Nakamura A, Inoue Y, Ishihara T, Matsunaga W, Ono T (1995) Acquired coccygeal nodule due to repeated stimulation by a bicycle saddle. J Dermatol 22:365–369

Richmond DR (1994) Handlebar problems in bicycling. Clin Sports Med 13:165–173

Variable Neighborhood Search with Path-Relinking for the Capacitated Location Routing Problem

Meilinda F. N. Maghfiroh, A. A. N. Perwira Redi and Vincent F. Yu

Abstract The Location Routing Problem (LRP) integrates strategic decisions (facility location) and tactical decisions (vehicle routing) aimed at minimizing the total cost associated with location opening cost and routing cost. It belongs to the class of NP-hard problems. In this study, we present a variable neighborhood search with path-relinking (VNSPR) for solving the CLRP. The path-relinking procedure is integrated into the variable neighborhood search (VNS) framework. We tested our heuristic approach on three well-know CLRP data sets and the results were compared with those reported in the literature. Computational results indicate that the proposed VNSPR heuristic is competitive with existing approaches for the CLRP.

Keywords Location routing problem · Variable neighborhood search · Path-relinking

1 Introduction

The Location Routing Problem (LRP) integrates strategic decisions (facility location), and tactical decisions (vehicle routing). This problem involves finding the optimal number and locations of the depots and simultaneously allocating customers to the depot and determining the routes to visit all customers. This

---

M. F. N. Maghfiroh (✉) · A. A. N. Perwira Redi · V. F. Yu
Department of Industrial Management, National Taiwan University of Science and Technology, Taipei, Taiwan
e-mail: meilinda.maghfiroh@gmail.com

A. A. N. Perwira Redi
e-mail: wira.redi@gmail.com

V. F. Yu
e-mail: vincent@mail.ntust.edu.tw

Y.-K. Lin et al. (eds.), *Proceedings of the Institute of Industrial Engineers Asian Conference 2013*, DOI: 10.1007/978-981-4451-98-7_34,
© Springer Science+Business Media Singapore 2013

problem belongs to the class of NP-hard problems since it combines two difficult sub-problems: the facility location problem (FLP) and the vehicle routing problem (VRP), where both are shown to be NP-hard (Lenstra and Kan 1981). In recent years, the attention given to the location routing problem has increased due to the interdependency of facility location problem and vehicle routing problem.

The LRP with capacity constraints on both depots and routes is called capacitated LRP (CLRP). This study focuses on solving the CLRP to minimize the total cost associated with depot opening cost, vehicle fixed cost and total travelling cost. The CLRP can be stated as follows. Given a graph $G = (V, E)$ where V and E represent the set of vertices and the set of edges of the graph, respectively. V consists of a subset I of m potential depot sites and a subset $J = \bigvee I$ of n customers. E contains edges connecting each pair of nodes in V. Associated with each edge (i, j) $\in E$ is a travelling cost c_{ij}. Each depot site $i \in I$ has a capacity W_i and opening cost O_i. Each customer $j \in J$ has a demand d_j which must be fulfilled by a single vehicle. A set K of vehicles with capacity Q is available. Each vehicle used by depot i incurs a depot dependent fixed cost F_i and performs a single route.

Many researchers have tried to solve CLRP using various solution methods in order to solve them. Laporte and Nobert (1981) proposed an exact algorithm for a single facility fixed fleet size LRP without tour length restrictions. They formulated the problem as an integer linear program and solved it by first relaxing integrality constraints and using a branch and bound technique to achieve integrality. Tuzun and Burke (1999) developed a two-phase tabu search (TS) for the LRP with capacitated routes and uncapacitated depots. The two phases of their TS algorithm are dedicated to vehicle routing problem and depot location decision. The algorithm iteratively adds a depot to the current solution until the solution degrades.

Prins et al. (2006) solved the CLRP with capacitated depots and routes by combining greedy randomized adaptive search procedure (GRASP) with a learning process and a path-relinking mechanism. Furthermore, Barreto et al. (2007) developed a class of three-phase heuristics based on clustering techniques. Clusters of customers fitting vehicle capacity are formed in the first phase. A travelling salesman problem (TSP) is solved for each cluster in the second phase. Finally in the third phase, the depots to be opened were determined by solving a facility location problem, where the TSP cycles were combined to form super nodes.

Yu et al. (2010) proposed an SALRP heuristic for the CLRP based on the popular simulated annealing heuristic. The proposed SALRP heuristic is tested on three sets of well-known benchmark instances and the results are compared with other heuristics in the literature. The computational study indicates that the SALRP heuristic is competitive with other well-known algorithms.

Escobar et al. (2013) proposed a two phase hybrid heuristic algorithm for the CLRP. The two phases are a construction phase and an improvement phase using a modified granular tabu search. Most recently, Ting and Chen (2013) developed a multiple ant colony optimization algorithm (MACO) for the CLRP. They decomposed the problem into a facility location problem and a multiple depot vehicle routing problem (MDVRP) where the second problem is treated as a sub-problem within the first problem.

Following the idea of variable neighborhood search (VNS) by systematically changing search neighborhood, this study proposed a variable neighborhood search with path-relinking (VNSPR) which combines with VNS path-relinking (PR) as a post optimization procedure after performing VNS algorithm. The proposed method was tested on a large number of instances and the results were compared with those reported in the literature. The rest of this study is organized as follows. Section 2 presents details of the proposed VNSPR. Section 3 discusses the computational results. Finally, Sect. 4 draws conclusions.

2 VNSPR Framework

In this section, the detail proposed VNSPR will be discussed. To avoid or at least alleviate the solution being trapped in the first local optimum found, the variable neighborhood search heuristic changes neighbourhood during the search process. Starting from an incumbent solution the VNS searches within a finite sequence of neighborhoods where the successive neighborhoods are explored using a local search algorithm. To enhance the performance of the VNS, our hybrid meta-heuristic includes VND as the local search component and uses PR as diversification strategy. Figure 1 illustrates the components and the general structure of the proposed VNSPR.

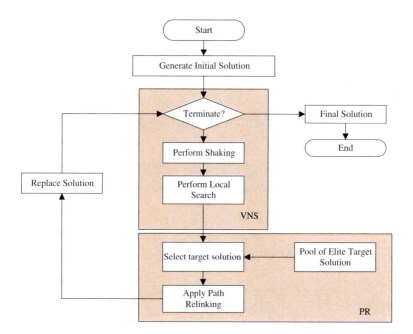

Fig. 1 VNSPR framework

The proposed VNSPR uses five parameters k_{max}, $N_{non\text{-}improving}$, m, EX, and I_{iter}. I_{iter} denotes the maximum number of iterations for the search procedure. k_{max} is the maximum number of pre-selected neighborhood structures, while m is the selected modulus number. The number of elite solutions to be put in the pool as target solutions is equal to EX. Finally, $N_{non\text{-}improving}$ is the maximum allowable number of iterations during which the best objective function value is not improved.

The current solution and objective value are set to be X and obj(X) respectively. Set the k value to be 1. The neighborhood will change depending on the value of k. At each iteration the next solution Y is obtained after the shaking phase. Then, Y' will be generated based on the local search result. The objective function is then evaluated. If obj(Y') < obj(X), then the current solution X will be replaced by Y' and the value of k remains the same. Otherwise, set $k = k + 1$ and the neighborhood will be changed based on the neighborhood sequence. This procedure is repeated until $k = k_{max}$. X_{best} will record the best solution found so far. To avoid being trapped at a local optimum, the current solution will be replaced with the best solution if k modulus m is equal to zero.

Further, the PR explores the path between a local optimum X obtained by VNS and an elite solution X' randomly chosen from EX. The PR uses a forward strategy to transform the initial solution X into the guiding solution X' by repairing from left to right the broken pairs in X, resulting in a path with non-increasing distance from X to $X'.X'$. The procedure looks for a node i with different positions in X and X'. Then node i exchanges positions with the node at the same position in X' to repair the difference. The best solution chosen will be a solution with arg $\min_{X \in EX} obj(X)$among the EX.

2.1 Initial Solution and Solution Representation

For solving the LRP problem, the initial solution is constructed randomly. The first position of the solution is the first selected open depot followed by the customers on the first route, customer on the second route and so forth. When all customers already assigned to the first depot, then the second depot will be selected and added into the solution representation. This whole process in continued until all customers are assigned in a depot. Finally, the closed depot is appended to the solution.

2.2 Neighborhood Structures

The proposed VNSPR algorithm incorporates five neighborhood structures to explore different possibilities of depot locations and to improve customer assignments to each depot. The detail of the neighborhoods used in this algorithm is present in Table 1. The two neighborhoods N_1 and N_2 correspond to insertion

Table 1 Neighborhood structures

N	k	Neighborhood
N_1	1–4	Random insertion
N_2	5–8	Random swap
N_3	9–12	2-opt
N_4	13–16	CROSS-exchange
N_5	17–20	iCROSS-exchange

and swap moves respectively. The classical 2-opt operator used in standard routing problem is a special case of our neighborhood N_3. The 2-opt procedure is done by replacing two node disjoint arcs by two other arcs. Another two neighborhoods applied are based on Taillard et al. (1997) whose proposed a neighborhood called CROSS-exchange. The main idea of this exchange is to take two segments of different routes and exchange them. The extension of CROSS-exchange called iCROSS-exchange is introduced by Braysy (2003). In this neighborhood, the sequences of the segment get inverted (Fig. 2).

The insertion (N_1) and swap (N_2) operations have several possibilities.

1. If i and j is both a customers, the customer i is reassigned to the depot that serves customer j. For swap operation, this possibility is worked in vice versa.
2. In the insertion operation, if the i and j is both a depots, the depot i will be closed and its customers will be assigned to the depot before depot i.
3. For swap operation, the depot i will be closed if depot j is closed before and depot i will be servicing depot j's customers and vice versa. This neighborhood will reassign the route from one to another depot. Thus, it can help to evaluate the savings occurred from the depot capacity, travel route and fixed cost.

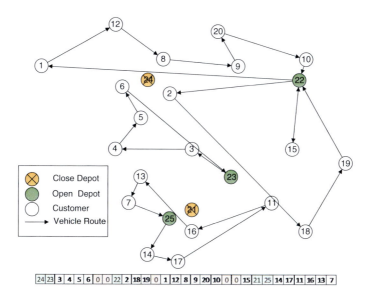

Fig. 2 An illustration of solution representation

2.3 Local Search

We choose neighborhoods by implementing the sequential deterministic order of neighborhoods within VND method. We call it sequential because the neighborhood structures are explored one by one in the given order. The algorithm goes to a neighborhood N_{k+1} unless it falls on a local optimum with respect to the neighborhood N_k. Otherwise, it is returned to the first one N_1. The effectiveness of VND is related to the order of neighborhoods considered and the order the solutions in each neighborhood are explored. To maintain the solution and computational time, the neighborhoods used for VND are limited to insertion (N_1) and swap (N_2) operation.

2.4 Post Optimization: Path Relinking

In order to intensification and divers the solution, path relinking (PR) algorithm is adopted. VNS with PR will maintains a pool of elite solutions EX generated using VNS. Each solution produce after local search is checked for inclusion in EX. This EX will be used as the initiating or guiding solutions. In this hybrid method the PR operator explores the paths between a local optimum obtained by VNS and a solution randomly chosen from EX.

The path from X to X' is then guided by the Hamming distance Δ, equal to the number of differences between the nodes at the same position in X and X', i.e., $\Delta(X, X') = \sum_{i=1,|X|} X_i \neq X'_i$.

3 Computational Result

In this section, we present the results of our computational experiments with the VNSPR described in previous section. The proposed VNSPR algorithm is coded in C++ Visual Studio 2008 and runs on a PC with a Dual Core (1.83 Ghz) processor and 2 GB RAM, under the Windows XP operating system. For the purpose of verification, VNSPR is applied to the benchmark instances used were designed by Barreto (2004). The proposed algorithm has been compared with three latest heuristic proposed to solve CLRP: SALRP (Yu et al. 2010), Modified Ant Colony Algorithm (MACO) (Ting and Chen 2013), and Two Phase Hybrid Heuristic (HH) (Escobar et al. 2013) (see Table 2).

Compared with the best know solutions, from 19 instances of this data set, the proposed VNSPR obtains 13 best solutions. The gap between the solution is vary from 0.00 to 1.60 % with average gap is 0.33 %. Compared with others algorithm, the result of VNSPR is competitive enough. The relative average gap of VNSPR is better than SALRP (0.43 %) and HH (0.78 %). In term of total number of BKS obtained, VNSPR result is outperforms MACO (12 BKS) and HH (7 BKS).

Table 2 Solutions cost for Prins et al.'s instances

Prob. ID	n	m	BKS	SALRP		MACO		HH		VNSPR	
				Cost	Gap (%)	Cost	Gap (%)	Cost	Gap (%)	Cost	Gap (%)
B1	21	5	424.9	424.9	**0.00**	424.9	**0.00**	424.9	**0.00**	424.9	**0.00**
B2	22	5	585.1	585.1	**0.00**	585.1	**0.00**	585.1	**0.00**	585.1	**0.00**
B3	29	5	512.1	512.1	**0.00**	512.1	**0.00**	512.1	**0.00**	512.1	**0.00**
B4	32	5	562.2	562.2	**0.00**	562.22	**0.00**	562.2	**0.00**	562.2	**0.00**
B5	32	5	504.3	504.3	**0.00**	504.3	**0.00**	504.3	**0.00**	504.3	**0.00**
B6	36	5	460.4	460.4	**0.00**	460.4	**0.00**	460.4	**0.00**	460.4	**0.00**
B7	50	5	565.6	565.6	**0.00**	565.6	**0.00**	580.4	2.62	565.6	**0.00**
B8	75	10	844.4	848	0.43	844.88	0.06	848.9	0.53	844.4	**0.00**
B9	100	10	833.4	838.3	0.59	836.75	0.40	838.6	0.62	833.43	**0.00**
B10	12	2	204	204	**0.00**	204	**0.00**	–	–	204	**0.00**
B11	55	15	1112.1	1112.8	0.06	1112.58	0.04	–	–	1112.1	**0.00**
B12	85	7	1622.5	1622.5	**0.00**	1623.14	0.04	–	–	1632.7	0.63
B13	318	4	557275	563493	1.12	560210.8	0.53	–	–	559224	0.35
B14	318	4	670119	684164	2.10	670118.5	**0.00**	–	–	678689	1.28
B15	27	5	3062	3062	**0.00**	3062	**0.00**	3062	**0.00**	3062	**0.00**
B16	134	8	5709	5709	**0.00**	5709	**0.00**	5890.6	3.18	5775.9	1.17
B17	88	8	355.8	355.8	**0.00**	355.8	**0.00**	362	1.74	355.8	**0.00**
B18	150	10	43919.9	45109.4	2.71	44131.02	0.48	44579	1.50	44314	0.90
B19	117	14	12290.3	12434.5	1.17	12355.91	0.53	–	–	12536	2.00
Average					*0.43*		*0.11*		*0.78*		*0.33*

4 Conclusions

In this study, both the location decision and routing problem in the LRP are tackled together. We propose a hybrid method, called variable neighborhood search with path-relinking, for the CLRP. The algorithm was tested on Barreto's instances to verify its performance. Computational results indicate the effectiveness and efficiency of the proposed algorithm. The proposed VNSPR is competitive with existing state-of-the-art algorithms in solving CLRP. Moreover, the relative percentage gap between best known solutions and the solutions obtained by the proposed VNSPR is relatively small, which shows the robustness of the algorithm.

References

Barreto SS (2004) Análise e Modelização de Problemas de localização-distribuição [Analysis and modelling of location-routing problems]. University of Aveiro, Portugal

Barreto S, Ferreira C, Paixão J, Santos BS (2007) Using clustering analysis in a capacitated location-routing problem. Eur J Oper Res 179(3):968–977

Braysy O (2003) A reactive variable neighborhood search for the vehicle-routing problem with time windows. INFORMS J Comput 15(4):347–368

Escobar JW, Linfati R, Toth P (2013) A two-phase hybrid heuristic algorithm for the capacitated location-routing problem. Comput Oper Res 40(1):70–79

Laporte G, Nobert Y (1981) An exact algorithm for minimizing routing and operating costs in depot location. Eur J Oper Res 6(2):224–226

Lenstra JK, Kan AHGR (1981) Complexity of vehicle routing and scheduling problems. Networks 11(2):221–227

Prins C, Prodhon C, Calvo R (2006) Solving the capacitated location-routing problem by a GRASP complemented by a learning process and a path relinking. 4OR 4(3):221–238

Taillard É, Badeau P, Gendreau M, Guertin F, Potvin JY (1997) A Tabu search heuristic for the vehicle routing problem with soft time windows. Transp Sci 31(2):170–186

Ting C-J, Chen C-H (2013) A multiple ant colony optimization algorithm for the capacitated location routing problem. Int J Prod Econ 141(1):34–44

Tuzun D, Burke LI (1999) A two-phase tabu search approach to the location routing problem. Eur J Oper Res 116(1):87–99

Yu VF, Lin S-W, Lee W, Ting C-J (2010) A simulated annealing heuristic for the capacitated location routing problem. Comput Ind Eng 58(2):288–299

Improving Optimization of Tool Path Planning in 5-Axis Flank Milling by Integrating Statistical Techniques

Chih-Hsing Chu and Chi-Lung Kuo

Abstract Optimization of the tool path planning in 5-axis flank milling of ruled surfaces involves search in an extremely high-dimensional solution space. The solutions obtained in previous studies suffer from lengthy computational time and suboptimal results. This paper proposes an optimization scheme by integrating statistical techniques to overcome these problems. The scheme first identified significant factors in the tool path planning that influence the machining error of a machined surface by a first sampling plan. We then conducted a series of simulation experiments designed by the two-level fractional factorial method to generate experimental data with various settings. A regression model was constructed with Response Surface Methodology (RSM) that approximates the machining error in terms of those identified factors. This simplified model accelerates estimation of the objective function, computed as a black-box function in previous studies, with less computation. Test results show that the proposed scheme outperforms PSO in both the computational efficiency and the solution quality.

Keywords 5-axis machining · Response surface methodology · Optimization

1 Introduction

With two rotational degrees of freedom in tool motion, 5-axis machining provides higher productivity and better shaping capability compared to traditional 3-axis machining. The 5-axis machining operation is categorized into two types: end

C.-H. Chu (✉) · C.-L. Kuo
Department of Industrial Engineering and Engineering Management,
National Tsing Hua University, Section 2, Kuang-Fu Road, Hsinchu, Taiwan
e-mail: chchu@ie.nthu.edu.tw

C.-L. Kuo
e-mail: u920827@gmail.com

Y.-K. Lin et al. (eds.), *Proceedings of the Institute of Industrial Engineers Asian Conference 2013*, DOI: 10.1007/978-981-4451-98-7_35,
© Springer Science+Business Media Singapore 2013

milling and flank milling. Tool path planning is a critical task in both milling operations, with avoidance of tool collision and machining error control as two major concerns (Rehsteiner 1993). To produce a machined surface exactly the same as its design specifications using a cylindrical cutter is highly difficult in 5-axis flank milling. The cutter cannot make a contact with a surface ruling without inducing overcut or undercut around the ruling due to local non-developability of a ruled surface (Chu and Chen 2006), except simple geometries like cylindrical and conical surfaces. A common method used in industry is to let the cutter follow the surface rulings, although extensive machining errors often occur on twisted surfaces (Tsay and Her 2001).

Previous studies (Chu et al. 2011; Hsieh and Chu 2011) have shown that the machining error can be effectively reduced through optimization of tool path planning in a global manner. Such an optimization approach works as a systematic mechanism for precise control of machining error. Wu and Chu (2008) transformed tool path planning in 5-axis flank milling into a curve matching problem and applied dynamic programming to solve for an optimal matching with the error of the machined surface as an objective function in the optimization. They solved the similar curve matching problem with Ant Colony Systems algorithm to reduce the lengthy time required by the dynamic programming approach. They allowed the cutter to freely make contact with the surface to be machined, rather than moving among pre-defined surface points in previous works. However, the PSO based search in the previous works (Hsieh and Chu 2011; Hsieh et al. 2013) suffers from unsatisfactory quality of sub-optimal solutions due to excessive nonlinearity inherited in the machining error estimation and high-dimensional solution space. The solutions obtained by PSO easily get trapped in local optima and are thus very far from global optima. The computational time in these works was lengthy, reducing the practicality of industrial applications. This study attempts to overcome these problems with an approximate optimization scheme that integrates various statistical techniques in the tool path planning. The scheme first identified significant factors that influence the machining error of a machined surface. We then conducted a series of simulation experiments designed by the two-level fractional factorial method to generate experimental data with various settings. A regression model was constructed with Response Surface Methodology (RSM) that approximates the machining error in terms of those identified factors. This simplified model accelerates estimation of the objective function with fewer computations compared to previous studies. Test results show that the proposed scheme outperforms PSO in both computational efficiency and solution quality.

2 Preliminaries

There are numerous variables in the tool path planning for 5-axis flank milling of a ruled surface. As illustrated in Table 1, some variables determine the precision of CNC machines and the others control the tool position at a cutter location.

Table 1 Possible factors in tool path planning for 5-axis flank milling

Factor	Notation
Number of error measuring points in the *u* direction	U^N
Number of error measuring points in the *v* direction	V^N
Tool radius	T^R
Tool cutting length	T^L
Number of cutter locations in a tool path	N^{CL}
Number of linear interpolation between two cutter locations	T^{IN}
Distance along the tangent direction on the first boundary curve	N^U
Distance along the normal direction on the first boundary curve	T^U
Distance along the bi-normal direction on the first boundary curve	B^U
Distance along the tangent direction on the second boundary curve	N^L
Distance along the normal direction on the second boundary curve	T^L
Distance along the bi-normal direction on the second boundary curve	B^L

The moving tri-hedron of the surface on two boundary curves is used to position a cylindrical cutter. Three parameters along the directions of the tangent, surface normal, and bi-normal specify the center point of each cutter end (see Fig. 1). We then categorize the parameters listed in Table 1 into two groups: normal factors and block factors. For example, users need to specify $U^N, V^N, T^R, T^L, N^{CL}, T^{IN}, T^U$ in a CNC machining operation and thus they are not variables to be optimized. We only need to consider $T^U, N^U, B^U, T^L, N^L, B^L$ in determining a cutter location.

The objective function used in optimization of tool path planning is the geometric error on the machined shape with respect to the design surface. Exact estimation of the machined geometry has no close-form solutions and involves highly non-linear equations. Most CNC tool path planning methods perform the error estimation approximately estimated with the z-buffer method. The estimation procedure consists of four steps as shown in Fig. 2. The design surface is sampled in a discrete manner. At each sampling point, two straight lines are extended along the positive and negative normal directions with a distance of the cutter radius. The lengths of these lines will get updated after the cutter sweeps across them along a given tool path. We approximate the tool swept surface by generating a finite

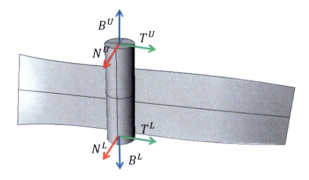

Fig. 1 Positioning a tool with the moving tri-hedron at a cutter location

Step	Illustration
Sample points from the design surface	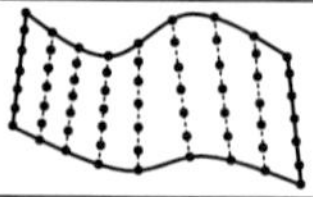
Create straight lines on each point	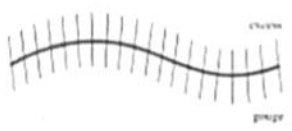
Intersect the lines with the cutter	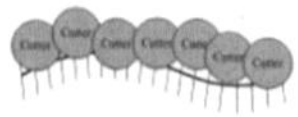
Calculatethe machining error	

Fig. 2 Approximate estimation of the machining error (Chu et al. 2011)

number of tool positions interpolated by two consecutive cutter locations. The next step is to intersect the lines with the peripheral surface of the cutter. The machining error is calculated as the sum of the lengths of the trimmed straight lines.

Previous studies (Hsieh and Chu 2011; Hsieh et al. 2013) reduced the machining error through optimization of the tool path planning based on PSO. Greedy approaches can only find sub-optimal solutions in the optimization, as interactions exist between cutter locations, i.e. cutter locations with minimized errors at an early stage may induce larger errors at later cutter locations. For machining complex geometries, a tool path normally consists of hundreds of cutter locations. Each cutter location contains 6 parameters to be specified when the moving tri-hedron is used for positioning the tool axis. PSO-based search suffers from several difficulties. Estimation of the machining error involves highly non-linear computation and has no closed-form solutions. This results in lengthy computational time in the optimization. The PSO-based search can quickly converge in local sub-optimal solutions in a high-dimensional solution space and thus produces unsatisfactory solution quality. This study proposes to solve the above mentioned difficulties by integrating statistical techniques.

3 Optimization Scheme

Optimization of tool path planning for 5-axis flank milling with the machining error as an objective function can be written as:

$$\text{Min} \sum_{i=0}^{N^{CL}-1} E_i$$

$$E_i = e\left(T_i^U, T_{i-1}^U, N_i^U, N_{i-1}^U, B_i^U, B_{i-1}^U\right) \quad i = 0, 1, \ldots, N^{CL} - 1$$

(1)

The machining error is the summation of the errors produced by each tool motion between two consecutive cutter locations. We then examined whether the 6 normal factors in Eq. (1) are significant ones using ANOVA techniques. A 2^{9-4} sampling plan was carried out to identify significant factors. In this plan, 2 means the number of levels of each factor investigated, 9 is the total number of factors, and 4 describes the size of the fraction of the full factorial used. This sampling design was chosen mainly because it has the minimum number of experiments in a resolution IV design. Based on the effect hierarchy principle, lower-order effects are more likely to be important than higher-order effects. A lower-resolution design has defining words of short length, which imply aliasing of low-order effects. Resolution R implies that no effect involving I factors is aliased with effects involving less than R-I factors (Wu and Hamada 2009). The p-values the 2^{9-4} sampling imply that $T^U, N^U, B^U, T^L, N^L, B^L$ are all significant factors. RSM is a strategy to achieve this goal that involves experimentation, modeling, data analysis, and optimization (Box and Wilson 1951). RSM-based optimization normally consists of two steps: response surface design and optimization. The response surface design constructs approximate functions for the independent variables and the response variable by systematically exploring the relationships between these two variables. RSM usually employs low-order polynomial approximation such as first-order regression models. Second-order or even higher-order polynomials can be used when the relationships to be investigated show the existence of a curvature. We applied least squares estimation to fit an approximate model and examined significance of the independent variables by regression analysis. The approximate model of the objective function is expressed as:

$$
\sum_{i=0}^{N^{CL}-1} |E_i| \approx \beta_0 + \sum_{j=1}^{N^{CL}-1} \sum_{k=1}^{r} \beta_{jk} X_{jk} + (n-1) \sum_{j=1}^{N^{CL}-1} \sum_{k=1}^{r} \beta_{jk} X_{jk}^2 + (n
$$
$$
-1) \sum_{i<k} \sum_{k=1}^{r} \sum_{j=1}^{N^{CL}-1} \beta_{ijk} X_{ik} X_{jk} + \varepsilon \tag{2}
$$

where X_i, X_j are significant factors, β_j are constants to be determined by RSM, and k is the number of significant factors. A second sampling plan was conducted to construct the above model based on RSM. The plan was designed by 5^{6-1} experiments. The upper- and lower-bounds of each factor were determined based on our previous study (Hsieh et al. 2013). A dicotomic search was applied to adjust individual parameter setting within the range. The final parameter settings are shown in Table 2. Constructing the RSM-model is equivalent to determining the β_i values from the experimental data generated from the 5^{6-1} sampling. Once the model had been constructed, the Steepest Descent method was applied to find its optimal solutions. The similar approximate model can be used as a greedy

Table 2 Parameter setting in constructing the RSM-based model

Factor	Upper bound	Lower bound
N^U	0.105	−0.105
T^U	0.5	−0.5
B^U	0.5	−0.5
N^L	0.105	−0.105
T^L	0.5	−0.5
B^L	0.5	−0.5

approach, i.e. the original problem is subdivided into ($N^{CL}-1$) regions, each corresponding to two consecutive cutter locations. The optimal solution of each region is computed sequentially and adds up to the final solution as:

$$y \approx \sum_{j=1}^{N^{CL}-1} \left(\beta_0 + \sum_{k=1}^{r} \beta_{jk} X_{jk} + (n-1) \sum_{k=1}^{r} \beta_{jk} X_{jk}^2 + (n-1) \sum_{k=1}^{r} \beta_{ijk} X_{ik} X_{jk} + \varepsilon \right). \tag{3}$$

4 Experimental Results

A ruled surface shown in Fig. 3 was used as test geometry to demonstrate the effectiveness of the proposed scheme. This surface has substantial twist near both ends, which induce excessive machining errors when a cylindrical cutter is used. We compare the computational efficiency by counting the total number of error estimations for a tool path. For the PSO-based search, the number is equal to the particle number (N) multiplying by the iteration number (G). The proposed scheme consists of two major steps: sampling and model construction. The total number of estimations conducted by the sampling step depends on the number of factors F, the level number L_F, and the fraction number P_F as $L_F^{F-P_F}$. The model construction requires the error estimation for $L_G^{G-P_G}$ times. Thus the scheme totally computes the machining error for ($L_F^{F-P_F} + L_G^{G-P_G}$) times.

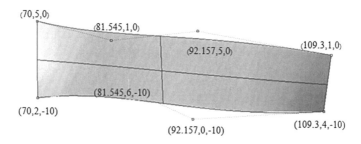

Fig. 3 Test ruled surface and its control points

Improving Optimization of Tool Path Planning

Table 3 Comparison between the solutions obtained by this work and PSO

Method	Machining error (mm)
RSM (global)	7.71
RSM (greedy)	9.15
PSO	11.55

Table 4 Parameter setting in PSO-based search

Parameter	Value
W	0.5
C1	0.5
C2	0.5
Iteration number	100
Particle number	100
Tool radius	2 mm
Tool length	30 mm
Number of cutter location	40

Table 5 Comparison of computational efficiency

Method	Number of error estimations
PSO (global)	10,000
RSM	3,157

As shown in Table 3, the tool path generated by the proposed scheme produces a smaller machining error than that of the PSO-based tool path planning. The parameter setting in the PSO is listed in Table 4. Both global search and greedy approaches outperform the previous work. The computational efficiency of the proposed scheme was also compared with the PSO-based tool path planning, as shown in Table 5. The comparison is based on the total number of error estimations for a given tool path. The error estimation conducted by the greedy RSM method is within two cutter locations rather than a complete tool path and thus not included in the comparison. The total number of computations required by our method is only one-third of the previous work. The proposed scheme not only produces better solution quality, it also requires fewer computations than the previous work.

5 Conclusions

This study described a new optimization scheme for the tool path planning in 5-axis flank milling of ruled surfaces. This scheme improves the solution quality and computation efficiency of previous PSO-based methods. It adopted various statistical techniques to approximate the total error on the machined surface. A first sampling plan was conducted to identify significant factors in the tool path planning. We carried out a series of simulation experiments designed by the two-level fractional factorial method to generate experimental data with systematic settings. The experimental data results in a second-order regression model constructed with the response surface methodology. This simplified model approximates the objective function in terms of those identified factors and enables quick estimation of the function. The test results on a ruled surface with excessive twist demonstrated the advantages of the proposed scheme. It enhanced the solution quality by reducing 33 % of the machining error produced by the PSO-based tool path planning with the number of computations required 31 % fewer. This work improves the practicality for optimization-based tool path planning in 5-axis flank milling.

References

Box GEP, Wilson KB (1951) On the experimental attainment of optimum conditions. J Royal Stat Soc B 13:1–45

Chu CH, Chen JT (2006) Automatic tool path generation for 5-axis flank milling based on developable surface approximation. Int J Adv Manuf Technol 29(7–8):707–713

Chu CH, Tien KW, Lee CT, Ting CJ (2011) Efficient tool path planning in 5-axis milling of ruled surfaces using ant colony system algorithms. Int J Prod Res 49(6):1557–1574

Hsieh ST, Chu CH (2011) PSO-based path planning for 5-axis flank milling accelerated by GPU. Int J Comput Integr Manuf 24(7):676–687

Hsieh ST, Tsai YC, Chu CH (2013) Multi-pass progressive tool path planning in five-axis flank milling by particle swarm optimization. Int J Comput Integr Manuf (in press)

Rehsteiner F (1993) Collision-free five-axis milling of twisted ruled surfaces. CIRP Ann 42(1):457–461

Tsay DM, Her MJ (2001) Accurate 5-axis machining of twisted ruled surfaces. ASME J Manuf Sci Eng 123:731–738

Wu CF, Hamada MS (2009) Experiments: planning, analysis and optimization. Wiley, New York

Wu PH, Li YW, Chu CH (2008) Tool path planning for 5-axis flank milling based on dynamic programming techniques. Int J Mach Tools Manuf 50:1224–1233

A Multiple Objectives Based DEA Model to Explore the Efficiency of Airports in Asia–Pacific

James J. H. Liou, Hsin-Yi Lee and Wen-Chein Yeh

Abstract Airport efficiency is an important issue for each country. The classical DEA models use different input and output weights in each decision making unit (DMU) that seems not reasonable. We present a multiple objectives based Data Envelopment Analysis (DEA) model which can be used to improve discriminating power of DEA method and generate a more reasonable input and out weights. The traditional DEA model is first replaced by a multiple objective linear program (MOLP) that a set of Pareto optimal solutions is obtained by genetic algorithm. We then choose a set of common weights for inputs and outputs within the Pareto solutions. A gap analysis is included in this study that can help airports understand their gaps of performances to aspiration levels. For this new proposed model based on MOLP it is observed that the number of efficient DUMs is reduced, improving the discrimination power. Numerical example from real-world airport data is provided to show some advantages of our method over the previous methods.

Keywords DEA · Airport · Multiple-criteria decision-making (MCDM) · Genetic algorithm

1 Introduction

Airport efficiency has been a central issue in cost control owing to reasons such as airport monopoly power, changing ownership structure, increasing competitive pressure from airlines and competing airports and government aspirations to

---

J. J. H. Liou (✉) · H.-Y. Lee
Department of Industrial Engineering and Management, National Taipei University
of Technology, Taipei, Taiwan
e-mail: jhliou@ntust.edu.tw

W.-C. Yeh
Department of Air Transportation, Kainan University, Luzhu Township,
Taoyuan County, Taiwan

Y.-K. Lin et al. (eds.), *Proceedings of the Institute of Industrial
Engineers Asian Conference 2013*, DOI: 10.1007/978-981-4451-98-7_36,
© Springer Science+Business Media Singapore 2013

develop their nations as air hub or logistics center (Lam et al. 2009). This paper examines the performance of major airports in Asia–Pacific region. The potential air transportation demand in the Asia–Pacific region is enormous due to the region's high population density, strong economic growth and the widespread adoption of open-skies policies. This region has been ranked the second largest air freight market, following only that of North America. The report by Airports Council International (ACI) indicated that if the current trend continues, the Asia–Pacific region will replace North America as the region with largest passenger volume. To meet the growing demands, airports in this region face a more severe demand for efficient and better service quality. Therefore, the accurate assessment of productive efficiencies has thus been one of the most pertinent issues in the unending quest towards global competitiveness within the international airline industry.

This paper contributes to the existing airport efficiency literature by presenting a new data envelopment analysis (DEA) model, which incorporates multiple objective linear programming (MOLP) and increases discriminating power. DEA has become a popular method often used in the literature to study the relative efficiencies of firms and it is a non-parametric, no statistical hypotheses and tests are required. DEA has been used to compare performance of airports within national boundaries, including the U.S. (Gillen and Lall 1997; Sarkis 2000), UK (Parker 1999), Spain (Martin and Roman 2001), Japan (Yoshida and Fujimoto 2004), Taiwan (Yu 2010), Italia (Gitto and Mancuso 2012), Portugal (Barros and Sampaio 2004) as well as airports around the world (Yang 2010; Adler and Berechman 2001). However, traditional DEA has been criticized by its weak discriminating power and unrealistic weight distribution (Li and Reeves 1999). To remedy the above shortcomings, Golany (1988) first integrated the DEA and MOLP by taking account the preference of the decision maker (DM). But the prior information could be biased or hard to obtain due to the complexity of problems. Another popular method, the cross-efficiency matrix, has been a widely used for improve discriminating power. However, the cross-efficiency matrix only provides cross and self-efficiency scores; it does not provide the new input and output weights corresponding to those new efficiency scores (Li and Reeves 1999). Our proposed MOLP based DEA model improves prior models in three ways. First, the proposed model is non-radial. All of the inputs and outputs will be projected on to the same efficient frontier. Thus, the weight distribution is more reasonable than traditional methods. Second, instead of solving n mathematical programming problems, it is required to solve only one. Finally, the efficiency measurement is based on aspiration levels, not relatively good DUMs from the existing DMUs. With the new concept, the decision maker sets an aspiration level as the benchmark, a DMU which might not exist in the current basket of apples, but decision-makers will understand the gaps between each DUM and the aspiration levels. Decision-makers can therefore devise and implement a strategy to reduce the gaps to aspiration levels. This paper improves prior studies and determines the most preferred solution and more reasonable weight distribution for decision makers.

2 The Multiple Objectives Based DEA Method

The classical DEA analysis considers individual DMUs separately and calculates a set of weights which brings maximal relative efficiency to each group. But this approach tends to obtain most of the DMUs are efficient, thus, hard to discriminate them. Current approach sets target values as the aspiration levels for inputs and outputs and calculates the multiple objectives programming to find a set of common based weights. We then compute the efficiency ratio of all DMUs and analyze their gaps to aspiration levels.

2.1 Classical Efficiency Measure

If the ath DMU uses m-dimension input variables x_{ia} $(i = 1, \ldots, m)$ to produce s-dimension output variables y_{ra} $(r = 1, \ldots, s)$, the efficiency of DMU h_a can be found from the following model:

$$\text{Max } h_a = \frac{\sum_{r=1}^{S} u_r y_{ra}}{\sum_{i=1}^{m} v_i x_{ia}} \tag{1}$$

Subject to:

$$\frac{\sum_{r=1}^{S} u_r y_{rk}}{\sum_{i=1}^{m} v_i x_{ik}} \leq 1, \quad k = 1, \ldots, n$$

$$0 < \varepsilon \leq u_r, 0 < \varepsilon \leq v_i, \quad i = 1, \ldots, m; \ r = 1, \ldots, s$$

where

x_{ik} Stands for the ith input of the kth DMU,
y_{rk} Stands for the rth output of kth DMU,
u_r, v_i Stand for the weight of the rth output and ith input respectively,
h_a Relative efficiency value.

2.2 Fuzzy Multiple Objective Programming

Based on Tsai et al. (2006), the Eq. (1) can be transformed as a MOLP problem. Multiple objective programming can be employed to find a set of common weight combinations so that the optimized efficiency value can be calculated for each

DMU in overall relative efficiency achievement. That means each DMU is analyzed on a same baseline rather than on different efficient frontiers. This above goal can be formulated in Eq. (2).

$$\text{Max } h_1 = \frac{\sum_{r=1}^{s} u_r y_{r1}}{\sum_{i=1}^{m} v_i x_{i1}} \tag{2}$$

$$\text{Max } h_2 = \frac{\sum_{r=1}^{s} u_r y_{r2}}{\sum_{i=1}^{m} v_i x_{i2}}$$

$$\vdots$$

$$\text{Max } h_n = \frac{\sum_{r=1}^{s} u_r y_{rn}}{\sum_{i=1}^{m} v_i x_{in}}$$

Subject to:

$$\frac{\sum_{r=1}^{s} u_r y_{rk}}{\sum_{i=1}^{m} v_i x_{ik}} \leq 1, \quad k = 1, 2, \ldots, n$$

$$u_r \geq \varepsilon > 0, \quad r = 1, 2, \ldots, s$$

$$v_i \geq \varepsilon > 0, \quad i = 1, 2, \ldots, m$$

2.3 Apply Genetic Algorithm NSGA-II to Obtain the Pareto Solutions

The presence of multiple objectives in a problem, in principle, gives rise to a set of optimal solutions (largely known as Pareto-optimal solutions), instead of a single optimal solution. Classical optimization methods suggest converting the MOLP to a single-objective optimization problem by emphasizing one particular Pareto-optimal solution at a time. Tsai et al. (2006) solves the MOLP by using the compromise solution as the optimal solution. But the compromise solution does not guarantee the solution is a Pareto solution. Furthermore, the obtained weight distribution from compromise solution sometimes produces extreme value in some inputs/outputs and zero values in some inputs/outputs that cannot reflect the real situation. We apply genetic algorithm NSGA II to solve Eq. (2) and to attain the Pareto solutions. Based on derived Pareto solutions, we select the solution according to airport managers' opinions. The common weights are then substituted into Eq. (1) to calculate the efficiency of each DMU.

A Multiple Objectives Based DEA Model 299

3 The Data and Variables for Empirical Case

To measure productive efficiency using DEA, one must first identify outputs that an airport produces and inputs it uses in producing those outputs. Based on prior studies (Parker 1999; Sarkis 2000; Martin and Roman 2001; Abbott and Wu 2002; Assaf et al. 2012), the characteristics of the selected DEA input/out variables are summarized in Table 1. On the input side, we consider employees, runways, terminal size and gates as the input variables. The most important inputs at airports are labor and capital. The easiest measure of the former is the number of employees. The size of the terminal and number of gates determine the airport's ability to load passengers and cargo into aircrafts and hence play a crucial role in airport operation activity. The number of runways decides the aircraft movements. An airport's primary function is to provide an interface between aircraft and passengers or freight. From this perspective, the selected outputs in our efficiency are the passengers, cargo and aircraft movements.

The selection of variables in DEA must be isotonic, i.e., the value of output variable cannot decrease when input variables are increase. This was tested with Pearson analysis for input and output variables. The results indicate the selected variables are positively correlated with significant p values. The data source for the input and output variables come from Airport Benchmarking Report (2010).

4 Results and Discussions of Empirical Case

The classical DEA model exist some problems, such as weak discriminating power and unrealistic weight distribution for inputs and outputs. The classical DEA solves n linear programming problems and uses different weights for inputs and outputs in each DMU calculation. These results imply the efficiency analysis is not on the same baseline. The unrealistic weight distribution refers to the situation where some DMUs can be treated as efficient by classical DEA model simply because they have extremely large weights in a single output and extremely small weights in a single input while these extreme weights are practically unreasonable.

Table 1 Characteristics of DEA input and output variables

	Variable	Definition
Input	Employee (persons)	Sum of employees
	Runway	Number of runways
	Terminal size (square meter)	Total area of passenger and cargo terminals
Output	Gate	Number of gates
	Passenger volume (persons)	Passenger throughput of the airport
	Cargo volume (tons)	Cargo throughput of the airport
	Aircraft movement	Number of flights at the airport

Table 2 The weights for inputs and outputs

	Input				Output		
	Employee	Runway	Terminal size	Gate	Passenger	Cargo	Aircraft movement
Weight	203.74	120.39	321.48	229.65	232.26	119.22	104.24

The reason is because the classical DEA model tends to pursue a favorable weight distribution when each DMU is evaluated. Another noted point is that the relative efficiency might not fit on today's competitive environment in the air transport markets. Enterprise should not pursue a relatively good firm as the benchmark but an aspired level to fulfill customers' needs. To remedy the above shortcomings, we introduce aspired levels for the inputs/outputs as the benchmark that is decided by airport managers based on their aspiration. After adding the aspired levels as a virtual DMU, we follow the procedures as described in Sect. 3 to formulate the problem as a MOLP model. The MOLP problem is then solved by genetic algorithm to find non-dominated solutions. Since the non-dominated solutions are not unique, we can select a most preferred one according to decision maker's opinions to reflect the real-world situation. In this study, we select the non-dominated solution based on the managers' input. Table 2 shows the relative weights in our analysis that is obtained from the MOLP problem solved by genetic algorithm. The results show there is no extreme weight distribution for input and output variables. It is evidence that the uneven weight distribution is fixed by our proposed method.

Using the obtained weight distribution (Table 2), we can derive the efficiencies for each DMU (Table 3). The results are clear and significant that no airport presents prefect efficiency except the aspired airport. The discrimination power has been significant improved by using the proposed model which efficient DUMs have reduced from 12 DMUs to only 1 DMU (the virtual UM). The leading airports are Hong Kong, followed by Tokyo Haneda and Sydney, respectively. Based on the obtained weight distribution for inputs and outputs, we further calculate the gaps to aspired levels for each airport. The gaps are analyzed by assuming that the under the current inputs of each airport, what are the gaps to reach the aspiration levels if the airports want to achieve perfect score. The gap numbers provide very useful information for airport management. The decision makers will not only know their relative position with their peers but also the gaps to the aspiration levels. Thus, airport managers can further set strategies to improve the volumes with respect to passengers, cargo and aircraft movements. Conversely, our model not only helps airports apprehend the gaps between current performance and aspiration levels but also provides them with a chance to surpass their leading competitors.

A Multiple Objectives Based DEA Model

Table 3 Analysis results of 22 airports in Asia–Pacific

No.	DUM	CCR	Tzeng's method	Proposed method	Gap to aspired levels		
					Passengers	Cargo (tons)	Aircraft movements
1	Auckland	1.000 (1)	1.000 (1)	0.533 (10)	34,158,403	1,164,722	1,079,839
2	Brisbane	0.962 (12)	0.767 (7)	0.548 (9)	35,604,473	328,236	824,535
3	Perth	1.000 (1)	0.527 (19)	0.421 (19)	28,526,491	416,003	593,230
4	Sydney	1.000 (1)	0.811 (6)	0.699 (3)	48,109,785	2,132,314	1,155,868
5	Guangzhou	1.000 (1)	1.000 (1)	0.662 (4)	63,582,220	3,793,511	1,447,186
6	Beijing	1.000 (1)	0.610 (12)	0.639 (5)	111,611,982	5,738,180	2,242,731
7	Hong Kong	0.926 (14)	1.000 (1)	0.926 (1)	54,204,438	10,815,252	986,354
8	Shanghai	0.750 (18)	0.556 (15)	0.458 (15)	133,322,384	18,691,234	2,461,311
9	Shenzhen	1.000 (1)	0.848 (5)	0.52 (12)	59,952,294	3,221,306	1,262,088
10	Xiamen	0.729(20)	0.661 (10)	0.436 (17)	36,909,224	1,253,180	789,816
11	Tokyo Haneda	1.000 (1)	0.742 (8)	0.878 (2)	42,302,792	1,607,590	954,948
12	Tokyo Narita	0.761 (16)	0.545 (16)	0.476 (14)	93,599,554	12,372,425	1,466,893
13	Kansai	0.726 (21)	0.542 (17)	0.456 (16)	46,405,109	4,642,605	1,091,671
14	Kuala Lumpur	0.751 (17)	0.589 (13)	0.422 (18)	93,871,998	3,798,086	1,658,757
15	Penang	1.000 (1)	0.527 (18)	0.415 (20)	14,599,629	1,071,098	346,418
16	Jakarta	1.000 (1)	0.978 (4)	0.636 (6)	59,539,996	1,543,514	1,151,265
17	New Delhi	0.606 (23)	0.435 (21)	0.377 (22)	100,637,825	3,728,368	1,998,137
18	Dubai	0.938 (13)	0.410 (22)	0.38 (21)	169,486,919	15,311,634	2,724,419
19	Incheon	0.738 (19)	0.615 (11)	0.586 (8)	77,089,326	12,568,962	1,264,853
20	Bangkok	0.837 (15)	0.665 (9)	0.527 (11)	84,039,895	4,695,710	1,363,302
21	Singapore	0.608 (22)	0.453 (20)	0.494 (13)	105,057,301	9,711,656	1,638,635
22	Taipei	1.000 (1)	0.588 (14)	0.607 (7)	49,333,100	6,483,728	781,319

5 Conclusion

This paper contributes to the existing literature by presenting a model to improve prior models in several ways. First, discriminating power has been improved with only aspired airport with perfect score and efficiencies of other airports are all less than 1. Second, the results can be solved by one MOLP model instead of n linear programming models. The obtained weight distribution is more reasonable than classical DEA models and other MOLP methods. Third, we introduce the concept of gaps to aspired levels that can help decision makers understanding of how airport can improve their performance to reach absolutely good levels, not a relatively good one. This gap analysis might be more suitable for today's competitive environment. Finally, little literature discusses the airport performance of Asia–Pacific region; we contribute to the existing research of airport efficiency. Our empirical example shows the effectiveness and usefulness of the proposed model. The results indicate Hong Kong is the leading airport in the Asia–Pacific region, followed by Tokyo Haneda and Sydney, respectively.

References

Abbott M, Wu S (2002) Total factor productivity and efficiency of Australian airports. Aust Econ Rev 35:244–260

Adler N, Berechman J (2001) Measuring airport quality from the airlines viewpoint: an application of data envelopment analysis. Transp Policy 8:171–181

Airport Benchmarking Report (2010) http://www.atrsworld.org/airportawards.html

Assaf AG, Gillen D, Barros C (2012) Performance assessment of UK airports: evidence from a Bayesian dynamic frontier model. Transp Res Part E: logistics Transp Rev 48:603–615

Barros CP, Sampaio A (2004) Technical and allocative efficiency of airports. Int J Transport Econ 31:355–377

Gillen D, Lall A (1997) Developing measures of airport productivity and performance: an application of data envelopment analysis. Transp Res E 33:261–273

Gitto S, Mancuso P (2012) Bootstrapping the Malmquist indexes for Italian airports. Int J Prod Econ 135:403–411

Golany B (1988) An interactive MOLP procedure for the extension of DEA to effectiveness analysis. J Oper Res Soc 39:725–734

Lam SW, Low JMW, Tang LC (2009) Operational efficiencies across Asia Pacific airports. Transp Res Part E 45:654–665

Li XB, Reeves GR (1999) A multiple criteria approach to data envelopment anaylsis. Eur J Oper Res 115:507–517

Martin JC, Roman C (2001) An application of DEA to measure the efficiency of Spanish airports prior to privatization. JAir Transp Manag 7:149–157

Parker D (1999) The performance of the BAA before and after privatization. J Transp Econ Policy 33:133–146

Sarkis J (2000) An analysis of the operational efficiency of major airports in the United States. J Oper Manag 18:335–351

Tsai HC, Chen CM, Tzeng GH (2006) The comparative productivity efficiency for global telecoms. Int J Prod Econ 103:509–526

Yang HH (2010) Measuring the efficiencies of Asia-Pacific international airports-Parametric and non-parametric evidence. Comput Ind Eng 59:697–702

Yoshida Y, Fujimoto H (2004) Japanese-airport benchmarking with the DEA and endogenous-weight TFP methods: testing the criticism of overinvestment in Japanese regional airports. Transp Res Part E 40:533–546

Yu MM (2010) Assessment of airport performance using the SBM-NDEA model. Omega 38:440–452

A Distributed Constraint Satisfaction Approach for Supply Chain Capable-to-Promise Coordination

Yeh-Chun Juan and Jyun-Rong Syu

Abstract Order promising starts with the available-to-promise (ATP) quantities. The short is then promised by capable-to-promise (CTP) quantities. Supply chain CTP coordination can be viewed as a distributed constraint satisfaction problem (DCSP) composed of a series of constraints about slack capacity, materials and orders distributed among supply chain members. To solve this problem, supply chain members should consider and resolve their intra- and inter-constraints via supply chain coordination. This research has proposed a DCSP approach for supply chain CTP coordination. With this approach, supply chain members can collaboratively determine a feasible integral supply chain CTP production plan.

Keywords Order promising · Capable-to-promise · Available-to-promise · Supply chain coordination · Distributed constraint satisfaction problem

1 Introduction

Order promising is a critical not only for the individual manufacturing companies but also for the entire supply chains (Min and Zhou 2002; Kim 2006). It starts with the available-to-promise (ATP) quantities, the short is then promised on the basis of capable-to-promise (CTP) quantities. CTP extends ATP by taking into account the slack capacity and the available materials (Stadtler 2005).

Y.-C. Juan (✉)
Department of Industrial Engineering and Management, Ming Chi University of Technology, New Taipei, Taiwan
e-mail: ycjuan@mail.mcut.edu.tw

J.-R. Syu
Department of Industrial Engineering and Management Information, Huafan University, New Taipei, Taiwan
e-mail: zone1230@hotmail.com

Y.-K. Lin et al. (eds.), *Proceedings of the Institute of Industrial Engineers Asian Conference 2013*, DOI: 10.1007/978-981-4451-98-7_37,
© Springer Science+Business Media Singapore 2013

In supply chain environment, CTP is viewed as a constraint satisfaction problem (CSP) composed of a series of constraints about the order (quantity and due date) and the availability of capacity and materials distributed among supply chain members (Lin and Chen 2005). Most of the existing related studies assumed that the supply chain is in a shared environment and all knowledge can be gathered from supply chain members to the leading company, so the CTP quantity are always calculated with a centralized CSP approach, e.g. the advanced planning and scheduling (APS), operation research (OR) and other optimization approaches.

However, considering the cost or security/privacy for individual supply chain members, collecting information from supply chain members and calculating CTP quantity with a centralized CSP approach may become infeasible for a distributed supply chain. In such case, the supply chain CTP planning becomes a distributed CSP (DCSP). To solve this problem, supply chain members should first consider and resolve their own (intra-) CTP constraints and then communicate with their upstream and downstream supply chain members to resolve their inter-CTP constraints with a distributed supply chain CTP coordination approach. This research has proposed a DCSP approach to support supply chain CTP coordination. With this approach, supply chain members can determine whether they can satisfy a new order with the coordinated supply chain CTP quantity.

2 Distributed Constraint Satisfaction Problem

A CSP is composed of n variables, $x_1, x_2, \ldots,$ and x_n which have their own finite and discrete domain, $D_1, D_2, \ldots,$ and D_n, and a set of constraints $(p_1, p_2, \ldots, p_k)$. A variable x_i should take its value from its own domain D_i. A constraint is defined by a predicate $p_k(x_{k1}, \ldots, x_{kj})$ which is defined on the Cartesian product $D_{k1} \times \ldots \times D_{kj}$. The predicate is true if and only if the values assigned to the variables satisfy the constraint (Yokoo and Hirayama 2000). Solving a CSP is to find the values assigned to all variables such that all constraints can be satisfied.

A DCSP is a CSP in which variables and constraints are distributed among multiple companies. Yokoo and Hirayama (2000) classified the existing DCSP algorithms according to the algorithm type (backtracking, iterative improvement, and hybrid) and the number of variables that must be solved in an agent (DCSPs with a single local variable, multiple local variables, and distributed partial CSPs).

The asynchronous backtracking (ABT) is an algorithm for agents with single local variable (Yokoo et al. 1992; Yokoo et al. 1998). It statically determines the priority order of variables/agents. Each agent communicates its current value to neighboring agents via *ok?* messages. The low-priority agents will change their value to fit high-priority agents. If high-priority agents select a bad value, low-priority agents need to perform an exhaustive search to their domain and communicate a new constraint to high-priority agents via *nogood* messages.

The asynchronous weak-commitment search (AWS) algorithm developed for agents with single local variable is similar to the ABT one. It dynamically changes the priority of variables/agents so that an exhaustive search can be avoided (Yokoo 1995; Yokoo et al. 1998). Besides, agents change their value and increase its priority value by using a min-conflict heuristic approach.

The distributed breakout (DB), an algorithm for single local variable, defines a weight for each constraint and uses the summation of weights of constraint violating pairs as an evaluation value (Yokoo and Hirayama 1996). To guarantee the evaluation value is improved, neighboring agents exchange the values of possible improvements by using *ok?* and *improve* messages, and only the agent that can maximally improve the evaluation value is given the right to change its value.

The agent-ordering AWS algorithm is based on AWS algorithm and introduces the prioritization among agents to handle multiple local variables (Yokoo and Hirayama 1998). Each agent first tries to find a local solution to fit the local solution of high-priority agents. If there is no such local solution, backtracking or modifying the agent priority by using various heuristics occurs.

The variable-ordering AWS is also an algorithm extending from AWS algorithm for handling multiple local variables (Yokoo and Hirayama 1998). Each agent creates multiple virtual agents, each of which corresponds to a local variable. An agent selects a variable with the highest priority from its local variables violating the constraints with high-priority variables and modifies its value so that the constraints can be satisfied. If there is no value can fit the value of high-priority agents, similar to AWS, the agent will increase its priority value.

For an over-constrained DCSP, i.e. a distributed partial CSP, most of the existing algorithms are to find a solution with a minimal number of violated constraints. The iterative DB algorithm modifies the DB one so that an agent can detect that the number of violated constraints is less than the current target number and propagate the new target number using *improve* messages. Agents continue this iteration until the target number becomes zero (Hirayama and Yokoo 1997). The asynchronous incremental relaxation (IR) algorithm applies the ABT one to solve the distributed partial CSPs (Yokoo 1993). Agents first try to solve an original DCSP by using ABT. If the problem is found to be over-constrained, agents give up constraints that are less important than a certain threshold. Next, agents apply ABT to find a solution to the relaxed DCSP.

3 The Proposed Approach

This section illustrates the DCSP model formulation and the proposed DCSP algorithm for supply chain CTP coordination.

Figure 1 is the formulated DCSP model for supply chain CTP coordination. The supply chain members are viewed as an agent and represented by a set of identifiers, $x_1, x_2, \ldots, x_n$, from downstream to upstream. Here, assume the supply chain has four members, including the end customer (x_1), assembler (x_2), manufacturer (x_3), and supplier (x_4).

The formulated model is a common variables/constraints model for DCSPs. Each agent x_i has four sets of variables.

1. $ATPQty_{x_i}$ and $ATPFD_{x_i}$ are the quantity and finish date of the scheduled ATP that x_i can immediately respond to customer order enquires. Their domains are $D\_ATPQty_{x_i}$ and $D\_ATPFD_{x_i}$ which can be derived by calculating ATP = MPS—(the sum of the actual orders) for each time period.
2. $SCap_{x_i}$ is x_i's slack capacity used to produce x_i's finished products for supply chain CTP production plan. Its domain $(D\_SCap_{x_i})$ is x_i's own capacity and x_i's subcontract capacity.
3. $MatQty_{x_i}$ and $MatDD_{x_i}$ are the quantity and due date of materials required for producing x_i's finished products for supply chain CTP production plan. Their domains are $D\_MatQty_{x_i}$ and $D\_MatDD_{x_i}$ which are determined by the CTP production plan of x_i's suppliers.
4. The *quantity* $(CTPQty_{x_i})$ and *due date* $(CTPFD_{x_i})$ are the quantity and finish date of the planned CTP that x_i can promise the customer order enquires. Their domains are $D\_CTPQty_{x_i}$ and $D\_CTPFD_{x_i}$ which can be calculated and planned by x_i's APS system.

Based on these variables, the intra- and inter-constraints for supply chain CTP coordination can be defined as follows.

1. Each agent x_i has two sets of intra-constraints, $p_{Capacity}$ and $p_{Material}$.

   (a) $p_{Capacity}(SCap_{x_i}, CTPQty_{x_i}, CTPFD_{x_i})$ is a slack capacity constraint in which $SCap_{x_i}$ must be sufficient to fulfill $CTPQty_{x_i}$ and $CTPFD_{x_i}$.
   (b) $p_{Material}(MatQty_{x_i}, MatDD_{x_i}, CTPQty_{x_i}, CTPFD_{x_i})$ is a material constraint in which the planned $MatQty_{x_i}$ and $MatDD_{x_i}$ must support the fulfillment of $CTPQty_{x_i}$ and $CTPFD_{x_i}$.

2. For agent x_i (supplier) and its downstream agent x_{i-1} (customer), there is a constraint p_{Order} between x_i and x_{i-1}

   (a) $p_{Order}(ATPQty_{x_i}, ATPFD_{x_i}, CTPQty_{x_i}, CTPFD_{x_i}, MatQty_{x_{i-1}}, MatDD_{x_{i-1}})$ is an order constraint in which $ATPQty_{x_i}$ and $CTPQty_{x_i}$ must satisfy $MatQty_{x_{i-1}}$, i.e. $MatQty_{x_i} \leq ATPQty_{x_i} + CTPQty_{x_i}$, and $ATPFD_{x_i}$ and $CTPFD_{x_i}$ must satisfy $MatDD_{x_i}$, i.e. $MatDD_{x_{i-1}} \geq APSFD_{x_i}$ and $CTPFD_{x_i}$.

From Fig. 1, the proposed DCSP algorithm for supply chain CTP coordination must be capable of handling multiple local variables and related constraints relaxation for slack capacity, materials and order. Since APS techniques and systems can solve the constraints of operation sequences, lead times and due dates to generate a feasible production plan (Chen and Ji 2007), this research use APS systems to solve the intra-constraints ($p_{Capacity}$ and $p_{Material}$). Besides, this research will modify and integrate the concepts of agent-ordering AWS algorithm and asynchronous IR algorithm to propose a coordination algorithm shown in Fig. 2 for solving the inter-constraints p_{Order}.

A Distributed Constraint Satisfaction Approach for Supply Chain 307

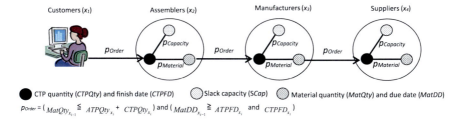

Fig. 1 DCSP model for supply chain CTP coordination

In the algorithm, agents can only communicate with their neighboring agents via *ok?* and *nogood* messages. The *ok?* message is formatted as (**ok?**, (x_i, (*Mat*, *MatQty*, *MatDD*))) which means agent x_i (customer) wishes its supplier to provide the quantity *MatQty* of product *Mat* by due date *MatDD*. The *nogood* message is formatted as (**nogood**, (x_i, (*order*, *relax_DD*, *relax_Qty*))) which means the agent x_i (supplier) cannot satisfy the original x_{i-1}'s (customer) order requirements and propose two constraint relaxation suggestions, a due-date constraint relaxation *relax_DD* and a quantity constraint relaxation *relax_Qty*. The procedures of the proposed algorithm are shown in Fig. 2 and explained as follows.

- In the initiation, the agent x_1 (end customer) proposes its demand by sending *ok?* message to agent x_2 (x_1's supplier) (Fig. 2a).
- When agent x_i receives an *ok?* message from downstream agent x_{i-1} (Fig. 2b), the order constraint p_{Order} among agent x_i and x_{i-1} is formed. If agent x_i can satisfy p_{Order} with scheduled ATP quantities, the integral CTP production plan is achieved and the CTP coordination process is terminated. Otherwise, the necessary CTP quantities are then calculated as ($MatQty_{x_{i-1}} - ATPQty_{x_i}$) and planned by APS system (Fig. 2d). APS system makes a local CTP production plan *CTP_PP* according to x_i's resources constraints ($p_{Capacity}$ and $p_{Material}$) to satisfy the *ctp* requirements for x_{i-1}'s order. If p_{Order} is satisfied and x_i has suppliers, agent x_i will propose its required material demand to agent x_{i+1} by sending *ok?* message. The order constraint p_{Order} is, therefore, further propagated upstream from agent x_i to agent x_{i+1}. If x_i doesn't have suppliers, the integral CTP production plan is achieved and the CTP coordination process is terminated. If x_i does not have enough resources for *ctp* requirements, the procedure **backtrack** is called.
- In **backtrack** procedure (Fig. 2e), APS system is called to provide two constraint relaxation suggestions, *relax_DD* and *relax_Qty*, via its simulation function. The *relax_DD* suggests to relax the originally requested due date ($MatDD_{x_{i-1}}$) to a possible earliest finish date ($CTPFD_{x_i}$) for the originally requested quantity ($MatQty_{x_{i-1}}$). The *relax_Qty* suggests to relax the originally requested quantity ($MatQty_{x_{i-1}}$) to a possible maximal finish quantity ($ATPQty_{x_i} + CTPQty_{x_i}$) before the originally requested due date ($MatDD_{x_{i-1}}$). Then, agent x_i sends a *nogood* message to inform agent x_{i-1} about constraint violations and relaxation suggestions.

when initialized do ------------------- **(a)**

   send (**ok?**, $(x_p, (Mat_{x_1} . MatQty_{x_1} , MatDD_{x_1})))$ to x_2;

end do;

when received (ok?, $(x_{i,i}, (Mat_{x_{i-1}} . MatQty_{x_{i-1}} , MatDD_{x_{i-1}}))$ **do** -------------- **(b)**

   $order \leftarrow (x_{i,i}, (Mat_{x_{i-1}} . MatQty_{x_{i-1}} , MatDD_{x_{i-1}})$

   if $(MatQty_{x_{-1}} \leq ATPQty_{x_i})$ and $(MatDD_{x_{i-1}} \geq ATPFD_{x_i})$ **then**

      broadcast to other agents that the integral CTP production plan has been achieved;

      terminate this CTP coordination process;

   end if;

   $ctp \leftarrow (x_{i,i}, (Mat_{x_{i-1}} . MatQty_{x_{i-1}} - ATPQty_{x_i} , MatDD_{x_{i-1}});$

   $CTP\_PP \cdot (SCap_{x_i} , MatQty_{x_i} , MatDD_{x_i} , CTPQty_{x_i} , CTPFD_{x_i}) \leftarrow$ **APS_planning** $(ctp);$

   if $(MatQty_{x_{i-1}} \leq ATPQty_{x_i} + CTP\_PP \cdot CTPQty_{x_i})$ and $(MatDD_{x_{i-1}} \geq ATPFD_{x_i}$ and $CTP\_PP \cdot CTPFD_{x_i})$

    then

     if x_i has supplier **then**

       send (**ok?**, $(x_r, (Mat_{x_i} , CTP\_PP \cdot MatQty_{x_i} , CTP\_PP \cdot MatDD_{x_i})))$ to x_{i+i};

     else

       broadcast to other agents that the integral CTP production plan has been achieved;

       terminate this CTP coordination process;

     end if;

    else

     backtrack;

    end if;

end do;

when received (nogood, $(x_{i,i}, (order, relax\_DD, relax\_Qty)))$ **do** --------- **(c)**

   add (**ok?**, $(x_r, (Mat_{x_i} , CTP\_PP \cdot MatQty_{x_i} , CTP\_PP \cdot MatDD_{x_i})))$ to $nogood\_list;$

   $CTP\_PP \cdot (SCap_{x_i} , MatQty_{x_i} , MatDD_{x_i} , CTPQty_{x_i} , CTPFD_{x_i}) \leftarrow$ **APS_planning** $(relaxDD);$

   if $(MatQty_{x_{i-1}} \leq ATPQty_{x_i} + CTP\_PP \cdot CTPQty_{x_i})$ and $(MatDD_{x_{i-1}} \geq ATPFD_{x_i}$ and $CTP\_PP \cdot CTPFD_{x_i})$

    then

     send (**ok?**, $(x_r, (Mat_{x_i} , CTP\_PP \cdot MatQty_{x_i} , CTP\_PP \cdot MatDD_{x_i})))$ to x_{i+i};

   else

    $CTP\_PP \cdot (SCap_{x_i} , MatQty_{x_i} , MatDD_{x_i} , CTPQty_x , CTPFD_{x_i}) \leftarrow$ **APS_planning** $(relaxQty);$

    if $(MatQty_{x_{i-1}} \leq ATPQty_{x_i} + CTP\_PP \cdot CTPQty_{x_i})$ and $(MatDD_{x_{i-1}} \geq ATPFD_{x_i}$ and $CTP\_PP \cdot CTPFD_{x_i})$

     then

      send (**ok?**, $(x_r, (Mat_{x_i} , CTP\_PP \cdot MatQty_{x_i} , CTP\_PP \cdot MatDD_{x_i})))$ to x_{i+i};

    else

     if x_i has customer **then**

      backtrack;

     else

      broadcast to other agents that there is no solution for the integral CTP production plan;

      terminate this CTP coordination process;

     end if;

    end if;

   end if;

end do;

procedure **APS_planning** ------------------ **(d)**

   Infinite Capacity Planning;

   Finite Capacity Planning;

   Detailed Scheduling;

Fig. 2 DCSP algorithm for supply chain CTP coordination

A Distributed Constraint Satisfaction Approach for Supply Chain

procedure **backtrack** ------------------- **(e)**

$(SCap_{x_i} , MatQty_{x_i} , MatDD_{x_i} , CTPQty_{x_i} = MatQty_{x_{i-1}} , CTPFD_{x_i} = \text{earliest}(MatDD_{x_{i-1}})) \leftarrow$ **APS_planning** (ctp);

$relax\_DD \leftarrow$ copy *order* and replace $MatDD_{x_{i-1}}$ with $CTPFD_{x_i}$;

$(SCap_{x_i} , MatQty_{x_i} , MatDD_{x_i} , CTPQty_{x_i} = \max(MatQty_{x_{i-1}} - ATPQty_{x_i}), CTPFD_{x_i} = MatDD_{x_{i-1}}) \leftarrow$ **APS_planning** (ctp);

$relax\_Qty \leftarrow$ copy *order* and replace $MatQty_{x_{i-1}}$ with $ATPQty_{x_i} + CTPQty_{x_i}$;

if *order* $\notin$ *nogood_sent* **then**
  send (**nogood**, $(x, (order, relax\_DD, relax\_Qty)))$ to $x_{i,j}$;
  add *order* to *nogood_sent*;
end if;

Fig. 2 continued

- When agent x_i receives a *nogood* message from its supplier (Fig. 2c), the original requirements become a new constraint and are recorded in a list *nogood_list* to avoid re-proposing. Besides, agent x_i uses APS system to evaluate the received two constraint relaxation suggestions. It first calls APS system to make a local CTP production plan according to the first relaxation suggestion *relax_DD*. If the planned quantity and finish date can satisfy x_i's customer requirements, agent x_i will accept the *relax_DD* suggestion and re-propose its material demand to agent x_{i+1}. Otherwise, it calls APS system to make a local production plan according to the second relaxation suggestion *relax_Qty*. If the planned quantity and finish date can satisfy x_i's customer requirements, agent x_i will accept the *relax_Qty* suggestion and send *ok?* message to agent x_{i+1} to re-propose the material demand. If both of *relax_DD* and *relax_Qty* are not accepted and x_i has customers, it will call procedure **backtrack** to send a *nogood* message to its customer. The *nogood* constraint is, therefore, propagated from upstream to downstream. If x_i is the end customer (x_1), it will broadcast to other agents that there is no solution for the integral CTP production plan and terminate the CTP coordination process.

4 The Completeness Analysis of the Proposed Algorithm

In the proposed algorithm, the constraint relaxation is suggested and backtracked to customer if a new p_{Order} constraint violation is found. Since the number of possible p_{Order} constraint violation is finite, the relaxation and backtracking cannot be done infinitely. Therefore, after a certain time point, the constraint violation will be diminished. Consequently, the completeness of the proposed coordination algorithm for solving the formulated DCSP is guaranteed.

5 Conclusions

Under the consideration of global manufacturing, we has formalized the supply chain CTP problem as a DCSP and proposed a coordination algorithm to solve it. The completeness of the proposed CTP coordination algorithm is also analyzed.

Acknowledgments The authors gratefully acknowledge the funding support by National Science Council, ROC through project No. NSC 99-2221-E-131-025.

References

Chen K, Ji P (2007) A mixed integer programming model for advanced planning and scheduling (APS). Eur J Oper Res 181:515–522

Hirayama K, Yokoo M (1997) Distributed partial constraint satisfaction problem. In: Proceedings of 3rd international conference on principles and practice of constraint programming, pp 222–336

Kim DS (2006) Process chain: a new paradigm of collaborative commerce and synchronized supply chain. Bus Horizons 49:359–367

Lin JT, Chen JH (2005) Enhance order promising with ATP allocation planning considering material and capacity constraints. J Chin Inst Ind Eng 22:282–292

Min H, Zhou G (2002) Supply chain modeling: past, present and future. Comput Ind Eng 43:231–249

Stadtler H (2005) Supply chain management and advanced planning: basics, overview and challenges. Eur J Oper Res 163:575–588

Yokoo M (1993) Constraint relaxation in distributed constraint satisfaction problem. In: Proceedings of 5th international conference on tools with artificial intelligence, pp 56–63

Yokoo M (1995) Asynchronous weak-commitment search for solving distributed constraint satisfaction problems. In: Proceedings of 1st international conference on principles and practice of constraint programming, pp 88–102

Yokoo M, Hirayama K (1996) Distributed breakout algorithm for solving distributed constraint satisfaction problems. In: Proceedings of 2nd international conference on Multiagent systems, pp 401–408

Yokoo M, Hirayama K (1998) Distributed constraint satisfaction algorithm for complex local problems. In: Proceedings of 3rd international conference on Multiagent systems, pp 372–379

Yokoo M, Hirayama K (2000) Algorithms for distributed constraint satisfaction: a review. Auton Agent Multi-Ag 3:185–207

Yokoo M, Durfee EH, Ishida T, Kuwabara K (1992) Distributed constraint satisfaction for formalizing distributed problem solving. In: Proceedings of 12th IEEE international conference on distributed computing systems, pp 614–621

Yokoo M, Durfee EH, Ishida T, Kuwabara K (1998) The distributed constraint satisfaction problem: formalization and algorithms. IEEE T Knowl Data En 10:673–685

Design and Selection of Plant Layout by Mean of Analytic Hierarchy Process: A Case Study of Medical Device Producer

Arthit Chaklang, Arnon Srisom and Chirakiat Saithong

Abstract The objectives of this research are to design the alternatives of plant layout and to select the most appropriate layout in which many criteria, both qualitative and quantitative criteria, are taken into account by mean of Analytic Hierarchy Process. Proximity requirements between each pair of facilities are determined by taking density of flow, harmful effect to nearby facilities and appropriateness into consideration. Furthermore, simulation approach is employed in order to evaluate the performance of quantitative criteria such as number of work in process and time in system of each alternative. The result of this research provides guideline to facility planner in order to select the most appropriate layout subject to a set of decision criteria.

Keywords Plant layout design · Plant layout selection · Analytic hierarchy process · Simulation

1 Introduction

In term of design problems, the unique answer for a given problem does not generally exist since an individual makes decision, which is selective, cumulative, and tentative, and this is contradictory to optimization problem (Simon 1975). In solving optimization problems, the derived answers are well defined and they are optimized to chosen criteria (Heragu 1997). Regarding to the characteristic of plant layout problem, plant layout problem should not be considered as neither pure design problem nor pure optimization problem (Heragu 1997). This is because there are many factors involved both qualitative factors and quantitative

---

A. Chaklang (✉) · A. Srisom · C. Saithong
Department of Industrial Engineering, Faculty of Engineering at Si Racha, Kasetsart University Si Racha Campus, Chon Buri 20230, Thailand
e-mail: sfengcrs@src.ku.ac.th

Y.-K. Lin et al. (eds.), *Proceedings of the Institute of Industrial Engineers Asian Conference 2013*, DOI: 10.1007/978-981-4451-98-7_38, © Springer Science+Business Media Singapore 2013

factors. Optimizing quantitative factors may sacrifice some quantitative factors whereas considering only qualitative factors may lead to poor efficiency of the plant (Heragu 1997). For illustration, a facility planner arranges the machines by their size; large, medium, and small machines, regardless to other factors. This causes highly flexible on allocating resources and it looks nice and tidy. However, this design would result in high cost of material handling especially in the case of high production volume (Tompkins et al 2010). It should be noted that material handling issue is considered as an integral part of plant layout design because it uses a number of working operators, utilizes more than half of floor space, and consumes majority of production time (Frazelle 1986). On the other hand, if the facility planner considers only minimizing material handling cost, the obtained design may be impractical because some important factors, which are not easy to be captured, are omitted (Heragu 1997). As a result, design of plant layout should regard not only quantitative factors but also qualitative factors as well.

In order to select the most appropriate plant layout, there are many factors involved with and the most appropriate one is selected subject to a set of decision criteria as well as the nature of business. For example, in Reconfigurable Manufacturing Systems (RMSs), quickly changing of market, not only in term of variety but also volume, results in requirement of flexibility of plant design in order to manipulate flow of material, material handling system, and other related issues as well. Thus, reconfigurable layout is a crucial factor of selection criteria. In addition, proximity requirement among facilities is also an important factor since a number of product families pass through facilities in accordance with their routing sequences (Abdi 2005). Furthermore, the other factors such as quality effect, maintainability, configuration cost, and productivity should be regarded as well (Abdi 2005; Azadeh and Izadbakhsh 2008; Ngampak and Phruksaphanrat 2011; Yang and Kuo 2003; Yanga and Shia 2002). Multi-criteria Decision Making (MCDM) should be taken employed in order to deal with the selection of plant layout process. Analytic Hierarchy Process (AHP) is a systematic approach dedicated to MCDM. It is based on pair-wise comparison and the relative scores represent how much an attribute dominate (is dominated) over attributes (Saaty 2008). Since this research considers multi-criteria in order to make decision, AHP is applied in order to select the most appropriate layout subject to potential criteria.

Regarding to the evaluation of quantitative factors, those factors must be evaluated according to the area of factors' type. Aguilar et al. (2009) considers for economic performance and net present value is used as evaluations' criteria. Yang and Kuo (2003) interests in flow density which is evaluated by the product between rectilinear distance and associated volume. Ngampak and Phruksaphanrat (2011) takes the distance moved between stations into account, however, they do not show the improved efficiency from adopting the selected layout.

In this research, a medical device producer is used as the case study to design and select plant layout. The problem is that the factory has insufficient production capacity due to the increase in demand. Thus, management decides to expand production capacity by establishing a new factory. At the existing factory, the factory arranges the machines according to the similarity of processing. This is because, according to the nature of business, there are various types of product with low in their volume. As a result, for type of layout in the new factory, the process layout should be continually employed and the following factors should be take into account; flexibility for reconfiguration, work in process, cycle time, traveled distance, and personal judgment on appropriateness.

2 Research Methodology

The following methodology, as can be seen in Fig. 1, is used in this research.

Fig. 1 Methodology for this research

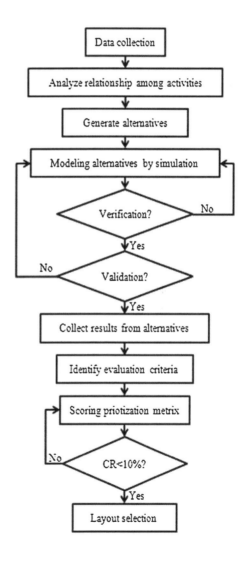

3 Analyze Activity Relationships and Construct Alternatives

After finishing data collection, the next step is to determine proximity requirement among facilities. All combinations of relationships are represented by using relationship chart, as can be seen in Fig. 2. The numerical values inside parenthesis and behind departments' name are the specified required space.

A number of alternatives are generated in accordance with activity relationship chart in Fig. 2. Then, the block diagrams of alternatives are constructed. After that, for each of block diagram, detail layout is designed. The arrangements of machines within department along with pickup and delivery point are specified. There are 15 alternatives generated from A1-A5, B1-B5, and C1-C5 and some examples of alternatives are shown in Fig. 3.

4 Modeling Alternatives by Simulation

Because of this research also considers for quantitative criteria, simulation approach is employed in order to evaluate performance quantitatively. However, there is some limitation about checking validation of the models. There lacks some

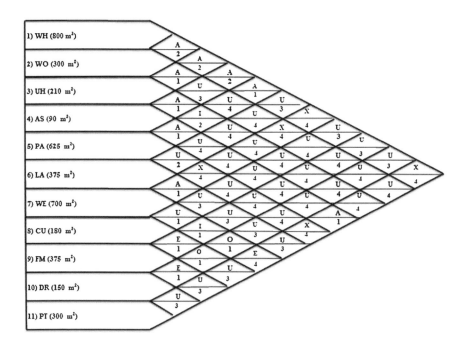

Fig. 2 Activity relationship chart

Design and Selection of Plant Layout 315

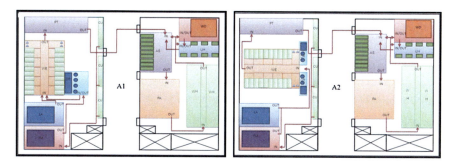

Fig. 3 Alternative layouts

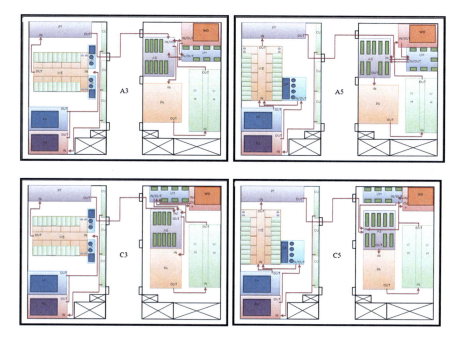

Fig. 3 continued

of important data in order to check validation of the models carefully because it deals with design a new factory. Thus, the number of produced products is used to check validation of the models only and the result shows that the error percentage is less than 3 % for all alternative models. Furthermore, the percentage of defective products from departments is assumed to be equal to the old factory. Figure 4 shows simulation model used in this research.

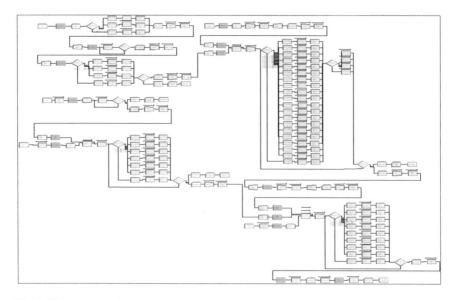

Fig. 4 Simulation model

Fig. 5 Comparison matrix

$$S_k = \begin{pmatrix} a_{11k} & a_{12k} & \ldots & a_{1jk} \\ \ldots & \ldots & a_{ijk} & \ldots \\ a_{i1k} & a_{i2k} & \ldots & a_{NPk} \end{pmatrix}$$

5 Analytic Hierarchy Process

After deriving performance of quantitative criteria from simulation, the next step is to evaluate total performance in which both quantitative and qualitative criteria are taken into account by mean of Analytic Hierarchy Process (AHP). AHP is based on pair wise comparison in which an attribute is compared to other attributes as can be seen in Fig. 5. The hierarchical structure of this problem is depicted in Fig. 6.

Where: a_{ijk} represents the relative important of attribute ith over attribute jth of individual kth.

The next step is to construct prioritization matrix, to calculate eigenvectors as well as consistency ratio (CR). The eigenvectors imply importance of attributes while CR values determine accuracy of the matrix evaluated. An example is illustrated in Fig. 7. After that, the most appropriate layout can be selected from ranking of layouts by considering all criteria as illustrated in Fig. 8.

Design and Selection of Plant Layout

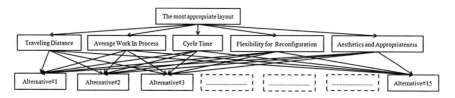

Fig. 6 Hierarchical structure

	Traveling Distance														Eigenvector
	A1	A2	A3	A4	A5	B1	B2	B3	B4	B5	C1	C2	C3	C4	C5	
A1	1	1	5	1	1	5	5	5	5	10	1	1	5	1	5	0.1215
A2	1	1	1	1	1	1	5	5	1	5	1	1	5	1	5	0.0875
A3	1/5	1	1	1/5	1	1	5	5	1	5	1/5	1/5	5	1	5	0.0655
A4	1	1	5	1	1	5	5	5	5	10	1	1	5	1	5	0.1215
A5	1	1	1	1	1	5	5	5	5	5	1	1	5	1	5	0.1025
B1	1/5	1	1	1/5	1/5	1	1	1	1	5	1/5	1/5	1	1/5	1	0.0293
B2	1/5	1/5	1/5	1/5	1/5	1	1	1	1	5	1/5	1/5	1	1/5	1	0.0222
B3	1/5	1/5	1/5	1/5	1/5	1	1	1	1	5	1/5	1/5	1	1/5	1	0.0222
B4	1/5	1	1	1/5	1/5	1	1	1	1	5	1/5	1/5	1	1/5	1	0.0293
B5	1/10	1/5	1/5	1/10	1/5	1/5	1/5	1/5	1/5	1	1/10	1/10	1/5	1/10	1/5	0.0088
C1	1	1	5	1	1	5	5	5	5	10	1	1	5	1	5	0.1215
C2	1	1	5	1	1	5	5	5	5	10	1	1	5	1	5	0.1215
C3	1/5	1/5	1/5	1/5	1/5	1	1	1	1	5	1/5	1/5	1	1/5	1	0.0222
C4	1	1	1	1	1	5	5	5	5	10	1	1	5	1	1	0.0976
C5	1/5	1/5	1/5	1/5	1/5	1	1	1	1	5	1/5	1/5	1	1	1	0.0269
	CR=6% Acceptable														

Fig. 7 An example of prioritization matrix, Eigenvector and CR value

	Criteria					Total weight	Ranking
	Distance	Ave. WIP	Cycle Time	Flexibility	Appropriateness		
A5	0.0287	0.0834	0.0043	0.0039	0.0149	0.1352	1
C4	0.0274	0.0149	0.0019	0.0131	0.0569	0.1142	2
A1	0.0341	0.0071	0.0012	0.0133	0.0536	0.1092	3
C2	0.0341	0.0440	0.0033	0.0030	0.0149	0.0993	4
C1	0.0341	0.0440	0.0021	0.0033	0.0146	0.0980	5
C3	0.0062	0.0584	0.0074	0.0030	0.0149	0.0899	6
A4	0.0341	0.0186	0.0022	0.0029	0.0149	0.0727	7
A2	0.0245	0.0054	0.0004	0.0031	0.0149	0.0483	8
C5	0.0075	0.0049	0.0004	0.0065	0.0282	0.0476	9
A3	0.0184	0.0054	0.0012	0.0030	0.0137	0.0417	10
B1	0.0082	0.0247	0.0007	0.0013	0.0053	0.0403	11
B4	0.0082	0.0149	0.0007	0.0027	0.0123	0.0389	12
B5	0.0025	0.0040	0.0002	0.0030	0.0149	0.0245	13
B3	0.0062	0.0095	0.0012	0.0010	0.0032	0.0211	14
B2	0.0062	0.0084	0.0003	0.0010	0.0032	0.0191	15

Fig. 8 Evaluation result

6 Conclusion and Recommendation

This research provides guideline to planner in order to design and select the most appropriate plant layout subject to both quantitative and qualitative criteria. Simulation approach is employed in order to assess quantitative criteria. Then, all of criteria are aggregated and the most appropriate layout is selected by mean of Analytic Hierarchy Process. In this case study, the alternative A5 is the most interesting layout. For recommendation, this research is conducted when some decisions are already made, e.g., facilities, structure, and required space. Thus, the alternatives generated are quite limited in order to explore more potential efficient designs.

Acknowledgments The authors would like to express their sincere appreciation to Faculty of Engineering at Si Racha, Kasetsart University Si Racha Campus for financial support. Lastly, this research cannot be carried out without greatly support from the case study company.

References

Abdi MR (2005) Selection of a layout configuration for reconfigurable manufacturing systems using the AHP. Paper presented at International Symposium on the Analytic Hierarchy Process 2005, Honolulu, Hawaii, 8–10 July 2005

Aguilar AA, Bautista MA, Ponsich A, Gonzalez MA (2009) An AHP-based decision-making tool for the solution of multiproduct batch plant design problem under imprecise demand. Comput Oper Res 36:711–736

Azadeh A, Izadbakhsh HR (2008) A multi-variate/multi-attribute approach for plant layout design. Int J Ind Eng 15:143–154

Frazelle EH (1986) Material handling: a technology for industrial competitiveness. Material handling research center technical report, Georgia Institute of Technology, Atlanta, Apr 1986

Ngampak N, Phruksaphanrat B (2011) cellular manufacturing layout design and selection: a case study of electronic manufacturing service plant. In: Proceedings of international multi conference of engineers and computer scientists 2011, Hong Kong, 16–18 Mar 2011

Saaty TL (2008) Decision making with the analytic hierarchy process. Int J Serv Sci 1:83–98

Simon HA (1975) Style in design. In: CM Eastman (ed) Spatial synthesis in computer-aid building design, Wiley, New York

Tompkins JA et al (2010) Facilities planning. Wiley, USA

Yang T, Kuo C (2003) A hierarchical AHP/DEA methodology for the facilities layout design problem. Eur J Oper Res 147:128–136

Yanga J, Shia P (2002) Applying analytic hierarchy process in firm's overall performance evaluation: a case study in China. Int J Bus 7:29–46

Using Taguchi Method for Coffee Cup Sleeve Design

Yiyo Kuo, Hsin-Yu Lin, Ying Chen Wu, Po-Hsi Kuo, Zhi-He Liang and Si Yong Wen

Abstract The present study aims to design the coffee cup sleeve which can detect the temperature of coffee and notify the consumers. When consumers find that the coffee is going to become cool, then they can drink the coffee quickly before it become sour. Due to the limitation of cost, thermal label is adopted to cohere on the coffee cup sleeve for detecting and notifying the important temperature. However, the temperature of coffee and the coffee cup sleeve are different which have to be taken into consideration by the thermal label. Different thermal labels are suitable for different temperature detection. Taguchi method is used for optimizing the selection of thermal labels for the importation temperature. Finally the surface of coffee cup sleeve with thermal labels on it is well designed to be a real product.

Y. Kuo (✉) · H.-Y. Lin
Department of Industrial Engineering and Management, Ming Chi University
of Technology, New Taipei, Taiwan
e-mail: yiyo@mail.mcut.edu.tw

H.-Y. Lin
e-mail: y801018@yahoo.com.tw

Y. C. Wu
Department of Visual Communication Design, Ming Chi University of Technology,
New Taipei, Taiwan
e-mail: Bohaha0725@yahoo.com.tw

P.-H. Kuo
Department of Industrial Design, Ming Chi University of Technology, New Taipei, Taiwan
e-mail: kevin8132@gmail.com

Z.-H. Liang · S. Y. Wen
Department of Business and Management, Ming Chi University of Technology,
New Taipei, Taiwan
e-mail: ktv8591@gmail.com

S. Y. Wen
e-mail: cool710186@gmail.com

Y.-K. Lin et al. (eds.), *Proceedings of the Institute of Industrial
Engineers Asian Conference 2013*, DOI: 10.1007/978-981-4451-98-7_39,
© Springer Science+Business Media Singapore 2013

320   Y. Kuo et al.

Keywords Taguchi method · Coffee cup sleeve · Product design · Thermal label

1 Introduction

Drinking coffee has become a routine behavior of many people in Taiwan. Usually consumers buy a cup of coffee on the way to office or school and then enjoy the coffee while they are working or studying. This life style can be view as enjoyment or fashion. However, after buying a cup of coffee, sometimes it is possible that consumer forget to drink the coffee due to they keep their mine on working or study. Then the coffee becomes cooler and cooler. Finally the coffee would be sour and hard to be drunk. Therefore, it would be a good ideal, if there is something that can remind consumers of the temperature of the coffee. Thus, the consumers can know the best timing to drink the coffee.

The coffee cup is the only thing that contacts the coffee. Therefore, coffee cup is the best choice for detecting the temperature of coffee. According to the survey for current product, one kind of cup made by ceramics can change the color when it detects different level of temperature. However, cup made by paper whose color can change while it detecting different level of temperature is not found. If a coffee bar sales coffee with paper cups which can remind the temperature of coffee. It would attract more consumers. However, sometimes coffee bars also provide coffee cup sleeve to keep consumers from scalding their hand. And the coffee cup sleeves would cover most part of the cup. Therefore, the present study aims to design a coffee cup sleeve which can remind the temperature of the coffee without affecting the total cost too much.

In this study, thermal labels are adopted to cohere on the coffee cup sleeve for detecting and notifying the important temperature. The thermal labels are designed for detecting certain level of temperature. There are two kind of thermal label, one is irreversible temperature indicators thermometer and the other one is reversible temperature indicators thermometer. When the temperature become higher and higher, the color of thermal labels are not changed until the temperature that they detect reach the certain level of temperature which they are designed to detect. However, when the temperature that they detect is lower than they are designed to detect, the color of irreversible temperature indicators thermometer will not return to its' original color but the reversible temperature indicators thermometer will. Ordinarily the temperature of coffee cup sleeve is room temperature. After the coffee cup fill with hot coffee, the temperature of coffee cup sleeve will become higher and higher and cause the thermal labels change color. When the temperature of coffee become lower and lower, this study expects the thermal labels will return to its' original color to remind consumer the temperatures of "not too hot" and "not too cold". Therefore, reversible temperature indicators thermometer is adopted for reminding the temperature of coffee.

Due to the coffee cup and coffee cup sleeve can reduce the detecting temperature of thermal labels. The temperature detected by thermal label is lower than the

Using Taguchi Method for Coffee Cup Sleeve Design

coffee. Moreover, there are several noise (uncontrollable) factors which can affect the detecting of the thermal label. Therefore, this study adopt Taguchi method for determining the optimal specification of thermal labels for detecting the temperatures of "not too hot" and "not too cold" of coffee.

The remainder of this article is organized a follows: a brief introduction to the case study problem is provided in Sect. 2. Section 3 shows the experimental structure can the corresponding results. Finally, Sect. 4 presents the conclusions.

2 Problem Description

This study aims to find the optimal specification of thermal labels for detecting the temperatures of "not too hot" and "not too cold" of coffee. Due to there are several factors which affect the coffee temperature detecting of thermal labels and should be taken into consideration in the experiment. These important factors are illustrated in Fig. 1 and discussed in the following.

2.1 Size of Coffee Cup

In Fig. 1, there are two size of coffee cup. Figure 1a is larger and Fig. 1b is smaller. According to the survey of market, the sleeve, cover and the bottom are all the same between larger and smaller cup. Therefore, the body of smaller cup is more oblique. If the sleeve and larger is totally touch with no gap between both of them, thus gap can be found between sleeve and smaller cup. The gap is illustrated in Fig. 1b. In this study only two sizes are considered.

2.2 Volume of Coffee

When the coffee is full of the cup, most part of cup is touched with the coffee, thus the temperature of coffee would have more opportunity transfer to the sleeve and

Fig. 1 Factors of the experiment

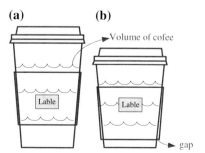

cause the temperature of the sleeve is higher. However, when there is less coffee in the cup, the opportunity of the temperature transferred to sleeve would be less. Therefore, the temperature detected by the thermal label would be lower. Thus, the volume of coffee in the cup will affect the temperature detected by the thermal label.

2.3 Location of Thermal Label

According to the gap introduced in Sect. 2.1, there is gap between small cup and cup sleeve. However, the gap is different between top and bottom of the cup sleeve. The gap is bigger in the bottom of the cup sleeve but no gap in the top of the cup sleeve. The temperature detected by thermal label would be different when the thermal label is located in the top, middle or bottom of cup sleeve. Therefore, the location that thermal labels cohered with cup sleeve could affect the detecting of coffee temperature.

2.4 Specification of Thermal Label

The specification of thermal label is the temperature that they change color. The color of thermal labels only change when the temperature that they detect pass their corresponding specification. For example, if a thermal label whose specification is 6 °C, when the temperature is rising and the temperature detected by thermal label is 6 °C, the color of the thermal label will change, vice versa.

However, the thermal label can only detect the surface temperature of cup sleeve, and the temperature between coffee and the surface temperature of cup sleeve are different. Determining the most suitable specification of thermal labels for detecting the "not too hot" and "not too cold" temperatures of coffee would be the most important job of the study.

3 Experiment

In this section, the temperature of "not too hot" and "not too cold" is introduced in Sect. 3.1. Based on the two temperatures, the experimental structure and results for optimizing the cup sleeve are provided in Sect. 3.2. Finally the optimal cup sleeve is presented in Sect. 3.3.

3.1 Target Value Determination

This study aims to optimize the cup sleeve design, when the temperature of coffee are "not too hot" and "not too cold", the thermal label can change the color to remind consumers to drink the coffee. Therefore, the temperature of "not too hot" and "not too cold" are the target values of the experiment. Lee and O'Mahony (2002) studied the preferred temperature for drinking coffee. The result show that the mean preferred temperature for drinking black coffee, coffee with creamer and sweetener, stronger black coffee and weaker black coffee are 61.5 °C (79.3–43.7 °C), 59.0 °C (73.7–44.3 °C), 59.3 °C (72.0–46.6 °C), 60.4 °C (73.5–47.3 °C) respectively. The range in the parentheses are 95 % confident interval for each group. Due to the function of the designed sleeve in this study is to remind the consumer that the temperature of coffee is accepted for drinking for wide-ranging consumer, 8 and 43 °C are chosen as the temperature of "not too hot" and "not too cold".

3.2 Experimental Structure

The factors which can affect the experiment are introduced in Sect. 2. The experimental levels for each factor are summarized in Table 1.

In Table 1, factor A and B are controllable but factor C and D are uncontrollable. However, according to the prior testing, when the volume is 4 °C %, the temperature is not easy to be detected. Therefore, only factor C is viewed as noise factors. The experimental structure generated by orthogonal array $L_8(2^7)$ with an outer orthogonal array $L_4(2^3)$ is illustrated in Table 2. The experimental results are shown in the Table 2.

3.3 Cup Sleeve Optimization

The signal-to-noise ratio (S/N ratio) is an effective way to find significant parameters by evaluating minimum variance. A higher S/N ratio means better

Table 1 Experimental levels of design factors

Factor	Description	Levels of factor			
		1	2	3	4
A	Specification of thermal label	30 °C	40 °C	50 °C	60 °C
B	Location of thermal label	Upper	Bottom		
C	Size of coffee cup	Large	Small		
D	Volume	80 %	40 %		

Table 2 Experimental structure and corresponding results

	Inner orthogonal array						Outer orthogonal array					
						C	1	2	2	1		
							1	2	1	2		
							1	1	2	2		
No.	A	B					Observations				S/N ratio	Mean
	(123)	4	5	6	7							
1	1	1	1	1	1		36	36	37	34	30.32	35.75
2	1	2	2	2	2		35	33	46	34	16.97	37.0
3	2	1	1	2	2		56	54	54	54	35.98	54.5
4	2	2	2	1	1		56	52	72	56	17.71	59.0
5	3	1	2	1	2		68	67	67	67	43.82	67.25
6	3	2	1	2	1		70	68	82	77	22.47	74.25
7	4	1	2	2	1		87	88	86	82	31.51	85.75
8	4	2	1	1	2		86	85	92[a]	92[a]	28.67	88.75

[a] the value should be higher than 92 and cannot be detected

performance for combinatorial parameters. The type of characteristic in this study is the nominal the better, Eq. (1) is used for calculating S/N ratio.

$$S/N \text{ ratio} = 10 \times \log(\bar{y}^2/s^2) \tag{1}$$

The S/N values of all treatments are shown in the last second column. It can be found that, if the thermal label is located in the top of the of cup sleeve, the S/N ratios are higher. Moreover, due to the factor A can be adjusted to approach the target values. This study selects thermal label with specification 35 °C for detecting coffee with 43 °C, and thermal label with specification 55 °C for detecting coffee with 80 °C.

4 Conclusion

This study aims to design a cup sleeve which can remind consumers the temperature state of coffee. Then the consumers can drink coffee when the temperature of coffee is preferred. Thermal labels are adopt for detecting the temperature and Taguchi method is used for optimizing the specification of thermal label and the location the it to be stuck. The design is a result of a project in a course named IDEA. It is developed by college of management and design. All sophomores in the college are divided into about 5 °C groups. The back ground of students of each group has to cover all department of the college. The course aims to Integrate students with the background of **D**esign, **E**ngineering and **A**dministration to develop a production. Therefore, the main contribution of this study is education rather than research.

Acknowledgments This work was supported, in part, by the National Science Council of Taiwan, Republic of China, under grant NSC-101-2221-E-131-043.

References

Lee HS, O'Mahony M (2002) At what temperatures do consumers like to drink coffee?: mixing method. J Food Sci 67:2774–2777

Ross PJ (1996) Taguchi techniques for quality engineering: loss function, orthogonal experiments, parameter and tolerance design, 2nd edn. McGraw-Hill, New York

Utilizing QFD and TRIZ Techniques to Design a Helmet Combined with the Wireless Camcorder

Shu-Jen Hu, Ling-Huey Su and Jhih-Hao Laio

Abstract The wireless transmission products become more important and popular in recent years because of their high efficiency and convenience. This study hopes to enhance its function and transform the product into a brand new one as well as to increase the added value of the wireless AV product substantially. In this study, a questionnaire designed with Likert scale is used to understand consumer's demand when wearing a construction site helmet which is combined with the wireless camcorder. With the application of quality function deployment (QFD), the key elements of product design can be more consistent with the voices of consumers. Next in the process of design-conducting and problem-solving, some TRIZ methods, such as selecting pairs of improving and worsening parameter, applying the contradiction matrix and 40 innovative principles, are applied to achieve a creative thinking and innovative approach for the product design. At last, the study provides an innovative design for the wireless video/audio transmission construction site helmet, in which the wide-angle lens camcorder is mounted at the front of the site helmet internally and the asymmetric ventilation holes are designed in the exterior part of the helmet. Finally, the Pro/Engineer drawing software is applied to finish the design drawing.

Keywords TRIZ · QFD · Construction site helmet · Wireless A/V transmission

---

S.-J. Hu (✉) · J.-H. Laio
Department of Industrial Management, Lunghwa University of Science and Technology, 300, Sec. 1, Wanshou Rd, Guishan, Taoyuan county 33306, Taiwan
e-mail: janicehu12@gmail.com

J.-H. Laio
e-mail: tonyha_717@yahoo.com.tw

L.-H. Su
Department of Industrial and Systems Engineering, Chung Yuan Christian University, 200 Chung Pei Rd, Chung Li 32023, Taiwan, Republic of China
e-mail: linghuey@cycu.edu.tw

Y.-K. Lin et al. (eds.), *Proceedings of the Institute of Industrial Engineers Asian Conference 2013*, DOI: 10.1007/978-981-4451-98-7_40,
© Springer Science+Business Media Singapore 2013

1 Introduction

Nowadays, advanced technology makes people pursue more convenient and efficient life as well as increasingly emphasize the importance of wireless transmission equipment and look forward to the real-time, highly efficiency wireless transmission device. However, it is found that the function and convenience of many wireless video/audio transmission products are incomplete and need to be improved. This study hopes to enhance this kind of product's function or even transform the product into a brand new one, furthermore to increase the added value of the wireless AV product substantially.

In this study, by applying the technology of quality functions deployment (abbr. as QFD) and the theory of inventive problem solving (abbreviated as TRIZ), a construction site helmet integrating of wireless transmission system for audio and video recording is designed to be valuable and practical as well as with better appearance. The objectives of this research are listed as follows:

a. Apply QFD onto the products to fit the consumer demand and improve market competitiveness of the products.
b. Apply TRIZ techniques to assist in product innovation and improvement.
c. Design a construction site helmet combined with a wireless Video/Audio transmission device and use Pro Engineering software to create 3D drawings of the designed product, and to make prototypes using a rapid prototype machine.

2 Literature Review

2.1 TRIZ

The TRIZ method was developed by Altshuller (1988), who analyzed over 400,000 patents and found that only 2 % of the patents are truly pioneering inventions, while the rest are only previous known ideas or concept added with some novel way. The invention process of TRIZ is based on the induction from summarization according to the technical information disclosed by the patent documents. Altshuller inferred that creativity can be systematic inducted and derived. TRIZ is a method to solve the problem by analyzing the problem and identifying the contradictions, then adopting different solutions according to their physical contradictions or technical contradictions. During the problem solving process, TRIZ applies some tools, including separation principle, substance-field analysis, 76 standard solutions, and ARIZ. The most famous and practical methods of TRIZ include 39 engineering parameters, a contradiction matrix, and 40 innovative TRIZ principles (Altshuller et al. 1999). TRIZ tools are used to obtain a general solution for lots of TRIZ problems, and that general solution is transformed into a solution applicable to engineering technology (Kim 2005; Li et al. 2007; Hsao 2011). It is vital for manufacturers and service companies to maintain their competitive power by utilizing TRIZ and relevant patents and continuing research and innovation on new products to differentiate themselves from competitors (Hsieh and Chen 2010; Hu et al. 2011a).

2.2 QFD

QFD was developed by Yoji Akao and Shigeru Mizuno who are Japanese quality management masters, and its main function is to ensure that the product design meets customer needs and customer desire. QFD helps transform customer needs (the voice of the customer [VOC]) into engineering characteristics for a product or service. QFD is a quality engineering management technology that involves multi-level interpretation and analysis for customer demand of the product. It transfers customer needs into product design requirements, then transfer to the characteristics of the components, and then the requirements of process design, finally transfer to the production requirements (Cheng 1996). QFD expresses the relationship matrix and assesses their important degree based on customer needs, quality characteristics, and engineering management measures. Quality Function Deployment can help identify the important quality characteristics and engineering management measures that have greatest influence on customer needs, letting the enterprise focus on the right place to ensure a good match between customer needs and real effectiveness. The QFD techniques have been applied to the field of automotive, home appliances, garments, integrated circuits, construction machinery, agricultural machinery and many other industries (Kuan 2004).

The inertia thinking mode, the depth of knowledge of the product design, and development personnel will affect the effectiveness of product innovation and R & D. And TRIZ theory can be used to assist developers in the conceptual design stage, remove psychological inertia of the engineers, and extend the field of knowledge. Systematic method of TRIZ help engineers correctly define the product and provide innovative approaches. In the part of market development, the use of quality function deployment can help businesses convert the voice of the customer to the product specifications. It is able to accurately grasp the market through QFD, which serves as a key point of making innovative products released earlier and increasing market share (Day and Chiang 2011).

You (2011) expected to find the processes of design management for the innovative product by combining QFD with innovative theoretical. In the conventional mold design process, the way of trial and error and intuitional approach is often used, so that a successful design often takes a long period of time and cost much. Product performance often depends on the designer's experience, ability and other factors, led to the design level is difficult to be controlled. Quality Function Deployment is used to convert customer needs into product features and specifications to improve the product design.

3 Methodology

3.1 Investigating the Voice of Customer

In this study, Likert-scaled questionnaire is applied to understand the requirement direction of consumers for the site safety helmet with wireless audio and video recording device. This part of the questionnaire is mainly targeting the construction industry practitioners or persons with experience using site helmets. 110 questionnaires were distributed, and 107 were valid. The questionnaire is reliable with Cronbach's Alpha value of 0.802. The results show that consumers consider important features as follows: durability, heat dissipation, overall weight (lighter the better), wearing comfortableness, camcorder screen coverage, videotaping resolution, video recording set location, as well as price.

3.2 Quality Elements Deployment

In order to identify factors affect the camcorder site helmet quality characteristics, several experts familiar with the site helmet construction are interviewed to provide the information about quality elements of this kind of site helmet. Then the table of quality elements deployment is completed as in Table 1.

3.3 Establish the Correlation Matrix of Customer Requirements and Design Requirements

The quality correlation matrix is established with the quality elements and the customer requirements in product features obtained from questionnaire analysis.

Table 1 Deployment of quality elements

	1st level	2nd level
Quality elements of the construction site helmet with wireless A/V transmitter	Material properties	Materials
		Corrosion resistance
		High temperature resistance
		Strength
	Design properties	Buckle
		Cap size
		Ventilation holes
		Construction
		Color
	Function properties	Wide-angle lens
		Pixels

Utilizing QFD and TRIZ Techniques to Design a Helmet

Design Requirements / Customer Requirements	Importance	Material properties				Design properties					Function properties			
		Materials	Corrosion resistance	High temperature resistance	Strength	Buckle	Cap size	Ventilation holes	Construction	Color	Wide-angle lens	Pixels		
High durability	5	7	7		9				5	1				
Good heat dissipation	4	5		9				9	7	5				
Lighter weight	3	9					1	1			2			
Wearing comfortable	3					9	9	1	9	1				
Wide camcorder coverage	2										9			
High videotaping resolution	2											9		
Adequate video recording allocation	2								1		7			
Low price	1	9	5	5	5		1	1	1		7	7		
Technical Importance — Absolute		91	40	41	50	27	31	43	83	28	45	25		
Technical Importance — Relative(%)		18	7.9	8.1	9.9	5.4	6.2	8.5	16	5.6	8.9	5		
Importance Ranking		A	B	B	A	C	C	B	A	C	B	C		

Fig. 1 Correlation matrix of customer requirements and design requirements for the site safety helmet

The strength of the correlation between the customer requirements and design requirements is explored in the matrix and given annotation as score 1–10. The higher the correlation, the higher the score, empty part means no correlation. Then it calculates the absolute weight and relative weight of each quality element. The whole matrix is shown in Fig. 1.

Weights of quality characteristics can be divided into three grades of A, B, and C. Wherein A is the percentage of the weighted scores which are greater than 9 %, B is from 7 to 9 %, and C is less than 7 %. In this study the quality features for the A and B grade items are taken into consideration by further using the method of TRIZ technical for contradiction removing and innovation of product improvement. They are in total 7 design elements include materials, corrosion resistance, high temperature resistance, strength, ventilation holes, the construction and wide-angle lens.

3.4 Applying TRIZ Contradiction Matrix and Inventive Principles

Further discussion of the design requirements for the seven quality features received in previous section is as following:

a. Materials: Lighter material should be used as much as possible so that consumers wearing site cap on the head will not feel the burden of heavy load.
b. Corrosion resistance: Site cap may be exposed to moisture or corrosive places, in order to prevent the site cap corrosion as well as prolong its usage life, a corrosion-resistant material is necessary.

Table 2 Summary of engineering parameters and innovation principles of this study

Problem	Improving parameters	Worsening parameters	Innovation principles
Temperature resistance	15: Durability of moving object	34: Ease of repair	10. Preliminary action
Materials	1: Weight of moving object	14: Strength	40. Composite materials
Ventilation holes	12: Shape	14: Strength	10. Preliminary action
	12: Shape	13: Stability of the object	4. Asymmetry
Construction	12: Shape	8: Volume of stationary	7. Nested doll
	12: Shape	11: Stress or pressure	10. Preliminary action
Wide-angle lens	27: Reliability	2: Weight of stationary	3. Local quality
	35: Adaptability or versatility	30: Object-affected harmful	11. Beforehand cushioning

c. High temperature resistance: When use the site cap in the field of no shade, it may cause cracking of the cap due to the strong sunshine, so that the feature of high temperature resistance is important.

d. Strength: People who work in dangerous places need to wear the site cap with high strength to protect their heads, in case some heavy objects falling from above and hit their heads.

e. Ventilation holes: The cap shell completely sealed site cap will let people feel very hot because of no ventilation, so that the site cap is better designed to set many ventilation holes to let heat dissipation.

f. Construction: Create an easy to wear and comfortable site cap.

g. Wide-angle lens: The video recording device is easily damaged, so that it should be allocated in a more secure place of the site cap, but still need to take into account its shooting angle range.

The improving and worsening parameters and their corresponding innovative solution principles are expressed in Table 2. Whereas only one of the practical innovation principles are chosen to solve the present contradiction problems and are indicated in Table 2.

4 Result

The main results of this study are analysis of the product in the following five aspects and make a structure and appearance design for this product (Hu et al. 2011b). The Pro/Engineer drawing software is applied to finish the design drawing as shown in Fig. 2.

a. The site helmet is produced with ABS plastic material because that this material has good strength and is corrosion resistance as well as high temperature resistance.

Fig. 2 Appearance of the designed helmet

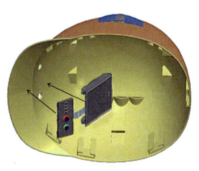

Fig. 3 Internal of the designed helmet

b. The site helmet is designed with many ventilation holes to enhance heat ventilation, while the asymmetric arrangement of the holes may increase the crashworthiness of the site helmet.
c. Design a rope net in the helmet interior, used as a compartment, to allow air circulation and be not hot. Additional sponge is set in the inside of the product to increase the space for heat dissipation, make the product more effectively ventilative with the circulation of the ventilation holes.
d. Cap with double D buckle design, this design can be based on each person's suitability to adjust its tightness.
e. Allocate the camcorder in helmets internal, can reduce collisions and increase service life. The camera is installed on the front of the site cap (as shown in Fig. 3), it will not lead to decline in range of photography shooting space.

5 Conclusion

In the present study, questionnaire survey is conducted to understand the consumer intentions and to identify consumer's expectation of combining the site cap with a wireless video/audio transmission device. Further questionnaire is used to identify consumer expectations and demands for the site cap camcorder, the eight demands with highest weighted score are then taken into QFD analysis. From the correlation matrix of customer needs and quality elements, the technical requirements for the development of the innovative product is obtained.

Next in the process of design-conducting and problem-solving, some TRIZ methods such as improving parameter and worsening parameter selection, application of contradiction matrix and 40 innovative principles are applied to achieve a creative thinking and innovative approach for the product design.

After then, the study provides an innovative design for the combination of wireless video/audio transmission device and a construction site helmet. For the design requirement of good material, corrosion resistance, high temperature resistance, and high strength, ABS material is used in the main part of the helmet. Amount of ventilation holes are added on the helmet to enhance ventilation. In addition, a thin sponge is installed inside served as a cushion to diminish the impact of falling objects. The double-D buckle design allows users to adjust the strap tightness more easily. Finally the wireless audio video AV sender transmitter is installed in the internal front part of the site cap to avoid being hit by heavy falling objects. Finally, the Pro/Engineer drawing software is applied to finish the design drawing of the helmet combined with wireless camcorder.

References

Altshuller G (1988) Creativity as an exact science. Translated by Anthony Williams. Gordon & Breach Science Publishers, New York

Altshuller G, Shulyak L, Rodman S (1999) The innovation algorithm: TRIZ, systematic innovation and technical creativity. Technical Innovation Center Inc, Worcester

Cheng CS (1996) Quality management. San Min Book Co., Taipei

Day JD, Chiang WC (2011) The research of the innovative design by QFD theories- on-road bicycle as an example. J Commercial Modernization 6(6):133–150

Hsao YC (2011) TRIZ technology for innovation (Original author: Isak Bukhman). Cubic Creativity Co, Taipei

Hsieh HT, Chen JL (2010) Using TRIZ methods in friction stir welding design. Int J Adv Manuf Technol 46:1085–1102

Hu SJ, Su LH, Lee KL, Chen JC, Chang CH (2011a) Applying TRIZ methodology to assist product improvement—take folding bicycle as an example. Key Eng Mater 450:27–30

Hu SJ, Su LH, Lee KL, Chen JC, Huang AC (2011b) Using TRIZ methods in product design and improvement of a presenter mouse. Key Eng Mater 486:13–16

Kim IC (2005) 40 principles as a problem finder. The TRIZ J, March 4. 2005. http://www.triz-journal.com/archives/2005/03/04.pdf

Kuan SP (2004) SPC continuous improvement simply. Chinese Society for Quality

Li Y, Wang J, Li X, Zhao W (2007) Design creativity in product innovation. Int J Adv Manuf Technol 33(3–4):213–222

You CS (2011) Integration of QFD, TRIZ and Taguchi method to develop disruptive innovation in bicycle sprocket process design, Master Thesis, National Chaughua U of Education. Interactive TRIZ matrix and 40 principles. http://www.triz40.com/

South East Asia Work Measurement Practices Challenges and Case Study

Thong Sze Yee, Zuraidah Mohd Zain and Bhuvenesh Rajamony

Abstract South-East-Asia has been producing wide range of products since decades ago, and hence it is often dubbed as the world's manufacturing-hub. In terms of operation, physical size and capital investment, there are family-owned businesses and Fortune 100 companies' biggest off-shore facilities co-existing. As for its workforce portfolio, the majority used to be of the kind that was non-skilled labor intensive. However, there had been workforce capability substantial upgrading, resulting in the niches and specialties development in automation. Nevertheless, the awareness of work measurement impact on productivity performance remained low. The literature shows that studies on work measurement practices in this region are very limited. It is an absolute waste if there have been tremendous improvements deployed in the machinery, systems, and tools, but they do not function to their maximum capacity because their interaction with the labor is not optimized, and if there is poor work measurement to understand the 'productivity-leak-factor' in the operation. This paper shares the literature on work measurement-related studies that are carried out in this region. It also discusses data collection and preliminary findings of the impact of work measurement method. It is found that much needs to be done to instill the appropriate awareness and understanding of work measurement.

Keywords Work measurement · Industrial engineering · Time study · Labor productivity · South East Asia manufacturing practices · Work sampling

---

T. S. Yee (✉) · Z. M. Zain · B. Rajamony
Universiti Malaysia Perlis (UniMAP), Perlis, Malaysia
e-mail: szeyee2@gmail.com; g1240510789@studentmail.unimap.edu.my

Z. M. Zain
e-mail: zuraidah@unimap.edu.my

B. Rajamony
e-mail: bhuvenesh@unimap.edu.my

Y.-K. Lin et al. (eds.), *Proceedings of the Institute of Industrial Engineers Asian Conference 2013*, DOI: 10.1007/978-981-4451-98-7_41, © Springer Science+Business Media Singapore 2013

1 Introduction

Method Study is the systematic recording and critical examination of ways of doing things in order to make improvements. It studies the steps and works for the task, identifies unnecessary/excess/non—optimized movements and eliminates ineffective time as a result of product/process shortcoming.

Work Measurement is the application of techniques designed to establish the time for a qualified worker to carry out a specific task at a defined rate of working. It can also be leveraged to compare the time consumed by different methods, and hence to pick the best methods to set challenging and yet achievable standards.

The combination of Method Study (to analyze, combine, simplify, eliminate unnecessary movement in order to make the job more economically to carry out) and Work Measurement (to evaluate, determine, eliminate ineffective time and set the standard to perform the task) work hand in hand to increase productivity by following a series of systematic steps and repeat the steps over and over again.

1.1 Introduction of Work Measurement Practices in South East Asia Industries

Research in Work Measurement started more than one century ago in the West. Among the famous ones are Frederick Taylor for his scientific management studies in the 19th century, and the Gilbreths' motion study research and development of "therbligs" that are based on the manufacturing environment. Subsequently, the concepts and models are included in university courses in Industrial Engineering (IE) during the pre-war period. Despite the fact that there were many scientific research and publications about work measurement, motion and time study in the next few decades, work measurement remained practices in the West only.

1960s was a period where the goods industry mushroomed after World War II. Over this period, the expansion of manufacturing boomed in SEA with the foreign investment. There were industrial parks established, of which many large-scale Americans' and Europeans' Original Equipment Manufacturer (OEM) set up their off-shore facilities. The entire economy was expanding, fuelled by large-scale labor workforce transformation from agriculture-base to industry-base.

During this time, foreign companies brought in their technical and scientific innovation as well as their production management systems. Work measurement was one of the production systems introduced. Some companies set up Industrial Engineering (IE) departments for work measurement and other basic IE functions. For smaller plants, the planning, production, or engineering departments were usually assigned to handle the work measurement-related functions.

1.2 Current Practices in South East Asia Industries

There had been steady and positive growth in the manufacturing sector in the past five decades in this region. As a result, there had been a substantial upgrading of human resource capability, resulting in the development of a range of niches and specialties, particularly in automation.

Nevertheless, the awareness of work measurement impact on productivity performance remains low. The literature shows that studies on work measurement practices in this region are very limited. It is an absolute waste if there have been tremendous improvements deployed in the machinery, systems and tools, but they do not function to their maximum capacity because their interaction with labor is not optimized. To comprehend further this region's work measurement practices, a survey is carried out. The findings will form part of the research on the impact of work measurement method on manufacturing productivity. The finding will enable South East Asian manufacturing industries to understand the trends and benefits of work measurement in order to set strategic moves in this direction so as to unleash maximum potential in this area.

2 Factors of Work Measurement Practices

According to *Work Measurement in Skilled Labor Environments* and the Industrial Engineering Publication *IE Terminology*, work measurement is "a generic term used to refer to the setting of a time standard by a recognized industrial engineering technique." While this definition may depict a simplistic image of work measurement, the process of determining a time standard in a complex labor setting is far from easy. Thus, the survey enlists all possible influencing factors on the work measurement practice as shown in Fig. 1.

2.1 Influencing Factors for Work Measurement Practices

There are four key influencing factors that contribute to the work measurement profile, and they are practitioner's knowledge, environmental circumstances, work measurement methods, and targets of measurement. For *practitioner's knowledge,* work measurement is about how humans interact with tools, processes, people, technologies and combinations of them. Knowledge learned in the books seldom suffix to cope with work measurement implementation. It has to be stated that a practitioner's skill set for analytical reasoning is critical to the success of the work measurement program. *Environmental* circumstances refer to the type of working environment. A wafer fabrication plant facility versus garment manufacturer impacts the workplace physical setting, work methods, capability, productivity

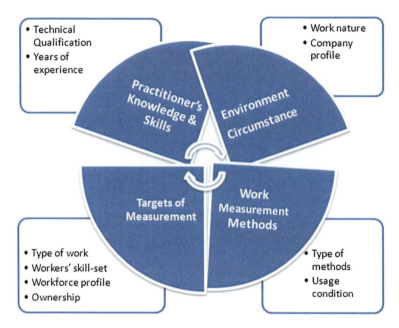

Fig. 1 Influencing factors for work measurement practices

rate setting, and workforce ratio differently, and these must certainly be taken into account. To effectively counter measure the factors, different *work measurement methods* must be applied. The availability and freedom to use a number of alternative methods is equally important. The *targets of measurement* is a set of complex raw data that must be considered in the work measurement process. This makes the analysis complicated and hence these become influencing factors.

2.2 The Approach of Data Collection

Surveys with questionnaires that cover the four influencing factors have been distributed to 400 manufacturers around the SEA countries, namely Indonesia, Malaysia, Singapore, Vietnam, Philippines, Thailand and Cambodia. There was no specific company selection criteria used. This is so as to widen the range of diversity in terms of service and product type. The questionnaires were written in English and designed in an e-survey format and Microsoft Excel for softcopy delivery, whilst hardcopies were delivered through postal service or door- to-door delivery to organizations located within the vicinity of where the study is carried out.

2.3 The Survey Response

The overall survey response rate is approximately 41 % (164/400) from all countries except Cambodia. Despite a good return rate, only 9 % (36 companies) provide complete or usable information. Out of those, 26 companies (6.5 %) report that their organizations have IE departments and confirm practicing work measurement. This percentage seems like a small population. Unfortunately, there is no track record to testify that this figure represents a healthy growth of work measurement practices in this region.

3 Case Study

The 36 companies which have completed the survey are categorized to 13 categories of industry sectors (see Table 1). 11 of them which are from the manufacturing sectors, fully or partially practice work measurement whilst the other two non-manufacturing industries from Finance and Research and Development (R&D) totally do not. There are a few different aspects observed between the groups of work measurement versus non-work measurement practice companies.

1. Work Measurement is used mainly in the manufacturing sectors with large headcount pool in workplace. Non manufacturing organizations which have small groups (<200 persons) of dedicated individuals with 100 % specialized skill who work in laboratory and office do not use work measurement.
2. Among the manufacturers, all (100 %) wafer fabrication, garment, and automobile sectors use work measurement systems in a variety of formats.
3. For small manufacturing plants ($\sim$200 workers), one or two types of work measurement systems are used. It is observed that tool types increase proportionally with worker population. Up to five types of work measurement methods are used in bigger plants.

The survey results also lead to the comprehension of how work measurement data is used. For example, despite the fact that approximately 50 % of the companies appoint Industrial Engineers to set the productivity rate (quantity of products produced within a period of time by a person) for their companies, only about 24 % of the companies use IE or work measurement method. Instead, the companies prefer to refer to "past records" and "estimations" as their main or cross reference method. In fact, all the manufacturing companies in this survey use more than one method for work measurement to set productivity rate.

Table 1 Comparison of company profile on work measurement practices

Company's work measurement practices	Environment	Workers' quantity range	Workers' skill set profile
Electronics manufacturer			
Wafer plant			
LED manufacturer			
Box-build assembly plant			
Heavy tooling manufacturer			
Petroleum			
Garment manufacturer			
Chemical production			
Automobile manufacturer			
Medical device manufacturer			
Finance organization			
R&D organization			
Food processing industry			
Legend			
All companies in this category practice work measurement	Manufacturing facilities	1,000 persons and above	50 % or more non skilled work
Some companies in this category do not practice or partially practice work measurement	Special type of work	Mixture of "1–200" and "1,000 and more"	Mixture of skilled and non skilled work
100 % companies in this category do not practice work measurement	Non manufacturing work	1–200 persons	100 % full skill or specialized work skill required

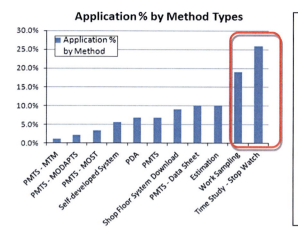

Fig. 2 Types of work measurement methods and percentage of use. *Note* MOST—Maynard Operation Sequence Technique; MODAPTS—Modular Arrangement of Predetermined Time Standards; PMTS—Predetermined motion time system; MTM—Methods Time Measurement

3.1 The Popular Work Measurement Method

The survey shows that work measurement using time study-stopwatch is the most popular method used (Fig. 2). Work measurement practitioners who use other methods (non 'time study—stopwatch') also rate 'time study—stopwatch' as the most preferred method (see Table 2 for details).

3.2 Relationship of Favorable Methods

There is no absolute conclusion drawn to the reason why time study-stopwatch is the favorable method, but the possible contributing factors could be:

- About 40 % of the work measurement practitioners pose Bachelor or Master Degree in IE, whilst 24 % learned work measurement-related subjects through internal company training programs.
- The top three work measurement courses that the practitioners have taken are, in order of popularity, Stopwatch, Work Sampling and PMTS.
- Majority of the companies use Time Study–stopwatch and the practitioners have two to five years of hands on experience using this method. Moreover, they work in small teams of five to ten persons. This is a good learning environment to continuously polish the work measurement skills.

The findings indicate that preference is driven from the users' academic background and the peers' interaction, which provides the technical support.

Table 2 Comparison of preferred work measurement method rated by users

Category of preference	Type of systems	Group of work measurement users who rate this	Remarks
Least preferred	MTM	Time study-stop watch, PTS in general, MOST	
Consider good	Time study—stop watch	Time study-stop watch, PTS in general, MOST, MODAPTS, estimation, PDA, self-developed systems	Stop watch is rated the most popular method among companies that use stop watch, work sampling, PTS, MOST, and all other methods listed
Best method	(1) Time study—stop watch (2) Work sampling	Time study-stop watch, PTS in general, MOST, MODAPTS, estimation, PDA, self-developed systems	Time study—stop watch remains as the popular one in the category of "best method". However, work sampling is also popular followed by that

Table 3 Influencing factors rating of switching work measurement type

Factor(s) and rate of influence level	The most	The least
Company's preference or standard	14	3
Follow majority staff's expertise	0	1
No or lack of certification	0	9
Constraint in budget, time or resources	2	3
Licensing and consultation fees is too high	1	7
Too many changes to manage to switch method	2	1
Others	1	0

3.3 Targets of Measurement

As discussed in Sect. 3.1, most companies use more than one work measurement method as the primary and/or cross reference/supporting function. Among the survey participants, only garment manufacturers are completely satisfied with their current method -GSD (A type of PMTS called General Sewing Data). The survey results also show that it is not easy to switch to other methods. For big corporation, there is usually a company standard that controls/influences the decision. On the other hand, small firms are most concerned about cost and the availability of resources to manage the transition. The least influencing factor is "certification" and this is probably due to the fact that the survey target population is corporate instead of individually-owned, whereby the companies are likely to afford paying for the training and certification (Table 3).

4 Conclusions

The research, at this stage, uncovers some fundamental work measurement facts in this region. It is a reflection of the current phenomenon. Despite the importance of determining productivity from an academic stand point, it is not a common practice among the industries. Even though there are many established work measurement tools available, the preferred and frequently used method narrowed down to the more traditional methods, namely, time study through stopwatch and work sampling, based on the practitioners' technical training, on-the-job experience and existing company practice. Except for the garment industry, which has already found the best fit method, others in the survey are neither satisfied nor unsatisfied with the present method. Switching of work measurement method is highly driven by the company rather than the individual's preference, technical relevance, or needs. There is no conclusion possibly drawn to the selection of method is based on the four key influencing factors as discussed in Sect. 2.1. It is also not conclusive that the presently-used methods contribute effectively to improvise manufacturing productivity in this region.

5 Future Work

The next phase of research will concentrate on the correlation factors of the choice of the labor work/motion study and work measurement methods based on actual practice in the workplace. A comparison between the productivity levels versus the work study methods applications will be carried out based on on-site observations.

Acknowledgments The authors would like to thank the companies which have completed the surveys and acknowledge the helpful comments. Intel Penang, B. Braun Medical Industries, Intel Kulim, Fairchild, B. Braun Vietnam, Agilent Penang, Philips Lumileds Penang, Osram Penang, Faeth Asia Pacific, Plexus Penang, Brady Penang, Smart Modular Technologies (M), Silterra (M), ST Microelectronics Singapore, Flextronics Senai, Talisman (M), Renesas Semiconductor (M), Sugihara Grand Industries, Pen Apparel, Flextronics Prai, Motorola Solutions (M), First Solar (M), Honeywell Aerospace Avionics, Square, Nokia, Hitachi Chemical, Perusahaan Otomobil Nasional, Pouchen Group, Flextronics Singapore, Tri Quint, KCK Garment, AUO Sunpower, Delphi Automotive System, AMD (M), CPF Public Company, Central Almenario de Tarlac and ProPlanner Asia.

References

Time and Motion Study, Wikipedia http://en.wikipedia.org/wiki/Time_and_motion_study
Tom Best, Work Measurement in Skilled Labor Environment. http://www.iienet2.org/uploaded Files/SHS_Community/Resources/Work%20Measurement%20in%20Skilled%20Labor%20 Environments.pdf

Decision Support System: Real-Time Dispatch of Manufacturing Processes

Chung-Wei Kan and An-Pin Chen

Abstract This research has highlighted the role of real-time dispatching (RTD) tools in the development of 300 mm manufacturing machinery systems. Dispatching production and distribution in the real-world 300 mm manufacturing environment is an extremely complex task requiring the consideration of numerous constraints and objectives. Decision support system (DSS) created for this purpose can potentially be used to provide support for related tasks, such as real-time optimization, operational planning, quality certificated, service and maintenance. The DSS comprises the ability to reinforce the RTD system which support both process operator and manager in the decision making process, allowing them to take full-scale of the physical system to implement it in a way where the optimized process control variables are under statistical control, resulting in optimized output that, in turn, secure higher productivity and improved quality.

Keywords Manufacturing systems · Industrial and systems engineering · Decision support system · Real-time dispatching

1 Introduction

In today's highly automated 300 mm fabs, operators would not make or validate scheduling decisions. Instead, real-time dispatching (RTD) system will decide what to do next, where the necessary materials are, and when they will arrive. Nor

---

C.-W. Kan (✉) · A.-P. Chen
Institute of Information Management, National Chiao Tung University,
1001 University Road, Hsinchu 300, Taiwan, Republic of China
e-mail: russntu@gmail.com

A.-P. Chen
e-mail: apc888888@gmail.com

Y.-K. Lin et al. (eds.), *Proceedings of the Institute of Industrial Engineers Asian Conference 2013*, DOI: 10.1007/978-981-4451-98-7_42,
© Springer Science+Business Media Singapore 2013

will operators move materials manually; instead, overhead hoist vehicles (OHVs), and rail-guided vehicles (RGVs) will transport materials and products as needed, with little or no human intervention.

This level of manufacturing sophistication can be archived only through the complete integration of other essential systems, such as: automated material handling system (AMHS), material control system (MCS), manufacturing execution system (MES) and real-time dispatching system.

Manufacturing system evaluation, design and selection issues have developed considerably over the part 3–4 decades. On the dispatching side, emphasis has shifted from dispatching conventional manufacturing systems to dispatching automated systems due to uncertainly in product demand and day to day technological developments. The spirit of these attempts is to justify the installation of RTD in phased manner or total.

2 Real Time Dispatching System

The dispatching system was implemented using the Real Time Dispatcher (RTD) application from Brooks Automation Inc. RTD accesses the status of the shop floor via the RTD repository and executes scheduling rules whenever signaled to do so by the MES. RTD performs the following functions:

- Allows detailed criteria to be used for dispatching lots.
- Uses the current state of the RTD repository to dispatch.
- Displays dispatch list in the fab (Gurshaman and Hoa 2001).

The dispatching application requires the detailed routing information for every product, which can be processed in the fab including the list of qualified tools for each operation, and theoretical and planned cycle times for each of the operation.

In addition, the fab produces low volume products. Rules were developed to support the manufacturing philosophy. RTD provides a critical link in the value chain by linking the manufacturing management directives and philosophy directly to the manufacturing floor.

2.1 Dispatching Architecture

It is very critical that the right information is displayed to the right people and that the information can be easily accessed. The dispatch architecture takes into account the information needs of all stakeholders and ensures that the right levels of detail are available to them. The stakeholders included operators of single tools and group of tools, manufacturing supervisors for different processing areas, manufacturing shift coordinators production planners, and fab management. In addition to the list of lots waiting for processing at a given tool, various other views of the ranked list of lots are available within the RTD system to meet the needs of different stakeholders. As an illustration, a ranked list of lots on hold is available by processing areas for the manufacturing supervisors. Another view includes a list of lots which cannot be processed or dispatched due to setup issues. These views facilitate managing the WIP which cannot be currently processed in the fab.

2.2 The Principle of "Real-Time" in Manufacturing

What is the real meaning of "real-time" in the context of manufacturing and operational processes? In a strict sense, real-time refers to application that have a time critical nature. A real-time process must perform a task in a determined length of time. For example, a typical "normal" program may be considered to perform correctly if it behaves as expected and gives the right answer. This means that data must be read and processed before the shaft rotates another revolution, otherwise the sampling rate will be compromised and inaccurate calculation may result.

A real-time program is considered to be correct only if it gives the right answer within a specified deadline. In other words, "real-time" adds a time constraint to the notion of a program being correct. The phrase "real-time" does not directly relate to how fast the program responds, even though many people believe that "real-time" means "real fast". This is a direct fall-out from the fact that it is easier to meet deadlines with a fast system. However, many operating systems now run on powerful hardware and are "fast", but that speed does not necessarily imply "determinism". Determinism is more important than average speed for real-time systems.

The research aims at developing an autonomous decision support system (DSS) for real-time dispatch of manufacturing systems, and this system is planned to be an integral part of the currently executed RTD system at all famous foundry companies, such as Taiwan Semiconductor Manufacturing Company Ltd. (TSMC), Powerchip Semiconductor Corp. (PSC) and others. The principle manifest in the development of the DSS, described in general terms below, and then designed tailored according to the requirements for practical control of the real-time dispatching process.

3 Decision Support System

The concept of a decision support system (DSS) is extremely broad and its definitions vary depending upon the author's point of view (Druzdzel and Flynn 1999). A DSS can take many different forms and the term can be used in many different ways (Alter 1980). On the one hand, Finlay (1994) and others define a DSS broadly as "a computer-based system that aids the process of decision making." In a more precise way, Turban (1995) defines it as "an interactive, flexible, and adaptable computer-based information system, especially developed for supporting the solution of a non-structured management problem for improved decision making. It utilizes data, provides an easy-to-use interface, and allows for the decision maker's own insights."

Other definitions fill the gap between these two extremes. For Keen and Scott Morton (1978), DSS couple the intellectual resources of individuals with the capabilities of the computer to improve the quality of decisions ("DSS are computer-based support for management decision makers who are dealing with semi-structured problems"). For Sprague and Carlson (1982), DSS are "interactive computer-based systems that help decision makers utilize data and models to solves unstructured problems."

On the other hand, Keen (1980) claims that it is impossible to give a precise definition including all the facets of DSS ("there can be no definition of decision support systems, only of decision support"). Nevertheless, according to Power (1977), the term decision support system remains a useful and inclusive term for many types of information systems that support decision making.

3.1 Design of the DSS

The decision support system (DSS) allows the application sponsors in the design, implementation and use of computerized systems that support business managers in the decision-making process. A DSS is a computer system that typically encompasses mathematical models as well as informational databases and a user interface in order to provide recommended decisions to manager-users. A DSS differs from a traditional information system (IS) or management information system (MIS) in that it not only provides the user with information or database as does an IS or MIS, but it also provides answers to user queries, i.e., decisions, through its modeling component. In essence a DSS is a computer system that helps managers makes decisions.

Users in the DSS option take a variety of specialized utilities in the information technology field that enhances the ability of their current developed computer systems. A completed set of DSS includes information systems, database management, networks and telecommunications, decision support system development and implementation, visual interface design, artificial intelligence, client/server systems, object-oriented programming, the internet, and simulation as well as various mathematical modeling techniques.

3.2 The DSS System Architecture of RTD

Sprague (1982) pointed out that a listing of the characteristic of DSS is more useful than definitions—the main features of DSS include these:

- Aimed at underspecified or less
- Well-structured problems
- Use models and analytic techniques as well as traditional data access and retrieval functions.
- Interactive and user-friendly
- Flexible and adaptive to environmental changes and accommodate the decision-making approach of the user.

Since part of the required system is already in place and some more are either in the process of upgrading or procurement, while the rest still to be designed, the overall design approach manifests in a combination of the above known approaches. Before we start building the DSS, it is essential to understand the detailed data flow, such as where information generated, how it is transferred to the operations managers or fab operators and how it is intended to incorporate the information in the decision making and control of the process. To simplify the operation of the DSS to be designed for the real-time dispatching system, the requirements for first, a real-time dispatch (RTD) system, second, the system approach of the DSS and third, control of the implemented RTD process, are detailed in the individual sections to follow.

3.3 DSS: An Alias for RTD

In the previous two sections, a brief review of the definitions, scopes and introduction of RTD and DSS were presented. Before defining the criteria by which RTD and DSS will be affiliated, a note about the RTD definitions is in order.

The RTD is a client and sever process. Dispatch client request dispatch list from the RTD, a server process on RTD system. At the meanwhile, the DSS should replicate with the same request coming from RTD as a secondary server process. These internal data comes from the RTD's micro-command calculation, the DSS will absorb them as a reference to its data generation. The DSS has to be an ongoing, dynamic system that continuously updates itself. Based on the information contained in the RTD, the solution-result from "what-if" analysis generated from DSS, the operators and operation manager are able to make decisions regarding process performance or improvement.

Both RTD and DSS are ultimately systems, hence making the system approach a suitable one. The basic premise of the system approach is that systems, regardless of their specific context, share a common set of elements, (Churchman 1998). The DSS has to assist the operators in making decisions, based on the real-time data (of the control variable) received from the RTD, while dispatching. It continuously monitors the process by comparing the real-time data with the corresponding benchmark data in the production configuration database, and alarms the operators when any control variable goes beyond the control limits. An additional aspect of the DSS is the inter-functional interaction between MES and RTD of the manufacturing systems.

DSS is an integrated decision support system with focus on the system-wide issues rather than being an individual or cloistered backup tool. It supports manufacturing decision activities by integrating model base with RTD database (Repository) and communication components (Adapter, Writer and Monitor). Originally, DSS was deemed only appropriate for clerical processing, i.e., for tasks that are well structured. However, the DSS proposed in this paper builds upon the industrial project and continue research. All the management levels (operational, tactical, and strategic) are targeted. In DSS, the strategic and tactical management level (top levels) were originally the primary targets, but more recent research suggest and provide evidence that DSS is appropriate at any level.

RTD renders most support in dispatch phase, and to a lesser extent in the intelligent and monitor phase. Exception reports (log files) generated by RTD itself give clues to existing problems thus aiding the intelligence phase. DSS too provides best support for the monitor phase. It is, however, appropriate for problem formulation and problem diagnosis to be collected and analyzed on the same system. Moreover, for DSS with the similar hardware specification towards RTD, it would not only for controlling the optimized process but a concomitant when RTD server is down.

Besides the ordinary RTD coupling to the MES database and MES Application, the DSS is connected hard-wire to the process control factors, i.e., only necessary data being collected for DSS to monitors the current activity and performance levels, anticipates future activity and forecasts the resources needed to provide desired levels of service and productivity. This control component of the real-time dispatch process is indicated by the secondary right block in Fig. 1. The DSS, a sub-system of RTD on the independent server, is clearly shown, indicating the information flow and the relationship MES and RTD.

4 Discussion

Research has indicated the importance of matching IT application or manufacturing systems with the competitive strategy of each company. Justification and implementation of advanced manufacturing technology involved decisions that are crucial for the practitioner regarding the survival of business in the present day uncertainties in the manufacturing world. Since advance systems require huge capital investment and offer large number of intangible benefits such as flexibility, quality, competitiveness, customer satisfaction etc.

Decision Support System: Real-Time Dispatch of Manufacturing Processes 351

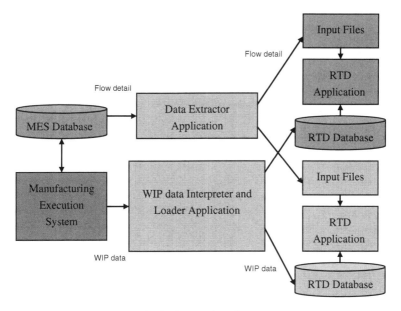

Fig. 1 A decision support system for RTD manufacturing process

In such a manner, this system will facilitate effective decision making in selecting appropriate applications that best match an organization's manufacturing strategy. The DSS may offer the user company the competitive edge it seeks. Furthermore, operators and operation managers can get authentic advice and alarm from the DSS in a timely, cost-effective manner.

5 Conclusions

The long-term goal for the research presented above is to develop a semi-autonomous decision support system. In most cases, decision making is a "man-in-the-loop" activity requiring adequate support. However, the DSS monitors the process control factors and compares the captured data on a real-time basis to against the production configuration database. Probably the most significant advantage of the "real-time" mode of DSS, is that the complex operating instructions and procedures are coded instead of having operators perform activities manually.

This prosperous implementation of the decision support system can be successfully expended to other areas of the manufacturing environment. Though the extended use of DSS, offers a powerful tool for decision support in the runtime phase of manufacturing system, decisions regarding which simulations to run, identification of abnormal conditions and so on, remains a human task. However, though the use of soft computing techniques such as artificial intelligence (AI) and genetic algorithms, these tasks can be supported as well, which will also be addressed in the future works.

References

Alter SL (1980) Decision support systems: current practice and continuing challenges. Addison-Wesley, Reading

Churcham CW (1998) The systems approach. Dell, New York

Druzdzel MJ, Flynn RR (1999) Decision support systems. Encyclopedia of library and information science. A Kent, Marcel Dekker, Inc

Finlay PN (1994) Introducing decision support systems. NCC Blackwell; Blackwell Publishers, Oxford, UK, Cambridge, Mass

Gurshaman SB, Hoa TL (2001) Real-time lot dispatching for semiconductor Fab, future Fab Intl. vol 11

Keen PGW, Scott Morton MS (1978) Decision support systems: an organizational perspective. Addision-Wesley, Reading

Keen PGW (1980) Decision support systems: a research perspective. Decision support systems: issue and challenges. In: Fick G, Sprague RH Pergamon Press, Oxford, New York

Power DJ (1977) What is a DSS? The on- line executive journal for data-intensive decision support 1(3)

Sprague RH (1982) A framework for the development of decision support system. MIS Q 4(4)

Sprague RH, Carlson ED (1982) Building effective decision support systems. Englewood Cliffs, Prentice-Hall

Turban E (1995) Decision support and expert systems: management support systems. Englewoood Cliffs, Prentice Hall. M. Young, The technical writer's handbook. University Science, Mill Valley, CA

The Application of MFCA Analysis in Process Improvement: A Case Study of Plastics Packaging Factory in Thailand

Chompoonoot Kasemset, Suchon Sasiopars and Sugun Suwiphat

Abstract This research aims to apply the Material Flow Cost Accounting (MFCA) for process improvement of the target product, 950 cc. plastic water bottles, a case study company in Thailand. The production line of this product consists of five processes, crushing, mixing, blow molding, printing, and packing. The data collection was carried out for all processes and analyzed based on MFCA procedure. During the process of MFCA, quantity of input and output material, material cost, system cost and energy cost were presented. Then, the cost of positive and negative products can be distinguished based on mass balancing for all processes. The results from MFCA calculation showed that the highest negative product cost occurred at blow molding process. Then, the operations flow at blow molding process was analyzed using motion study and ECRS concept in order to eliminate production defects. Finally, the improvement solution was proposed and the results showed that the defects were reduced 26.07 % from previous negative product cost.

Keywords Material flow cost accounting (MFCA) · Positive product · Negative product · Cost reduction

1 Introduction

The concept of manufacturing process improvement commonly concentrates on lead time reduction, waste or defect decreasing, and others which lead to increase productivity of any production line without interpreting the improvement in term

---

C. Kasemset (✉) · S. Sasiopars · S. Suwiphat
Department of Industrial Engineering, Faculty of Engineering, Chiang Mai University, Chiang Mai, Thailand
e-mail: chompook@gmail.com; chompoonoot.kasemset@cmu.ac.th

Y.-K. Lin et al. (eds.), *Proceedings of the Institute of Industrial Engineers Asian Conference 2013*, DOI: 10.1007/978-981-4451-98-7_43,
© Springer Science+Business Media Singapore 2013

of cost or monetary terms that is sometimes difficult for management persons to understanding the improvement results.

Material Flow Cost Accounting (MFCA) is developed to help organizations to better understand the effects of environment and finance of their used material and energy, and seek opportunities to gain both environmental and financial improvements.

This research paper aims to present the application of MFCA in manufacturing process improvement of one plastics packaging factory in Thailand.

2 Preliminaries

2.1 Material Flow Cost Accounting

MFCA is one of the environmental management accounting methods aimed to reduce both environmental impacts and costs. MFCA seeks to reduce costs through waste reduction, thereby improving business productivity. The detail of MFCA is addressed in many sources and also published as international standard ISO 14051:2011 as well.

The difference between MFCA and traditional cost accounting was presented in Nakajima (2004). Based on traditional cost accounting, the total production cost is put to products without considering waste production cost (i.e., defective parts, material losses, etc.). Product price is normally set from combination of total cost and profit. As long as companies satisfy with their profit, they will not care about how much of their losses in the production system.

Based on MFCA concept, cost can be classified as; material cost (cost of both direct and indirect material), system cost (cost of operating production system), energy cost (cost of energy used in production system) and waste management cost (cost of waste treatments). From those costs, each cost can be divided as positive and negative costs. Positive or product cost is the cost attached with the output of each process. Negative or material loss cost is the cost of loss from each process. Cost allocation between positive and negative cost is carried out based on the portion from material balancing between input and output materials. Finally, the operations with high negative cost are identified and improved in order to reduce negative cost.

The detail of MFCA implementation steps is explained in "Guide for Material Flow Cost Accounting" by Ministry of Economy, Trade and Industry, Japan (2011). There are seven steps as (1) preparation, (2) data collection and compilation, (3) MFCA calculation, (4) indentifying improvement requirement, (5) formulating improvement plans, (6) implementing Improvement and (7) evaluating improvement effects by re-calculating MFCA.

Many case studies in Japan applied MFCA for improving their production systems. The detail can be found in MFCA case examples 2011 published by Ministry of Economy, Trade and Industry, Japan.

For Thailand, MFCA is not widely known by Thai manufacturers. Some early MFCA applications in SMEs were presented in 2013. Kasemset et al. (2013) applied MFCA to reduced negative material cost in one small textile factory as a case study. Chompu-inwai et al. (2013) proposed to use MFCA to analyze the production of one type of wood product and applied the concept of design of experiment (DOE) to determine optimal parameters for wood cutting process in order to reduce defective wood sheets. Laosiritaworn et al. (2013) applied MFCA to analyze lost-wax casting process and proposed to recycle some material waste in order to reduce the cost of indirect materials. Another advanced study of MFCA was addressed in Chattinnawat (2013) when MFCA was combined with the concept of dynamic programming in order to identify the improvement plan that is economical when considering both cost of improvements and the benefit from increasing positive product cost or reducing negative product cost at the same time.

From those research works of MFCA application, MFCA is the effective tools in identifying the critical point that should be improved in the production system. Moreover, when MFCA is applied, the interpretation of the improvement in term of monetary is attractive and easy to be understood by management peoples as well.

2.2 Motion Study and ECRS

The purpose of motion study is to find the greatest economy of effort with due regard for safety and the human aspect. Through the use of motion study, the job can be broken down into steps, and each step can be analyzed to see if it is being done in the simplest, easiest, and safest possible manner. The improved systems based on motion study employed less number of workers while maintaining the same or extra amount of throughput.

ECRS is one of motion study technique used to improve production lines. E is to eliminate unnecessary work. C is to combine operations. R is to rearrange sequence of operations. Finally, S is to simplify the necessary operations.

Recently, there are many research works adopted ECRS as a tool for operations improvement. Lan (2010) applied ECRS to improve hands operations of electric motor assembly. Miranda (2011) proposed to adopt ECRS in increasing man efficiency of the clean room assembly process that can help in manpower cost reduction for one electronics manufacturer in Philippines. Wajanawichakon and Srimitee (2012) applied ECRS in drinking water production plant to increase the productivity of the case study factory by reducing the cycle time. Sindhuja et al. (2012) also applied ECRS to improve horn assembly line at the bottleneck process to increase the production rate of this assembly line.

Those research works of ECRS show that ECRS is the effective tool for manufacturing cycle time reduction as giving effect on production cost reduction as well.

3 Case Study

One plastics packaging factory in Chiang Mai, Thailand, was selected to be a case study. Main products of this factory are plastics baskets and bottles. To implement MFCA procedure (as the detail in Sect. 2.1), target product and process should be selected at firstly. Then, the data collection can be carried out. The detail of data collection can be addressed as follows.

The 950 cc. plastics bottle is selected to be the target product of this study (shown in Fig. 1). The production process of this product is shown in Table 1.

The data collection was carried out at this product/process. Data collected were all input materials and costs, all machines in this process, operating cost and labor cost and they were used in MFCA calculation.

4 Results

In this section, the results of MFCA implementation to the case study were presented as follows.

Fig. 1 Target product 950 cc. bottle

The Application of MFCA Analysis in Process Improvement

Table 1 Target process and details

Input	Process	Output
Recycled plastics from defective/ waste products	Plastics Crushing	Recycled plastics granules
New plastics granules and Recycled plastics granules	Mixing	Mixture plastics granules
Mixture plastics granules	Blow Molding	Plastics bottles
Printing Toner	Printing	Plastics bottles with label
Plastics bottles with label and plastics bags	Packing	Pack of plastics bottles

4.1 Material Flow Model

Figure 2 showed the material flow model of the target process. Wastes can be found only at blow molding and printing processes.

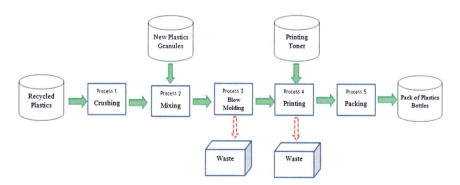

Fig. 2 Material flow model

4.2 Cost Allocation

Cost allocation of all processes is shown in Table 2.

There are only two processes generating wastes, blow molding and printing processes, and the results from cost calculation and allocation showed that blow molding process has the highest negative product cost and the largest portion of negative cost is material cost (shown in Table 3).

4.3 Identifying Improvement Requirement

At blow molding process, there are two types of material waste that are (1) defective bottles (5.54 %) and (2) normal waste from head and bottom cutting method (94.46 %) (shown in Fig. 3a, b).

Normal waste is the major of material waste at this process but the only way to reduce this waste is to invest new machine that need more technical detail in machine specification. Thus, only defective products are studied to find their root-cause using 7-QC tool. Major defective bottle is bottle with out-of-spec thickness that is 52.5 % from all defectives, so the cause-effect diagram is used to find the root cause of this defect (shown in Fig. 4).

The cause from inappropriate working method is basic way to improve, so motion study and ECRS were used to design new working procedure for reducing wastes from inappropriate working method.

Table 2 Cost allocation of all processes (in Thai Baht)

Process	Positive product (%)	Negative product (%)	MC	SC	EC
Crushing	100	0	160,569	4,286	3,823
Mixing	100	0	982,800	5,811.75	3,524
Blow molding	66.03	33.97	895,488	182,598	113,625
Printing	99.82	0.18	540,068	15,224	9,668.99
Packing	100	0	570,717.90	10,040	0

Table 3 Cost allocation of blow molding process (in Thai Baht)

	MC	SC	EC	WC	Total
Positive product	591,318.00	120,576.58	75,031.02	0.00	786,925.60
	49.62 %	10.12 %	6.30 %	0.00 %	66.03 %
Negative product	304,170.00	62,021.42	38,593.98	0.00	404,785.40
	25.52 %	5.20 %	3.24 %	0.00 %	33.97 %
Total	895,488.00	182,598.00	113,625.00	0.00	1,191,711.00
	75.14 %	15.32 %	9.53 %	0.00 %	

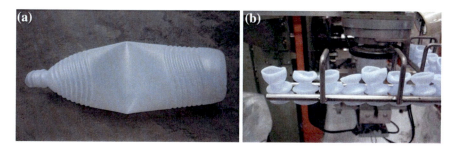

Fig. 3 **a** Defective bottle **b** Normal waste

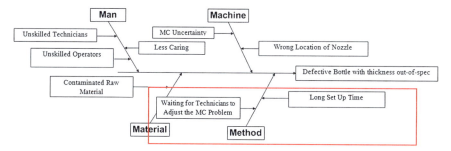

Fig. 4 Cause-effect diagram

Currently, when the operator at this process detects out-of spec product, he/she will walk to maintenance section and ask some technician to correct the problem without stopping machine so if he/she spend more time to find any technician, the more defective is continually produced. The solution procedures are (1) training operators to stop machine by him/herself, (2) introducing fast communication among operators and technician using radio communication devices and (3) reducing set up time by investing new digital weight scales. The first improvement solution is developed based on R as rearranging operations step of work and the second and third methods are developed based on E as eliminating unnecessary movements.

4.4 Evaluating Improvement Effects

After three solutions were implemented, MFCA calculation is carried out again to see how improvement in waste reduction as shown in Table 4.

After the improvement, positive material cost is increased to 49.81 from 49.62 % and negative material cost is reduced to 25.24 from 25.52 %.

Considering weight of material loss from defectives at blow molding process, the weight of material loss is reduced from 469 to 346.74 kg that is 26.07 %.

Table 4 Cost allocation of blow molding process (in Thai Baht) (improved)

	MC	SC	EC	WC	Total
Positive product	591,318.00	121,183.76	75,408.85	0	787,910.60
	49.81 %	10.21 %	6.35 %	0.00 %	66.37 %
Negative product	299,671.62	61,414.24	38,216.15	0.00	399,302.02
	25.24 %	5.17 %	3.22 %	0.00 %	33.63 %
Total	890,989.62	182,598.00	113,625.00	0.00	1,187,212.62
	75.05 %	15.38 %	9.57 %	0.00 %	

5 Conclusion and Discussion

This study aims to present the application of MFCA in process improvement of one plastics packaging factory as a case study. While MFCA was applied, the highest negative cost was identified at blow molding process. From data collection and observation, material loss can be classified as defective products and normal loss. The improvement procedure for work operations is introduced to reduced defective products. The results after improvement are increasing in positive material cost from 49.62 to 49.81 % and negative material cost reducing from 25.52 to 25.24 %. The effect on reducing in material loss weight is 26.07 %.

Although, normal loss from this process is larger portion of blow molding process than defective products, to reduce normal loss, some detail in machine specification and some investment are required. When investment is needed, the return on investment should be considered as well. However, the advantage of MFCA is material loss identification pointing not only defectives but also normal loss as well. Without MFCA application, producers will not care too much on normal loss because they think that it is a behavior of process that cannot be improved.

Acknowledgments The authors gratefully acknowledge support from National Science and Technology Development Agency, Thailand.

References

Chattinnawat W (2013) Identification of improvement for multistage serial processes with respect to material flow cost accounting via dynamic programming. In: EMAN-EU 2013 conference on material flow cost accounting conference proceedings, pp 30–33

Chompu-inwai R, Jaimjit B, Premsuriyanunt P (2013) Gainning competitive advantage in an SME using integration of material flow cost accounting and design of experiments: the case of a wood products manufacturing company in northern Thailand. In: EMAN-EU 2013 conference on material flow cost accounting conference proceedings, pp 141–144

Kasemset C, Chernsupornchai J, Pala-ud W (2013) The application of MFCA in textile factory: a case study. In: EMAN-EU 2013 conference on material flow cost accounting conference proceedings, pp 84–88

The Application of MFCA Analysis in Process Improvement

Lan S (2010) Optimization of electric motor assembly operation with work study. In: 2010 international conference on logistics systems and intelligent management. doi:10.1109/ICLSIM.2010.5461128

Laosiritaworn W, Kasemset C, Tara C, Poovilai W (2013) Application of material flow cost accounting technique in lost-wax casting process. In: EMAN-EU 2013 conference on material flow cost accounting conference proceedings, pp 80–83

Ministry of Economy, Trade and Industry, Japan (2011) MFCA case examples 2011. Ministry of Economy, Trade and Industry, Japan

Miranda FAA (2011) Application of work sampling and ECRS (Eliminate, Combine, Re-lay out and Simplify) principles of improvement at TO1 assembly. 21st ASEMEP National Technical Symposium. http://www.onsemi.com/ Accessed 24 April 2013

Nakajima M (2004) On the differences between material flow cost accounting and traditional cost accounting—reply to the questions and misunderstandings on material flow cost accounting. Kansai Univ Rev Bus Commerce 6:1–20

Sindhuja D, Mohandas GN, Madhumathi P (2012) Redesigning of horn assembly line using ECRS principles. Int J Eng Innovative Technol 1(3):214–217

Wajanawichakon K, Srimitee C (2012) ECRS's principles for a drinking water production plant. IOSR J Eng 2(5):956–960

Discussion of Water Footprint in Industrial Applications

Chung Chia Chiu, Wei-Jung Shiang and Chiuhsiang Joe Lin

Abstract Economic growth in the past half century brought an unprecedented comfortable life, but also had over-consumed Earth's natural resources. Species extinction and global warming caused by CO_2 emission make us start thinking highly of the surrounding environment. Therefore, the concepts of ecological footprint and carbon footprint have been proposed for assessing the extent of destruction on global environment. In year 2002, Dr. Hoekstra put forward the concept of water footprint for water consumption issues. The main concern is the freshwater used directly and indirectly by consumers or producers, including tracing the three key constituents as blue-, green-, and gray-water. Past studies of the water footprint have gathered a lot of information about agricultural water consumptions, but relatively few were studied for industrial applications. To face the possible shortage of water resources in the future, industries should take a serious attitude to water footprint issues. This study suggests that the water footprint in industrial applications can be used as a basis for improving process water usage, sewage treatment method, water cycle reuse and factories design.

C. C. Chiu (✉) · W.-J. Shiang
Department of Industrial and System Engineering, Chung Yuan Christian University, Chungli, Taiwan, Republic of China
e-mail: ccchiu@iner.gov.tw

W.-J. Shiang
e-mail: Wjs001@cycu.edu.tw

C. C. Chiu
Institute of Nuclear Energy Research, Atomic Energy Council, Executive Yuan, Taoyuan, Taiwan, Republic of China

C. J. Lin
Department of Industrial Management, National Taiwan University of Science and Technology, Taipei, Taiwan, Republic of China
e-mail: cjoelin@mail.ntust.edu.tw

Y.-K. Lin et al. (eds.), *Proceedings of the Institute of Industrial Engineers Asian Conference 2013*, DOI: 10.1007/978-981-4451-98-7_44,
© Springer Science+Business Media Singapore 2013

It will eventually help reach the objectives of saving water, reducing manufacturing costs, complying with international environmental protection requirements, and enhancing the corporate image and visibility.

Keywords Water footprint · Greenhouse effect · Water cycle

1 Introduction

Twentieth century, especially after mid-stage, is the most developed period of human science and technology. In order to boost the economy and improve the living standard, the world is continuously mining the earth's natural resources (for example: biomass, fossil fuels, metal ores and other minerals), and the exploitation has been increased nearly 45 % in the past 25 years (Giljum et al. 2009; Krausmann et al. 2009). These changes have already sacrificed the planet's ecosystems (Haberl 2006; Nelson et al. 2006; Rockstrom et al. 2009). The world population has increased more than 2 times at the end of twentieth century, and the consumption rate of global resources is far beyond Earth's renewed speed (Haberl et al. 2007; Hoekstra 2009). Among all the resources, the greatest impact on human living is due to the requirement for freshwater.

In the past half-century, the demand for freshwater has increased more than four times (Uitto and Schneider 1997). Due to population growth and lifestyle changes, agriculture and some industries also have been increasing the demand for freshwater besides household consumers. Freshwater has quickly become an important global resource, which is resulted from the rapid growing of the global trades on water-intensive products (Hoekstra et al. 2011), as well as a scarce and over-used resource (Bartram 2008; Falkenmark 2008; Vörösmarty et al. 2010). This has caused earthshaking effect upon aquatic ecosystems and livelihood. In recent years, nearly 80 % of the world's population is under a significant threat of water security (Vörösmarty et al. 2010). Climate change and increasing energy-crops consumptions have led to a greater demand for water (Dominguez-Faus et al. 2009; Liu et al. 2008). Regarding the application and management of water resource, in addition to water supply, the measurement of water needs can allocate water resource more effectively. Water footprint was recently promoted as an important indicator for water consumptions (Chapagain and Hoekstra 2004; Hoekstra and Chapagain 2007, 2008). In the past 10 years, quite a few studies about water footprint have been presented, but are mostly related with the water consumptions in agriculture and its relevant products. There lacks research in industrial and operational fields, which leave a new developing opportunity and space for the industrial engineering.

The purpose of this study is to investigate the water footprint for applying in the industrial field. The result of water footprint assessment can be the basis for improvement of operating procedures or a part of the evaluation on new engineering projects. It also can be used to establish international standards. The promotion and implementation of water footprint should have a positive meaning, and can be a supplement to the assessment on product supply chain as well.

2 The Concept of Water Footprint

The idea of considering water use along supply chains has gained interest after the introduction of the 'water footprints' concept by Hoekstra in 2002 (Hoekstra 2003). The water footprint thus offers a better and wider perspective on how a consumer or producer relates to the use of freshwater systems. It is a volumetric measure of water consumption and pollution. Water footprint accounts give spatiotemporally explicit information regarding how water is appropriated for various human purposes. The water footprint is an indicator of freshwater use that looks not only at direct water use of a consumer or producer, but also at the indirect water use. The water footprint can be regarded as a comprehensive indicator of freshwater resources appropriation, next to the traditional and restricted measure of water withdrawal (Hoekstra et al. 2011). The WF consists of three components: blue, green and grey water footprint (Fig. 1).

- Blue water footprint: The blue water footprint is an indicator of consumptive use of so-called blue water, in other words, fresh surface or groundwater.
- Green water footprint: The green water footprint is an indicator of the human use of so-called green water. Green water refers to the precipitation on land that

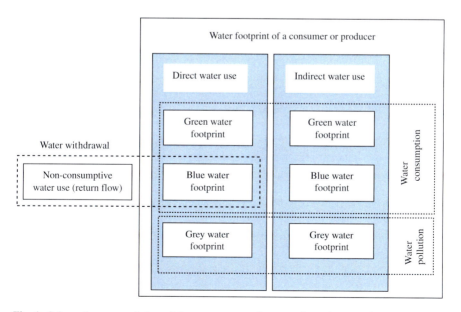

Fig. 1 Schematic representation of the components of a water footprint. (Hoekstra et al. 2011)

does not run off or recharge the groundwater but is stored in the soil or temporarily stays on top of the soil or vegetation.

- Grey water footprint: The grey water footprint of a process step is an indicator of the degree of freshwater pollution that can be associated with the process step. It is defined as the volume of freshwater that is required to assimilate the load of pollutants based on natural background concentrations and existing ambient water quality standards.

One can assess the water footprint of different entities, so it is most important to start specifying in which water footprint one is interested. One can be interested, for instance, in the:

- Water footprint of a process step.
- Water footprint of a product.
- Water footprint of a consumer.
- Water footprint a geographically delineated area.
- National water footprint accounting.
- Water footprint of catchments and river basin.
- Water footprint accounting for municipalities, provinces or other administrative units.
- Water footprint of a business.

3 Paper Survey of Water Footprint

This study has surveyed 61 water footprint related journal papers published in Water Footprint Network (WFN) and Science Direct Onsite (SDOS) during year 2005 to January 2013. The research fields and amounts are shown in Table 1.

Nearly 30 % of these papers were related to researches on water footprint in agriculture and its products, including plants such as: coffee and tea (Chapagain and Hoekstra 2007), cotton (Chapagain et al. 2006), tomato (Page et al. 2012), sweet carbonated drinks (Ercin et al. 2011), rice (Chapagain and Hoekstra 2011), sugar cane and cassava (Kongboon and Sampattagul 2012), spaghetti and pizza (Aldaya and Hoekstra 2010), paste and M&M's peanut (Ridoutt and Pfister 2010). For adapting to the development of renewable energy, quite attention has been paid to biomass such as the water footprint information of bioethanol (Winnie and Hoekstra 2012, Chiu and Wu 2012), and water footprint research for sweetener and bio-ethanol (Leenes and Hoekstra 2012). Almost all these papers investigated water footprint consumed for growing bio-crops, like corn, sugar cane and beet in different regions. These researches' field and methodology are the same as that for agricultural products, but different in final uses. There are 13.1 % of the total journal papers which are in the fields of water resource management and application; five of them are related to management. Studies on water resource applications have been started in recent 2 years. Investigations of virtual water flow direction are commonly seen in water footprint research, and actually involved in

Discussion of Water Footprint in Industrial Applications

Table 1 Research fields and amount of the journal papers related to water footprint

Research field	Numbers presented	Percentage	Main content
Management and utilization	8	13.1	Management and utilization of water resource
Virtual water (*Water footprint assessment*)	7	11.5	Virtual water flow of international trade
Agriculture and its products	18	29.5	Water footprint assessment for animals, crops and their products
Biomass crops	9	14.8	Water footprint assessment for bio-ethanol or bio-diesel crops
Energy carriers	3	4.9	Water footprint assessment for primary energy carrier, hydropower and microalgae
Industries applications	3	4.9	Applications of water footprint assessment results including: sweet carbonated drinks, paper, industries
Assessment technology	7	11.5	Development of water footprint assessment technologies: LCA, remote sensing, space coordinates
Others	6	9.8	Various water footprint assessments such as: footprint family, water conservation policy, national water footprint, water footprint for livelihood consumption, river water footprint

many agriculture water footprint studies mostly before year 2010. It is worth notice that literature on water footprint in relevant applications has been presented since year 2011. Although only three essays were surveyed in this study, it should be thought as a beginning of this kind of research anyhow.

From current literature, but most researches only emphasized the import and export flow directions of virtual water of agriculture or agricultural food products. The trade-off of water resource applications between agriculture and industry based on comparative advantage has not been studied. This study considers such a lack shall be filled with industrial engineering concept and technique.

4 Application and Development of Water Footprint

Mekonnen and Hoekstra (2012) thought water footprint assessment shall be an item for evaluating new and existing equipments. This suggestion is a very good footnote for water footprint application. In the past, industry development stressed how to make the best and the most products in the shortest time and how much resources were consumed or wasted during manufacture was not an important issue. After entering the 21st century, people paid more attention to earth's environmental and resources problems. How to reduce the harm to environment and conserve resources have become a new subject and a new aspect to strive for. Water footprint assessment can identify the hot-spot of water consumptions, and the corresponding improvement can be carried out on that hot-spot then. For example, Ercin et al. (2011) chose raw materials based on water footprint in the study of sweet carbonated drinks. Other researches such as: understanding the applications of water resource through computer games (Hoekstra 2012) and blue water footprint of hydroelectricity (Mekonnen and Hoekstra 2012) all utilized the results of water footprint assessment as the basis for improving existing facilities and building new facilities.

Although applications of water footprint in industries have been quietly implemented in the industrial circles, but not many papers related to industrial applications were presented in the academia. This may result in the lack of supply-chain products information for industrial promotion, because most industries use only blue water and grey water. Moreover, all the water consumption data can be acquired from production line except that of the products provided by supply-chain; this makes the calculation of water footprint easier than that for agriculture. Water is an important factor for the growth of agricultural crops. It needs irrigation (blue water) to satisfy the demand for crop growth in water shortage (green water) area, and will then inevitably increase the stress of water shortage. Though fertilization could increase crop's yield, it also consumes more grey water. All these are conflicts among agricultural water uses. The biggest difference from that of agriculture is the industrial water consumption hot-spots can be improved by many physical measures such as: changing the process and operating procedures, change of water-use habit, decreasing the concentration and amount of wastewater, water-cycle reuse and steam-recovery reuse so as to accomplish the purpose of reducing products' water footprint. These measures can achieve greater effects on lowering production cost and environmental protection compliance if they were implemented during design stage.

5 Conclusions

Freshwater resource has become an important global resource as well as a scarce and over-used resource due to world population growth and change of life style, and the resulted increase of water demand. Hoekstra proposed a water footprint concept of using actual water consumption instead of water taken in year 2002 and thus has raised a new wave of research in the field of water resource management application. But most of the abundant research literatures were focused on the water consumptions during agricultural crops growth and the extended virtual water trade flow amounts. This study collected the journal papers about water footprint in recent 8 years, 30 % of which were assessment or discussion on the water footprint of agriculture and relevant products, a total of more than 5 C % were related to agriculture. This shows water footprint in industrial applications is severely neglected.

Industries provide various kinds of products required for human living, large quantity of freshwater is consumed and different wastewaters are generated during manufacturing process. If we could adequately apply physical measures of industrial engineering discipline such as: changing the process and operating procedures, change of water-use habit, decreasing the concentration and amount of wastewater, water-cycle reuse and steam-recovery reuse so as to accomplish the purpose of reducing products' water footprint. The direct benefit is to lower production cost, reduce freshwater consumption and decrease the competing pressure with domestic water. The indirect benefit is to be in compliance with international environment protection requirement and promote enterprise's image and reputation.

The contribution of this study is to sort a direction for the research of water footprint in industrial applications and provide a clear target for follow-up researchers. Meanwhile through the establishment of water footprint information for various products, so that industries could choose materials accordingly and achieve the goal of efficient use of water resource.

References

Aldaya MM, Hoekstra AY (2010) The water needed for Italians to eat pasta and pizza. Agric Syst 103:351–360

Bartram J (2008) Improving on haves and have-nots. Nature 452(7185):283–284

Chapagain AK, Hoekstra AY (2011) The blue, green and grey water footprint of rice from production and consumption perspectives. Ecol Econ 70:749–758

Chapagain AK, Hoekstra AY (2004) Water footprints of nations. Research report series no. 16, UNESCO-IHE, Netherlands

Chapagain AK, Hoekstra AY (2007) The water footprint of coffee and tea consumption in the Netherlands. Ecol Econ 64(1):109–118

Chapagain AK, Hoekstra AY, Savenije HHG, Gautam R (2006) The water footprint of cotton consumption: an assessment of the impact of worldwide consumption of cotton products on the water resources in the cotton producing countries. Ecol Econ 60(1):186–203

Chiu YW, Wu M (2012) Assessing county-level water footprints of different cellulosic-biofuel feedstock pathways. Environ Sci Technol 46:9155–9162

Dominguez-Faus R, Powers SE, Burken JG, Alvarez PJ (2009) The water footprint of biofuels: a drink or drive issue? Environ Sci Technol 43(9):3005–3010

Ercin EA, Aldaya MM, Hoekstra AY (2011) Corporate water footprint accounting and impact assessment: The case of the water footprint of a sugar-containing carbonated beverage. Water Resour Manage 25:721–741. doi:10.1007/s11269-010-9723-8

Falkenmark M (2008) Water and sustainability: a reappraisal. Environment. http://www.environmentmagazine.org/Archives/Back%20Issues/March-April%202008/Falkenmark-full.html. Accessed 30 Mar 2013

Giljum S, Hinterberger F, Bruckner M, Burger E, Fruhmann J, Lutter S, Pirgmaier E, Polzin C, Waxwender H, Kernegger L, Warhurst M (2009) Overconsumption? Our use of the world's natural resources. Seri, Global 2000, Friends of the Earth Europe

Haberl H (2006) The global socioeconomic energetic metabolism as a sustainability problem. Energy 31:87–99

Haberl H, Erb KH, Krausmann F, Gaube V, Bondeau A, Plutzar C, Gingrich S, Lucht W, Fischer-Kowalski M (2007) Quantifying and mapping the human appropriation of net primary production in earth's terrestrial ecosystems. Proc Nat Acad Sci 104:12942–12947

Hoekstra AY (ed) (2003) Virtual Water trade. In: Proceedings of the international expert meeting on virtual water trade. Delft, The Netherlands, 12–13 December 2002, Value of water research report series no. 12, UNESCO-IHE, Delft, The Netherlands. www.waterfootprint.org/Reports/Report12.pdf. Accessed 10 Oct 2010

Hoekstra AY (2009) Human appropriation of natural capital: a comparison of ecological footprint and water footprint analysis. Ecol Econ 68:1963–1974

Hoekstra AY, Chapagain AK (2008) Globalization of water: sharing the planet's freshwater resources. Blackwell Publishing, Oxford

Hoekstra AY, Chapagain AK, Aldaya MM, Mekonnen MM (2011) The water footprint assessment manual. Earthscan, Washingtong

Hoekstra AY (2012) Computer-supported games and role plays in teaching water Management. Hydrol Earth Syst Sci 16:2985–2994. doi:10.5194/hess-16-2985-2012

Kongboon R, Sampattagul S (2012) The water footprint of sugarcane and cassava in northern Thailand. Procedia—Soc Behav Sci 40:451–460

Krausmann F, Gingrich S, Eisenmenger N, Erb KH, Haberl H, Fischer-Kowalski M (2009) Growth in global materials use GDP and population during the 20th century. Ecol Econ 68(10):2696–2705

Leenes WG, Hoekstra AY (2012) The water footprint of sweeteners and bio-ethanol. Environ Int 40:202–211

Liu JG, Yang H, Savenije HHG (2008) China's move to higher-meat diet hits water security. Nature 454(7203):397

Mekonnen MM, Hoekstra AY (2012) The blue water footprint of electricity from hydropower. Hydrol Earth Syst Sci 16:179–187. doi:10.5194/hess-16-179-2012

Nelson GC, Bennett E, Berhe AA, Cassman K, DeFries R, Dietz T, Dobermann A, Dobson A, Janetos A, Levy M, Marco D, Nakicenovic N, O'Neill B, Norgaard R, Petschel-Held G, Ojima D, Pingali P, Watson R, Zurek M (2006) Anthropogenic drivers of ecosystem change: an overview. Ecol Soc 11(2):29. http://www.ecologyandsociety.org/vol11/iss2/art29/. Accessed 10 Oct 2010

Page G, Ridoutt B, Bellotti B (2012) Carbon and water footprint tradeoffs in fresh tomato production. J Cleaner Prod 32:219–226

Ridoutt BG, Pfister S (2010) A revised approach to water footprinting to make transparent the impacts of consumption and production on global freshwater scarcity. Glob Environ Change 20:113–120

Rockstrom R, Steffen W, Noone K, Persson A, Chapin FS, Lambin EF, Lenton TM, Scheffer M, Folke C, Schellnhuber HJ, Nykvist B, de Wit CA, Hughes T, van der Leeuw S, Rodhe H, Sorlin S, Snyder PK, Costanza R, Svedin U, Falkenmark M, Karlberg L, Corell RW, Fabry VJ, Hansen J, Walker B, Liverman D, Richardson K, Crutzen P, Foley JA (2009) A safe operating space for humanity. Nature 461:472–475

Uitto JI, Schneider J (1997) Fresh resources in Arid Lands. United Nations University Press, Tokyo

Vörösmarty CJ, McIntyre PB, Gessner MO, Dudgeon D, Prusevich A, Green P, Glidden S, Bunn SE, Sullivan CA, Reidy Liermann C, Davies PM (2010) Global threats to human water security and river biodiversity. Nature 467(7315):555–561

Winnie GL, Hoekstra AY (2012) The water footprint of sweeteners and bio-ethanol. Environ Int 40:202–211

Mitigating Uncertainty Risks Through Inventory Management: A Case Study for an Automobile Company

Amy Chen, H. M. Wee, Chih-Ying Hsieh and Paul Wee

Abstract In recent years, global environment has changed dramatically due to unpredictable operational risks, disruption risks, natural and man-made disasters, global financial and European debt crisis. This greatly increases the complexity of the automotive supply chain. In this paper we investigate the inventory policy of the aftermarket parts for an automotive company. The key findings and insights from this study are: (1) to mitigate the risk of disruptive supply chain, enterprises need to reduce the monthly supplies of high priced products, (2) to improve profit, cash flow and fill rate, the use of A, B and C inventory management system is critical, (3) the case study provides managerial insights for other industries to develop an efficient inventory management system in a competitive and uncertain environment.

Keywords Supply chain risk management · ABC inventory management system · Fill rate · Sensitivity analysis · Uncertainty

---

A. Chen (✉) · H. M. Wee · P. Wee
Department of Industrial and System Engineering, Chung Yuan Christian University, Chung-Li, Taiwan, Republic of China
e-mail: chenamy999@gmail.com

H. M. Wee
e-mail: weehm@cycu.edu.tw

P. Wee
e-mail: pwee@ford.com

C.-Y. Hsieh
Department of Business, Vanung University, Chung-Li, Taiwan, Republic of China
e-mail: cishieh@vnu.edu.tw

Y.-K. Lin et al. (eds.), *Proceedings of the Institute of Industrial Engineers Asian Conference 2013*, DOI: 10.1007/978-981-4451-98-7_45, © Springer Science+Business Media Singapore 2013

1 Introduction

In an elaborate automobile network, aftermarket part is a major if not the biggest contributor to a company's profits. In the aftermarket part business, custom satisfaction is critical; and a quick response to customers' needs have become a basic requirements. Therefore, a good fill rate is crucial for a company to maintain a reasonable profit and cash flow. In order for a company to maintain an efficient fill rate, inventory management is critical.

In recent years, crisis that result in supply chain disruption are prevalent, for example the Japanese earthquake in March 2011, the Turkey earthquake in October 2011 and the Thailand floods in May 2012, these uncertain events have made the inventory control of aftermarket parts more complicated than production. Since aftermarket parts do not have a controllable volume and timing schedule, to quickly response to the unpredictable risks and maintain customer satisfaction, we need to identify an optimal inventory and fill rate with an automobile company as an example.

The aftermarket parts of the automotive company under study are managed by the PANDA system (Parts and Accessory logistic system). PANDA is an integrated system of aftermarket parts to run the complete parts operation. The system supports critical aspects of the aftermarket parts supply chain process including parts' ordering, demand forecast, delivering, storage and inventory management, the PANDA system is not only parts supply flow but also accounting flow to link dealers, service part's center and part's sources. Figure 1 shows an automotive network and Table 1 provides the definition of the ABC aftermarket parts' inventory system.

2 Literature Review

2.1 Supply Chain Risk Management

Wee et al. (2009) identified the supply chain risks in the automotive and electronic industries in Brazil and highlighted the urgency of the supply chain risk management (SCRM) and implementation. Sabio et al. (2010) presented an efficient decomposition method in order to expedite the solution of the underlying multi-objective model by exploiting its specification. They illustrated the capabilities of the proposed model framework and solution strategy through the application of a real case study in Spain. Blome and Schoenherr (2011) used in-depth case studies conducted among eight European enterprises and highlighted their approach to risk management and how they are related to Enterprise Risk Management.

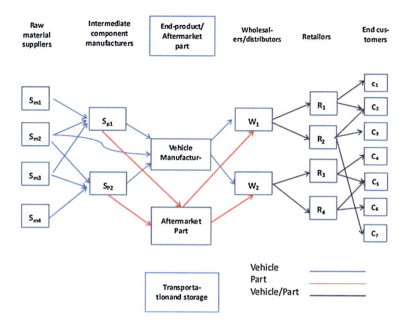

Fig. 1 The network of automobile industry

Table 1 The definition of ABCD inventory system

Categories	Annual sales forecast	Dealer demand
A Type	Sales forecast > 20 pieces	Average 1 piece per month
B Type	3.3 ≤ Sales forecast ≤ 20	Average 1 piece per 3 month
C Type	0 < Sales forecast < 3.3	Average 1 piece more than 3 month
D Type	Sales forecast = 0	• No demand
		• Obsolete parts—no order in over 2 years
		• New parts—last order within 2 years

2.2 Inventory Management

Schmitt (2011) modeled a multi-echelon system and evaluate multiple strategies for protecting customer service, also demonstrated that the greatest service level improvements can be made by providing both proactive inventory placement to cover short disruptions or the start of long disruptions, and reactive back-up methods to help the supply chain recover after long disruptions. Gupta et al. (2000) utilizes the framework of mid-term, multisite supply chain planning under demand uncertainty to safeguard against inventory depletion at the production sites and excessive shortage using a change constraint programming approach in conjunction with a two-stage stochastic programming methodology. Yu (2011) compared artificial-intelligence (AI)-based classification techniques with traditional multiple discriminant analysis (MDA), support vector machines (SVMs), backpropagation networks (BPNs), and the k-nearest neighbor (k-NN) algorithm and suggested that ERP systems can be used to implement AI-based classification techniques with multiple criteria in ABC analysis. The use of these techniques will improve the effectiveness and efficiency of inventory management.

2.3 Fill Rate

Silver et al. (2011) presented a news derivation of the fill rate and (R, S) system under normally distributed demand. A commonly used approximate method for selecting the value of the safety factor can lead to an appropriate service level. Axsater (2006) provided an alternative approximation technique for determining the order quantities for (R, Q) policy under a fill rate constraint and normal lead-time demand.

3 Model Development and Analyze

3.1 Data Source and Definition

The purpose of this model is to analyze inventory, monthly supply, inventory turnover and fill-rate of the aftermarket parts by Sensitivity Analysis so as to improve cash flow and profit. Aftermarket parts are sourced mainly from European and US markets, some parts are sourced from Taiwan and other Asian suppliers. This automotive company holds an average aftermarket part inventory of \$340 million to support about \$2,958.5 million in annual sales with a fill-rate of 92.8 %. **Fill-rate**:

> The first pick availability rate at the warehouse to meet dealer orders; including stock orders and urgent orders. That is critical for customer's satisfaction.

Monthly supply:

> Inventory turnover: Cost of sales divided by Inventory amount.
> Profit per part item is \$500.
> Fill-rate change of 1 % is around 13,500 items.
> Carrying cost is 20 % of inventory.

Inventory ABC categories are defined in Table 1. This study combines both C and D inventory types into C inventory category.

Mitigating Uncertainty Risks Through Inventory Management

3.2 Model

3.2.1 Tool: Excel Solver—Simple Linear Programs

3.2.2 Modeling

Propose: Using an Excel Solver to identify the optimal inventory level to drive the best practice in the after sales service parts business including cash flow, profit and customer' satisfaction.

Step 1 Define variances

Known variances (F): F_{ij}—fill-rates for each inventory categories by sources, i inventory category, j supply source

Unknown variances (S): S_{ij}—each inventory categories, i inventory category, j supplier source

Step 2 Functional target—an optimal solution to get reasonable inventory combination for a higher fill-rate Maximum of

$$\sum_{i \geq 1}^{n} (F_{ij} \times S_{ij}) / \sum_{i \geq 1}^{n} S_{ij}$$

Step 3 Constraint

Each source has a lead-time constraint considering the logistics of shipment. Therefore monthly supplies from each source are different.

$$M_{ij} > \, = 0$$

3.3 Data Analysis

3.3.1 Sensitivity Analysis Report

From the Sensitivity Analysis report generated from the Excel Solver, we found 4 top shadow prices; S_{A-LC} (82.6), S_{A-JA} (22.8), S_{A-EU} (14) then S_{B-LC} (8.4). If we plan to improve the service level in the short term, an increase in the monthly supply of S_{A-LC} (82.6) is necessary. The impact of fill-rate includes customer satisfaction, cash flow, profit and visible cash improvement. The improvements include S_{A-JA} (22.8), S_{A-EU} (14) then S_{B-LC} (8.4) where uncertain environment and limited resources are present. The details are shown in Table 2.

Table 2 Simulation report

Cell	Name	Final value	Reduced cost	Objezctive coefficient	Allowable increase	Allowable decrease	Minumun	Name	Maximun
Adjustable cells									
B8	A EU	57.1	0.0	0.98	1E + 30	0.98	0.00	A EU	1E + 30
C8	A JA	58.1	0.0	0.98	1E + 30	0.98	0.00	A JA	1E + 30
D8	A US	25.9	0.0	0.98	1E + 30	0.98	0.00	A US	1E + 30
E8	A LC	126.4	0.0	0.98	1E + 30	0.98	0.00	A LC	1E + 30
B9	B EU	10.7	0.0	0.85	1E + 30	0.85	0.00	B EU	1E + 30
C9	B JA	11.6	0.0	0.85	1E + 30	0.85	0.00	B JA	1E + 30
D9	B US	4.9	0.0	0.85	1E + 30	0.85	0.00	B US	1E + 30
E9	B LC	9.9	0.0	0.85	1E + 30	0.85	0.00	B LC	1E + 30
B10	C EU	9.9	0.0	0.60	1E + 30	0.60	0.00	C EU	1E + 30
C10	C JA	16.1	0.0	0.60	1E + 30	0.60	0.00	C JA	1E + 30
D10	C US	4.5	0.0	0.60	1E + 30	0.60	0.00	C US	1E + 30
E10	C LC	5.0	0.0	0.75	1E + 30	0.75	0.00	C LC	1E + 30

Cell	Name	Final value	Shadow price	Constraint R. H. side	Allowable increase	Allowable decrease	Minumun	Name	Maximun
Constraints									
B15	A EU	4.0	14.0	4.0	1E + 30	4.0	0.00	A EU	1E + 30
C15	A JA	2.5	22.8	2.5	1E + 30	2.5	0.00	A JA	1E + 30
D15	A US	4.0	6.4	4.0	1E + 30	4.0	0.00	A US	1E + 30
E15	A LC	1.5	82.6	1.5	1E + 30	1.5	0.00	A LC	1E + 30
B16	B EU	3.0	3.0	3.0	1E + 30	3.0	0.00	B EU	1E + 30
C16	B JA	2.0	4.9	2.0	1E + 30	2.0	0.00	B JA	1E + 30
D16	B US	3.0	1.4	3.0	1E + 30	3.0	0.00	B US	1E + 30
E16	B LC	1.0	8.4	1.0	1E + 30	1.0	0.00	B LC	1E + 30
B17	C EU	5.0	1.2	5.0	1E + 30	5.0	0.00	C EU	1E + 30
C17	C JA	5.0	1.9	5.0	1E + 30	5.0	0.00	C JA	1E + 30
D17	C US	5.0	0.5	5.0	1E + 30	5.0	0.00	C US	1E + 30
E17	C LC	1.0	3.7	1.0	1E + 30	1.0	0.00	C LC	1E + 30

3.3.2 From the Simulation Reports Below, a Summary is Listed

Figure 2, we can find the sensitivity for 0.1 month-supply interval for A/B/C inventory categories.

Category A, the profit impact is getting better as monthly supply is decreasing, also cash flow is improved significantly resulted from lower inventory levels, but the fill- rate becomes worse. That is because Category A is fast moving parts, month-supply change impacts fill-rate, profit, cash flow and fill-rate a lot.

Category B and C's profit and inventory impact are not significantly due to slow moving parts. Also the fill-rate of category B and C worsen as monthly supply is increased.

This study examines two cases, Case A to reduce 1 month supply for category A, B and C, Case B to increase 1 month supply for category A, B and C. The

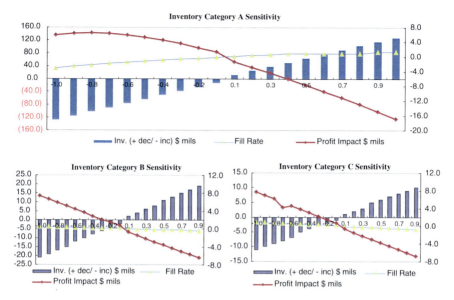

Fig. 2 Sensitivity analysis report

results are shown in Table 3: Case A has a higher profit impact with approximately $24.69 million but lower fill rate at 91.7 %; which means we will lose some loyal customers in the long term. Case B achieves a higher fill-rate 93.3 %, but the profit impact is the worst because it keeps the highest inventory levels. Therefore, we need to consider the optimal solution of monthly supply for category A, B and C pondering fill-rate and profit impact. Other cases are considered after simulation.

3.3.3 More Cases are Simulated Here

- Case C (Table 3)

    A category: 0.5 month decrease
    B category: 1 month decrease
    C category: 1 month decrease
    Fill-rate of 93.3 %, profit impact of $22.37 million at $244 inventory level
- Case D (Table 3)

    A category: no change
    B category: 1 month decrease
    C category: 1 month decrease
    Fill-rate of 94.3 %, and a profit impact of $15.67 million at $308 inventory level

Although Case C's profit impact is $22.37 million; $6.7 million higher than Case D, its fill-rate is 1 point lower than that of Case D. Other than profits; a

Table 3 The summary of 4 cases

Monthly adjustment	Case A A(−1) B(−1) C(−1)	Case B A(+1) B(+1) C(+1)	Case C A(−0.5) B(−1) C(−1)	Case D A(−1) B(−1) C(−1)
Inventory $ million	180	500	244	308
Month supply	1.1	3.1	1.5	1.9
Inventory turn	10.69	3.84	7.88	6.24
Fill rate (%)	91.7	93.3	93.3	94.3
Compared with original level				
Inventory (+dec/− inc) $ million	−160	160	−96	−32
Month supply	−1.0	1.0	−0.6	−0.2
Inventory turn	4.99	−1.86	2.18	0.54
Fill rate (%)	−1.1	0.5	0.5	1.5
Profit Impact ($ million)	24.69	−29.41	22.37	15.67

company should consider customer satisfaction (higher fill-rate) and cash flow (inventory level) to maintain a high profit.

4 Conclusion and Contribution

In this study, we optimize the ABC inventory management system to mitigate the risks of supply chain disruption. The case example provides the proposed enterprise with a strategy to improve profit, cash flow and fill-rate during the times of uncertainty and global financial economic pressures. Key findings in this study provide useful insights to enterprises:

1. To manage and control inventory, enterprises need to reduce the monthly supplies of high priced products as priority, the impact on the fill-rate (customer satisfaction) and profit can quickly be observed.
2. To improve profit, cash flow and fill rate, inventory management using A, B and C system is very critical for uncertain environment with limited resources.
3. The model developed in this study is user friendly and can be applied to other enterprises. It provides instantaneous fill-rate and inventory monitoring for senior management.

Finally, the case study provides managerial insights for other industries to develop an efficient and effective inventory management system in a competitive and uncertain environment.

References

Axsater S (2006) A simple procedure for determining order quantities under a fill rate constraint and normally distributed lead-time demand. Eur J Oper Res 174:480–491

Blome C, Schoenherr T (2011) Supply chain risk management in financial crises-A multiple case-study approach. Int J Prod Econ 10:1016

Gupta A, Maranas CD, McDonald CM (2000) Mid-term supply chain planning under demand uncertainty: customer demand satisfaction and inventory management. Computers and Chemical Engineering 24:2613–2621

Sabio N, Gadalla M, Guillen-Gosalbez G, Jimenezl L (2010) Strategic planning with risk control of hydrogen supply chains for vehicle use under uncertainty in operating costs: a case study of Spain. Int J Hydrogen Energy 35:6836–6852

Schmitt AJ (2011) Strategies for customer service level protection under multi-echelon supply chain disruption risk. Transp Res Part B 45:1266–1283

Silver EA, Bischak DP (2011) The exact fill rate in a periodic review base stock system under normally distributed demand. Omega 39:346–349

Wee HM, Blos MF, Quaddus M, Watanabe K (2009) Supply chain risk management (SCRM): a case study on the automotive and electronic industries in Brazil. Supply Chain Manage: Int J 14(4):247–252

Yu M-C (2011) Multi-criteria ABC analysis using artificial-intelligence-based classification techniques. Expert Syst Appl 38:3416–3421

Service Quality for the YouBike System in Taipei

Jung-Wei Chang, Xin-Yi Jiang, Xiu-Ru Chen, Chia-Chen Lin and Shih-Che Lo

Abstract In this study, we focused on the service quality for the public bicycle system, YouBike System, in Taipei, and used the station setting at the National Taiwan University of Science and Technology (NTUST) for the case study. YouBike System refers to the "rent-it-here, leave-it-there" bike sharing service provided by the Taipei City Government. We adopted the service quality models developed by Parasuraman, Zeithaml and Berry in 1970s to conduct the research. In the first stage before launching the station at the NTUST, a pre-using questionnaire was designed and distributed to the students at the NTUST to collect the opinions and their expectations about the YouBike System. Then, an after-using questionnaire was designed and distributed to the students at the NTUST to investigate whether the service quality of the YouBike System meet their expectations and what is the service level provided by the YouBike system. The after-using survey was conducted after one month of launching the station at the NTUST. The results analyzed from both sets of surveys would provide valuable information for the Taipei City Government to continuing improves their public transportation policy.

Keywords Service quality · Likert scale · Bike sharing service · Bike sharing network · Service science

1 Introduction

Due to the trend of saving power to protect environment and riding bicycles for citizens' health, big cities over the world established their public bicycle systems for many years. In addition to reducing traffic congestion during peak hours, riding

---

J.-W. Chang · X.-Y. Jiang · X.-R. Chen · C.-C. Lin · S.-C. Lo (✉)
Department of Industrial Management, National Taiwan University of Science
and Technology, No. 43, Keelung Road, Taipei 106, Taiwan, Republic of China
e-mail: sclo@mail.ntust.edu.tw

Y.-K. Lin et al. (eds.), *Proceedings of the Institute of Industrial
Engineers Asian Conference 2013*, DOI: 10.1007/978-981-4451-98-7_46,
© Springer Science+Business Media Singapore 2013

Table 1 Ten cities having bicycle sharing systems

Country	City	System	No. stations	No. bikes
Belgium	Brussels	Villo!	1,800	2,500
Canada	Montreal	Bixi	405	5,050
China	Hangzhou	Hangzou public bicycle	2,700	66,500
Denmark	Copenhagen	Bycylken	125	1,300
France	Lyon	Vélo'v	340	4,000
France	Paris	Vélib	1,450	20,600
Netherlands	Amsterdam	OV-fiets	240	6,000
South Korea	Changwon	NUBIJA	235	3,300
Spanish	Barcelona	BICING	420	6,000
UK	London	Barclays cycle hire	507	8,000

bicycles for commuters can also reduce gas emissions from vehicles (Frederick et al. 1959). Moreover, establishing public bicycle system has become a prosperous target to know the internationalism, the freedom, and the respect of environmental issue for big cities around the world (Brandt and Scharioth 1998).

Table 1 summarized 10 cities providing public bicycle sharing systems that are currently active. All systems listed in Table 1 allow users to pick up and drop off bicycles at any of the kiosk stations within the bicycle networks.

Taiwan is famous for manufacturing bicycles. Brand names, such as Giant Bicycles and Merida Bikes, are the companies making public bikes and the Taipei City Government is optimistic to the development of the public bike system in Taipei City and the vicinity. With the cooperation between enterprises and government officials, the YouBike system in Taipei had been established since 2009. For more than 7 years, Taipei City Government is in charge of supervising the engineering process, network expansion, and operations management of the YouBike. Moreover, the contractor is responsible for setting up, operation and maintenance of the YouBike. They hope that the YouBike can not only reducing the air pollution level and heavy traffic situation in Taipei City, but also providing a convenient travel method for the citizens and tourists (Huang 2010). There are twelve administrative districts in the Taipei City. In each district, the YouBike begins with setting stations near MRT stations, bus stops, markets, and residential area because of the stream of people. There are seventy stations in the Taipei City now and there will be 160 stations total at the end of this year.

In order to know users' considerations about service quality of the YouBike system in Taipei City, we adopted the service quality models developed by Parasuraman, Zeithaml and Berry in 1970s to conduct the research (Parasurman et al. 1984, 1985) in this paper. In the first stage before launching the station at the NTUST, a pre-using questionnaire was designed and distributed to the students at the NTUST to collect the opinions and their expectations about the YouBike System. Then, an after-using questionnaire was designed and distributed to the students at the NTUST to investigate whether the service quality of the YouBike System meet their expectations and what is the service level provided by the YouBike system (Noriaki et al. 1984).

Service Quality for the YouBike System in Taipei 383

The remainder of the paper is organized as follows. Section 2 is the design of both questionnaires and initial analysis of the data collection quality. Section 3 presents the experimental results and findings from the questionnaires. Section 4 offers conclusions, following the references in the final section.

2 Research Methodology

A before-using the YouBike system questionnaire was designed and a total of 868 questionnaires were collected to research expectations from students with/without using the YouBike service in NTUST before the kiosk station setting at the NTUST launched. Then the after-using questionnaire was designed and conducted to collect opinions from 162 students who have experience using the YouBike service at the NTUST station.

2.1 Reliability

Reliability is a way to test if a method is reliable by showing consistency and stability in the result. We use Cronbach α in five-point Likert scale to identify reliability of our questionnaire (Carman 1990). Cronbach's α is ranged between zero and one to represent its consistency. Generally, Cronbach's α falls at least larger than 0.5 and smaller than or equal to 0.7. If Cronbach's α larger than 0.7, it represents that the results from the questionnaire is very reliable. Table 2 summarized Cronbach's α values for both sets of questionnaires for the reliability analysis. The calculation of Cronbach's α is shown in the following equation.

$$\alpha = \frac{n}{n-1}\left(1 - \frac{\sum S_i^2}{S_H^2}\right),\qquad(1)$$

where S_i^2 is the deviation of every question and S_H^2 is the deviation of total questionnaire.

Table 2 Cronbach α from questionnaires

Questions	α (before-using)	α (after-using)
1 I am glad to see that the YouBike station setting at the NTUST	0.9031	0.8022
2 It is appropriate to set the YouBike station at the NTUST	0.9015	0.7842
3 I want to use the YouBike at the NTUST	0.8927	0.7757
4 It is convenient to set the YouBike kiosk station at the NTUST	0.8918	0.7775
5 It is helpful to set the YouBike station at the NTUST	0.8912	0.7671
6 If the YouBike is set at my destination, I will go there by the YouBike	0.9272	0.8695

Table 3 Cronbach α in users' opinions from Kano's model

Questions		α
1	Renting a well-functioned bike	0.8783
2	There is an available parking place to return the bike	0.8840
3	Instructions from the kiosk machine are in detail and clear	0.8776
4	If the kiosk machine can inform you the condition at every YouBike station	0.8720
5	When your easy card is locked with unknown reasons, the kiosk machine can teach you how to solve the problem	0.8776
6	You can rent a bike for free within 30 min	0.8810
7	You can apply the association member by both kiosk machine and on Internet	0.8849
8	The kiosk can inform you that your easy card has insufficient money	0.8794
9	The sensor can inform you when easy card touch the sensor	0.8759
10	If staff can arrive within 10 min when the YouBike station ran out of bikes or parking place	0.8792
11	You can return a bike by the kiosk machine when the station is full loaded	0.8970
12	You can get 1 h more for free when the parking space is full loaded	0.8856
13	You can get a transfer discount after using the YouBike service	0.8808

As shown in Table 2, α values from before-using questionnaire are around 0.9 and α values from after-using questionnaire are around 0.8. All values are bigger than 0.7 which confirmed that the data collected for our analysis is reliable.

Table 3 shows α values of after-using questionnaire designed from Kano model are around 0.88, which also confirmed that the data collected for our analysis is reliable.

2.2 Questionnaire Design

The before-using questionnaire was designed and focused on the users' expectation about the soon-to-be-launched YouBike station at NTUST. There are two sections in our questionnaire: (1) user information and (2) user's expectation about the YouBike service. Some of the questions in the first questionnaire are specially designed in order to compare the results from the second questionnaire. Next, the after-using questionnaire was designed and focused on the users' satisfaction and the service quality of the YouBike service. There are three sections in the second questionnaire: (1) user information, (2) users' satisfaction, and (3) users' opinion about the YouBike service provided by the YouBike station at NTUST. We integrated the Kano's model (Matzler and Hinterhuber 1998; Chang 2011) to design the second questionnaire in order to acquire more information for students after using the YouBike station at NTUST. Since we aim at the students having the YouBike service experience, so number of samples from the second questionnaire that we collected is smaller than number of samples from the first questionnaire. We analyzed the before-using results and after-using results to analysis if the

Service Quality for the YouBike System in Taipei

Table 4 Two-dimensional quality key elements

Service Requirements		Dysfunctional form of questions				
		I like it that way	It must be that way	I am neutral	I can live without it that way	I dislike it that way
Functional form of the questions	I like it that way	Q	A	A	A	O
	It must be that way	R	I	I	I	M
	I am neutral	R	I	I	I	M
	I can live without it that way	R	I	I	I	M
	I dislike it that way	R	R	R	R	Q

establishment of the YouBike station at NTUST fits in with students' expectations and the factors that they pay more attention to the YouBike service. A Two-dimensional quality key elements and quality improvement index (Kano 1984) were also used to calculate level of satisfaction and level of dissatisfaction from users, as shown in the following equations, where A: Attractive, O: One-dimensional, M: Must-be, I: Indifferent, R: Reverse, and Q: Questionable (Table 4).

$$\text{Satisfaction increment index} = (A + O)/(A + O + M + I). \quad (2)$$

$$\text{Dissatisfaction decrement index} = (O + M)/(A + O + M + I).$$

3 The Experimental Results

Table 5 shows the comparison results of before-using questionnaire and after-using questionnaire, where 5: very agree, 4: agree, 3: no comments, 2: disagree, and 1: very disagree.

Table 6 shows the before-using questionnaire results related to users' expectations before launching the new station of the YouBike service at NTUST with ranking these conditions. The first priority column means users believed that statements should be pay more attention, and the least priority means users usually do not pay more attention to these statements.

The results show that before launching new station at NTUST, up to 30% students pay much attention to "canceling 30 min free rental policy," and the second place in the first priority column is "no bike to be rent when you need a bike." Moreover, the first place in second priority column falls in "no place to

Table 5 The results of before-using questionnaire and after-using questionnaire

Questions	Before-using (first)					After-using (second)				
	5	4	3	2	1	5	4	3	2	1
1. Glad to see the YouBike station at NTUST	0.72	0.25	0.03	0	0	0.84	0.15	0	0.01	0
2. It is appropriate to set the YouBike station at NTUST	0.69	0.25	0.05	0.01	0	0.80	0.19	0.01	0	0
3. I want to use the YouBike at NTUST	0.65	0.23	0.11	0.01	0	0.83	0.16	0.01	0	0
4. It is convenient to have the station at NTUST	0.64	0.24	0.11	0.01	0	0.79	0.17	0.03	0.01	0
5. It is helpful to have the station at NTUST	0.57	0.25	0.15	0.02	0.01	0.76	0.19	0.04	0.01	0
6. If my destination has YouBike service, I will go there by the YouBike	0.33	0.26	0.32	0.08	0.01	0.54	0	0.17	0.04	0

Table 6 The before-using questionnaire results with users' expectations

Situations	First priority	Second priority	Least priority
No bike to rent when I need	0.21	0.19	0.08
No place to park bike at the station when returning	0.18	0.30	0.03
The YouBike station occupies more space of sidewalk	0.05	0.06	0.39
Renting a fault bike	0.09	0.13	0.06
Inconvenient to inform staff of the YouBike service	0.05	0.12	0.08
The kiosk machine cannot help you to solve your problem	0.09	0.13	0.07
30 min free rental policy cancelled	0.33	0.07	0.29

park a bike when returning," and the second place in second priority column is "no bike to be rent when you need a bike." Therefore, "canceling 30 min free rental policy," "no bike to be rent when you need a bike" and "no place to park a bike when returning" are the most important factors for students at NTUST. Also, there are approximately 40 % students do not care that the YouBike station occupies more space of sidewalk in the surrounding area. However, there are approximately 30 % of students do not care about "canceling 30 min free rental policy."

Table 7 summarizes the service quality analysis about users' considerations by the Kano's model. Moreover, the level of satisfaction and level of dissatisfaction analysis from the Kano model is shown in the following Table 8.

Service Quality for the YouBike System in Taipei

Table 7 The service quality analysis from users' considerations by Kano's model

Questions	Quality key elements (%)				Kano's model
	(A)	(O)	(M)	(I)	
1. Renting a well-functioned bike	20.3	48.4	17.6	13.7	O
2. There is an available parking place to return the bike	14.8	47.2	27.5	10.6	O
3. Instructions from the kiosk machine are in detail and clear	21.1	30.3	21.8	26.8	O
4. If the kiosk machine can inform you the condition at every YouBike station	26.1	37.3	15.5	21.1	O
5. When your easy card is locked with unknown reasons, the kiosk machine can teach you how to solve the problem	13.4	42.3	27.5	16.9	O
6. You can rent a bike for free within 30 min	31.3	54.2	8.3	6.2	O
7. You can apply the association member by both kiosk machine and on Internet	34.3	24.5	7.7	33.6	A
8. The kiosk can inform you that your easy card has insufficient money	20.6	29.8	18.4	31.2	I
9. The sensor can inform you when easy card touch the sensor	20.8	36.8	20.1	22.2	O
10. If staff can arrive within 10 min when the YouBike station ran out of bikes or parking place	29.9	43.8	12.5	13.9	O
11. You can return a bike by the kiosk machine when the station is full loaded	30.8	50	11.9	8.4	O
12. You can get 1 h more for free when the parking space is full loaded	32.8	43.1	13.1	10.9	O
13. You can get a transfer discount after using the YouBike service	45.8	37.5	6.9	9.7	A

Table 8 The level of satisfaction and level of dissatisfaction

	1	2	3	4	5	6	7
Satisfaction increment	0.68	0.62	0.51	0.63	0.56	0.85	0.59
Dissatisfaction decrement	−0.66	−0.75	−0.52	−0.53	−0.7	−0.63	−0.32
	8	9	10	11	12	13	
Satisfaction increment	0.50	0.58	0.74	0.8	0.76	0.83	
Dissatisfaction decrement	−0.48	−0.57	−0.56	−0.61	−0.56	−0.44	

4 Conclusions

After using the new YouBike station at NTUST, the satisfaction level from students increased with the YouBike system in Taipei City. Among 13 service criteria that we investigated, 10 of the service criteria fall in one-dimensional requirements, leading to the conclusions that these services remain in steady service quality. Users' satisfaction level about these services is easy to achieve and easy to increase.

From the analysis and results of quality improvement index, "canceling 30 min free rental policy," "no bike to be rent when you need a bike" and "no place to park a bike when returning" are the most important factors for students at NTUST. Both Taipei City Government and contractor of the YouBike system should consider providing hardware and software improvement plans to continue upgrade service level for users to provide the world class bicycle sharing system to the commuters and visitors in the Taipei City.

Acknowledgments The authors thank the Department of Transportation, Taipei City Government, for providing useful information about the YouBike System in Taipei, Taiwan.

References

Brandt DR, Scharioth J (1998) Attribute life cycle analysis. Alternatives to the Kanomethod in 51. ESOMAR-Congress, pp 413–429

Carman JM (1990) Consumer perceptions of service quality: an assessment of the SERVQUAL dimensions. J Retail 66(1):33–55

Chang YC (2011) The enhancement of customers' satisfaction for security industries using Kano and PZB Models: using P Company in Taiwan as an example. National Cheng Kung University, Dissertation

Frederick H, Mausner B, Snyderman BB (1959) The motivation to work, 2nd edn. Wiley, New York

Huang HJ (2010) The relationship among riding characteristics, service convenience and riding satisfaction of public bicycle—the case study of Taipei City YouBike. Dissertation, Chaoyang University of Technology

Kano N (1984) Attractive quality and must-be quality. Hinshitsu (Quality) 14(2):147–156

Matzler K, Hinterhuber HH (1998) How to make product development projects more successful by integrating Kano's model of customer satisfaction into quality function deployment. Technovation 18(1):25–38

Noriaki K, Seraku N, Takahashi F, Tsuji S (1984) Attractive quality and must-be quality. J Jpn Soc Qual Control 14(2):39–48

Parasurman A, Zeithaml VA, Berry LL (1984) A conceptual model of service quality and its implications for future research. J Mark 49:41–50

Parasurman A, Zeithaml VA, Berry LL (1985) Problems and strategies in services marketing. J Mark 49:33–40

Replenishment Strategies for the YouBike System in Taipei

Chia-Chen Lin, Xiu-Ru Chen, Jung-Wei Chang, Xin-Yi Jiang and Shih-Che Lo

Abstract In this study, we focused on the bike replenishment strategies for the public bicycle system, YouBike System, in Taipei. YouBike System refers to the "rent-it-here, leave-it-there" bike sharing service provided by the Taipei City Government. Recently, the bicycle system has become popular and has been used by over one million riders. During the rush hours, when people go on or off duty, there would be: (1) no bicycle for renting at the particular rental stations; or (2) no space to park bicycles at the rental stations near schools or MRT stations. These problems can be troublesome to many users/members of the YouBike System. In order to mimic the YouBike System in Taipei, we used computer simulation software to simulate the movement of the bicycles from one bicycle station to other bicycle station. As a result, our goal is to build an on-line monitoring system to provide real-time usage of the bikes and parking space of all YouBike stations. Feasible solutions and optimal strategies were proposed in this study to move bicycles between bicycle rental stations to balance: (1) number of bicycles in the rental station; and (2) number of available parking space for the bicycles in the bicycle rental station.

Keywords Inventory replenishment · Simulation · Bike sharing service · Bike sharing network · Service science

1 Introduction

As the conscience of environmental protection rises, many people ride bicycles for a short travel distance instead of other transportation modes such as private cars or motorcycles. In response to this trend, the Department of Transportation, Taipei

---

C.-C. Lin · X.-R. Chen · J.-W. Chang · X.-Y. Jiang · S.-C. Lo (✉)
Department of Industrial Management, National Taiwan University of Science and Technology, No. 43, Keelung Road, Taipei 106, Taiwan, Republic of China
e-mail: sclo@mail.ntust.edu.tw

Y.-K. Lin et al. (eds.), *Proceedings of the Institute of Industrial Engineers Asian Conference 2013*, DOI: 10.1007/978-981-4451-98-7_47,
© Springer Science+Business Media Singapore 2013

City Government, has carried out a public bicycle sharing system since 2008 (Chang 2000; Tzeng 2013). With the help from bicycle manufacturing company, Giant Bicycles, eleven bicycle rental stations with 500 bikes were firstly built in Xinyi District in Taipei City. The newly started public bicycle sharing system, the YouBike system, is established and operated by Giant Bicycles. Currently, there are more than 160 stations scattered in Taipei City and its vicinity, and the system has been used by more than one million riders.

The "rent-it-here, leave-it-there" bike sharing service of the YouBike brings much convenience to the citizens in Taipei City. Firstly, the YouBike system is tightly connected with mass rapid transit (MRT) system serving Taipei metropolitan area while the rental stations are often built near the entrance of the MRT stations. Riding the public system from MRT stations to people's home can greatly save time. Secondly, the well-designed rental stations and colorful vehicles with led lights can not only provide users a safety riding environment but also add energy into the busy city. The reduction of using private vehicles or motorcycles can also keep the air of the city clean and tidy.

However, many problem rises as the system become more and more popular. By asking the expectations from the users, major problems of the YouBike come out inevitably. During rush hours when people go on and off duty, there would be no bicycle for renting or no parking space at several stations, especially those near MRT stations or schools. The feasible solution to improve the YouBike system is a big issue for the Department of Transportation, Taipei City Government. One possible solution is to prepare a set of spare vehicles arranged besides existing rental stations in order to ensure that there would always bikes available to the riders. This method comes from double queue concept from inventory management study (Wang 2006). Another possible solution is that if renters bumped into a full rental station when returning bikes, they would be granted a free 1 h of time to ride to the nearest station which has enough empty parking spaces.

We found out that these problems happened mainly because the people in charge cannot be instantly informed of when and which stations need to add vehicles or to take the superfluous bikes away. Real-time monitoring system can be implemented by current information technology and traditional safety stock polity from inventory management can be applied to solve the situations (Li 1999; Graves and Willems 2003; Thomopoulos 2006; Humair and Willems 2011).

In this paper, we proposed several inventory replenishment strategies to provide a more convenient and feasible solution for solving this complicate problem. However, the problem itself is quite different from traditional inventory management studies since we need to move bikes from near full stations (no parking space available) to almost empty stations (no bike to rent). The inventory property in the problem is not only to control safety stock level, but also to avoid full queue situation. In order to mimic the situation, we used computer simulation software to simulate the movement of the bicycles from one bicycle station to other bicycle station. Moreover, several bikes inventory replenishment strategies were proposed and simulated through the simulation software. The replenishment strategies that we proposed can be implemented into an on-line monitoring system.

Table 1 Ten cities having bicycle sharing systems

Country	City	System	No. stations	No. bikes
Belgium	Brussels	Villo!	1,800	2,500
Canada	Montreal	Bixi	405	5,050
China	Hangzhou	Hangzou public bicycle	2,700	66,500
Denmark	Copenhagen	Bycylken	125	1,300
France	Lyon	Vélo'v	340	4,000
France	Paris	Vélib	1,450	20,600
Netherlands	Amsterdam	OV-fiets	240	6,000
South Korea	Changwon	NUBIJA	235	3,300
Spanish	Barcelona	BICING	420	6,000
UK	London	Barclays cycle hire	507	8,000

Therefore, the goal of our research is to balance the number of bikes and the available parking space for all rental stations in the YouBike network in Taipei City.

The public bicycle systems have been set up for a long time in other countries. It is common to see people riding bicycles through the city. We compared 10 cities known by bicycle sharing systems that are currently active as shown in Table 1. The remainder of the paper is organized as follows. Section 2 is the research methodology for the inventory replenishment strategies. Section 3 presents the experimental results and findings from the simulation software. Section 4 offers conclusions, following the references in the final section.

2 Research Methodology

In the first step of the research, we have to build the basic model to simulate real world YouBike System, and then we can develop several strategies to solve the problems by modifying the basic model with strategies applied. Finally, we compare the strategies from performance index, and propose the optimal solution.

The objective of the study is to find several feasible solutions to deal with the lack or surplus of bikes by simulation software. In order to mimic the operation of the YouBike System into our simulation model, there are four properties in our simulation models: (1) each station has different usage rate due to the location factors; (2) each station has different maximal capacity and different initiative amount of bikes (data obtained from the ratio of the total amount of bikes to the total parking space for every station from on-line website); (3) the travel distance between two stations must be considered and a proper distribution were used; (4) because of the "rent-it-here, leave-it-there" bike sharing service, it is possible for bicycles from all stations to reach each other.

Firstly, the basic model of the YouBike System is built and helps us to discover the bottleneck. We choose the 17 rental stations in Xinyi District in Taipei City to build our basic simulation model because the YouBike system in Xinyi District is the most mature one. For example, the lack of bikes is mostly happened in Citizen Square Station, Songde Station, as well as Wuchang Park Station, and the lack of parking space is mostly happened in MRT Taipei City Hall Station as well as MRT S.Y.S Memorial Hall Station. Secondly, we establish several models of different policy and use the statistical distributions to show the individual performance. In order to balance the quantity of bikes in all stations, we need to allocate the bicycles according to the statistical output data. One of our major concept of building the strategies for building simulation model is to monitor the operation of the YouBike System by a "double-queue" method. That is, prepare a set of bicycles as a second inventory queue and when original queue's inventory level reduce to 0, replenish the bikes to the station immediately.

Moreover, each station has a two-sided queue level to consider: (1) one is how many bikes waiting for the parking space when the parking space is full, and (2) the other condition is how many people currently waiting for bikes to rent. Hence, we use the concept of safety stock for both ends. Therefore, when the parking spaces are filled up over a particular level, the surplus bikes will be moved to other station based on some particular rules, such as moving to the nearest station. Also, if the available bicycles are less than a certain safety level, we will dispatch the appropriate amount of bikes from the other station to the target station.

To build an on-line monitoring system, it is necessary to use computer simulation software. Any Logic is a set of system simulation software providing both coding (by Java) and visual elements. It can be used on dynamic simulation analyzing, discrete simulation analyzing, etc. The users can take advantage of it to build up basic models with several fields.

In this paper, two sets of bicycles inventory replenishment policies were proposed: (1) policies focused on no bike to rent situation and (2) policies focused on no space to park the bikes situation. In order to shorten the length of the paper, all proposed inventory replenishment strategies are listed in Tables 3 and 4.

3 Experiment Results

All simulation models were running 30 days and two performance indexes were used to evaluate the proposed strategies. Table 2 shows the performance of the basic model which served as benchmark for number of average available bikes and number of average lack of parking space from basic model (without any strategy).

Table 2 Benchmark from the basic model

Benchmark (from basic model)
a. number of average available bikes: 75.13
b. number of average lack of parking space: 180.74

Replenishment Strategies for the YouBike System in Taipei

Table 3 Policies for no bike to rent situation

The description of the policy	Performance (30-day average)		Improvement from basic model	
	Bike	Space	Bike (%)	Space (%)
1. If the available bikes in one station are less than 20 % of the maximal capacity of the station, one bike is replenished in the order of the supply cycle which is developed according to the usage rate of each station	223.44	12.04	197.40	93.34
2. If the available bikes in one station are less than 10 % of the maximal capacity of the station, one bike is replenished in the order of the supply cycle which is developed according to the usage rate of each station	221.22	14.35	197.40	92.06
3. If the available bikes in one station are less than 20 % of the maximal capacity of the station, 10 % of the station's maximal available bikes are replenished in the order of the supply cycle which is developed according to the usage rate of each station	223.44	12.00	197.40	93.36
4. If the available bikes in one station are less than 10 % of the maximal capacity of the station, 10 % of the station's maximal available bikes are replenished in the order of the supply cycle which is developed according to the usage rate of each station	221.30	14.29	194.56	92.09
5. If the available bikes in one station are less than 20 % of the maximal capacity of the station, 20 % of the station's maximal available bikes are replenished in the order of the supply cycle which is developed according to the usage rate of each station	223.90	11.86	198.00	93.44
6. If the available bikes in one station are less than 10 % of the maximal capacity of the station, 20 % of the station's maximal available bikes are replenished in the order of the supply cycle which is developed according to the usage rate of each station	221.28	14.28	194.52	92.1
7. If the available bikes in one station are less than 20 % of the maximal capacity of the station, to increase the available bikes to 50 % of the station's maximal available bikes, 30 % of the station's maximal available bikes are replenished in the order of the supply cycle which is developed according to the usage rate of each station	223.87	11.90	197.98	93.42
8. If the available bikes in one station are less than 10 % of the maximal capacity of the station, to increase the available bikes to 50 % of the station's maximal available bikes, 40 % of the station's maximal available bikes are replenished in the order of the supply cycle which is developed according to the usage rate of each station	221.27	14.31	194.51	92.08

Table 4 Policies for no space to park the bikes situation

The description of the policy	Performance (30-day average)		Improvement Rate from basic model	
	Bike	Space	Bike (%)	Space (%)
1. If the parking space is 80 % full, one bike is moved to another station which is nearest	86.61	168.06	15.28	7.02
2. If the parking space is 70 % full, one bike is moved to another station which is nearest	170.68	83.97	127.17	53.54
3. If the parking space is 80 % full, one bike is moved to another two stations which are nearest	230.19	18.51	206.38	87.76
4. If the parking space is 70 % full, one bike is moved to another two stations which are nearest	214.66	31.83	185.70	82.40
5. If the parking space is 80 % full, 10 % of the station's maximal parking bikes are moved to another station which is nearest	86.31	168.47	14.88	6.79
6. If the parking space is 70 % full, 10 % of the station's maximal parking bikes are moved to another station which is nearest	83.62	171.06	11.3	5.36
7. If the parking space is 80 % full, 20 % of the station's maximal parking bikes are moved to another station which is nearest	87.01	167.53	15.82	7.31
8. If the parking space is 70 % full, 20 % of the station's maximal parking bikes are moved to another station which is nearest	233.52	64.00	210.82	97.61
9. If the parking space is 80 % full, 20 % of the station's maximal parking bikes are moved respectively to another two stations which are nearest. The moved bikes are in a total of 40 % of the station's maximal parking bikes	227.02	18.10	202.16	89.98
10. If the parking space is 70 % full, 20 % of the station's maximal parking bikes are moved respectively to another two stations which are nearest. The moved bikes are in a total of 40 % of the station's maximal parking bikes	213.33	31.79	183.94	82.41

Tables 3 and 4 show policies focused on no bike to rent situation and policies focused on no space to park the bikes situation, respectively.

The two performance indexes that we used to evaluate the performance from the proposed strategies have different meanings. For "number of average available bikes," the better policy will result in bigger value. However, for "number of average lack of parking space," the better policy will result in smaller value. Therefore, the improvement rates (IR) are defined and calculated by the following Eqs. (1) and (2).

$$IR_{\text{Bike}} = \frac{\overline{\text{available bike(with policy)}} - \overline{\text{available bike(benchmark)}}}{\overline{\text{available bike(benchmark)}}} \times 100\,\%.$$

$$(1)$$

$$IR_{\text{Space}} = \frac{\overline{\text{available space(with benchmark)}} - \overline{\text{available space(with policy)}}}{\overline{\text{available space(with policy)}}} \times 100\,\%.$$

$$(2)$$

4 Conclusions

In this study, through balancing the number of available bikes and parking space, our goal is to maximize the efficiency of the public bicycle sharing system, the YouBike system in Taipei. As a result, we use simulation software to build an on-line monitoring system which provides a real-time usage of bikes or the occupation of parking spaces of all rental stations. Moreover, according to this basic monitoring system, we developed several policies to solve the two main problems of the YouBike system: (1) no bike to rent and (2) no space to park the bikes in particular rental stations. Base on the result of simulation, we can obtain better performance from all strategies that we proposed and measure whether the policies could effectively solve the troublesome problems or not. Of all strategies that we proposed, the optimal policy that can increase 210.82 % available bikes and reduce 97.61 % lack of parking space strategy is "if the parking space is 70 % full, 20 % of the station's maximal parking bikes are moved to another station which is nearest." Further experiments can be done by expanding the model to the full size to verify that this strategy can efficiently solve the bicycle inventory replenishment problem.

Acknowledgments The authors thank the Department of Transportation, Taipei City Government, for providing useful information about the YouBike System in Taipei, Taiwan.

References

Chang L-C (2000) Design and management of urban bike sharing systems. Dissertation, National Cheng Kung University

Graves SC, Willems SP (2003) Optimizing strategic safety stock placement in supply chains. Manuf Serv Oper Manag 5:176–177

Humair S, Willems SP (2011) Optimizing Strategic Safety Stock Placement in General Acyclic Networks. Oper Res 59:781–787

Li J-R (1999) The study of optimal stock strategy under changeable lead time. Dissertation, National Taiwan University of Science and Technology

Thomopoulos NT (2006) Safety stock comparison with availability and service level. Paper presented at the international applied business research conference in Cancun, Mexico, 20–24 Mar 2006

Tzeng W-C (2013) Pricing of urban public bicycle-sharing system—Taipei YouBike, a case study. Dissertation, Tamkang University

Wang J-R (2006) Inventory policy for items with different demand patterns under continuous review. Dissertation, National Taiwan University of Science and Technology

A Tagging Mechanism for Solving the Capacitated Vehicle Routing Problem

Calvin K. Yu and Tsung-Chun Hsu

Abstract The Vehicle Routing Problem (VRP) is one of the difficult problems in combinatorial optimization and has wide applications in all aspects of our lives. In this research, a tag-based mechanism is proposed to prevent the formation of subtours in constructing the routing sequences, while each tour does not exceed the capacity of the vehicles and the total distance travelled is minimized. For each node, two tags are applied, one on each end, and to be assigned with different values. Two nodes can be connected, only if the tag value at one end of a node matches one of the end tag values of another node. The model is formulated as a mixed integer linear programming problem. Some variations of the vehicle routing problems can also be formulated in a similar manner. If the capacity of the vehicles is infinite, i.e., capacity restrictions are removed, the VRP reduces to the multiple Travelling Salesman Problem (mTSP). Computational results indicate that the proposed model is quite efficient for small sized problems.

Keywords Combinatorial optimization · Mixed Integer programming · Tagging mechanism · Traveling salesman problem · Vehicle routing problem

1 Introduction

Traveling Salesman Problem (TSP) is the basic type of Vehicle Routing Problems (VRPs) and is widely used in many industrial applications and academic researches. For example, applications can be found in the studies of production process,

---

C. K. Yu · T.-C. Hsu
Department of Industrial Engineering and Management, Ming Chi University of Technology, 84 Gungjuan Road, Taishan, New Taipei 24301, Taiwan, Republic of China
e-mail: calvinyu@mail.mcut.edu.tw

T.-C. Hsu
e-mail: darksnowii@gmail.com

Y.-K. Lin et al. (eds.), *Proceedings of the Institute of Industrial Engineers Asian Conference 2013*, DOI: 10.1007/978-981-4451-98-7_48,
© Springer Science+Business Media Singapore 2013

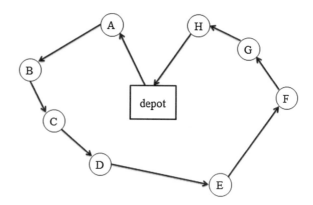

Fig. 1 The traveling salesman problem

transportation scheduling, biological engineering and electronic engineering, etc. Although TSP has proven its importance either in academy or in practice, however, it belongs to a large family of problems classified as NP-complete (Karp 1972).

The TSP contains a depot and several cities (nodes), and the distances between any two cities in the road network are known. A salesman departures from the depot and visits each city exactly once before returning back to the depot (Fig. 1). For economic consideration, the salesman has to find the shortest tour to visit all the cities.

The Multiple Traveling Salesman Problem (mTSP) is an extension of the well-known TSP, where more than one salesman is used in finding multiple tours. Moreover, the characteristics of the mTSP seem more appropriate for real life applications, and it is also possible to extend the problem to a wide variety of VRPs by incorporating some additional side constraints.

VRP is a derivative of the TSP (Lin 1965), and was first introduced in 1959 (Dantzig and Ramser 1959). The VRP can be described as the problem of designing optimal delivery or collection routes from one or several depots to a number of geographically scatted cities or customers, subject to side constrains. The VRP lies at the hearts of distribution management. There exist several versions of the problem and a wide variety of exact and approximate algorithms have been proposed for its solution. Exact algorithms can only solve relatively small problems (Laporte 1992).

Capacitated VRP (CVRP) is a VRP where vehicles have limited carrying capacity of the goods that must be delivered. Let one of the nodes be designated as the depot. With each node i, apart from the depot, is associated a demand Q_i that can be a delivery from or a pickup to the depot. The problem is to minimize the total distance traveled, such that the accumulated demand up to any node does not exceed a vehicle capacity. Each tour must start and end at the depot and all problem parameters are assumed to be known with certainty. Moreover, each customer must be served by exactly one vehicle (Fig. 2).

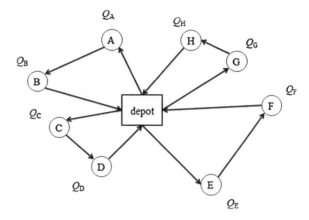

Fig. 2 The capacitated vehicle routing problem

The VRP is a NP-hard problem which makes it difficult to solve it to optimality. A survey by Laporte (1992) might be a good starting point to find out exact algorithms and heuristic algorithms in solving the VRPs.

In this paper, an exact algorithm is developed by applying tagging mechanism to each city to be visited, and a mixed integer program formulation is established in finding the optimal tours with minimum cost.

2 Analytical Approach

2.1 Notations

n	Number of cities
m	Number of salesman
c_{ij}	Travel cost from city i to city j
L_i	Left sequence tag of city i
R_i	Right sequence tag of city i
Q_i	Demand of city i
A_i	Cumulative quantity of arriving vehicle in city i
$vcap$	Capacity of the vehicles
x_{ij}	$\begin{cases} 1, & \text{if city } i \text{ precedes city } j \\ 0, & \text{otherwise} \end{cases}$

2.2 Assumptions

- Single depot with fixed location.
- Distances between nodes are known.

- Each node is served by a single salesman and each node is served exactly once.
- All salesmen shall start their tour from the depot and return to the depot to end their tour.
- There is no time limitation.

2.3 Tagging Mechanism

For each node, two tags are associated, namely, left and right, and let the value of the right tag always larger than the value the left tag by one. If two nodes i, j are connected to each other, and node i is visited prior to node j, then the right tag value of node i must equal to the left tag value of node j (Fig. 3). In this case, the subtours then can be eliminated.

To employ this mechanism on traveling sequence, first set $R_i = L_i + 1$ to each node i. If two nodes i and j are not directly connected (i.e., $x_{ij} = 0$), then the difference between R_i and L_j can be any value from $-(n-m)$ to $n-m$. If two nodes i and j are directly connected (i.e., $x_{ij} = 1$), then R_i must equal to L_j, which are:

when $x_{ij} = 0 : R_i - L_j \leq n - m$
when $x_{ij} = 1 : R_i - L_j = 0$

Rewriting these equations, we have:

$R_i - L_j \leq (n - m)(1 - x_{ij})$ where $x_{ij} = 0$
$R_i - L_j = (1 - x_{ij})$ where $x_{ij} = 0$

Combining these two equations, we have:

$R_i - L_j \leq (n - m)(1 - x_{ij})$ where $x_{ij} = 0$ or 1

The same procedure can be applied to the capacity restriction. If two nodes are not directly connected (i.e., $x_{ij} = 0$), then the difference between $Q_i + A_i$ and A_j

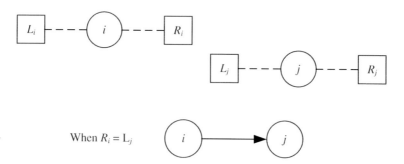

Fig. 3 The tagging mechanism

A Tagging Mechanism for Solving the Capacitated Vehicle   401

can be any value from $-vcap$ to $vcap$. If two nodes i and j are directly connected (i.e., $x_{ij} = 1$), then $Q_i + A_i$ must equal to A_j, which are:

when $x_{ij} = 0 : Q_i + A_i - A_j \leq vcap$
when $x_{ij} = 1 : Q_i + A_i - A_j = 0$

Rewriting these equations, we have:

$Q_i + A_i - A_j \leq vcap(1 - x_{ij})$ where $x_{ij} = 0$
$Q_i + A_i - A_j = (1 - x_{ij})$ where $x_{ij} = 1$

Combining these two equations, we have:

$Q_i + A_i - A_j \leq vcap(1 - x_{ij})$ where $x_{ij} = 0$ or 1

The complete formulation is:

$$\text{minimize} \sum_{i=1}^{n} \sum_{j=1}^{n} c_{ij} x_{ij} \tag{1}$$

$$\text{subject to} \sum_{i=1}^{n} x_{1j} = m \tag{2}$$

$$\sum_{i=1}^{n} x_{1j} = m \tag{3}$$

$$x_{ii} = 0 \quad (i = 1, 2, \ldots, n) \tag{4}$$

$$\sum_{j=1}^{n} x_{ij} = 1 \quad (i = 2, 3, \ldots, n) \tag{5}$$

$$\sum_{j=1}^{n} x_{ij} = 1 \quad (j = 2, 3, \ldots, n) \tag{6}$$

$$x_{ij} + x_{ji} \leq 1 \quad (i = 2, 3, \ldots, n; \ j = 2, 3, \ldots, n) \tag{7}$$

$$R_i = L_i + 1 \quad (i = 1, 2, \ldots, n) \tag{8}$$

$$R_i - L_j \leq (n - m)(1 - x_{ij}) \quad (i = 1, 2, \ldots, n; \ j = 2, 3, \ldots, n) \tag{9}$$

$$Q_i + A_i - A_j \leq vcap(1 - x_{ij}) \quad (i = 1, 2, \ldots, n; \ j = 2, 3, \ldots, n) \tag{10}$$

$$x_{ij} \in \{0, 1\} \tag{11}$$

In this formulation, Eq. (1) is the objective function, by summing up all distances between each city. Equations (2) and (3) ensure there is exactly m salesmen start from the depot and finally must return to the depot. Equation (4) ensures each node will not be connected to itself. Equations (5) and (6) ensure each node can only be visited once. Equation (7) ensures there is no loop occurs in each pairing nodes. Equations (8)–(10) are the tagging constraints discussed earlier.

3 Illustrative Example

Given an area, in which only 1 depot exists and the goods should be delivered to 6 cities and 3 vehicles are available. Each city shall be visited by one vehicle exactly once and each vehicle shall start its tour from the depot (starting point) and return to the depot to end its tour. Table 1 shows the distance matrix of city i (1–7) and city j (1–7) where node 1 represents the depot—the starting point.

The developed MIP formulation has been implemented in the LINGO software package and the results are shown in Table 2 and Fig. 4 respectively.

The optimal solution is $x_{15} = x_{16} = x_{17} = x_{23} = x_{31} = x_{41} = x_{54} = x_{61} = x_{72} = 1$ with minimum traveling distance 165.

By dropping constraint (10), the model becomes the mTSP, and the result is shown in Fig. 5. Further by setting the number of salesman $m = 1$, the model then becomes the TSP, and the result is shown in Fig. 6.

Table 1 Distance matrix for illustrative example

city j / city i	1	2	3	4	5	6	7
1	0	24	19	20	27	16	12
2	24	0	17	31	44	36	23
3	19	17	0	16	29	35	25
4	20	31	16	0	15	34	28
5	27	44	29	15	0	40	37
6	16	36	35	34	40	0	11
7	12	23	25	28	37	11	0

Table 2 Result for illustrative example

Li	Node i	Ri	Qi	Ai
0	1	1		
2	2	3	6	15
3	3	4	3	18
2	4	3	7	14
1	5	2	7	7
1	6	2	18	18
1	7	2	4	4

A Tagging Mechanism for Solving the Capacitated Vehicle

Fig. 4 Optimal tours with 3 vehicles

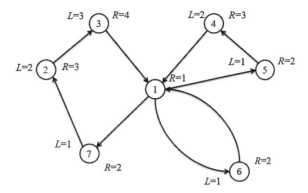

Fig. 5 Optimal mTSP tours

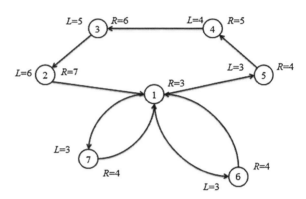

Fig. 6 Optimal TSP tour

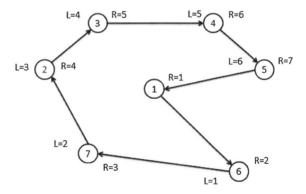

4 Conclusions

In this research, a MIP formulation of the CVRP is developed. By employing simple tagging mechanism to each node, the subtour elimination constraints are easy to implement and understand. For small size problem, this formulation is quite efficient, however, when the problem size gets larger, the time to obtain the optimal tours is also getting longer. Two variants of the CVRP are also presented in this paper. Simply by dropping or replacing value in the constraint, the formulation can easily be transformed into mTSP or classical TSP.

References

Dantzig GB, Ramser JH (1959) The truck dispatching problem. Manage Sci 6(1):80–91
Karp RM (1972) Reducibility among combinatorial problem. In: Miller R, Thatcher J (eds) Complexity of computer computations. Plenum Press, New York, pp 85–103
Laporte G (1992) The vehicle routing problem: an overview of exact and approximate algorithms. Eur J Oper Res 59(3):345–358
Lin S (1965) Computer solutions of the traveling salesman problem. Bell Syst Tech J 44(10):2245–2269

Two-Stage Multi-Project Scheduling with Minimum Makespan Under Limited Resource

Calvin K. Yu and Ching-Chin Liao

Abstract Project scheduling is one of the key components in the construction industry and has a significant effect on the overall cost. In business today, project managers have to manage several projects simultaneously and often face challenges with limited resources. In this paper, we propose a mixed integer linear programming formulation for the multiple projects scheduling problem with one set of limited resource to the identical activities within the same project or among different projects and minimum makespan objective. The proposed model is based on the Activity on Arrow (AOA) networks, a two-stage approach is applied for obtaining the optimal activities scheduling. At the first stage, the minimum makespan for completing all projects is computed. By fixing the minimum completion time obtained from the first stage, the second stage computation is then to maximize the slack time for each activity in order to obtain the effective start and finish time of each activity in constructing the multiple projects schedule. The experimental results have demonstrated practical viability of this approach.

Keywords Activity on arrow network · Mixed integer programming · Project management

---

C. K. Yu (✉) · C.-C. Liao
Department of Industrial Engineering and Management, Ming Chi University
of Technology, 84 Gungjuan Road, Taishan District, New Taipei City 24301, Taiwan
e-mail: calvinyu@mail.mcut.edu.tw

C.-C. Liao
e-mail: louis_liao@furong.com.tw

Y.-K. Lin et al. (eds.), *Proceedings of the Institute of Industrial Engineers Asian Conference 2013*, DOI: 10.1007/978-981-4451-98-7_49,
© Springer Science+Business Media Singapore 2013

1 Introduction

CPM and PERT have been successfully used in scheduling large complex projects that consist of many activities. When planning and evaluating the projects, project managers usually have to face the issue that the available resource is limited. Sometimes different projects have to share one set of valuable resource, which affects the effectiveness of project scheduling and timelines. Therefore, how to allocate these limited resources among the various projects and complete all projects in an acceptable time then becomes a big challenge to the project managers.

Demeulemeester et al. (2013) collects 11 carefully selected papers which deal with optimization or decision analysis problems in the field of project management and scheduling. This paper covers a considerable range of topics, including:

- solution methods for classical project scheduling problems
- mixed integer programming (MIP) formulations in order to solve scheduling problems with commercial MIP-solvers
- models and solution algorithms for multi-project scheduling
- extensions of the classical resource-constrained project scheduling problem
- models and solution approaches integrating decisions at different managerial levels such as selection and scheduling, or scheduling and control
- risk in project management and scheduling
- the assignment of resources to different stakeholders

Demeulemeester et al. also point out that the operations research techniques employed include common approaches such as metaheuristics and mixed-integer programming, but also specific approaches such as scenario-relaxation algorithms, Lagrangian-based algorithm, and Frank-Wolfe type algorithm.

In this paper, we propose a two stage approach by using the mixed integer programming models to generate optimal project schedules where all identical tasks have only one set of resource available and have to share among different projects. The objective is to complete all projects as early as possible.

2 Analytical Approach

Consider a multi-project scheduling problem where there is only one set of resource available for each task and the duration of each task is known with certainty. To avoid conflict in resource usage within the same project or among different projects, all identical tasks have to be scheduled either before or after each other. Therefore, a MIP model based on the AOA network is developed to help allocating valuable resources. The objective is to complete all projects as early as possible, i.e., minimize makespan.

2.1 Notations

t_{ij}	activity time of task j in project i
ns_{ij}	possible start time of task j in project i
nf_{ij}	possible finish time of task j in project i
et_{ij}	earliest start time of task j in project i
lt_{ij}	latest finish time of task j in project i
s_{ij}	slack time of task j in project i
$y_{ijj'}$	$\begin{cases} 1, \text{ if task } j \text{ finished before tase } j' \text{ within project } i \\ 0, \text{ otherwise} \end{cases}$
$z_{ikjk'}$	$\begin{cases} 1, \text{ if task } k \text{ in proj } i \text{ finished before tase } k' \text{ proj } j \\ 0, \text{ otherwise} \end{cases}$
M	a large positive number
T	finish time for all projects, i.e., makespan

2.2 Task Representation

Since the AOA network is used in developing the MIP model, the representation for each task j in project i is shown in Fig. 1.

The hollow circles in Fig. 1 are used to construct the project network, while the distance between the solid circles are the actual available time for completing a task. The dotted line represents the unusable time.

2.3 Stage 1 Computation

The first stage computation is to obtain the minimum makespan for all projects. The complete formulation is:

$$\text{minimize } T \quad (1)$$

subject to

Fig. 1 Task representation

(for each project i where task j precedes task k)

$$ns_{ik} \geq ns_{ij} + t_{ij} \tag{2}$$

$$nf_{ij} \leq nf_{ik} - t_{ik} \tag{3}$$

$$nf_{ij} = ns_{ik} \tag{4}$$

(for each project i and each task j)

$$et_{ij} \geq ns_{ij} \tag{5}$$

$$lt_{ij} \leq nf_{ij} \tag{6}$$

$$s_{ij} = lt_{ij} - et_{ij} - t_{ij} \tag{7}$$

(for each pair of identical tasks j and j' within the same project i)

$$et_{ij} + My_{ijj'} \geq lt_{ij'} \tag{8}$$

$$lt_{ij} + My_{ijj'} \geq et_{i'} \tag{9}$$

$$et_{ij'} + M(1 - y_{ijj'}) \geq lt_{ij} \tag{10}$$

$$lt_{ij'} + M(1 - y_{ijj'}) \geq et_{ij} \tag{11}$$

(for each pair of identical tasks k and k' between project i and j)

$$et_{ik} + Mz_{ikjk'} \geq lt_{jk'} \tag{12}$$

$$lt_{ik} + Mz_{ikjk'} \geq et_{jk'} \tag{13}$$

$$et_{jk'} + M(1 - z_{ikjk'}) \geq lt_{ik} \tag{14}$$

$$lt_{jk'} + M(1 - z_{ikjk'}) \geq et_{ik} \tag{15}$$

(for each project i)

$$lt_{i,finish} \leq T \tag{16}$$

$$y_{ijj'}, z_{ikjk'} \in \{0, 1\} \tag{17}$$

In this formulation, Eq. (1) is the objective function, by minimizing the total completion time T for all projects. (2–4) ensure the precedence relationship for each task j in project i. (5–7) ensure the actual time for each task falls in the available time range. (8–11) ensure the identical tasks within the same project to be staggered. (12–15) ensure the identical tasks among different projects to be staggered. (16) ensures all projects to be completed before time T.

2.4 Stage 2 Computation

At the completion of the first stage computation, the minimum makespan for all projects should be obtained without any question. However, in stage 1 computation, the possible finish times for tasks were pushed forward which causing most of the slack times for tasks are 0. To reclaim the slack times for tasks, therefore, the second stage computation is required. By fixing the minimum makespan obtained from stage 1, the stage 2 computation is simply by maximizing the sum of all slack times.

The MIP formulation for the second stage computation is basically the same as the first stage computation. The differences between these two formulations are:

- in objective function, Eq. (1) in stage 1 is to be replaced with:

$$\text{maximize} \sum_i \cdot \sum_j S_{ij} \tag{1'}$$

- in fixed makespan, Eq. (16) in stage 1 is to be replaced with:

$$lt_{i,finish} \leq minimum\ makespan\ obtained\ from\ stage\ 1 \tag{16'}$$

3 Illustrative Example

Consider three buildings construction projects where there is only one set of resource available for each task. The data is given in Table 1.

The AOA networks representation of these projects are shown in Fig. 2.

The developed MIP formulations have been implemented in the LINGO software package. After completing the first stage computation, the minimum makespan for all projects is 18 days, and the results are shown in Table 2. The Gantt chart representation of Table 2 is shown in Fig. 3.

Table 1 Data for illustrative example

		Project 1		Project 2		Project 3	
Task		Duration (days)	Predecessor(s)	Duration (days)	Predecessor(s)	Duration (days)	Predecessor(s)
A	Skeleton	2	×	3	×	2	×
B	Piping	3	A	4	A	3	A
C1	Wallboard	3	A	3	A	3	A
C2	Wallboard	2	B	–	–	2	B
D	Rock wool	2	C1	3	B	2	C1
E	Varnish	3	C2, D	2	D	3	C2, D

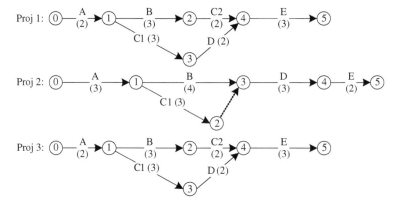

Fig. 2 Project network diagrams for illustrative example

Note that in Table 2, all tasks have zero slack time, which makes the finding of critical path somewhat difficult. Therefore, the second stage computation is applied, and the results are shown in Table 3. The Gantt chart representation of Table 3 is shown in Fig. 4. In Table 3, task E in project 1, task B in project 2 and task A in project 3 each has non-zero slack times, while the remaining tasks all with zero slack time. In this case, the critical path for each project is easy to find simply by taking the tasks with zero slack time.

Two-Stage Multi-Project Scheduling

Table 2 Solution from stage 1 computation

ns_{ij}		et_{ij}		et_{ij}		lt_{ij}		s_{ij}	
1A	0	1A	9	1A	7	1A	9	1A	0
1B	9	1B	13	1B	9	1B	12	1B	0
1C1	9	1C	13	1C	13	1C	13	1C	0
1C2	13	1C2	15	1C2	0	1C2	15	1C2	0
1D	13	1D	15	1D	13	1D	15	1D	0
1E	15	1E	18	1E	15	1E	18	1E	0
2A	0	2A	5	2A	2	2A	5	2A	0
2B	5	2B	10	2B	5	2B	9	2B	0
2C1	5	2C2	10	2C2	7	2C2	10	2C2	0
2D	0	2D	13	2D	10	2D	13	2D	0
2E	13	2E	18	2E	13	2E	15	2E	0
3A	0	3A	2	3A	0	3A	2	3A	0
3B	2	3B	5	3B	2	3B	5	3B	0
3C1	2	3C1	5	3C1	2	3C1	5	3C1	0
3C2	5	3C2	7	3C2	5	3C2	7	3C2	0
3D	5	3D	7	3D	5	3D	7	3D	0
3E	7	3E	18	3E	7	3E	10	3E	0

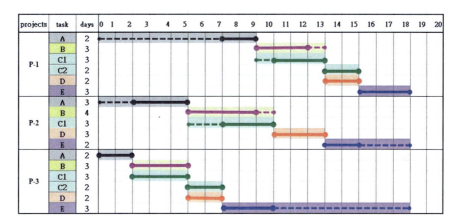

Fig. 3 Gantt chart representation for the stage 1 results

Table 3 Solution from stage 2 computation

ns_{ij}		et_{ij}		et_{ij}		lt_{ij}		s_{ij}	
1A	0	1A	2	1A	0	1A	2	1A	0
1B	2	1B	5	1B	2	1B	5	1B	0
1C1	2	1C1	5	1C1	2	1C1	5	1C	0
1C2	5	1C2	7	1C2	5	1C2	7	1C2	0
1D	5	1D	7	1D	5	1D	7	1D	0
1E	7	1E	18	1E	7	1E	13	1E	3
2A	0	2A	5	2A	2	2A	5	2A	0
2B	6	2B	10	2B	5	2B	10	2B	1
2C1	6	2C1	10	2C1	7	2C1	10	2C1	0
2D	10	2D	13	2D	10	2D	13	2D	0
2E	13	2E	18	2E	13	2E	15	2E	0
3A	0	3A	10	3A	5	3A	10	3A	3
3B	10	3B	13	3B	10	3B	13	3B	0
3C1	10	3C1	13	3C1	10	3C1	13	3C1	0
3C2	13	3C2	15	3C2	13	3C2	15	3C2	0
3D	13	3D	15	3D	13	3D	15	3D	0
3E	15	3E	18	3E	15	3E	18	3E	0

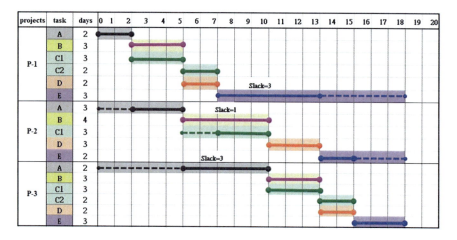

Fig. 4 Gantt chart representation for the stage 2 results

4 Conclusion

In this research, the MIP formulations of the multiple project networks scheduling with limited resource have been developed. In order to obtain the minimum makespan and construct the critical path, a two-stage computation approach is used. The first stage computation is to minimize the makespan for all projects, after that, the optimal solution obtained from the first stage is then used on the second stage to identify the critical path. The results show that this model is capable in allocating limited resources among several projects with minimum total completion time.

Reference

Demeulemeester E, Kolisch R, Salo A (2013) Project management and scheduling. Flex Serv Manuf J 25(1–2):1–5

A Weighting Approach for Scheduling Multi-Product Assembly Line with Multiple Objectives

Calvin K. Yu and Pei-Fang Lee

Abstract A flexible production scheduling in the assembly line is helpful for meeting the production demands and reducing the production costs. This paper considers the problem of scheduling multiple products on a single assembly line when multiple objectives exist. The specific objectives are to reduce the production changeovers and minimize overtime cost with the restrictions on production demands, due dates, and limited testing space. This system model derives from a real case of a company producing point of sale systems and a mixed integer programming model is developed. Different sets of weights that represent the priorities of specific objectives are applied to the model. By manipulating and adjusting these weights, it is capable to generate the desired production schedule that satisfies production managers' need. An example has been solved for illustrating the method. Preliminary findings suggest that the weighting approach is practicable and can provide valuable information for aggregate planning.

Keywords Assembly line · Mixed integer programming · Multi-objective optimization · Multi-product scheduling

1 Introduction

Business organizations continue to introduce new products, innovation and creativity to fulfill customer demands, and the control of cost is very necessary to the successful operation of business organizations. Scholars in the field of industrial

---

C. K. Yu (✉) · P.-F. Lee
Department of Industrial Engineering and Management, Ming Chi University of Technology, 84 Gungjuan Road, Taishan District, New Taipei City 24301, Taiwan
e-mail: calvinyu@mail.mcut.edu.tw

P.-F. Lee
e-mail: carollee68952003@yahoo.com.tw

Y.-K. Lin et al. (eds.), *Proceedings of the Institute of Industrial Engineers Asian Conference 2013*, DOI: 10.1007/978-981-4451-98-7_50,
© Springer Science+Business Media Singapore 2013

engineering have proposed many methods to reduce costs and improve quality which cover a wide range of business sizes and activities. In manufacturing organizations, product quality, delivery flexibility and cost of production are some important evaluation index of production management. Production management includes responsibility for product and process design, planning and control issues involving capacity and quality, and organization and supervision of the workforce. How to under limited resource to achieve effective production goal is testing the management ability of production managers. A good production schedule helps manufacturing organizations lower cost, reduce inventory, and improve customer service over all applicable time horizons.

Scheduling is the process of deciding how to commit resources between a variety of possible tasks. Multi-product scheduling can usually be treated as a multi-objective problem in which the planning schedule has multiple goals. Chang and Lo (2001) proposed an integrated approach to model the job shop scheduling problems, along with a genetic algorithm/tabu search mixture solution approach. The multiple objective functions modeled include both multiple quantitative (time and production) and multiple qualitative (marketing) objectives. Some illustrative examples are demonstrated using the genetic algorithm/tabu search solution approach. Lee (2001) evaluates artificial intelligence search methods for multi-machine two-stage scheduling problems with due date penalty, inventory, and machining costs. The results show that the two-phase tabu search is better in solution quality and computational time than the one-phase tabu search. Esquivel et al. (2002) shows how enhanced evolutionary approaches can solve the job shop scheduling problem in single and multi-objective optimization.

In this paper, we propose a weights method in determining the optimal production schedule for multi-product assembly line with multiple objectives. The objectives including consideration of production due date, number of time product types switched and necessity of overtime production.

2 Analytical Approach

2.1 Notations

n	number of product types
t	available production period (in days)
q_{ij}	production quantity of product type i on day j
dd_i	due date for product type i
d_i	demand for product type i
cap	capacity of testing room
chg_j	number of time product types switched on day j
ut_i	unit production time for product type i
rh	regular production hours per day

oh	overtime production hours per day
st	setup time for switching product types

$$y_{ij} \quad \begin{cases} 1, & \text{if product type } i \text{ is produced on day } j \\ 0, & \text{otherwise} \end{cases}$$

$$ot_j \quad \begin{cases} 1, & \text{if overtime is required on day } j \\ 0, & \text{otherwise} \end{cases}$$

$$shift_{ij} \quad \begin{cases} 1, & \text{if product type } i \text{ is getting on (off) line on day } j \\ 0, & \text{otherwise} \end{cases}$$

$$span_{ij} \quad \begin{cases} 1, & \text{if only product type } i \text{ can be produced on day } j \\ 0, & \text{otherwise} \end{cases}$$

wt_j	weight of production quantity on day j
wt_chg	weight on switching product types
wt_ot	weight on overtime
M	a large positive number

2.2 Assumptions

- All required material are ready prior to production
- Quality of raw material is stable
- Only one type of the product can be assembled at the same time
- Testing room has capacity restriction

2.3 Multi-objective Mixed Integer Programming Model

$$\text{Minimize} \sum_{i=1}^{n} \sum_{j=1}^{ddi} wt_j q_{ij} + \sum_{j=1}^{t} wt\_chg\left(chg_j\right) + \sum_{j=1}^{t} wt\_ot\left(ot_j\right) \tag{1}$$

$$\text{subject to} \sum_{j=1}^{ddi} q_{ij} \geq d_i \quad (i = 1, 2, \ldots, n) \tag{2}$$

$$\sum_{i=1}^{n} q_{ij} \leq cap \quad (j = 1, 2, \ldots, t) \tag{3}$$

$$\sum_{i=1}^{n} y_{ij} = chg_j \quad (j = 1, 2, \ldots, t) \tag{4}$$

$$\sum_{i=1}^{n} ut_i q_{ij} \leq rh + oh(ot_j) - st(chg_j) \quad (j = 1, 2, \ldots, t) \tag{5}$$

$$q_{ij} \leq M y_{ij} \quad (i = 1, 2, \ldots, n; \ j = 1, 2, \ldots, t) \tag{6}$$

$$shift_{i1} = y_{i1} \quad (i = 1, 2, \ldots, n) \tag{7}$$

$$y_{ij} - y_{ij+1} = s_{ij+1} - t_{ij+1} \quad (i = 1, 2, \ldots, n; \ j = 1, 2, \ldots, t-1) \tag{8}$$

$$shift_{ij+1} = s_{ij+1} + t_{ij+1} \quad (i = 1, 2, \ldots, n; \ j = 1, 2, \ldots, t-1) \tag{9}$$

$$\sum_{j-1}^{ddi} shift_{ij+1} \leq 2 \quad i = 1, 2, \ldots, n \tag{10}$$

$$y_{ij} + y_{ij+1} + y_{ij+2} \geq 3 span_{ij+1} \quad (i = 1, 2, \ldots, n; \ j = 1, 2, \ldots, t-2) \tag{11}$$

$$y_{ij} + y_{ij+1} + y_{ij+2} \leq 2 + span_{ij+1} \quad (i = 1, 2, \ldots, n; \ j = 1, 2, \ldots, t-2) \tag{12}$$

$$\sum_{i=1}^{n} span_{ij} \leq M z_j \quad (j = 1, 2, \ldots, t) \tag{13}$$

$$\sum_{i=1}^{n} y_{ij} \leq 1 + M(1 - z_j) \quad (j = 1, 2, \ldots, t) \tag{14}$$

$$q_{ij} \in integer \tag{15}$$

$$ot_j, y_{ij}, shift_{ij}, span_{ij}, s_{ij}, t_{ij}, z_j \in \{0, 1\} \tag{16}$$

Objective function (1) uses weights to control the production quantity, number of time product types switched and necessity of overtime production. The objective is minimizing the total weights gathered from all goals. The weight of production quantity on day j (wt_j), if setting it to a larger value, the production quantity on day j will be decreased, vice versa. If set relative large value to the weight on changing product types (wt_chg), it will make less product types switching during the production process. Increasing the weight of switching the assembly lines can avoid unnecessary switching. If set relative large value to the weight on overtime (wt_ot), it will make overtime unavailable.

Constraint (2) ensures the production quantity will meet the demand and due date. (3) ensures the daily production quantity will not exceed the capacity of testing facility. (4) counts the number of time product types switched. (5) ensures the daily production hours will not exceed the available working hours. (6) will force $y_{ij} = 1$, if $q_{ij} > 0$. (7–10) ensure per product type can only be on and off the production line one time, make it total less than 2 to fulfill batch processing requirement. (11–14) ensure the production will not be interrupted by another product type if the on line product type requires more than 3 days in production.

3 Illustrative Example

Consider a computer manufacturing factory carries out assembly operations during next 7 days and there are 6 operators working 8 h per day with additional available daily overtime of 2 h. A burn-in test should be run after the product is assembled. The capacity of burn-in test room is 400 computers. The time for switching product types is 20 min. Table 1 provides additional information on the product types, unit production times, demands and due dates.

The developed MIP formulations have been implemented in the LINGO software package. At first, overtime is not put into consideration (i.e., put heavy weight on overtime). By manipulating the weights of the production quantity, the resulting schedule is shown in Table 2. Since the weight on day 1 is much larger than the other days, we can expect a lower product quantity on day one. Due to the products due dates, the effect of day 2 is not obvious.

Table 1 Product demand and due date

Product type	Unit production time (min)	Demand	Due date
A	8	957	5
B	8	500	4
C	7	158	5
D	7	68	6
E	7	42	7
F	8	530	7
G	16	22	3
H	6	69	5

Table 2 Production schedule by assigning larger weights on the first two days

Day j	1	2	3	4	5	6	7	Total qty
Product i								
wt_j assigned	25	5	1	1	1	1	1	
A		81	357	357	162			957
B	266	274						500
C					158			158
D						68		68
E						42		42
F						256	274	530
G	22							22
H					69			69
Total qty	288	355	357	357	389	366	274	
Total time (min)	2,520	2,880	2,876	2,876	2,876	2,878	2,212	

Table 3 Production schedule by assigning larger weights on the last two days

Day j Product i	1	2	3	4	5	6	7	Total qty
wt_j assigned	1	1	1	1	1	5	25	
A			328	357	272			957
B	116	357	27					500
C	158							158
D					50	18		68
E					42			42
F						339	191	530
G	22							22
H	69							69
Total qty	365	357	355	357	364	357	191	
Total time (min)	2,880	2,876	2,880	2,876	2,880	2,878	1,548	

Table 4 Production schedule by assigning a smaller value to wt_chg

Day j Product i	1	2	3	4	5	6	7	Total qty
A		111	273	357	216			957
B	303	197						500
C					158			158
D						68		68
E						42		42
F						175	355	530
G		22						22
H	69							69
Total qty	372	330	273	357	374	285	355	
Total time (min)	2,878	2,876	2,204	2,876	2,874	2,230	2,860	

If the production is to be finished as early as possible, we can set larger weights on those days that near the end of the production period. The resulting schedule when we reverse the weights from the previous example is shown in Table 3.

By setting equal weighs on the production quantity, Table 4 show the result of putting smaller weight on the switching product types, while Table 5 uses a larger weight. The total switching times in Table 4 is 6, however, in Table 5, the switching times is 5 which is one time less than putting the smaller weight on the switching product types.

Table 6 shows the result of putting relatively smaller weight on overtime. As we can see, available working hours are fully used to fill up the capacity of the testing room most of the days.

A Weighting Approach for Scheduling Multi-Product Assembly Line

Table 5 Production schedule by assigning a larger value to *wt_chg*

Day j Product i	1	2	3	4	5	6	7	Total qty
A			357	357	243			957
B	170	330						500
C	158							158
D						68		68
E						42		42
F						173	357	530
G	22							22
H					69			69
Total qty	350	330	357	357	312	283	357	
Total time (min)	2,878	2,660	2,876	2,876	2,398	2,214	2,876	

Table 6 Production schedule by assigning a smaller value to *wt_ot*

Day j Product i	1	2	3	4	5	6	7	Total qty
wt_j assigned	1	1	1	1	1	5	25	
A			341	400	216			957
B	268	232						500
C		158						158
D	68							68
E	42							42
F					184	346		530
G	22							22
H		10	59					69
Total qty	400	400	400	400	400	346	0	
Total time (min)	3,346	3,082	3,122	3,220	3,240	2,788	0	
Over time (min)	466	202	242	340	360	0	0	

4 Conclusion

The model developed in this research can quickly obtain the optimal production schedule for each product type with the limitations of space, time and operators. Weights are used in this model, by manipulating and adjusting these weights, it is capable to generate the desired production schedule that satisfies production managers' need and make the production schedule more flexible.

References

Chang PT, Lo YT (2001) Modelling of job-shop scheduling with multiple quantitative and qualitative objectives and a GA/TS mixture approach. Int J Comput Integ M 14(4):367–384

Esquivel S, Ferrero S, Gallard R, Salto C, Alfonso H, Schütz M (2002) Enhanced evolutionary algorithms for single and multiobjective optimization in the job shop scheduling problem. Knowl-Based Syst 15(1):13–25

Lee I (2001) Artificial intelligence search methods for multi-machine two-stage scheduling with due date penalty, inventory, and machining costs. Comput Oper Res 28(9):835–852

Exploring Technology Feature with Patent Analysis

Ping Yu Hsu, Ming Shien Cheng, Kuo Yen Lu and Chen Yao Chung

Abstract The patent literature has documented 90 % of the world's technological achievements, which are protected by the patent law of each country. But with the increasingly competitive technology, enterprises have started the patent strategy research and attached great importance to patent analysis. The patent analysis uses statistics, data mining, and text mining to convert the information into a competitive intelligence that facilitates corporate decision making and prediction. Thus, the patent analysis has become a corporate weapon for long-term survival and protection of commercial technologies. The patent analysis in the past, compared with the trend analysis, mostly conducted the predictive analysis of a number of keywords and patents through the statistical analysis approach. However, the keywords found were limited to the already mature technology and could not locate the implicit emerging terms, so the patent analysis in the past could only find the words of obvious importance, but fail to find the emerging words that are unobvious yet will have a major impact on future technologies. Therefore, how to find these words of a low-frequency nature to make prediction of the correct trend is an important research topic. This study used the Chinese word segmentation system to find the words of the patent documents and extracted the words according to the probability model of the Cross Collection Mixture Model. This model targets the words under changes in the time series. The background model and the common theme in the model will eliminate frequent words without the meaning of identification and collect words persistently appearing across time. This method can quickly screen enormous volumes of patent documents, extract

P. Y. Hsu · K. Y. Lu · C. Y. Chung (✉)
Department of Business Administration, National Central University, No.300, Jhongda Road, Jhongli City, Taiwan (R.O.C.) 32001, Taoyuan County
e-mail: 984401019@cc.ncu.edu.tw

M. S. Cheng
Department of Industrial Engineering and Management, Ming Chi University of Technology, No.84, Gongzhuan Road, Taishan District, New Taipei City 24301, Taiwan (R.O.C)
e-mail: mscheng@mail.mcut.edu.tw

Y.-K. Lin et al. (eds.), *Proceedings of the Institute of Industrial Engineers Asian Conference 2013*, DOI: 10.1007/978-981-4451-98-7_51,
© Springer Science+Business Media Singapore 2013

from the patent summary emerging words of a low-frequency nature, successfully filter out the fashion words, and accurately detect the future trends of emerging technologies from the patent documents.

Keywords Patent analysis · Text mining · Cross collection mixture model · Technology feature

1 Introduction

Due to the rapidly evolving technology, various types of products are continually changing and increasing in their complexity. Under such a fast change and short product life cycle, the technological innovation of the product has become the main source of corporate profits. Therefore, how to master the key technology has become the main weapon of today's enterprises to maintain their corporate competitiveness. By virtue of a unique industrial technology, the enterprise will be able to improve the corporate profits from its products, so the enterprise should attach more importance to the patent layout for its own unique technology. A complete patent layout for the enterprise is like a big umbrella that protects the corporate from being harmed by foreign competitors, and by expanding this patent umbrella enterprises can keep flourishing.

The quantitative prediction analysis of existing patents is aimed at historical data, while the research data columns are aimed at countries, inventors, patent classification, calculation of the growth of literature category, and technological growth and decline for predicting possible future developments, the predictive results are all the known research areas and cannot accurately tell what will be a potential area of . Therefore, knowledge mining by virtue of the combination of data mining and text-mining techniques to identify the possible potential knowledge from the semi-structured data, such as the summary and documentation of the patent, has also become an increasingly important analytical tool in recent patent analysis.

The keyword search was based on the keywords as defined by the knowledge of experts and therefore was often unable to uncover the implicit emerging technical documents or new words. So, it has been suggested that text-mining be used to find the keywords of the patent documents. This method of information retrieval is important because it can accurately find out the keywords representative of the patent documents and pinpoint this keyword using the growth curve to conduct analysis in the hopes of finding the emerging keywords. But usually the enterprise deliberately substitutes other words to prevent competitors retrieving this keyword, leading to the finding of keywords relating to already mature technology, which can cause great difficulties for the growth curve forecasts. Therefore to find out the emerging technology words from patent documents to help enterprises quickly cut into the possible technical development is the main objective of this study.

This study will pinpoint the content of summary of the unstructured patent documents to conduct trend analysis of words and identify possible potential new words in the hopes of finding words of a low-frequency nature in the past but under the changes in the time series will pop up in the future to become emerging words.

This study is structured as follow: Sect. 2 is the Literature that views this study's related patent predictive analysis and Cross Collection Mixture Model. Section 3 is System Processes for deriving the algorithm proposed in this study. Section 4 is the Empirical Analysis, which used the results of the research model to conduct the words trend analysis. Section 5 is the Conclusion, and the future research direction derived by this study.

2 Related Work

As the patent document has a time stamp, we hope to find the words and the change of theme under the time change, so that we can see the variations of the patent keyword. In 2004, Morinage and Yamanishi (Morinage and Yamanishi 2004) proposed the Finite Mixture Model for the real-time analysis of CRM (Customer Relationship Management), knowledge management, or Web monitoring. This model can use dynamic monitoring variation on the theme and use theme features to facilitate document clustering (Wu et al. 2010).

Correlations exist among the time, authors, or locations in documents. For example, at a press event the articles written by the same reporter will have the same style of coverage, or in the coverage of the tsunami event the theme reported in the time change will usually be different. In 2006, Mei and Zhai (Mei and Zhai 2006) proposed the contextual probabilistic latent semantic analysis (CPLSA) to improve the old probabilistic latent semantic analysis (PLSA). The new model introduces a new environment variable, such as spatiotemporal mining, author-topic analysis, temporal text-mining, etc. The documents have different underlying theme background time, place, and author (Porter et al. 1991; Perkiom et al. 2004; Chen et al. 2005). Zhai et al. (2004) adopted the Temporal Text-mining of Cross Collection Mixture Model (Fig. 1) to identify the life cycle changes of the tsunami news events and the intensity changes of the theme in each time of the news events. This study quoted the Cross Collection Mixture Model and by the background model removed the words that appeared too often. Meanwhile, the words collected by the common theme are specific words existing persistently in the documents in time, so they are popular and representative. The specific theme will only collect particular words appearing at certain times. The probability of these words standing out represents its intensity at the specific time. The purpose of using the specific theme is to see the frequency of these words in these times and observe which words have a sudden appearance and a high likelihood to continue its emergence in the future.

Fig. 1 Cross collection mixture model

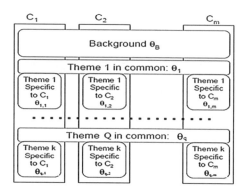

3 System Design

The past patent analysis shows the patent analysis only focuses on the structured data (patent applicant, the technical category) to conduct the statistical analysis by aiming at the statistical model from a variety of different angles, such as the number of inventors, the amount of patent applications, and patent citations for quantitative analysis. But the problem often faced is that the structured data of patent cannot accurately represent the contents of the patent, only find out the titled keywords-related documents or specifically classified documents, but this search method requires many professionals to conduct the analysis and judgment to digest the large number of messages. Therefore, the use of computer-aid to effectively find out the valuable words of the patent document is an important subject we need to discuss. Our laboratory aims at WEBPAT Taiwan to provide the patent summary, emerging words strength across time, identification of keywords, and difference of emerging degree to identify possible future emerging technology areas, with the hope of increasing the predictive accuracy.

In this study, the algorithmic steps include:

1. Data collection and extraction.
2. Chinese word segmentation system processing/initial value setting.
3. Processing Cross-Collection Mixture Model-SQL.
4. Parameters Estimation with EM Algorithm.
5. Word analysis under specific theme.

Figure 2 Patent content documents collection $C = \{d_1, d_2, \ldots, d_k\}$, in order to extract a theme evolution, this study made the time of a year as a time interval. The documents were divided into m time intervals, i.e., $C = C_1 \cup C_2 \cup C_3 \cup C_m$, and the summary content of each document was composed of words from the collection $d = \{w_1, w_2, \ldots, w_{|V|}\}$.

Themes (θ) were to collect words with the conditional probability as the theme or sub-theme of semantic coherence. And the collection of the theme was done by the distribution of $P(w|\theta)$ with the probability of appearance of each word.

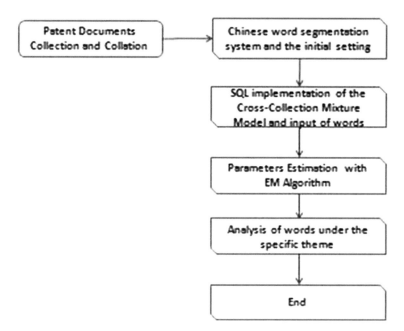

Fig. 2 System design flow

These words were sorted with probability, and the language model θ was composed of some words of higher $P(w|\theta)$ as a summary and definition of this theme. In which $\sum_{w \in v} P(w|\theta) = 1$, the extraction theme of this study makes use of the Cross Collection Mixture Model to divide the results of the conditional probability into the thematic cluster θ or background theme θ_B, common theme θ_j and specific theme θ_{ij}. Each patent document will be attributed to the Q theme group in the form of π_{ijd}

According to this study, as we hoped to find emerging words, we had to segment the words out of the patent documents. And in the natural language processing, the word is the most basic unit in processing, so we conducted the Chinese word processing for the collated summary. Through the Chinese word segmentation processing, the unstructured text can be collated into structured data to facilitate text-mining analysis.

Based on the source of information to set the value, the initial value may expressed as follows:

The initial value of $P^0(w|\theta_i)$ expressed as follows

$$P^0(w|\theta_i) = \lambda_c \times \frac{\sum_{j=1}^{m} \sum_{d \in c_j} \pi_{jdi} c(w, d)}{\sum_{j=1}^{m} \sum_{d \in c_j} \sum_{w' \in V} \pi_{jdi} c(w', d)} \quad (1)$$

The initial value of $P^0(w|\theta_{ij})$ expressed as follows

$$P^0\left(w|\theta_{ij}\right) = (1 - \lambda_c) \times \frac{\sum_{d\in c_j} \pi_{jdi}c(w,d)}{\sum_{d\in c_j}\sum_{w'\in V}\pi_{jdi}c(w',d)} \tag{2}$$

This study uses the Cross-Collection Mixture Model and conditional probability distribution to collect words. There are Q common themes and $Q \times |C|$ specific themes and background model in this model. The probability distribution of the words collection operation is as follows:

$$p_{d,C_j}(w) = \lambda_B p(w/\theta_B) + (1 - \lambda_B)\left(\sum_{\theta_i\in\Theta}\pi_{jdi}(1-\lambda_s)p(w/\theta_i) + \lambda_s p\left(w/\theta_{ij}\right)\right) \tag{3}$$

We express all word probability distributions as follows:

$$\log P\left(<C_1,..,C_m>\right) = \sum_{j=1}^{m}\sum_{d\in C_j}\sum_{w\in V}\left\{ c(w,d)k_{d,j}\log[\lambda_B p(w|\theta_B)\right.$$
$$\left. + (1 - \lambda_B)\left(\sum_{\theta_i\in\Theta}\pi_{jdi}((1-\lambda_s)p(w/\theta_i) + \lambda_s p\left(w/\theta_{ij}\right))\right)]\right\} \tag{4}$$

This study used the Expectation Maximization (EM) Algorithm to calculate the relative maximum probability value of each word falling into each theme, with the relative concept of probability to calculate the relative probability value of each sample falling into all the theme clusters.

The emerging term is defined in this study as the word that used to be of low-frequency in the past but has suddenly become a high-frequency word during this period. This study uses the Cross-Collection Mixture Model. After screening the words, those words collected from a specific theme in collection time. Based on strength of words, we look for the words of the top one percent strength in j period and analyze the words of low-frequency in the j-1 period. The words found will belong to the scope of our emerging words. And we will use the third period and the common theme to verify our results and see whether the emerging words continue to appear or have fallen into the common theme.

4 Empirical Analysis

In this study, first the patent documents of the WEBPAT Taiwan are collected based on IPC classification rules of patent documents to conduct research on the four categories of patent documents G06Q, H04B, H04L, and H04 N [8]. These four categories of patent documents classification are described as follows:

Exploring Technology Feature with Patent Analysis

1. G06Q: The data processing systems or methods applies specifically to the purpose of administration, management, commerce, operation, supervision, or prediction; and the same not included in other categories.
2. H04L: Transmission of digital information, such as telegraph communications.
3. H04 N: Image communication, such as TV.
4. H04B: Transmission.

In this study, content of the patent documents was retrieved from the patent summary, a total of 1,562 patent contents were selected from patent documents released from 2003 to 2008, and patent documents per year were taken as a collection. 2003: 378 cases $(C_1 = \{d_1, d_2, \ldots, d_{378}\})$; 2004: 253 cases $(C_2 = \{d_{379}, d_{380}, \ldots, d_{631}\})$; 2005: 324 cases $(C_3 = \{d_{632}, d_{633}, \ldots, d_{955}\})$; 2006: 275 cases $(C_4 = \{d_{956}, d_{957}, \ldots, d_{1230}\})$; 2007: 265 cases $(C_5 = \{d_{1231}, d_{1232}, \ldots, d_{1495}\})$; In 2008 because of the open time limit for patent information that is publicly available, there are only 67 cases $(C_6 = \{d_{1496}, d_{1498}, \ldots, d_{1562}\})$.

Cases of patent document under each category: G06Q had 379 cases, H04L: 448 cases, H04N: 396 cases, and H04B: 337 cases. The information column name after filtration and collation are as follows: the patent document names; collecting data for six years; the patent summary.

Data of this 1,562 cases were fed to the Chinese word segmentation system to analyze the summary of the patent documents about the words and calculate the number of times c (w, d) the words appeared in each document. The resulting number of times the word appeared and the patent documents formed a two-dimensional matrix. This word segmentation system based on the corpus to conduct a contrast collected a total of 3,901 words. $V = \{w_1, w_2, \ldots, w_{|V|}\}$. The following analysis was then conducted.

Based on our study, the sources of patent documents collected had four categories, so we default the four themes. With a total of six years of data, making a total of 24 specific themes, the default cluster 1 was the patent documents H04B with a total 337 cases, cluster 2 H04L with 448 cases, cluster 3 H04N with 396 cases, and cluster 4 with G06Q 379 cases.

The Cross-Collection Mixture Model used a probability distribution for words collection, which consisted of four groups of θ_i (common theme) and θ_B (background model). Whereas it must first set value, the size of the parameter value is set by the researcher. This study set the model to gather words in θ_B with the proportion or weight as $\lambda_B = 0.95$. As most of the data of the patent documents in this study were 300–500 words, with many words scattered, and the more frequent words were only concentrated in a few words, a larger λ_B value could be set. Because this study hopes to identify more specific themes, we set the λ_C value as $\lambda_C = 0.4$ with the hope of making a specific theme value of a larger difference. In addition, we also set the other two sets of parameters. One set was $\lambda_B = 0.95$, $\lambda_C = 0.4$ and the other set was $\lambda_B = 0.95$, $\lambda_C = 0.6$ to contrast whether the results were subject to change when the parameter was different and to compare with our experimental results.

Table 1 Top ten key words of each theme

keyid	Common1	keyid	Common2
偏振片 (Polarizer)	0.037229	緩衝區 (Buffer)	0.024247
廣視角 (Wide Perspective)	0.01986	最大數 (Maximum Number)	0.013471
視角膜 (Perspective Film)	0.01986	合會 (RCAs)	0.013344
靜態影像 (Static Image)	0.013374	管理資訊 (Management Information)	0.009685
替代性 (Alternative)	0.012145	界線 (Boundary)	0.00899
占卜 (Divination)	0.011866	薪資 (Salary)	0.008648
圖案 (Pattern)	0.01143	數位內容 (Digital Content)	0.007228
管理中心 (Management Center)	0.009016	通行碼 (Pass Code)	0.007136
安全模組 (Security Module)	0.008354	光罩 (Mask)	0.007071
色度 (Chroma)	0.007779	記憶體管理 (Memory Management)	0.006225
keyid	**Common3**	**keyid**	**Common4**
運動 (Movement)	0.103633	物件 (Object)	0.191586
柱形 (Columnar)	0.080316	串流 (Streaming)	0.131786
字體 (Font)	0.076037	像素 (Pixels)	0.121299
積分器 (Integrator)	0.065446	媒體 (Medium)	0.111169
端面 (End)	0.055025	像框 (Frame)	0.077681
預測值 (Predictive Value)	0.051258	參數 (Parameter)	0.067922
凹面鏡 (Concave Mirror)	0.028581	相機 (Camera)	0.065405
預測誤差 (Prediction Error)	0.02445	區段 (Section)	0.063112
集成 (Integrated)	0.020575	序列 (Sequence)	0.050709
射角 (Angle)	0.017725	鏡頭 (Shot)	0.046344

Table 1 shows $P(w|\theta_i)$ (i = 1, 2, 3, 4) of the top ten words. According to the original model common theme $P(w|\theta_i)$ can be used to describe the clustering of the themes, thereby locating the high strength words of the common theme to help describe the theme. The θ_1 keywords are the words related to the visual images, θ_2 are the words related to digital content management, θ_3 are the words for processing method for image perspective, and θ_4 are the words related to the digital content image data transmission method. All of these words continue to appear and have the power of identification. So we can based on the description of each theme filter out wanted keywords content

5 Conclusion and Future Research

From the study results, we can successfully find the emerging words of low-frequency in the past, which is different from the previous relevant studies with patent retrieval analysis focusing on the obvious and important words, while these obvious but important emerging words can help enterprises make accurate search to look for patent documents of emerging technologies or help enterprises by providing another possible direction for future research and development. We hope to take this approach to provide another keyword search method on the patent retrieval analysis. This method can help enterprises save a lot of labor costs in finding the answer they want from a large amount of patent documents content. The emerging words search is different from the past general data mining. The occurrences of emerging technologies are fast and rising rapidly so when carrying out trend analysis we should focus on words that first appeared but not obviously.

These keywords can help enterprises find the hidden patent documents of emerging technologies, which may also be where the enterprises' emerging technologies are located. Therefore, this study can help make more accurate results for the forecast of emerging technologies.

The literature of the technology gap mentions the use of the principal component analysis to reduce the dimension of the word. We hope the word search method of this study can be added to the empty hole diagram of the technology gap as discussed in the literature. This method can find the technology gap of the patent documents using the scatter diagram of the first principal component and the second principal component. We hope the word search will adopt the words of a common theme that continually appear. With a descriptive significance for the theme, the empty diagram can therefore be integrated to improve the shortcoming that the principal component analysis cannot effectively describe the field.

References

Chen CC, Chen MC, and Chen MS (2005) Liped: Hmm-based life profiles for adaptive event detection

Mei QZ, Zhai CX (2006) A mixture model for contextual text mining, ACM 1-59593-339-5/06/0008

Morinaga S, Yamanishi K (2004) Tracking dynamics of topic trends using a finite mixture model, ACM 1-58113-888-1/04/0008

Perkiom J, Buntine W, and Perttu S (2004) Exploring independent trends in a topic-based search engine

Porter AL, Roper AT, Mason TW et al (1991) Forecasting and management of technology. Wiley, New York

Wu YH, Ding YX, Wang XL et al (2010) Topic Detection by topic model induced distance using biased initiation. Advance in Computer Science and Information Technology

Zhai CX, Velivelli A, Yu B (2004) A cross-collection mixture model for comparative text mining, ACM 1-58113-888-1/04/0008

Making the MOST® Out of Economical Key-Tabbing Automation

P. A. Brenda Yap, S. L. Serene Choo and Thong Sze Yee

Abstract Industrial engineers observe and analyze movements and steps taken by a worker in completing a given task. The data obtained is defined and summed up through coded values to predetermine the motion time of process activities that occur within the production line. Although softwares have been developed to ease the recording and computation of such studies, they are expensive and require trainings to operate. The use of Maynard Operations Sequence Technique (MOST®) system has been recognized as one of the standards for predetermined time calculation. However, the sequential breakdown and movement analysis techniques of this system involving coded values can be complicated, time-consuming and tedious. This is worsened when the production line contains numerous activities with lengthy processes. The objective of this paper is thus to introduce an easy and cost-effective solution that uses simple Microsoft Excel macros key-in method which enables a trained analyst to record, determine and generate a MOST® time study within seconds. Case studies performed have shown that this method speeds up the calculation process at a 50–60 % rate faster than the ordinary individual code definition and tabulation recording method. As such, it has been proven to save time and eliminate the possibility of a miscalculation.

Keywords MOST · Time study · Industrial engineering · Cycle time · Work measurement · Predetermined time studies

---

P. A. B. Yap (✉) · S. L. S. Choo · T. S. Yee
A1 OPEX.No.25, JalanPuteri 11/8, Bandar PuteriPuchong 47100 Puchong, Malaysia
e-mail: brendayap@gmail.com

S. L. S. Choo
e-mail: slingchoo@hotmail.com

T. S. Yee
e-mail: g1240510789@studentmail.unimap.edu.my

Y.-K. Lin et al. (eds.), *Proceedings of the Institute of Industrial Engineers Asian Conference 2013*, DOI: 10.1007/978-981-4451-98-7_52,
© Springer Science+Business Media Singapore 2013

1 Introduction

Maynard Operations Sequence Technique (MOST®) is an index coded work measurement system where analyzed actions and processes are recorded down, code assigned and tabulated to provide a predetermined time study. It's a system designed to measure work through movement; an alternative method to using other systems such as MODAPTS (Modular Arrangement of Predetermined Time) or Methods-Time Measurement (MTM) Standards.Developed by H.B Maynard and Company,Maynard Operations Sequence Technique (MOST®) gained a wide reognition in Industrial Engineering practice since 1975 based on the fundamental statistical principles and work measurement data which evolved into a logical and natural way to measure work. Ithas since then, been applied in all sorts of industries ranging from production lines to offices, material handling and even finishing operations.

When basic MOST is applied into an analysis, the actions and movements are categorized according to 3 different sequence models. They are The General Move Sequence, The Controlled Move Sequence and the Tool Used Sequence Model. Each of these models has its sub activities broken down and represented by a series of alphabet indexes. A numbering value index will also be assigned to each alphabet depending on the steps and number of times a movement is repeated. Table 1 shows the basic MOST sequence models of the categorized activities.

2 Defining MOST

MOST uses time units that defined in TMU (Time Measurement Units) whereby 1 TMU is equivalent to 0.00001 h or 0.0006 min or 0.036 s. MOST® works by adding up the total index parameter values which an analyst will assign to the movement/actionand later adding it all up to get the predetermined time in Time Measurement Units (TMU). From there, the total TMU will be converted in order

Table 1 Basic MOST sequence models

Activity	Sequence model	Sub-activities
General move	ABG ABP A	A—action distance
		B—body motion
		C—gain control
		P—placement
Controlled move	ABG MXI A	M—move controlled
		X—process time
		I—alignment
Tool used	ABG ABP U ABP A	U—can be replaced with a specific action such as fasten, loosen, cut, surface treatment, measure, record and think.

Making the MOST® Out of Economical Key-Tabbing Automation   435

to get the final time in either minutes or seconds. The index value assigned to each sub-activity has already been set in accordance to a prefixed MOST reference data card which classifies and categorizes the number and types of movements with the index value that an analyst will base on. For example, in a scenario where a skilled operator walks 5 steps to turn on a simple tact button switch by pressing and remain standing there; the MOST®code sequence will be classified as a controlled move with the parameters A B G M X I A. The index value assigned for the parameters are demonstrated in Table 2.

3 Problem Description

While MOST is known to be an efficient method of getting a predetermined time study, it is also complex and tedious in the compilation of data analysis. Thus, by having a handful of procedures, one will tend to make mistakes in observation, recording or even in the tabulation of data.Common human errors that are bound to influence or affect the accuracy of the recordings and most importantly affecting the summation of time values are:

- Fatigue and tiredness
- Loss of focus due to concentration on the video and process analysis
- Complicated processes that require more time and attention of the analyst
- A repeated sequence or action can confuse the analyst in making calculations
- Negligence in calculation

4 Collecting Data and Applying MOST

A certified MOST analyst is able to breakdown the entire process that is being observed and classifying it with parameters as well as assigning index values to each code sequence. The process is long, tedious and can be troublesome at times especially when the process involves complicated and sophisticated movements or activities. In many cases, a video of the entire process is captured and recorded first in order to playback and analyze with the MOST system. The video analysis

Table 2 MOST analysis sample

Activity sequence:
Walk 5 steps to a machine and turn on the tact switch by pressing it and remain standing there. *Code defined:*
A10 B0 G1 M1 X0 I0 A0
$= 10 + 0 + 1 + 0 + 0 = 120$ TMU (1)   (1)
$= 120$ TMU $\times 0.036 = 4.32$ s   (1)

requires the analyst to playback the video in a reduced speed and may even have to repeat the video more than once to determine the significant action(s), eliminating non value added activities and even identifying steps that do not contribute to the efficiency of the process. However, the analyst has to weigh and ration between the identification of steps that are practical eventhough some of it may seem like a waste of time.In an experimental comparison of a certified analyst who used a manual recording method of MOST versus the usage of the MOST Excel sheet to analyze a 15 min video of a production operator at work, the application of the MOST Excel sheet used to produce the final cycle time result and tabulation was 60 % faster compared to the other manual tabulation analysis method.

4.1 Manual Versus Computerized Macros

The advantages of using MOST macros sheet is mainly speed, convenience and accuracy. It is clear that formanual pen-paper writing or individual key-in of parameters and a shortcut key generated parameter; the latter is a very much faster and efficient solution.With the Microsoft Excel macros sheet tabbing method, the shortcut keys assigned to generate the MOST parameter alphabets immediately allows the analyst to generate the string of action codes within seconds and only has to define the index code value to each alphabet. It is also cost-effective in terms of the need to purchase other work measurement software or licenses in the market that provides the same function of doing a time study. The comparisons between manual and macros recording methods these two are mapped out in Table 3 (Table 4).

Table 3 Comparison of MOST recording methods

Manual/single key in MOST	Macros MOST
• Analyst has to remember the alphabet codes to represent the correct action category and what it stands for.	• Sequence parameter alphabet codes are generated immediately by pressing two buttons. Saves time
• If written on paper, errors are harder to be corrected and checking back data will be difficult	• Analyst only requires to remember which shortcut generates which parameter
• Analyst is required to add up the time using a calculator and also make conversion for the unit of time measurement	• Computer generated data makes amendments and check backs easy
• Repeated processes or sequences can be confusing to the assigned analysis; sometimes affecting the tabulation accuracy	• The summation of time are added up and converted in different measurement units automatically
	• The number of a repeated action or sequence of an activity can be easily edited without having to make a recalculation of the data
	• Cost effective. No external/additional software or licensing is required

Making the MOST® Out of Economical Key-Tabbing Automation 437

Table 4 Example of MOST parameters generated using macros sheet

PREDETERMINED TIME STUDY

Customer: Standard Time (sec): 25.56   Prepared By:

Product: Assembly: UPH: 140   Date:

No	Method	L/R	MOST CODE	f	TMU	f	Total TMU	Time(sec)
1	Move Tray to workbench	R	A_3 B_0 G_1 A_3 B_0 P_1 A_0		80	1	80	2.88
2	Put screw to part	LR	A_1 B_0 G_1 A_1 B_0 P_1 A_0	2	60	1	60	2.16
3		L	A_0 B_0 G_1 A_0 B_0 P_0 A_0		10	1	10	0.36
4	Put Spring on fixture	R	A_1 B_0 G_1 A_3 B_0 P_1 A_0		60	1	60	2.16
5	Fasten screw	R	A_1 B_0 G_1 A_1 B_0 P_1 A_0		40	1	40	1.44
6		LR	A_0 B_0 G_0 A_1 B_0 P_1 A_0		20	1	20	0.72
7		L	A_0 B_0 G_1 A_0 B_0 P_0 A_0		10	1	10	0.36
8		R	A_0 B_0 G_0 A_0 B_0 P_0 F_{27} A_1 B_0 P_2 A_0		300	1	300	10.80
9	Bend part	R	A_1 B_0 G_1 M_1 X_0 I_0 A_0		40	1	40	1.44
10		R	A_0 B_0 G_1 M_1 X_0 I_0 A_0		20	1	20	0.72
11		L	A_0 B_0 G_0 A_1 B_0 P_1 A_0		20	1	20	0.72
12	Place assembled part to bin	L	A_0 B_0 G_1 A_3 B_0 P_1 A_0		50	1	50	1.80

5 Macros Predetermined Time Sheet In Detail

Microsoft Excel's Macros function allows the user to set certain functions as part of a 'pre-recorded' memory to be used so that the user does not have to retype the same data when needed over again. The key-tabbing shortcut method in Microsoft Excel's MOST sheet is programmed to generate a series of MOST parameter sequence according to the analyzed actions. It is a simple excel sheet that contains allocated columns for the recording of actions with assigned shortcut keys for the MOST parameter to be called out within seconds. After which, the user will key in the value codes for each alphabet and the total TMU and converted seconds will be automatically added up in another column respectively. In the event a particular alphabet is not used or does not have any value in its string of MOST alphabets sequence, the analyst may eliminate the inclusion of the alphabet just by keying in the assigned value as zero ('0'). The top part of the sheet contains indicators to fill in information such as the person in charge, customer, product, unit per hour (UPH), standard time, type of assembly and date. The second part of the sheet contains columns to be filled in with details and only the MOST code column preset with macros shortcut keys.

5.1 Colour Coded Indicator

Whenever a particular action or movement is repeated more than once; the analyst will key in the number of times an additional action is repeated below the alphabet code index and the value of the alphabet will automatically multiply itself with the additional number of times an action is made. The index colour will then turn yellow to indicate where the additional value is assigned from. This method also saves time for the analyst so that a repeated keying in of the whole action is not needed (Fig. 1).

5.2 Macros Shortcut Keys

The macros parameter shortcut keys that are assigned according to the appropriate parameters will generate in the MOST code column of the macros sheet by pressing two assigned keys at the same time. They are (Table 5):

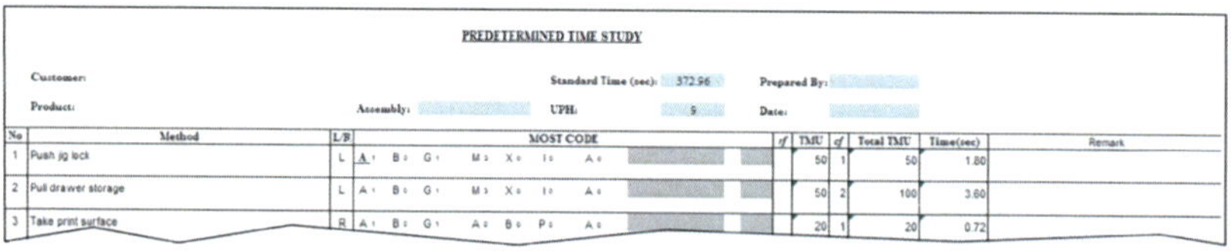

Columns	Definition
No.	To fill in the step or procedure numbering
Method	The methods or actions that are being observed are recorded here
L/R	To record down whether left or right hand or both hands are used in this process
MOST	The parameter and index value codes are keyed and defined here
rf	Abbreviation for 'repeat frequency'. The number of times a specific action or movement is keyedin here to multiply the repeated action(s).
TMU	Time Measurement Unit.
cf	Abbreviation for 'cycle frequency'. The number of times a full cycle of a repeated action or sequence is keyed in here to multiply the a whole repeated cycle.
Total TMU	Total Time Measurement Unit (after the inclusion of 'cf' or 'rf')
Seconds	Conversion of total time from TMU to seconds
Remarks	Additional notes or comments can be recorded in this coloumn

Fig. 1 Macros MOST excel sheet columns

Table 5 Sequence shortcut keys

Shortcut Keys	Sequence parameter generated
Ctrl + d	General move
Ctrl + q	Tool use sequence
Ctrl + l	Controlled move sequence

6 Conclusion

With the utilization and application of macros MOST predetermined sheet method, the work of an industrial engineer for a predetermine time study recording is made convenient and simplified. Doing a time study analysis requires focus and concentration that only becomes accurate if broken down and analyzed properly. As human mistakes are bound to happen, using ways that can minimize the tendency of carelessness helps to produce effective results in a short amount of time. In addition, the MOST predetermined sheet method is a tool that can help to save cost without the need to use a pen and paper or the requirement to purchase an additional software and license to produce the same time study result.

Integer Program Modeling of Portfolio Optimization with Mental Accounts Using Simulated Tail Distribution

Kuo-Hwa Chang, Yi Shou Shu and Michael Nayat Young

Abstract Since Markowitz introduced mean–variance portfolio theory, there have been many portfolio selection problems proposed. One of them is the safety-first portfolio optimization considering the downside risk. From behavioral portfolio theory, investors may not consider their portfolios as a whole. Instead, they may consider their portfolios as collections of subportfolios over many mental accounts. In this study, we present a mixed-integer programming model of portfolio optimization considering mental accounts (MAs). In this study, varied MAs are described by different level of risk-aversion. We measure the risk as the probability of a return failing to reach a threshold level, called the downside risk. An investor in each MA specifies the threshold level of return and the probability of failing to reach this return. Usually the portfolio's returns are assumed as normally distributed, but this move may underestimate the downside risks. Accordingly, we estimate the downside risk by using models utilizing extreme-value theory and copula. We generate scenarios of the tail distribution based on this model, on which the mixed-integer program is applied. In the end, we use historical data to back test our model and the results are consistent with what they expected. These actions result in a better understanding of the relation between investor goals and portfolio production, and portfolio optimization.

Keywords Integer programming · Simulation model · Portfolio optimization · VaR

K.-H. Chang (✉) · Y. S. Shu · M. N. Young
Department of ISE, Chung Yuan Christian University, Chung Li, 302, Taiwan, ROC
e-mail: kuohwa@cycu.edu.tw

Y. S. Shu
e-mail: g10074035@cycu.edu.tw

M. N. Young
e-mail: g10174053@cycu.edu.tw

Y.-K. Lin et al. (eds.), *Proceedings of the Institute of Industrial Engineers Asian Conference 2013*, DOI: 10.1007/978-981-4451-98-7_53, © Springer Science+Business Media Singapore 2013

441

1 Introduction

The behavior of people in terms of their investments has been a hot topic for research throughout the years. From the initial findings of Friedman and Savage (1948) Insurance-Lottery Framework to the (Das et al. 2010) Mental Accounting Framework (MA), investors tend to behave based on personal goals and aspirations (Friedman and Savage 1948). Insurance-Lottery Framework study started it all, they found out that certain people are risk seeking enough to buy lottery tickets for their financial aspiration and also risk averse enough to buy insurances for their financial security. Markowitz (1952) followed it with his Mean–Variance Theory (MVT) which is based from the expected utility theory and extended the Insurance-Lottery framework, wherein people who practice MVT don't buy lottery tickets as they are more risk averse. Shefrin and Statman (2000) followed suite with their goal based Behavioral Portfolio Theory (BPT), wherein investors divide their money into different mental layers with associated risk level tolerance within a portfolio. Each layer corresponds to a specific goal of the investor like retirement security, college education plans, and travel aspirations. Recently, Das et al. (2010) combined key aspects of MVT and the mental accounts of BPT to developed MA, but, similar to previous works, they assumed portfolio returns are normally distributed.

The foundation of BPT lies in (Lopes 1987) Safety First Portfolio Theory (SP/A Theory) and (Kahneman & Tversky 1979) Frame Dependence which are both developmental studies of the Insurance-Lottery Framework. SP/A Theory and Frame Dependence are all about how to deal with the risk threshold and framework of each investor to beat the inefficient market. These studies greatly impacted the Finance world in the 1990s which led to the formation of Behavioral Finance and BPT. In BPT, investors view their portfolio not as a whole but as a collection of several mental accounts that are associated with specific goals and varying risk attitude for each mental account. Investors can treat one mental account as their way to have protection from being poor, while other mental accounts as their tactic to have the opportunity to be rich.

In this study, we describe varied MAs by different level of risk-aversion. We measure the risk as the probability of a return failing to reach a threshold level, which is called the downside risk. An investor in each MA specifies the threshold level of return and the probability of failing to reach this return. Usually the portfolio's returns are assumed as normally distributed, but this move may underestimate the downside risks. Accordingly, we estimate the downside risk by using simulated tail distribution utilizing extreme value theory (EVT) and copula developed in (Chang and Wang 2012). We generate scenarios based on distribution, on which the mixed-integer program will be applied. In the end, we use historical data to back test our model.

2 Model Formulation

To further study BPT, we consider the following Telser's safety-first model model (Telser 1955):

$$Max \quad E_h(W)$$
$$s.t. \quad P(W < A) \le \alpha \tag{1}$$

where $E_h(W)$ is the expected wealth under the transformed decumulative function which is mentioned in Lopes and Oden (1999), A is the minimum acceptable rate of return, and α is the risk threshold or the probability that the occurring return fall on the lower end of the acceptable risk.

In BPT, risk is measured as the probability of a return failing to reach a threshold level, called the downside risk. There are a lot of ways to estimate the probability. In this study, we are going to use the scenario data to estimate the risk based on a simulation model. Usually the portfolio's returns are assumed as normally distributed, but in practice, findings from past studies say that this is not the case. The rate of returns usually has heavy-tailed properties, so using normal distribution will underestimate the risk of incidence. Accordingly, to handle this problem, we adopt the model in Chang and Wang (2012), we consider heavy-tailed distribution by incorporating the EVT and the copula structure to the Monte Carlo Simulation to generate scenarios to be used as the focal point of this study.

There are n assets and g mental accounts. There are m generated scenarios. We define the variables as follows:

$w_{i,k}$	denote the percentage of wealth invested on asset i in mental account k, $i = 1, 2, \ldots, n$, $k = 1, 2, \ldots, g$
$r_{ij,k}$	denote the return of asset i in simulated scenario j in mental account k, $i = 1, 2, \ldots, n$, $j = 1, 2, \ldots, m$, $k = 1, 2, \ldots, g$
$p_{j,k}$	denote the probability that simulated scenario j will occur in mental account k, $j = 1, 2, \ldots, m$, $k = 1, 2, \ldots, g$
$\bar{r}_i$	denote the mean return of asset i, $i = 1, 2, \ldots, n$
$R_{p,k} = \sum_{i=1}^{n} w_{i,k} r_{ij,k}$	denote the return of the portfolio in mental account k, $i = 1, 2, \ldots, n$, $j = 1, 2, \ldots, m$, $k = 1, 2, \ldots, g$
$\bar{R}_{p,k} = \sum_{i=1}^{n} w_{i,k} \bar{r}_i$	denote the expected return of the portfolio in mental account k, $i = 1, 2, \ldots, n$, $j = 1, 2, \ldots, m$, $k = 1, 2, \ldots, g$
$R_{L,k}$	denote the tolerance level on return rate that investor desires in mental account k, $k = 1, 2, \ldots, g$
α_k	denote the acceptable probability of the return failing to reach the threshold level in mental account k, $k = 1, 2, \ldots, g$

For mental account k, the corresponding (1) is

$$Max \quad \bar{R}_{p,k}$$

$$s.t. \quad P(R_{p,k} < R_{L,k}) \le \alpha_k \tag{2}$$

By Norkin and Boyko (2010), we can rewrite the constraints as follows:

$$Max \quad \overline{R}_{p,k}$$

$$s.t. \quad \sum_{j=1}^{m} p_{j,k} z_{j,k} \leq \alpha_k \tag{3}$$

$$R_{L,k} - R_{p,k} \leq (M + R_{L,k}) z_{j,k} \tag{4}$$

where $z_{j,k} \in \{0, 1\}$ and M is approaching to infinity.

To be more practical, the portfolio should be presented in the number of shares or units purchased on the assets. Let $x_{i,k}$ be the numbers of units of asset i purchased in portfolio in account k. If S_i is the initial price of asset i and we have initial capital b_k for account k, so the percentage of wealth on asset i in account k is

$$w_{i,k} = \frac{x_{i,k} \times S_i}{b_k} \tag{5}$$

We assume that the probabilities that simulated scenario j will occur, $p_{j,k}$, are equiprobable, so the constraint (3) can be written in

$$\sum_{j=1}^{m} z_{j,k} \leq \alpha_k \times m \tag{6}$$

where m is the total number of the simulated data.

In this setting, let $C_k = R_{L,k} \times b_k$ be the tolerance level on the loss from b_k. After writing $w_{i,k}$ in terms of $x_{i,k}$ and writing $R_{L,k}$ in term of C_k and adding the bounds on initial capital, our model is as follows.

$$Max \quad \sum_{i=1}^{n} x_{i,k} \overline{r}_i S_i \tag{7}$$

$$s.t. \quad \sum_{j=1}^{m} z_{j,k} \leq \alpha_k \times m$$

$$C_k - \sum_{i=1}^{n} x_{i,k} r_{ij,k} S_i \leq (M + C_k) z_{j,k} \tag{8}$$

$$\sum_{i=1}^{n} x_{i,k} S_i \leq b_k \tag{9}$$

$$\sum_{k=1}^{g} b_k \leq B \tag{10}$$

$$x_{i,k} \geq 0, integer, \quad i = 1, 2, \ldots, n, \ k = 1, 2, \ldots, g$$

$$z_{j,k} \in \{0, 1\}, \quad j = 1, 2, \ldots, m, \ k = 1, 2, \ldots, g$$

where B is the total initial capital.

We then use this integer programming model to solve for the optimal portfolios.

3 Results and Analysis

We use the stocks of the MSCI Taiwan Index Top 20 largest constituents, and through the use of the Taiwan Economic Journal (TEJ) we get those stocks that falls after a week of the announcement of dividends. We calculate 102 portfolios from October 2010 to September 2012, and each portfolio uses 200 returns from the historical data. We use the data collected from November 2006 to September of 2012 to adjust and balanced the effect of the rate of returns. We then integrated the 200 original data to the pair-copula structure and simulated 20,000 scenarios. With reference to the EVT and the scenarios numerical distribution we were able to develop the model to solve for the optimal portfolios.

We consider that investors have 3 mental accounts with the parameters shown below :

"Weight" represents the proportion of wealth invested in the respective account. From the Table 1, Accounts 1 and 2 have the same risk threshold level but different return expectations, accounts 2 and 3 have different risk threshold levels but have similar return expectations. We then use different methods to study the investment returns.

First, we look at the accounts' means and standard deviations and at the same time compare them with the Market values as shown below :

In Table 2, we can see that account 1 has the least standard deviation and also outperforms the market. Similarly, Account 1 has the best mean return and out-performs the market. We can then observe the distribution of the returns in Fig. 1.

From Fig. 1, we can see that there is smaller volatility in Account 1 and Aggregate account. Then we can analyze the cumulative distribution of the returns through Fig. 2.

From Fig. 2, with the set of parameters in Table 1 it is clear that each account has a smaller probability of getting bad returns compared to the Market, especially in Account 1 and Aggregate account which gave the best performances.

Similar to the study in Chang and Wang (2012), we use following 3 cases $(-,-)$, $(+,-)$ and $(-,+)$ to represent the returns of the accounts and the Market. $(-,-)$ indicates that the account and market return rates are both negative. $(+,-)$ indicates

Table 1 Parameter settings

	Account 1 (%)	Account 2 (%)	Account 3 (%)
R_L	−3	−5	−5
α	7.5	7.5	10
Weight	40	30	30

Table 2 Mean and standard deviation comparison

	Account 1	Account 2	Account 3	Aggregate	Market
Mean	0.002999	0.002412	0.00117	0.002273	−0.00017
Standard Deviation	0.018376	0.027671	0.02835	0.023487	0.025825

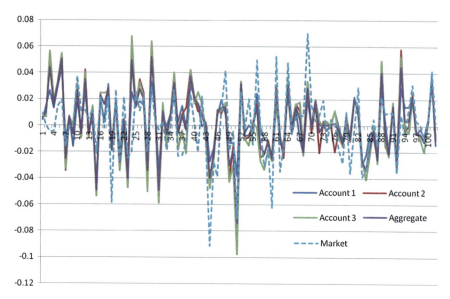

Fig. 1 Portfolio rate of return

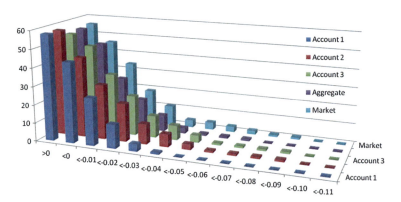

Fig. 2 Returns cumulative distribution map

that the account has positive returns, while the market has negative returns. (−,+) indicates that the account has negative returns, while the market has positive returns. The case (+,+) wherein both the market and the account have positive returns is not risky so it was excluded from the comparison. Using these 3 cases the summaries of the comparison are shown in Tables 3 and 4.

Finally we have geometric long term cumulative returns in Fig. 3:

From Fig. 3 it is clear that each account's rate of returns outperforms the Market and that each account has positive returns compared to the market with negative returns.

Table 3 Account 1 versus market & account 2 versus market

	Account 1	Market	Account 2	Market
(−,−)	25	8	17	15
(+,−)	13	0	14	0
(−,+)	0	11	0	12
Total	38	19	31	27

Table 4 Account 3 versus market & aggregate versus market

	Account 3	Market	Aggregate	Market
(−,−)	16	18	19	15
(+,−)	12	0	12	0
(−,+)	0	14	0	13
Total	28	32	31	28

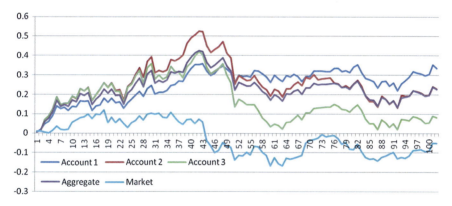

Fig. 3 Geometric long term cumulative returns

4 Conclusions

The main contribution of this study is to combine the mental accounting portfolio optimization with the safety-first model. We use mixed integer programming to find the optimal portfolios. Through the back test, it shows that the performances of the 3 accounts are within the pre-determined risk threshold and that the integrated portfolio performances are superior in all aspects of the market. These results supported our model in terms of obtaining optimal portfolios. To have a possibly more accurate portfolio, future studies should consider having varying occurrence probability for the returns and also allow short selling.

References

Chang KH, Wang ZJ (2012) Safety-first optimal portfolio with approximated bound on one-sided risk using simulated tail distribution. Working paper, Department of Industrial and Systems Engineering, Chung-Yuan Christian University

Das S, Markowitz H, Scheid J, Statman M (2010) Portfolio optimization with mental accounts. J Fin Quant Anal 45(2):311–334

Friedman M, Savage LJ (1948) The utility analysis of choices involving risk. J Polit Econ 56(4):279–304

Jiang CH, Ma YK, An YB (2013) International portfolio selection with exchange rate risk: a behavioral portfolio theory perspective. J Bank Finance 37:648–659

Kahneman D, Tversky A (1979) Prospect theory: an analysis of decision under risk. Econometrica 47(2):263–291

Lopes LL (1987) Between hope and fear: the psychology of risk. Adv Exp Soc Psychol 20:255–296

Lopes LL, Oden GC (1999) The role of aspiration level in risky choice: a comparison of cumulative prospect theory and SP/A theory. J Math Psychol 43:286–313

Markowitz H (1952) Portfolio selection. J Finance 7:77–91

Norkin V, Boyko S (2010) On the safety first portfolio selection. Working paper, Glushkov Institute of Cybernetics, National Academy of Science of Ukraine

Shefrin H, Statman M (2000) Behavioral portfolio theory. J Fin Quant Anal 35(2):127

Telser LG (1955) Safety-first and hedging. Rev Econ Stud 23:1–60

Simulated Annealing Algorithm for Berth Allocation Problems

Shih-Wei Line and Ching-Jung Ting

Abstract Maritime transport is the backbone of global supply chain. Around 80 % of global trade by volume is carried by sea and is handled by ports worldwide. The fierce competition among different ports forces the container port operators to improve the terminal operation efficiency and competitiveness. This research addresses the dynamic and discrete berth allocation problem (BAP) which is critical to the terminal operations. The objective is to minimize the total service times for vessels. To solve the NP-hard BAP problem, we develop a simulated annealing (SA) algorithm to obtain the near optimal solutions. Benchmark instances from the literature are tested for the effectiveness of the proposed SA and compared with other leading heuristics in the literature. Computational results show that our SA is competitive and find the optimal solutions in all instances.

Keywords Berth allocation problem · Simulated annealing · Container port

1 Introduction

Maritime transportation has been an important component of the international trade since the introduction of the container in the 1950s. There are more than 8.4 billion tons of goods carried by ships annually (UNCTAD 2011). Due to the growth of the ocean shipping demand, shippers and carriers would expect to speed

---

S.-W. Line (✉)
Department of Information Management, Chang Gung University, Kwei-Shan, Taoyuan, 33302, Taiwan, Republic of China
e-mail: swlin@mail.cgu.edu.tw

C.-J. Ting
Department of Industrial Engineering and Management, Yuan Ze University, Chung-Li, 32003, Taiwan, Republic of China
e-mail: ietingcj@saturn.yzu.edu.tw

Y.-K. Lin et al. (eds.), *Proceedings of the Institute of Industrial Engineers Asian Conference 2013*, DOI: 10.1007/978-981-4451-98-7_54,
© Springer Science+Business Media Singapore 2013

up their operations at the port stop. Thus, how to provide efficient and cost-effective services by using the limited port resources becomes an important issue for port authority. Among the port operations, berth allocation problem (BAP) that assign the berth positions to a set of vessels with the planning horizon to minimize the total service time catch both practical and academic attention.

The BAP has been tackled in both spatial and temporal variations: (1) discrete versus continuous; (2) static versus dynamic arrival times (Bierwirth and Meisel 2010). In the discrete case, the quay is divided into several berths and exactly one vessel can be served at a time in each berth. In the continuous case, there is no partition of the quay and vessels can moor at any position. A static BAP assumes that all vessels already in port before the berth allocation is planned, while the dynamic case allows vessels to arrive at any time during the planning horizon with known arrival information. The dynamic and discrete BAP that is known to be NP-hard (Cordeau et al. 2005) is the main focus of this paper. We propose a simulated algorithm to solve the dynamic and discrete BAP.

The remainder of the paper is organized as follows. Section 2 presents a brief literature review about the BAP. The proposed simulated annealing algorithm is presented in Sect. 3. In Sect. 4, computational experiments are performed and compared with the promised heuristics from the literature. Finally, our conclusion is summarized in Sect. 5.

2 Literature Review

Both the discrete and continuous BAP in various models have been proposed in the literature; however, most studies address the discrete BAP. Steenken et al. (2004), Stahlbock and Voß (2008) and Bierwirth and Meisel (2010) provided comprehensive overviews of application and optimization models in this field. In this paper, we will address the discrete BAP and only briefly review continuous BAP.

Imai et al. (1997) formulated a static BAP as a nonlinear integer programming model to minimize the weighted objectives which include berth performance and dissatisfaction. Imai et al. (2001) studied a dynamic BAP whose objective was to minimize the sum of waiting and handling times of all vessels. A Lagrangian relaxation based heuristic was proposed to solve the problem. Nishimura et al. (2001) extended the dynamic BAP with multi-water depth configuration. A genetic algorithm (GA) was developed to solve the problem. Imai et al. (2003) considered a dynamic BAP that vessels have different service priorities. The authors also proposed a GA to solve the problem.

Cordeau et al. (2005) addressed a dynamic BAP with time windows based on data from a terminal in the port of Gioia Tauro. The problem was formulated as a multiple depot vehicle routing problem with time windows (MDVRPTW), and solved by a tabu search algorithm. Imai et al. (2007) analyzed a two-objective BAP which minimizes service time and delay time. They proposed a Lagrangian relaxation with subgradient optimization and a GA to identify the non-inferior solutions. Hansen et al. (2008) extended Imai et al.'s (2003) model and developed a variable neighborhood search heuristic to solve it. Imai et al. (2008) studied a variant of the dynamic BAP in which an external terminal was available when the berth capacity is not enough.

Mauri et al. (2008) proposed a hybrid column generation algorithm to solve the MDVRPTW model in Cordeau et al. (2005). Barros et al. (2011) developed and analyzed a berth allocation model with tidal time windows, where vessels can only be served by berths during those time windows. Buhrkal et al. (2011) formulated the discrete BAP as a generalized set partitioning problem (GSPP). Their computational results provided the best optimal solutions for comparison in later research. de Oliveira et al. (2012b) applied a clustering search method with simulated annealing to solve the discrete BAP.

Lim (1998) was the first one to study the continuous BAP whose berths can be shared by different vessels. The problem was modeled as a restricted form of the 2-dimensional packing problem. Li et al. (1998) and Guan et al. (2002) modeled the continuous BAP as a machine scheduling problem with multiprocessor tasks. Park and Kim (2003) formulated a mixed integer programming model for the continuous to minimize the penalty cost associated with service delay and a non-preferred location. A Lagrangian relaxation with subgradient optimization method was developed to solve the problem. Kim and Moon (2003) formulated the continuous BAP as a mixed integer linear programming model and proposed a simulated annealing algorithm to solve it. Imai et al. (2005) enhanced their previous discrete BAP to a continuous BAP and proposed a heuristic for the problem. Wang and Lim (2007) proposed a stochastic beam search algorithm to solve the BAP in a multiple stage decision making procedure.

Tang et al. (2009) proposed two mathematical models and developed an improved Lagrangian relaxation algorithm to solve the continuous BAP at the raw material docks in an iron and steel complex. Lee et al. (2010) developed two greedy randomized adaptive search procedure (GRASP) heuristics for the continuous BAP. Seyedalizadeh Ganji et al. (2010) applied a GA to solve the problem proposed by Imai et al. (2005). Mauri et al. (2011) proposed a memetic algorithm to solve the continuous BAP. De Oliveira et al. (2012a) presented a clustering search method with simulated annealing heuristic to solve the continuous BAP.

3 Simulated Annealing Algorithm

The simulated annealing (SA) reaching a (near) global optimum during the search process mimics the crystallization cooling procedure (Metropolis et al. 1953; Kirkpatrick et al. 1983). SA first generates a random initial solution as an incumbent solution. The algorithm moves to a new solution from the predetermined neighborhood of the current solution. The objective function value of the new solution is compared to the current one to determine whether the new solution is better. If the objective function value of a new solution is better than that of the incumbent one, then the new solution is automatically accepted, and becomes the incumbent solution from which the search will continue. The procedure then continues with the next iteration. A worse objective function value for the new solution may also be accepted as the incumbent solution under certain conditions. By accepting a worse solution, the procedure may escape from the local optima.

A solution is represented by a string of numbers consisting of a permutation of n ships denoted by the set $\{1, 2, ..., n\}$ and $m - 1$ zeros for separating ships into m berths. The numbers between two zero are the sequence of the vessels that will be served by a berth. The completion time of each ship in the berth can be easily calculated according to its arrival time, the sequence in the berth, and the availability of the berth. If an infeasible solution is obtained, a penalty will be added to the objective function. We generate the initial solution based on first-come-first-serve of vessels' arrival times. Each ship is sequentially positioned to the berth which can provide the shortest completion time for the current assigned ship. After the initial solution X is obtained, a local search procedure is applied to improve X. The local search procedure applies swap moves and insertion moves to X sequentially.

To obtain a better solution, best-of-g-trial moves that choose the best solution among g neighborhood solutions as the next solution are also performed. At each iteration, a new solution Y is generated from the neighborhood of the current solution X, $N(X)$, and its objective function value is evaluated. Let $\Delta = \text{obj}(Y) - \text{obj}(X)$. If $\Delta \leq 0$, the probability of replacing X with Y is 1; otherwise it is based on another random generated probability and $T/(T^2 + \Delta^2)$, where T is the current temperature. T is reduced after running I_{iter} iterations from the previous decrease, according to the formula $T = \alpha T$, where $0 < \alpha < 1$. After each temperature reduction, a local search procedure is used to improve X_{best}, the best solution found so far. We use swap moves and insertion moves to X_{best} sequentially in the local search.

4 Computational Results

The proposed SA was implemented using the C language in Windows XP operating system, and run on a personal computer with an Intel Core 2 2.5 GHz CPU and 2G RAM. Each instance was solved using 10 runs of the proposed approach. The performance of the proposed SA_{RS} heuristic was compared with other existing algorithms for DBAP, namely tabu search (T^2S) (Cordeau et al. 2005), population training algorithm with linear programming (PTA/LP) (Mauri et al. 2008), clustering search (CS) approach (de Oliveira et al. 2012b). The benchmark instances of Cordeau et al. (2005) were used in this study. The instances were randomly generated based on data from the port of Gioia Tauro (Italy). The instances may be categorized as two sets, I2 and I3. The I2 set includes five instance sizes: 25 ships with 5, 7, and 10 berths; 35 ships with 7 and 10 berths and a set of 10 instances generated for each size. The I3 set includes 30 instances with 60 ships and 13 berths.

Simulated Annealing Algorithm for Berth Allocation Problems

Table 1 Computational result for I2 problem set

Instance	GSPP		T²S	SA		
	Opt.	Time	Best	Best	Avg.	Time
25 × 5_1	759	5.99	759	759	759.0	0.03
25 × 5_2	964	3.70	965	964	964.0	0.27
25 × 5_3	970	2.95	974	970	970.0	1.02
25 × 5_4	688	2.72	702	688	688.0	0.10
25 × 5_5	955	6.97	965	955	955.0	0.66
25 × 5_6	1,129	3.10	1,129	1,129	1,129.0	0.01
25 × 5_7	835	2.31	835	835	835.0	0.01
25 × 5_8	627	1.92	629	627	627.0	0.03
25 × 5_9	752	4.76	755	752	752.0	0.08
25 × 5_10	1,073	6.38	1,077	1,073	1,073.3	8.75
25 × 7_1	657	3.62	667	657	657.0	0.00
25 × 7_2	662	3.15	671	662	662.0	0.02
25 × 7_3	807	4.28	823	807	807.0	0.29
25 × 7_4	648	3.78	655	648	648.0	0.31
25 × 7_5	725	3.85	728	725	725.0	0.02
25 × 7_6	794	3.60	794	794	794.0	0.01
25 × 7_7	734	3.54	740	734	734.0	0.21
25 × 7_8	768	3.93	782	768	768.0	0.07
25 × 7_9	749	3.73	759	749	749.0	0.02
25 × 7_10	825	3.82	830	825	825.0	0.02
25 × 10_1	713	5.83	717	713	713.0	0.04
25 × 10_2	727	6.99	736	727	727.0	0.13
25 × 10_3	761	6.12	764	761	761.0	0.33
25 × 10_4	810	5.38	819	810	810.0	0.72
25 × 10_5	840	6.77	855	840	840.0	0.07
25 × 10_6	689	5.57	694	689	689.0	0.01
25 × 10_7	666	5.83	673	666	666.0	0.00
25 × 10_8	855	5.87	860	855	855.0	0.01
25 × 10_9	711	5.38	726	711	711.0	0.16
25 × 10_10	801	5.96	812	801	801.0	0.04
35 × 7_1	1,000	12.57	1,019	1,000	1,000.1	5.35
35 × 7_2	1,192	15.93	1,196	1,192	1,192.8	13.33
35 × 7_3	1,201	7.16	1,230	1,201	1,201.0	2.93
35 × 7_4	1,139	13.59	1,150	1,139	1,139.0	1.25
35 × 7_5	1,164	11.50	1,179	1,164	1,164.3	3.39
35 × 7_6	1,686	29.16	1,703	1,686	1,686.4	11.72
35 × 7_7	1,176	12.89	1,181	1,176	1,176.0	1.95
35 × 7_8	1,318	17.52	1330	1,318	1,318.1	4.44
35 × 7_9	1,245	8.41	1,245	1,245	1,245.0	0.76
35 × 7_10	1,109	14.39	1,130	1,109	1,109.1	4.00
35 × 10_1	1,124	19.98	1,128	1,124	1,124.0	0.33
35 × 10_2	1,189	11.37	1,197	1,189	1,189.0	5.64
35 × 10_3	938	8.97	953	938	938.0	0.13

(continued)

Table 1 (continued)

Instance	GSPP		T²S	SA		
	Opt.	Time	Best	Best	Avg.	Time
35 × 10_4	1,226	10.28	1,239	1,226	1,227.1	16.21
35 × 10_5	1,349	22.31	1,372	1,349	1,349.0	0.95
35 × 10_6	1,188	10.92	1,221	1,188	1,188.0	0.39
35 × 10_7	1,051	9.74	1,052	1,051	1,051.0	0.33
35 × 10_8	1,194	9.39	1,219	1,194	1,194.0	0.08
35 × 10_9	1,311	29.45	1,315	1,311	1,311.0	2.30
35 × 10_10	1,189	14.28	1,198	1,189	1,189.0	0.05
Average	953.7	8.60	963.0	953.7	953.7	1.78

Table 2 Computational result for I3 problem set

Inst.	GSPP		T²S	PTA/LP		CS		SA		
	Opt.	Time	Best	Best	Time	Best	Time	Best	Avg.	Time
i01	1,409	17.92	1,415	1,409	74.61	1,409	12.47	1,409	1,409.0	1.03
i02	1,261	15.77	1,263	1,261	60.75	1,261	12.59	1,261	1,261.0	0.05
i03	1,129	13.54	1,139	1,129	135.45	1,129	12.64	1,129	1,129.0	0.18
i04	1,302	14.48	1,303	1,302	110.17	1,302	12.59	1,302	1,302.0	0.09
i05	1,207	17.21	1,208	1,207	124.70	1,207	12.68	1,207	1,207.0	0.07
i06	1,261	13.85	1,262	1,261	78.34	1,261	12.56	1,261	1,261.0	0.00
i07	1,279	14.60	1,279	1,279	114.20	1,279	12.63	1,279	1,279.0	0.96
i08	1,299	14.21	1,299	1,299	57.06	1,299	12.57	1,299	1,299.0	0.30
i09	1,444	16.51	1,444	1,444	96.47	1,444	12.58	1,444	1,444.0	0.22
i10	1,213	14.16	1,213	1,213	99.41	1,213	12.61	1,213	1,213.0	0.11
i11	1,368	14.13	1,378	1,369	99.34	1,368	12.58	1,368	1,368.0	2.16
i12	1,325	15.60	1,325	1,325	80.69	1,325	12.56	1,325	1,325.0	2.51
i13	1,360	13.87	1,360	1,360	89.94	1,360	12.61	1,360	1,360.0	0.04
i14	1,233	15.60	1,233	1,233	73.95	1,233	12.67	1,233	1,233.0	0.05
i15	1,295	13.52	1,295	1,295	74.19	1,295	13.80	1,295	1,295.0	0.00
i16	1,364	13.68	1,375	1,365	170.36	1,364	14.46	1,364	1,364.0	3.15
i17	1,283	13.37	1,283	1,283	46.58	1,283	13.73	1,283	1,283.0	0.02
i18	1,345	13.51	1,346	1,345	84.02	1,345	12.72	1,345	1,345.0	0.00
i19	1,367	14.59	1,370	1,367	123.19	1,367	13.39	1,367	1,367.0	4.49
i20	1,328	16.64	1,328	1,328	82.30	1,328	12.82	1,328	1,328.0	3.31
i21	1,341	13.37	1,346	1,341	108.08	1,341	12.68	1,341	1,341.0	4.79
i22	1,326	15.24	1,332	1,326	105.38	1,326	12.62	1,326	1,326.0	1.16
i23	1,266	13.65	1,266	1,266	43.72	1,266	12.62	1,266	1,266.0	0.06
i24	1,260	15.58	1,261	1,260	78.91	1,260	12.64	1,260	1,260.0	0.07
i25	1,376	15.80	1,379	1,376	96.58	1,376	12.62	1,376	1,376.0	4.75
i26	1,318	15.38	1,330	1,318	101.11	1,318	12.62	1,318	1,318.0	0.46
i27	1,261	15.52	1,261	1,261	82.86	1,261	12.64	1,261	1,261.0	0.09
i28	1,359	16.22	1,365	1,360	52.91	1,359	12.71	1,359	1,359.4	13.53
i29	1,280	15.30	1,282	1,280	203.36	1,280	12.62	1,280	1,280.0	2.48
i30	1,344	16.52	1,351	1,344	71.02	1,344	12.58	1,344	1,344.0	3.80
Avg.	1,306.8	14.98	1,309.7	1,306.9	93.99	1,306.8	12.79	1,306.8	1,306.8	1.66

Simulated Annealing Algorithm for Berth Allocation Problems 455

The computational results for both set of instances are presented in Tables 1 and 2, respectively. The optimal solution was provided by the GSPP model using CPLEX 11 (Buhrkal et al. 2011). The proposed SA is compared with T^2S for the I2 problem set as shown in Table 1. The first three columns show the instances, the optimal solution and the computational time. The results of T^2S and the best and average solution and CPU time of our SA are presented in columns 4–7. Our SA can obtain the optimal solutions in all I2 instances, while T^2S can only find five optimal solutions. Optimal solutions can be obtained in all 10 runs except for only nine instances in which the largest gap is 0.090 %.

Table 2 lists the results of T^2S, PTA/LP, CS and the proposed SA_{WRS} and SA_{RS} heuristics for the I3 problem set. The first three columns show the instances, the optimal solution and the computational time. The results of T^2S and the best and CPU time of PTA/LP and CS are presented in columns 4–8 followed by our SA results. The SA and CS obtain all optimal solutions, whereas PTA/LP cannot reach the optimal solutions in three instances.

5 Conclusions

In this paper we have studied the berth allocation problem with dynamic arrival times. We propose a simulated annealing (SA) heuristic to solve the problem and test our algorithm with two sets of instances. The results are also compared with the optimal and best known solutions from the literature. Computational results indicate that the proposed SA algorithm is fairly effective. Our SA algorithm is able to find all the optimal solutions of the BAP instances. In the future, we can apply the proposed SA algorithm to the continuous BAP in which vessels can berth anywhere along the quayside.

References

Barros VH, Costa TS, Oliveira ACM and Lorena LAN (2011) Model and heuristic for berth allocation in tidal bulk ports with stock level constraints. Comput Ind Eng 60:606–613

Bierwirth C and Meisel F (2010) A survey of berth allocation and quay crane scheduling problems in container terminals. Eur J Oper Res 202:615–627

Buhrkal K, Zuglian S, Ropke S, Larsen J, Lusby R (2011) Models for the discrete berth allocation problem: a computational comparison. Transp Res Part E 47:461–473

Cordeau JF, Laporte G, Legato P, Moccia L (2005) Models and tabu search heuristics for the berth-allocation problem. Transp Sci 39:526–538

de Oliveira RM, Mauri GR, Lorena LAN (2012a) Clustering search heuristics for solving a continuous berth allocation problem. Lect Notes Comput Sci 7245:49–62

de Oliveira RM, Mauri GR, Lorena LAN (2012b) Clustering search for the berth allocation problem. Expert Syst Appl 39:5499–5505

Guan Y, Xiao WQ, Cheung RK and Li CL (2002) A multiprocessor task scheduling model for berth allocation: heuristic and worst-case analysis. Oper Res Lett 30:343–350

Hansen P, Oğuz C, Mladenovic N (2008) Variable neighborhood search for minimum cost berth allocation. Eur J Oper Res 191:636–649

Imai A, Nagaiwa K, Chan WT (1997) Efficient planning of berth allocation for container terminals in Asia. J Adv Transp 31:75–94

Imai A, Nishimura E, Papadimitriou S (2001) The dynamic berth allocation problem for a container port. Transp Res Part B 35:401–417

Imai A, Nishimura E, Papadimitriou S (2003) Berth allocation with service priority. Transp Res Part B 37:437–457

Imai A, Nishimura E, Papadimitriou S (2008) Berthing ships at a multi-user container terminal with a limited quay capacity. Transp Res Part E 44:136–151

Imai A, Sun X, Nishimura E, Papadimitriou S (2005) Berth allocation in a container port: using a continuous location space approach. Transp Res Part B 39:199–221

Imai A, Zhang JT, Nishimura E and Papadimitriou S (2007) The berth allocation problem with service time and delay time objectives. Marit Econ Logistics 9:269–290

Kim KH and Moon KC (2003) Berth scheduling by simulated annealing. Transp Res Part B 37:541–560

Kirkpatrick S, Gelatt CD and Vecch MP (1983) Optimization by simulated annealing. Science 220:671–680

Lee DH, Chen JH and Cao JX (2010) The continuous berth allocation problem: A greedy randomized adaptive search solution. Transp Res Part E 46:101–1029

Li CL, Cai X and Lee CY (1998) Scheduling with multiple-job-on-one-processor pattern. IIE Trans 30:433–445

Lim A (1998) The berth planning problem. Oper Res Lett 22:105–110

Mauri GR, Oliveira ACM and Lorena LAN (2008) A hybrid column generation approach for the berth allocation problem. Lect Notes Comput Sci 4972:110–122

Mauri GR, Andrade LN and Lorena LAN (2011) A memetic algorithm for a continuous case of the berth allocation problem. 2011 International conference on evolutionary computation theory and applications, Paris, France

Metropolis N, Rosenbluth A, Rosenbluth M, Teller A and Teller E (1953) Equation of state calculations by fast computing machines. J Chem Phys 21:1087–1090

Nishimura E, Imai A, Papadimitriou S (2001) Berth allocation planning in the public berth system by genetic algorithms. Eur J Oper Res 131:282–292

Park YM, Kim KH (2003) A scheduling method for berth and quay cranes. OR Spectrum 25:1–23

Stahlbock R and Voß S (2008) Operations research at container terminals: a literature update. OR Spectrum 30:1–52

Steenken D, Voß S, Stahlbock R (2004) Container terminal operation and operations research a classification and literature review. OR Spectrum 26:3–49

Seyedalizadeh Ganji SR, Babazadeh A and Arabshahi N (2010) Analysis of the continuous berth allocation problem in container ports using a genetic algorithm. J Mar Sci Technol 15:408–416

Tang L, Li S, Liu J (2009) Dynamically scheduling ships to multiple continuous berth spaces in an iron and steel complex. Int Trans Oper Res 16:87–107

UNCTAD (2011) Review of maritime transportation. United Nations Conference on Trade and Development

Wang F, Lim A (2007) A stochastic beam search for the berth allocation problem. Decis Support Syst 42:2186–2196

Using Hyperbolic Tangent Function for Nonlinear Profile Monitoring

Shu-Kai S. Fan and Tzu-Yi Lee

Abstract For most of the Statistical process control (SPC) applications, the quality of a process or product is measured by one or multiple quality characteristics. In some particular circumstances, quality characteristics depend on the relationship between the response variable and one and/or explanatory variables. Therefore, such a quality characteristic is represented by a function or a curve, which is called a 'profile'. In this paper, a new method of using the hyperbolic tangent function will be addressed for modeling the vacuum heat treatment process data. The hyperbolic tangent function approach is compared to the smoothing spline approach when modeling the nonlinear profiles. The vector of parameter estimates is monitored by using the Hotelling's T^2 for the parametric approach and by the metrics method for the nonparametric approach. In Phase I, the proposed hyperbolic tangent approach is able to correctly identify the outlying profiles.

Keywords Nonlinear profile · Hyperbolic tangent function · Smoothing spline · Hotelling's T^2

1 Introduction

Statistical process control (SPC) has been widely recognized for quality and productivity improvement in many domains, especially in manufacturing industries. It was pioneered by Walter A. Shewhart in the early 1920s. For most SPC

S.-K. S. Fan (✉) · T.-Y. Lee
Department of Industrial Engineering and Management, National Taipei University of Technology, Taipei City 10608, Taiwan, Republic of China
e-mail: morrisfan@ntut.edu.tw

T.-Y. Lee
e-mail: ritatzu1007@gmail.com

Y.-K. Lin et al. (eds.), *Proceedings of the Institute of Industrial Engineers Asian Conference 2013*, DOI: 10.1007/978-981-4451-98-7_55,
© Springer Science+Business Media Singapore 2013

applications, the quality of a process or product is measured by one or more quality characteristics. However, in some practical circumstances, the quality characteristic is better characterized by a curve. Such a functional relationship is called a profile. A new economic age is beginning in which the demand for quality is rapidly increasing, with a resulting global competition of companies striving to provide quality products or services. In practice, the aluminum alloy heat treatments are quite difficult to link to an SPC program. Due to its nonlinear data type, some researchers attempted to develop thermal monitoring hardware and SPC software that makes it possible to automatically gather data on the thermal portions on the heat treatment of the aluminum alloy rim process in real time. In this kind of applications, profile monitoring intends to keep watch on the relationship between the response variable and the explanatory variables and then draw inferences about the profile of quality characteristics. There are two distinct phases in profile monitoring. The purpose of phase I analysis is to evaluate the stability of a process, to find and remove any outliers attributed to some potential assignable cause, and to estimate the in-control values of the process parameters. Based on the parameter estimates obtained from phase I, the goal of phase II is to monitor the online data in order to quickly detect any change in the process parameter as possible. To date, the majority of profile monitoring works focus on phase II analysis.

A wide variety of literature has appeared in recent years, demonstrating a growing popularity in profile monitoring. Woodall et al. (2004) gave an excellent overview of the SPC literature regarding profile monitoring. For further discussions on linear profiles, interested readers can be referred to this article. A few references have directly addressed the issues of nonlinear profiles. Basically, nonlinear profile monitoring approaches can be classified two categories: parametric and nonparametric. For example, the parametric approaches include that Williams et al. (2007) addressed the monitoring of nonlinear profiles by considering a parametric nonlinear regression model. Jensen and Birch (2009) proposed a nonlinear mixed model to monitor nonlinear profiles to account for the correlation structure. Fan et al. (2011) proposed a piecewise linear approximation approach to model nonlinear profile data. Chen and Nembhard (2011) used the adaptive Neyman test and the discrete Fourier transform to develop a monitoring procedure for linear and nonlinear profile data. For the nonparametric approaches, Ding et al. (2006) utilized several types of reduction techniques for nonlinear profile data. Moguerza et al. (2007) used support vector machines to monitor the fitted curves instead of monitoring the parameters of models that fit the curves. Zou et al. (2007) discussed profile monitoring via nonparametric regression methods. Qiu et al. (2010) proposed a novel control chart for phase II analysis, which incorporates a local linear kernel smoothing technique into the exponentially weighted moving average (EWMA) control scheme when within-profile data are correlated.

The SPC problem with a nonlinear profile is particularly challenging by nature. Unlike the linear profile, the nonlinear models are different from one case to another. Different models can be used to fit different nonlinear profiles. Jin and Shi (1999) proposed monitoring the tonnage stamping profile by the use of dimension reduction techniques. Ding et al. (2006) proposed using nonparametric procedures

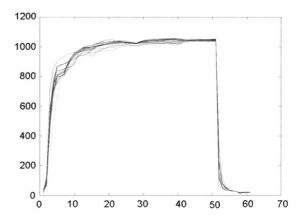

Fig. 1 A typical aluminum alloy rim heat treatment profile

to perform phase I analysis for multivariate nonlinear profiles. They adopted data reduction components that projected the original data onto a subspace of lower dimension while preserving the original data clustering structure. The main goal of this paper is to monitor the heat treatment of aluminum alloy rim process. A set of profile data is obtained from an industrial company in Taiwan. As demonstrated in Fig. 1, the sample data depicts nicely the typical thermal profile graph. It should be set up to monitor oven stability and to identify any process change before it affects product quality. In the first part, the temperature of the data is recorded by the thermal barrier through the production line, then analytical tool gains essential information that allows the process engineer to collect statistics and perform the monitoring.

In this paper, the hyperbolic tangent function is proposed to model this nonlinear profile data set. We evaluate the effectiveness of the proposed approach by comparing it with the smoothing spline approach. In phase I, the Hotelling T^2 and metric charts will be used for the outlier detection and the estimation of the in-control parameters. The rest of the article is organized as follows. The definitions of the hyperbolic tangent function and the parameters used in the function are explained in Sect. 2. Phase I analysis is performed in Sect. 3. Conclusions are drawn in Sect. 4.

2 Hyperbolic Tangent Function

The proposed approach in this article is to monitor nonlinear aluminum alloy heat treatment profiles which are fitted using the "modified" hyperbolic tangent function in comparison to the smoothing spline approach based on a model-building viewpoint. The first step in monitoring nonlinear profiles is to fit a parametric model that characterizes the relationship between the response variables and the explanatory variables. In essence, various modeling approaches should be tried in order to achieve the best fitting result. Here, we use the small sample Akaike Information Criterion (AIC) and Schwarz Information Criterion (SIC) for model selection. The aluminum alloy heat treatment data is used to illustrate the proposed approach.

The hyperbolic tangent function is the major model-building approach investigated in this paper. It is adopted for periodic functions and is a linear combination that consists of the hyperbolic tangent function with constant multipliers. With the hyperbolic tangent function, a single profile can be represented by.

$$ y_j = \sum_{i=1}^{s} a_i (x_j - b_i) \tanh \left(\frac{x_j - b_i}{w_i} \right) + \varepsilon_j, \quad i = 1, \ldots, s; \quad j = 1, \ldots, n, \quad (1) $$

where a_i is the strength of curve, b_i is the transition time position for going into/leaving the state, w_i is the width at each tangent wave term, s is the number of terms in the series, and n is the number of observations in the profile. The parameters are estimated through the nonlinear least squares estimation method. In this paper, the Marguardt's algorithm is utilized. The error term is assumed to have a zero mean and a constant variance and to be uncorrelated with each other. The hyperbolic tangent function also takes into account the variation among the different points of the data. The criterion of minimizing the squared errors is opted to decide the most appropriate model. Typically, the coefficient of determination should be high enough to indicate a good fitting result. According to Kang and Albin (2000), profile data consist of a set of measurements with a response variable y and one or more explanatory variables $(x's)$, which are used to evaluate the quality of manufactured items. Assume in this paper that n observations in the i-th random sample (i.e., profile) collected over time are available, indicated by (x_{ij}, y_{ij}) for $j = 1, \ldots, n$ and $i = 1, \ldots, m$. If the case of $s = 2$ is considered, the relationship between the paired observations as the process is in statistical control can be expressed by

$$ f(x_{ij}) = a_{1j}(x_{ij} - b_{1j}) \tanh \left(\frac{x_{ij} - b_{1j}}{w_{1j}} \right) + a_{2j}(x_{ij} - b_{2j}) \tanh \left(\frac{x_{ij} - b_{2j}}{w_{2j}} \right) + \varepsilon_{ij}. \quad (2) $$

Without losing generality, adding the intercept to Eq. (2) gives

$$ f(x_{ij}) = a_{0j} + a_{1j}(x_{ij} - b_{1j}) \tanh \left(\frac{x_{ij} - b_{1j}}{w_{1j}} \right) + a_{2j}(x_{ij} - b_{2j}) \tanh \left(\frac{x_{ij} - b_{2j}}{w_{2j}} \right) + \varepsilon_{ij}. $$

$$ (3) $$

where the fitted profiles can be constructed via the estimation of the unknown parameters $\beta_j = (a_{0j}, a_{1j}, a_{2j}, b_{1j}, b_{2j}, w_{1j}, w_{2j})'$.

3 Phase I Analysis

Akaike information criterion (AIC) is a statistical indicator for the selection of several models that can best explain the data. This criterion not only considers the fitting result but also considers the number of parameters. As the parameters increase, the model would become more complex accordingly and then affecting the effectiveness of fitting result. Akaike information criterion mainly consists of accurate fitting result and the parameters number. A minimum AIC value is desirable.

At the Fig. 2 shows that the modeling function using hyperbolic triple tangent functions exhibits a better fit than hyperbolic double tangent functions. $\overline{AIC}$ is used to decide the best hyperbolic tangent function. We developed double, triple, and quadruple hyperbolic tangent function by hyperbolic tangent function. Table 1 shows that using hyperbolic triple tangent function yields is the smallest $\overline{AIC}$, so the choice of the hyperbolic triple tangent function is chosen for this case study of the vacuum heat treatment data.

In order to find the model near that best fits profiles, extending to hyperbolic triple and quadruple tangent functions are attempted. The hyperbolic triple tangent function can be described by

$$f(X_{ij}, \beta_i) = a_{0i} + a_{1i}(x_{ij} - b_{1i}) \tanh\left(\frac{x_{ij} - b_{1i}}{w_1}\right) + a_{2i}(x_{ij} - b_{2i}) \tanh\left(\frac{x_{ij} - b_{2i}}{w_2}\right) \\ + a_{3i}(x_{ij} - b_{3i}) \tanh\left(\frac{x_{ij} - b_{3i}}{w_3}\right) + \varepsilon_j, \tag{7}$$

Figure 3 shows the T^2 control charts by using the hyperbolic double tangent function method from the control chart shows the 9th profile appears to be an outlier as its parameter estimates are quite different from those of remaining profiles.

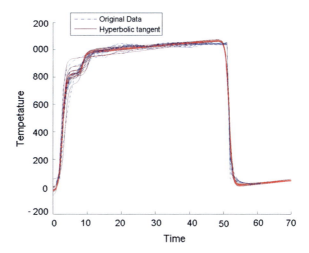

Fig. 2 Profile fitting by hyperbolic triple tangent

Table 1 The fitting results for the hyperbolic tangent functions

Knots	Double	Triple	Quadruple
$\overline{R^2}$	0.9841	0.99197	0.992097
$\overline{R^2_{adj}}$	0.9823	0.98521	0.990121
$\overline{AIC}$	25.8799	**23.7199**	29.7199

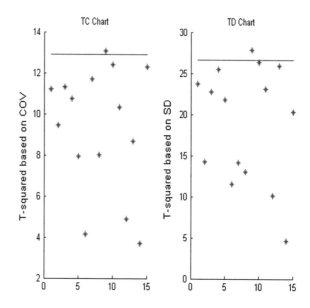

Fig. 3 T^2 control charts for 15 profiles data

4 Conclusion

The field of SPC is becoming more and more promising as the modern technology advances. As products and process become more competitive and complex, there is an increasing demand for the development of statistical methodologies to meet these needs. This objective of this paper is to develop a new approach that can properly characterize the aluminum alloy vacuum heat treatment process data and also provide competitive phase I monitoring performances. The specific shape of the profile addressed in this thesis is nonlinear. In Phase I, we propose using the modified hyperbolic tangent function to fit the aluminum alloy heat treatment process data. The proposed approach provides a good fit to the studied data. By means of the two T^2 statistics, the outlying profiles can be correctly identified through this monitoring framework.

Studying the fundamental of manufacturing processes could always be an interesting area in SPC. It helps understand whether the profiles are being over-fitted or under-fitted. Also, a comparison study of proposed approach to the other nonparametric methods, appearing in literature may be a good idea for further study. Building upon the research of the aluminum vacuum heat treatment process monitoring, how to create an effective combination of non-parametric charts in phase I analysis also deserves a future study.

References

Chen S, Nembhard HB (2011) A high-dimensional control chart for profile monitoring. Qual Reliab Eng Int 27(4):451–464

Ding Y, Zeng L, Zhou S (2006) Phase I analysis for monitoring nonlinear profiles in manufacturing processes. J Qual Technol 38(3):199–216

Fan SKS, Yao NC, Chang YJ, Jen CH (2011) Statistical monitoring of nonlinear profiles by using piecewise linear approximation. J Process Control 21:1217–1229

Jensen AJ, Birch JB (2009) Profile monitoring via nonlinear mixed models. J Qual Technol 41(4):18–34

Jin J, Shi J (1999) Feature-preserving data compression of stamping tonnage information using wavelets. Technometrics 41(4):327–339

Kang L, Albin SL (2000) On-line monitoring when the process yields a linear profile. J Qual Technol 32(4):418–426

Moguerza JM, Munoz A, Psarakis S (2007) Monitoring nonlinear profiles using support vector machines. Lecture notes in computer science 4789, Springer, Berlin, pp 574–583

Qiu P, Zou C, Wang Z (2010) Nonparametric profile monitoring by mixed effects modeling. Technometrics 52(3):265–293

Williams JD, Woodall WH, Birch JB (2007) Statistical monitoring of nonlinear product and process quality profiles. Qual Reliab Eng Int 23(8):925–941

Woodall WH, Spitzner DJ, Montgomery DC, Gutpa S (2004) Using control charts to monitor process and product quality profiles. J Qual Technol 36:309–320

Zou C, Tsung F, Wang Z (2007) Monitoring profiles based on nonparametric regression models. Technometrics 49:395–408

Full Fault Detection for Semiconductor Processes Using Independent Component Analysis

Shu-Kai S. Fan and Shih-Han Huang

Abstract Nowadays, semiconductor industry has been marching toward an increasingly automated, ubiquitous data gathering production system that is full of manufacturing complexity and environmental uncertainty. Hence, developing an effective fault detection system is virtually essential for the semiconductor camp. This paper focuses on the physical vapor deposition (PVD) process. In order to rectify the aforementioned difficulties that could realistically take place, an independent component analysis approach is proposed that decomposes every process parameter of interest into the basis data. A fault detection method is presented to identify the faults of the process and construct a process monitoring model by means of the obtained basis data.

Keywords Process monitoring · Independent component analysis (ICA) · Semiconductor manufacturing · Fault detection

1 Introduction

Due to the high complexity of semiconductor manufacturing, monitoring and diagnosis are gaining importance in manufacturing system. In semiconductor process, the common process variations can be due to process disturbances such as process mean shift, drift, autocorrelation, etc. For variance reduction, the critical source of process variation needs to be first identified. The effective fault detection

---

S.-K. S. Fan (✉) · S.-H. Huang
Department of Industrial Engineering and Management, National Taipei University of Technology, Taipei 10608, Taiwan, Republic of China
e-mail: morrisfan@ntut.edu.tw

S.-H. Huang
e-mail: e2813260@hotmail.com

Y.-K. Lin et al. (eds.), *Proceedings of the Institute of Industrial Engineers Asian Conference 2013*, DOI: 10.1007/978-981-4451-98-7_56, © Springer Science+Business Media Singapore 2013

techniques can help semiconductors reduce scrap, increase equipment uptime, and reduce the usage of test wafers. In semiconductor industry, a massive amount of trace or machine data is generated and recorded. Traditional univariate statistical process control charts have long been used for fault detection, such as the Shewhart, CUSUM (cumulative sum), and EWMA (exponentially weighted moving average) charts have long been applied to reducing process variablility. Although univariate statistical techniques are easy to implement, they often lead to a significant number of false alarms on multivariate processes where the sensor measurements are highly correlated because of physical and chemical principles governing the process operation, such as mass and energy balances. Multivariate statistical fault detection methods such principal component analysis (PCA) and partial least squares (PLS) have drawn increasing interest in semiconductor manufacturing industry recently. PCA and PLS-based methods have been tremendously successful in continuous process applications such as petrochemical processes and its application to traditional chemical batch processes has been extensively studied in the last decade.

2 Independent Component Analysis

To rigorously define ICA (Hyvärinen et al. 2001), a statistical "latent variables" model is considered. Assume that we observe n linear mixtures $x_1, \ldots, x_n$ of n independent components, as expressed by

$$x_j = a_{j1}s_1 + a_{j2}s_2 + \cdots + a_{jn}s_n. \tag{1}$$

In the ICA model, we assume that each mixture $\mathbf{x}$ as well as each independent component s_n is a random variable. It is convenient to use vector-matrix notation instead of the sums as in equation. Let us denote by $\mathbf{x}$ the random vector whose elements are the mixtures $x_1, \ldots, x_n$, and likewise, by s the random vector with elements $s_1, \ldots, s_n$. Let us also denote by $\mathbf{A}$ the mixing matrix with elements a_{ij}. In terms of vector–matrix notation, the above mixing model becomes

$$\mathbf{X}_{m \times k} = \mathbf{A}_{m \times n} \cdot \mathbf{S}_{n \times k}. \tag{2}$$

The statistical model in equation is called independent component analysis, or ICA model. The ICA model is a generative model, implying that it describes how the observed data are generated by a process of mixing the component $\mathbf{S}$. The independent components are termed latent variables, meaning that they cannot be directly observed. Thus, the mixing matrix is assumed to be unknown. The ICA aims to find a demixing matrix $\mathbf{S}$ by means of the random vector $\mathbf{x}$, such that

$$\mathbf{Y}_{n \times k} = \hat{\mathbf{S}}_{n \times k} = \mathbf{W}_{n \times m} \cdot \mathbf{X}_{m \times k}. \tag{3}$$

Full Fault Detection for Semiconductor Processes 467

The fundamental restriction imposed on ICA is that the independent components are assumed statically independent and they must have non-Gaussian distributions. These two conditions are also critical techniques that make ICA different from any other methods.

3 Proposed Fault Detection Method

We use the Independent Component Analysis (ICA) method to perform the fault detection in the process variable and information that are obtained from the sensor. Wu have proposed an ICA method that will be borrowed for use in Wu (2011). In his thesis, they use the ICA method to detect the defects of solar cells/modules in electroluminescence images. Here, we utilize their ICA method to design our fault detection procedure in semiconductor process and we find the ICA method can easily help us to make distinction between the normal data and abnormal data in the process. Here, we focus on the sputter deposition that is a physical vapor deposition (PVD) method of depositing thin films by sputtering. Sputter deposition is a physical vapor deposition (PVD) method of depositing thin films by sputtering. Resputtering is re-emission of the deposited material during the deposition process by ion or atom bombardment. Sputtered atoms ejected from the target have a wide energy distribution, typically up to tens of eV (100,000 K). The sputtered ions can ballistically fly from the target in straight lines and impact energetically on the substrates or vacuum chamber (causing resputtering).

In what follows, the proposed fault detection method will be described. We have to firstly choose the normal data to be trained by using the ICA and then we can obtain the basis data. Therefore, using the linear combination of the basis data to reconstruct each testing data.

Step 1: Choose several normal data from the sensor to be the training data.

$\mathbf{X}$ is the matrix of the training data. Each normal data is in each column then the sample matrix is shown below:

$$\mathbf{X} = \begin{bmatrix} x_{11} & x_{12} & \cdots & x_{1N} \\ x_{21} & x_{22} & \cdots & x_{2N} \\ \vdots & \vdots & \ddots & \vdots \\ x_{M1} & x_{M2} & \cdots & x_{MN} \end{bmatrix}. \tag{4}$$

M Number of the normal data
N Number of observations in each normal data

Step 2: Obtain the source matrix by using the FastICA to train the training data.

In this step, we can take the input data ($\mathbf{X}$) into FastICA process, and then we can obtain the basis data $\mathbf{U}$. The size of $\mathbf{U}$ is $M \times N$. ICA model is as below:

$$U = W \cdot X, \quad (5)$$

where **W** is the demixing data that is obtained by using FastICA, and the size of **W** is $H \times H$. The basic data **U** ($U = [u_1, u_2, \cdots, u_H]^T$) is the source matrix that stands for the linear combination of basis data. Due to the page limitation, for the procedure of reconstructing the test data, interested readers can be referred to Huang.

4 Experimental Results

We will discuss the experimental results of ICA for each process parameters (see Fig. 1) and we will find the key step for each process parameter. Because we focus on the sputter deposition, then we have four important process parameters to be investigated that include Gas2, Gas3, PRESSURE and PWR. Gas2 and Gas3 are the reactive gas, PRESSURE is the pressure measured inside in the chamber and PWR is the pressure of the Ar. We collect 227 parameters that are known as normal data serving as the training data. There are 68 observations in Gas2, 69 observations in Gas3, 64 observations in PRESSURE and 68 observations in PWR. We have 20 test data that include 10 normal data (good wafers) and 10 abnormal data (bad wafers). We collect 20 sets of the testing data. Here we use FastICA

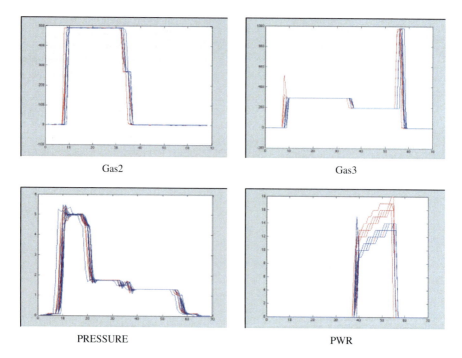

Fig..1 Process parameter

Full Fault Detection for Semiconductor Processes 469

algorithm to retrieve the basis data from the training data in each process parameters. The flow of FastICA algorithm has been described in Sect. 3.3.3.1. By using FasICA algorithm we obtained 67 basis data from Gas2, 68 basis data from Gas3, 63 basis data from PRESSURE and 23 basis data from PWR.

When we use ICA to analyze the data, we have to set up the threshold for each process parameter to help us to distinguish the good and bad wafers in the testing data. Therefore, we consider the empirical rule to find the suitable threshold after we analyze the testing data by using ICA via a comprehensive preliminary test. The thresholds for each parameter are in Table 1.

After we set up the empirical rule for the threshold, then we can start to detect the faults of the data that has been analyzed (i.e., the distance between the testing data and reconstructed data has been calculated). The success rate along with its standard deviation for each process parameter are displayed in Table 2.

In order to further investigate the accuracy rate, then we compute the type I error and type II error for each parameter. Here, the type I error indicates the rate of the normal data that is mistakenly identified as the abnormal data; the type II error indicates the rate of the abnormal data that is not correctly identified as abnormal data. The results of the type I error and type II error are in Tables 3 and 4.

Table 1 Threshold of each process parameter

Threshold	Gas2	Gas3	PRESSURE	PWR
Model 1	1,555	1,620	13.95	28
Model 2	1,648	1,760	15.3	29

Table 2 Experimental results of each parameter monitored

Threshold (Std)	Gas2	Gas3	PRESSURE	PWR
Model 1	57 % (9.5026)	55 % (10.083)	62 % (9.5867)	77 % (6.0886)
Model 2	56 % (4.3457)	61 % (4.759)	65 % (12.3886)	70 % (9.0238)

Table 3 Type I error

Type I error	Gas2 (%)	Gas3 (%)	PRESSURE (%)	PWR (%)
Model 1	44	49	43	31
Model 2	47	41	36	26

Table 4 Type II error

Type II error	Gas2 (%)	Gas3 (%)	PRESSURE (%)	PWR (%)
Model 1	52	53	41	33
Model 2	39	28	32	24

5 Conclusions

In this study of sputter deposition, the observations for each process parameter in the semiconductor process can be always divided to steps 0–10. We take steps 0–10 for each observations in each process parameter to perform the simulation by using the ICA method. For experimental results of ICA model1, the success rates of Gas2, Gas3, PRESSURE and PWR are 57, 55, 62 and 77 %, respectively. However, if we only take the key (step 2–7 in Gas2, step 1–4 in Gas3, step 0–2 in PRESSURE and step 8–9 in PWR) to perform the simulation in experimental result of ICA model 1, the success rates of Gas2, Gas3, PRESSURE and PWR are 64, 62, 68 and 79 %, respectively. From these experimental results, we can obviously see that the success rate of only considering the key step is completely higher than taking steps 0–10. It means that if we take steps 0–10 to perform ICA analysis, there will produce much useless information that could influence the efficiency of ICA in the observations. Therefore, in order to reduce the influence of this useless information, we can only take the key step to perform ICA. For time delay and missing values, we have several methods to deal with these problems. For time delay data, after we correct the time delay data according to the difference in time, we find that using no insert data is a suitable method from the experimental results. For missing values, using a half standard deviation control chart is suitable method to deal this problem.

References

Hyvärinen A, Karhunen J, Oja E (2001) Independent component analysis. Wiley, New York
Wu SH (2011) Machine vision-based defect detection of solar cells/modules in electroluminescence images, Master thesis. Yuan-Ze university. Department of Industrial Engineering, Taoyuan County, Taiwan

Multi-Objective Optimal Placement of Automatic Line Switches in Power Distribution Networks

Diego Orlando Logrono, Wen-Fang Wu and Yi-An Lu

Abstract The installation of automatic line switches in distribution networks provides major benefits to the reliability of power distribution systems. However, it involves an increased investment cost. For distribution utilities, obtaining a high level of reliability while minimizing investment costs constitutes an optimization problem. In order to solve this problem, the present paper introduces a computational procedure based on Non-dominated Sorting Genetic Algorithm (NSGA-II). The proposed methodology is able to obtain a set of optimal trade-off solutions identifying the number and placement of automatic switches in distribution networks for which we can obtain the most reliability benefit out of the utility investment. To determine the effectiveness of the procedure, an actual power distribution system was considered as an example. The system belongs to Taiwan Power Company, and it was selected to drive comparisons with a previous study. The result indicates improvements in system reliability indices due to the addition of automatic switching devices in a distribution network, and demonstrates the present methodology satisfies the system requirements in a better way than the mentioned previous study.

Keywords Automatic line switches · Genetic algorithm · Power distribution networks · Multi-objective optimization

---

D. O. Logrono (✉)
Graduate Institute of Industrial Engineering, National Taiwan University,
Taipei 10617, Taiwan, Republic of China
e-mail: r99546042@ntu.edu.tw

W.-F. Wu
Graduate Institute of Industrial Engineering and Department of Mechanical Engineering,
National Taiwan University, Taipei 10617, Taiwan, Republic of China
e-mail: wfwu@ntu.edu.tw

Y.-A. Lu
Department of Mechanical Engineering, National Taiwan University, Taipei 10617,
Taiwan, Republic of China
e-mail: Skippy_lu11@hotmail.com

Y.-K. Lin et al. (eds.), *Proceedings of the Institute of Industrial Engineers Asian Conference 2013*, DOI: 10.1007/978-981-4451-98-7_57,
© Springer Science+Business Media Singapore 2013

1 Introduction

For modern power distribution utilities, service quality and continuity are the two most important demands. To attend to this demand, distribution automation is used for improving the system response in the event of outages. On distribution automation projects, the biggest challenge is how to achieve the most possible benefit while minimizing the investment costs (Abiri-Jahromi et al. 2012). The optimal placement and number of automatic line switches in distribution networks is essential for utilities in order to reduce power outages. However this optimization is a combinatorial constrained problem described by nonlinear and nondifferential objective functions and their solution can be challenging to solve (Tippachonl and Rerkpreedapong 2009). Many studies provided insight on the subject, for example, Abiri-Jahromi et al. (2012) and Chen et al. (2006). In papers such as Conti et al. (2011), different versions of NSGA-II have been applied for the allocation of protective and switching devices. This paper proposes the development of a computational algorithm to address the optimal placement of automatic line switches in distribution networks that simultaneously minimizes cost expenditures and maximizes system reliability. A version of NSGA-II is employed.

2 Problem Formulation

2.1 Multi-objective Optimization

The multi-objective optimization problem (MOOP), in its general form, can be expressed using the following structure (Deb 2001):

$$
\begin{aligned}
&\text{Minimize/Maximize } f_m(\mathbf{x}), \quad m = 1, 2, \ldots, M; \\
&\text{subject to } g_i(\mathbf{x}) \geq 0, \quad j = 1, 2, \ldots, J; \\
&\qquad\qquad h_k(\mathbf{x}) \geq 0, \quad k = 1, 2, \ldots, K; \\
&\qquad x_i^{Lower} \leq x_i \leq x_i^{Upper}, \quad i = 1, 2, \ldots, n.
\end{aligned}
\tag{1}
$$

where $f_m(\mathbf{x}) = (f_1(\mathbf{x}), f_2(\mathbf{x}), \ldots, f_M(\mathbf{x}))^T$ is a vector of M objective functions to be optimized. A solution $\mathbf{x}$ is a vector of n decision variables $\mathbf{x} = (x_1, x_2, \ldots, x_n)^T$. When a solution $\mathbf{x}$ satisfies all the $(J + K)$ constraints and is allocated within the $2n$ variable bounds, it is known as a feasible solution for the optimization problem and can be mapped into the M objective functions to obtain the objective space. We can perform all possible pair-comparisons for a finite set of solutions, and find which solutions dominate. The solution set of a multi-objective optimization consists of all non-dominated solutions, and it is known as Pareto-optimal set or Pareto-optimal front.

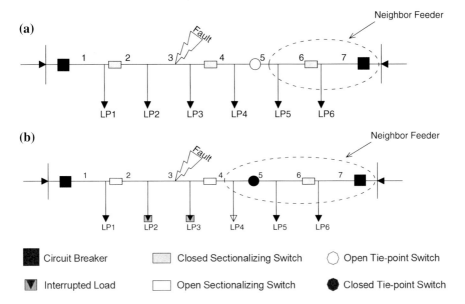

Fig. 1 Automatic sectionalizing and tie-point switches in distribution networks

2.2 Distribution Feeder Model

An illustration of a radial feeder of a distribution system is presented in Fig. 1. There are two types of line switches normally installed along distribution feeders: sectionalizing switch (D) and tie-point switch (TP). For Fig. 1a, the sets of switching devices can be expressed as: D = {1,4,6} and TP = {5}.

When a failure occurs in Sect. 3, the configuration becomes as that of Fig. 1b. Customers in load point LP1 and LP4 will experience an interruption equal to the switching time of the devices, while loads between the two automatic switches (LP2 and LP3) will experience longer outage duration equal to the repair time of feeder Sect. 3.

2.3 Objective Functions

We select three objectives to be minimized simultaneously: System Average Interruption Frequency Index (SAIFI), System Average Interruption Duration Index (SAIDI) and TCOST (total investment cost). We use a method based on a modification of the study performed by Tippachonl and Rerkpreedapong (2009), defined as follows:

- *SAIFI*, $f_1(\mathbf{x})$: system average interruption frequency index.

$$SAIFI = \frac{\sum_{i=1}^{n} \left(\sum_{s=1}^{m} \lambda_{is} \right) N_i}{\sum_{i=1}^{n} N_i} \quad \text{(int./cust. - year)} \tag{2}$$

where N_i is the number of customers at load point i, n identifies the number of load points and m the number of sections. λ_{is} is the permanent failure rate of load point i due to failure in section s.

- SAIDI, $f_2(\mathbf{x})$: system average interruption duration index. It is referred to as the average time that a customer is interrupted per year.

$$SAIDI = \frac{\sum_{i=1}^{n} \left(\sum_{s=1}^{m} \lambda_{is} r_{is} \right) N_i}{\sum_{i=1}^{n} N_i} \quad \text{(min/cust. - year)} \tag{3}$$

where r_{is} is the average time per interruption of load point i due to outages in section s.

- TCOST, $f_3(\mathbf{x})$: total cost.

$$TCOST = Num_D \times C_D + Num_{TP} \times C_{TP} \text{ (US\$ - year)} \tag{4}$$

where Num_D accounts for the number of sectionalizing switches and Num_{TP} for the number of tie-point switches to be installed. C_D and C_{TP} are the total costs including purchase and installation of sectionalizing and tie-points, respectively. For this task, a decision variable will be associated to every section for computation purposes and its value represents the cases in which: 0—no device, 1—sectionalizing switch, or 2—a tie-point switch.

2.4 Constraints

The following operation constraints are considered in this study:

- Each decision variable can only take integer values 0, 1 or 2.
- Only one automatic tie-point switch can connect two neighbor feeders.
- When performing the load transfer for service restoration, no overloading should be introduced to the power transformers.

3 Proposed Integer Version of NSGA-II

Genetic Algorithms (GAs) handle a population of solutions that is modified over the course of a number of generations using genetic operators and are able to work with a wide range of types and number of objective functions making them suitable for our multi-objective optimization problem. A modified version of NSGA-II was selected as the search mechanism for the optimal solutions to the proposed problem.

Feasible Solution =	x_1	x_2	x_3	x_4	x_5	x_6	...	x_{m-3}	x_{m-2}
	0	0	1	0	1	2		1	0
section	2	3	4	5	6	7		m-2	m-1

Fig. 2 Chromosomal representation for a feasible solution

3.1 Solution Codification

The optimal placement of automatic line switches in distribution networks can be considered as an integer optimization problem (integer phenotype). In this paper, integer-coding is used.

A decision variable has been associated to every section of the distribution feeder where a switch can be allocated (0 → no switch, 1 → auto sectionalizing switch, 2 → auto tie-point switch). An illustration of the chromosomal representation for this feeder is shown in Fig. 2.

3.2 Algorithm Procedure

The optimal solution searching process performed by the proposed NSGA-II follows the procedures shown in the Fig. 3. The steps are presented as follows: (1) *Generate the Initial Population;* (2) *Fitness Assignment;* (3) *Elitist Selection;* (4) *Tournament Selection;* (5) *Crossover;* (6) *Mutation.*

4 Numerical Results

An actual distribution system has been considered for comparison. The system is part of Taiwan Power Company and it is located in the Fengshan area. The system is described in Chen et al. (2006).

The NSGA-II settings for both case studies are determined in Table 1. Other data have been retrieved from Chen et al. (2006) in Table 2. For this case study, we run the proposed NSGA-II using a population size and number of generations of 100 and 500, respectively, with a total of 50,000 computational evaluations. The scatter of the non-dominated solutions and the Pareto-optimal solutions for the switch placement optimization problem is found as that of Fig. 4.

In this study, a max–min approach has been used to select a final solution for the multi-objective problem. Each solution in the nondominated set is first normalized and then a max–min operator is applied to them using the following expression as in Tippachonl and Rerkpreedapong (2009):

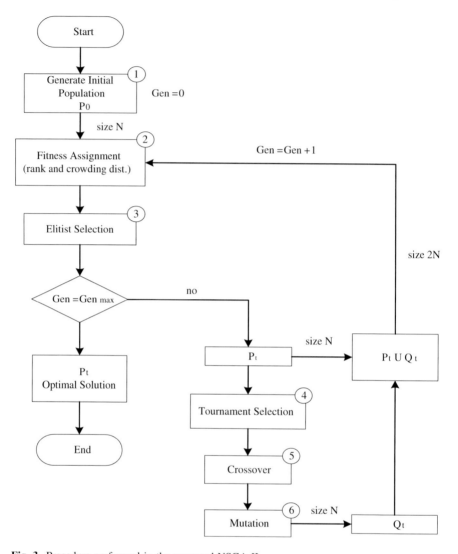

Fig. 3 Procedure performed in the proposed NSGA-II

$$\max\left[\min\left(\frac{SAIFI_{\max} - SAIFI_i}{SAIFI_{\max} - SAIFI_{\min}}, \frac{SAIDI_{\max} - SAIDI_i}{SAIDI_{\max} - SAIDI_{\min}}, \frac{TCOST_{\max} - TCOST_i}{TCOST_{\max} - TCOST_{\min}}\right)\right]. \tag{5}$$

In Table 3, we present a comparison of the reliability indexes in the original system; after the partial automation study; and finally our proposed study for the optimal placement of automatic line switches.

Multi-Objective Optimal Placement of Automatic Line Switches

Table 1 NSGA-II parameter settings

	Parameter	Value
Crossover	Crossover fraction	0.9
	Ratio	0.1
Mutation	Mutation fraction	0.4
	Scale	0.5
	Shrink	0

Table 2 Distribution feeder parameters

Parameter		Rate/Duration time
Average permanent failure rate, λ_s		0.132 failures/year-km
Average repair time, r_{rs}		240 min.
Switching time, r_{sw}	Upstream the failure	5 min
(automatic switches)	Downstream the failure	0.33 min

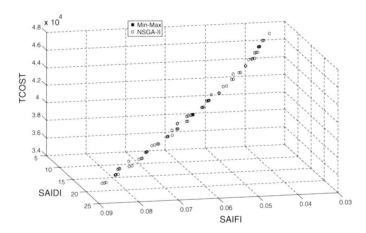

Fig. 4 Pareto-optimal solutions and max–min solution for the case study

The solution obtained by the max–min approach represents an improvement of 62 % in SAIFI and 32 % in SAIDI, when compared to the previous study. Consequently, SAIFI has been reduced from 0.157 to 0.059 int./cust. − year, and SAIDI from 20.872 to 14.212 (min./cust. − year) An investment 2.8 times higher than the later study is required for gaining the above mentioned reliability benefit.

Despite the higher TCOST, the max–min solution not only excels in providing a remarkable reduction in the current reliability indices, but also the reliability values obtained by its possible implementation can be still useful in the future since Taipower Company has now set a new goal of 15.5 min./cust. − year for SAIDI by the year 2030 (Runte 2012).

Table 3 Results of the proposed study

	Original system manual switching	Previous study partial automation	Our proposed solution fully automated
SAIFI (int./cust. − yr.)	0.231	0.157	0.059
SAIDI (min./cust. − yr.)	32.233	20.872	14.212
TCOST (US$ − yr.)	–	14,326	40,516.9

5 Conclusion

This paper provides a methodology to solve the multi-objective optimal placement of automatic line switches in distribution networks. The proposed integer version of NSGA-II has showed that this methodology guarantees a very good approximation to the true Pareto-front. The decision maker can select a final solution from the Pareto-optimal according to his/her professional experience. However, a selection approach has also been presented in this study in order to choose the final solution based on Max–min method. The methodology was tested using an actual distribution system. The proposed solution for the system allows obtaining notable improvements for reliability indices but involves higher investment cost for the utility.

References

Abiri-Jahrom A, Fotuhi-Firuzabad M, Parvania M, Mosleh M (2012) Optimized sectionalizing switch placement strategy in distribution systems. Power Delivery, IEEE Trans 99:1–7
Chen CS, Lin CH, Chuang HJ, Li CS, Huang MY, Huang CW (2006) Optimal placement of line switches for distribution automation systems using immune algorithm. Power Syst, IEEE Trans 21(3):1209–1217
Conti S, Nicolosi R, Rizzo SA (2011) Optimal investment assessment for distribution reliability through a multi-objective evolutionary algorithm. Paper presented at 2011 International conference on clean electrical power (ICCEP)
Deb K. (2001) Multi-objective optimization using evolutionary algorithms. Chichester New York Weinheim [etc.]: J. Wiley
Runte G (2012) Taiwan power: quietly getting the smart grid, Power Generation
Tippachon W, Rerkpreedapong D (2009) Multiobjective optimal placement of switches and protective devices in electric power distribution systems using ant colony optimization. Electr Power Syst Res 79(7):1171–1178. doi:10.1016/j.epsr.2009.02.006

The Design of Combing Hair Assistive Device to Increase the Upper Limb Activities for Female Hemiplegia

Jo-Han Chang

Abstract Many researches show that the progress of the upper limb function of patients with more than one year apoplexy appears to be Learned Nonuse. This research takes concepts of the Constraint-induced Movement Therapy and User-Centered Design and develops the combing hair assistive device to increase the upper limb activities for female Hemiplegia. We adopt the AD-TOWS (Assistive Devices-Threats, Opportunities, Weaknesses, and Strengths) matrix to develop 33 design concepts among which 4 concepts are screened to make models, and then invite 5 participators in the experiment. By analyzing the results of upper limb lifting angle, the upper limb movement angle forced by the "Joint Adjustable Device" is the biggest, which is followed by the "Comb Convertible Device", and then is the "Comb Convertible Lengthening Device". The upper limb average angle of operating the above mentioned three assistive devices are bigger than that of operating the existing devices. By the result of the part unable to be combed, we find that the most difficult action to users who use the existing long-handled comb is to comb their sutural bone and occipital bone, however, the "Double-Handled Device" is good at improving the action unable to be done.

Keywords Activities · Constraint-induced movement therapy · Hemiplegia · Upper limb · User-centered design

1 Introduction

For the patients with cerebral apoplexy, besides suffering from physical dysfunction, they also suffer from great psychological impact (Qiu 2008; Mitchell and Moore 2004). Patients will have Body Image Disturbance, and even hold a

---

J.-H. Chang (✉)
Department of Industrial Design, National Taipei University of Technology, 1, Section 3, Chung-hsiao E. Road, Taipei 10608, Taiwan, Republic of China
e-mail: johan@ntut.edu.tw

Y.-K. Lin et al. (eds.), *Proceedings of the Institute of Industrial Engineers Asian Conference 2013*, DOI: 10.1007/978-981-4451-98-7_58,
© Springer Science+Business Media Singapore 2013

negative attitude towards the disease. Many researches show that the upper limb function of the patients with more than one year history of cerebral apoplexy gradually present the symptom of Learned Nonuse (Taub et al. 1994).

In recent years, Constraint-Induced Movement Therapy (CIMT) has been used clinically, called Forced Use. Many researches also certified that the limb function of the affected side of 50 % patients is significantly improved after a lot of appropriate training (Liepert et al. 1998).

The proportions of items in which the female has difficulties in self-care are higher than those of male, and the proportion of complete loss of self-care ability in walking up and down stairs, washing hair and outdoor walking for female is over 5 % higher than that for male (Census and Statistics Department of Ministry of the Inferior 2006). For many females, it is very important that they can tidy personal appearance and hygiene independently. The goal of medical team devoted is to help the patients live independently and restore their self-esteem. This paper proposes an assumption that if we can design an assistive device based on the concept of CIMT by the love of beauty in female's nature, and the assistive device can not only help the patients live independently, but also can promote their synchronous motion, so as to help the patients achieve the purpose of rehabilitation at an early date, even return to normal life.

2 Literature Review

The biggest difference between disabled people and general users is the physiological function of the disabled people is divided into normal side and affected side, and the conditions of normal side and affected size are different, as a result, when they use an assistive device, it can bring opportunities to achieve the purpose, but also may cause danger due to inapplicability. Ma et al. (2007) proposed AD-SWOT (Assistive Devices-Strength, Weakness, Opportunity, Threat), imported the Business Management Strategy and Method-SWOT (Strength, Weakness, Opportunity, Threat) into the design procedure of assistive devices, analyzed the physiological condition of users by AD-SWOT, and developed the design of assistive devices according to the AD-SWOT analysis results. Then, Wu et al. (2009) developed the assistive devices using the method of AD-TOWS (Assistive Devices-Threat, Opportunity, Weakness, Strength) matrix to take full advantage of patients' physiological conditions, therefore both the affected side and normal can be moved during the process of using assistive devices.

The purpose of this paper is to develop an assistive device which can promote movements. This paper try to develop the assistive devices based on the concept of CIMT and develop a new assistive device through AD-SWOT and AD-TOWS procedures, so as to make good use of patients' limbs to help the patients complete tasks with their own strength, and they can also make limb movement in the process.

3 New Assistive Devices Design

In this stage, a participator was invited to have a hemiplegia condition analysis, so as to design new assistive devices according to the research results.

3.1 Physiological Condition Analysis of Female Hemiplegia

3.1.1 Ability Measurement of Participator

To learn about the consciousness, cognitive function, communication condition, motor coordination ability and motor function of the participator.

1. Consciousness, cognitive function and communication condition: Diagnosed by doctor, the participator was a left hemiplegia apoplexy patient caused by right cerebral artery occlusion, whose consciousness, cognitive function and communication condition were normal.
2. Motor coordination ability: The participator had left upper limb hemiplegia, myasthenia and poor upward activity of shoulder joint.
3. Motor function: The participator can overcome the basic synergistic effects and easily perform actions such as Hand to Sacrum, Raise Arm Forward to Horizontal, Pronation (Elbow Flexed) and Supination (Elbow Flexed).

3.1.2 AD-SWOT of Participator

AD-SWOT of the participator was summarized and analyzed according to the observation on actions and activity of upper limbs (Table 1).

3.2 Combing Hair Assistive Device Design for Female Hemiplegia

In this stage, the assistive device design was developed according to the steps of AD-SWOT and AD-TOWS, and AD-WO developed 9 concepts, AD-SO and AD-ST developed 12 concepts respectively. In order to screen the best design, 19 evaluation items containing W (Weakness) of user and T (Threat) of assistive device as well as WT (Weakness/Threat) were proposed to screen the best design from the 33 design concepts. Four design concepts with the highest score are shown in Table 2, named as Comb Convertible Device, Joint Adjust-able Device, Double-Handled Device and Comb Convertible Lengthening Device.

Table 1 AD-SWOT of participator

SWOT	Description
Strength	1 The shoulder joint of left upper limb can bend forward and lift to the height of shoulder
	2 The ROM and myodynamia of forward bending the right upper limb shoulder joint are complete
	3 The shoulder joint of right upper limb can lift to the height of shoulder
	4 The ROM and myodynamia of the elbow, wrist and fingers of right upper limb are complete
	5 Shoulder joints of right and left upper limbs can rotate inward and outward
Weakness	1 The shoulder joint of left upper limb can't bend forward, lift and abduct beyond the shoulder
	2 Myodynamia of moving the shoulder, elbow, wrist and finger of left upper limb are insufficient
	3 The shoulder joint of right upper limb can't lift and abduct beyond ear, and without insufficient myodynamia
	4 Myodynamia of the shoulder joint of right upper limb to rotate inward and outward is insufficient
Opportunity	1 The user can complete the whole task using assistive devices
	2 The user's physiological functions can be induced by using the assistive devices and the users utilize their own abilities
	3 The assistive devices can help rehabilitation
Threat	1 The operational motion makes the user lose balance
	2 The pain and change of blood flow volume are caused when the user exerts strength suddenly and violently
	3 Bad compensatory postures cause other pains

4 Research Methods

In order to evaluate whether the assistive devices developed by AD-SWOT and AD-TOWS meet the concept of CIMT, this program used mock-up to perform experiment of testing and evaluation. In the evaluation, 5 left hemiplegia patients were invited to actually operate the combing hair assistive devices to learn their level of enforcement; the existing combing hair assistive devices (long-handled comb) and the four assistive devices in this program were given to participators; instead of being asked to perform typical tasks, the participators were allowed to operate assistive devices freely to test the applicability of the assistive devices. The content of evaluation contained two aspects:

4.1 Combing Completion

Parts able to be combed were analyzed, including parietal bone, right parietal bone, left parietal bone, sutural bone, occipital bone, right temporal bone and left temporal bone.

The Design of Combing Hair Assistive Device 483

Table 2 New combing hair devices for female hemiplegia

Assistive device	Picture	Design description
Comb convertible device		1. The area of comb bed is increased 2. The contact point of comb bed and handle is arranged with a rotational joint, so that the comb bed can rotate to different angles so as to help the users with myodynamia insufficiency 3. The handle can be pressed and the surface of handle is granular, which can stimulate user's sensation of touch
Joint adjustable device		1. The area of comb bed is increased 2. The contact point of comb bed and handle is arranged with a rotational joint, so that the comb bed can rotate to different angles and adjust its length to help the users with myodynamia insufficiency as well as those who are unable to lift up upper limbs 3. The handle can be pressed and the surface of handle is granular, which can stimulate user's sensation of touch
Double-handled device		1. With two handles and the handles are rotatable (All of the four joints can rotate smoothly) 2. The comb bed appears to be curved 3. The comb bed is widened
Comb convertible lengthening device		1. The comb bed is rotatable (The joint can rotate smoothly) 2. The handle is lengthened 3. The comb bed is widened

4.2 Angle of Upper Limb Activity

Observing and measuring the upper limb movement angle of operating the assistive devices, so as to learn whether the new assistive devices are effective on improving upper limb activity; recording the maximum lifting angle and analyzing the average lifting angle respectively.

5 Results

Table 3 is combing completion in the experimental. Table 4 is the angle of upper limb activity in the experimental.

By analyzing the combing completion, we found that it was more difficult for participators to comb sutural bone and occipital the combing completion pital bone by using the existing long-handled comb. As for the four new type assistive devices, the parts unable to be combed of the participators were different; the Double-Handled Device could best improve the action unable to be done. Moreover, two participators could complete the whole combing task by using the Comb Convertible Device and Double-Handled Device.

By analyzing the average angle of upper limb activity, we found that the upper limb movement angle forced by Joint Adjustable Device was the biggest, followed by "Comb Convertible Device", and after that is "Comb Convertible Lengthening Device"; and the upper limb average angle of operating the mentioned three assistive devices was bigger than that of operating the existing devices.

By analyzing the maximum lifting angle of upper limb activity, we found that, both the existing and new combs could promote the upper limb lifting when users comb the parietal bone, while the Double-Handled Device can lift two upper limbs up to the sutural bone. In addition, two participators could lift upper limbs when combing the left and right temporal bones with Comb Convertible Device; and another two participators lifted upper limbs when combing the left temporal bone with Joint Adjustable Device.

Table 3 The combing completion in the experimental

Assistive devices	Left parietal bone	Parietal bone	Right parietal bone	Occipital bone	Sutural bone	Left temporal bone	Right temporal bone	Total
Long-handled comb	0	0	0	4/5	4/5	0	1/5	1 4/5
Comb convertible device	0	1/5	0	3/5	4/5	0	0	1 3/5
Comb convertible lengthening device	2/5	0	1/5	3/5	3/5	4/5	1/5	2 4/5
Joint adjustable device	2/5	2/5	2/5	4/5	2/5	2/5	1/5	3
Double-handled device	0	0	0	1/5	0	1/5	1/5	3/5
Total	4/5	3/5	3/5	3	2 3/5	1 2/5	4/5	

Table 4 The angle of upper limb activity in the experimental

Assistive devices	Maximum angle	Minimal angle	Average angle	Standard deviation
Long-handled comb	105	15	46	35.1
Comb convertible device	25	125	66	37.5
Comb convertible lengthening device	15	110	53	36.8
Joint adjustable device	20	165	83	56.0
Double-handled device	40	15	28	17.7

6 Discussion

This experiment shows that the new combing hair assistive devices for female hemiplegia can improve the upper limb activity of the users. The Double-Handled Device can best improve the action unable to be done, and it can help users to lift bilateral upper limbs to sutural bone so as to achieve the purpose of forcing the movement of the affected part. The products developed by AD-SWOT and AD-TOWS can utilize the user's limbs, so as to successfully promote the movement of the affected part.

By analyzing the experimental record, we found that when the comb bed was perpendicular to the handle, the contact area of comb bed and hair was reduced, so that the participators must lift their upper limbs to complete the task. The results showed that the upper limb lifting can be promoted by reducing the contact area of comb bed and hair, which can serve as a reference for future design.

Previous researches showed that for the patients with more than one year history of cerebral apoplexy, if we limit their the upper limb activity of normal side for 2 weeks and give enough appropriate training on the affected side, about 50 % patients will obviously improve their motor function of the affected upper limb (Liepert et al. 1998); and this improvement can be continued and obviously observed during the application of the affected upper limb in daily life. In this research, the assistive device design evaluation was mainly focused on the promotion of upper limb activity and validity of using the products. In the future, we will continue to observe whether this Double-Handled Device can achieve rehabilitation effects.

Acknowledgments Special thanks to National Science Council (NSC 98-2218-E-033-005-) for its support for this study.

References

Census and Statistics Department of Ministry of the Inferior (2006) The report of people with disabilities living needs. Census and Statistics Department of Ministry of the Inferior, Taipei

Liepert J, Miltner WH, Bauder H, Sommer M, Dettmers C, Taub E, Weiller C (1998) Motor cortex plasticity during constraint-induced movement therapy in stroke patients. Neurosci Lett 250(1):5–8

Ma MY, Wu FG, Chang RH (2007) A new design approach of user-centered design on a personal assistive bathing device for hemiplegia. Disabil Rehabil 29(14):1077–1089

Mitchell E, Moore K (2004) Stroke: holistic care and management. Nursing Standard 33(18):43–52

Qiu HY (2008) The current status of stroke and epidemiological characteristics. Nao Zhong Feng Hui Xun 15(3):2–4

Taub E, Crago JE, Burgio LD, Groomes TE, Cook EW, DeLuca SC (1994) An operant approach to rehabilitation medicine: overcoming learned nonuse by shaping. J Exp Anal Behav 61:281–293

Wu FG, Ma MY, Chang RH (2009) A new design approach of user-centered design on hair washing assistive device design for users with shoulder restriction. Appl Ergon 40(5):878–886

Surgical Suites Scheduling with Integrating Upstream and Downstream Operations

Huang Kwei-Long, Lin Yu-Chien and Chen Hao-Huai

Abstract Surgical operations are a critical function for hospitals because they are responsible for almost two-thirds of all profits. Hence, medical resources associated with surgical operations and operation rooms (OR) should be utilized efficiently. Conducting an overall plan for the allocation of medical resources and capacities used in surgical operations is complicated because numerous types of resources are involved in an OR scheduling problem. Furthermore, the upstream and downstream operations of a surgery, such as the number of beds for preoperative examination and intensive care unit, also significantly affect OR schedule performance. The objective of OR scheduling is to minimize the overtime cost of surgical operations and the idle cost of ORs. Using the concept of OR suites and modes, we construct a mixed integer linear programming model by considering surgical resources and the corresponding upstream and downstream operations. A five-stage heuristic method is proposed to solve large-scale problems effectively and efficiently. A numerical study is conducted, and the results show that the proposed method can reduce the total cost of managing operation rooms and improve the quality of surgical services.

Keywords Operation room · Scheduling · Resource capacity constraint · Heuristic algorithm

---

H. Kwei-Long (✉) · L. Yu-Chien · C. Hao-Huai
Institute of Industrial Engineering College of Engineering, National Taiwan University,
No. 1, Sec. 4, Roosevelt Road, Taipei, Taiwan, Republic of China
e-mail: craighuang@ntu.edu.tw

L. Yu-Chien
e-mail: r99546044@ntu.edu.tw

C. Hao-Huai
e-mail: r01546025@ntu.edu.tw

Y.-K. Lin et al. (eds.), *Proceedings of the Institute of Industrial Engineers Asian Conference 2013*, DOI: 10.1007/978-981-4451-98-7_59,
© Springer Science+Business Media Singapore 2013

1 Introduction

The managerial aspect of providing health services to hospital patients is becoming increasingly important. Among all hospital operations, surgery is one of the most critical because the revenue derived from surgeries is about two-thirds of the total income of the hospital (Jackson 2002). Denton et al. (2010) indicated that operating rooms (ORs) constitute about 40 % of a hospital's total expenses. ORs are therefore simultaneously the most expensive center and the greatest source of revenue for most hospitals. To lower the cost of ORs and run ORs efficiently, OR scheduling must be implemented. If the schedule is too tight, ORs will be less flexible in handling emergencies. Conversely, if the schedule is too loose, ORs may become idle. Thus, methods on how to improve the efficiency of OR scheduling have received considerable attention in recent years.

So far, most OR scheduling problems only consider how surgeries can be scheduled to increase the performance of ORs. However, to improve the efficiency of ORs, the periods when the OR is available should be considered along with the coordination of other resources. Weinbroum et al. (2003) showed that the better coordinated the upstream and downstream resources, the higher the efficiency of ORs in operation room scheduling. Pham and Klinkert (2008) indicated that a complete surgery is divided into three stages: preoperative, perioperative, and postoperative. They also proposed the concept of the OR suite to illustrate that considering other support facilities when scheduling surgeries is important. In the present study, we considered the simultaneous limitation of upstream and downstream OR resources and used the concept of the OR suite to integrate and plan resources. This method prevents ORs from becoming idle or occupied by a surgery for an extended period, which causes a lack of resources upstream and downstream. We propose a mixed integer linear programming (MILP) model to produce a multi-day OR schedule. The resulting model can help hospital managers decide the required number of ORs, how to assign surgeries to the OR, and the time each medical team should conduct operations.

The remainder of this article is organized as follows. The next section gives a brief review of related literature. In Sect. 3, we describe the problem and definitions. Section 4 proposes a heuristic algorithm and shows the numerical results of algorithm. In the final section, we summarize our main conclusions.

2 Literature Review

Pham and Klinkert (2008) indicated that scheduling priority for different cases and predictability should be considered separately in the scheduling problems of surgical cases. They also divided surgical cases into four types: emergency, urgent, elective, and add-elective cases.

Surgical Suites Scheduling with Integrating Upstream and Downstream Operations   489

In a hospital, the proper functioning of an OR depends upon the doctors, the number of operating and recovery rooms, anesthesiologists, and nurses. Moreover, the crucial points of different factors should be focused for generating an effective OR schedule, which can improve healthcare quality and reduce costs in numerous restrictions. Denton et al. (2007) divided OR scheduling into two systems: the block system and the open system. The open system is designed to arrange the patients to be served during the scheduled period. The surgery team will apply for an appointment to use the OR after a surgery case is held, then the supervisor will schedule the time when the ORs may be used. By contrast, Patterson simplified the open system into a first-come, first-served policy. Furthermore, Dexter et al. (2003) defined the "any workday strategy," in which the surgical team is free to make an appointment for a surgery date and an OR. Gable indicated that the open system is rarely utilized and that surgery cases are often cancelled because surgery teams prefer to complete surgery cases during fixed days. However, the open system forces them to complete all surgery cases in a day or a separated group of days, thereby reducing the satisfaction of the surgery team. In practice, the open system is rarely adopted in hospitals.

In the OR scheduling problem, upstream and downstream resource capacity constraints, including caregivers, surgery teams, ORs, and equipment, must be considered. Insufficient capacity may result in surgical delays and OR/personnel scheduling changes, leading to additional healthcare costs and reduced patient satisfaction. Jonnalagadda et al. (2005) showed that 15 % of surgery cases in hospitals are cancelled because of the lack of available recovery beds. About 24 % of appointed surgery cases are refused because of a full intensive care unit (ICU); moreover, the patients' length of stay in the ICU and the number of ICU beds significantly affect surgery scheduling (Sobolev et al. 2005; Kim 2000).

3 Problem Description and Definition

We consider an OR scheduling problem with the limited upstream and downstream capacities and integrate the demand of resources, including staff and facilities in every stage of surgery. The objective is to reduce the total overtime cost and total idle cost of ORs. Moreover, a stable schedule can reduce the number of delayed or cancelled surgeries and the extra costs caused by schedule adjustments.

Our problem is a scheduling problem without a buffer. Using the OR suite concept, we define a surgery as a job that can be separated into three stages: preoperative, perioperative, and postoperative. In each stage, resources are needed when conducting the operation, including human resources (anesthesiologists, nurses, and surgeons) and facilities (OR, recovery room, and other medical equipment). The combination of resources in every stage is called a mode. Different operations require different modes. Therefore, for any job, two concurrent operations will never use the same resource. In this study, different modes may process the same operation. Some assumptions for the mode are given as follows:

- The available mode for any operation is known. The number of available modes is limited by the capacity of the hospital.
- When a mode is chosen for an operation, all the resources of that mode are occupied for that period of time, which consists of processing time, setup time, and cleanup time.
- Other modes that include the same resources as the chosen mode cannot provide services when the chosen mode is in progress.
- The lengths of various modes are slightly different. Every mode has an available time limit, and the available time interval of every mode is decided at the beginning of a problem setting.

In the early stages of planning, we will verify the modes available to process every stage of every job and to indicate that the available path of every job is different and known. If we consider the mode of every stage as different functions of a machine, we can turn the scheduling problem of each stage into a flexible scheduling model for the unrelated parallel machine scheduling problem and formulate a mixed integer linear programming model (MILP).

4 Heuristic Algorithm and Numerical Analysis

4.1 Heuristics

This study investigates the scheduling problem with the flexible unrelated parallel machine, which is an NP-hard problem. Therefore, we need to develop a heuristic algorithm to solve large-scale problems. This heuristic algorithm is divided into five steps to obtain a near optimal solution in a short period.

Initial step: Dividing surgical cases into three types

The first type of surgical case is the critical surgical case. It has the highest priority in scheduling. Next, the second type of surgical case has a higher priority than the third type because it requires more time to be accomplished.

Step 1: Planning critical surgical cases

Planning the perioperative and postoperative stages and scheduling OR suites are the main objective of this step. In addition, the first type of case that requires downstream resources is a high priority for planning. To avoid the chance of overtime and overtime costs and increase the efficiency of scheduling surgical cases, a preliminary plan for the order of processing for the first type of cases and OR suites is created in this step.

Step 2: Calculating the adjustment factor and deciding the adjustment range

Surgical Suites Scheduling with Integrating Upstream and Downstream Operations

Fig. 1 Maximum possible idling gap

If the first type of cases occupied downstream resources for a long period, then other cases must wait for the release of downstream resources, causing other surgery cases to become blocked and upstream resources to become idle. The blocking of the critical surgical case with the longest time on ICU and the surgical case with the shortest processing time result in the maximum possible idling gap (denoted as $G_{\max}$) as shown in Fig. 1.

In this study, we can derive an adjustment factor from the maximum possible idling gap, idle costs, and overtime costs of the OR suites to decide the maximum possible adjustment range by this factor. When considering placing surgical cases into the maximum possible idling gap, if the processing time of the surgical case is less than the maximum possible idling gap, the surgical case can be placed into the time interval. Conversely, it will result in overtime costs for the OR suite.

We define the unit cost of the processing time of the mode r as C_r^p and the unit overtime cost of the mode r as C_o^r. P_r is the processing time of the surgical case, which is most likely to be scheduled into the maximum possible idling gap. The idle time cost is C_{idle}. If we want to use another mode r' to deal with this surgical case, the unit cost of time and processing time are defined as $C_{r'}^p$ and $P_{r'}$. If the processing and overtime costs of the mode r are less than or equal to the processing cost of the mode r' and the idle cost of $G_{\max}$, then placing this surgical case into the maximum possible idling gap can reduce the total cost. Moreover, the critical value can be derived from Eqs. (1) and (2):

$$C_r^p \times P_r + C_o^r \times (P_r - G_{\max}) = C_{r'}^p \times P_{r'} + G_{\max} \times C_{idle}. \tag{1}$$

$$P_r = \frac{C_{r'}^p \times P_{r'} + G_{\max} \times (C_{idle} + C_o^r)}{C_o^r + C_r^p}. \tag{2}$$

More critical surgical cases require the use of the ICU and are usually operated by an experienced medical team. The processing and overtime costs of the combination of resources are therefore almost the same. We can simplify Eq. (2) by assuming $C_r^p \cong C_{r'}^p$ and $P_r \cong P_{r'}$ to derive Eq. (3):

$$P_r = \frac{G_{\max} \times (C_{idle} + C_o^r)}{C_o^r}. \tag{3}$$

If the processing time of the surgical case exceeds the critical value, placing the surgical case into the maximum possible idling gap is not advisable.

Step 3: Planning the surgical case not within the adjustment range

After finding the maximum adjustment range in Eq. (3), a small- or medium-sized surgical case with a processing time larger than the adjustment range will be planned first. The processing time of these surgical cases is relatively longer than that of other cases; in addition, these cases are expected to use more downstream resources which may also result in the idling gap. Therefore, these surgical cases should be given higher priority in planning than other cases.

Step 4: Adjusting the upper and lower bounds of the start time of planned surgeries

To improve the idle situation caused by system congestion, small and medium-sized surgical cases can be placed into the idle time interval in this step. However, not all idle time intervals can accommodate small and medium-sized surgical cases. Thus, we aim to place very small cases into the interval by a slight adjustment of the start time or end time of planned cases to increase the utilization of the OR. Given the processing time of unplanned surgical cases, an adjustment threshold (G_{adjust}) can be derived as Eq. (4).

$$G_{\text{adjust}} = \frac{C_o^r \times P_r}{\left(C_{idle}^k + C_o^r\right)}. \tag{4}$$

Step 5: Planning all surgeries

The start time and end time of surgical cases planned in Steps 1 and 3 are adjusted based on G_{adjust} and considered as constraints for planning these surgical cases. Further, the resource modes for these planned surgical cases are recorded as the input data. Because of the reduction of the problem size, all surgical cases and resource allocations are scheduled by the mixed integer linear programming model.

4.2 Parameter Analysis and Comparison

The objective of this study is to minimize the total fixed costs, overtime costs, and total cost of idle ORs. In the parameter analysis, we compared the example with five types of surgical cases and three ORs with 13 combinations of resources within a 3-day planning period solved by an MILP model and the proposed heuristic algorithm. Three factors are studied: overtime cost, idle time cost, and ability represented by O, I, and A, respectively. These factors are set to high and low standards, which are represented by subscripts H and L, respectively. Therefore, eight scenarios are studies and the results are summarized in Table 1.

Surgical Suites Scheduling with Integrating Upstream and Downstream Operations

Table 1 Comparison of MILP and Heuristic results

	MILP result				Heuristic result			
Scenario	Total cost	Overtime cost	Idle time cost	OR utility (%)	Total cost	Overtime cost	Idle time cost	OR utility (%)
1(O_H, I_H, A_H)	$161,632	$9,142	$1,200	96.30	$169,115	$22,855	$2,800	96.30
2(O_H, I_H, A_L)	$192,216	$54,976	$800	97.04	$203,347	$45,207	$100	99.63
3(O_H, I_L, A_H)	$166,230	$0	$360	90.48	$172,680	$2,150	$360	90.48
4(O_H, I_L, A_L)	$187,380	$13,060	$620	83.60	$197,070	$0	$960	78.24
5(O_L, I_H, A_H)	$163,400	$32,000	$0	100	$164,650	$33,250	$0	100
6(O_L, I_H, A_L)	$164,270	$18,500	$1,500	94.44	$169,280	$24,050	$1,500	100.00
7(O_L, I_L, A_H)	$139,780	$19,000	$0	100	$143,030	$22,250	$0	100
8(O_L, I_L, A_L)	$151,160	$29,300	$0	100	$151,160	$29,300	$0	100

Figure 2 shows that the total medical cost has roughly the same pattern as the fixed costs and that the fixed costs of opening or not opening OR suites significantly affect the total cost of the hospital. The difference between the total medical cost and the fixed cost in Scenario 2 is slightly large because the overtime work involved in the situation is more serious than the others, thereby leading to higher overtime costs.

Although the result of MILP is similar to that of the heuristic algorithm, they differ in overtime costs. The total cost in the heuristic algorithm is higher than the approximate optimal solution by only about 3.2 % in average, but the time required by the MILP to reach a solution is about 443 times longer than that by the heuristic algorithm. Therefore, we suggest using the heuristic algorithm to save time.

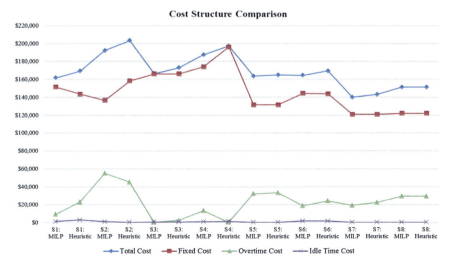

Fig. 2 Costs comparison for each scenario

5 Conclusion

In this study, we integrate resources and plan the schedule using the concept of the OR suite while considering constraints of upstream and downstream resources of ORs in a hospital to avoid ORs being occupied or idle because of a lack of upstream and downstream capacities. Using the OR scheduling method proposed in this study, the forced cancellation or delay of surgeries for the maintenance of the quality of medical and surgical services can be avoided. Hence, the satisfaction of staff and patients is also improved.

This problem can be formulated as an MILP model, and the objective is to minimize the fixed costs, overtime costs, and idle costs of ORs. Given the high degree of complexity of this problem, we propose a five-step heuristic algorithm to increase the efficiency of solving this large-scale problem. We also compare and discuss the results of MILP and the heuristic algorithm to identify the applicable conditions for the problems faced by the different scenarios.

In conclusion, the differences in total costs between MILP and the heuristic algorithm when solving the different scenarios are within 6 %. However, the heuristic algorithm is more efficient, and it can be used to solve not only different scenarios within this problem but also large-scale practical problems.

Acknowledgments This research was supported by National Science Council of Taiwan (NSC 99-2221-E-002-155-MY3).

References

Denton B, Viapiano J, Vogl A (2007) Optimization of surgery sequencing and scheduling decisions under uncertainty. Health Care Manage Sci 10(1):13–24

Denton BT, Miller AJ, Balasubramanian HJ, Huschka TR (2010) Optimal allocation of surgery blocks to operating rooms under uncertainty. Oper Res 58(41):802–816

Dexter F, Traub RD, Macario A (2003) How the release allocated operating room time to increase efficiency: predicting which surgical service will have the most underutilized operating room time. Anesth Analg 96:507–512

Jackson RL (2002) The business of surgery: managing the OR as a profit center requires more than just IT. It requires a profit-making mindset, too. Oper Room Inf Syst, Health Manage Technol

Jonnalagadda R, Walrond ER, Hariharan S, Walrond M, Prasad C (2005) Evaluation of the reasons for cancellation and delays of surgical procedures in a developing country. Int J Clin Pract 59(6):716–720

Kim SC (2000) Flexible bed allocation and performance in the intensive care unit. J Oper Manage 18:427–443

Pham DN, Klinkert A (2008) Surgical case scheduling as a generalized job shop scheduling problem. Eur J Oper Res 185:1011–1025

Sobolev BG, Brown PM, Zelt D, FitzGerald M (2005) Priority waiting lists: is there a clinically ordered queue? J Eval Clin Pract 11(4):408–410

Weinbroum AA, Ekstein P, Ezri T (2003) Efficiency of the operating room suite. Am J Surg 185:244–250

Research on Culture Supply Chain Intension and Its Operation Models

Xiaojing Li and Qian Zhang

Abstract In recent years, China has become the cultural industry toward highly centralized, with the rapid development of international direction. Between the competitions of the 21st century, not a business enterprise competition, but competition in the supply chain and supply chain, supply chain is the development trend of industry chain as the value chain of cultural products. Supply chain functions, like integration, and optimization, are gradually reflected in the cultural industries. The cultural characteristics of the supply chain are proposed in this paper. From the perspective of the cultural industry and supply chain's intension, several cultural supply chain operating models are given in this paper. From the angle of the supply chain finance and capital flow, the supply chain operation mechanism is presented for how to promoting the culture of the supply chain, which is provided a reference for increasing the cultural development of the supply chain.

Keywords Culture industry · Culture supply chain · Operation mode · Supply chain finance · Operating mechanism

In 2010, the overall size of Chinese cultural industry market total transactions amount reached to 169.4 billion, an increase of 41 % from 2009. In 2011, Chinese cultural industry market transactions amount reached to 210.8 billion. In 2012, the first time China surpassed United States to become the world's largest art market; Chinese cultural industry output value is expected to exceed 4 trillion Yuan, make further enhance the proportion of GDP. These series growth dates indicated that cultural industry market as a special market have tremendous development potential and cultural industry is gradually mature in china.

X. Li (✉) · Q. Zhang
School of Economics and Finance, Huaqiao University, Quanzhou, Fujian 362021, People's Republic of China
e-mail: 1017126988@qq.com

Q. Zhang
e-mail: zhangyl@hqu.edu.cn

1 The Connotation of Cultural Supply Chain

What is "culture"? In 1871 American anthropologist Edward Taylor put forward to a concept, the cultural is defined as "a completed system including knowledge, faith, art, morality, custom, and all the abilities and habitats from which a social member would acquire". Cultural products are created by a person or department, and cultural or artistic as the main content, which the goals are meet the needs of the human spirit, reflect the social ideology and become mass entertainment culture carrier. UNESCO United Nations defined cultural industry is "According to industry standards, cultural industry includes production, reproduction, storage, and distribution of cultural products and a series of activities." Cultural industry market is different from other material economy markets. When mining existing cultural resources to create a new cultural products or services, instead of bring consumption and loss of original cultural resources; we will increase the content of the original cultural resources. Supply chain is building around the core business, starting from the procurement of raw materials, intermediate products and final products, then deliver to consumers by sales network, through control information flow, logistics, capital flow, together suppliers, manufacturers, distributors, retailers, until end-user into an overall functional network chain structure. Now Chinese cultural industry is constantly maturity, development, and gradually forms a complete industrial chain. Putting supply chain theory into cultural industry, using supply chain basic ideas to achieve these goals that include integrate the cultural industry, achieve efficient operation of cultural industry, create time value, place value, and other value-added; achieve cultural industry restructuring, optimize the industrial structure, improve the functional organization of urban land use, improve the investment environment, the protection of the urban environment and bring about other important social functions. Through anglicizing and combing the concept of cultural product and supply chain, this paper present cultural connotation of product supply chain and it unlike other material goods supply chain, put forward to the definition of cultural products supply chain and features that differ from other products. Taking the market as the orientation, Cultural supply chain is based on leading enterprises, linked with interest, put the cultural product creation, production, processed, sales, as a whole, be integrated with flow of material, information and capital, finally form a network structure with industrial chain. The cultural flow chart of the supply chain as follows in Fig. 1.

2 Basic Characteristics of Cultural Supply Chain

The key to form a complete supply chain is to increase the added value of cultural industry chain. If you do not use supply chain theory to optimize logistics, information, capital on the whole industry chain flow, to reduce unnecessary links, the china is just a cultural industry chain, not a cultural industry supply chain.

Fig. 1 The cultural supply chain flow chart

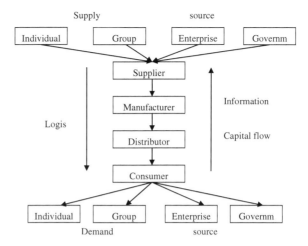

In the cultural industry, around the core enterprise, optimize the characteristics of cultural products and supply chain management, control the logistics flow, information flow, capital flow, to form a supply chain network structure from the creative source to the final consumers. According to the analysis of cultural industry and supply chain, cultural industry supply chain has characteristics of general industry supply chain, also has its own characteristics. The cultural supply chain basic characteristics are as follows:

1. Higher complexity. Compare with other products, Cultural products supply chain more complexity, this kind of complexity is composed of several aspects. First, the supply chain upstream—cultural product creation, cultural product features: symbolic, invisible, ideology and so on, determine the diversity of cultural product types, and sources of product creators. The product creators including individuals, social groups, enterprises and government. Cultural upstream supply chain show its complexity, so that we should pay attention to upstream supply chain. Second, compare with other general products, the manufacturing process of cultural product become more complexity, cultural manufacture process is not a one-time processing, but through industrial chain, take advantage of cultural resources, put in-depth development to repeatedly output on the content, including offer rich additional market value for related industries. This special complexity is determined by cultural features like the eternal value and advanced. Finally, the last thing is cultural product sales, because of cultural product consumption has a lot of uncertainty, compared with the general sales, cultural products become more complexity. These uncertainties are directly or indirectly connect with the policy changes, the demand of social media, and not simplify for supply and demand theory determine the entire marketing process. At the same time, cultural products consumers may also be new cultural product creators, therefore cultural products of service object may become to cultural products provider.

2. Better agility. Compared with general industry, the ideology and intangibility of cultural products represent cultural products in constant spread, demand is also constantly changing, so better agility of whole cultural supply chain to be required. As a creative of cultural industry, cultural industry supply chain need dynamic more, so its agility requires higher. The agility of cultural supply chain is mainly reflected in three aspects: Firstly, source of cultural supply chain— cultural products suppliers, Cultural product innovation and mining maximum cultural value directly affects entire cultural industry supply chain from source of manufacture to sales. Secondly, manufacturing process of cultural product, because of culture features: innovative frequent, the entire manufacturing operation require to make quick response to meet supply chain requirements, to shorten the operating cycle, to reduce unnecessary work link, can be reflect to supply chain upstream in time. Finally, cultural products marketing process, that is to maximize product speed which is transferred to the consumer market, shorten the operating cycle of cultural products.
3. More network nodes. Cultural supply chain is a network structure which consists of core business of supplier, the supplier's suppliers and users, end-users. Because of the special attributes of cultural products, suppliers, intermediaries and users of cultural supply chain appear diversification, while government and special assessment agencies are also involved in the whole cultural product supply chain process. Compare with other products, network nodes of cultural supply chain with more hierarchical, more integration use to entire cultural supply chain. A reasonable network node has a vital role to improve entire cultural products supply chain efficiency and integration.

3 Cultural Supply Chain Operation Modes

Supply chain theory in the application of cultural industry are mainly reflected in two aspects: **Integrated Supply Chain (ISC).** The use of ISC in cultural industry mainly centres on product-orientation, geographical area, various enterprises and Industry internal and external systems etc. In general, ISC deem supply chain as integration which include integration of supply demand relations, integration of logistics, information and management. **Green Chain.** Green Chain is that considering environmental factors and regeneration influence factors on the industrial chain integration. In the development of cultural industries, it contains Green Chain idea. For example, ecological park is a kind of green tourism, green is an important factor to consider in the development of park, need to ensure minimum damage to the environment, to realize the green supply chain operation, reduce unnecessary waste. Again for instance, recovery based on knowledge of culture product, also embodies the idea of Green Chain. Nowadays the competition between enterprises is the competition between supply chains. Through analysis the characteristics and connotation of cultural supply chain, combined with the actual situation of China's cultural industry development, according to the development of cultural industry relying on the land space needs, gradually formed two types of supply chain development models: one is the entity of the industrial park and industrial integration mode, second is the virtual network level mode.

3.1 Physical Space: The Cultural Industry Cluster Supply Chain Operation Mode

At present, innovation is an important role in cultural industry development, innovation make all-round, large-scale influence on cultural industry development. According to innovation and degree of dependence on tangible resources, mainly have the following several kinds of supply chain development path: (1) Relying on resources mode. (2) Creative guidance modes. (3) Non inheritance innovation mode.

3.2 Virtual: Network Level Cultural Supply Chain Operations Mode

With the development of international information, Science and technology, e-commerce in various industries have achieved a certain degree of development. Cultural supply chain network modes consist of Cultural Assets and Equity Exchange, namely establish a network platform that realize seamless connectivity between service providers and customers, reduce unnecessary operational flow. This mode has the following features:

1. Supply chain management functions. Through integrated information, technology to the entire operation platform, controlled logistics, information, capital flow from suppliers and customers, reduced unnecessary operation process, such as suppliers, etc., to achieve the whole chain integration and optimization.
2. Integration of capital flows. From the view point of capital flow, currently chinese cultural property rights exchange mainly to solve cultural products pricing, settlement, delivery, financing and other capital flows problems. Because cultural property trading platform can put individual, group, government, corporate and other participants together, take advantage of information technology as soon as possible to achieve circulation of entire supply chain flow and to get rid of the unnecessary capital flows operation, to achieve capital flow integration of entire supply chain and value-added of industrial chain, to achieve integration of capital flow operation (Fig. 2).

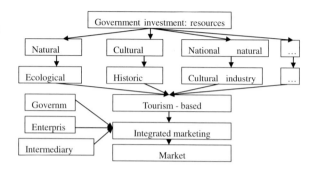

Fig. 2 Resources type culture industry supply chain mode of operation

4 Cultural Supply Chain Operation Mechanism

Cultural supply chain operating mechanism is use to optimize and integrate cultural industry in logistics, information and capital flow. Logistics: cultural supply chain logistics related to cultural products such as procurement, production, inventory, package, transportation, sales process. Information flow: through information sharing, cultural supply chain information flow mainly realize seamless connectivity to reduce the bullwhip effect, achieve to fast response and entire supply chain agility. Capital flow: cultural supply chain capital flow mainly is reflected in financing activities, settlement payment, procurement activities and risk control.

Supply chain finance embedded industrial clusters can get more new resources "blood" to promote cultural industry cluster rapid development. Industrial clusters also can make supply chain finance more widely used, and then provide a good credit environment to improve income levels and reduce the risk for financial institutions. From point view of supply chain finance, optimizing integration of cultural supply chain we can from following aspects:

1. Establish clear cultural industry supply chain finance subject. Depending on the supply chain model, subjects have following categories: (1) government, macro managers. As planners and investors of Chinese cultural industry parks, ecological parks and other gathering place, in the early stages development of cultural supply chain government is a very important finance body. (2) Core enterprises. As cultural supply chain dominant enterprises, core enterprises open up cultural supply chain upstream and downstream financing needs to stabilize entire supply chain cash flow and to promote supply chain integration and flexible development. (3) Other enterprises. As other supply chain node enterprise, be able to reduce capital flow pressure, to some extent, promoting supply chain stable and orderly development. (4) other organizations. Other organizations such as bank play an important payment and settlement role which affecting whole cultural supply chain capital flow. Making a clear cultural supply chain financial body category, and then proposing different financial strategies for financial body have significant significance.

2. Coordination mechanisms. From point view of core enterprise capital flow, Supply chain finance collaborative management is base on collaborative theory. At strategic, motivation, business level, through financing, procurement, payment, we wish settlement and risk control activities can achieve whole cultural supply chain financial body coordination, achieve cultural supply chain capital flow integration, shorten current cycle of cultural products, improve the operating efficiency of entire supply chain, achieve whole culture supply chain coordination, agility and flexibility.
3. Establish partnership. Supply chain partnership is the node enterprises for the common goal to establish a benefit-sharing, risk-sharing partnership in the supply chain. Cultural supply chain operations have certain risks, such as credit risk. Establishing partnerships is an effective means to avoid credit risk, help to improve the reputation of the business-to-business, to reduce unnecessary approval operational aspects, to control supply chain finance risks, to achieve a higher level of integration of supply chain, to improve the operation of the entire supply chain finance.

Acknowledgments This paper was supported in part by National Natural Science Foundation of China under grant number 71040009, in part by Program for New Century Excellent Talents in University under grant number NCET-10-0118, in part by Youth Foundation of Chinese Education by Huo Ying-dong under grant number 104009.

References

Fang Z, Zhang X (2010) Cluster-based supply chain study of cultural creative industry. In: E-business and information system security (EBISS)
Sheng W (2008) Cultural industry supply chain, industrial chain and value chain—in order to Dafen Industrial Park as an example of cultural characteristics. Urban Probl 12:2008
Song H (2012) Relationship of supply chain finance and the development of industrial clusters. North Finance 01:2012
Wu Z (2008) Cluster-based supply chain, the development of cultural industry garden path. J Nanjing Univ Finance and Econ 05:2008
Yu P (2013) Chinese cultural innovation report. Social Science Culture Press, Beijing
Zang X, You T (2011) Cultural product: further understanding of the characteristics and attributes. Exploration No. 05 2011
Zhu Z, Zhang S (2012) Cultural creative industry concept and form analysis. Northeastern University News No.01 2012

Power System by Variable Scaling Hybrid Differential Evolution

Ji-Pyng Chiou, Chong-Wei Lo and Chung-Fu Chang

Abstract In this paper, the variable scaling hybrid differential evolution (VSHDE) used to solve the large-scale static economic dispatch problem. Different from the hybrid differential evolution (HDE), the concept of a variable scaling factor is used in the VSHDE method. The variable scaling factor based on the 1/5 success rule of evolution strategies (ESs) is embedded in the original HDE to accelerate the search for the global solution. The use of the variable scaling factor in the VSHDE can overcome the drawback of the fixed and random scaling factor used in HDE. To alleviate the drawback of the penalty method for equality constraints, the repair method is proposed. One 40-unit practical static economic dispatch (SED) system of Taiwan Power Company is used to compare the performance of the proposed method with HDE. Numerical results show that the performance of the proposed method combining with the repair method is better than the other methods.

Keywords VSHDE · Economic dispatch · Repair method

J.-P. Chiou (✉) · C.-W. Lo
Department of Electrical Engineering, Ming Chi University of Technology, New Taipei, Taiwan
e-mail: jipyng@mail.mcut.edu.tw

C.-W. Lo
e-mail: milk_tea666666@hotmail.com

C.-F. Chang
Department of Electrical Engineering, WuFeng University, Chiayi, Taiwan
e-mail: cfchang@wfc.edu.tw

Y.-K. Lin et al. (eds.), *Proceedings of the Institute of Industrial Engineers Asian Conference 2013*, DOI: 10.1007/978-981-4451-98-7_61,
© Springer Science+Business Media Singapore 2013

1 Introduction

The aim of the economic dispatch (ED) is to obtain the great benefits in power system. So, many mathematic programming methods have been researched for ED problems. Su and Chiou (1995) applied Hopfield network approach to solve the economic dispatch problems. However, the Hopfield network method requires two phase computations. Sewtohul et al. (2004) proposed genetic algorithms (GAs) to solve the economic dispatch problem. Gaing (2003) proposed a practical swarm optimization (PSO) method for solving the economic dispatch problems in power systems. The population size is set to $10 \times b$, where b, is the number of decision parameters. In so doing, much more computation time is required to evaluate the fitness function. Sinha et al. (2003) used an evolutionary programming (EP) method to solve economic dispatch problems.

Hybrid differential evolution (HDE) (Chiou and Wang 1999, 2001) is a stochastic search and optimization method. The fittest of an offspring competes one by one with that of the corresponding parent. This competition implies that the parent is replaced by its offspring if the fitness of the offspring is better than that of its parent. On the other hand, the parent is retained in the next generation if the fitness of the offspring is worse than that of its parent. This one by one competition gives rise to a faster convergence rate. However, this faster convergence also leads to a higher probability of obtaining a local optimum because the diversity of the population descends faster during the solution process. To overcome this drawback, migrating operator and accelerated operator act as a trade-off operator for the diversity of population and convergence property in HDE. However, a fixed scaling factor is used in HDE. Using a smaller scaling factor, HDE becomes increasingly robust. However, much computational time should be expanded to evaluate the objective function. HDE with a larger scaling factor generally produces a local solution or misconvergence. Lin et al. (2000) used a random number that its value is between zero and one as a scaling factor. However, a random scaling factor could not guarantee the fast convergence.

In this study, a variable scaling hybrid differential evolution (VSHDE) (Chiou et al. 2005) for solving the large-scale economic dispatch problems is proposed. Different from the HDE, the scaling factor based on the 1/5 success rule of evolution strategies (ESs) (Back and Schwefel 1991, 1993) is used in VSHDE method to accelerate searching out the global solution. The repair method is used to alleviate the drawback of the penalty method for equality constraints. To illustrate the convergence property of the proposed method, one 40-unit economic dispatch system of Taiwan power system is used to compare the performance of the proposed method with HDE.

2 Problem Formulation

The economic dispatch problem can be mathematically described as follows:

$$min \sum_{i \in \Psi} P_i = P_D + P_L \tag{1}$$

where i is index of dispatchable units, P_i is power generation of unit i, Ψ is a set of all dispatchable units. Subject to the following constraints:

1. Power balance constraint

$$\sum_{i \in \Psi} P_i = P_D + P_L \tag{2}$$

$$P_L = \sum_i \sum_j B_{ij} P_i P_j \tag{3}$$

where P_D is total load demand, P_L is power losses and B_{ij} is power loss coefficient.

2. Generation limits of units

$$P_{i\,min} \leq P_i \leq P_{i\,max} \tag{4}$$

where $P_{i\,min}$ and $P_{i\,max}$ are the minimum and maximum of unit i, respectively.

3 VSHDE Method

The VSHDE method is briefly described in the following.

Step 1: Initialization

The initial population is chosen randomly in an attempt to cover the entire parameter space uniformly as (5).

$$Z_i^0 = Z_{i,min} + \left(\sigma_i \cdot \left(Z_{i,max} - Z_{i,min} \right) \right), \quad i = 1, 2, \ldots, N_p \tag{5}$$

where $\sigma_i \in (0, 1]$ is a random number. The initial process can produce N_P individuals of Z_i^0 randomly.

Step 2: Mutation operation

The essential ingredient in the mutation operation is the difference vector. Each individual pair in a population at the G-th generation defines a difference vector D_{jk} as

$$D_{jk} = Z_j^G - Z_k^G \tag{6}$$

The mutation process at the G-th generation begins by randomly selecting either two or four population individuals Z_j^G, Z_k^G, Z_j^G and Z_m^G for any j, k, l and m. These four individuals are then combined to form a difference vector D_{jklm} as

$$D_{jklm} = D_{jk} + D_{lm} = \left(Z_j^G - Z_k^G\right) + \left(Z_l^G - Z_m^G\right) \tag{7}$$

A mutant vector is then generated based on the present individual in the mutation process by

$$\hat{Z}_i^{G+1} = Z_p^G + \left(F \cdot D_{jklm}\right), \quad i = 1, 2, \ldots, N_p \tag{8}$$

where F is the scaling factor. Further more, j, k, l and m are randomly selected.

Step 3: Crossover operation

The perturbed individual of $\hat{Z}_i^{G+1}$ and the present individual of Z_i^G are chosen by a binomial distribution to progress the crossover operation to generate the offspring. Each gene of i-th individual is reproduced from the mutant vectors $\hat{Z}_i^{G+1} = \left[\hat{Z}_{1i}^{G+1}, \hat{Z}_{2i}^{G+1}, \ldots, \hat{Z}_{ni}^{G+1}\right]$ and the present individual $Z_i^G = \left[Z_{1i}^G, Z_{2i}^G, \ldots, Z_{ni}^G\right]$. That

$$\hat{Z}_i^{G+1} = \begin{cases} Z_{gi}^G & \text{if a random number} > C_r \\ \hat{Z}_{gi}^{G+1} & \text{otherwise} \end{cases} \tag{9}$$

where $i = 1, \ldots, N_p; g = 1, \ldots, n$; and the crossover factor $C_r \in [0, 1]$ is assigned by the user.

Step 4: Estimation and selection

The evaluation function of a child is one-to-one competed to that of its parent. This competition means that the parent is replaced by its child if the fitness of the child is better than that of its parent. On the other hand, the parent is retained in the next generation if the fitness of the child is worse than that of its parent, i.e.

$$Z_i^{G+1} = \arg \min\left\{f\left(Z_i^G\right), f\left(\hat{Z}_i^{G+1}\right)\right\} \tag{10}$$

$$Z_b^{G+1} = \arg \min\left\{f\left(Z_i^G\right)\right\} \tag{11}$$

where $\arg\ min$ means the argument of the minimum.

Step 5: Migrating operation if necessary

In order to effectively enhance the investigation of the search space and reduce the choice pressure of a small population, a migration phase is introduced to regenerate a new diverse population of individuals. The new population is yielded

Power System 509

based on the best individual Z_b^{G+1}. The g-th gene of the i-th individual is as follows:

$$
Z_{ig}^{G+1} = \begin{cases} Z_{bg}^{G+1} + \left(\sigma_i \cdot \left(Z_{g\,min} - Z_{bg}^{G+1}\right)\right), & \text{if } \delta \frac{Z_{bg}^{G+1} - Z_{g\,min}}{Z_{g\,max} - Z_{g\,min}} \\ Z_{bg}^{G+1} + \left(\sigma_i \cdot \left(Z_{g\,max} - Z_{bg}^{G+1}\right)\right), & \text{otherwise} \\ i = 1, 2, \dots, N_p; \quad g = 1, 2, \dots, n \end{cases} \tag{12}
$$

where σ_i and δ are randomly generated numbers uniformly distributed in the range of $[0,1]$; $i = 1, \dots, N_p$; and $g = 1, \dots, n$.

The migrating operation is executed only if a measure fails to match the desired tolerance of population diversity. The measure is defined as follows:

$$
\varepsilon = \sum_{\substack{i=1 \\ i \neq b}}^{N_p} \sum_{g=1}^{n} \frac{\eta_z}{\left(n \cdot (N_p - 1)\right)} < \varepsilon_1 \tag{13}
$$

where

$$
\eta_z = \begin{cases} 0, & \text{if } \varepsilon_2 < \left\| \frac{Z_{gi}^{G+1} - Z_{bi}^{G+1}}{Z_{bi}^{G+1}} \right\| \\ 1, & \text{otherwise} \end{cases} \tag{14}
$$

Parameter $\varepsilon_1, \varepsilon_2 \in [0, 1]$ expresses the desired tolerance for the population diversity and the gene diversity with respect to the best individual.

Step 6: Accelerated operation if necessary

When the best individual at the present generation is not improved any longer by the mutation and crossover operations. A decent method is then employed to push the present best individual toward attaining a better point. Thus, the accelerated phase is expressed as follows

$$
Z_b^{G+1} = \begin{cases} Z_b^{G+1}, & \text{if } f\left(Z_b^{G+1}\right) < f\left(Z_b^{G}\right) \\ Z_b^{G+1} - \alpha \nabla f, & \text{otherwise} \end{cases} \tag{15}
$$

where Z_b^G denotes the best individual, as obtained from Eq. (15). The step size $\alpha \in (0, 1]$ in (15) is determined by the descent property. Initially, α is set to one to obtain the new individual.

Step 7: Updating the scaling factor if necessary

The scaling factor should be updated in every q iterations as follow:

$$
F^{t+1} = \begin{cases} c_d \times F^t & \text{if } p_s^t < \frac{1}{5} \\ c_j \times F^t & \text{if } p_s^t > \frac{1}{5} \\ F^t & \text{if } p_s^t = \frac{1}{5} \end{cases} \tag{16}
$$

where p'_s is the frequency of successful mutations measured. The initial value of the scaling factor, F, is set to 2 (Storn and Price 1996; Price 1997). The factors of $c_d = 0.82$ and $c_j = 1/0.82$ (Michalewicz 1999) are used for adjustment, which should be taken place for every q iterations.

When the migrating operation performed or the scaling factor is too small to find the better solution, the scaling factor is reset as follow:

$$F = 1 - \frac{iter}{itermax} \qquad (17)$$

where *iter* and *itermax* are the number of current iteration and the maximum iteration, respectively.

Step 8: Repeat step 2 to step 7 until the maximum iteration quantity or the desired fitness is accomplished.

4 Example

To investigate the convergence property of the VSHDE, a 40-unit practical ED system of Taiwan Power Company (TPC) is employed as an example. The power loss is released in this example. The total demand is 10,500 MW. The penalty method and repair method are used for the equality constraints, respectively. To compare the performance of the VSHDE and HDE methods, this system is repeated 20 independent trials. The best convergence property among this 20 runs of the HDE and VSHDE methods are lists in Fig. 1. From the Fig. 1, the convergence property of the VSHDE combining the repair method is better than the other methods. The largest and smallest values among the best solutions of the 20

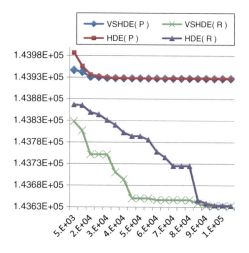

Fig. 1 The best convergence property of the HDE and VSHDE methods

Power System   511

Table 1 Computational result for 20 runs of example

| | Penalty method | | Repair method | |
	VSHDE	HDE	VSHDE	HDE
STD	0.1598983	0.1673993	41.9851067	58.0260301
MAX. ($)	143926.9138	143927.2861	143778.2017	143831.3298
MIN. ($)	143926.4272	143926.7005	143632.7084	143633.0706
AVE. ($)	143926.6289	143927.0029	143724.7618	143757.7223

runs are, respectively, expressed in Table 1. The average for the best solutions of the 20 runs and the standard deviation with respect to the average are also shown in this table. A smaller standard deviation implies that almost all the best solutions are close to the average best solution. The standard deviation for the VSHDE method is smaller than the HDE method. From the above discussion, the convergence property of the VSHDE method is better than the HDE method.

5 Conclusion

This paper uses the VSHDE and HDE methods is solved the large-scale economic dispatch systems, respectively. The VSHDE method utilized the 1/5 success rule of the evolution strategies (ESs) to adjust the scaling factor to accelerate searching out the global solution. The variable scaling factor is used to overcome the drawback of fixed and random scaling factor used in HDE. To alleviate the drawback of the penalty method for the equality constraints, the repair method is used in this paper. The computational results obtained of solving one 40-unit practical ED system of Taiwan Power Company are investigated. From the computation results, the VSHDE method combing the repair method is better than the other methods.

Acknowledgments Financial research support from the National Science Council of the R. O. C. under grant NSC 100-2632-E-131-001-MY3 2/3 is greatly appreciated.

References

Back T, Schwefel HP (1993) An overview of evolutionary algorithms for parameter optimization. Evol Comput 1:1–23

Chiou JP, Wang FS (1999) Hybrid method of evolutionary algorithms for static and dynamic optimization problems with application to fed-batch fermentation process. Comput Chem Eng 23:1277–1291

Chiou JP, Wang FS (2001) Estimation of monod parameters by hybrid differential evolution. Bioprocess Biosyst Eng 24(2):109–113

Chiou JP, Chang CF, Su CT (2005) Variable scaling hybrid differential evolution for solving network reconfiguration of distribution systems. IEEE Trans Power Syst 20(2):668–674

Gaing ZL (2003) Particle swarm optimization to solving the economic dispatch considering the generator constraints. IEEE Trans Power Syst 8(3):1187–1195

Lin YC, Hwang KS, Wang FS (2000) Plant scheduling and planning using mixed-integer hybrid differential evolution with multiplier updating. Congr Evol Comput 1:593–600

Michalewicz Z (1999) Genetic algorithms + data structures = evolution programs, 3rd edn. Springer, New York

Price KV (1997) Differential evolution vs. functions of the 2nd ICEC. IEEE Conf Evol Comput 1:153–157

Sewtohul LG, Ah K, Rufhooputh HCS (2004) Genetic algorithms for economic dispatch with valve point effect. IEEE Int Conf Network, Sens Control 2:1358–1363

Sinha N, Chakrabarti R, Chattopadhyay PK (2003) Evolutionary programming techniques for economic load dispatch. IEEE Trans Evol Comput 7(1):83–94

Storn R, Price KV (1996) Minimizing the real functions of the ICEC'96 contest by differential evolution. IEEE Conference on Evolutionary Computation pp 842–844

Su CT, Chiou GJ (1995) Hopfield network approach to economic dispatch with prohibited operating zones. In: Proceedings of IEEE on energy management and power delivery conference, pp 382–387

Investigating the Replenishment Policy for Retail Industries Under VMI Strategic Alliance Using Simulation

Ping-Yu Chang

Abstract Recently, retail industries are booming rapidly because of the convenience and low price. However, as the competition increases, inventory shortage becomes a serious problem for both retailer and supplier. Therefore, developing a suitable inventory management for retailer and supplier has been a pertinent area of study in recent years. Some strategies such as Quick Response (QP), Continuous replenishment (CR), and Vendor Managed Inventory (VMI) have been proven to have impact on retailer-supplier inventory. However, only qualitative research are devoted in these strategies which might not provide detailed insights of the usefulness for using these strategies. Therefore, this paper investigates VMI strategies using simulation models and develops the replenishment policy for implementing VMI. Furthermore, this research uses Automatic Pipeline Inventory Order Based Production Control System (APIOBPCS) to identify the factors of using VMI in retailer-supplier inventory management and a simulation model based on these factors is developed to achieve the replenishment policy. The results show that total cost will be reduced and inventory turnover can be increased using this replenishment policy in VMI.

Keywords Retailer · VMI · Strategic alliance · Replenishment policy

1 Introduction

Recently, retail industries are booming rapidly because of the convenience and low price. The major function of retailers is being the middleman between suppliers and customers. For the suppliers, retailers provide exihibition place for their

P.-Y. Chang (✉)
Department of Industrial Engineering and Management, Ming Chi University of Technology, #84 Gungjuan Rd, Taishan District, New Taipei City 243, Taiwan
e-mail: pchang@mail.mcut.edu.tw

Y.-K. Lin et al. (eds.), *Proceedings of the Institute of Industrial Engineers Asian Conference 2013*, DOI: 10.1007/978-981-4451-98-7_62, © Springer Science+Business Media Singapore 2013

product while retailers maintain various product and reduce transaction cost for customers. Simchi-Levi et al. (2001) pointed out the reason of bullwhip effect and showed that reducing uncertainty and constructingstrategic alliance will reduce the bullwhip effect. For strategic alliance, Quick Response (QR), Continuous Replenishment (CR), and Vendor Management Inventory (VMI) are usually implemented and discussed. All the strategic alliances are proved to reduce inventory and lead time while increasing efficiency. Furthermore, Tyan and Wee (2003) analyzed VMI and demonstrated that VMI can solve the problem of inventory shortage. VMI is realized to have the most significant effect on reducing inventory and lead time than QR and CR. However, its effectiveness will be different among various strategies executions. To realize the impact, further analysis should be performed with appropriate measurements. Computer simulation has become a very popular tool in the analysis and design of manufacturing systems and has been done for a variety of purposes from assessing the flexibility of the system to comparing different system designs. Therefore, this research investigates the effect of implementing VMI into industries using simulation.

2 Literature Reviews

VMI is known as the suppliers (vendors) control the quantity and the date of the shipment so that inventory levels of suppliers and retailers can be reduced. Alberto compared the situations of traditional supply chain and VMI supply chain when orders occur. His research proved that VMI can be effective for inventory control because of the just-in-time information flow between suppliers and retailers. Tyan and Wee (2003) investigated the impact of implementing VMI in retail industries in Taiwan. The results showed that great advantages can be achieved using VMI in inventory control for retail industries. The results also indicated that VMI can be applied in electronic industries to reduce inventory level and lead time. Disney and Towill (2003a, b) described VMI as the special collaborative strategy between manufacturers and distributors based on literatures. Distributors will collect and pass customers' purchasing information to manufacturers and suppliers so that forecast estimation accuracy and reorder point can be improved. Furthermore, Disney and Towill (2003a, b) proved that VMI can solve bullwhip effect within traditional supply chain while inventory cost was reduced and inventory turnover is increased. Disney et al. (2004) also evaluated the impact of using VMI in different echelons of supply chain and concluded that VMI was appropriate for dynamic supply chain. Hong (2002) used IOBPCS (Inventory and Order Based Production Control System) to present the problems within the production systems and provided strategies to solve the proposed problems. IOBPCS was first mentioned by Coyle in 1977 and strengthened by Naim and Towill in 1993. IOBPCS represents the simplify model of the relations within the real production system and was introduced based on observation of British industries. Berry discussed the variables that were appropriate for IOBPCS. These variables are production completion rate, inventory level, lead time, and fill rate.

John introduced WIP (work in progress) into IOBPCS to enhanced the stability of the system and proposed APIOBPCS(Automatic Pipeline Inventory and Order Based Production Control System,APIOBPCS) for modeling the production system. Disney and Towill (2002a, b, c) then applied APIOBPCS to increase the stability of VMI supply chain and the results showed that the production system can be more stable with WIP as the performance measurement.

3 Methodology and Results

This research develops an APIOBPCS model of VMI for retailers and suppliers based on the IPBPCS model created by Disney. The proposed model can be applied to evaluate the impact of implementing VMI into the relationship of suppliers and retailers. Figure 1 demonstrates the developed APIOBPCS model of suppliers and retailers. In Fig. 1, the direction with a positive sign represents the decision making process will add value to the supply chain. For example, a positive sign existed on the direction from warehousing to suppliers safety stock. The reason of positive sign is that inventory is moved from warehouse to suppliers to reduce the total inventory level of the supply chain. The negative sign represents the decision making process will deduct the value to the supply chain. The reason

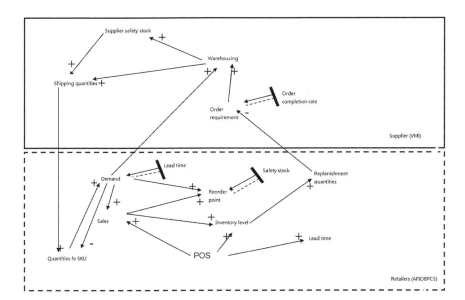

Fig. 1 Retailer and supplier APIOBPCS

of a negative sign between replenishment quantities and order requirement is that the order requirement will have negative impact on the replenishment quantities.

In the VMI supply chain, suppliers will obtain demand information from retailer's database using information technology. The suppliers will determine shipping quantities and time to deliver product to retailers. Through APIOBPCS, the flow of information and the value of the decision making process are evaluated. Based on APIOBPCS, decisions that can provide positive value to the supply chain are mentioned. To realize the significance effect of using VMI and APIOBPCS, simulation models are constructed and factors with different levels are tested in the simulation models.

3.1 Simulation Results

The simulation models are developed using ARENA with order processing time and supplier replenishing time as the performance measurements. Table 1 shows the factors and levels of the simulation experiments. In Table 1, three factors, product variety, inventory levels, and replenishment quantities are determined based on literatures for the simulation experiments.

Table 2 demonstrates the results of order processing time using simulation for VMI supply chain and traditional supply chain. The average order processing time of VMI is significant shorter than the average order processing time of traditional supply chain. The decrement of the order processing time indicates the reduction of the lead time. The most significant reduction is when inventory level is low and replenishment quantities is high. The average processing time for traditional supply chain and VMI are 105.098 and 73.843, respectively. Furthermore, the variances of traditional supply chain and VMI in this experiment are 126.817 and 23.953, respectively. This result indicates that implementing VMI not only reduce the average order processing time but stabilize the system performance.

Table 3 demonstrates the replenishing time for the traditional supply chain and VMI. It takes more time for the traditional supply chain to replenish inventory than the time for VMI. The product varieties do not have significant effect on the average replenishing time. However, the variances of the replenishing time between traditional supply chain and VMI do not have significant difference.

Based on the results showed in Tables 2 and 3, VMI is a better supply chain strategy than the traditional supply chain with the performance measures of order processing time, order waiting time, and replenishing time. The variance of order

Table 1 Factors and levels

Factors	Levels
Product variety	Uniform (10, 20), Uniform (20, 40)
Inventory levels	50, 60
Replenishment quantities	50, 70, 80

Investigating the Replenishment Policy

Table 2 Order processing time

	Product variety	Inventory level	Replenishment quantities	Average	Variance
Traditional	Uniform (20, 40)	60	50	85.175	58.457
			70	94.725	106.84
			80	98.292	97.904
		50	50	88.342	76.864
			70	96.256	86.626
			80	105.098	126.817
	Uniform (10, 20)	60	50	42.435	12.239
			70	47.104	24.716
			80	49.720	23.973
		50	50	44.987	11.364
			70	50.004	15.408
			80	52.486	19.133
VMI	Uniform (20, 40)	60	50	74.004	21.124
			70	73.638	27.918
			80	73.843	23.953
		50	50	74.004	21.124
			70	73.638	27.918
			80	73.843	23.953
	Uniform (10, 20)	60	50	37.677	4.061
			70	37.629	2.550
			80	37.428	1.992
		50	50	37.916	5.706
			70	37.536	5.190
			80	39.265	19.507

processing time and order waiting time for VMI outperform traditional supply chain. However, no difference of the variance of replenishing time between VMI and the traditional supply chain is found.

4 Conclusions

This research uses simulation to prove the advantages of adapting VMI into supply chain. Three factors, product varieties, inventory levels, and replenishment quantities with different levels are tested in the models to realize the impact of using VMI in the supply chain. The results of the simulation show that VMI can be implemented to shorten order processing time, order waiting time, and replenishing time. With the reduction in these three categories, order lead time can be decreased. Furthermore, VMI outperform the traditional supply chain in different experiments. The results also indicate that VMI can be extended to be implemented in retailers-suppliers supply chain. Although strengths of using VMI are addressed in this research, more topics can be discussed as future research. Topics

Table 3 Replenishing time

	Product variety	Inventory level	Replenishment quantities	Average	Variance
Traditional	Uniform (20, 40)	60	50	379.272	446.817
			70	396.646	681.108
			80	409.999	1098.48
		50	50	382.717	482.991
			70	389.307	1291.476
			80	404.483	997.099
	Uniform (10, 20)	60	50	375.289	626.856
			70	399.523	953.422
			80	412.295	1038.913
		50	50	381.951	464.437
			70	407.785	1002.174
			80	411.624	776.065
VMI	Uniform (20, 40)	60	50	287.896	542.757
			70	266.594	534.132
			80	272.476	1422.782
		50	50	287.896	542.757
			70	266.594	534.132
			80	272.476	1422.782
	Uniform (10, 20)	60	50	311.294	459.662
			70	274.125	772.330
			80	272.476	1422.782
		50	50	313.158	332.286
			70	285.631	663.091
			80	288.429	768.464

such as safety stocks, adapting optimized retailer-supplier relations in the supply chain, and determining optimal levels for the factors can be discuss further to identify the advantages of using VMI.

References

Disney SM, Naim MM, Potter A (2004) Assessing the impact of E-business on supply chain dynamics. Inter J prod Econ 89(2):109–118

Disney SM, Towill DR (2002a) A procedure for the optimization of the dynamic response of a vendor managed inventory system. Com Ind Eng 43(1–2):27–58

Disney SM, Towill DR (2003a) The effect of vendor managed inventory(VMI) dynamics on the bullwhip effect in supply chains. Inter J Prod Econ 85(2):199–215

Disney SM, Towill DR (2002b) A discrete transfer function model to determine the dynamic stability of a vendor managed inventory supply chain. Inter J Prod Res 40(1):179–204

Disney SM, Towill DR (2003b) On the bullwhip and inventory variance produced by an ordering policy. Omega: The Inter J Manage Sci 31(3):157–167

Disney SM, Towill DR (2002c) A robust and stable analytical solution to the production and inventory control problem via a z-transform approach. Logistics Systems Dynamics Group,

Cardiff Business School, Cardiff University. Retrieved Feb 2007, from http://www.ipe.liu.se/rwg/igls/igls2002/Paper039.pdf

Hong-Ming S (2002) Re-engineering the UK private house building supply chain. University of Wales, U.K., degree of Doctor of Philosophy

Simchi-Levi D, Kaminsky P, Simchi-Levi E (2001) Designing and managing the supply chain: concepts, strategies, and case studies. Irwin/McGraw-Hill, Boston

Tyan J, Wee HM (2003) Vendor managed inventory : a survey of the taiwanese grocery industry. J Purchasing Sup Manage 9(1):11–18

Applying RFID in Picker's Positioning in a Warehouse

Kai Ying Chen, Mei Xiu Wu and Shih Min Chen

Abstract As the RFID technology gradually matures, the research on the RFID technology-based positioning system has attracted more attention. The RFID technology applied to positioning can be used for orientation recognition, tracking moving trajectories, and optimal path analysis, as well as information related to the picker's position. According to previous literatures, warehousing management consumes high cost and time in business operation, more specifically; picking is one of the most costly operations in warehousing management, which accounts for 55 % of total warehousing cost. Therefore, this study constructed an actual picking environment based on the RFID technology. This system uses the back-propagation network method to analyze the received signal strength indicator (RSSI), so as to obtain the position of the picker. In addition, this study applied this locating device to picking activities, and discussed the effects of the positioning device on different picking situations.

Keywords RFID · Back-propagation network · Positioning · Warehouse management

---

K. Y. Chen (✉) · M. X. Wu · S. M. Chen
Department of Industrial Engineering and Management, National Taipei University
of Technology, No. 1, Chunghsiao East Road, Sec 3, Taipei, Taiwan, Republic of China
e-mail: kychen@ntut.edu.tw

M. X. Wu
e-mail: fly6616@gmail.com

S. M. Chen
e-mail: teaorme555@msn.com

Y.-K. Lin et al. (eds.), *Proceedings of the Institute of Industrial Engineers Asian Conference 2013*, DOI: 10.1007/978-981-4451-98-7_63,
© Springer Science+Business Media Singapore 2013

1 Introduction

In recent years, the vigorous development and continuous improvement of wireless communications technology have been promoting the gradual growth of RFID technology. From 2000 onward, RFID has been widely applied in transportation management, logistics management, livestock management and personal ID recognition. With increasingly maturing RFID technology, studies on RFID technology-based positioning system are gaining more and more attention. The application of RFID technology in positioning can provide direction recognition, movement tracking and optimal route analysis as well as information relating to the user location. To reduce the redundant labor cost in picking and maximize the benefits, this study uses RFID technology to build a real application environment and judge the location of the order picking personnel in real time with the help of the RSSI reading results. Moreover, this study also applies the positioning device in order picking activities to explore the overall effectiveness of the positioning devices in order picking activities to reduce unnecessary paperwork and respond quickly to temporary orders or changes.

2 Literature Review

2.1 Introduction to RFID

RFID is mainly originated in the technology for military purpose of identifying the friend or foe aircraft in WWII in 1940. The one gives rise to the wave of RFID application is the largest chain store of the United States, the Wal-Mart. In a retailer exhibition held in Chicago in 2003, it announced to request the top 100 suppliers to place RFID tags on the pallets and packaging cartons before 2005 (Glover and Bhatt 2006).

RFID is a "non-contact" automatic identification technology, mainly consisting of transponder, reader and middleware system. Its automatic identification technology can complete management operations without manual labor by mainly using the wireless wave to transmit the identification data. The fixed or hand-held reader then automatically receives the identification data contained in the tag chip to realize the purpose of identity recognition. The received data will then be filtered in real time and sent back to the back end application systems to generate data with added-value after summarization. RFID can be integrated with production, logistics, marketing and financial management information for combined applications (Lahiri 2006).

2.2 Discussions on Positioning Approaches

The principle of triangulation is mainly to use the known geometric relationship of lengths or angles of three points in the space as the reference for positioning to calculate the object position. With distance as the reference basis, it mainly measures the relative distances in between three fixed points of known locations. With the distances between fixed points and the center of the round as the radius, the location of the point to be measured can be inferred by the intersections of triangular geometry. If using the angle as the reference positioning basis, location is identified by the angle of the signals.

The positioning concepts of Scene Analysis are mainly to use the RSSI and mathematical models to construct database of the collected parameters regarding specific place or location. In the positioning of certain object, the measured parameters can be matched and compared with the database to find out similar data to infer the location of the object (Bouet and Santos 2008).

Proximity approach is also known as the connectivity approach, which mainly uses the proximity to characteristics in the judgment of location. It can be mainly divided into the centralized and the distributed method. The former is to set up antenna in specific location. When the point to be positioned approaches a certain location, the antenna will receive the information about the point to be positioned and uses the location of strongest RSSI as the coordinates. The latter uses a large amount of reference points of coordinates. When a certain reference point approaches, it will read its coordinates to realize positioning (Ward et al. 1997).

Spot ON positioning system is the first indoor positioning system using the RFID technology. Spot ON uses self-developed RFID equipment to realize the function of indoor positioning. It mainly uses the RFID reader and a number of sensing tags to construct an indoor wireless sensing network environment covering a certain range (Hinckley and Sinclair 1999).

Location Identification based on dynamic Active RFID Calibration (LANDMARC) is a recently developed positioning system using the RFID technology. LANDMARC system mainly follows the method of the Spot ON positioning system to predict the location of the unknown object by relative RSSI estimated distance. The main idea is to use the additional reference tags of fixed positions to enable the LANDMARC to considerably improve the accuracy rate of location determination.

Neural network is a mathematical model. It is an in-formation processing system composed of biological simulation and neural network system. It can conduct a series of actions including storage, learning and feedback to the input signals. The neural network has high capability of information processing with excellent non-linear mapping capability for non-linear calculation. At present, the commonly used neural networks can be divided by network connection architecture into the feed-forward neural network and feedback neural network (Basheer and Hajmeer 2000).

3 Research Method

This study is mainly to develop a real picking positioning environment on the basis of the RFID technology. Since it does not need too many RFID readers and the environmental equipment for the experiment is relatively simple, coupled with the built-in chips of the RFID tags that can store object information, it can be a system for warehouse management and positioning to save costs considerably. The RSSI can be sent back to the back end application systems via readers for analysis by using the BPN (Back-Propagation Network) method to realize the effects of positioning.

In this study, we mainly use the RFID readers, antennas and a number of RFID tags developed by the Omron Company to build the actual positioning environment for warehouse picking. The reader type is Omron V750-series UHF RFID System, the communications frequency is 922–928 MHz, and is in line with the standards of EPC global Class 1 Gen2, it is suitable for the process management in manufacturing site, the order delivery of logistics units and material incoming inspections.

The experimental situation of this study is: the environment is assumedly set up in an open space. A RFID Reader and a number of receiving antennas fixed in different places are used in the system. According to different distances between the antenna and the personnel, when positioning, it can read K RSSI values in one reading. In this experiment, objects of different impacts will be placed on the shelf (such as water, metal etc.). By the RSSI values the reader receives, we can calculate all the possible locations of the order picking personnel. The experimental scenarios are as shown in Fig. 1.

In this study, we use Matlab software as the programming tool to build the BPN. Among all neural net-work learning models, BPN model is the so far the most representative and most widely applied model. It is an architecture of multiple-layer

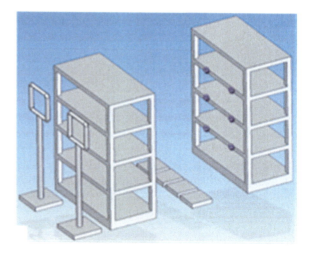

Fig. 1 Experimental scenarios

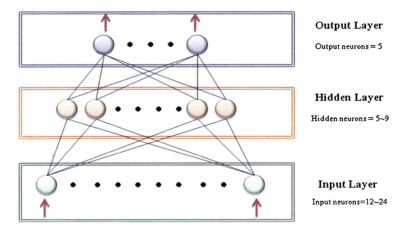

Fig. 2 Architecture of neural network

sensors. It is a multiple layers of feed-forward neural network using the super-vising learning algorithm. It is suitable for applications including classification, identification, prediction, system control, noise filtering and data compression.

During the learning process, BPN will give training examples to the neural network. Each training example contains input item and target output value, and the target output value will continuously urge the network to modify the weight value of the transmission connectivity, and the repeatedly adjust the strengths of the network links by training to reduce the difference between the network output value and target output values until the difference is below the critical value.

The network used in this study contains three layers: input layer, hidden layer and output layer, and its architecture are as shown in Fig. 2.

Input layer: the number of neurons is determined by the corresponding number of input vectors. In this study, the experimental design has 12–24 features, hence the number of the neurons of the input layer is set in the range of 12–24 neurons.

Output layer: in this study, the output value is the location of the personnel. The location of the order picking personnel is assumed in the experiment in five cases including Position 0 representing that the picking personnel is not at the site of the warehouse, Position 1–4 representing that the picking personnel is at the No. 1, No. 2 … location, respectively. Hence, the number of neurons of the output layer in this experiment is set as five.

Hidden layer: generally, the unit number of the hidden layer is set as (input layer unit number + output layer unit number) $\div 2$. To improve the accuracy of the experiment, in this study, we set the hidden layer unit number in the range of 5–9 in search of the optimal number of units.

The BPN training process is to input the training examples in the network to allow the network to slowly adjust the weighted value matrix and partial weight vectors to comply with the requirements of training example target value T.

4 Experiment Design and Analysis

Since the RFID operation and positioning are subject to the impact of external environmental factors such as object properties, blocking, multiple routes, and dispersion. Hence, this study tests the system in case of empty shelf and shelf with objects of different properties to simulate the positioning in actual picking environment. This study designs four scenarios in experiment. Scenario 1 is the shelf without any object. Scenario 2 is the shelf with paper products, Scenario 3 is the shelf with liquid objects and Scenario 4 is the shelf with metal objects. Regarding five locations of Position 0–4, we collect the RSSI values when the personnel standing at different locations in case of four scenarios and input them into the BPN model for prediction to determine the location of the personnel. The photos of experimental scenarios are as shown in Figs. 3, 4, 5 and 6.

Before the analysis RSSI value, the data should be processed. The preprocessing steps in this study can be summarized as shown below:

Step 1: list the RSSI values read by tags of same number in a same column
Step 2: make up for the omitted value with the minimum value of the receiving level
Step 3: set the target output vector of each location (T)
Step 4: data normalization

In this study, we divide the variable data into the training set and testing set by ratio of 7: 3. The collected data are divided into two parts: the first part consists of 1,440 data of 1,000 training samples and 440 testing samples; the second part consists of 2,860 samples of 2,000 training samples and 860 testing samples. Compare the data of the two parts to analyze the relationship be-tween the accuracy rate and Mean Square Error (MSE) value in case of different scenarios.

Fig. 3 Scenario 1: without any object on the shelf

Applying RFID in Picker's Positioning in a Warehouse 527

Fig. 4 Scenario 2: paper products on the shelf

Fig. 5 Scenario 3: liquid objects on the shelf

Since there is no definite standard regarding the se-lection of neural network architecture and relevant parameter setting, we have to use the trial and error method. Therefore, this study will explore the impact of different parameter setting on the network accuracy (for example: Hidden layer, Hidden node, Learning rate, Epoch and Momentum). After the comparison of the tests of different parameters, the BPN optimal parameter settings of this study are as shown in Table 1:

After the network analysis of the four scenarios, the accuracy rate and MSE value of the analysis results are as shown in Fig. 4.

Fig. 6 Scenario 4: metal objects on the shelf

Table 1 Parameter setting

Parameters	Value
Input Node	12
Hidden Node	7
Output Node	5
Learning Rate	0.3
Momentum	0.7
Epoch	2,000

Fig. 7 Accuracy rate of the analysis results

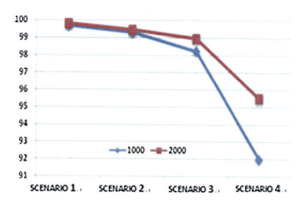

According to the analysis results in this study, no matter 1,000 or 2,000 training samples, it can be learnt from Fig. 7, the accuracy rate in case of four scenarios gradually reduces from that of the empty shelf (99.7 %) → paper products (99.3 %) → liquids (98.22 %) → metals (92 %). It can be learnt from Fig. 8, the MSE value gradually rises from that of the empty shelf (0.00080418) → paper

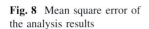

Fig. 8 Mean square error of the analysis results

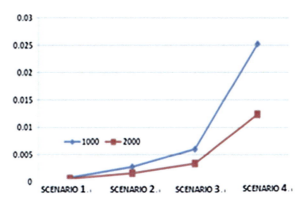

products (0.0027778) → liquids (0.0060043) → metals (0.025204). Hence, according to the experimental results, the RSSI will be more severely interfered in case of being blocked by metal products. As a result, the ac-curacy rate and MSE value will be poorer consequently.

5 Conclusion

According to the experimental results, more training samples can lead to better accuracy rate and MSE value, and the accuracy rate and MSE value in case of picking metal products is the worst. Namely, when positioning and reading, the receiving level of RSSI will be more affected in case of the blocking of metal objects than other objects. In this study, we compare the predicted location and the original data and the result can effectively predict the location of the picking personnel.

Hence, the proposed picking positioning system is expected to reduce unnecessary labor and equipment cost in picking activities to maximize effectiveness. Moreover, the system is expected to respond quickly to temporary orders or emergent orders to increase picking efficiency.

References

Basheer IA, Hajmeer M (2000) Artificial neural networks: fundamentals, computing, de-sign, and application. J Microbiol Methods 43:3–31
Bouet M, Santos ALD (2008) RFID tags: positioning principles and localization techniques. Paper presented at the wireless days, Dubai, 24–27 Nov 2008
Glover B, Bhatt H (2006) RFID essential. O'Reilly Media, California

Hinckley K, Sinclair M (1999) Touch-sensing input devices. Paper presented at the 1999 conference on human factors in computing systems, Pittsburgh, 18–20 May 1999

Lahiri S (2006) RFID sourcebook. IBM Press, Indiana, USA

Ward A, Jones A, Hopper A (1997) A new location technique for the active office. IEEE Wireless Commun 4:42–47

An Innovation Planning Approach Based on Combining Technology Progress Trends and Market Price Trends

Wen-Chieh Chuang and Guan-Ling Lin

Abstract Innovation planning is important for manufactures to maximize the payoffs of their limited research and development expenditures. The key to the success of such innovation planning relies on a valid approach to estimate the value of each R&D project can produce. The purpose of this research is to build a model that considers a broad spectrum of trends in technology development and market price trends. Base on the assumption that the market prices can be a good indication of customer values, this modeling method collects historical product feature data, as well as their historical market prices and trains a neural network to track how electronic product specifications evolutions affect market prices. Predictions can be made from the behavior of the trained model to evaluate the value of each product improvement and, therefore, effective innovation plan can be made based on this model. The structure of this paper is threefold. First, it describes the evolutionary patterns of electronic products and their market prices. Second, it proposes artificial neural network methods to model the evolutionary processes; predictions concerning digital cameras to prove the validity of the model. Third, this research discusses the implications of these findings and draws conclusions.

Keywords Innovation planning · Technology trends · Market trends · Neural network

---

W.-C. Chuang (✉) · G.-L. Lin
Department of Industrial Engineering and Systems Management, Feng-Chia University, Taichung, Taiwan
e-mail: wcchuang@fcu.edu.tw

Y.-K. Lin et al. (eds.), *Proceedings of the Institute of Industrial Engineers Asian Conference 2013*, DOI: 10.1007/978-981-4451-98-7_64,
© Springer Science+Business Media Singapore 2013

1 Introduction

Every year, market competition forces electronics manufacturers to develop products with new specifications. A new model of any electronic product typically offers new features or improved specifications. Manufacturers compete by frequent introductions of new products with slightly better specifications than previous products. For example, Canon, one of the largest digital camera manufactures of the world, has brought in at least 7 series, 165 in total compact camera models for the U.S. market along since 1998. The resolutions of these cameras have been evolving from 810 k pixel, for the early models, to 12.1 megapixel for the latest products (Canon 2013).

Most products can be improved in many ways. For example, a digital camera manufacturer might need to decide whether to invest on making its new model to capture 200 more pixels, or to weigh 20 g less, or to increase optical zooming by 20 %. Omae (1982) referred to the various possible directions of product improvement as the strategic degrees of freedom (SDFs). For the new product strategic developers, with the fact that each of these improvements might require separate product improvement projects to achieve, each cost different R&D resources, it is important for them to use innovation planning to maximize the payoffs of their limited research and development (R&D) expenditures;

An innovation planning, therefore, can be simply defined as the determination of the value of each product improvement project in this paper. Such value can be complicated in many ways. In fact, to determine the valuation of a technology development has long been described as an art (Park and Park 2004) rather than a science. Davenport et al. (1996), have pointed out that any study on knowledge work improvement should focus on making products/services more attractive (to customers) in order to increase value. Although how attractive a product improvement is can be somehow derived using some market survey techniques, these results are subjective to how far the survey reached. An alternative approach is therefore proposed in this paper to evaluate product improvement values. The historical data of the technology evolving trend, as well as market price evolving trend is collected to form a model. Based on the assumption that the value of each product improvement can be measured by the amount of money that customers are willing to pay, which can be indicated by their market price, a neural network is trained to model how specification improvement as well as time factors to affect product prices. The price of future product can be therefore predicted by projecting; as a result the value of each product improvement can be estimated.

The structure of this paper is threefold. First, it describes the evolutionary patterns of electronic products and their market prices. Second, it proposes artificial neural network methods to model the evolutionary processes; predictions concerning digital cameras to prove the validity of the model. Third, this research discusses the implications of these findings and draws conclusions.

2 Technology Progress Trends and Market Price Trends for Electronic

Ever-shortening product life cycles (PLCs) and technology s-curves are two important notions for electronic product evolution trends. The first notion, PLC, can be defined as "the period of a product in the market" (Abernathy and Utterback 1975). It has been widely used to represent the sales pattern of a product from its introduction to termination. Rink and Swan (1979) presented some typical product life cycle patterns and suggested that businesses might improve their planning processes by changing their product patterns. Due to rapid developments in technology and brisk competition, electronic products often have life cycles that can be measured in months. According to data collected from Intel's website, Intel's microprocessors have evolved at least 72 times between the company's inception in 1971 and 2013 (Intel 2013).

The second idea, the s-curve, is often used to depict technological progress over time (Abernathy and Clark 1988). Movement along a given s-curve generally results from incremental improvements within an existing technological approach. When the progress reaches its limit and starts to slow down, one expects a new s-curve to replace older curves as a driver of technological growth. Theoretical as well as empirical discussion of s-curve models can be found in Nieto et al. (1998).

By combining the two notions above, an evolutionary model for electronic products can be built. Figure 1 illustrates this model. S1 and S1 in Fig. 1 represent two technological s-curves. As the technology progresses overtime, new products (P1, P2, ..., etc.), each holds better specifications than previous models, are introduced to market and replace the market share of the old models quickly.

It is more interesting when to add the information of product price into the model. Although the relationship between advances in performance and price has long been recognized (Porter 1980), price does not usually move in lockstep with technological innovations. Generally, new products with higher specifications are expected to be more valuable and deserve higher price tags. However, empirical studies do not support this notion. For example, a study to Google's computing infrastructure equipments showed that, in compared with three successive

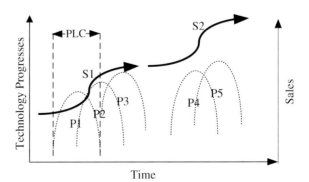

Fig. 1 Typical technology progress trends for electronics

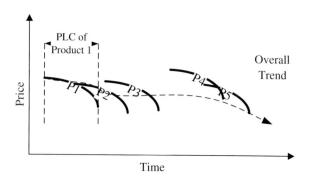

Fig. 2 Typical market price trends for electronics

generations of Google hardware of servers, the performance trend goes up with a much steeper pattern than the performance/server price trend (Barroso 2005). One likely cause of this divergence between price and performance is that technological advances reduce manufacturing costs while improving product features. Other factors might include the competition effects or even strategic considerations. While the patterns of new product pricing are hard to track, a typical price pattern can be easily observed for many electronics; Companies cutting the prices of old products while some new products with higher specifications are introduced to market. A typical new product price pattern can be therefore summarized as Fig. 2.

Note that the curves in Fig. 2 overlap; at any given time it is likely that multiple products with different features and prices will be offered to the same consumers. Foster (1982) and McGrath (1998) have emphasized that products with similar price/performance ratio tend to share the market. For example, in January, 2010, there were more than 23 CPU products on the market from Intel only, ranging from $51 to $998 in pricing (Sharky Extreme 2013). The wide spread of prices range is mainly derived from their technical features differences.

It can be concluded that, in spite of non-technical factors which are hard to be predict, the prices of electronics are affected by two major factors. First, how attractive their features are in compared with other competition product in the market; Second, the time factor itself. Based on this conclusion, an evolutionary model that takes specification evolution as input and the market prices as output can be built. Using this model, the price of future product can be therefore predicted by projecting; as a result the value of each product innovation plan can be estimated.

3 Artificial Neural Network for Electronics Price Trend Modeling

The nonlinear interactions and non-numeric data of this domain can be modeled by an artificial neural network. An artificial neural network (ANN) is an interconnected collection of artificial neurons that uses a mathematical or computational model for information processing (Smith 1993). Neural networks have long been applied as practical non-linear statistical data modeling tools. A predetermined "neural network" is built and a set of known examples "trains" the correct linkage strengths, or weights, between input and output neurons. Once the "correct" weights have been obtained from training, the model can be used for prediction.

To illustrate the proposed methodology and to verify its validity, a prediction model is constructed for a real-world case—namely, point-and-shoot digital cameras.

3.1 Model Construction and Testing

In order to construct the model, 269 records, describing a 58-month period, were collected from the point-and-shoot digital camera market in Taiwan. Six variables were selected as inputs: effective pixels, optical zooms, ISO, camera size, weight, and data collection time. The market price of each record at the data collection time was assigned as the model output.

Various network structures were tested; a simple multi-layer neural network (1 hidden layer with 3 neurons) was selected because of the training results. Twenty records that are after the date of 2010/7/1 were selected as the verification set. The other 249 records were randomly divided into three groups: a training set (180 records); a cross-validation set (25 records); and a testing set (44 records). Data in the cross-validation set was used in the training process to prevent possible over-training.

After the weights had been obtained, data from the testing set were used to verify that the trained ANN model would be valid with respect to new data. Figure 3 compares the results of model prediction with actual prices of the testing set data. As shown in Fig. 3, the average differences between predicted data to actual data is about 18 %, which were satisfactory in this case.

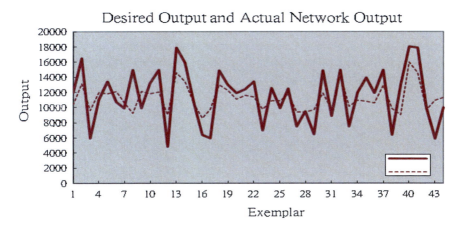

Fig. 3 Comparison between predicted prices and actual prices for the testing set data

Table 1 Comparison between predicted prices to actual prices

No.	Time	Effective Pixel	Optical Zoom	Max ISO	Volume (cm3)	Weight (gram)	Prices (predicted)	Prices (Actual)	Diff. (%)
1	2010/7/1	1,410	2.3	3200	88,536	105	8,727	7,980	9
2	2010/7/1	1,200	2.3	3200	101,834	133	9,188	13,980	34
3	2010/7/1	1,220	2.3	3200	102,323	130	9,121	9,980	9
4	2010/7/1	600	2.3	6400	658,419	341	11,827	8,800	34
5	2010/7/1	504	2	3200	117,494	141	8,215	8,000	3
6	2010/7/1	1,510	24	6400	954,180	496	16,506	20,450	19
7	2010/7/1	1,410	2.3	3200	88,536	105	8,727	8,800	1
8	2010/8/1	1,410	2.3	6400	152,712	180	11,383	11,150	2
9	2010/8/1	1,410	2.3	3200	98,532	116	8,867	9,100	3
10	2010/8/1	1,410	2.3	3200	92,806	103	8,529	6,200	38
11	2010/8/1	1,000	3.8	3200	132,151	190	10,201	11,700	13
12	2010/8/1	1,000	3.8	3200	171,935	170	9,292	14,300	35
13	2010/8/1	1,200	3	1600	130,014	135	8,044	7,980	1
14	2010/8/19	1,210	4	1600	379,845	260	9,985	8,500	17
15	2010/9/1	1,010	1.7	12800	395,010	360	12,728	17,900	29
16	2010/9/1	1,210	2.3	6400	106,029	133	10,267	9,900	4
17	2010/9/1	1,210	2.3	3200	184,080	209	10,643	9,900	8
18	2010/9/2	600	2	12800	202,488	195	11,845	10,000	18
19	2010/10/10	1,410	35	1600	1223,037	552	14,293	15,600	8
20	2010/10/20	1,000	5	3200	412,580	351	12,858	17,900	28

3.2 Model Application for Innovation Planning

The trends in the data were extrapolated to predict future market behavior. The inputs from the verification data set were applied to the model to test the performance of the model to extrapolated data. Table 1 shows the testing results.

As shown in Table 1, the differences between actual prices and predicated prices are ranging from 1 to 38 % with an average 15 %. The accuracy of the prediction model is within acceptable range. Further test the accuracy trend of the prediction model. The average accuracy decreases gradually from 15 % for the first projected month to 18 % for the fourth projected month. This information can be use to evaluate how far the model can be projected.

4 Conclusions

Traditional technology models have often applied s-curves to explain how market prices arrived at their present states. This paper has combined PLCs and s-curves to predict how feature advances will affect market prices. This research trained a neural network from data about product features and market prices over a period of

An Innovation Planning Approach

time. Price trends affected by product features were identified. The trained model was used to predict digital camera prices as a proof of concept. Corporations can use the methods explained in this paper to plan future product innovations.

References

Abernathy WJ, Clark KB (1988) Innovation: mapping the winds of creative destruction. In: Tushman ML, Moore WL (eds) Readings in the management of innovation. Harper, New York, pp 55–78

Abernathy W, Utterback J (1975) A dynamic model of process and product innovation. Omega 3:639–656

Barroso LA (2005) The price of performance, queue—multiprocessors. Queue, 3(7):48–53

Canon (2013) Canon home page. http://www.cannon.com/

Davenport T, Javenpaa S, Beers M (1996) Improving knowledge work processes. Sloan Manage Rev 37:53–65

Foster R (1982) Boosting the payoff from R&D. Research Manage 26:22

Intel (2013) Intel website, http://www.intel.com/

McGrath RN (1998) Technological discontinuities and media patterns: assessing electric vehicle batteries. Technovation 18(11):677–687

Nieto M, Lopez F, Cruz F (1998) Performance analysis of technology using the s curve model: the case of digital signal processing (DSP) technologies. Technovation 18(6/7):439–457

Omae K (1982) The mind of the strategist: the art of Japanese business. McGraw-Hill, New York

Park Y, Park P (2004) A new method for technology valuation in monetary value: procedure and application. Technovation 24:387–394

Porter ME (1980) Competitive strategy. The Free Press, New York

Rink D, Swan J (1979) Product life cycle research: a literature review. J Bus Res, 40, 219–243

Sharky Extreme (2013) Sharky extreme website. http://www.sharkyextreme.com

Smith M (1993) Neural networks for statistical modeling. Van Nostrand Reinhold, New York

Preemptive Two-Agent Scheduling in Open Shops Subject to Machine Availability and Eligibility Constraints

Ming-Chih Hsiao and Ling-Huey Su

Abstract We address the scheduling problem in which two agents, each with a set of preemptive jobs with release dates, compete to perform their jobs on open shop with machine availability and eligibility constraints. The objective of this study is to minimize make span, given that one agent will accept a schedule of time up to Q. We proposed a heuristic and a network based linear programming to solve the problem. Computational experiments show that the heuristic generates a good quality schedule with a deviation from the optimum of 0.25 % on average and the network based linear programming model can solve problems up to 110 jobs combined with 10 machines.

Keywords Scheduling · Two-agent · Open shop · Preemptive · Machine availability constraint · Machine eligibility

1 Introduction

Consider two agents who have to schedule two sets of jobs on an m-machine open shop. Each job has k operations, $k \leq m$, and each operation must be performed on the corresponding specialized machine. The order in which the operations of each job are performed is irrelevant. Operation preemption is allowed and the machine availability and eligibility constraints are considered. The machine availability arises when machines are subject to breakdowns, maintenance, or perhaps high

---

M.-C. Hsiao (✉) · L.-H. Su
Department of Industrial and System Engineering, Chung-Yuan Christian University,
Chung-Li, Taiwan, Republic of China
e-mail: Miller_Hsiao@aseglobal.com

L.-H. Su
e-mail: linghuey@cycu.edu.tw

Y.-K. Lin et al. (eds.), *Proceedings of the Institute of Industrial Engineers Asian Conference 2013*, DOI: 10.1007/978-981-4451-98-7_65,
© Springer Science+Business Media Singapore 2013

priority tasks are prescheduled in certain time intervals. The machine eligibility constraints are imposed when the number of operations of each job i can be less than m. Each machine can handle at most one operation at a time and each operation can be processed on at most one machine at a time. Two agents are called Agent A and B. Agents A and B have n_a (n_b) jobs. Let n denote the total number of jobs, i.e., $n = n_a + n_b$, and each job j of agent A (B) is denoted by $J_j^a \left(J_j^b \right)$. The processing time of a job j of agent A (B) on machine i is denoted by $P_{ji}^a \left(P_{ji}^b \right)$. The release time of a job j of agent A (B) is denoted by $r_j^a \left(r_j^b \right)$. Each machine i is available for processing in the given $N(i)$ intervals, which are $[b_i^k, f_i^k]$, $i = 1, \ldots, m$, $k = 1, \ldots, N(i)$, and $b_i^{k+1} > f_i^k$, where b_i^k and f_i^k are the start time and end time of the kth availability interval of machine i, respectively. The operations of job j of agent A (B) are processed on a specified subset $M_j^a \left(M_j^b \right)$ of the machines in an arbitrary order. We use C_j^a (C_j^b) to denote the completion time of J_j for agent A (B) and $C_{\max} = \max\{ \max_{1 \le j \le n_a} C_j^a, \max_{1 \le j \le n_b} C_j^b \}$ to denote the makespan. The objective is to minimize makespan, given that agent B will accept a schedule of cost up to Q. According to the notation for machine scheduling, the problem is denoted as $O, NC_{win} | pmtn^a, r_j^a, M_j^a : pmtn^b, r_j^b, M_j^b | C_{\max} : C_{\max}^b \le Q$, where O indicates open machines, NC_{win} means that the machines are not available in certain time intervals, $pmtn^{a(b)}$ signifies job preemption, $r_j^{a(b)}$ implies that each job has a release date, $M_j^{a(b)}$ denotes the specific subset of machines to process job j, and $C_{\max}^b \le Q$ denotes agent B will accept a schedule of time up to Q.

The multi-agent scheduling problems have received increasing attention recently. However, most of the research focuses on the single machine problem. The Baker and Smith study (2003) was perhaps the first to consider the problem in which two agents compete on the use of a single machine. They demonstrated that although determining a minimum cost schedule according to any of three criteria: makespan, minimizing maximum lateness, and minimizing total weighted completion time for a single machine, is polynomial, the problem of minimizing a mix of these criteria is NP-hard. Agnetis et al. (2004) studied a two-agent setting for a single machine, two-machine flowshop and two-machine open shop environments. The objective function value of the primary customer is minimized subject to the requirement that the objective function value of the second customer cannot exceed a given number. The objective functions are the maximum of regular functions, the number of late jobs, and the total weighted completion times. The problem in a similar two-agent single machine were further studied by Cheng et al. (2006), Ng et al. (2006), Agnetis et al. (2007), Cheng et al. (2008), Agnetis et al. (2009), and Leung et al. (2010). When release times are considered, Lee et al. (2012) proposed three genetic algorithms to minimize the total tardiness of jobs for the first agent given that the second agent will accept a schedule with maximum tardiness up to Q. Yin et al. (2012a) address the same problem while using the approaches of mixed integer programming, branch and bound, and marriage in honey-bees optimization to solve the problem.

The multi-agent problems are extended by considering variations in job processing time such as controllable processing time (Wan et al. 2010), deteriorating job processing time (Cheng et al. 2011a; Liu et al. 2011), learning effect (Cheng et al. 2011b; Lee and Hsu 2012; Wu et al. 2011; Yin et al. 2012b; Cheng 2012), and setup time (Ding and Sun 2011). Mor and Mosheiov (2010) considered minimizing the maximum earliness cost or total weighted earliness cost of one agent, subject to an upper bound on the maximum earliness cost of the other agent. They showed that both minimax and minsum cases are polynomially solvable while the weighted minsum case is NP-hard.

Lee et al. (2011) extended the single machine two-agent problem to the two-machine flowshop problem where the objective is to minimize the total completion time of the first agent with no tardy jobs for the second agent. A branch-and-bound and simulated annealing heuristic were proposed to find the optimal and near-optimal solutions, respectively.

As for the preemptive open shop, Gonzalez and Sahni (1976) proposed a polynomial time algorithm to obtain the minimum makespan. Breit et al. (2001) studied a two-machine open shop where one machine is not available for processing during a given time interval. The objective is to minimize the makespan. They showed that the problem is NP-hard and presented a heuristic with a worst-case ratio of 4/3. When time-windows is considered for each job on an open shop and the objective is to minimize makespan, Sedeno-Noda et al. (2006) introduced a network flow procedure to check feasibility and a max-flow parametrical algorithm to minimize the makespan. Sedeno-Noda et al. (2009) extended the same problem by considering performance costs including resource and personnel involvement.

So far as we know, there is no result related to a preemptive open shop wherein two-agents and machine availability and eligibility, as well as job release times, are considered.

2 Computational Results

The objective of the computational experiments described in this section is to evaluate both the performances of the heuristic and the exact algorithms. All experimental tests were run on a personal computer with AMD 2.91 GHz CPU. The heuristic algorithm was coded in Visual Basic, and the linear programming model and the constraint programming model were solved by LINGO11.0 and ILOG OPL 6.1, respectively. The experiment involves the instances with the number of jobs n = 40, 60, 80, 100 and 110 in which three pairs of n_a and n_b are set as

$\{(n_a = 20, n_b = 20), (n_a = 25, n_b = 15), \text{ and } (n_a = 15, n_b = 25)\}$
$\{(n_a = 30, n_b = 30), (n_a = 35, n_b = 25), \text{ and } (n_a = 25, n_b = 35)\}$
$\{(n_a = 40, n_b = 40), (n_a = 45, n_b = 35), \text{ and } (n_a = 35, n_b = 45)\}$
$\{(n_a = 50, n_b = 50), (n_a = 60, n_b = 40), \text{ and } (n_a = 40, n_b = 60)\}, \text{ and}$
$\{(n_a = 55, n_b = 55), (n_a = 60, n_b = 50), \text{ and } (n_a = 50, n_b = 60)\}, \text{respectively.}$

The processing time of p_{ji}^x are randomly generated from a uniform distribution $U[0, 10]$. The job arrival time r_j^x refers to Chu (1992) and are generated from

$$U[0, \bar{P}], \text{ where } \bar{P} = \left(\sum_{i=1}^{m} \sum_{j=1}^{n_a} p_{ji}^a + \sum_{i=1}^{m} \sum_{j=1}^{n_b} p_{ji}^b \right) \Big/ m \text{ is the mean processing time on}$$

each machine. The number of machine m is set as m = 6, 8 and 10. To generate the upper bound of makespan for agent b, we refer to Bank and Werner (2001)

$$\text{with} \quad Q = \beta \left[\frac{n}{m} \left(p + \frac{\sum_{j=1}^{n_a} r_j^a + \sum_{j=1}^{n_b} r_j^b}{n} \right) \right] \frac{n_b}{n}, \quad \text{where} \quad \bar{P} = \frac{\sum_{i=1}^{m} \sum_{j=1}^{n_a} p_{ji}^a + \sum_{i=1}^{m} \sum_{j=1}^{n_b} p_{ji}^b}{n} \quad \text{and}$$

$\beta \in \{1.2, 1.5, 2, 2.5\}$. The rate of machine availability θ is set as 0.7, 0.9 and 1.0. Five instances are generated for each combination of n, m, p_{ji}^x, r_j^x, θ, Q, yielding 900 instances. In comparing heuristic performance, the following formula is used to determine the deviation of the heuristic solution over the optimal solution. Deviation (%) = [(heuristic-optimum)/optimum] $\times$ 100 %.

Table 1 shows the solution quality of the heuristic. The influences of m, n, n_a/n_b and θ on the solution quality of the heuristic are analyzed.

Table 1 illustrates that the deviation of the heuristic from the optimal solution appears in descending trend as the value of m increases. The reason is that the heuristic selects the maximum total remaining processing time of each job, instead of that of each machine, to be processed on the corresponding machine. For the rate of machine availability θ, the value of 1 showing that all machines are available at any times. Since none of $[b_i^k, f_i^k]$, $i = 1, \ldots, m, k = 1, \ldots, N(i)$ incurred, the number of time intervals decreases and the span of time interval increases, therefore more jobs are competing to be scheduled in each time interval and hence the error increases. As to the number of jobs, the higher value that n is, the better the performance of the heuristic. One reason is that the higher value for n implies more time epochs r_j^x incurred and a smaller time span for each T_l making the heuristic easier to assigning the operations correctly. Another reason is that the denominator increases, whereas the deviation of the heuristic solution from the optimal solution may not increase in proportion to the denominator. There is no significant difference in the performance of the heuristic on the value of n_a/n_b, but $n_a/n_b = 0.5$ gives the best performance. When n_a/n_b increases, the deviation increases due to the fact that the heuristic gives priority to agent B and thus more operations of agent A should compete to schedule in each time interval.

Table 1 The average solution quality of the heuristic algorithm

θ	m	β	40			60			80			100			110		
			na,nb 20,20 Dev (%)	na,nb 25,15 Dev (%)	na,nb 15,25 Dev (%)	na,nb 30,30 Dev (%)	na,nb 35,25 Dev (%)	na,nb 25,35 Dev (%)	na,nb 40,40 Dev (%)	na,nb 45,35 Dev (%)	na,nb 35,45 Dev (%)	na,nb 50,50 Dev (%)	na,nb 60,40 Dev (%)	na,nb 40,60 Dev (%)	na,nb 55,55 Dev (%)	na,nb 60,50 Dev (%)	na,nb 50,60 Dev (%)
0.7	6	1.2	0.00	4.78	0.28	0.31	0.00	0.00	0.00	0.00	0.13	0.00	0.00	0.00	0.00	0.00	0.00
		1.5	0.00	4.52	0.00	0.00	0.36	0.00	0.00	0.00	0.00	0.00	0.00	0.00	0.00	0.00	0.00
		2	0.39	0.48	0.90	0.00	0.00	0.00	0.00	0.00	0.00	0.00	0.00	0.00	0.00	0.00	0.00
		2.5	0.00	0.00	0.00	0.00	0.00	0.00	0.00	0.00	0.00	0.00	0.00	0.00	0.00	0.00	0.00
	8	1.2	0.62	–	0.78	2.04	0.18	0.00	0.00	0.73	0.00	0.00	0.00	0.00	0.00	0.00	0.00
		1.5	0.00	–	0.00	0.00	0.57	0.81	0.00	0.00	0.00	0.00	0.00	0.00	0.00	0.00	0.00
		2	0.00	0.65	0.00	0.00	0.00	0.51	0.00	0.00	0.00	0.00	0.00	0.00	0.00	0.00	0.00
		2.5	0.00	0.00	0.60	0.00	0.00	0.00	0.00	0.00	0.00	0.00	0.00	0.00	0.00	0.00	0.00
	10	1.2	2.09	–	0.00	2.88	–	1.19	1.00	1.49	2.70	0.63	0.75	0.00	0.00	0.10	0.00
		1.5	0.57	–	1.31	0.25	0.00	0.00	0.00	0.00	0.00	0.00	0.00	0.00	0.00	0.00	0.00
		2	0.25	0.42	0.61	0.00	0.00	0.00	0.00	0.00	0.00	0.00	0.00	0.00	0.00	0.00	0.00
		2.5	0.00	1.81	0.00	0.17	0.00	0.00	0.00	0.00	0.00	0.00	0.00	0.00	0.00	0.00	0.00
0.9	6	1.2	0.00	0.75	0.00	1.12	0.00	0.48	0.00	0.28	0.00	0.00	0.00	0.00	0.00	0.00	0.00
		1.5	0.00	0.00	0.00	0.00	0.00	0.62	0.00	0.00	0.00	0.00	0.00	0.00	0.00	0.00	0.00
		2	0.00	0.00	0.00	0.00	0.00	0.00	0.39	0.00	0.00	0.00	0.00	0.00	0.00	0.00	0.00
		2.5	0.18	0.00	0.00	0.00	0.00	0.00	0.00	0.00	0.00	0.00	0.00	0.00	0.00	0.00	0.00
	8	1.2	0.00	–	0.00	0.38	0.00	0.00	0.00	0.61	0.00	0.00	0.00	0.00	0.00	0.00	0.00
		1.5	0.00	0.00	0.00	0.00	0.00	0.00	0.56	0.00	0.00	0.00	1.67	0.00	0.00	0.00	0.43
		2	0.00	0.00	0.00	0.00	0.00	0.00	0.00	0.00	0.00	0.00	0.00	0.00	0.00	0.00	0.00
		2.5	0.78	1.28	0.00	0.00	0.00	0.00	0.00	0.00	0.00	0.00	0.00	0.00	0.00	0.00	1.64
	10	1.2	0.00	0.00	0.00	0.00	0.00	0.00	0.00	0.00	0.00	0.00	0.00	0.00	0.00	0.00	0.00
		1.5	3.28	0.00	0.00	0.00	0.12	0.00	0.00	0.00	0.00	0.00	0.00	0.00	0.00	0.00	0.00

(continued)

Table 1 (continued)

θ	m	β	40 na,nb 20,20 Dev (%)	na,nb 25,15 Dev (%)	na,nb 15,25 Dev (%)	60 na,nb 30,30 Dev (%)	na,nb 35,25 Dev (%)	na,nb 25,35 Dev (%)	80 na,nb 40,40 Dev (%)	na,nb 45,35 Dev (%)	na,nb 35,45 Dev (%)	100 na,nb 50,50 Dev (%)	na,nb 60,40 Dev (%)	na,nb 40,60 Dev (%)	110 na,nb 55,55 Dev (%)	na,nb 60,50 Dev (%)	na,nb 50,60 Dev (%)
		2	0.00	0.92	0.00	0.00	0.00	0.00	0.00	0.00	0.00	0.00	0.00	0.00	0.00	0.00	0.00
		2.5	0.73	0.00	0.00	0.00	0.00	0.00	0.00	0.00	0.00	0.00	0.00	0.00	0.00	0.00	0.00
1	6	1.2	0.00	1.64	0.00	0.00	0.00	1.99	0.00	0.00	0.00	0.00	0.00	0.00	0.00	0.50	0.00
		1.5	0.00	1.45	0.00	0.00	0.00	0.00	0.00	0.97	0.00	0.00	0.84	1.47	1.38	0.00	0.00
		2	1.56	1.91	1.06	0.00	0.00	0.00	0.00	0.48	0.11	0.00	0.00	0.00	0.00	0.82	0.00
		2.5	0.00	0.00	0.00	0.00	0.00	0.00	0.00	0.00	0.00	0.00	0.00	0.00	0.93	0.00	0.00
	8	1.2	2.92	–	0.00	0.14	3.16	1.85	1.31	0.00	0.33	0.00	0.00	0.00	0.00	0.00	0.00
		1.5	0.00	1.58	2.01	0.00	0.00	0.00	0.65	0.00	0.87	0.00	0.00	0.7	0.00	0.00	0.00
		2	1.40	0.00	0.00	1.15	0.00	1.67	0.00	0.00	0.00	0.00	2.43	0.00	0.00	0.00	0.00
		2.5	4.64	1.40	0.00	2.56	0.00	0.00	0.00	1.03	1.41	0.00	0.00	0.00	0.00	0.00	0.00
	10	1.2	4.30	–	1.37	0.00	2.36	3.22	1.05	0.00	1.35	0.00	0.00	0.00	2.89	0.00	2.53
		1.5	0.00	–	0.00	1.20	1.39	0.00	0.86	0.00	1.37	0.00	0.00	0.00	0.00	0.00	0.00
		2	0.00	2.72	0.00	0.00	0.00	0.00	0.00	0.00	0.00	1.70	0.00	0.00	0.00	0.00	0.00
		2.5	0.00	0.00	0.00	0.00	0.00	0.00	0.00	0.00	0.00	0.00	0.00	0.00	0.00	0.00	0.00

Total average deviation:
0.25 %

"–" no feasible solution is incurred due to that the time limit Q for the second agent is too small.

Preemptive Two-Agent Scheduling in Open Shops

Table 2 Execution times of Lingo and Heuristic for small-size instances

m	N	Lingo(ss)	Heuristic(ss)
6	90	19.25	1.25
6	100	37.75	1.75
6	110	30.25	1.75
8	90	32.25	1.75
8	100	165.25	2.25
8	110	71.25	2.50
10	90	181.25	2.00
10	100	629.00	3.00
10	110	1002.75	3.00

Based on this analysis, we found that decreasing machine number m and the value of n_a/n_b reduced the deviation of the heuristic, while decreasing job number n increased the deviation of the heuristic. The heuristic generates a good quality schedule with a deviation from the optimum of 0.25 % on average.

As to the execution time of the heuristic, Table 2 shows the average execution time of the network based linear programming and the heuristic. The average execution time of the heuristic is small compared to that of the linear programming.

In Table 3, the average execution times in seconds for Linear Programming model, Constraint Programming model, and Combined model (Linear programming and Constraint Programming) are shown in columns lingo, OPL, and Total, respectively. For $n = 80$ and $m = 10$, the average execution time for the combined model exceeds two hours. Therefore, an efficient heuristic algorithm is much required.

3 Conclusions

In this paper, we have analyzed the preemptive open-shop with machine availability and eligibility constraints for two-agent scheduling problem. The objective is to minimize makespan, given that one agent will accept a schedule of time up to Q. This problem arises in environments where both TFT-LCD and E-Paper are manufactured and units go through a series of diagnostic tests that do not have to be performed in any specified order. We proposed an effective heuristic to find a nearly optimal solution and a linear programming model that based on minimum cost flow network to optimally solve the problem. Computational experiments show that the heuristic generates a good quality schedule with a deviation from the optimum of 0.25 % on average.

Table 3 Execution times of L P model, OPL model and those of both models

θ	m	β	n								
			40			60			80		
			lingo	OPL	Total	Lingo	OPL	Total	Lingo	OPL	Total
0.7	6	1.2	1.67	3471.55	3473.22	5.83	5632.6	5638.43	12.50	5383.2	5395.70
		1.5	2.50	1621.4	1623.90	5.83	4866.14	4871 .97	12.33	6217.37	6229.70
		2	1.83	4443.77	4445.60	4.83	3658.6	3663.43	12.83	8374.42	8387.25
		2.5	1.67	2189.61	2191.28	5.50	4569.55	4575.05	11.00	7941.41	7952.41
	8	1.2	4.83	3563.08	3567.91	21.50	4424.8	4446.03	78.67	6268.49	6347.16
		1.5	7.00	1443.82	1450.82	18.50	3954.75	3973.25	38.17	7704.92	7743.09
		2	7.67	5066.59	5074.26	18.83	6614.74	6633.57	27.33	2103.56	2130.89
		2.5	3.33	4791.44	4794.77	27.33	5982.09	6009.42	40.17	7497.72	7537.89
	10	1.2	6.00	4189.73	4195.73	54.83	6148.1	6202.93	143.00	5235.4	5378.40
		1.5	8.50	3821.1	3829.60	49.50	5743.97	5793.47	114.33	6818.86	6933.19
		2	19.00	4693.21	4712.21	49.17	7330.26	7379.43	183.17	8758.87	8942.04
		2.5	15.50	2043.46	2058.96	46.17	6236.85	6283.02	62.17	8695.65	8757.82
0.9	6	1.2	3.67	3039.84	3043.51	6.67	2850.65	2857.32	26.33	5551.2	5577.53
		1.5	1.67	2232.66	2234.33	9.50	3644.31	3653.81	20.17	5946.28	5966.45
		2	3.00	2767.03	2770.03	8.50	1418.79	1427.29	15.83	7841.27	7857.10
		2.5	3.00	1283.03	1286.03	13.67	3345.55	3359.22	17.50	6982.52	7000.02
	8	1.2	4.33	2918.39	2922.72	44.17	3 86.79	430.96	121.17	6002.08	6123.25
		1.5	3.83	1051.23	1055.06	1 1 .67	2515.51	2527.18	50.83	6147.72	6198.55
		2	12.33	3845.41	3857.74	51.50	4309.71	4361.21	84.00	6349.53	6433.53
		2.5	15.50	2504.07	2519.57	17.17	4862.58	4879.75	45.83	6600.61	6646.44
	10	1.2	9.17	3579.19	3588.36	25.50	6039.01	6064.51	168.33	7074.55	7242.88
		1.5	14.83	298.89	313.72	58.00	5946.64	6004.64	183.00	6661.49	6844.49
		2	14.83	4352.03	4366.86	51.67	7150.68	7202.35	123.00	8079.32	8202.32
		2.5	11.67	2143.27	2154.94	25.33	6270.04	6295.37	99.00	7973.84	8072.84
1	6	1.2	1.50	3215.97	3217.47	15.83	3695.52	371 1.35	12.17	6414.66	6426.83
		1.5	1.83	3013.4	3015.23	4.50	5124.51	5129.01	16.83	6301.5	6318.33
		2	2.17	2111.1	2113.27	4.50	4288.64	4293.14	25.50	6042.42	6067.92
		2.5	1.67	1390.48	1392.15	5.50	5092.22	5097.72	12.17	1745.63	1757.80
	8	1.2	2.67	2791.07	2793.74	17.00	2167.23	2184.23	48.50	5986.16	6034.66
		1.5	4.50	2551.8	2556.30	12.17	2216.43	2228.60	53.00	6284.06	6337.06
		2	2.67	3868.56	3871.23	15.00	4533.6	4548.60	20.00	5285.74	5305.74
		2.5	9.17	4052.98	4062.15	13.67	5139.19	5152.86	86.33	6349.87	6436.20
	10	1.2	6.50	3989.02	3995.52	35.17	6148.1	6183.27	101.17	9280.7	9381.87
		1.5	6.33	3484	3490.33	39.50	5743.97	5783.47	139.83	6818.86	6958.69
		2	4.83	3693.21	3698.04	13.00	7330.26	7343.26	36.17	8758.87	8795.04
		2.5	3.33	3043.1	3046.43	14.67	6236.85	6251.52	54.00	8695.65	8749.65

Acknowledgements We acknowledge the support given to this project by the National Science Council of Taiwan, R.O.C. under NSC 101-2221-E-033-022.

References

Agnetis A, Mirchandani PB, Pacciarelli D, Pacifici A (2004) Scheduling problems with two competing agents. Oper Res 52:229–242

Agnetis A, Pacciarelli D, Pacifici A (2007) Multi-agent single machine scheduling. Ann Oper Res 150:3–15

Agnetis A, de Pascale G, Pacciarelli D (2009) A Lagrangian approach to single-machine scheduling problems with two competing agents. J Sched 12:401–415

Bank J, Werner F (2001) Heuristic algorithms for unrelated parallel machine scheduling with a common due date, release dates, and linear earliness and tardiness penalties. Math Comput Model 33:363–383

Breit J, Schmidt G, Strusevich VA (2001) Two-machine open shop scheduling with an availability constraint. Oper Res Lett 29:65–77

Baker KR, Smith JC (2003) A multiple-criterion model for machine scheduling. J Sched 6:7–16

Chu CB (1992) A branch-and-bound algorithm to minimize total tardiness with different release dates. Naval Res Logistics 39:265–283

Cheng TCE, Ng CT, Yuan JJ (2006) Multi-agent scheduling on a single machine to minimize total weighted number of tardy jobs. Theoret Comput Sci 362:273–281

Cheng TCE, Ng CT, Yuan JJ (2008) Multi-agent scheduling on a single machine with max-form criteria. Eur J Oper Res 188:603–609

Cheng TCE, Cheng SR, Wu WH, Hsu PH, Wu CC (2011) A two-agent single-machine scheduling problem with truncated sum-of-processing-times-based learning considerations. Comput Ind Eng 60:534–541

Cheng TCE, Wu WH, Cheng SR, Wu CC (2011b) Two-agent scheduling with position-based deteriorating jobs and learning effects. Appl Math Comput 217(21):8804–8824

Cheng SR (2012) A single-machine two-agent scheduling problem by GA approach. Asia-Pacific J Oper Res 29:1250013 (22 pages)

Ding G, Sun S (2011) Single machine family scheduling with two cometing agents to minimize makespan. Asia-Pacific J Oper Res 28:773–785

Gonzalez T, Sahni S (1976) Open shop scheduling to minimize finish time. Ann Oper Res 23:665–679

Leung JYT, Pinedo M, Wan GH (2010) Competitive two-agent scheduling and its applications. Oper Res 58:458–469

Lee WC, Chen SK, Chen CW, Wu CC (2011) A two-machine flowshop problem with two agents. Comput Oper Res 38:98–104

Liu P, Yi N, Zhou X (2011) Two-agent single-machine scheduling problems under increasing linear deterioration. Appl Math Model 35:2290–2296

Lee DC, Hsu PH (2012) Solving a two-agent single-machine scheduling problem considering learning effect. Comput Oper Res 39:1644–1651

Lee WC, Chung YH, Hu MC (2012) Genetic algorithms for a two-agent single-machine problem with release time. Comput Oper Res. doi:10.1016/j.asoc.2012.06.015

Mor B, Mosheiov G (2010) Scheduling problems with two competing agents to minimize minmax and minsum earliness measures. European J Oper Res 206:540–546

Ng CT, Cheng TCE, Yuan JJ (2006) A note on the complexity of the problem of two-agent scheduling on a single machine. J Comb Optim 12:386–393

Sedeno-Noda A, Alcaide D, Gonzalez-Martin C (2006) Network flow approaches to pre-emptive open-shop scheduling problems with time-windows. Eur J Oper Res 174:1501–1518

Sedeno-Noda A, Pablo de DAL, Gonzalez-Martin C (2009) A network flow-based method to solve performance cost and makespan open-shop scheduling problems with time-windows. Eur J Oper Res 196:140–154

Wan GH, Vakati SR, Leung JYT, Pinedo M (2010) Scheduling two agents with controllable processing times. Eur J Oper Res 205:528–539

Wu CC, Huang SK, Lee WC (2011) Two-agent scheduling with learning consideration. Comput Ind Eng 61:1324–1335

Yin Y, Cheng SR, Wu CC (2012a) Scheduling problems with two agents and a linear non-increasing deterioration to minimize earliness penalties. Inf Sci 189:282–292

Yin Y, Wu WH, Cheng SR, Wu CC (2012b) An investigation on a two-agent single-machine scheduling problem with unequal release dates. Comput Oper Res 39:3062–3073

Supply Risk Management via Social Capital Theory and Its Impact on Buyer's Performance Improvement and Innovation

Yugowati Praharsi, Maffie Linda Araos Dioquino and Hui-Ming Wee

Abstract Today's supply chain managers are facing plenty of risks due to uncertainties of inbound supplies. Buyer must learn how to mitigate those unexpected risks. In this study, we explore supply risk management via structural, relational, and cognitive approaches from a buying firm's perspective based on the social capital theory. We also propose that the three forms of social capitals are positively related to buyer–supplier performance improvements. Consequently, the performance improvement of buyer–supplier will positively influence the innovation performance.

Keywords Supply risk · Social capital theory · Buyer performance improvement · Supplier performance improvement · Innovation performance

1 Introduction

Modern supply chain managers have to deal with uncertainties of inbound supplies such as changes of demand volume, on-time delivery, competitive pricing, technologically behind competitors, and quality standards. Buyer must learn how to mitigate those unexpected risks.

Firms establish networking relationships to obtain resources, valuable information, and knowledge to overcome uncertainty in the business environment. Buyers and suppliers that work closely with one another tend to form social

Y. Praharsi (✉) · H.-M. Wee
Department of Industrial and Systems Engineering, Chung Yuan Christian University, Chung Li, Taiwan
e-mail: yugowati.praharsi@staff.uksw.edu

M. L. A. Dioquino
Department of Information Technology, Satya Wacana Christian University, Salatiga, Indonesia

Y.-K. Lin et al. (eds.), *Proceedings of the Institute of Industrial Engineers Asian Conference 2013*, DOI: 10.1007/978-981-4451-98-7_66,
© Springer Science+Business Media Singapore 2013

networks and develop relationships in uncertain situations to manage supply risk (Acquaah 2006). Social capital is an important risk management strategy because social relationships are one of the ways to deal with uncertainty (Woolcock 2001).

Nahapiet and Ghoshal (1998) proposed 3 dimensions of social capital, i.e.: the relational, the cognitive, and the structural dimensions. Research that has examined buyer–supplier relationship effects on buyer–supplier performance has primarily focused on relational capital (Johnston et al. 2004; Cousins et al. 2006). More recently researchers have begun to consider other dimensions of social capital. Krause et al. (2007) considered the structural and cognitive aspects of social capital and their effects on various aspects of buyer and supplier performance. Lawson et al. (2008) considered the effects of relational and structural capital, resulting from relational and structural embeddedness. Embeddedness refers to the degree to which economic activity is constrained by non-economic institutions. The term was created by economic historian Karl Polanyi. Carey et al. (2011) proposed an integrative model examining the relationships among relational, structural, and cognitive dimensions of social capital. Hughes and Perrons (2011) explored the evolution of social capital dimensions. Tsai et al. (2012) postulated that innovation performance is indirectly affected by 3 forms of social capital.

There are few studies that discuss on risk management in the relationship with social capital. Cheng et al. (2012) explored supply risk management via the relational approach in the Chinese business environment (i.e. guanxi). The current study jointly examines three forms of social capital as a form to perceive supply risk. Also, there are few studies explored the dimensions of social capital embeddedness. Lawson et al. (2008) considered relational embeddedness such as supplier integration and supplier closeness and structural embeddedness such as managerial communication and technical exchange. Our study thus considers transaction-specific supplier development in addition to relational embeddedness and shared norms in addition to cognitive embeddedness. Either social capital theory or social capital theory embeddedness has been respectively included in the previous research as form to perceive supply risk, but none of the previous research has tried to simultaneously take both into account for explaining performance outcomes. Hence, this study is motivated by their initiatives.

In answering these gaps, we make four key contributions to the supply chain literature. First, we show that when a buying firm faces supply risk, it tends to form social networks with its key supplier to reduce risk in 3 forms of social capital. Secondly, we extend the application of social capital theory in SC research by explicitly recognizing relational aspect of embeddedness (transaction-specific supplier development) and cognitive aspect of embeddedness (shared norms). Thirdly, we propose that the dimensions of social capital theory (social interaction ties, trust in supplier, shared vision) and the dimensions of social capital embeddedness (managerial communication, technical exchange, supplier integration, transaction-specific supplier development, shared norms) influence buyer and supplier performance improvements. Finally, we propose that the performance improvement of buyer and supplier will positively influence the innovation performance.

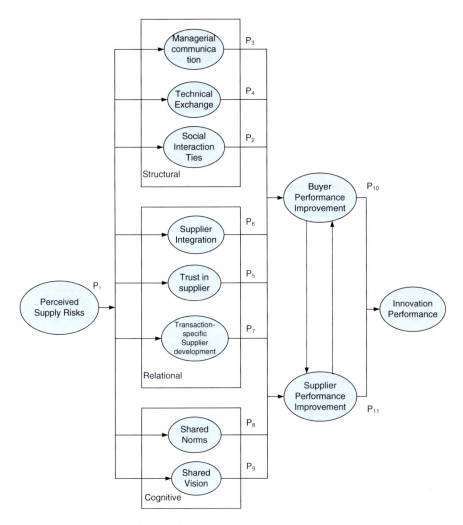

Fig. 1 Proposed research framework

2 Literature Review and Model Formulation

In this study, we explore supply risk management via structural, relational, and cognitive approaches from a buying firm's perspective. Figure 1 shows a major framework in this study.

2.1 Supply Risk Management

Risk exists in supply chains and its occurrence can detrimentally affect the provision of products and services to the final consumer. Supply risk has been defined as the probability of an incident associated with inbound supply from individual supplier failures or the supply market, in which its outcomes result in the inability of the purchasing firm to meet customer demand or cause threats to customer life and safety (Zsidisin and Ellram 2003). Cheng et al. (2012) proposed that when buyers perceive a supply risk situation with their key supplier, they amass social capital to deal with uncertainty and risk. Social capital can be used as a resource to reduce uncertainty and risk in different forms above. This leads to:

Proposition 1 *Perceived supply risk has a positive relationship with social capital development*

2.2 Social Capital Theory

The social capital theory is used to explain the relationship between the buyer and the supplier. Social capital theory (SCT) has become an important perspective for theorizing the nature of connection and cooperation between organizations (Adler and Kwon 2002). As interactions within the linkage between the firms increase, social capital is improved, thereby potentially increasing the flow of benefits. These benefits can include access to knowledge, resources, technologies, markets, and business opportunities (Inkpen and Tsang 2005).

Nahapiet and Ghoshal (1998) proposed 3 dimensions of social capital: (1) the relational dimension (e.g. trust, friendship, respect, reciprocity, identification and obligation), referring to the strength of social relationships developed between buyers and suppliers in the network that is developed through a history of prior interactions among these people and that influences their subsequent behaviors in the network; (2) the cognitive dimension (e.g. shared ambition, goals, vision, culture, and values), referring to rules and expectations of behaviors between buyers and suppliers in a network that define how a community or society will perform; (3) the structural dimension (e.g. strength and number of ties between actors, social interaction ties), referring to structural links or interactions between buyers and suppliers in social relationship.

Structural social capital refers to the configuration of linkages between parties that is whom you know and how you reach them (Nahapiet and Ghoshal 1998). Conceptualizing structural capital as the strength of the social interaction ties exists between buyer–supplier (Tsai and Ghoshal 1998). Social interaction ties facilitate cooperation in dyadic buyer–supplier relationships, and are defined as purposefully designed, specialized processes or events, implemented to coordinate and structurally embed the relationship between buyer and supplier (Cousins et al. 2006; Nahapiet and Ghoshal 1998).

Social interaction ties have also been linked to performance improvements and value creation in buyer–supplier relationships (Cousins et al. 2006), because they provide a forum whereby buyers and suppliers can share information and identify gaps that may exist in current work practices. This leads to:

Proposition 2 *Social interaction ties positively influence buyer and supplier performance improvements*

Structural embeddedness creates the opportunity for future structural capital benefits. There are two factors in the structural embeddedness, i.e.: managerial communication and technical exchange (Lawson et al. 2008). In order to achieve the benefits of collaboration, effective communication between partners' personnel is essential (Cummings 1984). Quality performance was superior between buyers and suppliers when communication occurred among design, engineering, quality control, purchasing and other functions. This leads to:

Proposition 3 *Managerial communication positively influence buyer and supplier performance improvements*

Simple technology exchanges can enhance supplier performance and are independent of whether a buyer and supplier have established familiarity through long-term relationship. Others have argued that technical exchanges help to improve buyer performance (Lamming 1993). This leads to:

Proposition 4 *Technical exchange positively influence buyer and supplier performance improvements*

Relational capital dimension refers to personal relationships that develop through a history of interactions, i.e. the extent to which trust, friendship, respect, obligation, identification, and reciprocity exist between parties (Nahapiet and Ghoshal 1998; Villena et al. 2011). These networks encourage buyer and supplier to act according to one another's expectations and to the commonly held values, beliefs, and norms of reciprocity. The latter maintain mutual trust between supply chain partners, which reduces the negative consequences of uncertainty and risk.

Trust is an essential element of relationships and one of the key aspects of relational social capital. Trust reduces the risk of opportunistic behavior and brings partnering firms closer together to collaborate more richly and to withhold potentially relevant resources (Inkpen and Tsang 2005). Improved relationships and trust can lead to improved buyer and supplier performances (Johnston et al. 2004). This leads to:

Proposition 5 *Trust in supplier positively influence buyer and supplier performance improvements*

Relational embeddedness between buyers and their key suppliers can be defined as the range of activities integrated, the direct investments between both parties, and their relational capital. As a relationship evolves with a key strategic supplier, the relationship becomes embedded as the supplier becomes more integrated and the buyer gives more efforts in supplier development. There are two factors in the relational embeddedness, i.e.: supplier integration (Lawson et al. 2008) and transaction-specific supplier development.

When buying firms are committed to full supplier integration, they are arguably prepared to help their key suppliers through information sharing, technical assistance, training, process control, and direct investment in supplier operations, in return for the benefits of improved performance and joint value creation (Frohlich and Westbrook 2001). More strategic suppliers become fully integrated and the more positive experiences draw them closer to a buying firm, the richer the information exchanged (Koka and Prescott 2002). This leads to:

Proposition 6 *Supplier integration positively influence the buyer and supplier performance improvements*

As buying firms increasingly realize that supplier performance is crucial to their establishing and maintaining competitive advantage, supplier development has been a subject of considerable research in supply chain management (Govindan et al. 2010). Supplier development is defined as any effort of a buying firm to increase the performance and capabilities of the supplier and to meet the buying firm's short and/or long term supply needs (Krause and Ellram 1997).

Transaction-specific supplier development significantly correlated with and had direct effects on the performance of both purchasing and supplier organizations in terms of buyer and supplier performance improvement (Krause 1997). Transaction-specific supplier development leads to closer cooperation between manufacturers and their suppliers. This leads to:

Proposition 7 *Transaction-specific supplier development positively influence the buyer and supplier performance improvements*

The cognitive dimension refers to the resources that provide parties with shared expectations, interpretations, representations and systems of meaning (Nahapiet and Ghoshal 1998). The cognitive capital is also defined as symbolic of shared goals, vision and values among parties in a social system (Tsai and Ghoshal 1998). Congruent goals represent the degree to which parties share a common understanding and approach to the achievement of common tasks and outcomes. The establishment of congruent goals can guide the nature, direction, and magnitude of the efforts of the parties (Jap and Anderson 2003).

Shared culture refers to the degree to which norms of behavior govern relationships. Shared norms can be created by expectation that govern appropriate behavior and affect the nature and degree of cooperation among firms. Strong social norms associated with a closed social network encourage compliance with local rules and customs can reduce the need for formal control (Adler and Kwon 2002).

A lack of norms similarities and compatible goals may trigger conflicts that result in frustration and have negative effects on performance (Inkpen and Tsang 2005). Krause et al. (2007) found support for the positive effect of shared norms and goals on cost reduction. This leads to:

Proposition 8 *Shared norms positively influence the buyer and supplier performance improvements*

Proposition 9 *Shared goals positively influence the buyer and supplier performance improvements*

2.3 Buyer and Supplier Performance Outcomes

Performance improvements sought by buying firms are often only possible when they commit to long-term relationships with key suppliers. Long term commitment means that the buyer regards its suppliers as partners (Krause and Ellram 1997). In the global market where buyer's competitive advantage can be rapidly initiated by competitors, a commitment to innovation is inevitable to sustain competitive advantages and innovation performance. Commitment to innovation refers to employee's duty such as pledge or obligation to work on innovation. By boosting commitment to the innovation, great performance on innovation can be primarily achieved. This leads to:

Proposition 10 *Buyer performance improvement positively influence innovation performance*

Supplier performance improvement is defined as upgrading existing suppliers' performance and capabilities has been recognized as one of the initiations of supplier development to meet the changing competitive requirements (Hahn et al. 1990). Rewards for supplier's improvement are also a stimulating tool that indicates buyer's recognition and provides incentive to supplier for further outstanding achievement. This leads to:

Proposition 11 *Supplier performance improvement positively influence innovation performance*

The performance improvement in essence also comes from promoting buyer and supplier cooperative behavior that increases the creativity of their actions (Nahapiet and Ghoshal 1998). The creativity encourages the accomplishment of innovation performance such as the development of new products and markets. More recently, some studies suggested pursuing innovation performance besides the traditional operational improvement. A set of innovation performance shows that it takes longer to reach the threshold of innovation performance compared with operational benefits (Villena et al. 2011). Therefore, in this study we pose innovation performance as the effect of buyer and supplier performance improvements.

3 Conclusion and Future Research

This study developed an integrative framework of structural, relational, and cognitive approaches to supply risk management grounded in the social capital theory and proposed its impact on buyer performance improvement and innovation. Our study contributed to the literature on a number of fronts.

First, we showed that when a buying firm faces supply risk, it tends to form social networks with its key supplier to reduce risk in three dimensions of social capital. Secondly, we extend the application of social capital theory in supply chain research by explicitly recognizing relational aspect of embeddedness and cognitive aspect of embeddedness. Thirdly, we analyze the three dimensions of social capital and the social capital embeddedness in a single model, which has rarely been done in previous studies. Finally, we use a complete set of performance measures to develop a more complete view of how social capital facilitates a value creation. For future works, a more comprehensive study to develop performance criteria and indicators can be done.

References

Acquaah M (2006) The impact of managerial networking relationships on organizational performance in Sub-Saharan Africa: evidence from ghana. Organiz Manage J 3:115–138

Adler PS, Kwon SW (2002) Social capital: prospects for a new concept. Acad Manage Rev 27:17–40

Carey S, Lawson B, Krause DR (2011) Social capital configuration, legal bonds and performance in buyer-supplier relationships. J Oper Manage 29:277–288

Cheng TCE, Yip FK, Yeung ACL (2012) Supply risk management via guanxi in the Chinese business context: The buyer's perspective. Int J Prod Econ 139:3–13

Cousins PD, Handfield RB, Lawson B, Petersen KJ (2006) Creating supply chain relational capital: the impact of formal and informal socialization processes. J Oper Manage 24:851–863

Cummings T (1984) Transorganizational development. Res Organiz Behav 6:367–422

Frohlich MT, Westbrook R (2001) Arcs of integration: an international study of supply chain strategies. J Oper Manage 19:185–200

Govindan K, Kannan D, Haq AN (2010) Analyzing supplier development criteria for an automobile industry. Ind Manage Data Syst 110:43–62

Hahn CK, Watts CA, Kim KY (1990) The supplier development program: a conceptual model. Int J Mater Manage 26:2–7

Hughes M, Perrons RK (2011) Shaping and re-shaping social capital in buyer-supplier relationships. J Bus Res 64:164–171

Inkpen AC, Tsang EWK (2005) Social capital, networks and knowledge transfer. Acad Manag Rev 30:146–165

Jap SD, Anderson E (2003) Safeguarding interorganizational performance and continuity under ex post opportunism. Manage Sci 49:1684–1701

Johnston DA, Mccutcheon DM, Stuart FI, Kerwood H (2004) Effects of supplier trust on performance of cooperative supplier relationships. J Oper Manage 22:23–28

Koka BR, Prescott JE (2002) Strategic alliances as social capital: a multidimensional view. Strat Manage J 23:795–816

Krause DR (1997) Supplier development: current practices and outcomes. Int J Purchasing and Mater Manage 33:2–19

Krause DR, Ellram LM (1997) Critical elements of supplier development. Eur J Purchasing and Supply Manage 3:21–31

Krause DR, Handfield RB, Tyler BB (2007) The relationships between supplier development, commitment, social capital accumulation and performance improvement. J Oper Manage 25:528–545

Lamming RC (1993) Beyond partnership: strategies for innovation and lean supply. Prentice-Hall, London

Lawson B, Tyler BB, Cousins PD (2008) Antecedents and consequences of social capital on buyer performance improvement. J Oper Manage 26:446–460

Nahapiet J, Ghoshal S (1998) Social capital, intellectual capital, and the organizational advantage. Acad Manage Rev 23:242–266

Tsai W, Ghoshal S (1998) Social capital and value creation: the role of intra firm networks. Acad Manage J 41:464–476

Tsai YH, Joe SW, Ding CG, Lin CP (2012) Modeling technological innovation performance and its determinants: an aspect of buyer-seller social capital. Technological forecasting and social change. http://dx.doi.org/10.1016/j.techfore.2012.10.028. Accessed 24 March 2013

Villena VH, Revilla E, Choi TY (2011) The dark side of buyer-supplier relationships: a social capital perspective. J Oper Manage 29:561–576

Woolcock M (2001) The place of social capital in understanding social and economic outcomes. Canadian J Policy Res 2:11–17

Zsidisin GA, Ellram LM (2003) Managerial perceptions of supply risk. J Supply Chain Manage 39:14–25

Variable Time Windows-Based Three-Phase Combined Algorithm for On-Line Batch Processing Machine Scheduling with Limited Waiting Time Constraints

Dongwei Yang, Wenyou Jia, Zhibin Jiang and You Li

Abstract In this paper, a variable time windows-based three-phase combined algorithm is proposed to address the scheduling problem of on-line batch processing machine for minimizing total tardiness with limited waiting time constraints and dynamic arrivals in the semiconductor wafer fabrication system (SWFS). This problem is known to be NP-hard. In the first phase, the on-line information of scheduling parameters is preserved and sent. In the second phase, the computational results of reforming and sequencing are obtained. In the third phase, the super-hot batch is loaded. With the rolling horizon control strategy, the three-phase combined algorithm can update solution continually. Each interval of rolling horizon is a time window. The length of each time window is variable. The experiments are implemented on the real-time scheduling simulation platform of SWFS and ILOG CPLEX to demonstrate the effectiveness of our proposed algorithm.

Keywords Three-phase combined algorithm · Batch processing machine · Variable time windows

---

D. Yang
Department of Economics & Management, Shanghai University of Electric Power, Shanghai, China

W. Jia · Z. Jiang (✉) · Y. Li
Department of Industrial Engineering and Logistics Engineering, Shanghai Jiao Tong University, Shanghai, China
e-mail: zbjiang@sjtu.edu.cn

W. Jia
e-mail: jiawy@sjtu.edu.cn

Y. Li
e-mail: liyoustar@sjtu.edu.cn

W. Jia
School of Mechanical and Automotive Engineering, Anhui Polytechnic University, Wuhu, China

Y.-K. Lin et al. (eds.), *Proceedings of the Institute of Industrial Engineers Asian Conference 2013*, DOI: 10.1007/978-981-4451-98-7_67,
© Springer Science+Business Media Singapore 2013

1 Introduction

Batch processing machines (BPM) are frequently encountered in the semiconductor wafer fabrication system (SWFS) such as furnace operations. For an overview of batching and scheduling problems, we invite the reader to refer to Potts and Kovalyov (2000). Johnson (1954) first proposes polynomial time algorithms the items minimizing the total elapsed time. Ahmadi et al. (1992) consider the batching and scheduling problems in a flow-shop scenario with a batch-processing machine and a discrete-processing machine. Lin and Cheng (2001) consider scheduling a set of jobs in two batch machines. Su (2003) formulates a hybrid two-stage model comprising of a BPM in the first stage and a discrete machine in the next stage with limited waiting time constraints. In order to consider on-line scheduling, the rule decomposition-based is a good strategy. Rolling horizon control strategy, a kind of time-sequence-based decomposition method, is developed for dynamic scheduling problems (Ovacikt and Uzsoy 1994, 1995). According to rolling horizon control strategy, a scheduling problem can be divided into several sub-problems along time-axis. Each sub-problem corresponds to a time window of the whole schedule. This approach is extended and applied to schedule in SWFS (Klemmt et al. 2011).

In this paper, we extend time-sequence-based decomposition method to address the problem of scheduling on-line BPM for minimizing total tardiness with limited waiting time constraints and dynamic arrivals in SWFS. Du and Leung (1990) prove that a special case with single machines is NP-hard. Definitely, our problem is also NP-hard. According to time-sequence-based decomposition rule, the large BPM is divided into smaller ones. Each smaller one includes three-phase.

The remaining sections of this paper are organized as follows. In Sect. 2, we present the problem and notations. The methodology of proposed variable time windows-based three-phase combined algorithm is defined in Sect. 3. The computational experiments are designed and conducted in Sect. 4. Section 5 concludes the paper with future research directions.

2 Problem Definition and Notations

In this research, we model two BPMs with limited waiting time constraints and dynamic arrivals. As illustrated by Fig. 1, the definitions are stated as follows.

This model includes many machine groups (MG) such as MG^1, MG^2, and MG^3. The process flow is $IN \rightarrow MG^1 \rightarrow MG^2 \rightarrow MG^1 \rightarrow MG^3 \rightarrow OUT$, where re-entrant flow is allowed. The information of jobs in Buffer 3 is obtained in the first phase. Once any MG is started, it can't be interrupted. Assume MG^3 group has the same processing time, and is not starved.

For convenience, we use the following notations in the rest of the paper.

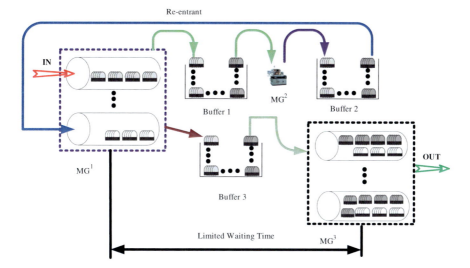

Fig. 1 The process flow of the manufacturing system

Let I be number of different products, J number of different batches coming from MG^1 group in Buffer 3, N number of reforming batches waiting to be scheduled on MG^3 group and S capacity of MG^3 group. t denotes current time of simulation clock. P_{MG}^3 presents processing time of MG^3 group. For batch j, q_j, s_j, r_j, d_j, ZS_j, DS_j, PR_j, and P_{kj} stand for limited waiting time between MG^1 group and MG^3 group, size, arriving time, due date time, total number of processing steps, number of processing step on MG^3 group, pure remaining processing time after MG^3 group, and processing time of the kth processing step, respectively. For the reforming batch n, dr_n, Cr_n and PRr_n denote due date time, completion time, and pure remaining processing time after MG^3 group, respectively. M is an extremely huge positive integer.

3 Methodology

In this section, a variable time windows-based three-phase combined algorithm is proposed. We begin with introducing the overall algorithm structure. Then we analyze the slack-based mixed integer linear programming (MILP) model.

Based on decomposition rule and rolling horizon control strategy, the problem can be decomposed into many time windows. The length of each time window is variable, which is equivalent to the interval of between two adjacent time points of the super-hot batch to be loaded. At each time window, the problem includes three sub-problems (i.e., three phases): to get information of scheduling parameters, to reform and sequence batches, and to load super-hot reforming batch and update the state of manufacturing system. The first and third sub-problems are resolved by real-time scheduling simulation (ReS^2) platform of SWFS (Liu et al. 2005). The second sub-problem is resolved by slack-based MILP model which is executed by ILOG CPLEX and the program developed by ourselves. The proposed algorithm is illustrated in Fig. 2 and implemented as follows.

step 1 The last idle and available BPM (i.e., MG^3 group) is loaded. Initialize the time window and start to run the first phase strategy. Read and preserve the on-line information of batches in Buffer 3. Check the on-line status of MG^3 group.
step 2 When a condition that one MG^3 group is idle and available is met, the event of loading batch is registered.
step 3 Because no reformed and sequenced batch can be loaded, the idle and available MG^3 group must be a status of waiting.
step 4 Compute the total number (N) of reformed batches which are waiting to be scheduled on MG^3 group. If N > 1, then go to *Step 5*; else go to *Step 9*.

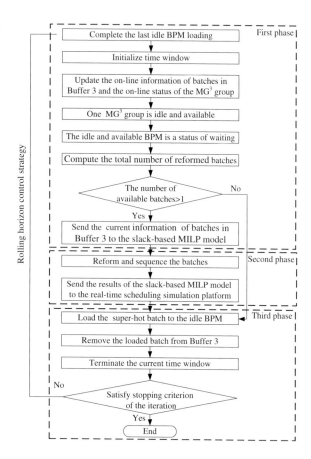

Fig. 2 The overall flow chart of three-phase combined algorithm

step 5 Output the on-line information of batches in Buffer 3 and set up a database. The database is the inputs of the slack-based MILP model.

step 6 The event of reforming and sequencing batch is registered. That is, start to run the ILOG CPLEX and the program developed by ourselves.

step 7 Generate an array of the optimal batches sequencing and output database of this position array. This database is returned to ReS^2 platform.

step 8 The event of loading the super-batch is registered.

step 9 Start to load the super-batch to idle and available MG^3 group.

step 10 Remove the loaded batch from Buffer 3. Update the status of MG^3 group. Terminate the current time window.

step 11 If the stopping criterion of the iteration is reached, the procedure stops. Otherwise, it returns to *Step 1*. Where, the stopping criterion of the iteration is controlled by the total running time of the ReS^2 platform of SWFS.

For the event of reforming and sequencing batch, we consider the optimal sequence positions indices that can be found efficiently by slack-based MILP model.

The reforming batches binary parameters and optimal sequencing positions indices are the decision variables. The processing time requirements need to be satisfied, which are the constraints. The total weighted tardiness is to be minimized, which is the objective function. We define the optimal sequencing positions indices as the one-dimensional array which are noted the [Position (1), Position (2), ..., Position (N)]. Such problems are conventionally formulated as follows.

Minimize

$$\sum T = \sum_{n=1}^{N} \max[(Cr_n - dr_n), 0] \tag{1}$$

Subject to

$$Cr_n = t + \text{Position}(n) \cdot P_{MG^3} + PRr_n \tag{2}$$

$$\text{Position}(n) \in \{1, 2, \cdots, N\}; \forall n \in \{1, 2, \cdots, N\} \tag{3}$$

$$\text{If } n \neq l \text{ then Position}(n) \neq \text{Position}(l); \ \forall n \text{ and } l \in \{1, 2, \cdots, N\} \tag{4}$$

$$x_{jn} = \begin{cases} 1 & \text{If the batch } j \text{ in Buffer3 is assigned the reforming batch } n \\ 0 & \text{Otherwise} \end{cases}$$

$$\forall n \in \{1, 2, \cdots, N\}; \ \forall j \in \{1, 2, \cdots, J\} \tag{5}$$

$$\sum_{n=1}^{N} x_{jn} = 1; \forall j \in \{1, 2, \cdots, J\} \tag{6}$$

$$PRr_n = \sum_{j=1}^{J} x_{jn} \cdot PR_j; \forall n \in \{1, 2, \cdots, N\} \tag{7}$$

$$PR_j = \sum_{k=DS_j+1}^{ZS_j} P_{kj}; \forall j \in \{1, 2, \cdots, J\} \tag{8}$$

$$dr_n = \sum_{j=1}^{J} x_{jn} \cdot d_j; \forall n \in \{1, 2, \cdots, N\} \tag{9}$$

$$\sum_{j=1}^{J} x_{jn} \cdot s_j \leq S; \forall n \in \{1, 2, \cdots, N\} \tag{10}$$

$$t - r_j + \text{Position}(n) \cdot P_{GM^3} - M(1 - x_{jn}) \leq q_j; \forall n \in \{1, 2, \cdots, N\}; \\ \forall j \in \{1, 2, \cdots, J\} \tag{11}$$

$$N = \left\lceil \left(\sum_{j=1}^{J} s_j \right) / S \right\rceil \tag{12}$$

Objective function (1) minimizes total weighted tardiness for each time window. Constraint (2) is used to compute the completion time of the nth reforming batch. Constraints (3) and (4) ensure that each reforming batch is sequenced to a position and each position has only one reforming batch to be assigned. Constraint (5) imposes the reforming batches binary restrictions. Constraint (6) ensures that each batch in Buffer 3 is assigned to exactly one reforming batch. Constraint (7) is used to compute the pure remaining processing time of the nth reforming batch after MG^3 group. Constraint (8) is used to compute the pure remaining processing time of the batch j after MG^3 group. Constraint (9) is used to compute the due date time of the nth reforming batch. Constraint (10) ensures that the total size of all batches assigned to a reforming batch does not exceed MG^3 group capacity. Constraint (11) ensures that the total waiting time of the jth batch in Buffer 3 does not exceed the limited waiting time. Constraint (12) is used to compute the total number of reformed batches waiting to be scheduled on MG^3 group. Constraints (1–12) conduct slack-based mathematical linear model, we call it slack-based MILP model which can be fit for being solved by ILOG CPLEX .

4 Simulation Experiments

In this section, a computational experiment is invested to evaluate the proposed three-phase combined algorithm, so a virtual semiconductor manufacturing system (VSMS) is modeled and set to run. As shown in Fig. 3, there are 8 different

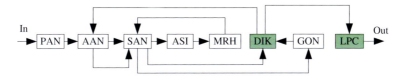

Fig. 3 The typical process flow of VSMS with batch processors

Fig. 4 The processing time of the different processing steps

equipment areas. Among the overflow of VSMS, DIK (i.e., GM^1 group) and LPC (i.e., GM^3 group) belong to BPMs with limited waiting time constraints and dynamic arrivals which are our object of research.

Four different types of wafer products are assumed to be processed simultaneously (I = 4). We assume that each product has the same processing steps, i.e., $ZS_j = DS_j = 12$. The processing time of each processing step is shown in Fig. 4, where $PT_{i, DIK}2$ represents processing time of product i which is processed in DIK group for the second time (i.e., re-entrant). The values of $PT_{i, DIK}2$ (i = 1, 2, 3, 4) are 180, 263, 410, and 300. The maximum capacity of LPC is 12 jobs (S = 12). For each batch, let q_j be 11360. Because LPC is the last processing step, $PR_j = 0$.

In order to evaluate efficiency of the proposed three-phase combined algorithm, genetic algorithm (GA) is used as reference heuristics to assess the performance of slack-based MILP method. The computational experiments are programmed and implemented in Visual Basic. NET 2008 and ILOG CPLEX 12.2 on a Intel Core 2 Duo 2.0 GHz with 2 GB DDR-2 RAM. We consider four different performance measures and two cases. The four different performance measures include CPU time, improvement rate, objective function (optimal) value and deviation.

Whenever one machine in LPC is idle, the information of scheduling parameters of LPC is sent to the slack-based MILP model. Two cases are Case 1 and Case 2. With respect to Case 1 the total number of batches waiting to be scheduled for LPC is 8 (J = 8). The detailed database is that s_j = [3, 4, 5, 6, 6, 7, 8, 9], r_j = [100, 150, 200, 300, 180, 263, 410, 300] and d_j = [1,211, 1,211, 1,227, 1,227, 1,236, 1,236, 1,245, 1,245]. While Case 2, J = 16, the detailed database is that s_j = [3, 4, 5, 6, 6, 7, 8, 9, 3, 4, 5, 6, 6, 7, 8, 9], r_j = [100, 150, 200, 300, 180, 263, 410, 300, 100, 150, 200, 300, 180, 263, 410, 300] and d_j = [1,211, 1,211, 1,227, 1,227, 1,236, 1,236, 1,245, 1,245, 1,211, 1,211, 1,227, 1,227, 1,236, 1,236, 1,245, 1,245].

For the main parameters of GA, let population size, crossover probability, mutation probability, and number of generations be 80, 0.8, 0.1 and 120, respectively.

Table 1 The comparisons between slack-based MILP method and GA method

Case	CPU time (s)		Percentage improvement (%)	$\min(\sum T)$		Deviation (%)
	Slack-based MILP	GA		Slack-based MILP	GA	
Case 1	0.06	55.83	+99.87	1,500	1,542	+2.72
Case 2	77.58	77.46	−0.15	15,762	15,870	+0.68

The experimental results of slack-based MILP method and GA are listed in Table 1. In Case 1, the objective function value obtained from slack-based MILP is better than GA by 2.72 %, the CPU time of slack-based MILP is better than GA by 99.87 %. In Case 2, although the objective function value obtained from slack-based MILP is better than GA by 0.68 %, the CPU time of slack-based MILP is longer than GA by 0.15 %. From Table 1, slack-based MILP method performs quite well, especially for smaller number of scheduling batches. Table 1 also shows that our proposed slack-based MILP method can provide the tradeoff between CPU time and solution quality.

5 Conclusions

In this paper, we have discussed a scheduling problem of on-line BPM for minimizing total tardiness with limited waiting time constraints and dynamic arrivals in SWFS. To resolve this NP-hard problem, we have proposed a variable time windows-based three-phase combined algorithm. According to the VSMS experiment, we have evaluated the performance of the proposed algorithm. The results have demonstrated the effectiveness. Our ongoing work is on where real SWFS industry BPM scheduling can be considered.

Acknowledgments This work is supported by National Science and Technology Major Special Projects (2011ZX02501-005) and Research Grant from National Natural Science Foundation of China (No.50475027). We also acknowledge the helpful comments and suggestions of the reviewers, which have improved the presentation.

References

Ahmadi JH, Ahmadi RH, Dasu S, Tang CS (1992) Batching and scheduling jobs on batch and discrete processors. Oper Res, 40 (4):750–763

Du J, Leung JYT (1990) Minimizing total tardiness on one machine is np-hard. Math Oper Res, 15 (3):483–495

Johnson SM (1954) Optimal two-and three-stage production schedules with setup times included. Naval Res Logistics Q 1(1):61–68

Klemmt A, Weigert G, Werner S (2011) Optimisation approaches for batch scheduling in semiconductor manufacturing. Eur J Ind Eng 5(3):338–359

Lin BMT, Cheng T (2001) Batch scheduling in the no-wait two-machine flow shop to minimize the make span. Comput Oper Res 28(7):613–624

Liu H, Fung RYK, Jiang Z (2005) Modelling of semiconductor wafer fabrication systems by extended object-oriented petri nets. Int J Prod Res 43(3):471–495

Ovacik IM, Uzsoy R (1995) Rolling horizon procedures for dynamic parallel machine scheduling with sequence-dependent setup times. Int J Prod Res 33(11):3173–3192

Ovacikt I, Uzsoy R (1994) Rolling horizon algorithms for a single-machine dynamic scheduling problem with sequence-dependent setup times. Int J Prod Res 32(6):1243–1263

Potts CN, Kovalyov MY (2000) Scheduling with batching: a review. Eur J Oper Res, 120 (2):228–249

Su LH (2003) A hybrid two-stage flow shop with limited waiting time constraints. Comput Ind Eng 44(3):409–424

Optimal Organic Rankine Cycle Installation Planning for Factory Waste Heat Recovery

Yu-Lin Chen and Chun-Wei Lin

Abstract As Taiwan's industry developed rapidly, the energy demand also rises simultaneously. In the production process, there's a lot of energy consumed in the process. Formally, the energy used in generating the heat in the production process. In the total energy consumption, 40 % of the heat was used in process heat, mechanical work, chemical energy and electricity. The remaining 50 % were released into the environment. It will cause energy waste and environment pollution. There are many ways for recovering the waste heat in factory. Organic Rankine Cycle (ORC) system can produce electricity and reduce energy costs by recovering the waste of low temperature heat in the factory. In addition, ORC is the technology with the highest power generating efficiency in low-temperature heat recycling. However, most of factories are still hesitated because of the implementation cost of ORC system, even they generate a lot of waste heat. Therefore, this study constructed a nonlinear mathematical model of waste heat recovery equipment configuration to maximize profits, and generated the most desirable model and number of ORC system installed by using the particle swarm optimization method.

Keywords Waste heat · Low temperature · Organic rankine cycle · Particle swarm optimization

---

Y.-L. Chen (✉) · C.-W. Lin
Industrial Engineering and Management, National Yunlin University of Science
& Technology, 123 University Road, Sect. 3, Douliou, Yunlin 64002, Taiwan,
Republic of China
e-mail: m10021048@yuntech.edu.tw

C.-W. Lin
e-mail: lincwr@yuntech.edu.tw

Y.-K. Lin et al. (eds.), *Proceedings of the Institute of Industrial Engineers Asian Conference 2013*, DOI: 10.1007/978-981-4451-98-7_68,
© Springer Science+Business Media Singapore 2013

1 Introduction

It is required large amounts of energy and fuel consumption when factories in the process of manufacturing. According to Taiwan's department of energy distribution, the industrial department accounted 38.56 % with total energy, and 50 % of carbon dioxide emissions with total national emissions (Bureau of Energy 2011a). However, the energy is used in the form of heat energy, thermal energy is accounted for the total energy consumption by more than 90 %, only 40 % of heat energy is converted into process heat, mechanical work, chemical energy, and electricity, other 50 % heat released as waste heat form to environment, it is causing of energy waste and environmental pollution (Kuo et al. 2012). According to statistics, the temperature of most Industry waste heat discharge is between 130 °C and 650 °C, and about 2 million (KLOE) of waste heat below 250 °C, accounting for 62.72 % of the total waste heat, belongs to the low temperature of waste heat discharge. Then, the temperature between 251 °C and 650 °C (middle temperature waste heat) and above 651 °C (high temperature waste heat) account for 19.08 and 18.20 % of total waste heat (Bureau of Energy 2011b).

Today for low-temperature waste heat recovery power generation system, the organic rankine cycle (ORC) system is the highest technology with power generation efficiency, it is widely used in industrial waste heat, geothermal hot springs, biomass heat and waste heat power generation purposes (Kuo and Luo 2012). In addition, ORC system almost has no fuel consumption during it running period, it can also reduce carbon dioxide and sulfur dioxide emissions and other pollutants (Wei et al. 2007). ORC system use organic working fluid as a medium of pick up thermal power generation (Lee et al. 2011). It can choose suitable low boiling point substances as the working fluid (such as refrigerant, ammonia, etc.) by different heat source temperature range, and making low temperature heat energy converted into electricity or brake horsepower output. For the medium and the low temperature heat source, converted to electricity, is more efficient than water (Drescher and Brüggemann 2007). The loop circuit of ORC system, the main components include: pump, evaporator, expander, generator and the condenser. Working fluid circulating in the circuit model is started form (1) booster pump, (2) evaporated into vapor, (3) promote the expansion machine and generators, (4) then condensing to liquid, to complete the cycle (Durma et al. 2012; Somayaji et al. 2007; Chang et al. 2011; Kuo and Luo 2012).

However, in Taiwan the environment of tariffs is generally low, invest of ORC generator have high risk of too long investment recovery period and low return on investment, so the industry still maintain hesitated, even if the factory has lots of waste heat but still lack of invest desire, only few factories are willing to invest in ORC generator to conduct waste heat recovery power generation (Lee et al. 2011).

Therefore, the goal of this study is using ORC power generation units to construct a nonlinear mathematical model of waste heat recovery equipment configuration to maximize profits, will let the low temperature waste heat into

Optimal Organic Rankine Cycle Installation Planning 571

electric power via ORC system, and using particle swarm algorithm to find out the suitable generator set number and model, that will be an effectively assist of factory in their waste heat recovery.

2 Model Construction

This study focus on the factory's low temperature waste heat with recycling value, and through ORC system for it recovery. With the condition of maximum output power, only considering heat source temperature and discharge temperature, heat source mass flow rate and specific heat, the working fluid evaporation temperature, and the condensing temperature, than obtained by waste heat of heat capacity (Fenn 2003). It is include the loss of power transmission and the generator efficiency, therefore, without discussing each element thermal efficiency, achieve the overall optimization cycle thermal efficiency (Lee 2009; Lee et al. 2011). In the cost analysis of investment of ORC generator, the most important is the purchased equipment costs of unit itself. Besides, have to consider generating set surrounding the installation costs (Lukawski 2009), and taking into consideration of the time value of money, use levelization concept. The times change might also change the money value, transform to annuity. In addition, the operation and maintenance costs of waste heat recovery equipment operate will increase year by year, must through constant-escalation-levelization-factor to the cost of the time value of money considerations (Lukawski 2009; Meng and Jacobi 2011).

In this study, the objective function is maximize the annual net profit, include the annual sell electricity income, annual investment cost, annual operation and maintenance costs, government subsidies and annual salvage value. At last, it can figure out the suitable number of units from the waste heat recovery equipment configuration optimization. Then can be learned that all of the waste heat source in factory should be parallel configuration of those generators model and number.

2.1 Parameters

I	Number of heat source in factory
J	Number of generators model of ORC
HT_i	The temperature of the heat source i (°C)
LT_i	The discharge temperature of the factory requirements (°C)
M_i	Total mass of the heat source i (kg/h)
S_i	Specific heat of the heat source i [cal/g(°C)]
ET_j	The evaporation temperature of the working fluid in the model j of ORC (kelvin, K)
CT_j	The condensing temperature of the working fluid in the model j of ORC (kelvin, K)
RH	The running hours of the model j of ORC in a year (h)
EP	Electricity price per kWh

EGR	Electricity growth rate
PEC_j	Purchased equipment cost of the model j of ORC
SV_j	Salvage value of the model j of ORC
A_j	Each generator government grants of the model j of ORC
CI_j	Cost of installation of the model j of ORC
CP_j	Cost of piping of the model j of ORC
CEE_j	Cost of electrical equipment of the model j of ORC
CC_j	Cost of civil and structural work of the model j of ORC
CS_j	Cold source of the model j of ORC
DPC_j	Design and planning costs of the model j of ORC
MC_{ij}	Maintenance costs of the model j of ORC installation at the heat source i
PRC_{ij}	Plant repair costs of the model j of ORC installation at the heat source i
$GRMC$	Growth rate of the maintenance costs
B	Factory budget
FPB	Payback required by the factory
$MARR$	Minimum acceptable rate of return
n	Assessment of useful life

2.2 Variables

x_{ij}	The number of the model j of ORC installation at the heat source i
m_{ij}	The mass diverted to the model j of ORC from the heat source i (kg/h)

$$y_{ij} = \begin{cases} 1, & \text{the heat source } i \text{ has installed the model } j \text{ of ORC ;} \\ 0, & \text{the heat source } i \text{ has not installed the model } j \text{ of ORC} \end{cases}$$

$$Y_j = \begin{cases} 1, & \text{the factory has installed the model } j \text{ of ORC} \left(\sum_{i=1}^{I} y_{ij} \geq 1 \right); \\ 0, & \text{the factory has not installed the model } j \text{ of ORC} \left(\sum_{i=1}^{I} y_{ij} = 0 \right) \end{cases}$$

2.3 Mathematical Model

Objective function for waste heat recovery equipment configuration profits (annual sell electricity income-annual investment cost-annual operation and maintenance costs + government subsidies + annual salvage value) maximize is:

Optimal Organic Rankine Cycle Installation Planning

$$P = \left\{ \begin{array}{l} \left\{ \sum_{i=1}^{I} \sum_{j=1}^{J} \left[m_{ij} \cdot S_i \cdot (HT_i - LT_i) \div 860 \right] \times \left(1 - \sqrt{CT_j \div ET_j} \right) \times RH \right\} \\ \times EP \times \left[\dfrac{(1+EGR)^n - 1}{(1+EGR) - 1} \right] \times \left[\dfrac{MARR \times (1+MARR)^n}{(1+MARR)^n - 1} \right] \end{array} \right\}$$

$$- \left\{ \left\{ \sum_{j=1}^{J} \left[\sum_{i=1}^{I} \left(\begin{array}{c} PEC_j + CI_j + CP_j \\ +CEE_j + CC_j + CS_j \\ +DPC_j \times Y_j \end{array} \right) \times x_{ij} \right] \right\} \times \left[\dfrac{MARR \times (1+MARR)^n}{(1+MARR)^n - 1} \right] \right\}$$

$$- \left\{ \begin{array}{l} \left[\sum_{i=1}^{I} \sum_{j=1}^{J} (MC_{ij} + PRC_{ij}) \times x_{ij} \right] \times \left(\dfrac{1+GRMC}{1+MARR} \right) \\ \times \left\{ \left[1 - \left(\dfrac{1+GRMC}{1+MARR} \right)^n \right] \div \left[1 - \left(\dfrac{1+GRMC}{1+MARR} \right) \right] \right\} \times \left[\dfrac{MARR \times (1+MARR)^n}{(1+MARR)^n - 1} \right] \end{array} \right\}$$

$$+ \left\{ \sum_{i=1}^{I} \sum_{j=1}^{J} A_j ij \div n \right\} + \left\{ \left[\sum_{i=1}^{I} \sum_{j=1}^{J} (SV_j \times x_{ij}) \right] \times \{ MARR \div [(1+MARR)^n - 1] \} \right\}. \tag{1}$$

s.t.

$$ET_j \times y_{ij} \leq HT_i \quad \forall i, \forall j. \tag{2}$$

$$y_{ij} \leq x_{ij} \leq M \cdot y_{ij} \quad \forall i, \forall j. \tag{3}$$

$$Y_j \leq \sum_{i=1}^{I} y_{ij} \leq M \cdot Y_j \quad \forall j. \tag{4}$$

$$\sum_{j=1}^{J} m_{ij} \leq M_i \quad \forall i. \tag{5}$$

$$\left\{ \left[m_{ij} \cdot S_i \cdot (HT_i - LT_i) \div 860 \right] \times \left(1 - \sqrt{CT_j \div ET_j} \right) \right\} \div x_{ij} \leq GC_j \quad \forall i, \forall j. \tag{6}$$

$$\left\{ \sum_{j=1}^{J} \left[\sum_{i=1}^{I} \left(\begin{array}{c} PEC_j + CI_j + CP_j \\ +CEE_j + CC_j + CS_j \end{array} \right) \times x_{ij} + DPC_j \times Y_j \right] \right\} \div P \leq FPB \quad \forall i, \forall j. \tag{7}$$

$$\sum_{j=1}^{J} \left[\sum_{i=1}^{I} \left(\begin{array}{c} PEC_j + CI_j + CP_j \\ +CEE_j + CC_j + CS_j \end{array} \right) \times x_{ij} + DPC_j \times Y_j \right] \leq B \quad \forall i, \forall j. \tag{8}$$

$$P \div \left\{ \sum_{j=1}^{J} \left[\sum_{i=1}^{I} \left(\begin{array}{c} PEC_j + CI_j + CP_j \\ +CEE_j + CC_j + CS_j \end{array} \right) \times x_{ij} + DPC_j \times Y_j \right] \right\} \geq MARR \quad \forall i, \forall j. \tag{9}$$

$$x_{ij} \geq 0, x_{ij} \in Z, m_{ij} \geq 0, y_{ij} \geq 0, y_{ij} \in \{0, 1\}, Y_j \geq 0 \quad \forall i, \forall j. \tag{10}$$

Equation (1) is the objective function of this study. Equation (2) as the heat source i with model j of ORC, the evaporation temperature of ORC can't higher than the temperature of that source. Equation (3) for the generator heat source have no install, the installation number is 0. If factory installed the j generator, then Eq. (4) represent at least one heat source will installed the generators, which must consider the generator design and planning costs. M of Eqs. (3) and (4) as an infinite value. Equation (5) as the sum of mass diverted to the model j of ORC from heat source i, those mass can't more than the total mass of that source. Equation (6) as available generating capacity by the heat source i installed the model j of ORC, it can't more than the generating capacity of that model. Equation (7) is restriction of payback period. Equation (8) is restriction of budget. Equation (9) is restriction of return on investment. Equation (10) tells all variables must be greater than or equal to zero.

3 Validation

This study discusses the case for a steel works within the two color coating line, in the process of production will produce waste heat. The heat source data is below Table 1, the conditions of ORC generator power generation, and the relative of investment cost, according to the literature and the case data which company provided, the simulation parameters set as Table 2 (Lukawski 2009; Lee et al. 2011; Chang et al. 2011; Kuo et al. 2012). This case company for investment restrictions: $MARR$ is 10 (%), FPB is 5(year), B is 50,000,000 (NT). And assuming RH is 8,000 (h/year), EP is 3.01 (NT/kWh), EGR is 2 (%), n is 20 (year), $GRMC$ is 1 (%).

This study use Particle Swarm Optimization of simulate the flock foraging, through global search and particle search, then find the best solution after iterative

Table 1 Heat source data

Heat source	M_i (kg/h)	S_i (cal/g °C)	HT_i (°C)	LT_i (°C)
$i = 1$	23,760	0.31	320	170
$i = 2$	23,280	0.31	332	170

Table 2 Parameters of the ORC

Model of ORC	GC_j (kW)	ET_j (K)	CT_j (K)	PEC_j (NT)	SV_j (NT)	A_j (NT)	CI_j (NT)
$j = 1$	50	381.15	313.15	4,500,000	450,000	2,000,000	270,000
$j = 2$	125	394.15	294.15	11,250,000	1,125,000	5,000,000	675,000

Model of ORC	CP_j (NT)	CEE_j (NT)	CC_j (NT)	CS_j (NT)	DPC_j (NT)	MC_j (NT)	PRC_j (NT)
$j = 1$	405,000	180,000	135,000	225,000	225,000	225,000	225,000
$j = 2$	1,012,500	450,000	337,500	562,500	562,500	562,500	562,500

Optimal Organic Rankine Cycle Installation Planning

Table 3 PSO parameter table

Parameters	Generations	Particle number	Maximum weight (w_{max})	Minimum weight (w_{min})	Studying factor 1 c_1	Studying factor 2 C_2
Value	500	100	0.9	0.4	2	2

Table 4 Results of the validation

Heat source i with model j of ORC	$i=1, j=1$	$i=1, j=2$	$i=2, j=1$	$i=2, j=2$
Number of installation x_{ij}	1	1	1	1
Mass m_{ij} (kg/h)	6,715	16,920	7,494	15,662
Annual net profit (NT/year)	14,845,764			

update (Xia and Wu 2005). This research use linear decreasing weighting method (Kennedy and Eberhart 1995), set the related parameters as Table 3. Use MAT-LAB 7.10 to write a program, get the results in Table 4.

4 Conclusions

This study constructed a nonlinear mathematical model of waste heat recovery equipment configuration to maximize profits of ORC system, with the problem of painting process of waste heat emissions in steel company as a case, using particle swarm algorithm for validation. Then find out the best configuration of company's color coating process for optimal of waste heat recovery equipment, to achieve the objective of the waste heat recovery.

Acknowledgments Thanks for my professor and the assistant manager of steel company, Mr. Lin, Thank you for helping this study complete with smoothly way. Also sincerely thanks for the seminar last review committee members and all the coworkers.

References

Bureau of energy, ministry of economic affairs (2011a) Energy statistics handbook. Taiwan, Republic of China

Bureau of energy, ministry of economic affairs (2011b) 能源局採取「創新管理」方式推動能源回收再利用 http://www.moea.gov.tw/Mns/populace/news/News.aspx?kind=1&menu_id=40&news_id=22745

Chang K-H, Kuo C-R, Hsu S-W, Wang C-C (2011) Analysis of a 50 kW organic Rankine cycle system and its heat exchangers. Energ HVAC Eng 72:45–56

Drescher U, Brüggemann D (2007) Fluid selection for the organic rankine cycle (ORC) in biomass power and heat plants. Therm Eng 27:223–228

Durma A, Pugh R, Yazici S, Erdogan K, Kosan A (2012) Novel application of organic rankine cycle (ORC) technology for waste heat recovery from reheat furnace evaporative cooling system. Papers presented at the AISTech conference, Atlanta, Georgia, pp 7–10

Fenn JB (2003) Engines, energy, and entropy: a thermodynamics primer. USA

Kennedy J, Eberhart R (1995) Particle swarm optimization. Proc IEEE Int Conf Neural Networks 4:1942–1948

Kuo C-R, Luo S-T (2012) 低溫熱能發電系統現況與展望. Energ Monthly 7:31–34

Kuo C-R, Li Y-R, Hsu S-W (2012) Product development and applications of organic Rankone cycle power units. Mechatronic Ind 355:93–100

Lee Jhe-Yu (2009) Finite time endoreversible maximum useful energy rate analysis of thermodynamics cogeneration cycles. National Cheng Kung University, Dissertation

Lee Yu-Ren, Kuo Chi-Ron, Hsu Sung-Wei, Kuo Y-L (2011) Development of ORC for low grade thermal energy conversion. Mechatronic Ind 343:138–148

Lukawski M (2009) Design and optimization of standardized organic rankine cycle power plant for European conditions. Dissertation, University of Akureyri

Meng L, Jacobi AM (2011) Optimization of polymer tube-bundle heat exchangers using a genetic algorithm. papers presented at the ASME 2011 International Mechanical Engineering Congress And Exposition, Colorado, USA

Somayaji C, Chamra LM, Mago PJ (2007) Performance analysis of different working fluids for use in organic Rankine cycles. Power and Energ 221:255–263

Wei D, Lu X, Lu Z, Gu J (2007) Performance analysis and optimization of organic Rankine cycle (ORC) for waste heat recovery. Energy Convers Manage 48:1113–1119

Xia Wu, Wu Z (2005) An effective hybrid optimization approach for multi-objective flexible job-shop scheduling problems. Comput Ind Eng 48:409–425

Evaluation of Risky Driving Performance in Lighting Transition Zones Near Tunnel Portals

Ying-Yin Huang and Marino Menozzi

Abstract Driving behavior is affected by rapidly varying lighting conditions that frequently occur in transition zones near tunnel portals. When entering a tunnel in daytime, car drivers might encounter various difficulties in keeping high performance for safety concerns: (1) Physiological issues caused by high-level glare—the adaptation of the eye requires recovery time, during which time there will be impaired vision and reduced visibility of on-road objects; (2) Psychological issues caused by visual discomfort and distraction—limited resources for performing information processing are shared by the distracting and disturbing sensation; (3) Behavioral issues caused by driving patterns of the driver and other road users—many drivers reduce their speeds when entering the tunnel; thus a car driver must react to the sudden speed change. This study investigated the records of traffic accidents on Zurich highways in tunnel areas and conducted experiments using a driving simulator. The analysis of accident records shows that the frequency of accidents increases near tunnel portals. Experimental results show that discomfort glare impairs both peripheral visual attention and motion discrimination in simulated driving tasks. In conclusion, we suggest considering tunnel portals as a key factor causing elevated risk in traffic safety. Lighting designs and road layout near tunnel portal areas should be carefully defined.

Keywords Driving performance · Lighting transition · Tunnel portal · Risky event · Visual attention · Motion discrimination

---

Y.-Y. Huang (✉)
Department of Management, Technology and Economics and Department of Health Sciences and Technology, Ergonomics of Information Media Research Group, ETH Zurich, Scheuchzerstrasse 7, CH-8092 Zurich, Switzerland
e-mail: yingyinhuang@ethz.ch

M. Menozzi
Department of Health Sciences and Technology, Ergonomics of Information Media Research Group, ETH Zurich, Scheuchzerstrasse 7, CH-8092 Zurich, Switzerland
e-mail: mmenozzi@ethz.ch

Y.-K. Lin et al. (eds.), *Proceedings of the Institute of Industrial Engineers Asian Conference 2013*, DOI: 10.1007/978-981-4451-98-7_69,
© Springer Science+Business Media Singapore 2013

1 Introduction

Modern development and applications of the tunnel construction technology have enabled fast and convenient transportation in our daily life. With the increasing number of travelers using tunnel connections, efficient management of traffic flows and traffic safety have become an important topic. Continuously improved design concepts regarding tunnel constructions have been proposed in order to fulfill various aspects of major importance, such as safety concerns and optimized traffic volumes, as well as economic and energy considerations. For example, an Austrian study has suggested that traffic safety in uni-directional traffic tunnels is significantly higher than in bi-directional traffic tunnels (Nussbaumer 2007). Other suggestions have also been discussed and made for the layout of tunnels, lighting systems in tunnels (DIN 2008), traffic organization and fast control of tunnels, and so on, regarding safety. In fact, studies have shown that road tunnels are as safe as (or even safer than) other high standard open stretches of roads (Amundsen and Ranes 2000; Nussbaumer 2007) in terms of the probability of accidents occurring and road users being injured. However, when an accident happens in a tunnel, the cost and severity are higher than on open stretches of roads. Besides, higher accident rates occurred in the entrance zone of tunnels.

Several issues arise when one is driving into a tunnel. Car drivers encounter various difficulties in the transition zones of tunnels, i.e. from outside to inside of tunnels, which may result in risky driving performance. Firstly, rapidly changing lighting conditions may perturb one's visual system. Depending on lighting conditions and on individual susceptibility, glare may cause visual disability, discomfort or a combination of both (Boyce 2003). On a sunny day, for instance, a driver may be exposed to an extremely high-level luminance of $10,000 \text{ cd/m}^2$ when gazing on the tunnel entrance wall from the outside of the tunnel; as soon as the driver enters the dark environment inside the tunnel, the eye must get adapted to a very low-level luminance of 5 cd/m^2 within a short time. Such an abrupt and great variation of luminance causes so called disability glare and affects one's visual performance because of the inertia of the light adaptation mechanism of the eye. As visibility of objects depends on the light level one's eye is adapted to, and the light adaptation process is rather slow, a car driver may therefore face a risky driving event during the transition period due to reduced visibility of on-road objects. In addition to disability glare, the changing lighting conditions may result in visual discomfort without affecting acute visual performance, known as discomfort glare. Discomfort glare, which causes visual distraction, and/or vice versa (Lynes 1977), has been discussed mainly for some work-place related issues (Osterhaus 2005; Sheedy et al. 2005; Hemphälä and Eklund 2012). Little is known about its potential effects on traffic safety in tunnel portal areas. When discomfort glare acts as a distracter, our cognitive mechanism decides how to proceed with our visual attention for incoming driving events. As a result, while we are suffering from sensations of discomfort, discomfort glare shares our perception resources and increases our mental loads. As the amount of available resources for information processing are limited (Wickens et al. 2005), discomfort glare may bind resources therefore reducing the amount of available resources during the transition phase. Secondly, changes in the driving environment in the transition areas require car drivers' attention resources. When entering a tunnel, drivers have to proceed and rebuild the radical image changes of the visual environment from an open road to a closed and narrower area. Besides, some events may take place during the transition period such as switching the radio channel, adapting speed, changing lane, adjusting the head-lamp, attending to traffic signs and indicators, etc. More resources are occupied and less is left for handling any unexpected or sudden critical driving event. Thirdly, driving behaviors and patterns of road users towards a tunnel may affect other drivers. It has been noticed that drivers tend to reduce the car speed when approaching a tunnel entrance. This means that a driver must keep alert to any sudden speed changes of other cars and react simultaneously, especially to keep a safe distance with the front car. Again, such a task uses the limited resources and increases the mental load of a driver.

In conclusion, to drive in lighting transition zones near tunnel portals is a heavy driving task of a road user. Factors among physiological issues of light adaptation, psychological and behavioral issues of increased loads and reduced resources being available, may raise the probability of a risky driving event. In this study we aimed to analyze some car accident records near highway tunnel portals in Switzerland and support the hypothesis that an elevated potential of risk of car accidents may be expected in the transition zones near tunnel portals. In addition, by summarizing results of several experiments carried out in our driving simulator, we aimed to investigate certain abovementioned effects which could result in a risky driving performance caused by visual discomfort and distraction. Finally we aimed to conclude with some suggestions which may help in preventing risky driving performance near tunnel portals.

2 Method

2.1 Data Analysis: Records of Accidents on Zurich Highway Within the Last Decade

Support to the hypothesis of an elevated risk for accidents in proximity of tunnel portals is investigated by analyzing records of accidents on Zurich highways, traffic data etc., which were collected within the last decade. Data from 1,110 accidents were included and analyzed in this study, the factors bring considered are lighting, meteorological and other environmental factors, traffic parameters and individual factors of drivers involved in the accidents.

2.2 Visual Attention Tests in Simulated Driving Tasks: Evaluation of Effects on Visual and Mental Performance Caused by Discomfort Glare

Several experiments were carried out in a driving simulator in virtual reality. First, we investigated the effect of discomfort glare on peripheral visual attention in a realistic driving scenario. Participants performed the attention test (Fig. 1a) which required detection of simultaneously presented visual information both in the central visual field and in periphery. In a two alternative forced choice task, participants reported whether the orientations of two arrows, one in the central visual field (0°) and the other in the peripheral visual field (18°), were the same or not under different glare conditions. In 50 % of the total trials, a white frame with a luminance of 25 cd/m^2 was flashed prior to the presentation of the arrow set. In the other 50 %, no glare stimuli were applied.

In another experiment we investigated the effect of discomfort glare on the speed discrimination of a front car. Participants performed the motion detection test in a simulated city-scenario (Fig. 1b) under different glare conditions. Participants reported whether the perceived speed of a front car was faster or slower than own car speed. The car speed of the participants was fixed at 50 km per hour. The speed of the front car was given pseudo-randomly among the total trials, varying from 45 to 54 km per hour. One-third of the total trials were applied a flash of a white frame with a luminance of 19 cd/m^2 before the stimuli were presented. In another one-third of the trials, a grey glare mask with a luminance of 3.5 cd/m^2 was flashed before the front car presentation. In the rest trials no glare was applied.

Fig. 1 a Template scene of the attention test—*two arrows* were simultaneously presented and participants were asked to report whether the orientations of two arrows were the same **b** Template scene of the motion detection test—a front car with different speed settings was presented and participants were asked to report whether the speed of the front car was faster or slower than his/her own speed

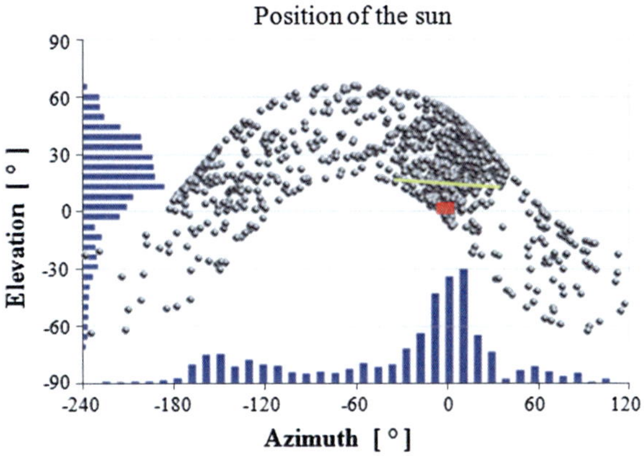

Evaluation of Risky Driving Performance

Fig. 2 Accidents with elevation and azimuth of the position of the sun. The tunnel portal is marked with a *red box* (indicative, not true to scale) and the horizon with a *yellow line* (indicative). Columns on the axes represent the frequency distribution of the elevation and the azimuth (Mauch 2012)

Fig. 3 Box-plot diagram presenting the distribution of detectability d′ in the glare condition and the no-glare condition of the total 56 participants

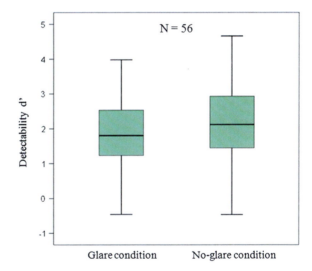

After analyzing the data, it was found that there is a strong relation of accidents frequency increase with near tunnel portal areas (Mauch 2012). In extreme cases, it was found that there is about seven times higher accident frequency occurred at the tunnel entrance compare to the inner section of the tunnel. An elevated accident frequency was also found at the exit portal of tunnels. One interest finding is that the luminance level before the tunnel portal area does not affect the accident frequency, however, the position of the sun at the time of accident occurrence shows significantly correlation with the accident frequency (Fig. 2).

Results of the attention test were evaluated based on the theory of signal detection (Gescheider 1985). Participants performed significantly better when no glare stimuli were presented (two-tailed t test, $t(55) = -2.614$, $p = 0.01$) in detectability d′ ($d'_{glare} = 1.87$; $d'_{no\ glare} = 2.11$; $\Delta d' = 0.24$) as illustrated in Fig. 3. Discomfort glare was shown to cause peripheral visual attention

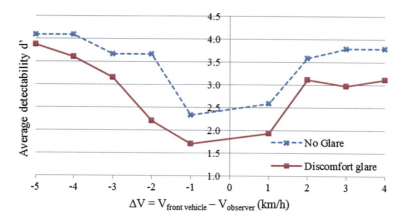

Fig. 4 Graphical representation of the results of average detectability d' versus the speed difference between the participant and the front car under different glare conditions

impairment in a complex and temporarily varying visual environment. As attention and in particular peripheral attention have been identified as a predominant factor directly related to driving skills (Ball et al. 1993), we therefore propose to consider discomfort glare as an important cause of risky driving performance in lighting transition zones of tunnels.

Results of the motion detection test revealed that participants performed the speed discrimination task significantly better in the no glare condition (Fig. 4). Discomfort glare interfered in visual perception of motion and caused reduced detectability in speed discrimination of a front car, which could be a major concern of safety issues during the transition period.

Above-mentioned effects on drivers' attention of abruptly varying lighting situations, as can be found in the transition zones, seem to play a role on traffic safety. Inattentiveness while entering a tunnel was named as one of the most important causes leading to impact with the front car in the transition zones. Possibly a lack of attention might have prevented some drivers in correctly estimating the speed of the front car, despite good visibility of the car ahead in terms of luminance contrast.

3 Conclusions

We suggest that tunnel portal areas may contain several risky factors which increase the probability of risky driving performance regarding traffic safety. Car drivers face various difficulties in the lighting transition zones when driving into a tunnel. Discomfort glare impairs drivers' attention to visual information on the road and impair their ability to estimate a front car's speed. Therefore, we suggest considering discomfort glare as a major importance in driving safety in addition to effects caused by a disturbed light adaptation of the eye. Among measures for improving driving safety in tunnel portal areas, we suggest increasing the lighting level at tunnel entrances with respect to current regulations by an important amount, e.g. a factor of 5–10, and/or reducing the lighting level before tunnel entrances by introducing some penumbra using roofs, windows, etc. for instance. Additionally, we recommend that the coefficient of reflection of the tunnel portal surface be reduced by changing the painting material or using dark-green vegetation, or smaller surface areas of tunnel portals, Furthermore, we suggest avoiding complex driving environments before tunnel portal areas, such as by decreasing the amount of road indicators and traffic signs, avoiding junction-connections, and so on.

Acknowledgments The authors thank the Federal Road Office of Switzerland (ASTRA) for supporting our research work on traffic safety.

References

Amundsen FH, Ranes G (2000) Studies on traffic accidents in Norwegian road tunnels. Tunn Undergr Space Technol 15(1):3–11

Ball K, Owsley C, Sloane ME, Roenker DL, Brun JR (1993) Visual attention problems as a predictor of vehicle crashes in older drivers. Invest Ophthalmol Vis Sci 34(11):3110–3123

Boyce PR (2003) Human factors in lighting, 2nd edn. Taylor & Francis, New York

DIN 67524 (2008) Lighting of street tunnels and underpasses

Gescheider GA (1985) Psychophysics: method, theory, and application, 2nd edn. Lawrence Earlbaum Associated, Houndmills

Hemphälä H, Eklund J (2012) A visual ergonomic intervention in mail sorting facilities: Effects on eyes, muscles and productivity. Applied Ergonomics 43:217–229

Lynes JA (1977) Discomfort glare and visual distraction. Light Res Technol 9:51–52

Mauch D (2012) Accidents in the Gubrist Tunnel: an analysis on the momentary inattention. Master Thesis, MSc MTEC, ETH Zürich

Nussbaumer C (2007) Comparative analysis of safety in tunnels. Paper presented at the Young Researchers Seminar, Austrian Road Safety Board, Brno, 2007

Osterhaus WKE (2005) Discomfort glare assessment and prevention for daylight applications in office environments. Solar Energy 79:140–158

Sheedy JE, Smith R, Hayes J (2005) Visual effects of the luminance surrounding a computer display. Ergonomics 48:1114–1128

Wickens CD, McCarley JS, Alexander AL, Thomas LC, Ambinder M, Zheng S (2005) Attention-Situation Awareness (A-SA) Model of Pilot Error. Technical Report AHFD-05-15/NASA-04-5

Application of Maple on Solving Some Differential Problems

Chii-Huei Yu

Abstract This article takes the mathematical software Maple as the auxiliary tool to study the differential problem of some types of functions. We can obtain the infinite series forms of any order derivatives of these functions by using differentiation term by term theorem, and hence greatly reduce the difficulty of calculating their higher order derivative values. On the other hand, we propose some functions to evaluate their any order derivatives, and calculate some of their higher order derivative values practically. The research methods adopted in this study involved finding solutions through manual calculations and verifying these solutions by using Maple. This type of research method not only allows the discovery of calculation errors, but also helps modify the original directions of thinking from manual and Maple calculations. For this reason, Maple provides insights and guidance regarding problem-solving methods.

Keywords Derivatives · Differentiation term by term theorem · Infinite series forms · Maple

1 Introduction

As information technology advances, whether computers can become comparable with human brains to perform abstract tasks, such as abstract art similar to the paintings of Picasso and musical compositions similar to those of Mozart, is a natural question. Currently, this appears unattainable. In addition, whether computers can solve abstract and difficult mathematical problems and develop abstract mathematical theories such as those of mathematicians also appears unfeasible.

C.-H. Yu (✉)
Department of Management and Information, Nan Jeon Institute of Technology, No.178,
Chaoqin Rd., Yanshui Dist, Tainan City, 73746 Taiwan, ROC
e-mail: chiihuei@mail.njtc.edu.tw

Y.-K. Lin et al. (eds.), *Proceedings of the Institute of Industrial Engineers Asian Conference 2013*, DOI: 10.1007/978-981-4451-98-7_70,
© Springer Science+Business Media Singapore 2013

Nevertheless, in seeking for alternatives, we can study what assistance mathematical software can provide. This study introduces how to conduct mathematical research using the mathematical software Maple. The main reasons of using Maple in this study are its simple instructions and ease of use, which enable beginners to learn the operating techniques in a short period. By employing the powerful computing capabilities of Maple, difficult problems can be easily solved. Even when Maple cannot determine the solution, problem-solving hints can be identified and inferred from the approximate values calculated and solutions to similar problems, as determined by Maple. For this reason, Maple can provide insights into scientific research. Inquiring through an online support system provided by Maple or browsing the Maple website (www.maplesoft.com) can facilitate further understanding of Maple and might provide unexpected insights. For the instructions and operations of Maple, we can refer to (Abell 2005; Dodson and Gonzalez 1995; Garvan 2001; Richards 2002; Robertson 1996; Stroeker and Kaashoek 1999).

In calculus courses, determining the nth order derivative value $f^{(n)}(c)$ of a function $f(x)$ at $x = c$, in general, needs two procedures: firstly finding the nth order derivative $f^{(n)}(x)$ of $f(x)$, and secondly taking $x = c$ into $f^{(n)}(x)$. These two procedures will make us face with increasingly complex calculations when calculating higher order derivative values of a function (i.e. n is larger), Therefore, to obtain the answers by manual calculations is not easy. For the study of differential problems can refer to (Edwards and Penney 1986, Chap. 3; Grossman 1992, Chap. 2; Larson et al. 2006, Chap. 2; Flatto 1976, Chap. 3; Yu 2013a, b, c, d, e, 2012). In this paper, we mainly studied the differential problems of the following two types of functions

$$f(x) = \frac{x^p}{\left(1 + x^q + x^{2q} + \cdots + x^{nq}\right)^m} \tag{1}$$

$$g(x) = \frac{x^p}{\left[1 - x^q + x^{2q} - \cdots + (-1)^n x^{nq}\right]^m} \tag{2}$$

where m, n, p, q are positive integers. We can determine any order derivatives of these two types of functions by using differentiation term by term theorem; these are the main results of this study (i.e., Theorems 1 and 2), and hence greatly reduce the difficulty of evaluating their higher order derivative values. Additionally, two examples in which Theorems 1 and 2 were practically employed to determine any order derivatives of these two functions and calculate some of their higher order derivative values. The research methods adopted in this study involved finding solutions through manual calculations and verifying these solutions by using Maple. This type of research method not only allows the discovery of calculation errors, but also helps modify the original directions of thinking from manual and Maple calculations. For this reason, Maple provides insights and guidance regarding problem-solving methods.

Application of Maple on Solving Some Differential Problems

2 Main Results

Firstly, we introduce some notations and theorems used in this study.

Notations.
Suppose r is any real number, n is any positive integer. Define
$$(r)_n = r(r-1)\cdots(r-n+1), \text{ and } (r)_0 = 1.$$

Finite geometric series.
Suppose n is a positive integer, y is a real number, $y \neq 1$. Then
$$1 + y + y^2 + \cdots + y^n = \frac{1-y^{n+1}}{1-y}, \text{ where } y \neq 1.$$
$$1 - y + y^2 - \cdots + (-1)^n y^n = \frac{1-(-1)^{n+1}y^{n+1}}{1+y}, \text{ where } y \neq -1.$$

Binomial theorem.
Assume u is a real number, m is a positive integer. Then $(1+u)^m = \sum_{k=0}^{m} \frac{(m)_k}{k!} u^k$.

Binomial series (Apostol 1975, p. 244).
If w, a are real numbers, $|w| < 1$. Then $(1+w)^a = \sum_{q=0}^{\infty} \frac{(a)_q}{q!} w^q$.

Differentiation term by term theorem (Apostol 1975, p. 230).
If, for all non-negative integer k, the functions $g_k : (a,b) \to R$ satisfy the following three conditions: (1) there exists a point $x_0 \in (a,b)$ such that $\sum_{k=0}^{\infty} g_k(x_0)$ is convergent, (2) all functions $g_k(x)$ are differentiable on open interval $(a,\ b)$, (3) $\sum_{k=0}^{\infty} \frac{d}{dx} g_k(x)$ is uniformly convergent on (a,b). Then $\sum_{k=0}^{\infty} g_k(x)$ is uniformly convergent and differentiable on $(a,\ b)$. Moreover, its derivative $\frac{d}{dx} \sum_{k=0}^{\infty} g_k(x) = \sum_{k=0}^{\infty} \frac{d}{dx} g_k(x)$.

Firstly, we determined the infinite series forms of any order derivatives of function (1).

Theorem 1 *Suppose m, n, p, q are positive integers and let the domain of the function*

$$f(x) = \frac{x^p}{(1 + x^q + x^{2q} + \cdots + x^{nq})^m}$$

be $\{x \in R | x \neq 0, \pm 1\}$.

(1) *If $|x| < 1$ and $x \neq 0$, then the kth order derivative of $f(x)$,*

$$f^{(k)}(x) = \sum_{r=0}^{m} \sum_{s=0}^{\infty} \frac{(-1)^{r+s}(m)_r(-m)_s(qr+p+nqs+qs)_k}{r!s!} x^{qr+p+nqs+qs-k} \quad (3)$$

(2) If $|x| > 1$, then

$$f^{(k)}(x) = \sum_{r=0}^{m} \sum_{s=0}^{\infty} \frac{(-1)^{m+r+s}(m)_r(-m)_s(qr+p-nqm-nqs-qm-qs)_k}{r!s!} x^{qr+p-nqm-nqs-qm-qs-k}$$

$$(4)$$

Proof

(1) If $|x| < 1$ and $x \neq 0$, then

$$\begin{aligned}
f(x) &= \frac{x^p}{(1+x^q+x^{2q}+\cdots+x^{nq})^m} \\
&= \frac{x^p(1-x^q)^m}{[1-x^{(n+1)q}]^m} \quad \text{(by finite geometric series)}
\end{aligned}$$

$$(5)$$

$$= x^p \cdot \sum_{r=0}^{m} \frac{(m)_r}{r!}(-x^q)^r \cdot \sum_{s=0}^{\infty} \frac{(-m)_s}{s!}[-x^{(n+1)q}]^s$$

(by binomial theorem and binomial series)

$$= \sum_{r=0}^{m} \frac{(-1)^{r+s}(m)_r}{r!} x^{qr+p} \cdot \sum_{s=0}^{\infty} \frac{(-m)_s}{s!} x^{(n+1)qs}$$

$$= \sum_{r=0}^{m} \sum_{s=0}^{\infty} \frac{(-1)^{r+s}(m)_r(-m)_s}{r!s!} x^{qr+p+nqs+qs}$$

$$(6)$$

By differentiation term by term theorem, differentiating x by k times on both sides of (6), we obtained the kth order derivative of $f(x)$,

$$f^{(k)}(x) = \sum_{r=0}^{m} \sum_{s=0}^{\infty} \frac{(-1)^{r+s}(m)_r(-m)_s(qr+p+nqs+qs)_k}{r!s!} x^{qr+p+nqs+qs-k}$$

(2) If $|x| > 1$, then by (5) we obtained

$$\begin{aligned}
f(x) &= x^p(1-x^q)^m \cdot \frac{1}{[1-x^{(n+1)q}]^m} \\
&= x^p(1-x^q)^m \cdot \frac{(-1)^m}{x^{(n+1)qm}[1-x^{-(n+1)q}]^m} \\
&= \sum_{r=0}^{m} \frac{(-1)^r(m)_r}{r!} x^{qr+p-nqm-qm} \cdot \sum_{s=0}^{\infty} \frac{(-1)^{m+s}(-m)_s}{s!} x^{-(n+1)qs} \\
&= \sum_{r=0}^{m} \sum_{s=0}^{\infty} \frac{(-1)^{m+r+s}(m)_r(-m)_s}{r!s!} x^{qr+p-nqm-nqs-qm-qs}
\end{aligned}$$

$$(7)$$

Application of Maple on Solving Some Differential Problems 589

Also, using differentiation term by term theorem, differentiating x by k times on both sides of (7), we determined $f^{(k)}(x)$

$$= \sum_{r=0}^{m} \sum_{s=0}^{\infty} \frac{(-1)^{m+r+s}(m)_r(-m)_s(qr+p-nqm-nqs-qm-qs)_k}{r!s!} x^{qr+p-nqm-nqs-qm-qs-k}$$

The similar proof of Theorem 1, we can easily determine the infinite series forms of any order derivatives of function (2).

Theorem 2 *If the assumptions are the same as Theorem 1, and let the domain of function* $g(x) = \frac{x^p}{[1-x^q+x^{2q}-\cdots+(-1)^n x^{nq}]^m}$ *be* $\{x \in R | x \neq 0, \pm 1\}$.

(1) *If* $|x| < 1$ *and* $x \neq 0$, *then the kth order derivative of* $g(x)$,

$$g^{(k)}(x) = \sum_{r=0}^{m} \sum_{s=0}^{\infty} \frac{(-1)^{ns}(m)_r(-m)_s(qr+p+nqs+qs)_k}{r!s!} x^{qr+p+nqs+qs-k} \qquad (8)$$

(2) *If* $|x| > 1$, *then* $g^{(k)}(x)$

$$= \sum_{r=0}^{m} \sum_{s=0}^{\infty} \frac{(-1)^{ns+nm}(m)_r(-m)_s(qr+p-nqm-nqs-qm-qs)_k}{r!s!} x^{qr+p-nqm-nqs-qm-qs-k}$$

$$(9)$$

3 Examples

In the following, we provide two examples, aimed at the differential problem of two types of rational functions to demonstrate our results. We determined their any order derivatives and some of their higher-order derivative values by using Theorems 1 and 2. On the other hand, we employ Maple to calculate the approximations of these higher-order derivative values and their infinite series forms for verifying our answers.

Example 1 Suppose the domain of function

$$f(x) = \frac{x^3}{(1+x^2+x^4+x^6)^4} \qquad (10)$$

is $\{x \in R | x \neq 0, \pm 1\}$. By (i) of Theorem 1, we determined the 7th order derivative value of $f(x)$ at $x = 1/2$,

$$f^{(7)}\left(\frac{1}{2}\right) = \sum_{r=0}^{4}\sum_{s=0}^{\infty}\frac{(-1)^{r+s}(4)_r(-4)_s(8s+2r+3)_7}{r!s!}\cdot\left(\frac{1}{2}\right)^{8s+2r-4} \tag{11}$$

We use Maple to verify our answer.
>f: = x → x^3/(1 + x^2 + x^4 + x^6)^4;

$$f := x \longrightarrow \frac{x^3}{\left(1+x^2+x^4+x^6\right)^4}$$

>evalf((D@@7)(f)(1/2), 14);
−35551.764617819
>evalf(sum(sum((−1)^(r + s)*product(4−i,   i = 0···(r − 1))*product(−4−
j,j = 0···(s − 1))*   product(8*s + 2*r + 3 − k,   k = 0···6)/(r!*s!)*(1/
2)^(8*s + 2*r − 4),s = 0···infinity), r = 0···4),14);
−35551.764617819
On the other hand, using (2) of Theorem 1, we obtained the 10th order derivative value of $f(x)$ at $x = -2$,

$$f^{(10)}(-2) = \sum_{r=0}^{4}\sum_{s=0}^{\infty}\frac{(-1)^{r+s}(4)_r(-4)_s(2r-8s-29)_{10}}{r!s!}\cdot(-2)^{2r-8s-39} \tag{12}$$

Using Maple to verify the correctness of (12).
>evalf((D@@10)(f)(−2), 14);
−2255.1259338446
>evalf(sum(sum((−1)^(r + s)*product(4 − i,i = 0..(r − 1))*product(−4 −
j,j = 0···(s − 1))* product(2*r − 8*s − 29 − k, k = 0..9)/(r!*s!)*(−2)^(2*r −
8*s − 39), s = 0···infinity), r = 0···4), 14);
−2255.1259338446

Example 2 Suppose the domain of function

$$g(x) = \frac{x^5}{\left(1 - x^3 + x^6 - x^9 + x^{12}\right)^3} \tag{13}$$

is $\{x \in R | x \neq 0, \pm1\}$. Using (i) of Theorem 2, we obtained the 6th order derivative value of $g(x)$ at $x = 1/3$,

$$g^{(6)}\left(\frac{1}{3}\right) = \sum_{r=0}^{3}\sum_{s=0}^{\infty}\frac{(3)_r(-3)_s(3r+15s+5)_6}{r!s!}\cdot\left(\frac{1}{3}\right)^{3r+15s-1} \tag{14}$$

We can use Maple to verify the correctness of (14).
>g: = x → x^5/(1 − x^3 + x^6 − x^9 + x^12)^3;

$$g := x \longrightarrow \frac{x^5}{\left(1 - x^3 + x^6 - x^9 + x^{12}\right)^3}$$

Application of Maple on Solving Some Differential Problems

>evalf((D@@6)(g)(1/3), 14);
11133.205916540
>evalf(sum(sum(product(3 − i,i = 0···(r − 1))*product(−3 − j, j = 0···(s − 1))*product(3*r + 15*s + 5 − k,   k = 0···5)/(r!*s!)*(1/3)^(3*r + 15*s − 1), s = 0···infinity), r = 0···3), 14);
11133.205916540

Next, by (ii) of Theorem 2, we obtained the 9th order derivative value of $g(x)$ at $x = 3$,

$$g^{(9)}(3) = \sum_{r=0}^{3} \sum_{s=0}^{\infty} \frac{(3)_r(-3)_s(3r - 15s - 40)_9}{r!s!} \cdot 3^{3r-15s-49} \tag{15}$$

Using Maple to verify the correctness of (15).
>evalf((D@ @9)(g)(3), 14);
−0.0000079153266419116
>evalf(sum(sum(product(3 − i, i = 0···(r − 1))*product(−3 − j, j = 0···(s − 1))*product(3*r − 15*s − 40 − k,   k = 0···8)/(r!*s!)*3^(3*r − 15*s − 49), s = 0···infinity), r = 0···3), 14);
−0.0000079153266419116

4 Conclusions

As mentioned, the differentiation term by term theorem plays a significant role in the theoretical inferences of this study. In fact, the application of this theorem is extensive, and can be used to easily solve many difficult problems; we endeavor to conduct further studies on related applications.

On the other hand, Maple also plays a vital assistive role in problem-solving. In the future, we will extend the research topic to other calculus and engineering mathematics problems and solve these problems by using Maple. These results will be used as teaching materials for Maple on education and research to enhance the connotations of calculus and engineering mathematics.

References

Abell ML, Braselton JP (2005) Maple by example, 3rd edn. Elsevier, Amsterdam
Apostol TM (1975) Mathematical analysis, 2nd edn. Addison-Wesley Publishing Co., Inc, Reading, MA
Dodson CTJ, Gonzalez EA (1995) Experiments in mathematics using Maple, Springer, Berlin
Edwards CH Jr, Penney DE (1986) Calculus and analytic geometry, 2nd edn. Prentice-Hall, Inc, Englewood Cliffs, NJ
Flatto L (1976) Advanced calculus. The Williams & Wilkins Co, Baltimore, MD
Garvan F (2001) The Maple book, Chapman & Hall/CRC, London

Grossman SI (1992) Calculus, 5th edn. Saunders College Publishing, London

Larson R, Hostetler RP, Edwards BH (2006) Calculus with analytic geometry, 8th edn. Houghton Mifflin, Boston

Richards D (2002) Advanced mathematical methods with Maple, Cambridge University Press, Cambridge

Robertson JS (1996) Engineering mathematics with Maple, McGraw-Hill, New York

Stroeker RJ, Kaashoek JF (1999) Discovering mathematics with Maple: an interactive exploration for mathematicians, engineers and econometricians, Birkhauser Verlag, Basel

Yu CH (2013a) The differential problems of fractional type and exponential type of functions. De Lin J 26:33–42

Yu CH (2013b) The differential problem of two types of rational functions. Meiho J, 32(1) (accepted)

Yu CH (2013c) The differential problem of four types of functions, Kang-Ning J 15 (accepted)

Yu CH (2013d) A study on the differential problem of some trigonometric functions. Jen-Teh J 10 (accepted)

Yu CH (2013e) The differential problem of two types of exponential functions. Nan Jeon J 16 (accepted)

Yu CH (2012) Application of Maple on the differential problem of hyperbolic functions. International Conference on Safety & Security Management and Engineering Technology 2012, WuFeng University, Taiwan, pp 481–484

Six Sigma Approach Applied to LCD Photolithography Process Improvement

Yung-Tsan Jou and Yih-Chuan Wu

Abstract Liquid Crystal Display (LCD) makes to pursue light and thin trend, and towards high resolution development. To promote the LCD high resolution, the process requires more precise micromachining. The major key is lithography, but also generally known as the photolithography process. This research uses Six Sigma DMAIC steps (Define, Measure, Analyze, Improve, Control) to an empirical study of the pattern pitch machining process in a domestic optoelectronic manufacturer. Constructed in the photolithography process, the pattern pitch machining process capability improves and a process optimization prediction mode. Taguchi method is used to explore the parameters combination of photolithography process optimization and to understand the impact of various parameters. The findings of this research show that C_{pk} can upgrade from 0.85 to 1.56 which achieve quality improvement goals and to enhance the LCD photolithography process capability.

Keywords DMAIC · LCD · Photolithography · Six sigma · Taguchi method

1 Introduction

The liquid crystal display (LCD) has major advantages over its competitors. It is thinner, smaller and lighter than other displays and also has low power consumption, low-radiation, high-contrast and high-dpi. Hence, the LCD panel is

---

Y.-T. Jou (✉) · Y.-C. Wu
Department of Industrial & Systems Engineering, Chung Yuan Christian University,
No.200, Chung-Pei Rd, Chung Li, Taoyuan, Taiwan
e-mail: ytjou@cycu.edu.tw

Y.-C. Wu
e-mail: jason_wu1102@yahoo.com.tw

Y.-K. Lin et al. (eds.), *Proceedings of the Institute of Industrial
Engineers Asian Conference 2013*, DOI: 10.1007/978-981-4451-98-7_71,
© Springer Science+Business Media Singapore 2013

widely applied in daily electronic products, and the demand for the LCD panel increases. LCD production process consists of the following three main processes: TFT Array process, Cell assembly process, and Module assembly process, and each process consists of some sub-processes (Chen et al. 2006). The most critical modules that include mask process numbers and LCD resolution quality are photolithography technology. Significantly reduction in manufacturing costs result if there are fewer a mask process number during photolithography process. In addition, the higher resolution of the photolithography process will improve the LCD quality. This study focuses on the LCD array process which requires 4–8 mask process numbers and each mask process needs handling after photolithography process. It hopes to find a suitable model for the LCD photolithography process improvement by actual machining experiments of one domestic opto-electronic manufacturer.

In recent years, Six Sigma DMAIC (Define-Measure-Analyze-Improve-Control) approach has been widely used in many research areas (Breyfogle et al. 2001). Through Six Sigma approach, Analysis of variance (ANOVA), and Taguchi method (TM), the study aims to construct a suitable LCD photolithography process improvement mode for the optoelectronics industry. Therefore, the study has the following objectives: (1) Six Sigma DMAIC approach is used to establish a project process, and combined with Taguchi method and ANOVA analysis to design a new process analysis and prediction mode. (2) Taguchi method is used to identify the various design parameters and key factors of the photolithography process to find out the optimal process conditions and to verify the results of the experiments.

2 Literature Review

2.1 Six Sigma

Dr. Mike J. Harry developed Six Sigma for management practices in 1980s. Six Sigma's most common and well-known methodology is its problem-solving DMAIC approach. The 5-step DMAIC method is often called the process improvement methodology. Traditionally, this approach is to be applied to a problem with an existing, steady-state process or product and service offering. DMAIC resolves issues of defects or failures, deviation from a target, excess cost or time, and deterioration (Pete and Larry 2002). DMAIC identifies key requirements, deliverables, tasks, and standard tools for a project team to utilize when tackling a problem. The DMAIC process and key steps are shown as in Fig. 1.

Fig. 1 DMAIC process and key steps

2.2 Taguchi Method

Taguchi technique provides a simple, efficient, and systematic approach to optimize design for performance, quality, and cost. Taguchi quality control is divided into three stages, system design, parameter design, and tolerance design (Su 2008). This methodology is valuable when design parameters are qualitative and discrete. Taguchi parameter design can optimize the performance characteristics through the setting of design parameters and reduce the sensitivity of the system performance to the source of variation (Chen and Chuang 2008).

This study focuses on the photolithography process of LCD Array and adopts nominal-the-best, NTB characteristic. The closer to the original mask design value, the better the quality characteristics after developing pattern pitch. It means that the quality characteristics measured value close to the target is better. The Signal-to-Noise ratio, S/N for each design parameter level is computed based on S/N analysis. The nominal S/N ratio corresponds to a better performance characteristic. Hence, the optimal design parameter level is the level with the nominal S/N ratio. The nominal S/N ratio can be written as given in Eq. (1). The confidence interval, CI then can be calculated as Eqs. (2) and (3) (Su 2008).

$$\eta = -10\log_{10}\left(\frac{u^2}{\sigma^2}\right) = -10\log_{10}(MSD) = -10\log_{10}\left[(u-m)^2 + \sigma^2\right] \quad (1)$$

$$CI_1 = \sqrt{F_{\alpha;1,2} \times V_e \times \left[\frac{1}{n_{\mathit{eff}}}\right]} \qquad (2)$$

$$CI_2 = \sqrt{F_{\alpha;1,2} \times V_e \times \left[\frac{1}{n_{\mathit{eff}}} + \frac{1}{r}\right]} \qquad (3)$$

where, represents S/N ratio; *MSD* stands for Mean Square Deviation. F_α represents critical values of the F-distribution for the significance level α; V_e represents the pooled error variance.

3 Methodology

3.1 Define

A TFT-LCD array process consists of three main modules—Thin Film, Photolithography, and Etching (Chen et al. 2006). Photolithography process includes some sub-processes such as dehydration bake, HMDS, photoresist coating, soft/pre-bake, exposure, development, and post-exposure (as in Fig. 2). The high complexity and precision processing, means that every sub-process directly influences the success or failure to the entire photolithography process. LCD manufacturing is a glass substrate which must go through photolithography process 4–8 times to achieve complete circuit. Therefore photolithography process plays an important role of the entire LCD manufacturing process.

Patterns after developing the photolithography process can be classified into normal developing, incomplete developing, under-development and over-development etc. This study focuses on the pattern pitch after developing of the

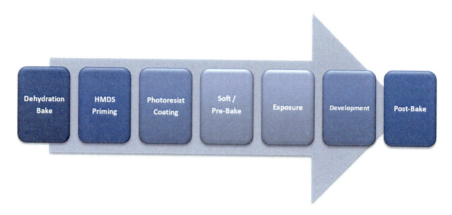

Fig. 2 Photolithography process

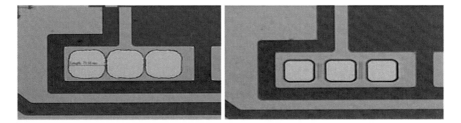

Fig. 3 Actual pattern pitch from developing

photolithography process to meet the design target value 60 ± 5 (nm). The closer to the original mask design value, the better it is. Developing from the pattern pitch the left is incomplete developing, and the right is developing normally, as shown in Fig. 3.

3.2 Measurement

Process capability indices (PCIs) are practical and powerful tools for measuring process performance. Process capability indices have been widely used in the manufacturing industry to provide numerical measures on whether a process is capable of reproducing items meeting the manufacturing quality requirement preset in the factory. Numerous capability indices have been proposed to measure process potential and performance. The two most commonly used indices C_p and C_{pk} discussed in Kane (1974), and the two more-advanced indices C_{pm} and C_{pmk} are developed by Chan et al. (1990), and Pearn et al. (1992). C_{pk} measures how close one is to one's target and how consistent one is to one's average performance (Wright 1995). The larger the index, the less likely it is that any item will be outside the specifications. Of this study before improvement the C_{pk} 0.85 < 1.33 presents the process is substandard, unstable and insufficient process. Process improvements must be given high priority and documented in a corrective action plan. The measurement data of the pattern pitch by 30 randomly selected developing products are shown in Table 1.

Table 1 Measure data of pattern pitch (nm)

57.90	58.97	58.34	60.39	57.36	62.39
59.99	62.39	61.80	59.64	62.90	58.90
57.91	59.64	60.50	62.84	63.20	63.28
59.45	61.32	62.32	60.30	58.00	57.65
56.00	63.75	63.90	61.09	61.50	60.12

Note 1 pitch = 130 nm

Table 2 Control factors and fixed factors

Factors		Range	Conditions
Control factors	Developing time (sec)	45 ~ 80	50
	Exposure energy (mj)	200 ~ 700	300
	Photoresist thickness (Å)	25,000 ~ 29,000	26,000
	Temperature (°C)	110 ~ 125	115
Fixed factors	Ingredients	TMAH	
	Photoresist developers	2.5 %	

3.3 Analyze

Three senior engineers with an average engineering experience of more than six years in the photolithography process, participated in this project. These experts identified control factors and fixed factors. The control factor chosen for this study had 4 control factors, including developing time, exposure energy, photoresist developers, and temperature. Table 2 presents the chosen control factors and the developer ingredients are Tetra Methyl Ammonium Hydride (TMAH).

3.4 Improve

Design of experiments (DOE) is a discipline that has very broad application across all the natural and social sciences and engineering. An excellent solution to this project is an approach known as Taguchi Parameter Design. As a type of fractional factorial design, Taguchi Parameter Design is similar to traditional DOE methods in that multiple input parameters can be considered for a given response. There are, however, some key differences, for which Taguchi Parameter Design lends itself well to optimizing a production process (Muthukrishnan et al. 2012). Taguchi Parameter Design methodology includes selection of parameters, utilizing an orthogonal array (OA), conducting experimental runs, data analysis, determining the optimum combination, and verification. In this experiment, the pattern pitch 60 ± 5 (nm) after the developing be used to obtain a set of best combination of experimental parameters. The use of orthogonal array can effectively reduce the number of experiment necessary. The experiment layout using a Taguchi's L_9'OA, as shown in Table 3, was used to design the Taguchi experiment in this study.

The greater the S/N ratio, the better quality has. The largest S/N ratio can get the best parameter level combinations. The experimental data shows that the S/N ratios for optimal factors combination are $A_2B_3C_2D_2$. Figure 4 presents main effects plot for S/N ratios.

Perform ANOVA to identify significant parameters. ANOVA establishes the relative significance of parameters. Factors A, C, and D are significant in this experiment. The percentage of the experimental error of 2.42 %, less than 15 %

Six Sigma Approach Applied To LCD Photolithography

Table 3 Experimental layout using $L'_9(3^4)$ OA

A	B	C	D	Developing time	Exposure energy	PR thickness	Temp.	SN
1	1	1	1	60	300	25,000	115	40.869
1	2	2	2	60	400	27,000	120	75.607
2	3	3	3	60	500	29,000	125	70.692
2	1	2	3	70	300	29,000	125	80.281
2	2	3	1	70	400	25,000	115	68.871
2	3	1	2	70	500	27,000	120	71.406
3	1	3	2	80	300	27,000	120	65.550
3	2	1	3	80	400	25,000	125	41.147
3	3	2	1	80	500	29,000	115	57.332

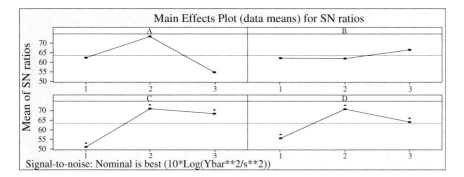

Fig. 4 Response graph for SN ratios

Table 4 Results of confirmation experiment

Factors				Pattern pitch values after developing			SN
A	B	C	D	Observed value 1	Observed value 2	Observed value 3	
2	3	2	2	60.3	60.31	60.32	75.6078
2	3	2	2	59.64	59.63	59.64	80.2815
2	3	2	2	60.31	60.29	60.32	71.9275

can be identified in this experiment did not ignore some important factors. The F value of experimental factors A, C, D is greater than 4, which means that the factors' effect is considerable. The calculated optimal expectation S/N ratios and CI_1 are 88.39 and 16.82. According to S/N ratios and CI_1, the range of the optimal conditions is $[88.39 \pm 16.82] = [71.57, 105.21]$. This experiment performed three trials. Table 4 presents the results of the confirmation experiment.

Further, CI_2 is 20.103, the S/N ratio confidence interval of confirmation experiment is $[88.39 \pm 20.103] = [68.287, 105.50]$. Through the confirmation experiment, S/N ratios fall within the 95 % confidence interval which indicates the success of the experimental results. Using the patterns pitch after developing and

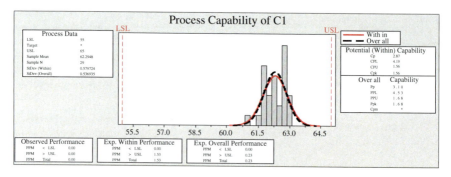

Fig. 5 Process capability after improvement

optimal factors combination of the experiment perform 30 times, process capability indices C_{pk} is improved from 0.85 to 1.56 in experimentation (as in Fig. 5). This study clearly shows that through optimal factors combination $A_2B_3C_2D_2$, including the developing time of 70 s, the exposure energy of 500 mj, the photoresist thickness of 27,000 Å, and temperature of 115 °C, machining process is greatly improved for photolithography process.

3.5 Control

After confirmation experiments and process capability analysis, this study thus can verify conclusions and has two-point suggestions for control. (1) Changing the factors combination, the control factors of the developing time from 50 to 70 s, exposure energy from 300 to 500 mj, the photoresist thickness from 26,000 to 27,000 Å, and maintaining the temperature at 115 °C. (2) Keeping observation the photolithography process records and monitoring the process capability analysis, the products quality and process stability will be improved.

4 Conclusions

This study uses Six Sigma DMAIC steps, technical data of one domestic optoelectronics manufacturer, to do investigate the potential improvement of photolithography process. Combined with Taguchi DOE and ANOVA analysis, the chosen key factors are use to identify the pattern pitch 60 ± 5 (nm) process optimization. Taguchi method was applied to this experiment to get the optimal processing factors including the developing time of 70 s, the exposure energy of 500 mj, the photoresist thickness of 27,000 Å, the temperature of 115 °C, and its process capability analysis results show that C_{pk} is upgraded from 0.85 to 1.56 in experimentation. It will obviously improve the products quality of precision and process capability. In addition to increasing the competitiveness of products, the experiment results of photolithography process can also provide technical references for the domestic optoelectronic manufacturers and related industries.

References

Breyfogle FW, Cupello JM, Meadows B (2001) Managing six-sigma. John Wiley & Sons, New York

Chan LK, Xiong Z, Zhang D (1990) On the asymptotic distributions of some process capability indices. Commun Stat-Theor Methods 19(1):11–18

Chen CC, Chuang MC (2008) Integrating the Kano model into a robust design approach to enhance customer satisfaction with product design. Int J Prod Econ 114:667–681

Chen KS, Wang CH, Chen HT (2006) A MAIC approach to TFT-LCD panel quality improvement. Microelectron Reliab 46:1189–1198

Kane VE (1974) Process capability indices. J Qual Technol 18(1):41–52

Muthukrishnan N, Mahesh Babu TS, Ramanujam R (2012) Fabrication and turning of A1/S$_i$C/ B$_4$C hybrid metal matrix composites optimization using desirability analysis. J Chin Inst Ind Eng 29(8):515–525

Pearn WL, Kotz S, Johnson NL (1992) Distributional and inferential properties of process capability indices. J Qual Technol 24:216–231

Pete P, Larry H (2002) A brief introduction to Six Sigma for employees. McGraw-Hill, Taiwan

Su CT (2008) Quality engineering. Chin Soc Qual, Taiwan

Wright PA (1995) A process capacity index sensitive to skewness. Commun Stat-Simul Comput 52:195–203

A Study of the Integrals of Trigonometric Functions with Maple

Chii-Huei Yu

Abstract This paper uses Maple for the auxiliary tool to evaluate the integrals of some types of trigonometric functions. We can obtain the Fourier series expansions of these integrals by using integration term by term theorem. On the other hand, we propose some related integrals to do calculation practically. Our research way is to count the answers by hand, and then use Maple to verify our results. This research way can not only let us find the calculation errors but also help us to revise the original thinking direction because we can verify the correctness of our theory from the consistency of hand count and Maple calculations.

Keywords Integrals · Trigonometric functions · Integration term by term theorem · Fourier series expansions · Maple

1 Introduction

The computer algebra system (CAS) has been widely employed in mathematical and scientific studies. The rapid computations and the visually appealing graphical interface of the program render creative research possible. Maple possesses significance among mathematical calculation systems and can be considered a leading tool in the CAS field. The superiority of Maple lies in its simple instructions and ease of use, which enable beginners to learn the operating techniques in a short period. In addition, through the numerical and symbolic computations performed by Maple, the logic of thinking can be converted into a series of instructions. The computation results of Maple can be used to modify previous thinking directions, thereby forming direct and constructive feedback that

---

C.-H. Yu (✉)
Department of Management and Information, Nan Jeon Institute of Technology,
No. 178, Chaoqin Road, Yanshui, Tainan 73746, Taiwan, Republic of China
e-mail: chiihuei@mail.njtc.edu.tw

can aid in improving understanding of problems and cultivating research interests. Inquiring through an online support system provided by Maple or browsing the Maple website (www.maplesoft.com) can facilitate further understanding of Maple and might provide unexpected insights. For the instructions and operations of Maple, we can refer to (Abell 2005; Dodson and Gonzalez 1995; Garvan 2001; Richards 2002; Robertson 1996; Stroeker and Kaashoek 1999).

In calculus courses, we learnt many methods to solve the integral problems, including change of variables method, integration by parts method, partial fractions method, trigonometric substitution method, and so on. The introduction of these methods can refer to (Edwards and Penney 1986, Chap. 9; Grossman, 1992, Chap. 7; Larson et al. 2006, Chap. 8). This paper mainly studies the integrals of the following two types of trigonometric functions, which are not easy to obtain their answers by using the methods mentioned above.

$$f(x) = \frac{r^{m+1} \cos(m-1)\lambda x - ar^m \cos m\lambda x}{r^2 - 2ar \cos \lambda x + a^2} \tag{1}$$

$$g(x) = \frac{r^{m+1} \sin(m-1)\lambda x - ar^m \sin m\lambda x}{r^2 - 2ar \cos \lambda x + a^2} \tag{2}$$

where a, λ, r are real numbers, a, λ, $r \neq 0$, m is any integer, and $|r| \neq |a|$. The integrals of these two types of trigonometric functions are different from the integrals studied in Yu (2011, 2012a, b, c). We can obtain the Fourier series expansions of the integrals of trigonometric functions (1) and (2) by using integration term by term theorem; these are the main results of this study (i.e., Theorems 1 and 2). On the other hand, we propose two integrals to do calculation practically. The research methods adopted in this study involved finding solutions through manual calculations and verifying these solutions by using Maple. This type of research method not only allows the discovery of calculation errors, but also helps modify the original directions of thinking from manual and Maple calculations. Therefore, Maple provides insights and guidance regarding problem-solving methods.

2 Main Results

Firstly, we introduce some theorems used in this study.

Geometric series.

Suppose z is a complex number, $|z| < 1$. Then $\frac{1}{1-z} = \sum_{k=0}^{\infty} z^k$.

Integration term by term theorem (Apostol 1975, p. 269).

Suppose $\{g_n\}_{n=0}^{\infty}$ is a sequence of Lebesgue integrable functions defined on inteval I. If $\sum_{n=0}^{\infty} \int_I |g_n|$ is convergent, then $\int_I \sum_{n=0}^{\infty} g_n = \sum_{n=0}^{\infty} \int_I g_n$.

A Study of the Integrals of Trigonometric Functions

Before deriving our main results, we need a lemma.

Lemma *Suppose z is a complex number, m is any integer, a is a real number, $a \neq 0$ and $|z| \neq |a|$. Then*

$$\frac{z^m}{z-a} = -\sum_{k=0}^{\infty} \frac{1}{a^{k+1}} z^{k+m} \quad \text{if } |z| < |a| \tag{3}$$

$$= \sum_{k=0}^{\infty} a^k z^{-k-1+m} \quad \text{if } |z| > |a| \tag{4}$$

Proof If $|z| < |a|$, then

$$\frac{z^m}{z-a} = -z^m \cdot \frac{1}{a} \cdot \frac{1}{1 - \frac{z}{a}}$$

$$= -z^m \cdot \frac{1}{a} \cdot \sum_{k=0}^{\infty} \left(\frac{z}{a}\right)^k \quad \text{(by geometric series)}$$

$$= -\sum_{k=0}^{\infty} \frac{1}{a^{k+1}} z^{k+m}.$$

If $|z| > |a|$, then

$$\frac{z^m}{z-a} = z^m \cdot \frac{1}{z} \cdot \frac{1}{1 - \frac{a}{z}}$$

$$= z^m \cdot \frac{1}{z} \cdot \sum_{k=0}^{\infty} \left(\frac{a}{z}\right)^k \quad \text{(by geometric series)}$$

$$= \sum_{k=0}^{\infty} a^k z^{-k-1+m}$$

□

Next, we determine the Fourier series expansion of the integral of trigonometric function (1).

Theorem 1 *Suppose a, λ, r are real numbers, a, λ, $r \neq 0$, m is any integer, and $|r| \neq |a|$. Let $\delta(m) = \begin{cases} 1 & m \leq 0 \\ 0 & m > 0 \end{cases}$, $\sigma(m) = \begin{cases} 1 & m \geq 1 \\ 0 & m < 1 \end{cases}$. Then the integral*

$$\int \frac{r^{m+1} \cos(m-1)\lambda x - a r^m \cos m\lambda x}{r^2 - 2ar\cos\lambda x + a^2} dx$$

$$= -\delta(m) \cdot a^{m-1} x - \frac{1}{\lambda} \cdot \sum_{\substack{k=0 \\ k \neq -m}}^{\infty} \frac{r^{k+m}}{a^{k+1}(k+m)} \sin(k+m)\lambda x + C \quad \text{if } |r| < |a| \tag{5}$$

$$= \sigma(m) \cdot a^{m-1}x + \frac{1}{\lambda} \cdot \sum_{\substack{k=0 \\ k \neq m-1}}^{\infty} \frac{a^k}{r^{k+1-m}(k+1-m)} \sin(k+1-m)\lambda x + C \tag{6}$$

if $|r| > |a|$

Proof Because

$$f(x) = \frac{r^{m+1}\cos(m-1)\lambda x - ar^m \cos m\lambda x}{r^2 - 2ar\cos\lambda x + a^2}$$

$$= \mathrm{Re}\left[\frac{r^{m+1}e^{i(m-1)\lambda x} - ar^m e^{im\lambda x}}{(re^{i\lambda x} - a)(re^{-i\lambda x} - a)}\right] \tag{7}$$

$$= \mathrm{Re}\left(\frac{z^m}{z-a}\right) \quad (\text{where } z = re^{i\lambda x})$$

We obtained $\int \frac{r^{m+1}\cos(m-1)\lambda x - ar^m \cos m\lambda x}{r^2 - 2ar\cos\lambda x + a^2}dx$

$$= \int \mathrm{Re}\left(\frac{z^m}{z-a}\right)dx$$

[by(7)]

$$= \mathrm{Re}\left(\int \frac{z^m}{z-a}dx\right) \tag{8}$$

If $|r| < |a|$, then $\int \frac{r^{m+1}\cos(m-1)\lambda x - ar^m \cos m\lambda x}{r^2 - 2ar\cos\lambda x + a^2}dx$

$$= \mathrm{Re}\left(\int -\sum_{k=0}^{\infty}\frac{1}{a^{k+1}}z^{k+m}dx\right) \quad [\text{by (3), where } z = re^{i\lambda x}]$$

$$= \int\left[-\sum_{k=0}^{\infty}\frac{r^{k+m}}{a^{k+1}}\cos(k+m)\lambda x\right]dx \quad (\text{because } z = re^{i\lambda x})$$

$$= -\delta(m) \cdot a^{m-1}x - \frac{1}{\lambda} \cdot \sum_{\substack{k=0 \\ k \neq -m}}^{\infty} \frac{r^{k+m}}{a^{k+1}(k+m)}\sin(k+m)\lambda x + C.$$

(by integration term by term theorem)

If $|r| > |a|$, then $\int \frac{r^{m+1}\cos(m-1)\lambda x - ar^m \cos m\lambda x}{r^2 - 2ar\cos\lambda x + a^2}dx$

$$= \mathrm{Re}\left(\int \sum_{k=0}^{\infty} a^k z^{-k-1+m}dx\right) \quad [\text{by (4)}]$$

$$= \int\left[\sum_{k=0}^{\infty}\frac{a^k}{r^{k+1-m}}\cos(k+1-m)\lambda x\right]dx \quad (\text{because } z = re^{i\lambda x})$$

A Study of the Integrals of Trigonometric Functions 607

$$= \sigma(m) \cdot a^{m-1}x + \frac{1}{\lambda} \cdot \sum_{\substack{k=0 \\ k \neq m-1}}^{\infty} \frac{a^k}{r^{k+1-m}(k+1-m)} \sin(k+1-m)\lambda x + C$$

(by integration term by term theorem) □

The same proof as Theorem 1, we can easily determine the Fourier series expansion of the integral of trigonometric function (2).

Theorem 2 *If the assumptions are the same as Theorem 1, then the integral*

$$\int \frac{r^{m+1}\sin(m-1)\lambda x - ar^m \sin m\lambda x}{r^2 - 2ar\cos\lambda x + a^2}dx$$

$$= \frac{1}{\lambda} \cdot \sum_{\substack{k=0 \\ k \neq -m}}^{\infty} \frac{r^{k+m}}{a^{k+1}(k+m)}\cos(k+m)\lambda x + C \quad \text{if } |r| < |a| \qquad (9)$$

$$= \frac{1}{\lambda} \cdot \sum_{\substack{k=0 \\ k \neq m-1}}^{\infty} \frac{a^k}{r^{k+1-m}(k+1-m)}\cos(k+1-m)\lambda x + C \quad \text{if } |r| > |a| \qquad (10)$$

3 Examples

In the following, aimed at the integrals of these two types of trigonometric functions, we provide two examples and use Theorems 1 and 2 to determine their Fourier series expansions. On the other hand, we use Maple to calculate the approximations of some definite integrals and their solutions for verifying our answers.

Example 1 In Theorem 1, we take $r = 2, a = 3, \lambda = 4, m = 5$. Then by (5), we obtained the Fourier series expansion of the integral

$$\int \frac{64\cos 16x - 96\cos 20x}{13 - 12\cos 4x}dx = -\frac{1}{4} \cdot \sum_{k=0}^{\infty} \frac{2^{k+5}}{3^{k+1}(k+5)}\sin(4k+20)x + C \qquad (11)$$

Therefore, we can determine the following definite integral

$$\int_{\pi/8}^{\pi/4} \frac{64\cos 16x - 96\cos 20x}{13 - 12\cos 4x}dx = \frac{1}{4} \cdot \sum_{k=0}^{\infty} \frac{2^{k+5}}{3^{k+1}(k+5)}\sin\frac{(k+5)\pi}{2} \qquad (12)$$

We use Maple to verify the correctness of (12) as follows:

>evalf(int((64*cos(16*x)−96*cos(20*x))/(13−12*cos(4*x)), x = Pi/8..Pi/4),18);

0.4070527218382429

>evalf(1/4*sum(2^(k + 5)/(3^(k + 1)*(k + 5))*sin((k + 5)*Pi/2), k = 0...∞),
18);

$$0.4070527218382429$$

Also, in Theorem 1, if we take $r = 5$, $a = 4$, $\lambda = 2$, $m = 3$. Then using (6), the Fourier series expansion of the integral

$$\int \frac{625\cos 4x - 500\cos 6x}{41 - 40\cos 2x}dx = 16x + \frac{1}{2} \cdot \sum_{\substack{k=0 \\ k\neq 2}}^{\infty} \frac{4^k}{5^{k-2}(k-2)}\sin(2k-4)x + C \quad (3)$$

Thus, we can determine the definite integral

$$\int_{\pi/4}^{\pi/2} \frac{625\cos 4x - 500\cos 6x}{41 - 40\cos 2x}dx = 4\pi - \frac{1}{2} \cdot \sum_{\substack{k=0 \\ k\neq 2}}^{\infty} \frac{4^k}{5^{k-2}(k-2)}\sin\frac{(k+2)\pi}{2} \quad (14)$$

Using Maple to verify the correctness of (14).

>evalf(int((625*cos(4*x)-500*cos(6*x))/(41−40*cos(2*x)), x = Pi/4...Pi/2),18);

$$-2.8315569234292484$$

>evalf(4*Pi−1/2*sum(4^k/(5^(k − 2)*(k − 2))*sin((k + 2)*Pi/2),k = 0...1)
−1/2*sum(4^k/(5^(k − 2)*(k − 2))*sin((k + 2)*Pi/2), k = 3...∞),18);

$$-2.83155692342924830$$

Example 2 In Theorem 2, taking $r = 3, a = 5, \lambda = 2, m = -4$. Then by (9), we obtained the Fourier series expansion of the integral

$$\int \frac{-\frac{1}{27}\sin 10x + \frac{5}{81}\sin 8x}{34 - 30\cos 2x}dx = \frac{1}{2} \cdot \sum_{\substack{k=0 \\ k\neq 4}}^{\infty} \frac{3^{k-4}}{5^{k+1}(k-4)}\cos(2k-8)x + C \quad (15)$$

Hence, we determined the following definite integral

$$\int_{\pi/4}^{\pi} \frac{-\frac{1}{27}\sin 10x + \frac{5}{81}\sin 8x}{34 - 30\cos 2x}dx = \frac{1}{2} \cdot \sum_{\substack{k=0 \\ k\neq 4}}^{\infty} \frac{3^{k-4}}{5^{k+1}(k-4)}\left(1 - \cos\frac{k\pi}{2}\right) \quad (16)$$

A Study of the Integrals of Trigonometric Functions   609

Also, we employ Maple to verify the correctness of (16).

\>evalf(int((−1/27*sin(10*x) + 5/81*sin(8*x))/(34 − 30*cos(2*x)),x = Pi/4...Pi),18);

$$−0.000786819398278323030$$

\>evalf(1/2*sum(3^(k-4)/(5^(k + 1)*(k-4))*(1-cos(k*Pi/2)),k = 0..3) + 1/2*sum(3^(k-4)/(5^(k + 1)*(k-4))*(1-cos(k*Pi/2)),k = 5..infinity),18);

$$−0.000786819398278323030$$

On the other hand, in Theorem 2, if we take $r = 2, a = 1, \lambda = 3, m = 6$. By (10), the Fourier series expansion of the following integral

$$\int \frac{128 \sin 15x − 64 \sin 18x}{5 − 4 \cos 3x} dx = \frac{1}{3} \cdot \sum_{\substack{k=0 \\ k \neq 5}}^{\infty} \frac{1}{2^{k-5}(k − 5)} \cos(3k − 15)x + C \quad (17)$$

Therefore, we can determine the following definite integral

$$\int_{\pi/3}^{2\pi/3} \frac{128 \sin 15x − 64 \sin 18x}{5 − 4 \cos 3x} dx = \frac{1}{3} \cdot \sum_{\substack{k=0 \\ k \neq 5}}^{\infty} \frac{1}{2^{k-5}(k − 5)} [1 − \cos(k + 1)\pi] \quad (18)$$

We employ Maple to verify the correctness of (16) as follows.

\>evalf(int((128*sin(15*x)−64*sin(18*x))/(5 − 4*cos(3*x)), x = Pi/3...2*Pi/3),18);

$$−7.01157368155507455$$

\>evalf(1/3*sum(1/(2^(k − 5)*(k − 5))*(1 − cos((k + 1)*Pi)),k = 0...4) + 1/3*sum(1/(2^(k − 5)*(k − 5))*(1 − cos((k + 1)*Pi)),k = 6...∞),18);

$$−7.01157368155507455$$

4 Conclusions

As mentioned, the integration term by term theorem plays a significant role in the theoretical inferences of this study. In fact, the application of this theorem is extensive, and can be used to easily solve many difficult problems; we endeavor to conduct further studies on related applications.

On the other hand, Maple also plays a vital assistive role in problem-solving. In the future, we will extend the research topic to other calculus and engineering mathematics problems and solve these problems by using Maple. These results will be used as teaching materials for Maple on education and research to enhance the connotations of calculus and engineering mathematics.

References

Abell ML, Braselton JP (2005) Maple by example, 3rd ed. Elsevier Academic Press

Apostol TM (1975) Mathematical analysis, 2nd ed. Addison-Wesley Publishing Co., Inc. Available at http://en.wikipedia.org/wiki/Inverse_trigonometric_functions

Dodson CTJ, Gonzalez EA (1995) Experiments in mathematics using Maple. Springer

Edwards CH Jr, Penney DE (1986) Calculus and analytic geometry, 2nd ed. Prentice-Hall, Inc., New Jersey

Garvan F (2001) The Maple book. Chapman & Hall/CRC

Grossman SI (1992) Calculus, 5th ed. Saunders College Publishing

Larson R, Hostetler RP, Edwards BH (2006) Calculus with analytic geometry, 8th ed. Houghton Mifflin

Richards D (2002) Advanced mathematical methods with Maple. Cambridge University Press

Robertson JS (1996) Engineering mathematics with Maple. McGraw-Hill

Stroeker RJ, Kaashoek JF (1999) Discovering mathematics with Maple: an interactive exploration for mathematicians, engineers and econometricians. Birkhauser Verlag

Yu CH (2011) Application of Fourier series on some integral problems. In: 100-Year general education symposium, National Pingtung University of Science and Technology, Taiwan, pp 307–320

Yu CH (2012a) Application of Maple: taking the integral problems as examples. 2012 digital technology and innovation management seminar, Huafan University, Taiwan, p A74

Yu CH (2012b) Application of Maple on some integral problems. In: 2012 international conference on safety and security management and engineering technology, WuFeng University, Taiwan, pp 290–294

Yu CH (2012c) Application of Maple: taking two special integral problems as examples. In: The 8th international conference on knowledge community, Chinese Culture University, Taiwan, pp 803–811

A Study of Optimization on Mainland Tourist Souvenir Shops Service Reliability

Kang-Hung Yang, Li-Peng Fang and Z-John Liu

Abstract Recently, Taiwan government launched tourism for mainland tourists, who have become the largest inbound tourists. Business opportunities burst, especially a souvenir shop. With rapid growths of souvenir shops, they compete with each other intensely. In order to survive in the competition market, a shop has to promote its service system. The best practice is to assess and optimize the current status of service quality to maintain a high reliability on its service systems. Reliability is a key indicator for the service quality, which is often applied to manufacturing and service. Failure Mode and Effects Analysis (FMEA) is an important tool for analyzing the reliability to forecast a system failure risk. This study applies FMEA to establish a failure risk evaluation model for a mainland tourist souvenir shop. Store H is validated by this approach, which can prioritize the potential failure items of Store H to improve the service quality and suggest persuasive corrective actions. The proposed model essentially ameliorates service level, effectively reduces the mainland tourist souvenir shops failure risks, and enhance the reliability of service. This approach is not only for the mainland tourist souvenir shops, but also be widely used in the general souvenir shops.

Keywords Mainland tourists · Souvenir shops · Service reliability · FMEA

K.-H. Yang (✉) · L.-P. Fang
Department of Industrial and Systems Engineering, Chung Yang Christian University,
Taiwan, China
e-mail: kanghungyang@cycu.edu.tw

L.-P. Fang
e-mail: g9902406@cycu.edu.tw

Z.-J. Liu
Department of International Business, Ling Tung University, Taiwan, China
e-mail: b91069025@yahoo.com.tw

Y.-K. Lin et al. (eds.), *Proceedings of the Institute of Industrial Engineers Asian Conference 2013*, DOI: 10.1007/978-981-4451-98-7_73,
© Springer Science+Business Media Singapore 2013

1 Introduction

Since 2008 Taiwan launched tourism for mainland tourists, mainland tourists has become the largest inbound tourists in Taiwan, the growth trend of tourists to Taiwan by country in the past years can be illustrated as Fig. 1 (Taiwan Tourism Bureau 2012), which generated opportunities for businesses, especially like a souvenir shop. While the domestic souvenir shops for mainland tourists increase year by year, those shops compete with each other a lot and operate in difficult with little marginal profits. The quality of service is not only the most important competitive weapon for a shop (Stafford 1996), but also a key feature to reveal differentiation of the horizontal competition (Morrall 1994). Customers will not only lose their loyalty and lead to their leaving, but affect the overall image of the shops if they are not satisfied with the shops' service. Therefore, how to evaluate high reliability, optimize the quality of service for maintaining the service system is the primary key for the mainland tourists souvenir shops to win in a competitive operation.

Reliability is an important indicator of the measurement in quality, which is often applied in the manufacturing industry and service sectors. To establish or maintain a high reliability depends on the application of analysis tools which include Failure Mode and Effects Analysis (FMEA). In addition, the tool can prevent from the risk of system failure. For the mainland tourist souvenir shops' service system in the process, each service will likely cause failures and lead to customer complaints. The service projects' failure situation, their effect and cause are all essential and as observed improvement factors to pursue a high quality of service. If the service failure occurs during the process of service, the causes of failure must be analyzed and be improved to reduce the risk of failure. With view of above reasons, this study investigates the improvement of the reliability in mainland tourist souvenir shops service system and attempts to apply them in FMEA. In addition, this study also analyzes the impact of the mainland tourist souvenir shops' service system failure. The projects for the system are given a weighting score for severity of failure, occurrence of degree and difficulties checking level. Finally, this study constructs FMEA assessment model for flow of

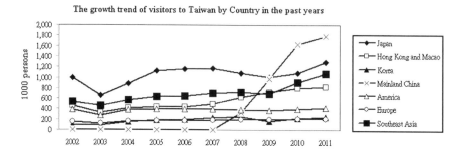

Fig. 1 The growth trend of tourists to Taiwan by country in the past years

the shops' service system and propose improvements to reduce the risk of failure of the service system and further optimize the reliability of the shops' service system.

2 Literature Review

2.1 Quality of Service

Service quality is the main form of customer satisfaction (Bitner 1990). Positive quality of service will improve service value and customer satisfaction (Rust and Oliver 1994). Customers would be willing to continue its transaction if the enterprise is providing good service (Keaveney 1995). Store atmosphere, the behavior of service staff, and merchandise display skills, it will affect the purchasing intention of tourists (Swanson and Horridge 2006). Chih et al. (2009) investigation gas station customer consumption behavior found that customer satisfaction can build up customer loyalty and customer satisfaction has a direct effect on customer loyalty.

2.2 Reliability

Weng (2011) pointed out that the "quality" and "reliability" is often confused, but in fact they are not the same. "Quality" means the production process, through the measure of actual quality performance with established quality standards for comparison, and to take the necessary production quality management decisions for which the quality difference. "Reliability" is on time considerations, in the setting of the use of environmental conditions or time conditions, and product or service can reach the required functional standard. In short, reliability is a product or service, whether in life or cycle process is normal, will be influenced by the reliability of the level of satisfaction on the quality of goods or services. It is increasing that uses reliability analysis to improve the quality of cases in service industry, failure mode and effects analysis (FMEA), and Fault Tree Analysis method is more used frequently.

2.3 Failure Mode and Effects Analysis

FMEA is a systematic method of reliability analysis, it is pointed out that the design and manufacturing of potential failure modes. It cans analysis the impact of failure for the system not only to give their assessment but take the necessary measures for system reliability problems (Hsu 2001). FMEA emphasis on prevention rather than corrective action after, can be applied in the system and at any stage of the product life cycle, such as the research and development stage of the system design, improve product reliability, maintainability and safety, as well as in the manufacturing of quality improvement (Hu 2008).

FMEA technology originated in the 1950s, the Grumman Aircraft Corporation, is mainly applied to the aircraft's main control system failure analysis. The United States Space Agency (NASA) used in the space program successfully in the 1960s. In 1970s, U.S. military also began to apply the FMEA technology, the FMEA operating procedures of the military standard MIL-STD-1692 and published in 1974. In 1980s, this standard was revised to MIL-STD-1692A, become important FMEA reference index (Hsu 2001). FMEA technology has already been widely used in the development of high-tech industries and traditional manufacturing or manufacturing stage. To the automotive industry, in 1993, General Motors, Ford and Chrysler, under the auspices of the U.S. quality control automobile branch of the Association and the Automotive Industry Action Group (AIAG), compile an effective operation manual "Potential Failure Mode and Effects Analysis Reference Manual", and can master the reliability. This technology with the QS 9000 quality system requirements as product development and process design tools in major automobile manufacturing center and its satellite factories (Chang et al. 2004).

FMEA is a tool for the analysis of safety and reliability, but also a proactive management technology. Because of its systemic thinking, simple and easy to use, it extended to the service system design and failure prevention services, even among the international standard ISO14001, OHSAS18001, TS16949 or technical specification in recent years (Hu 2008).

3 Research Methods

Study design is the primary work in research, the design process including decisions on the following: which variables to study, how to measure these variables, the method used to study, which is the object of the study, how to collect and analyze data, through the analysis of information, identify the best solution of the research problem (Jung 2008).

How to optimize the reliability of mainland tourist souvenir shops service is the main purpose of this study. Weng and Lin (2011) proposed FMEA execution steps are used for reference to construct FMEA assessment model in study. The following steps:

1. Understand the flow of mainland tourist souvenir shops service system.
2. Investigate all possible failure of the project on the system's degree of influence over the mainland tourist souvenir shops service system. Grading estimates and giving weights to the seriousness effects (severity, S), the likelihood that failure

associated with those effects will occur (occurrence, O), and an ability to detect the failure (detection, D). The assessment criteria are as follows:

(a) Severity (S): severity by failure to set the 10 level, the highest score of 10 points (very serious); level descending lowest score 1 point (non-hazardous).
(b) Occurrence (O): 10 level according to the frequency of failure occurs, a maximum score of 10 points (very high frequency); level descending, the lowest score 1 point (low frequency extremely).
(c) Detection (D): According the failure occurs be perceived of difficulty level set 10 rating, the highest score of 10 points (almost cannot be detected); level descending, the lowest scores 1 point (easily detected).

3. Using the assessment of risk priority number (RPN), RPN is a mathematical product of severity, occurrence, and detection, which can be expressed as follows:

$$RPN = S \times O \times D \qquad (1)$$

The highest failure factor of RPN is the highest risk of failure cause to the system, it must be improvement priority.

4. Proposed recommendations to optimize the mainland tourist souvenir shops' service system reliability according to the findings.

4 Empirical Analysis

4.1 The Collection of Factors in Failure Service

The empirical Store H is one of the first five big mainland tourist souvenir shops. This study selects Store H for mainland tourists that have purchased. The research methods were conducted by simple random sampling and random interview on service failure to understand the reasons in September 2012. The reasons of service failure are summarized into two systems at Store H. The 10 failures of the projects are as follows:

1. Purchase system

A1 It's prone to chaotic situations under un-planning route for customer
A2 The stores few service personnel can not be in response to the majority of customers at the same time questions
A3 The service personnel are not cordial expression on face
A4 Customers have doubts about that try to eat merchandise stores provide few

A5   When the shelves selling merchandise sold out, failed to make up to provide customers with purchase quickly.

2. Checkout system

B1   Checkout must wait in line (takes more than 5 min)
B2   The cashier operates slowly to delay customer time
B3   The cashier is not cordial expression on face
B4   The cashier chats with other colleagues to poor customer perception
B5   Checkout barcode does not scan interpretation and must type the artificial verbatim.

4.2 The Statistics of Risk in Failure Service

The 10 failures of the projects are arranged as question items. The 'simple random sampling' is applied for collecting data on Store H over the mainland tourists that have purchased by depth interview in November 2012. Before the interview, the 10 failure items and the scoring criteria on the severity and occurrence are explained to customers. The mainland tourist rating information is collected by question-and-answer format immediately. The interview samples are 250 in total. The incomplete answers are excluded as invalid samples. Eventually, 213 samples are valid. In addition, the depth interview is also conducted on the manager and 12 employees at Store H to collect the data of score in 10 failure difficult detection score. Finally, the average value of severity, occurrence and detection are obtained, and the RPN of 10 failure items are also calculated. The statistics by RPN from large to small rearrange as shown in Table 1. The failure project No. 6 [Checkout must wait in line (takes more than 5 min)] RPN = 81.84 is the highest, it must be of the highest priority improvement project. Failure project No. 5 (When the shelves selling merchandise sold out, failed to make up to provide customers with purchase quickly) RPN = 71.38 is the second highest, it must be classified as the second priority improvement project, the order of other failure projects for improvement shown in Table 1.

FMEA assessment results via constructed in this study can be screened to the failure of the project must be the first priority to improve for No. 6 [Checkout must wait in line (takes more than 5 min)]. The improvement practices can be adopted to increase the number of POS and train the cashiers more skillful on operation to reduce customers waiting time, and avoid mainland tourists are unsatisfied due to long waiting periods. Improvement method proposed for other failure projects shown in Table 2.

A Study of Optimization on Mainland Tourist

Table 1 The order of priority to improve in failure projects

Priority	No	Failure project	RPN
1	6	Checkout must wait in line (takes more than 5 min)	81.84
2	5	When the shelves selling merchandise sold out, failed to make up to provide customers with purchase quickly	71.38
3	7	The cashier operates slowly to delay customer time	64.80
4	2	The stores few service personnel can not be in response to the majority of customers at the same time questions	60.59
5	3	The service personnel are not cordial expression on face	59.94
6	8	The cashier is not cordial expression on face	54.67
7	1	It's prone to chaotic situations under un-planning route for customer	54.12
8	9	The cashier chats with other colleagues to poor customer perception	51.83
9	10	Checkout barcode does not scan interpretation and must type the artificial verbatim	43.55
10	4	Customers have doubts about that try to eat merchandise stores provide few	28.70

Table 2 The improvement suggestions for failure projects

Priority	No	RPN	Improvement method
1	6	81.84	To increase the number of POS and train the cashiers more skillful on operation
2	5	71.38	To take the hot-selling goods enough, and designate staffs to fill up
3	7	64.80	To train the cashiers more skillful on operation
4	2	60.59	Can provide enough service staffs to respond the questions of customers at any time
5	3	59.94	To train the service staffs more courtesy and polite
6	8	54.67	To train the cashiers more courtesy and polite
7	1	54.12	To re-plan the purchasing route for customer and designate service staffs to guide
8	9	51.83	To set penalties and fines for the cashiers chat with colleagues during the operation
9	10	43.55	To repair scanning bar code machines and control barcode manufactured process
10	4	28.70	Increase the types of goods to try eating

5 Conclusions

The empirical result of assessment model shows that the service system failure of Store H must be prioritized to improve in project No. 6 [Checkout must wait in line (takes more than 5 min)]. This failure of the project is alike with the most of people in the process of buying behavior "impatient" and produces the same effect as unsatisfactory. It is often one of the main reasons to refuse to purchase again. This shows that the proposed model essentially ameliorate service level and effectively reduce the mainland tourist souvenir shops failure risks and enhance the reliability of service.

The FMEA has systemic thinking and with advantage of easily use. This evaluation model not only can provide reference for the mainland tourist souvenir shops, but also be widely used in the general souvenir shops.

References

Bitner MJ (1990) Evaluating service encounters: the effects of physical surroundings and employee responses. J Mark 54(2):69–82

Chang TM, Wu YW, Ho TP, Chiang JK (2004) The Application of "PFMEA" in manufacturing of automobile part. Hsiuping J 9:137–156

Chih WH, Chen JL, Pan MH (2009) Brand equity, service quality, perceived value, customer satisfaction and customer loyalty-an empirical study of Chinese petroleum corporation, Taiwan service stations. J Qual 16(4):291–309

Hsu SP (2001) A QFD and FMEA integrated model. Master dissertation, Department of Industrial Engineering and Management, Yuan Ze University, Taiwan

Hu KY (2008) Application of FMEA to reduce service failure. Tajen J 33:35–50

Jung TS (2008) Business research methods, 3rd edn. Wu Nan, Taipei

Keaveney SM (1995) Customer switching behavior in service industries: an exploratory study. J Mark 59(2):71–82

Morrall K (1994) Service quality: the ultimate differentiator. Bank Mark 26(10):33–38

Rust RT, Oliver RL (1994) Service quality: insights and managerial implications from the frontier, in theory and practice. In: Rust RT, Oliver RL (eds) Service quality: new directions. Thousand Oaks, Sage, pp 1–19

Sightseeing statistics chart (Taiwan Tourism Bureau) (2012) http://admin.taiwan.net.tw/public/public.aspx?no=315. Accessed 13 Aug 2012

Stafford MR (1996) Demographic discriminators of service quality in the banking industry. J Serv Mark 10(4):6–22

Swanson KK, Horridge PE (2006) Travel motivations as souvenir purchase indicators. Tourism Manag 27(4):671–683

Weng SJ (2011) Short introduction of reliability engineering. Qual Mag 47(1):44–47

Weng SJ, Lin KP (2011) Reliability engineering-introduction of FMEA. Qual Mag 47(2):32–34

The Effects of Background Music Style on Study Performance

An-Che Chen and Chen-Shun Wen

Abstract Due to the increasing popularity of personal digital devices, many students listen to music while they study. It is however a controversial issue whether music listening is helpful to study performance. This study investigates the effects of different types of background music on study performance among college students through lab experiments. Two major categories of study activities (i.e., reading comprehension and mathematical computation) are examined for four different treatments of background music style (i.e., soft music, rock music, heavy metal music, and no music). For each student subject, objective measures, such as test scores and heart rates, were recorded for all conditions of the experiment design. Subjective measures concerning treatment evaluations along with personal preference and behaviors on music listening were instrumented in the individual interviews after the experiments. Data analysis on the objective measures indicates that neither test scores nor heart rates of reading comprehension and mathematic computation for different styles of background music are with statistical significance. By further cross-referencing with the subjective measures, our results suggest that, for a better studying performance, college students may choose to listen to background music with preferred music for reading activities but non-preferred music for mathematic computation.

Keywords Background music · Study performance

A.-C. Chen (✉)
Ming Chi University of Technology, New Taipei City, Taiwan, Republic of China
e-mail: anche@mail.mcut.edu.tw

C.-S. Wen
Nanya Technology, New Taipei City, Taiwan, Republic of China
e-mail: s114202001@hotmail.com

Y.-K. Lin et al. (eds.), *Proceedings of the Institute of Industrial Engineers Asian Conference 2013*, DOI: 10.1007/978-981-4451-98-7_74,
© Springer Science+Business Media Singapore 2013

1 Introduction and Background

Due to the increasing popularity of personal digital devices, many students listen to music while they study. It is however a controversial issue whether music listening is helpful to cognitive memory or study performance (Bellezza 1996; Pietschnig et al. 2010). Deems (2001) both found that students who normally listened to music while studying scored higher on reading comprehension tests compared to those who usually studied without any background music. Nittono et al. (2000) tested on 24 undergraduates performing a self-paced line tracing task with different tempos of background music and found that fast music accelerated performance compared with slow music. The study of Haynes indicated that studying to background music did reduce the math anxiety of college students.

On the other hand, Kiger (1989) reported that, for high school students, reading comprehension was best when material is learned in silence and worst in presence of high information-load music. Tucker and Bushman (1991) found that rock and roll background music decreased performance of undergraduate students on math and verbal tests, but not their scores on reading comprehension. Manthei and Kelly (1999) reported that the music had no statistically significant effect on the math test scores. Burns et al. (2002) suggested that different types of music have different effects on stress. While the analysis does not indicate that listening to classical or relaxing music decreases anxiety, it does suggest that hard rock music may compromise one's ability to relax. However, the test scores were not affected by the background music for students with different levels of anxiety. These studies suggest that background music act as a distracter to students trying to focus on studying.

Upon past research the influences of background music to learning performance may still be unclear and seems to have certain connections with personal preference. This present study therefore seeks to re-investigate the effects of different types of background music on study performance in terms of reading comprehension and math calculation through lab experiments and further linked with personal preference in music.

2 Method

This study investigates the effects of different types of background music on study performance through lab experiments. College students with non-music major are our target experiment subjects. Prior to the experiment, each student subject was requested to finish a questionnaire regarding demographic information (e.g., age, gender, and major), frequency of music listening while studying, and the preferred background music type.

In the experiment, two major categories of study activities (i.e., reading comprehension and mathematical computation) are examined for four different treatments of background music style (i.e., soft music, rock music, heavy metal music, and no music). In the reading comprehension sessions, subjects are asked to read several short essays (in Chinese) and answer the quiz questions following each essay. Mathematical questions on algebra and equation solving in fundamental high-school levels are instrumented for the math computation sessions. For each of the three background music types, two songs sampled from the pop music market in Taiwan are alternatively played to cover the entire experiment session. A within-subject completed randomized experiment design is therefore instrumented. That is, each student subject was tested on a set of all eight conditions (i.e., the combination of the two study categories with the four background music types) in random orders. Each experiment condition lasted for 30 min to ensure the sensitivity and validity of performance measures. These performance or objective measures, such as test scores (the percentages of the correct answers) and heart rate variation, were recorded for all conditions of the experiment design.

Subjective measures concerning treatment evaluations and cognitive influences were instrumented in the individual interviews after the experiments. The primary interests of these semi-structured post-experiment debriefings include possible distractions by lyrics or singer's image, positive or negative influences on attentions with different music types, and the major cognitive passageways behind those influences.

3 Results and Discussion

Twenty university students with non-music majors in Taiwan voluntarily participated in this study. Sixty percent were male (n = 12) and 40 % female (n = 8). The ANOVA results of gender and music type to test scores for reading comprehension and math computation show no significant effects except the gender effect to reading comprehension ($p = 0.008$). Figure 1 depicts the detailed data of gender differences in test performance under various types of background music. It is obvious that the test performance of female subjects is usually higher than that of male subjects for reading comprehension. This plot also suggests that, for reading comprehension in particular, the performance of females be less sensitive to music types while males perform better in listening to rock music or without any background music.

By further comparing the effects of music type across two study categories, as shown in Fig. 2, it is obvious that very limited variation in test performance was found across different types of background music for the reading comprehension sessions but the effect of music type to math computation showed a different pattern. For the math computation sessions in particular, the performance without background music was relatively poor while the performance of providing soft music was the best.

As to the measures of heart rate variation, which were the differences (in BPM) between the average HR of the last 5 min in each experiment session to that of the

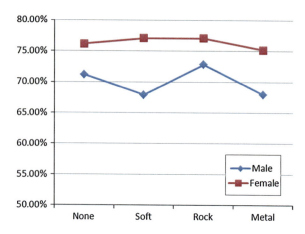

Fig. 1 Test performance of reading comprehension under different background music types

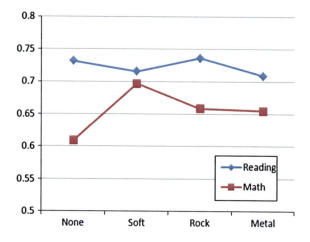

Fig. 2 Test performance under different types of background music

first five minutes, the ANOVA results show no statistical significance except the gender effect. By further examining the interaction plot shown in Fig. 3, it is obvious that male subjects generally demonstrated greater HR elevation than the females did. These data also suggest, mostly, the insignificant effects to heart rate variations across different background music types. But for male subjects in the math computation sessions, however, soft music showed a prominent HR elevation effect in contrast to other types of background music.

Therefore, in general, our analysis suggests that music type show any statistically significant effect on neither test performance nor heart rate elevation. This result rather concurs with both the findings reported in Manthei and Kelly (1999) and Haynes (2003). The analysis regarding the discrepancy between reading comprehension and math computation in our study, which is depicted in Fig. 2, shows similar patterns reported in Tucker and Bushman (1991). Our analysis on the post-experiment debriefings indicates that the lyrics in all music types and the

Fig. 3 The interaction plot for gender and music type to heart rate variation

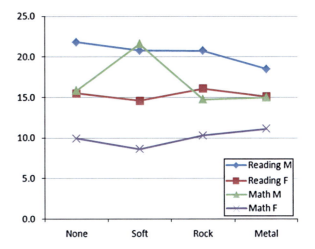

noisy strong beats in rock and roll and heavy metal music acted as distractors to focused attentions in study activities. This result seems to well reflect the findings in Kiger (1989) and Burns et al. (2002) as well.

In order to investigate the individual preference effects demonstrated in Etaugh and Michals (1975) and Deems (2001), we cross-referenced the personal preference data from the pre-test questionnaires with the actual performance data collected in the experiments. Figure 4 shows the actual performance rankings of the preferred music type indicated by the individual subjects before the experiments. It is apparent that the results of two study categories did not share similar patterns. For reading comprehension, more than half (55 %) of subjects performed better (i.e., ranked best or second) when the type of background music provided was their personal preference. For math computation, in a reverse fashion, more than two-third of the test performance of the preferred music provided actually came in worse (i.e., third or worst ranking), compared to their non-preferred music. In the

Fig. 4 Test performance rankings for individual preferred music types

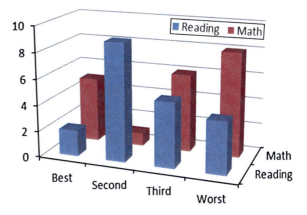

post-experiment debriefings, several subjects reported that non-preferred music were in fact less distracting and therefore made the focused attention easier for study activities.

4 Conclusions

Data analysis on the objective measures indicates that neither test scores nor heart rates of reading comprehension and mathematic computation for different styles of background music are with statistical significance. Gender differences were found and the females seem to be less sensitive to the changes of background music types. By further cross-referencing with the subjective measures, our results suggest that, for a better studying performance, college students may choose to listen to background music with preferred music for reading activities but non-preferred music for mathematic computation.

References

Bellezza F (1996) Mnemonic methods to enhance storage and retrieval. In: Bjork E, Bjork R (eds) Memory. Academic Press, New York, pp 345–380

Burns JL, Labbé E, Arke B, Capeless K, Cooksey B, Steadman A, Gonzales C (2002) The effects of different types of music on perceived and physiological measures of stress. J Music Ther XXXIX(2):101–116

Deems DA (2001) The effects of sound on reading comprehension and short-term memory. Department of Psychology, MWSC

Kiger DM (1989) Effects of music information load on a reading comprehension task. Percept Mot Skills 69:531–534

Manthei M, Kelly SN (1999) Effects of popular and classical background music on undergraduate math test scores. Res Perspect Music Educ 1:38–42

Nittono H, Tsuda A, Nakajima Y (2000) Tempo of background sound and performance speed. Percept Mot Skills 90(3/2):1122

Pietschnig J, Voracek M, Formann AK (2010) Mozart effect- a meta-analysis. Intelligence 38(3):314–323

Tucker A, Bushman BJ (1991) Effects of rock and roll music on mathematical, verbal, and reading comprehension performance. Percept Mot Skills 72:942

Using Maple to Study the Multiple Improper Integral Problem

Chii-Huei Yu

Abstract Multiple improper integral problem is closely related with probability theory and quantum field theory. Therefore, the evaluation and numerical calculation of multiple improper integrals is an important issue. This paper takes the mathematical software Maple as the auxiliary tool to study some types of multiple improper integrals. We can obtain the infinite series forms of these types of multiple improper integrals by using integration term by term theorem. On the other hand, we propose some examples to do calculation practically. Our research way is to count the answers by hand, and then use Maple to verify our results. This research way can not only let us find the calculation errors but also help us to revise the original thinking direction because we can verify the correctness of our theory from the consistency of hand count and Maple calculations. Therefore, Maple can bring us inspiration and guide us to find the problem-solving method, this is not an exaggeration.

Keywords Multiple improper integrals · Integration term by term theorem · Infinite series forms · Maple

1 Introduction

As information technology advances, whether computers can become comparable with human brains to perform abstract tasks, such as abstract art similar to the paintings of Picasso and musical compositions similar to those of Mozart, is a natural question. Currently, this appears unattainable. In addition, whether computers can solve abstract and difficult mathematical problems and develop abstract

---

C.-H. Yu (✉)
Department of Management and Information, Nan Jeon Institute of Technology, No. 178, Chaoqin Road, Yanshui District, Tainan City 73746, Taiwan, Republic of China
e-mail: chiihuei@mail.njtc.edu.tw

626   C.-H. Yu

mathematical theories such as those of mathematicians also appears unfeasible. Nevertheless, in seeking for alternatives, we can study what assistance mathematical software can provide. This study introduces how to conduct mathematical research using the mathematical software Maple. The main reasons of using Maple in this study are its simple instructions and ease of use, which enable beginners to learn the operating techniques in a short period. By employing the powerful computing capabilities of Maple, difficult problems can be easily solved. Even when Maple cannot determine the solution, problem-solving hints can be identified and inferred from the approximate values calculated and solutions to similar problems, as determined by Maple. For this reason, Maple can provide insights into scientific research. Inquiring through an online support system provided by Maple or browsing the Maple website (www.maplesoft.com) can facilitate further understanding of Maple and might provide unexpected insights. For the instructions and operations of Maple, we can refer to (Abell 2005; Dodson and Gonzalez 1995; Garvan 2001; Richards 2002; Robertson 1996; Stroeker and Kaashoek 1999).

In this paper, we mainly study the multiple improper integral problem. This problem is closely related with probability theory and quantum field theory, and can refer to (Ryder 1996; Streit 1970). Therefore, the evaluation and numerical calculation of multiple improper integrals is an important issue, and can be studied based on (Edwards and Penney 1986, Chap. 16; Grossman 1992, Chap. 14; Larson et al. 2006, Chap. 14; Widder 1961, 6; Yu 2012a, b, c, 2013). In this paper, we determined the following two types of multiple improper integrals

$$\int_0^1 \cdots \int_0^1 \prod_{m=1}^n (\ln x_m)^{p_m} \cdot \prod_{m=1}^n x_m^{a_m} \cdot \sin^{-1}\left(\prod_{m=1}^n x_m^{b_m}\right) dx_1 \ldots dx_n \tag{1}$$

$$\int_0^1 \cdots \int_0^1 \prod_{m=1}^n (\ln x_m)^{p_m} \cdot \prod_{m=1}^n x_m^{a_m} \cdot \cos^{-1}\left(\prod_{m=1}^n x_m^{b_m}\right) dx_1 \ldots dx_n \tag{2}$$

where n is any positive integer, and $a_m, b_m > 0$, p_m are positive integers for all $m = 1, \ldots, n$. We can find the infinite series forms of these two types of multiple improper integrals; these are the main results of this study (i.e., Theorems 1 and 2). At the same time, we obtained two corollaries from these two theorems. Additionally, in several examples in which Theorems 1 and 2 were practically employed to determine the infinite series forms of these multiple improper integrals. The research methods adopted in this study involved finding solutions through manual calculations and verifying these solutions by using Maple. This type of research method not only allows the discovery of calculation errors, but also helps modify the original directions of thinking from manual and Maple calculations. Therefore, Maple provides insights and guidance regarding problem-solving methods.

Using Maple to Study the Multiple Improper Integral Problem

2 Main Results

Firstly, we introduce some formulas and theorems used in this study.

Formulas:
(Available at http://en.wikipedia.org/wiki/Inverse_trigonometric_functions)

1. $\sin^{-1} y = \sum_{k=0}^{\infty} \frac{\binom{2k}{k}}{4^k(2k+1)} y^{2k+1}$, where $y \in R, |y| \le 1$.

2. $\cos^{-1} y = \frac{\pi}{2} - \sum_{k=0}^{\infty} \frac{\binom{2k}{k}}{4^k(2k+1)} y^{2k+1}$, where $y \in R, |y| \le 1$.

Differentiation term by term theorem (Apostol 1975, p 230). If, for all non-negative integer k, the functions $g_k : (a,b) \to R$ satisfy the following three conditions: (1) there exists a point $x_0 \in (a,b)$ such that $\sum_{k=0}^{\infty} g_k(x_0)$ is convergent, (2) all functions $g_k(x)$ are differentiable on open interval (a,b), (3) $\sum_{k=0}^{\infty} \frac{d}{dx} g_k(x)$ is uniformly convergent on (a,b). Then $\sum_{k=0}^{\infty} g_k(x)$ is uniformly convergent and differentiable on (a,b). Moreover, its derivative $\frac{d}{dx} \sum_{k=0}^{\infty} g_k(x) = \sum_{k=0}^{\infty} \frac{d}{dx} g_k(x)$.

Integration term by term theorem (Apostol 1975, p 269). Suppose that $\{g_n\}_{n=0}^{\infty}$ is a sequence of Lebesgue integrable functions defined on interval I. If $\sum_{n=0}^{\infty} \int_I |g_n|$ is convergent, then $\int_I \sum_{n=0}^{\infty} g_n = \sum_{n=0}^{\infty} \int_I g_n$.

Differentiation with respect to a parameter (Flatto 1976, p 405). Suppose $I = [a_{11}, a_{12}] \times [a_{21}, a_{22}] \times \ldots \times [a_{n1}, a_{n2}]$, and n-variables function $f(x_1, x_2, \ldots, x_n)$ is defined on I. If $f(x_1, x_2, \ldots, x_n)$ and its partial derivative $\frac{\partial f}{\partial x_1}(x_1, x_2, \ldots, x_n)$ are continuous functions on I. Then

$$F(x_1) = \int_{a_{21}}^{a_{22}} \int_{a_{31}}^{a_{32}} \cdots \int_{a_{n1}}^{a_{n2}} f(x_1, x_2, x_3, \ldots, x_n) dx_2 dx_3 \ldots dx_n$$

is differentiable on the open interval (a_{11}, a_{12}), and its derivative

$$\frac{d}{dx_1} F(x_1) = \int_{a_{21}}^{a_{22}} \int_{a_{31}}^{a_{32}} \cdots \int_{a_{n1}}^{a_{n2}} \frac{\partial f}{\partial x_1}(x_1, x_2, x_3, \ldots, x_n) dx_2 dx_3 \ldots dx_n$$

Firstly, we determined the infinite series form of multiple improper integral (1).

628   C.-H. Yu

Theorem 1 Assume n is any positive integer, and $a_m, b_m > 0, p_m$ are positive integers for all $m = 1, \ldots, n$. Then the n-tiple improper integral

$$
\int_0^1 \cdots \int_0^1 \prod_{m=1}^n (\ln x_m)^{p_m} \cdot \prod_{m=1}^n x_m^{a_m} \cdot \sin^{-1}\left(\prod_{m=1}^n x_m^{b_m}\right) dx_1 \ldots dx_n
$$

$$
= (-1)^{\sum_{m=1}^n p_m} \cdot \prod_{m=1}^n p_m! \cdot \sum_{k=0}^\infty \frac{\binom{2k}{k}}{4^k(2k+1) \cdot \prod_{m=1}^n (2b_m k + a_m + b_m + 1)^{p_m+1}}
$$

$$\tag{3}$$

Proof If $0 \le x_m \le 1$, for all $m = 1, \ldots, n$, then

$$
\prod_{m=1}^n x_m^{a_m} \sin^{-1}\left(\prod_{m=1}^n x_m^{b_m}\right) = \prod_{m=1}^n x_m^{a_m} \cdot \sum_{k=0}^\infty \frac{\binom{2k}{k}}{4^k(2k+1)} \left(\prod_{m=1}^n x_m^{b_m}\right)^{2k+1} \quad \text{(by formula (1))}
$$

$$
= \sum_{k=0}^\infty \frac{\binom{2k}{k}}{4^k(2k+1)} \cdot \prod_{m=1}^n x_m^{2b_m k + a_m + b_m}
$$

$$\tag{4}$$

Therefore,

$$
\int_0^1 \cdots \int_0^1 \prod_{m=1}^n x_m^{a_m} \cdot \sin^{-1}\left(\prod_{m=1}^n x_m^{b_m}\right) dx_1 \ldots dx_n
$$

$$
= \int_0^1 \cdots \int_0^1 \sum_{k=0}^\infty \frac{\binom{2k}{k}}{4^k(2k+1)} \cdot \prod_{m=1}^n x_m^{2b_m k + a_m + b_m} dx_1 \ldots dx_n
$$

$$
= \sum_{k=0}^\infty \frac{\binom{2k}{k}}{4^k(2k+1)} \cdot \int_0^1 \cdots \int_0^1 \prod_{m=1}^n x_m^{2b_m k + a_m + b_m} dx_1 \ldots dx_n.
$$

$$\tag{5}$$

$$
\text{(by integration term by term theorem)}
$$

$$
= \sum_{k=0}^\infty \frac{\binom{2k}{k}}{4^k(2k+1)} \cdot \prod_{m=1}^n \left(\int_0^1 x_m^{2b_m k + a_m + b_m} dx_m\right)
$$

$$
= \sum_{k=0}^\infty \frac{\binom{2k}{k}}{4^k(2k+1) \cdot \prod_{m=1}^n (2b_m k + a_m + b_m + 1)}
$$

Using Maple to Study the Multiple Improper Integral Problem

By differentiation with respect to a parameter and differentiation term by term theorem, differentiating each a_m by p_m times on both sides of Eq. (5), we obtained

$$\int_0^1 \cdots \int_0^1 \prod_{m=1}^n (\ln x_m)^{p_m} \cdot \prod_{m=1}^n x_m^{a_m} \cdot \sin^{-1}\left(\prod_{m=1}^n x_m^{b_m}\right) dx_1 \ldots dx_n$$

$$=(-1)^{\sum_{m=1}^n p_m} \cdot \prod_{m=1}^n p_m! \cdot \sum_{k=0}^\infty \frac{\binom{2k}{k}}{4^k(2k+1) \cdot \prod_{m=1}^n (2b_m k + a_m + b_m + 1)^{p_m+1}}$$

In Theorem 1, substituting $x_m = e^{-t_m}$ (where $t_m > 0$ for all $m = 1, \ldots, n$), we obtained the following result.

Corollary 1 Suppose the assumptions are the same as Theorem 1. Then the n-tiple multiple improper integral

$$\int_1^\infty \cdots \int_1^\infty \prod_{m=1}^n t_m^{p_m} \cdot \exp-\left(\sum_{m=1}^n a_m t_m\right) \cdot \sin^{-1}\left[\exp-\left(\sum_{m=1}^n b_m t_m\right)\right] dx_1 \ldots dx_n$$

$$=\prod_{m=1}^n p_m! \cdot \sum_{k=0}^\infty \frac{\binom{2k}{k}}{4^k(2k+1) \cdot \prod_{m=1}^n (2b_m k + a_m + b_m + 1)^{p_m+1}}$$

$$(6)$$

Using Theorem 1 and formula (2), we can easily determine the infinite series form of the multiple improper integral (2).

Theorem 2 If the assumptions are the same as Theorem 1, then the n-tiple multiple improper integral

$$\int_0^1 \cdots \int_0^1 \prod_{m=1}^n (\ln x_m)^{p_m} \cdot \prod_{m=1}^n x_m^{a_m} \cdot \cos^{-1}\left(\prod_{m=1}^n x_m^{b_m}\right) dx_1 \ldots dx_n$$

$$=(-1)^{\sum_{m=1}^n p_m} \cdot \prod_{m=1}^n p_m! \cdot \left[\frac{\pi}{2 \cdot \prod_{m=1}^n (a_m + 1)^{p_m+1}} - \sum_{k=0}^\infty \frac{\binom{2k}{k}}{4^k(2k+1) \prod_{m=1}^n (2b_m k + a_m + b_m + 1)^{p_m+1}}\right]$$

$$(7)$$

In Theorem 2, substituting $x_m = e^{-t_m}$ (where $t_m > 0$ for all $m = 1, \ldots, n$), we obtained the following result.

Corollary 2 Suppose the assumptions are the same as Theorem 1. Then the n-tiple multiple improper integral

$$\int_1^\infty \cdots \int_1^\infty \prod_{m=1}^n t_m^{p_m} \cdot \exp-\left(\sum_{m=1}^n a_m t_m\right) \cdot \cos^{-1}\left[\exp-\left(\sum_{m=1}^n b_m t_m\right)\right] dx_1 \ldots dx_n$$

$$= \prod_{m=1}^n p_m! \cdot \left[\frac{\pi}{2 \cdot \prod_{m=1}^n (a_m+1)^{p_m+1}} - \sum_{k=0}^\infty \frac{\binom{2k}{k}}{4^k(2k+1)\prod_{m=1}^n (2b_mk + a_m + b_m + 1)^{p_m+1}}\right]$$

$$(8)$$

3 Examples

In the following, aimed at the two types of multiple improper integrals, we propose two examples and use Theorems 1 and 2 to determine their solutions. On the other hand, we employ Maple to calculate the approximations of these multiple improper integrals and their infinite series forms for verifying our answers.

Example 1 By theorem 1, we determined the following double improper integral

$$\int_0^1 \int_0^1 (\ln x_1)^2 (\ln x_2) \cdot x_1 x_2^3 \cdot \sin^{-1}(x_1^2 x_2) dx_1 dx_2$$

$$= -2 \cdot \sum_{k=0}^\infty \frac{\binom{2k}{k}}{4^k(2k+1) \cdot (4k+4)^3 (2k+5)^2} \tag{9}$$

We use Maple to verify our answer.

>evalf(Doubleint((ln(x1))^2*(ln(x2))*x1*x2^3*arcsin(x1^2*x2),x1=0..1,x2= 0..1),18);

$$-0.00126460806826573483$$

>evalf(-2*sum((2*k)!/(k!*k!*4^k*(2*k+1)*(4*k+4)^3*(2*k+5)^2), k=0..infinity),18);

$$-0.00126460806826573482$$

Example 2 Using theorem 2, we obtained the following triple improper integral

$$\int_0^1 \int_0^1 \int_0^1 (\ln x_1)^3 (\ln x_2)^2 (\ln x_3) \cdot x_1^2 x_2 x_3^3 \cdot \cos^{-1}\left(x_1 x_2 x_3^2\right) dx_1 dx_2 dx_3$$

$$= 12 \cdot \left[\frac{\pi}{20736} - \sum_{k=0}^{\infty} \frac{\binom{2k}{k}}{4^k = (2k+1)(2k+4)^4 (2k+3)^3 (4k+6)^2} \right]$$

$$(10)$$

Also, we employ Maple to verify our answer.

>evalf(Tripleint(((ln(x1))^3*(ln(x2))^2*(ln(x3))*x1^2*x2*x3^3*arccos(x1*x2*x3^2),x1=0..1,x2=0..1,x3=0..1),14);

$$0.0017696990100762$$

>evalf(12*(Pi/20736-sum((2*k)!/(k!*k!*4^k*(2*k+1)*(2*k+4)^4*(2*k+3)^3*(4*k+6)^2),k=0..infinity)),14);

$$0.0017696990100763$$

4 Conclusions

As mentioned, the integration term by term theorem, the differentiation term by term theorem, and the differentiation with respect to a parameter play significant roles in the theoretical inferences of this study. In fact, the applications of these theorems are extensive, and can be used to easily solve many difficult problems; we endeavor to conduct further studies on related applications.

On the other hand, Maple also plays a vital assistive role in problem-solving. In the future, we will extend the research topic to other calculus and engineering mathematics problems and solve these problems by using Maple. These results will be used as teaching materials for Maple on education and research to enhance the connotations of calculus and engineering mathematics.

References

Abell ML, Braselton JP (2005) Maple by example, 3rd edn. Elsevier Academic Press, Waltham

Apostol TM (1975) Mathematical analysis, 2nd edn. Addison-Wesley Publishing Co., Inc, Botson. Available at http://en.wikipedia.org/wiki/Inverse_trigonometric_functions

Dodson CTJ, Gonzalez EA (1995) Experiments in mathematics using Maple. Springer, New York

Edwards CH Jr, Penney DE (1986) Calculus and analytic geometry, 2nd edn. Prentice-Hall Inc, Upper Saddle River

Flatto L (1976) Advanced calculus. The Williams & Wilkins Co, Baltimore

Garvan F (2001) The maple book. Chapman & Hall/CRC, London

Grossman SI (1992) Calculus, 5th edn. Saunders College Publishing, Orlando

Larson R, Hostetler RP, Edwards BH (2006) Calculus with analytic geometry, 8th edn. Houghton Mifflin Harcourt, Botson

Richards D (2002) Advanced mathematical methods with maple. Cambridge University Press, Cambridge

Robertson JS (1996) Engineering mathematics with maple. McGraw-Hill, New York

Ryder LH (1996) Quantum field theory, 2nd edn. Cambridge University Press, Cambridge

Streit F (1970) On multiple integral geometric integrals and their applications to probability theory. Can J Math 22:151–163

Stroeker RJ, Kaashoek JF (1999) Discovering mathematics with maple: an interactive exploration for mathematicians, engineers and econometricians. Birkhauser Verlag, Berlin

Widder DV (1961) Advanced calculus. Prentice-Hall, Upper Saddle River

Yu CH (2012a) Evaluation of two types of multiple improper integrals, 2012 Changhua, Yunlin and Chiayi Colleges Union Symposium, Da-Yeh University, Taiwan

Yu CH (2012b) Application of maple on multiple improper integral problems, 2012 optoelectronics and communication engineering workshop. National Kaohsiung University of Applied Sciences, Taiwan

Yu CH (2012c) Evaluating multiple improper integral problems, 101 year general education symposium. National Pingtung University of Science and Technology, Taiwan

Yu CH (2013) Application of maple: taking the double improper integrals as examples, 2013 information education and technology application seminar. Overseas Chinese University, Taiwan

On Reformulation of a Berth Allocation Model

Yun-Chia Liang, Angela Hsiang-Ling Chen and Horacio Yamil Lovo Gutierrezmil

Abstract Over the last decade, Ports' operations have been a focal point for global supply chain management and logistics network structures around the world. The berth allocation problem (BAP) is closely related to the operational performance of any port. BAP consists of optimally assigning ships to berthing areas along the quay in a port. A good allocation of ships to berths has a positive impact on terminal's productivity and customer's satisfaction. Therefore, finding valid formulations which captures the nature of the BAP and accounts for the interest of ports operation management is imperative for practitioners and researchers in this field. In this study, various arrival times of ships are embedded in a real time scheduling system to address the berth allocation planning problem in a dynamic environment. A new mix integer mathematical formulation (MBAP) that accounts for two objectives—total waiting time and total handling time is proposed. The MBAP model is further evaluated in terms of computational time, and has demonstrated to be competitive compared to another well-known BAP formulation in the literature.

Keywords Berth allocation problem · Supply chain management · Ports operation management

---

Y.-C. Liang (✉) · H. Y. L. Gutierrezmil
Department of Industrial Engineering and Management, Yuan Ze University,
No. 135, Yuan-Tung Rd., Chungli 320 Taoyuan County, Taiwan, Republic of China
e-mail: ycliang@saturn.yzu.edu.tw

H. Y. L. Gutierrezmil
e-mail: s985458@mail.yzu.edu.tw

A. H.-L. Chen
Department of Marketing and Distribution Management, Taoyuan Innovation Institute
of Technology, No. 141, Sec. 3, Jhongshan Rd., Chungli 320 Taoyuan County,
Taiwan, Republic of China
e-mail: achen@tiit.edu.tw

Y.-K. Lin et al. (eds.), *Proceedings of the Institute of Industrial
Engineers Asian Conference 2013*, DOI: 10.1007/978-981-4451-98-7_76,
© Springer Science+Business Media Singapore 2013

1 Introduction

In the past few years, the containerized maritime transportation has gained a crucial role in the trade of goods all over the world. The transshipment ports handle a large number of containers and require high levels of efficiency to stay its competitiveness. This has made the operation managers to endeavor their efforts in the stream linearization of the port's terminal activities and the efficient utilization of different resources. Generally, when a container ship arrives at a port, a part of the quay (a berth) is arranged to take care of the container. The quay cranes are in charge of loading and unloading the cargo which is moved from and to the yard by straddle carriers. The overall time a ship spends in port depends on the waiting time and the handling time at the berth where the ship is moored. Berths, therefore, seem one of the most important resources to be utilized for a cost-effective container terminal management. More so, a good berth allocation enhances not only terminal productivity and ship owners' satisfaction, but also produces a positive impact in any company's revenue levels.

The berth allocation problem (BAP) is an NP-Hard as noticed by Imai et al. (2001). The complexity of the problem from an optimization point of view has captured the attention of researchers who pursue to develop valid mathematical formulations and find efficient algorithms which provide good quality solutions in a practical amount of time. One of the most cited researches by Imai et al. (2001) presented the dynamic berth allocation problem where the arrival time of ships was already known; however, the arrival time assigned to a berth might be later than the time the berth became available. While the waiting time of a particular ship was represented by the departure of its predecessor minus the arrival time of the ship, the objective was to minimize the total of waiting and handling times.

Monaco and Sammarra (2007) reformulated the Imai et al. (2001) model, where the idle times per berth depends only on the berth and the ship sequence associated to. However, such idle times did not depend on the ship eventually scheduled to the position. The effectiveness of the Monaco and Sammarra (2007) model was proven by Buhrkal et al. (2011). Buhrkal et al. (2011) compared four mathematical models by three categories. The first category was based on Parallel Machine Scheduling; models proposed by Imai et al. (2001) and Monaco and Sammarra (2007) are in this category. The second category was based on the Multi-Depot Vehicle Routing Problem with Time Windows (MDVRPTW) and the third category was for the BAP formulated as a Generalized Set Partitioning problem (GSSP).

This research attempts to model the berth allocation problem in a way to balance the interest of the parties involved in the system. In other words, this study aims to minimize the handling time of ships served in a given time horizon and the waiting time for those ships have to spend before being moored in a berth. Different from other researches that have addressed the discrete berth allocation problem with dynamic arrival time of the ships (DBAP), the handling time of the vessels are not assumed to be integer.

On Reformulation of a Berth Allocation Model 635

The remaining paper is divided into four parts. Section 2 reviews previous approaches and models that addressed the berth allocation problem, and proposes a new mathematical model. Section 3 presents comparisons with existing models in the literature, and Sect. 4 draws the conclusions from the findings.

2 BAP Models

Efficiently allocating ships to berths and finding the best position for ships to be served in the service sequence of each berth has a positive impact in the operations of a port. Therefore, this section is devoted to review a main approach addressing the BAP and discusses a modified berth allocation problem (MBAP) model proposed in this study.

2.1 Monaco and Sammarra (2007) DBAP$^+$ Model

When it comes to the berth allocation problem one of the most cited research is the one presented by Imai et al. (2001). The objective function pursues to minimize the total of waiting and handling times. The waiting time of a particular ship is represented by the departure of its predecessor minus the arrival time of the ship. Later Monaco and Sammarra (2007) presented a reformulation of the Imai et al. (2001) model. In their study, they assumed the idle times per berth depended only on the berth and the kth index of the ship sequence associated. However, such idle time did not depend on the ship eventually scheduled in the kth position. Taking that into consideration, proper changes in the objective function and constraints were made.

Furthermore, in Monaco and Sammarra (2007), the total quay dimension can be split into a finite number of berths (segment of the total quay) available for incoming ships to load or unload cargo. While having dynamic arrival times and different quay crane capacity, not all the vessels will be in the port at the beginning of the period; in fact, they will often arrive over time. As a result, the handling time of a ship is affected by its assigned berth which can serve only one ship at the time. In addition, not all berths might be ready for berthing activities at the beginning of the period; yet, they will eventually start operations. The former mentioned system is known as Discrete Berth Allocation Problem with Dynamic Arrival times (DBAP$^+$). The mathematical model of the DBAP$^+$ is as follows:

$$Min \sum_{i \in B} \sum_{j \in V} \sum_{k \in K} \left[(n - k + 1)c_{ij} + s_i - a_j \right] x_{jki} + \sum_{i \in B} \sum_{k \in K} (n - k + 1)y_{ki} \qquad (1)$$

subject to

$$\sum_{i \in B} \sum_{k \in K} x_{jki} = 1, \forall j \in V \tag{2}$$

$$\sum_{j \in V} x_{jki} \leq 1, \forall i \in B, \forall k \in K \tag{3}$$

$$\sum_{h < k \in K} \left(y_{hi} + \sum_{l \in V} c_{li} x_{lhi} \right) - \sum_{j \in d} (a_j - s_i) x_{jki} + y_{ki} \geq 0, \forall i \in B, \forall k \in K \tag{4}$$

$$x_{jki} \in \{0, 1\}, \forall i \in B, \forall j \in V, \forall k \in K \tag{5}$$

$$y_{ki} \geq 0, \forall i \in B, \forall k \in K \tag{6}$$

Though the above model is similar to the one proposed by Imai et al. (2001), the number of constraints and variables related with the idle time are reduced because the idle time is handled as a two index variable. Therefore, when it comes to calculating the idle time in the objective function there is no need to account for the ships index anymore. Also, constraint (4) does not need the constraints that accounts for the ships index of the idle time.

2.2 MBAP

This study introduces a Modified Berth Allocation Problem model, which deals with the minimization of the total handling time of ships attended within a time horizon and the total waiting time those ships spend before being berthed. This objective allows schedules that are not only operationally efficient, but also capable of reducing an activity that does not add value to the customers (e.g., waiting for being served).

In MBAP, the main decision to be made is "where" and "when" the ships should be berthed. This naturally lends itself in a two-dimensional assumption. That is, all berths can handle any incoming ship; thus, the number of ships to be served, the arrival and handling time of the ships are known in advance. The handling time is defined to be the time when the ship is at the berth, while the service time is the total time the ship spends at the port (that includes the handling time plus any waiting time the ship experiences before being moored). Since all berths can handle any incoming ship, the time when a berth starts operation is given as well. The mathematical model introduced in this study share some characteristics from previous models from Monaco and Sammarra (2007) and Golias et al. (2009).

$$Min \sum_{i \in B} \sum_{j \in V} \sum_{k \in K} (n - k + 1)(h_{ij}) x_{ijk} + \sum_{i \in B} \sum_{k \in K} w_{ik} \tag{7}$$

subject to

$$\sum_{i \in B} \sum_{k \in K} x_{ijk} = 1, \forall j \in V \tag{8}$$

$$\sum_{j \in V} x_{ijk} \leq 1, \forall i \in B, \forall k \in K \tag{9}$$

$$y_{i1} \geq \sum_{j \in V} (a_j - s_i) x_{ij1}, \forall i \in B \tag{10}$$

$$y_{ik} \geq \sum_{j \in V} (a_j - s_i) x_{ijk} - \sum_{m < k \in K} \left(y_{im} + \sum_{l \in V} (h_{il}) x_{ilm} \right), \forall i \in B, \forall k \in K - \{1\} \tag{11}$$

$$w_{i1} \geq \sum_{j \in V} (s_i - a_j) x_{ij1}, \forall i \in B \tag{12}$$

$$w_{ik} \geq \sum_{m < k \in K} \left(y_{im} + \sum_{l \in V} (h_{il}) x_{ilm} \right) - \sum_{j \in V} (a_j - s_i) x_{ijk}, \forall i \in B, \forall k \in K - \{1\} \tag{13}$$

$$x_{ijk} \in \{0, 1\}, \forall i \in B, \forall j \in V, \forall k \in K \tag{14}$$

$$y_{ik} \geq 0, \forall i \in B, \forall k \in K \tag{15}$$

$$w_{ik} \geq 0, \forall i \in B, \forall k \in K \tag{16}$$

The objective is to minimize the sum of total handling and waiting time. Constraint (8) makes sure that every ship be assigned to a berth and a position in the service order. Constraint (9) does not allow a ship to be assigned to more than one berth and one position. Constraints (10) and (12) define the idle and waiting time respectively for ships scheduled in the first service position in every berth. Constraint (11) identifies the idle time of a given ship assigned to a specific berth and a position to be dependent of the completion time, which is equal to the handling time plus the time the berth is idle, of previous ships assigned to the same berth and the arrival time of the ship assigned to the current position. Constraint (13) defines the waiting time for any given ship scheduled at a determined berth and position. Constraint (14) denotes x as a binary variable which takes the value of 1 if a ship is assigned to a berth and a position in the service order and takes the value of zero otherwise. Constraints (15) and (16) define the idle and waiting time as continuous variables greater or equal than zero.

3 Model Comparisons

In this section the proposed MBAP model is compared with the DBAP$^+$ model proposed by Monaco and Sammarra (2007) which has been proven to be one of the most efficient formulations. Besides, the MBAP model proposed in this research share some similarities with the DBAP$^+$ model proposed by Monaco and Sammarra (2007) such as working with two index idle time variables. Therefore it will be interesting to test if the performance of both models will have similar performance in terms of computational effort. The mathematical models related to the instances of the problem were solved using the commercial package Gurobi optimizer 4.5. It was run on a computer Intel Core Duo 3.00 GHz and 2.00 GB of RAM.

The generation of the test instances was based on the characteristics of the ports of Singapore (Singapore), Kaohsiung (Taiwan) and Rotterdan (Netherlands). The information related to their number of berths and number of ships handled per month is available to the public online (World port source 2011). Consequently, the size of the instances was inspired by information gathered from those ports' terminals. It is worth to mention that those ports are some of the busiest ports in the world (Rosenberg 2011). Therefore, they provide valuable information to this research.

The inter-arrival time is exponentially distributed with arrival rate equal to 23.6, equivalent to 0.0422 ships per minute, or 2.53 ships per hour. That distribution was obtained from real data, inter-arrival rate, published online by the port of Kaohsiung in Taiwan (Kaohsiung harbor bureau 2011). In order to make a thorough analysis of the algorithms employed, it is intended to replicate different levels of operational pace. For that matter, the service rate (which is related to the handling time of the ships) is set proportionally higher or lower than the arrival rate (see Table 1 where LB and UB stand for the lower and upper bound parameters of the uniform distribution, respectively). On the other hand, the opening time of the berths was set based on the observation made in Cordeau et al. (2005). In the research it is pointed out that realistic opening times for the berths are 1/7–1/21 (expected proportion of ships present in the port) the amount of time from the arrival of the last ship with respect to the arrival time of the first ship taken into account in the time horizon plan. However, instead of setting a fix proportion for all berths, in this research the proportion varies uniformly

Table 1 Distribution and parameters of service rate	Uniform service rate as a percentage of the inter-arrival time with parameters [LB, UB]
	[90, 150]
	[85, 130]
	[80, 110]
	[60, 90]
	[50, 80]

distributed from 1/7 to 1/21 which is more realistic yet the berths does not become all available at the same time.

In Table 2, it can be noticed that MBAP model outperforms the DBAP$^+$ formulation in terms of computational time. As the instance becomes more complex the MBAP formulation presented in this research becomes more attractive. For example for instances of size 3×8 the difference in terms of performance is barely apparent between the two models. However, for instance 7×21 the MBAP formulation proposed is almost 30 times faster than the DBAP$^+$ formulation over 5 instances. Furthermore, the DBAP$^+$ model is not able to find an optimal solution within the 3 h run time limit set to Gurobi optimizer for instances 2, 3, 4, and 5 of size 9×30 while the MBAP formulation is able to solve the instances to optimality in less than 270 s for all instances of size 9×30.

The reason behind the superior performance of MBAP model in terms of CPU time is most likely related to the objective function taken into account and the constraints proposed. Given the similarities of the MBAP model and DBAP$^+$ formulation, if both mathematical models had the same objective function then one model would converge to another. For instance, if it is pursued to minimize the total handling and waiting times with the DBAP$^+$ model then constraint (11) has to be introduced yet that set of constraints defines the waiting times. Then it will be noticed that the DBAP$^+$ has converged to the MBAP formulation. On the other hand, if the MBAP model employs the objective function of the DBAP$^+$ model then constraint (11) becomes superfluous and therefore could be eliminated from the model. In that case the MBAP formulation reduces to the DBAP$^+$ model.

Table 2 Comparison between MBAP and DBAP$^+$ in terms of CPU time

Instance size	Instance ID	CPU time (s)	
		MBAP	DBAP$^+$
3×8	1	0.156	0.375
	2	0.188	0.250
	3	0.188	0.219
	4	0.203	0.250
	5	0.313	0.265
7×21	1	5.359	273.516
	2	4.781	176.953
	3	3.594	100.969
	4	16.219	321.109
	5	8.453	304.094
9×30	1	51.875	3,465.484
	2	36.672	[a]
	3	61.703	[a]
	4	21.031	[a]
	5	263.438	[a]

[a] Instance not solved to optimality under pre-specified 3 h run time limit

4 Conclusions

This research focuses on the discrete berth allocation problem with dynamic arrival times, and proposes a new mixed integer formulation, named modified berth allocation problem, which pursues to deliver solutions with minimal total service time and total waiting times for the ships to be served during the horizon plan. The MBAP model not only presents a valid formulation of the discrete berth allocation problem with dynamic arrival times but also presents a worthy option for those terminals interested in having a low total handling time while offering a service with minimum waiting times for the ships to be served. The model was further evaluated in terms of computational time. The MBAP model has demonstrated to be competitive compared to another well-known formulation in the literature.

The objective function proposed in this research tries to account for two indicators, total waiting time and total handling time that are of special interest for any service provider. Without a doubt, every port is aware of its interest and priorities. Therefore, the decision maker would select an objective function which accounts for those metrics operations recognize as important. Furthermore, the decision maker has available optimal solutions in a relative short time for small instances of the problem. Since a port might consist of terminals of different sizes, exact methods are still viable (although the problem is NP-Hard) approaches for small enough terminals within a port.

References

Buhrkal K, Zuglian S, Ropke S, Larsen J, Lusby R (2011) Models for the discrete berth allocation problem: a computational comparison. Transport Res Part E 47:461–473

Cordeau J-F, Laporte G, Legato P, Moccia L (2005) Models and tabu search heuristics for the berth-allocation problem. Transport Sci 39:526–538

Golias M, Boile M, Theofanis S (2009) A lambda-optimal based heuristic for the berth scheduling problem. Transport Res Part C 18:794–806

Imai A, Nishimura E, Papadimitriou S (2001) The dynamic berth allocation problem for a container port. Transport Res Part B 35:401–417

Kaohsiung harbor bureau (2011) Vessel actual time of arrival, http://www.khb.gov.tw/english/. Accessed Sept 2011

Monaco MF, Sammarra M (2007) The berth allocation problem: a strong formulation solved by a Lagrangian approach. Transport Sci 41:265–280

Rosenberg M (2011) Busiest ports in the world. http://geography.about.com/cs/transportation/a/aa061603.htm. Accessed 5 Oct 2011

World port source (2011) Ports. http://www.worldportsource.com/. Accessed Sept 2011

Forecast of Development Trends in Cloud Computing Industry

Wei-Hsiu Weng, Woo-Tsong Lin and Wei-Tai Weng

Abstract This paper presents a study on the future development of Taiwan's Cloud Computing industry. The forecast of development trends was made through interviews and focus group discussions of industry professionals. We construct an analysis framework for analyzing value chain and production value for the emerging Cloud Computing industry. Based on the analysis of recent Cloud Computing business models and the value activities of Cloud Computing vendors, the result provides a reference for IT business developers and innovative vendors interested in entering the emerging Cloud Computing market.

Keywords Cloud computing · Industry development · Forecasting · Value chain · SWOT

1 Introduction

The Cloud Computing concept has made a major impact on the products, services and business models of the IT software and hardware industries (Armbrust et al. 2009; Buyya et al. 2008, Keahey and Freeman 2008). Cloud computing has therefore become the emerging concept and technology that has drawn the most

---

W.-H. Weng (✉) · W.-T. Lin
Department of Management Information Systems, National Chengchi University,
Taipei city, Taiwan, Republic of China
e-mail: wh.weng@msa.hinet.net

W.-T. Lin
e-mail: lin@mis.nccu.edu.tw

W.-T. Weng
Department of Industrial Engineering and Management, Ming Chi University
of Technology, Guangzhou, Taiwan, Republic of China
e-mail: wtweng@mail.mcut.edu.tw

Y.-K. Lin et al. (eds.), *Proceedings of the Institute of Industrial
Engineers Asian Conference 2013*, DOI: 10.1007/978-981-4451-98-7_77,
© Springer Science+Business Media Singapore 2013

attention from the IT software and hardware industries in the period of the 2008 global financial crisis. The sheer scope of the industry as well as the fact that it spans both the enterprise and consumer markets has led to much discussion on its future business potential as well (Katzan 2009; Foster et al. 2008). Nevertheless, Cloud Computing technologies and business models as well as the new products, services, competition and alliances that arise as a result offer an emerging market that is well worth monitoring.

Currently, major IT firms are exploiting possible business opportunity into Cloud Computing market (Vouk 2008; Sotomayor et al. 2008). The Taiwanese IT vendors are strong players worldwide in the manufacturing and integration of IT devices and services. To assist the Taiwanese IT vendors moving forward towards the emerging Cloud Computing market, this research aims to address the question of deriving future trends for the Taiwanese Cloud Computing industry.

2 Research Method

Qualitative analysis is employed instead of quantitative analysis, by way of expert panel, vendor interviews and focus groups.

2.1 Expert Panel

The expert panel from industry experts is to assist the convergence process of data analysis. To this objective, industrial experts panel of eleven people were selected. The panel consists of CEO, CIO and line of business managers from various domains of Taiwanese IT industry. All of them are from publicly listed firms. Their business domains include System Integration (SI), Independent Software Vendor (ISV), Internet Service Provider (ISV), device manufacturer, and data center operator. These are the major participants in the Cloud Computing industry. The main function of the Expert Panel is to help determining the research framework and deriving strategy. In particular, the following questions are discussed.

1. What are the possible Cloud Computing business models within the context of the Taiwanese IT industry and environment?
2. With regards to these business models, what is the value chain or value system of the Taiwanese Cloud Computing industry?
3. Within this value chain or value system, what are the business environments in terms of internal strengths and weaknesses, as well as external opportunities and threats of the IT firms?

Forecast of Development Trends in Cloud Computing Industry 643

4. Recognizing the internal and external business environments, what is the possible production value that could be estimated for the Cloud Computing industry?

2.2 Vendor Interviews and Focus Groups

Representative IT firms from Taiwan are selected as the objects of this study. The selection process is based on the rank of the revenue of the firms as well as their reputation in terms of technology innovation and market visibility. IT vendors of Taiwan enjoy high market share worldwide in the sectors such as computer, communication and consumer electronics manufacturing. Currently the Taiwanese vendors participate in Cloud Computing include IT device manufacturers, IT service providers, and Internet datacenter operators. We collect and analyze business proposal data of 66 Taiwanese IT vendors in cloud computing. The selection criteria are as follows.

1. Revenue of the firm is among the top five in its industry domain.
2. The firm has announced in public its vision, strategy, products or service toward Cloud Computing market.

Based on these criteria, representatives from 38 IT firms are selected for vendor interviews and focus groups. These firms are summarized in the following Table 1.

3 Development of Cloud Computing Industry

3.1 Development of Cloud Computing Business Models

Business model is a term often used to describe the key components of a given business. It is particularly popular among e-businesses and within research on e-businesses (Afuah and Tucci 2001). Despite of its importance, no generally accepted definition of the term "business model" has emerged. Diversity in the

Table 1 Selected cases for vendor interviews and focus groups

Business domain	Number of firms
Independent software vendor (ISV)	8
System integration provider (SI)	10
Telecom operator	4
Server and storage device manufacture	6
Networking device manufacture	4
Mobile device manufacture	6
Total	38

Table 2 Cloud computing business models

Service/product	Delivery model	Revenue model
Cloud user technology	IT vendor provides hardware and software products and technology for Cloud Computing client device, including client OS, user interface, middleware and applications	License
Software as a Service (SaaS)	Commercial application software are hosted by data center and accessed via internet service on a pay-as-you-go basis	Subscription
Platform as a Service (PaaS)	Software vendors provide APIs or development/deployment environment platforms for ISVs to develop cloud version software	Subscription
Infrastructure as a Service (IaaS)	Service providers provide the virtual computing environments for users, including virtual CPU, OS, storage and network	Subscription
Cloud infrastructure technology	IT service vendors provide technology, product or service to support enterprise or government agencies for implementing public or private cloud service	License

available definitions poses substantive challenges for delimiting the nature and components of a model and determining what constitutes a good model. It also leads to confusion in terminology, as business model, strategy, business concept, revenue model, and economic model are often used interchangeably. Moreover, the business model has been referred to as architecture, design, pattern, plan, method, assumption, and statement (Morris et al. 2003).

Several researches have proposed frameworks to identify business models. Morris et al. (2003) proposed a set of six questions for defining basic components of business models. In the context of Cloud Computing, Rappa (2004) described a framework to classify business models using the customer relationship as the primary dimension for defining categories. Using this approach, nine major categories are used to classify a number of different types of business models that have been identified in practice among web-based enterprises and also utility computing firms.

Business models of Cloud Computing are identified and summarized in Table 2.

3.2 Development of Cloud Computing Value Chain

Porter (1980, 1985, 1991) discusses the concept and framework of value chain analysis. The value chain provides a template for understanding cost position, because activities are the elemental unit of cost behavior. The value chain also provides a means to systematically understand the sources of customer value and conduct differentiation. Customer value is created when a firm lowers its customer's cost or enhances its customer's benefit. The term "value system" is

Forecast of Development Trends in Cloud Computing Industry 645

Table 3 Cloud computing vendors' value activities

Cloud computing sub-industries	Main value activities	Taiwanese vendor examples
Cloud user device	Entry into the cloud supply chain through smartphones, tablet, netbook PC, network communication equipment, and other products	Mainly smart mobile device manufactures, such as: HTC, Asus, Acer, D-Link, Gemtek
Cloud infrastructure equipment	Entry into the cloud supply chain through servers, storage equipment, power suppliers, and other branded products of OEM operations	Mainly server and storage device manufactures, such as: quanta, inventec, gigabyte, Delta electronics, infortrend, promise
Cloud service and data center operation	Provision of broadband service, data center services, and various XaaS services needed in cloud computing	Mainly cloud service provider and internet data center operators, such as: Chunghwa telecom, taiwan fixed network, FETnet, acer eDC, ASUS webstorage, GSS
Cloud infrastructure software and IT service	Assistance offered to cloud service and mobile device setup through system integration, software development, consultancy services, and other operations	Mainly system integrators and independent software vendors, such as: Systex, Data systems, the Syscom Group, Fortune Information Sysetms Corp., Tatung System Technologies Inc., Genesis Technology, Inc., Stark Technology Inc

sometimes employed when the concept of value chain is extended from intra-firm value activities to inter-firm value activities.

By examining the Cloud Computing business models obtained from the above sections, value activities can be identified and the Cloud Computing value chain can be extracted. IT vendors participating in the Cloud Computing value activities are classified into four clusters, which are identified as the major sub-industries of cloud computing. The value activities of these four clusters are summarized in the following Table 3.

4 Analysis on Forecasting Results

4.1 SWOT Analysis of Taiwan's Cloud Computing Industry

The SWOT analysis is an established method for assisting the formulation of strategy. SWOT analysis aims to identify the strengths and weaknesses of an organization and the opportunities and threats in the environment (Pickton and Wright 1998). The strengths and weaknesses are identified by an internal appraisal of the organization and the opportunities and threats by an external appraisal

Table 4 Competitiveness analysis of Taiwan's cloud computing industry

Strengths	Weaknesses
A. Possesses both hardware and software solutions and provides professional consultancy services experience	A. Customers are mostly large enterprises or government agencies; fewer dealings with SMEs and consumers
B. Cooperated with the global leading companies for many years, and prices are flexible	B. Weak research and development of key software technologies such as virtualization and Big Data analytics
C. Possesses in-depth vertical domain knowledge, and localized Know-How as well	
Opportunities	*Threats*
A. Open up the market of private cloud deployment	A. Global leading companies lead technologies and standards
B. Develop the SaaS model of software to attract new customers	B. Part of the business is replaced by emerging cloud services
C. Cloud services governance, including security, auditing, and quality control	C. Industries rise in the emerging markets such as Mainland China, India, and others
D. Enterprise mobile application software and services	

(Dyson 2004). Having identified these factors strategies are developed which may build on the strengths, eliminate the weaknesses, exploit the opportunities or counter the threats (Weihrich 1982).

By applying the analysis method of Weihrich (1982), the SWOT matrix of Taiwan's Cloud Computing industry is derived as follows (Table 4).

4.2 Forecast of Taiwan's Cloud Computing Production Value

With the gradual development of the cloud market, the number of Taiwanese companies in the cloud computing industry and their scale has gradually expanded, while production value has increased annually. At present, vendors of smartphones, tablets, netbooks, servers, network communication equipments, data centers, system integration, and software solution have actively participated in the development. With the continued cloud service development and the combined market effects brought about by the trend of digital conversion, the production value is expected to achieve steady growth through 2018, as shown in Table 5.

Table 5 Production value of Taiwan's cloud computing industry (Unit: USD Million)

Cloud computing sub-industries	2013	2014	2015	2016	2017	2018
Cloud user device	12,025	14,312	16,785	18,937	21,278	23,786
Cloud infrastructure equipment	1,937	2,052	2,188	2,305	2,487	2,650
Cloud service and data center operation	365	453	536	665	745	842
Cloud infrastructure software and IT service	144	166	180	205	223	248
Total production value	14,471	16,983	19,689	22,112	24,733	27,526
YoY growth (%)	–	17.36	15.93	12.31	11.85	11.29

5 Conclusions

The cloud computing industry in accordance with the value activities are divided into four sub-industries, namely, the cloud infrastructure software and IT service industries, the cloud service and data center operation industry, the cloud infrastructure equipment industry, as well as the cloud user device industry. The internal strength and weakness, the external opportunity and threat, as well as the production value forecast, are analyzed and presented.

Vendors interested in exploring the market opportunities of Cloud Computing can use this analysis process and outcome of this research as a reference for their strategic planning, and avoid many unnecessary trial and error efforts. In particular, with a clear picture of the Cloud Computing business model and value chain, vendors can position themselves more precisely for a market sector of their competitive advantage.

Acknowledgments The authors gratefully acknowledge the helpful comments and suggestions of the reviewers, which have improved the presentation.

References

Afuah A, Tucci CL (2001) Internet business models and strategies: text and cases. McGraw-Hill, Boston

Armbrust M, Fox A, Griffith R, Joseph A, Katz, R, Konwinski A, Lee G, Patterson D, Rabkin A, Stoica I, Zaharia M (2009) Above the clouds: a Berkeley view of cloud computing, Technical Report. No.UCB/EECS-2009-28, University of California at Berkley, USA

Buyya R, Venugopal S, Yeo CS (2008) Market oriented cloud computing: vision, hype and reality for delivering IT services as computing utilities. In: Proceedings of the 10th IEEE international conference on high performance computing and communications

Dyson RG (2004) Strategic development and SWOT analysis at the University of Warwick. Eur J Oper Res 152:631–640

Foster I, Zhao Y, Raicu I, Lu S (2008) Cloud computing and grid computing 360-degree compared. In: Proceedings of the IEEE grid computing environments workshop 1–10

Katzan H Jr (2009) Cloud software service: concepts, technology, Economics. Service Sci 1(4):256–269

Keahey K, Freeman T (2008) Science clouds: early experiences in cloud computing for scientific applications. Workshop on cloud computing and its applications 2008 (CCA08), Chicago, Illinois, USA

Morris M, Schindehutte M, Allen J (2003) The entrepreneur's business model: toward a unified perspective. J Bus Res 58:726–735

Pickton DW, Wright S (1998) What's SWOT in strategic analysis? Strategic Change 7(2):101–109

Porter ME (1980) Competitive Strategy. Free Press, New York

Porter ME (1985) Competitive Advantage. Free Press, New York

Porter ME (1991) Towards dynamic theory of strategy. Strateg Manag J 12(Winter):95–117

Rappa MA (2004) The utility business model and the future of computing services. IBM Syst J 43(1):32–42

Sotomayor B, Montero RS, Llorente IM, Foster I (2008) Capacity leasing in cloud systems using the Opennebula engine, workshop on cloud computing and its applications 2008 (CCA08), Chicago, Illinois, USA

Vouk MA (2008) Cloud computing–issues, research and implementations. J Comput Inf Technol 16(4):235–246

Weihrich H (1982) The TOWS matrix: a tool for situational analysis. Long Range Plan 15(2):54–66

Self-Organizing Maps with Support Vector Regression for Sales Forecasting: A Case Study in Fresh Food Data

Annisa Uswatun Khasanah, Wan-Hsien Lin and Ren-Jieh Kuo

Abstract Many food stores face the same problem, "how many products should we make and how much ingredients should we order?" For most managers, they cannot predict a specific quantity of sales for upcoming week. If their prediction is not accurate, it will cause lots of products waste or the opposite, products shortage. Fresh food products have time limit. When consumers buy food, they would first consider if the foods are fresh or has been expired. As a result, customer demand forecasting is an important issue in food product market. With the recent development of artificial intelligence models, several methods have been employed in order to conduct forecasting model to be more effective than the conventional one. This research presents a two-stage forecasting model. The noise detecting and the removing will be considered first, and then all data will be clustered to increase the accuracy and practicability of the model.

Keywords Clustering · Forecasting · Self-organizing maps · Support vector regression

---

A. U. Khasanah (✉) · W.-H. Lin · R.-J. Kuo
Department of Industrial Management, National Taiwan University of Science and Technology, Taipei city, Taiwan, Republic of China
e-mail: M10101816@mail.ntust.edu.tw

W.-H. Lin
e-mail: M10101004@mail.ntust.edu.tw

R.-J. Kuo
e-mail: rjkuo@mail.ntust.edu.tw

Y.-K. Lin et al. (eds.), *Proceedings of the Institute of Industrial Engineers Asian Conference 2013*, DOI: 10.1007/978-981-4451-98-7_78,
© Springer Science+Business Media Singapore 2013

1 Introduction

Food products have time limit. When consumers buy food products, they would consider whether the foods are fresh or not for the first time, and the other factors will be considered next. So, this is why expiry dates become important issues in food industry, especially in fresh food business. Products can be categorized by its expiry dates. They can be categorized into long-term products (can be kept for over one year) and perishable products (can only be kept for less than 15 days).

As sales forecasting can predict consumer demand before the sale begins, it can be used to determine the required inventory level to meet this demand and avoid the problem of under stocking. In addition, sales forecasting can have implications on corporate financial planning, marketing, client management and other areas. Therefore, improving the accuracy of sales forecasting has become an important issue in business operation.

During the last few years, many scholars have developed different kinds of forecasting techniques to increase the forecasting accuracy. Among these methods, regression models and autoregressive moving average model techniques are classified as traditional methods, which are criticized by researchers for their weakness of non-linear fitting capability.

Differential evolution (DE), a recent optimization technique, has been considered as a novel evolutionary computation technique, and outperformed other evolutionary computation techniques such as genetic algorithm (GA), particle swarm optimization (PSO). The advantages of DE are not only it is easy to be implemented but also it requires few parameters. DE has been successfully employed in many real work applications such as pattern recognition, classification and multi-objective optimization. But, only few researchers use DE algorithm for SVR parameters optimization in forecasting problem.

In this study, we apply DE algorithm to select the appropriate parameters in support vector regression model for improving the model's forecasting accuracy, preceded by clustering the data using self-organizing maps (SOM) before performing the forecasting step. We apply both of those methods to compare the forecasting result between clustered data and the non-clustered one.

2 Literature Review

Artificial neural networks (ANNs), such as support vector regression (SVR), have been found to be useful techniques for sales/demand forecasting due to their ability to capture subtle functional relationships among the empirical data even though the underlying relationships are unknown or hard to be described, and unlike traditional time series forecasting model, such as ARIMA and multivariate regression analysis (MARS), they do not require strong model assumptions and can map any nonlinear function without a priori assumption about the properties of the data. Fildes et al. in (2008), Alahakoon and Halgamuge (2000) proves coupled with superior performance in constructing nonlinear models.

However, Chang et al. in (2006) and Wang et al. in (2009) have mentioned that no matter what kind of data, some noise may influence the forecast result a lot. It seems data preprocessing become more and more important. Dash et al. in (1995) and Chang et al. in (2000) have applied different methods to select key factors in their forecasting system.

Furthermore, in the recent years, hybrid system is widely developed in different areas and has many positive performances as shown by Chen in (2003) and Marx-Gómez et al. (2002). Chang et al. in (2006) have developed various hybrid methods in dealing with the sales forecasting problems in different industrial sectors.

Cheng et al. (2009) have developed a hybrid model by integrating K-mean cluster and fuzzy neural network (KFNN) to forecast the future sales of a printed circuit board factory. This sales forecasting model is designed with the purpose of improving the forecasting accuracy and providing timely information to help managers make better decisions.

Wang and Lu have developed a demand forecasting model which noise detecting and removing task will be considered first, and then all data will be clustered to increase the accuracy and the practicability of the model. In their research, a hybrid forecasting model which combines ICA, GHSOM, and SVR algorithm is proposed. The GHSOM clusters input data into several disjoined clusters and each cluster contains similar objects. Next, an individual SVR model for each cluster is constructed and the final forecasting results can be obtained.

3 Methodology

3.1 Self-organizing Maps

The SOM architecture was originally motivated by the topological maps and the self organization of sensory pathways in the brain. It is a kind of unsupervised learning neural network. The main focus of SOM is to summarize information while preserving topological relationship. The objective of SOM is to represent high dimensional input patterns with weight vectors that can be visualized in a usually two dimensional (2D) lattice structure (Kuo et al. 2012). Each unit in the lattice structure is called neuron and adjust neuron are connect to each other, which gives the clear topology of how the network fits itself to the input space. Input layer and the output layer are connected with weight and this weight will be update during the training. A cluster can be defined as a group of neurons with short distances between them and long distance to the other neurons.

There are several procedures that must be followed to apply this method (Alahakoon and Halgamuge 2000).

1. Set up network parameters (learning rate, neighborhood radius).
2. Set up connecting weight matrix, w, randomly.
3. Input a training sample's input vector, x.
4. Select the winning output node using Euclidean distance. The winning neuron is denoted as wc.

$$|x - wc| = \min|x - wi|$$ (1)

5. where i is the position for ith weight.
6. Updating weight

$$wij(new) = wij(old) + a[xi - wij(old)]$$ (2)

7. Repeat step 3–5 until all training samples have been presented.
8. Shrink learning rate and neighborhood radius.
9. Repeat step 3–7 until the termination criteria is satisfied.
10. Error measurement

$$\sum_p \left(\min_j d_j^p \right) d_j^p = \sqrt{\sum_i \left(X_i^p - W_{ij} \right)^2}$$ (3)

3.2 Support Vector Regression

The basic concept of SVR is introduced. A nonlinear mapping $\varphi(.) : \Re^n \to \Re^{n_h}$ is defined to map the input data (training data set) $\{(x_i, y_i)\}_{i=1}^N$ into a so-called high dimensional space (which may have infinite dimensions), $\Re^{n_h}$. Then, in the high dimensional feature space, a linear function, f, is used to formulate the nonlinear relationship between input data and output data. This linear function, called SVR function, is defined as Eq. (4),

$$f(x) = w^T \varphi(x) + b$$ (4)

The purpose of SVR method is minimizing the empirical risk through Eq. (5),

$$R_{emp}(f) = \frac{1}{N} \sum_{i=1}^N \Theta_\varepsilon(y_i, w^T \varphi(x_i) + b)$$ (5)

Where $\Theta_\varepsilon(y_i, w^T \varphi(x_i) + b)$ is the ε-insensitive loss function and it can be explain more as shown in Eq. (6),

$$\Theta_\varepsilon(y_i, w^T \varphi(x_i) + b) = \begin{cases} |w^T \varphi(x_i) + b - y_i| - \varepsilon, & \text{if } |w^T \varphi(x_i) + b - y_i| \geq \varepsilon \\ 0, & \text{otherwise} \end{cases}$$

(6)

Self-Organizing Maps with Support Vector Regression for Sales Forecasting

in addition, $\Theta_\varepsilon(y_i, w^T\varphi(x_i) + b)$ is employed to find out the optimum hyper plane on the high dimensional feature space to maximize the distance separating the training data into two subsets. Thus, the SVR focuses on finding the optimum hyper plane and minimizing the training error between the training data and the ε-insensitive loss function. Then, the SVR minimize the overall errors through Eq. (7)

$$\underset{w,b,\xi^*,\xi}{Min} \ R_\varepsilon(w, \xi^*, \xi) = \frac{1}{2}w^Tw + C\sum_{i=1}^{N}(\xi_i^* + \xi_i) \tag{7}$$

with the constraints

$$y_i - w^T\varphi(x_i) - b \leq \varepsilon + \xi_i^*, \ i = 1, 2, \ldots, N$$
$$-y_i + w^T\varphi(x_i) + b \leq \varepsilon + \xi_i, \ i = 1, 2, \ldots, N$$
$$\xi_i^* \geq 0, \ i = 1, 2, \ldots, N$$
$$\xi_i \geq 0, \ i = 1, 2, \ldots, N$$

Training errors above ε are denoted as ξ_i^*, whereas training errors below ε are denoted as ξ_i. After the quadratic optimization problem with inequality constraints is solved, the parameter vector w in Eq. (8) is obtained,

$$w = \sum_{i=1}^{N}(\beta_i^* - \beta_i)\varphi(x_i) \tag{8}$$

where β_i^*, β_i are obtained by solving a quadratic program and are the Lagrangian multipliers. Finally, the SVR function is obtained as Eq. (9) in the dual space,

$$f(x) = \sum_{i=1}^{N}(\beta_i^* - \beta_i)k(x_i, x_j) + b \tag{9}$$

where $k(x_i, x_j)$ is called the kernel function, and the value of the kernel equals the inner product of two vectors, x_i and x_j in the feature space $\varphi(x_i)$ and $\varphi(x_j)$, respectively; so, $K(x_i, x_j) = \varphi(x_i) \circ \varphi(x_j)$. Any function that meets Mercer's condition can be used as the kernel function. The most used kernel functions are the Gaussian radial basis functions (RBF) with a width of $\sigma : K(x_i, x_j) = \exp(-0.5||x_i - x_j||^2/\sigma^2)$.

4 Results

The first step in this study is to cluster the data with SOM. The number of cluster can be represented by the topological size. Topology size is an important factor to be decided, however, there is no theory which can exactly determine and solve this

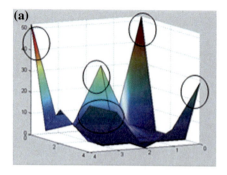 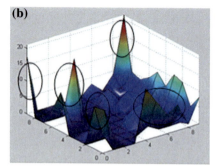

Fig. 1 **a** 5 × 5, **b** 10 × 10 topological clustering results

problem. In the preliminary test, different sizes of topology are tested and compared to see the effect of this difference in the result.

The data is nine months "mochi" sales data, and it includes 262 days. There are several factors to be considered: holiday, temperature, humidity, indicator for the occurrence of rain or drizzle, snow or ice pellets, thunder, and fog.

Figure 1 represents the clustering result when 5 × 5 and 10 × 10 topology sizes are applied. The number of clusters can be determined by the number of peaks than can be seen. The bigger the topology size, there will be more peaks appear. Just like what it has been shown in Fig. 1 there is more peaks appear in 10 × 10 than in 5 × 5 topology. But, overall it can be concluded that there are five high peaks. For several topology sizes that have been tested, in general they all show five peaks.

The error measurement for each topology also computes through Eq. (3). Figure 2 represents the MAD comparisons for each topology. It can be seen that the bigger the topology size, the smaller the MAD. High topology size will produce small MAD, but sometime the number of cluster is more difficult to determine because there will be so many peaks.

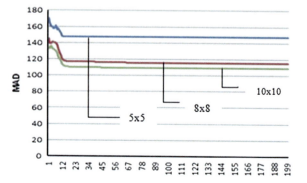

Fig. 2 MAD comparison

Fig. 3 Forecasting result comparing

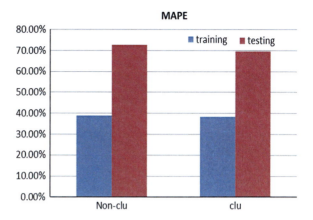

The second step is to forecasting step using SVR. In the forecasting method, two models are compared. The first model is conducted without clustering the data and the forecasting model is constructed by considering all of the factors. The eight day sale is predicted by using the previous seven days. The second model is constructed by applying five clustered data. In this model, the forecasting is constructed for each cluster.

Performance evaluation using MAPE is conducted to show the forecasting result. The MAPE value for non-clustered data and clustered data respectively are 38.8 and 38.5 % for training, and 72.7 and 69.84 % for testing. From this result, it can be concluded that clustered data has smaller percentage error compare with non-clustered one even the differences is not so much. It is because, when we clustered the data, the variance error within the cluster is smaller than the non clustered one, and the variance error will become smaller when more topology size is applied. In the other word, for example if there are 10 high sales value among the 262 data, in non clustered model those 10 data will be compared with the other 252, so the 10 high sales data will be not very significant. But, in clustered model those 10 data will be compared with smaller number data in a cluster. Because, it has been said before that in clustered model, the forecasting method is conducted in each cluster.

In this case, the error testing is still very big. This problem is caused by the daily data is quite difficult to predict. By considering several factors that may be influenced the mochi's sales, make this kind of forecasting is more difficult to be done. There are lots of uncertainties Fig. 3.

5 Conclusion

This study constructs a forecasting model by comparing the non clustered data and the clustered one use SOM and SVR. From the experimental result, the clustered data gives better forecasting result than non clustered data. It is proved by the MAPE value.

In the clustering issues using SOM, it is important to define the topology size. There is no rule of thumbs to decide the topology size. Finally, it must be decide by the analyzer. And in the forecasting issues, it is important to do preprocessing data. Data clustering is one this step. It is important to know how the data looks like, and it is also important to recognize what kinds of factor that can influence the data and the forecasting result due to minimize noise. So, finally accurate forecasting result can be obtained. To forecast daily data is more difficult to conduct, because there will be more uncertainties to be considered.

References

Alahakoon D, Halgamuge SK (2000) Dynamics self-organizing maps with controlled growth for knowledge discovery. IEEE Trans Neural Netw 11(3):601–604

Chang PT, Huang LC, Lin HJ (2000) The fuzzy Delphi method via fuzzy statistic sand membership function fitting and an application to the human resources. Fuzzy Sets Syst 112:511–520

Chang PC, Wang YW, Liu CH (2006) Combining SOM and GA-CBR for flow time prediction in semiconductor manufacturing factory. Lect Notes Comput Sci, pp 777–767

Chang PC, Liu CH, Fan CY (2009) Data clustering and fuzzy neural network for sales forecasting: a case study in printed circuit board industry. Knowl Based Syst J

Chen T (2003) A fuzzy back propagation network for output time prediction in a wafer fab. Applied Soft Comput 2(3):211–222

Dash PK, Liew AC, Rahman S (1995) Peak load forecasted using a fuzzy neural network. Electric Power Syst Res 32:19–23

Fildes R, Nikolopoulos K, Crone SF, Syntetos AA (2008) Forecasting and operational research: a review. J Oper Res Soc 59(9):1150–1172

Kuo RJ, Wang CF, Chen ZY (2012) Integration of growing self-organizing map and continuous genetic algorithm for grading lithium-ion battery cells. Appl Soft Comput J

Marx-Gómez J, Rautenstrauch C, Nürnberger A, Kruse R (2002) Neural-fuzzy approach to forecast returns of scrapped products to recycling and remanufacturing. Knowl-Based Syst 15:119–128

Wang YW, Liu CH, Fan CY (2009) The hybrid model development of clustering and back propagation network in printed circuit board sales forecasting. Opportunities Challenges Next-Generation Appl Intell, pp 213–218

State of Charge Estimation for Lithium-Ion Batteries Using a Temperature-Based Equivalent Circuit Model

Yinjiao Xing and Kwok-Leung Tsui

Abstract This study investigates battery state-of-charge (SOC) estimation under different temperature conditions. A battery modeling approach is developed aiming to improve the accuracy of the SOC estimation when ambient temperature is taken into account. Firstly, a widely used equivalent circuit model with the one-order resistance-capacitor (RC) network is modified to capture battery dynamics at different temperatures. Secondly, since the open-circuit voltage verse SOC (OCV-SOC) incorporated into the battery model is also influenced by the temperature, OCV-SOC-Temperature (OCV-SOC-T) table is constructed to replace the original table based on our experimental data. The experiments with two dynamic load tests, dynamic stress test (DST) and federal urban driving schedule (FUDS) are run on the battery. The purpose of DST profile is to identify the battery model, while FUDS data is used to emulate the operation conditions and evaluate the performance of our proposed model by unscented Kalman filtering. Finally, the comparative results indicate that our temperature-based model provide more accurate SOC estimation with root mean square estimated errors than the original model without regard to temperature dependence.

Keywords Lithium-ion batteries · State-of-charge estimation · Equivalent circuit model · Unscented Kalman filtering · Dynamic load tests

---

Y. Xing (✉) · K.-L. Tsui
Department of Systems Engineering and Engineering Management,
City University of Hong Kong, Tat Chee Avenue, Kowloon, Hong Kong SAR
e-mail: yxing3@student.cityu.edu.hk

K.-L. Tsui
e-mail: kltsui@cityu.edu.hk

Y.-K. Lin et al. (eds.), *Proceedings of the Institute of Industrial Engineers Asian Conference 2013*, DOI: 10.1007/978-981-4451-98-7_79,
© Springer Science+Business Media Singapore 2013

1 Introduction

Batteries are receiving a large amount attention as the major energy storage device in electric vehicles (EVs) and uninterruptible power supplies (UPS). In order to guarantee a safe, reliable and effective operation of a battery-powered system, the key information of a battery should be monitored, and provided to the end users through the battery management system (BMS). The major functions in BMS include state-of-charge (SOC), state-of-health (SOH) and remaining useful performance (RUP) (Xing et al. 2011). For EV drivers, one of the top concerns is running out of the cars on the road. In other words, an accurate battery gauge indicator is significant for users on residual range of EV. However, battery SOC cannot be measured directly but must be estimated according to measurable parameters, such as current, terminal voltage, ambient temperature, internal resistance etc. In particular, when the battery is operated under different conditions, for example, dynamic loadings and different ambient temperatures, the performance of the battery SOC estimator is subjected to challenge. Thus, it is necessary to develop an effective and robust method to estimate battery SOC under different working conditions.

The existing methods for SOC estimation can be categorized into three types, including Coulomb counting, black box modeling, and a combination of coulomb counting method with model-based estimation. The Coulomb counting approach is a definition-based method for SOC estimation. It accumulates the charge flowing in and out of a battery directly from the initial time. However, the initial SOC is difficult to determine. In addition, this method is an open-loop estimator. The measurement errors due to current sensors and noise disturbances will cause accumulating errors (Charkhgard and Farrokhi 2010). Thus, the recalibration methods were widely employed for offline adjustment, such as open-circuit voltage (OCV) table. Secondly, the black box modeling methods were discussed in Hansen and Wang (2005), Singh et al. (2006), Li et al. (2007), Cheng et al. (2011). They mainly refer to neural networks-based, fuzzy-based, support vector-based estimators. This modeling method is time consuming. The robustness of the model will suffer from the uncertainty of the new dataset. Thirdly, a model-based filtering method combined with Coulomb counting approach is being popular and studied in Plett (2004), Junping et al. (2009), Roscher and Sauer (2011), Sun et al. (2011), Dai et al. (2012). The models refer to electrochemical models (first principle model) and equivalent circuit models. Equivalent circuit model is preferred in today's BMSs because it is easy to implement on the online system.

However, the existing methods did not take into account temperature dependence of battery modeling. There are two facets involved. One is that model parameters vary with temperature; the other is that the OCV-SOC function in the equivalent circuit model is temperature dependent. In this paper, we put forward a temperature-based equivalent circuit model incorporated with OCV-SOC-Temperature (OCV-SOC-T) to estimate the SOC of lithium-ion batteries. Our work proceeds as follow. Experiments are introduced in Sect. 2. Our methodology combining the proposed model with unscented Kalman filtering is followed by Sect. 3. The results of SOC estimation are shown in Sect. 4 to demonstrate the performance of our proposed model in comparison with that of the original model without regard to temperature effects.

2 Experiments

Lithium-ion batteries (LiFePO4) with the nominal capacity of 1.1 Amp-hours were run in our lab to simulate the practical operation. Battery monitoring system was used to control battery charge/discharge, measure the battery data. Each test as follows was tested from 0 to 50 °C at an interval of 10 °C. Three kinds of tests were run on our test samples.

Driving stress test (DST) profile was tested for model identification. It is similar to a pulse discharge regime that simulates the expected demands of an EV battery (Hunt 1996). Since a standard DST cycle only lasts for 360 s, the discharge process includes several cycles when our test sample discharge from a fully charge at 3.6 V (100 % SOC) to empty at 2 V (0 % SOC). FUDS profile was tested for evaluating the performance of our proposed method because it is a more complicated dynamic current profile to simulate the practical operation data. DST and FUDS data tested at 20 °C are shown in Fig. 1.

The OCV-SOC relationship for is significant for SOC recalibration. However, it is temperature dependent (Johnson et al. 2001; Plett 2004; Roscher and Sauer 2011). In other words, a normal OCV-SOC tested at room temperature will cause large errors for the SOC inference. Figure 2 shows OCV curves form 30 % SOC to 80 % SOC at different temperatures. It can be seen that the same OCV i.e. 3.3v

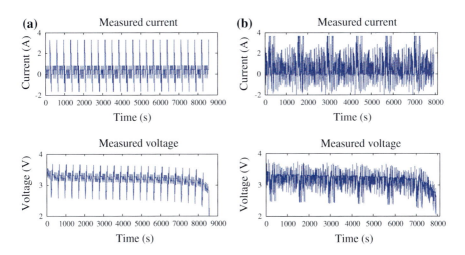

Fig. 1 Dynamic load test profile tested at 20 °C

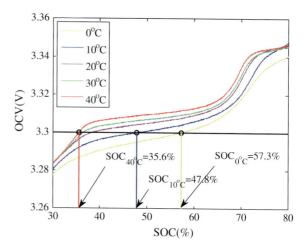

Fig. 2 SOC differences at the OCV of 3.3 V at different temperatures

corresponds to different SOC values at different temperatures. Thus, to minimize the deviation of the OCV-SOC at different temperatures, we tested the OCV curve from 0 to 50 °C at an interval of 10 °C. OCV-SOC-Temperature (OCV-SOC-T) will replace the original OCV-SOC relationship.

3 Methodology

3.1 Battery RC Model and Model Identification

A typical one-order resistance–capacitance (RC) model was employed in our study. It is a well-behaved model, and has been widely used to capture LiFePO4 dynamics (Hu et al. 2011; Roscher and Sauer 2011; He et al. 2012). Its schematic diagram is shown in Fig. 3.

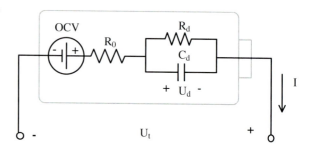

Fig. 3 Schematic diagram of LiFePO4 model

State of Charge Estimation for Lithium-Ion Batteries

Table 1 Statistical list of the battery model

Temperature (°C)	R_0	R_p	C_p	MAE	RMSE
0	0.263	0.115	400.2	0.00132	3.4E-03
10	0.229	0.090	618.5	0.00083	2.1E-03
20	0.217	0.076	820.4	0.00061	1.5E-03
25	0.194	0.070	889.8	0.00047	8.6E-04
30	0.177	0.061	986.4	0.00040	6.0E-04
40	0.151	0.050	1139.1	0.00037	5.6E-04
50	0.176	0.0426	1340.1	0.00036	5.2E-04

$$\begin{cases} \dot{U}_d = \dfrac{I}{C_d} - \dfrac{U_d}{C_d R_d} \\ U_t = OCV(SOC) - U_d - I \times R_0 \end{cases} \tag{1}$$

where U_t is the measured terminal voltage of the battery, I is the load current with a positive value at discharge and a negative value at charge, R_0 is the ohmic resistance, C_d and R_d are the polarization capacitance and resistance, respectively. U_d is the voltage across the C_d.

As mentioned in Sect. 2, the sequence of current and terminal voltage of DST at the given temperature was used to fit model using least square. The specific OCV-SOC- T i.e. OCV-SOC-40 °C was selected. For the OCV inference in Eq. (1), the accurate SOC at time k should be obtained. Since the battery was discharged from 100 % SOC, the accumulative SOC can be calculated. Accordingly, model parameters were fitted as shown in Table 1.

Mean absolute modeling error (MAE) and root mean square modeling error (RMSE) are used to evaluate the goodness of fit of the model. Here, modeling error refers to the deviation between the measured terminal voltage ($U_{t,k}$) and the estimated terminal voltage ($\widehat{U}_{t,k}$). Table 1 demonstrates that the model can fit the measured data well with small MAEs and RMSEs taking into account temperature. Additionally, model parameters vary with temperature. Thus, a temperature-based model, TRC model, was proposed as follows.

$$\begin{cases} \dot{U}_d = \dfrac{I}{C_d(T)} - \dfrac{U_d}{C_d(T)R_d(T)} \\ U_t = OCV(SOC, T) - U_d - I \times R_0(T) \end{cases} \tag{2}$$

3.2 Unscented Kalman filtering for State Estimation

The online SOC estimation has strong nonlinearity. In order to improve the estimation accuracy, unscented Kalman filtering (UKF) was employed for battery SOC estimation. The characteristics of UKF is based on unscented transformation, which could avoid the weakness using Taylor series expansion like extended Kalman filtering (EKF) (Wan and Van Der Merwe 2000, 2001). Incorporating with the discrete form of Eq. (2) and Coulomb counting principle, a state-space model for UKF estimation can be formulated in Eqs. (3) and (4).

State function:

$$\mathbf{x}_{k+1} = \begin{bmatrix} SOC_{k+1} \\ U_{d,k+1} \\ R_{0,k+1} \end{bmatrix} = \mathbf{A}_k \begin{bmatrix} SOC_k \\ U_{d,k} \\ R_{0,k} \end{bmatrix} + \mathbf{B}_k I_k + \begin{bmatrix} \omega_{SOC,k} \\ \omega_{U_d,k} \\ \omega_{R_0,k} \end{bmatrix} \quad (3)$$

Measurement function:

$$\mathbf{y}_k = U_{t,k} = OCV(SOC_k, T) - I_k R_{0,k} - U_{d,k} + \xi_k \quad (4)$$

where $\mathbf{A}_k = \begin{bmatrix} 1 & 0 & 0 \\ 0 & \exp(-\Delta T/C_p R_p) & 0 \\ 0 & 0 & 1 \end{bmatrix}, \mathbf{B}_k = \begin{bmatrix} -\Delta T/C_n \\ R_d[1 - \exp(-\Delta T/C_p R_p)] \\ 0 \end{bmatrix}, k$

is the time, C_n is the nominal capacity of the battery. ΔT is the sampling period of 1 s in our test, $\omega_{SOC}, \omega_{U_d}, \omega_{R_0}, \xi$ are zero man while stochastic process.

4 Results

Based on our proposed temperature-based model as Eq. (2), the sequence of FUDS data tested at different temperatures was used to evaluate the performance of our developed method. FUDS profile is applied to emulate the operation conditions. A comparison was made by using RC and TRC models at different temperatures. Initialized parameters, including the initial guess of SOC_0, measurement covariance and process noise, were set the same for these two estimators. Figure 4a and b compare the estimated results based on these two models. For both of these two figures, the initial guess of SOC was set at 30 %, while the initial SOC was 80 %. Figure 4a is the estimated results when the battery operated at a relatively high temperature i.e. 40 °C. Figure 4b is the results at a low temperature i.e. 10 °C.

According to the estimated results, several conclusions can be obtained as follows. Firstly, the estimated SOC based on TRC model was much more close to the true SOC than RC model. Without the knowledge of the true initial SOC value, our method still can quickly track and converge to the true SOC. Secondly, there were large estimated errors at low temperature, especially using RC model. That means, the estimation based on RC model would not be able to capture the battery dynamics. Finally, Fig. 4b show that the estimated errors based on TRC model would be relatively large when the SOC is between 40 and 60 %. The reason is that the OCV curve is flat in this stage. Thus, a small deviation from the OCV inference will cause the fluctuation of the estimated SOC.

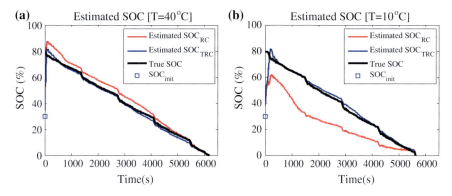

Fig. 4 True and estimated SOC at two temperatures

5 Conclusions

State of charge (SOC) estimation is one of significant functions in a battery management system. It does not only indicate residual range for battery recharge, but also provides information to avoid overcharge and over-discharge for a long battery life. However, when the battery is operated under different working conditions i.e. different temperatures, the existing methods for SOC estimation is subjected to challenge.

In this paper, we discussed temperature effects on SOC estimation based on our test data. To address this problem, a temperature-based battery model (TRC model) incorporated with the OCV-SOC-Temperature table was put forward to improve the accuracy of the SOC estimation by unscented Kalman filtering. Experimental results demonstrated that our model could provide more accurate SOC estimation with smaller root mean square estimated errors than the original model (RC model) without regard to temperature dependence. In addition, TRC model would perform better at high temperature than at low temperature. Therefore, in order to improve the SOC estimation at low temperature, a more accurate model to capture battery dynamics is worthy to investigate for further study.

References

Charkhgard M, Farrokhi M (2010) State-of-charge estimation for lithium-ion batteries using neural networks and EKF. IEEE Trans Ind Electron 57:4178–4187

Cheng C-S, Chen P-W, Huang K-K (2011) Estimating the shift size in the process mean with support vector regression and neural networks. Expert Syst Appl 38:10624–10630

Dai H, Wei X, Sun Z, Wang J, Gu W (2012) Online cell SOC estimation of Li-ion battery packs using a dual time-scale Kalman filtering for EV applications. Appl Energ 95:227–237

Hansen T, Wang C-J (2005) Support vector based battery state of charge estimator. J Power Sources 141:351–358

He H, Xiong R, Guo H (2012) Online estimation of model parameters and state-of-charge of LiFePO4 batteries in electric vehicles. Appl Energ 89:413–420

Hu X, Li S, Peng H (2011) A comparative study of equivalent circuit models for Li-ion batteries. J Power Sources 198:359–367

Hunt G (1996) USABC electric vehicle battery test procedures manual.

Johnson VH, Pesaran AA, Sack T, America S (2001) Temperature-dependent battery models for high-power lithium-ion batteries. National Renewable Energy Laboratory, City of Golden

Junping W, Jingang G, Lei D (2009) An adaptive Kalman filtering based state of charge combined estimator for electric vehicle battery pack. Energ Convers Manage 50:3182–3186

Li IH, Wang WY, Su SF, Lee YS (2007) A merged fuzzy neural network and its applications in battery state-of-charge estimation. IEEE Trans Energ Convers 22:697–708

Plett GL (2004) Extended Kalman filtering for battery management systems of LiPB-based HEV battery packs: Part 2. Modeling and identification. J Power Sources 134:262–276

Roscher MA, Sauer DU (2011) Dynamic electric behavior and open-circuit-voltage modeling of LiFePO4-based lithium ion secondary batteries. J Power Sources 196:331–336

Singh P, Vinjamuri R, Wang X, Reisner D (2006) Design and implementation of a fuzzy logic-based state-of-charge meter for Li-ion batteries used in portable defibrillators. J Power Sources 162:829–836

Sun F, Hu X, Zou Y, Li S (2011) Adaptive unscented Kalman filtering for state of charge estimation of a lithium-ion battery for electric vehicles. Energy 36:3531–3540

Wan EA, Van Der Merwe R (2000) The unscented Kalman filter for nonlinear estimation. In: Adaptive systems for signal processing, communications, and control symposium 2000. AS-SPCC. The IEEE 2000. pp 153–158, IEEE

Wan EA, Van Der Merwe R (2001) The unscented Kalman filter. Kalman Filtering Neural Netw, Wiley, 221–80

Xing Y, Ma EWM, Tsui KL, Pecht M (2011) Battery management systems in electric and hybrid vehicles. Energies 4:1840–1857

Linking Individual Investors' Preferences to a Portfolio Optimization Model

Angela Hsiang-Ling Chen, Yun-Chia Liang and Chieh Chiang

Abstract When optimizing a portfolio, individual investors today already knew potential returns from the capital are often offset by the amount of risk willing to take; hence, to lower the risk associated with the investment, they'll have to diversify through a pool of portfolio from different asset classes (such as stocks, bonds, mutual funds, and cash, etc.). Since most studies have disregarded preferences of individual investors and selections of different asset classes in model formulations, this study provides a systematic approach to set priorities among multi-criteria and trade-off among objectives for Taiwanese individual investors. For that, the Analytic Network Process (ANP) is suggested to determine an asset allocation scheme tailored to the specific requirements of individual investors. Such scheme is then applied to the Markowitz model of portfolio optimization. The Variable Neighborhood Search (VNS) algorithm is constructed to build an efficient frontier of multiple portfolios which offer investors more alternatives on asset selections.

Keywords Asset allocation · Portfolio optimization · Analytic network process · Variable neighborhood search

---

A. H.-L. Chen (✉)
Department of Marketing and Distribution Management, Taoyuan Innovation Institute of Technology, No. 414, Sec. 3, Jhongshan E. Rd., Chungli 320, Taoyuan County, Taiwan, Republic of China
e-mail: achen@tiit.edu.tw

Y.-C. Liang · C. Chiang
Department of Industrial Engineering and Management, Yuan Ze University, No 135 Yuan-tung Rd., Chungli 320, Taoyuan County, Taiwan, Republic of China
e-mail: ycliang@saturn.yzue.du.tw

C. Chiang
e-mail: s995433@mail.yzu.edu.tw

Y.-K. Lin et al. (eds.), *Proceedings of the Institute of Industrial Engineers Asian Conference 2013*, DOI: 10.1007/978-981-4451-98-7_80,
© Springer Science+Business Media Singapore 2013

1 Introduction

Long-term investors have known to diversify investment through a pool of portfolio amongst different asset classes in order to minimize the unacceptable risk and maximize the expected profit. Academic interests on diversification began and have grown since Markowitz (1952) first proposed a mean–variance model. Today, many attempts have been made to linearize Markowitz's model (Chang et al. 2000) and to include other various realistic constraints (Markowitz et al. 2000; Soleimani et al. 2009; Anagnostopoulos and Mamanis 2011; Golmakani and Fazel 2011; Lwin and Qu 2013).

According to Markowitz model, investors are better off to buy a large number of different securities, but to manage a portfolio with a large number of different securities is rather weary and expensive for any investors. Moreover, in Matarazzo (1979) individuals with extensive financial and mental capabilities often make different investment decisions from those with fewer capabilities. Konno (1990) observed that most investors actually buy portfolios apart from the efficient frontier as a result of different views on attributes, e.g. the number of securities in a portfolio, holding periods, and profit growth, etc. Several papers have provided insight into the preferences of the investors and the characteristics of the investment opportunities (Ballestero 1998; Bronson et al. 2007; Charouz and Ramik 2010; Chen et al. 2011; Le 2011).

Since both investor preferences and portfolio optimality are desirable attributes, this study proposes three distinct and useful contributions to the problem of portfolio selection and optimization. First, the risk and diversification of various asset classes match investor preferences by developing a set of suitable and diversified weights using the Analytical Network Process (ANP). Second, portfolio selection is optimized with a cardinality-constrained mean–variance (CCMV) model in Chang et al. (2000) and its computation has been proven to be NP-hard (Moral-Escudero et al. 2006). As well, the realistic constraints are introduced, its computational complexity elevates as well. For that, the Variable Neighborhood Search (VNS) used to overcome challenges faced by deterministic optimization methods. Finally, this study concludes by validating the model with empirically available data and deriving the best combination of portfolios assigned to different financial assets. We hope, with evidence mounting, our study can better meet the modeling needs of investors other than of the standard variety.

2 Problem Formulation

The model of portfolio optimization is worked out in light of the recommendation of financial advisors and finance theory (Khaksari et al. 1989; Puelz and Puelz 1992; Bolster et al. 1995; Bodie and Crane 1997). We consider a four-level hierarchical decision system (Fig. 1). Such hierarchy aims to find investor

Linking Individual Investors' Preferences to a Portfolio Optimization Model 667

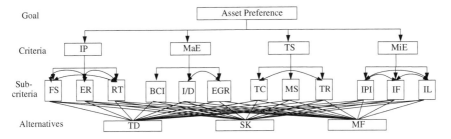

Fig. 1 A four-level hierarchical decision system

preferences among different asset classes—*fix income securities* (e.g. term deposit, TD), *mutual funds* (MF), and *company equities* (e.g. stocks, SK). Major criteria and attributes are listed in Table 1. The array of possible outcomes is the weightings of the tree assets held in investor's portfolio.

Table 1 Definitions for attributes and sub-elements of investment preferences

Criterion/attribute	Definition
Investor's profile (IP)	
Financial status (FS)	The financial position and goals that an investor maintains
Expected return (ER)	The rate of return that an investor desires to achieve for his/her investment
Risk tolerance (RT)	The level of risk that an investor is willing to take for the possibility of incurring a loss
Macroeconomic environment (MaE)	
Business cycle indicator (BCI)	An economic indicator that shows the recurring and fluctuating levels of economic activity over a long period of time
Inflation/deflation (I/D)	The overall general price movement (upward/downward) of goods and services in an economy
Economic growth rate (EGR)	A measure of economic change that a nation's GDP goes from one period to another in percentage terms
Trade situation (TS)	
Transaction costs (TC)	Expenses incurred in the process of carrying out a transaction
Market scale (MS)	The marketability of the financial securities that is relevant
Trade regulation (TR)	Certain financial requirements, restrictions and guidelines, handled by either a government or non-government organization
Microeconomic environment (MiE)	
Industrial production index (IPI)	An economic indicator, highly sensitive to interest rates and consumer demand, measures changes in output for the manufacturing, mining, and utilities
Industrial forecasting (IF)	The trend in which the demographics, socio-cultural, economic, technological, regulatory, and natural environments of an industry will continue to grow
Industrial lifecycle (IL)	The distinct stages of an industry include introduction, growth, maturity, and decline

While the target preferences (i.e. weightings) of the asset classes are defined, we use a CCMV model to find optimal portfolios for each asset class for a single period of investment. Due to the page limitation, the details of the model can be found in Chang et al. (2000). The two objectives (i.e. risk and return) are interdependent and formulated as follows:

$$\min\left(\lambda \sum_{i=1}^{N}\sum_{j=1}^{N} w_i w_j \sigma_{ij} - (1-\lambda)\sum_{i=1}^{N} w_i \mu_i\right). \tag{1}$$

In Eq. (1), each asset has associated a real valued expected return (per period) (μ_i) and each pair o assets has real valued covariance (σ_{ij}). Values of $\lambda (0 < \lambda < 1)$ represent an explicit trade-off between risk and return. At $\lambda = 0$, the optimal solution consists of maximum expected return (μ_i) regardless of the risk (σ_{ij}) involved; whereas, at $\lambda = 1$, the optimal solution becomes minimum risk regardless of the return involved. Also, three additional constraints were adopted: the cardinality constraint imposes a limit K on a desired number of assets in the portfolio (i.e. $\sum_{i=1}^{N} z_i = K$); the budget constraint insures the proportions held of asset i in the portfolio add up to one (i.e. $\sum_{i=1}^{N} w_i = 1, 0 \le w_i \le 1$); the quantity constraint defines upper and lower limits for each asset in the portfolio (i.e. $\varepsilon_i z_i \le w_i \le \delta_i z_i$).

3 Methodology

3.1 Analytical Network Process

When it comes to solve multi-criteria decision problems, the well-known Analytical Hierarch Process (AHP) by Saaty assumes its functional independence on upper and lower parts of alternatives or criteria. Nevertheless, many real-life decision-making involves the interaction and dependence of several alternatives and criteria; thus, the ANP becomes more appropriate. The process utilizes pairwise comparisons of the "n" alternatives and/or criteria (C_n), forming a 'supermatrix (W_{ij})' to deal with component dependence and feedback. Within this supermatrix, the criteria is further decomposed into the sub-elements $(C_i = \{e_{i1}, \ldots, e_{in}\})$, where $\{e_{i1}, \ldots, e_{in}\}$ denotes the components of sub-elements and eigenvector solutions. The final priority weights are derived by the limiting matrix, multiplying the supermatrix by itself until Cesaro summability occurs. Due to the page limitation, the details of this method can be found in several Saaty's works (Niemira and Saaty 2004).

3.2 Variable Neighborhood Search

VNS, one of metaheuristics for solving combinatorial optimization problems, is based on the strategy of a systematic change of neighborhood structure (Mladenovic and Hansen 1997). Applications have been rapidly pertained to many fields, and references can be found in (Hansen et al. 2010). This study contains a single weighted objective so the single objective VNS algorithm is employed here, but the archive of Pareto front, i.e. the efficient frontier, will be built up to store the non-dominated solutions. The procedures are as follows:

Stage 1. Set up a set of predefined neighborhood structures (U_k, $k = 1, 2, 3$)

Stage 2. (Initialization) Randomly generate 1,000 solutions, and each owns K assets. For each asset K, determine the corresponding s_i value, which is used for weight calculation, and the weight w_i, which represents the proportion held of asset i, according to Chang et al. (2000).

Stage 3. (Local Search) Within the chosen neighborhood, randomly generate 50 neighboring solutions and select one among them. For its associated s_i value, add a random number ranging between $(-0.1, 0.1)$ in the first neighborhood, $(-0.2, 0.2)$ in the second neighborhood, and $(-0.3, 0.3)$ in the third neighborhood. For s_i value less than zero will be replaced by any unselected asset; otherwise, update the value. All neighboring solutions will be used to update the efficient frontier.

Stage 4. Accept the solution when its objective function value at *stage* 3 is better than its base solution. Note that whenever a neighborhood structures generates a better solution, the search start over; otherwise, the same base solution will be used for next neighborhood.

Stage 5. (Termination) At the stopping criterion equal to the total number of evaluations in this study), stop the algorithm; otherwise, return to *stage* 3.

4 Computational Results

To illustrate the use and advantages of the combined ANP and CCMV model in portfolio optimization, we present empirical data obtained from Taiwanese investors and Taiwan financial markets. For determining the relative importance between these elements in Table 1, a series of pair-wise comparisons with Saaty's nine-point scale is made. Then the geometric mean method is used to aggregate their assessments, expressed in the form as the unweighted supermatrix. The unweighted supermatrix is then multiplied by the priority weights from the clusters, yielding the weighted supermatrix.

Next, in order to evaluate the weights of elements, the limiting process method of the powers of the supermatrix was employed. The solution is derived by the power method process generating the limiting matrix (shown in Fig. 2). Indeed, the

$$W = \begin{array}{c} \\ IP \\ MaE \\ TS \\ MiE \\ FS \\ ER \\ RT \\ BCI \\ I/D \\ EGR \\ TC \\ MS \\ TR \\ IPI \\ IF \\ IL \\ TD \\ SK \\ MF \end{array}$$

	IP	MaE	TS	MiE	FS	ER	RT	BCI	I/D	EGR	TC	MS	TR	IPI	IF	IL	TD	SK	MF
IP	0.000	0.000	0.000	0.000	0.000	0.000	0.000	0.000	0.000	0.000	0.000	0.000	0.000	0.000	0.000	0.000	0.126	0.224	0.096
MaE	0.000	0.000	0.000	0.000	0.000	0.000	0.000	0.000	0.000	0.000	0.000	0.000	0.000	0.000	0.000	0.000	0.066	0.313	0.106
TS	0.000	0.000	0.000	0.000	0.000	0.000	0.000	0.000	0.000	0.000	0.000	0.000	0.000	0.000	0.000	0.000	0.063	0.248	0.169
MiE	0.000	0.000	0.000	0.000	0.000	0.000	0.000	0.000	0.000	0.000	0.000	0.000	0.000	0.000	0.000	0.000	0.053	0.311	0.124
FS	0.236	0.000	0.000	0.000	0.236	0.236	0.236	0.000	0.000	0.000	0.000	0.000	0.000	0.000	0.000	0.000	0.000	0.000	0.000
ER	0.043	0.000	0.000	0.000	0.043	0.043	0.043	0.000	0.000	0.000	0.000	0.000	0.000	0.000	0.000	0.000	0.000	0.000	0.000
RT	0.206	0.000	0.000	0.000	0.206	0.206	0.206	0.000	0.000	0.000	0.000	0.000	0.000	0.000	0.000	0.000	0.000	0.000	0.000
BCI	0.000	0.149	0.000	0.000	0.000	0.000	0.000	0.500	0.000	0.000	0.000	0.000	0.000	0.000	0.000	0.000	0.000	0.000	0.000
I/D	0.000	0.157	0.000	0.000	0.000	0.000	0.000	0.000	0.250	0.250	0.000	0.000	0.000	0.000	0.000	0.000	0.000	0.000	0.000
EGR	0.000	0.157	0.000	0.000	0.000	0.000	0.000	0.000	0.250	0.250	0.000	0.000	0.000	0.000	0.000	0.000	0.000	0.000	0.000
TC	0.000	0.000	0.183	0.000	0.000	0.000	0.000	0.000	0.000	0.000	0.234	0.234	0.000	0.000	0.000	0.000	0.000	0.000	0.000
MS	0.000	0.000	0.203	0.000	0.000	0.000	0.000	0.000	0.000	0.000	0.259	0.259	0.000	0.000	0.000	0.000	0.000	0.000	0.000
TR	0.000	0.000	0.051	0.000	0.000	0.000	0.000	0.000	0.000	0.000	0.000	0.000	0.500	0.000	0.000	0.000	0.000	0.000	0.000
IPI	0.000	0.000	0.000	0.133	0.000	0.000	0.000	0.000	0.000	0.000	0.000	0.000	0.000	0.500	0.000	0.000	0.000	0.000	0.000
IF	0.000	0.000	0.000	0.167	0.000	0.000	0.000	0.000	0.000	0.000	0.000	0.000	0.000	0.000	0.250	0.250	0.000	0.000	0.000
IL	0.000	0.000	0.000	0.167	0.000	0.000	0.000	0.000	0.000	0.000	0.000	0.000	0.000	0.000	0.250	0.250	0.000	0.000	0.000
TD	0.000	0.000	0.000	0.000	0.126	0.126	0.126	0.069	0.063	0.063	0.061	0.061	0.052	0.056	0.051	0.051	0.000	0.000	0.000
SK	0.000	0.000	0.000	0.000	0.224	0.224	0.224	0.284	0.314	0.314	0.243	0.243	0.310	0.284	0.306	0.306	0.000	0.000	0.000
MF	0.000	0.000	0.000	0.000	0.096	0.096	0.096	0.096	0.108	0.108	0.178	0.178	0.124	0.118	0.126	0.126	0.000	0.000	0.000

Fig. 2 The limiting supermatrix

calculations of the supermatrix can be easily solved by using the professional software named "super decisions," and then the overall normalized priorities were obtained: $W = \{IP, MaE, TS, MiE\} = \{0.439, 0.173, 0.184, 0.204\}$, $W_{IP} = \{FS, ER, RT\} = \{0.487, 0.089, 0.424\}$, $W_{MaE} = \{BCI, I/D, EGR\} = \{0.321, 0.340, 0.339\}$, $W_{TS} = \{TC, MS, TR\} = \{0.418, 0.464, 0.117\}$, and $W_{MiE} = \{IPI, IF, IL\} = \{0.285, 0.357, 0.357\}$, and $A = \{TD, SK, MF\} = \{0.173, 0.574, 0.253\}$. These ANP results provide the relative importance weights for every factor in the model. The most considered factor is investor's profile (IP) due to the highest priority of 0.439. Among the alternatives, the rank is SK (i.e. stocks), MF (mutual fund), and TD (term deposit). Their weights are used as priorities in final portfolio optimization formulation. $(TD, SK, MF) = (w_1, w_2, w_3) = (0.173, 0.574, 0.253)$, where w_j are the values of the three asset classes in our investment portfolio.

Now that the investor's preferences have been determined, the VNS algorithm coded in Borland C++ 6.0 is applied to optimize a structure-judgmental portfolio for each asset classes (i.e. stocks and mutual funds). Our sample consists of 731 stocks from all the sectors listed in the TSE and 258 mutual funds currently traded in the market. The study period includes the years 2007–2011, and the closing prices were recorded in a weekly basis. The stopping criteria are set to 1,000 N, where N represents the total number of assets in each class. The λ value is increased from 0 with increment of 0.02; thus, a total number of 51 λ values is considered. For both asset classes, we set different values of K (i.e., 5, 10, 15, and 20), references to the desired number of assets in the portfolio. The upper (δ_i) and lower (ε_i) limits of the weights are 1.0 and 0.01, respectively.

To illustrate our proposed model in action, we assume a base-line investor who prefers a more conservative allocation, lower level of risk and would like to maintain a portfolio of 10 assets in each of asset classes. Therefore, via the VNS, we generate a recommended portfolio for both stocks and mutual funds (Tables 2 and 3).

Table 2 A recommendation of a stock portfolio at K = 10

Stock no.	Weights (%)	Stock no.	Weights (%)	Stock no.	Weights (%)
512	1.057	502	1.005	418	1.014
28	26.785	385	1.006	619	1.009
657	22.763	466	21.174	215	1.013
45	23.174				

Table 3 A recommendation of a mutual fund portfolio at K = 10

MF no.	Weights (%)	MF no.	Weights (%)	MF no.	Weights (%)
233	1.000	201	1.004	108	43.604
166	12.797	251	1.015	208	1.055
203	2.055	150	1.018	205	34.637
91	2.815				

5 Conclusions

In this work, we have proposed an investor's preference framework in portfolio selection problem. Moreover, we consider selections of different investment instruments (i.e. asset classes) in formulating portfolio selection model. As few studies consider preferences of individual investors, and even fewer investigate portfolio selection with different investment instruments, our study shows potential to meet the needs of investors.

References

Anagnostopoulos KP, Mamanis G (2011) The mean-variance cardinality constrained portfolio optimization problem: an experimental evaluation of five multiobjective evolutionary algorithms. Expert Syst Appl 38:14208–14217

Ballestero E (1998) Approximating the optimum portfolio for an investor with particular preferences. J Oper Res Soc 49:998–1000

Bodie Z, Crane DB (1997) Personal investing: advice, theory, and evidence from a survey of TIAA–CREF participants. Financ Anal J 53:13–23

Bolster PJ, Janjigian V, Trahan EA (1995) Determining investor suitability using the analytic hierarchy process. Financ Anal J 51:63–75

Bronson J, Scanlan M, Squires J (2007) Managing individual investor portfolios, CFA level III candidate body of knowledge. CFA Institute, Charlottesville

Chang TJ, Meade N, Beasley JE, Sharaiha YM (2000) Heuristics for cardinality constrained portfolio optimization. Compu Oper Res 27:1271–1302

Charouz J, Ramík J (2010) A multicriteria decision making at portfolio management. Ekonomika A Manage 2:44–52

Chen AHL, Cheng K, Lee ZH (2011) The behavior of Taiwanese investors in asset allocation. Asia-Pacific J Bus Admin 3:62–74

Golmakani HR, Fazel M (2011) Constrained portfolio selection using particle swarm optimization. Expert Syst Appl 38:8327–8335

Hansen P, Mladenović N, Pérez JAM (2010) Variable neighborhood search: methods and applications. Annals Oper Res 175:367–407

Khaksari S, Kamath R, Grieves R (1989) A new approach to determining optimum portfolio mix. J Portfolio Manage 15:43–49

Konno H (1990) Piecewise linear risk function and portfolio optimization. J Oper Res Soc Jpn 33:139–156

Le SV (2011) Asset allocation: an application of the analytic hierarchy process. J Bus Econ Res (JBER) 6:87–94

Lwin K, Qu R (2013) A hybrid algorithm for constrained portfolio selection problems. Appl Intell 1–16. doi:10.1007/s10489-012-0411-7

Markowitz HM (1952) Portfolio selection. J Financ 7:77–91

Markowitz HM, Todd GP, Sharpe WF (2000) Mean-variance analysis in Portfolio choice and capital markets. Frank J. Fabozzi Series: book, vol 66. Wiley, Hoboken

Matarazzo J (1979) Wechsler's measurement and appraisal of adult intelligence, 5th edn. Oxford University Press, New York, pp 449–454

Mladenović N, Hansen P (1997) Variable neighborhood search. Comput Oper Res 24:1097–1100

Moral-Escudero R, Ruiz-Torrubiano R, Suárez A (2006) Selection of optimal investment portfolios with cardinality constraints. In: Evol Comput 2006 (CEC 2006) IEEE Congress on. July 2006, 2382–2388

Niemira MP, Saaty TL (2004) An analytic network process model for financial-crisis forecasting. Intl J Forecast 20:573–587

Puelz AV, Puelz R (1992) Personal financial planning and the allocation of disposable wealth. Financ Serv Rev 1:87–99

Soleimani H, Golmakani HR, Salimi MH (2009) Markowitz-based portfolio selection with minimum transaction lots, cardinality constraints and regarding section capitalization using genetic algorithm. Expert Syst Appl 36:5058–5063

Models and Partial Re-Optimization Heuristics for Dynamic Hub-and-Spoke Transferring Route Problems

Ming-Der May

Abstract The major advantages of hub-and-spoke network are the reduction of the number of routes and the effect of economies of scale, which can effectively save the shipping cost of the transportation industry. In the landside pick-up services of the international express industry, the application of such a model has achieved high efficient transit operations. The entire operational area is divided into numbers of partitions. Each partition sets up a station as the cargo collection place (meeting point). Vehicle routes meet here to consolidate goods that collected from customers into truckload shipment and transfer those to regional centers by larger trucks. In this study, the dynamic vehicle routing problem are extended to this type of hub-and-spoke network architecture, which make it different from the Vehicle Routing Problem (VRP), and is formulated as the Dynamic Hub-and-Spoke Problem with Transferring Route (DHSPTR). Test problems with dynamic pickup flows of export goods under H-S networks are designed to evaluate the proposed hybrid ACO solution methods. Partial re-optimization heuristics are realized by ACO for its ability to keep the solution status while new demands keep coming during the process and are inserted to existing routes dynamically.

Keywords Dynamic vehicle routing · Hub-and-spoke · Partial re-optimization · Ant colony optimization

1 Introduction

Generally, international express carriers set up a regional warehouse in the airport of some given market area, and dividing the large region into several smaller sub-regions. For example, if Taiwan is one of the major market regions of some

---

M.-D. May (✉)
Lung Hwa University of Science and Technology, No.300, Section1, Wanshou Road, Guishan Shiang, Taoyuan County, Taiwan, Republic of China
e-mail: mdmay@mail.lhu.edu.tw

Y.-K. Lin et al. (eds.), *Proceedings of the Institute of Industrial Engineers Asian Conference 2013*, DOI: 10.1007/978-981-4451-98-7_81, © Springer Science+Business Media Singapore 2013

international express carrier, thus one or some sub-regions may be clustered within this area. Each sub-region might be served by one or several transshipping points (TP) to handle the pickup or delivery of goods. In the end of every work days or some given deadline, the carrier has to transfer all collected cargos to regional warehouse in the airport for the followed international transportation processes. This operational framework is shown as in the Fig. 1.

In this framework, there are three participants within it, which are the regional warehouses in the airport, TPs, and customers. These three players are interacted in two stages. In the first stage, customer orders are collected to TPs. One or more vehicles (ex. pickup truck or van) route to customers to collect parcels and back to the TP. In the second stage, express carriers transfer collected freights at TPs to the airport warehouse. Such two stages collecting and transferring networks are quite common for LTL or express carriers, and are able to save the operation cost dramatically when the customer demands are large enough.

In this paper, we develop transshipment route models that consider three major decisions simultaneously. Phase (I) is planning interior operation paths which connect carrier's regional hub to every TPs, and phases (II) is the selection of TPs, finally, the phase (III) is the routing to every customers assigned to each TP. The transferring routes path from the single central hub to every TP is also considered. One dynamic solution framework is proposed for these models to select TPs and routing paths dynamically based on the arrival of new orders. Afterward the numerical testing results would be described and followed with some conclusion remarks in the end.

The concrete definition of DVRP was first addressed by Psaraftis (1988), who was discussion the meaning, characteristic of DVRP and solving strategy of it. A new review study by Pillac et al. (2013) is recommended for the further knowledge about DVRP. The MDVRP consists of constructing a set of vehicle routes in such a way that: (1) each route starts and ends at the same depot, (2) each customer is visited exactly once by a vehicle, (3) the total demand of each route does not exceed the vehicle capacity Q, (4) the total duration of each route (including travel and service time) does not exceed a preset limit L and (5) the total routing cost is minimized (Renaud et al. 1996). Generally, the objective of the MDVRP is to

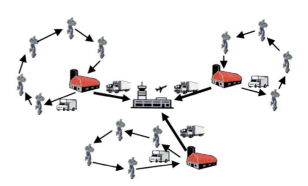

Fig. 1 Operational framework of pickup service for express carriers

minimize the total delivery distance or time spent in serving all customers. Every customer is visited by a vehicle based at one of several depots.

Irnich (2000) examined a 'Multi-Depot Pickup and Delivery Problem with a Single Hub and Heterogeneous Vehicles' (MDPDPSH), where pickup always means to load something at a location and deliver it to the hub, and delivery is always defined as loading some goods at the hub and deliver it to a location. Their work is inspired by a practical problem at the Deutsche Post AG, Germany's post service. The MDPDPSH is one subproblem, the ground feeding problem, in the design process of the global area transportation network. Crevier et al. (2007) consider a 'Multi-Depot Vehicle Routing Problem with Inter-Depot Routes' (MDVRPI) that arises from a real-life grocery distribution problem in the Montreal area. They mention that several similar applications are encountered in the context where the route of a vehicle can be composed of multiple stops at intermediate depots in order for the vehicle to be replenished.

Transshipping problem is gathering more and more attention in recent years. The similar and classical problem is Hub-and-Spoke (H/S) problem, which can reduce transportation cost by transshipping with hubs. But Zäpfel and Wasner (2002) argued that hybrid direct shipment and transshipping will be more efficient than only direct shipment or transshipping. Wasner and Zapfel (2004) further pointed out that using transshipping operation can reduce the number of depots and operating cost.

Above literatures show related topics like MDVRP and H/S problem have been studied well and have obvious results in each area, but either DVRP are limited to single depot or MDVRP are limited on static scenario. Therefore, this study will focus on integrating DVRP, MDVRP and transshipping problem.

The ant colony optimization (ACO) algorithm is one of the metaheuristics for combinatorial problem, and was proposed by Dorigo (1992). ACO can be seen as multiagent systems in which each single agent is inspired by the behavior of a real ant. The dynamic memory structure, which incorporates information on the effectiveness of previously obtained results, guide the construction process of initial solution, and resume the re-optimization process if the problems is interrupted and changed during the running process.

2 Operational Framework of Express Pick Service

In this section, we introduce the operational processes in pickup procedure of local express carriers. Generally, express carriers set up a regional warehouse in their market region, and dividing the large region into several smaller sub-regions. For example, if the Taipei city area is the major market region of one local express carrier, thus one or some administrative districts may be cluster into the sub-regions. Each sub-region will be served by a small transshipping post to pickup or delivery. However, they will transfer the collected cargos to the regional warehouse for the followed transportation processes. The operational framework has been shown in the Fig. 1.

In this framework, there are three levels, which are the warehouses, transshipping points, and customers. These three levels are operated in two stages. First, customer orders are collected to transshipping points in stage 1. One or more vehicles route to every customers to collect parcels and back to the transshipping point. In the stage two, express carriers transfer collected freight from transshipping points to regional warehouse. Some local carriers dispatch a truck especially for direct shipping between transshipping point and regional warehouse. Although this method is easy to implement, the efficiency and flexibility for serving customers would be getting worse if the demand is varied. Because a large truck is often adopted to do line haul from transshipping post to the regional warehouse. That usually leads to low utilization rate and increased operational cost. Moreover, the transshipping operation processed in the end of every business day that cumulate the cargos into truck load will defer customers collected before the midday to the next day. So it seems to have some problems in existing transshipping framework, that is the purpose of this study to try to improve and solve in our proposed models and dynamic transferring framework. These will be discussed in next section.

3 Models and Formulation

As mentioned above, this study focus on the pickup operation of local express service providers, and consider the customer demand appeared in real time. There are three levels and two stages in the considered serving framework. Briefly, the first level is the customers, the transshipping point and warehouse (or hub) is at the second and the third level respectively. The stage (I) is routing network from the transfer post to every customers that have been assigned to this transshipping point, and the stage (II) is interior operation, i.e. no customers involved, for paths connecting the hub and every transshipping point. The proposed models are based on Warehouse Location Routing Problem (WLRP) which presented by Perl and Daskin (1985) and Ambrosino and Scutella (2005) respectively, and DVRPTW which presented by Shieh and May(1998). Some new variables and constrain are created to formulate this problem and some assumptions in these two models are made and will be described later.

The purpose of this model is to reduce transshipping cost. For this reason, only one big truck moves back and forth between the regional hub and transship point, and may route several times a day. In this way, we have assumed that: (1) Customers can assign to any transshipping point, and the customers appear in real time; (2) Cargos are transferred from a small vehicle to a big truck at the transshipping points; (3) The small vehicles site in transshipping points and the big trucks site in regional hub; and (4) Vehicles do pickup operations to collect demand to accumulated to the vehicle's capacity, if so, the vehicle will be back to

Models and Partial Re-Optimization Heuristics

the transshipping point. Again, the big truck travel to all transshipping points to collect the cargos and move back to the hub.

3.1 Multi-Depot and Routing Transshipment

The route starts from the hub to each transshipping point with big truck. For this reason, we need to adjust and define additional notation as follows.

$V_1 = \{1,\ldots,v_1\}$: set of vehicles which route form the hub to transshipping points;

$V_2 = \{1,\ldots,v_2\}$: set of vehicles which route form transshipping points to customers;

$R = \{1,\ldots,m + H\}$: $TP \cup H$;

$TCgh$: the cost of routing from node g to node h, $\forall g, h \in R; g \neq h$

$Qk1$: capacity of vehicle $k1$, $\forall k_1 \in V_1$;

$Qk2$: capacity of vehicle $k2$, $\forall k_2 \in V_2$;

$sgk1$: start service time of node g by vehicle $k1$; $\forall g \in R$;

$shk1$: start service time of node h by vehicle $k1$; $\forall h \in R$;

$sgk2$: start service time of node g by vehicle $k2$; $\forall g \in Y$;

$shk2$: start service time of node h by vehicle $k2$; $\forall h \in Y$;

$t1gh$: travel time from node g to node h including service time of node g, $\forall g, h \in R; g \neq h$;

$t2gh$: travel time from node g to node h including service time of node g, $\forall g, h \in Y; g \neq h$;

$[a_{1g}, b_{1g}]$: time window of routing from hub;

$[a_{2g}, b_{2g}]$: time window of routing from transshipping poitn;

The DHSPTR is formulated in the following.

$$Min \sum_{j \in TP} VC_j \left(\sum_{i \in C} D_i y_{ij} \right) + \sum_{k \in V_1} \sum_{g \in R} \sum_{h \in R} TC_{gh} r_{ghk1} + \sum_{k \in V_2} \sum_{g \in Y} \sum_{h \in Y} C_{gh} x_{ghk2} \quad (1)$$

$$\sum_{k \in V_1} \sum_{h \in Y} r_{ihk} = 1, \quad \forall i \in TP, \quad (2)$$

$$\sum_{k \in V_2} \sum_{h \in Y} x_{ihk} = 1, \quad \forall i \in C, \quad (3)$$

$$\sum_{g \in TP} r_{hgk_1} - \sum_{g \in TP} r_{ghk_1} = 0, \quad \forall h \in R, \forall k_1 \in V_2, \quad (4)$$

$$\sum_{g \in C} x_{hgk_2} - \sum_{g \in C} x_{ghk_2} = 0, \quad \forall h \in Y, \forall k_2 \in V_2, \quad (5)$$

$$\sum_{g \in TP} \sum_{h \in H} r_{ghk1} \leq 1, \quad \forall k_1 \in V_1, \quad (6)$$

$$\sum_{g \in C} \sum_{h \in TP} x_{ghk_2} \leq 1, \quad \forall k_2 \in V_2, \tag{7}$$

$$\sum_{h \in H} \sum_{g \in R} r_{ghk_1} = 1, \quad \forall k_1 \in V_1, \tag{8}$$

$$\sum_{h \in R} \sum_{g \in H} r_{hgk_1} = 1, \quad \forall k_1 \in V_1, \tag{9}$$

$$\sum_{g \in TP} \sum_{h \in Y} x_{ghk_2} = 1, \quad \forall k_2 \in V_2, \tag{10}$$

$$\sum_{h \in Y} \sum_{g \in TP} x_{hgk_2} = 1, \quad \forall k_2 \in V_2, \tag{11}$$

$$\sum_{i \in C} \sum_{j \in TP} D_i y_{ij} \sum_{h \in R} r_{jhk_1} - Q_{k1} \leq 0, \quad \forall k_1 \in V_1, \tag{12}$$

$$\sum_{i \in C} D_i \sum_{h \in Y} x_{ihk_2} - Q_{k2} \leq 0, \quad \forall k_2 \in V_2, \tag{13}$$

$$\sum_{h \in Y} x_{ihk_2} + \sum_{h \in Y} x_{jhk_2} - y_{ij} \leq 1, \quad \forall i \in C, \forall j \in TP, \forall k_2 \in V_2, \tag{14}$$

$$s_{1gk} + t_{1gh} - K(1 - r_{ghk_1}) \leq s_{1hk}, \quad \forall g, h \in R, \forall k_1 \in V_1, \tag{15}$$

$$s_{2gk} + t_{2gh} - K(1 - x_{ghk_2}) \leq s_{2hk}, \quad \forall g, h \in Y, \forall k_2 \in V_2, \tag{16}$$

$$a_{1g} \leq s_{1gk_1} \leq b_{1g}, \quad \forall g \in R, \forall k_2 \in V_2, \tag{17}$$

$$a_{2g} \leq s_{2gk_2} \leq b_{2g}, \quad \forall g \in Y, \forall k_1 \in V_1, \tag{18}$$

$$r_{ghk_1} \in \{0, 1\}, \quad \forall g, h \in R, \forall k_1 \in V_1, \tag{19}$$

$$x_{ghk_2} \in \{0, 1\}, \quad \forall g, h \in Y, \forall k_2 \in V_2, \tag{20}$$

$$y_{ij} \in \{0, 1\}, \quad \forall i \in C, \forall j \in TP, \tag{21}$$

The objective function (1) is minimizing total cost of pickup and transshipping operation, and the meaning of constrains are also similar to the related constrains in the above DHSPTR formulation. The framework of this DHSPTR is depicted as Fig. 2.

Fig. 2 Operation framework of DHSPTR

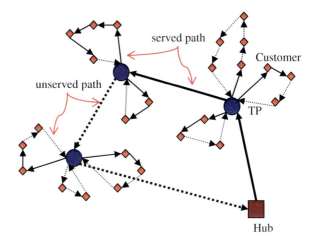

4 Computational Test

In order to investigate the performance of our models, we do some computational tests in some instances as shown in Table 1. To test our two models, we randomly generate 6 instances, and try to find how the transshipping point (TP) numbers affect their results, so we set TP about 2–3, and number of customers are given from 3 to 10; the vehicles are given by 3 and 5. The demand of the customers set form 1 to 25; the routing cost is generated randomly, transshipping cost is also generated randomly. Finally, time window is generated in interval [0,180] for early time and [10, 190] for later time. To ensure both the arrival and departure to be at TP within reasonable time, we set time window of every TP as [0,200].

Table 1 The instances data for computational test

Problem number	Number of TP	Number of customer	Number of vehicle	Number of constraints	Number of variables
1	2	5	3	216	167
2	2	7	5	542	431
3	2	10	5	908	755
4	3	5	3	279	219
5	3	7	5	674	534
6	3	10	5	1086	892

5 Conclusions

In this paper, we consider the actual operation of local express carriers. In order to solve the dynamic customer demand in real time, we have presented mathematical programming formulations that determine customer's assignment, routing, and transshipping in the same time. We develop ACO with dynamic framework to carry out solution. However, this research has provided stable basis for further development, and how to solve instants efficiently will be the major subjects for the future work.

References

Ambrosino D, Scutella MG (2005) Distribution network design: new problems and related models. Eur J Oper Res 165(3):610–624

Crevier B, Cordeau JF, Laporte G (2007) Multi-depot vehicle routing problem with inter-depot routes. Eur J Oper Res 176:756–773

Dorigo M (1992) Optimization, learning and natural algorithms (in Italian), PhD Thesis, Department of Electronics, Politecnico di Milano, Italy

Irnich S (2000) A multi-depot pickup and delivery problem with a single hub and heterogeneous vehicles. Eur J Oper Res 122(2):310–328

Pillac V, Gendreau M, Guéret C, Medaglia AL (2013) A review of dynamic vehicle routing problems. Eur J Oper Res 225:1–11

Psaraftis HN (1988) Dynamic vehicle routing problems. In: Golden BL, Assad AA (eds) Vehicle routing: methods and studies. Elsevier, North-Holland, pp 223–248

Renaud J, Laporte G, Boctor FF (1996) A tabu search heuristic for the multi-depot vehicle routing problem. Comput Oper Res 23:229–235

Wasner M, Zapfel G (2004) An integrated multi-depot hub-location vehicle routing model for network planning of parcel service. Int J Prod Econ 90(3):403–419

Zäpfel G, Wasner M (2002) Planning and optimization of hub-and-spoke transportation networks of cooperative third-party logistics providers. Int J Prod Econ 78(2):207–220

Hazards and Risks Associated with Warehouse Workers: A Field Study

Ren-Liu Jang and An-Che Chen

Abstract This study evaluated risk factors of manual material jobs performed in the field using questionnaire and NIOSH 1991 lifting guide. Discomfort assessment survey was used to investigate the risk of musculoskeletal injury in the warehousing operations. The questionnaire contains basic personal information, medical history, working hour and discomfort symptoms of the body such as pain, soreness, numbness at the end of the working day. Work analysis and ergonomic evaluation was also conducted to understand how workers engaged in their work. The results found the musculoskeletal risk factors for workers in warehouses were the overweight of objects lifted, awkward postures, static working postures, and long working duration. Awkward working postures such as lifting involved forward bending or twisting would increase the risk of lower back pain. Static working postures such as prolonged sitting or standing were also associated with the occurrence of low back pain and other discomfort.

Keywords Manual material handling · Biomechanics · Warehouse

1 Introduction

The main operations of warehouses are to manipulate the flow of goods and quick response to customer demand. Although most operations in warehouses are designed toward automation considering on cost and efficiency, it is difficult to run warehouses fully automation without involving manual handling. For example,

R.-L. Jang (✉) · A.-C. Chen
Department of Industrial Engineering and Management, Ming Chi University of Technology, Hsinchu County, Taiwan, Republic of China
e-mail: renliuj@mail.mcut.edu.tw

A.-C. Chen
e-mail: anche@mail.mcut.edu.tw

picking up the goods from shelves to pallets, manual handling is still the most efficient way.

The manual material handling was the main cause for workers to develop musculoskeletal injury. The National Association of Wholesale Grocers of America and the International Foodservice Distribution Association found that 30 % of occupational injuries of the warehousing personnel were back sprain and strain (Waters 1993). Studies showed that manual lifting jobs were related to back injuries and was the primary cause of back injuries in 54 % of all cases studied. Compared with other workers in warehouses, order selectors had a very high incidence rate of lower back injuries, particular when handling frequency was high (Waters 1993; Waters et al. 1995) .

Order selectors had a higher risk in musculoskeletal disorders because of many factors which resulted in fatigue and high metabolic rate and hardly keeping up with the pace of work. The study found that order selectors' workload exceeded the metabolism and biomechanical standards, mainly due to shelf height, workplace layout and long working hours (Waters et al. 1993). There were 50 % of 38 retail pickers reported at least injured once within 12 months in which at least 18 % of them were back injuries. Another similar report, 63 % of the workers reported at least injured once within 12 months and 47 % of them report at least injured once on back (Waters 1993; Waters et al. 1995).

Work-related low-back disorders (LBDs) are still the major cause of compensation costs. The relative contributions between personal, workplace, organizational and environmental variables to the development and severity of LBDs remain unclear. Dempsey et al. (1997) investigated the personal variables associated with LBDs in industrial populations. The results suggested that age, gender, injury history, relative strength, smoking, and psychosocial variables be studied further. The personal variables: height, weight, pathologies, genetic factors, maximum oxygen uptake, and absolute strength were unlikely to produce significant effects.

Xu et al. (1997) found that physically hard work, frequently twisting or bending, standing up, and concentration demands were the risk factors for the occurrence of low back pain, even after the following variables: age, sex, educational level, and duration of employment in a specific occupation were controlled.

Macfarlane et al. (1997) found those whose jobs involved lifting/pulling/pushing objects of at least 25 lbs, or whose jobs involved prolonged periods of standing or walking had an increased risk of a new episode of low back pain. Occupational activities such as working with heavy weights or lengthy periods of standing or walking, were associated with the occurrence of low back pain.

Bongers et al. (1993) studied psychosocial factors at work and musculoskeletal disease, and concluded that monotonous work, high perceived work load, and time pressure were related to musculoskeletal symptoms. The data also suggested that low control on the job and lack of social support by colleagues were positively associated with musculoskeletal disease. In addition, stress symptoms were often associated with musculoskeletal disease, and some studies indicated that stress symptoms did contribute to the development of this disease.

Besides the weight of object to be lifted, the locations of stacking items, and the sizes of the items would affect how order selectors appropriately handled the items. Awkward postures would result in high compressing forces on spinal disc (Marras et al. 1999).

2 Methods

This study evaluated risk factors of manual material jobs performed in the field using questionnaire and NIOSH 1991 lifting guide. Discomfort assessment survey was used to investigate the risk of musculoskeletal injury in the warehousing operations. The questionnaire contains: basic personal information, medical history, working hour and discomfort symptoms of the body such as pain, soreness, numbness at the end of the working day. Work Analysis and ergonomic evaluation was also conducted to understand how workers engaged in their work. In ergonomic evaluation, there was a videotaping on the order fulfillment process, from activities performed at the warehouse to activities performed at the shipping sites. Biomechanical evaluations of compressing force on low back were performed using the revised NIOSH lifting equation.

2.1 Respondents

95 workers participated in this survey from 8 local logistic centers. There were 29 female and 66 male. 62 % of female worked in the warehouse sector and 34 % worked in the management department; 67 % of male worked in the warehouse sector and 33 % worked in the shipping sector.

The female workers' average age ranged from 26 to 40 years, with an average of 33. Height ranged from 149 to 168 cm, with an average of 157. Weight ranged from 49 to 60 kg, with an average of 55. The male workers' average age ranged from 26 to 35 years, with an average of 32.7. Height ranged from 164 to 177 cm, with an average of 171. Weight ranged from 67 to 74 kg, with an average of 69.

3 Results

3.1 Respondents Background

Female respondents' seniority in laborious work was 8.5 years on average. The average working time was 8.6 h and the average sleeping time 7 h. 31 % of respondents were engaged in labor work more than 10 years and 38 % between

5 and 10 years. 59 % of respondents working at their current positions for more than five years and 38 % less than three years.

Male respondents' seniority in laborious work was 7.5 years on average. The average working time was 9.3 h and the average sleeping time 6.5 h. 38 % of respondents were engaged in labor work more than 10 years and 26 % between 5 and 10 years. 33 % of respondents working at their current positions for more than five years and 56 % less than three years.

3.2 Discomfort Assessment Survey

Medical history showed that eight female respondents (27 %) had been injured, and four had fully recovered. 29 male respondents (44 %) had been injured, and fifteen had fully recovered.

The response of three major questions in DAS was analyzed.

Q 1: In the past 12 months, have you experienced any discomfort, fatigue, numbness, or pain that relates to your job by body location?

65.5 % of female respondents experienced discomfort on shoulder and hand and 62 % on low back; 65.2 % of male respondents experienced discomfort on shoulder and 57.6 % on neck and low back.

Q 2: How often (daily, weekly, monthly) do you experience discomfort, fatigue, numbness, or pain in the region of the body?

41.7 % of female respondents experienced discomfort daily on legs and feet and 38.9 % on low back; 42.2 % of male respondents experienced discomfort daily on feet and 42.1 % on low back. 70 % of female respondents experienced discomfort weekly on low leg and 69.2 % on feet; 71.4 % of male respondents experienced discomfort weekly on low leg and 68.4 % on neck.

Q 3: On average, how severe (affecting work and daily living) is the discomfort, fatigue, numbness, or pain in this region of the body?

72.2 % of female respondents experienced discomfort affecting work and daily living on low back and 70 % on arms; 73.7 % of male respondents experienced discomfort affecting work and daily living on low back and 69.8 % on shoulder.

Regardless of gender, low back, followed by shoulder and neck were the body parts most frequently affected and experienced discomfort, fatigue, numbness or pain within 12 months. Low back, hand/wrist, and feet were the most responded locations for experienced discomfort daily. Above 50 % of respondents reported every discomfort of body part would affect work and daily living no matter what location being experienced discomfort.

3.3 Work Analysis

The workers in the warehouse sector were mainly order selectors. When picking a case from shelves, they normally would slide or rolled cases to the front edge of pallets before lifting to avoid reaching across pallets to lift and always pick up the cases in front and at waist height first. The workers in the management department were data-entry workers, which was sedentary work. The workers in the shipping sector were drivers who used pallet jacks to move the pallet of goods directly to delivery trucks.

The hazards and risks associated with workers were found that:

1. The NIOSH lifting criteria showed that most of the case lifting tasks exceeded the recommended weight limit (RWL).
2. The depth on the shelf exceeded the normal reach limit of the order selectors. Extended reaches for heavy cases may significantly increase the risk for musculoskeletal injuries.
3. Awkward positions of the hand, wrist, forearm, shoulder and low back increased the risk for experiencing discomfort, fatigue, numbness and pain during and after the work.
4. Awkward postures such as lifting involved forward bending or twisting increased the risk of lower back injuries.
5. Long work hours with only few short breaks resulted in more discomforts in body parts, particularly on legs and feet.

4 Conclusions

This study evaluated risk factors of manual material jobs performed in the field using questionnaire and NIOSH 1991 lifting guide. The results found the musculoskeletal risk factors for workers in warehouse were the overweight of objects lifted, awkward postures, static work postures, and long work duration. To reduce workers' exposures to hazards and risks, the following guidelines are recommended:

1. Reduce the weight of loads such as putting fewer items in the container or using a smaller and/or lighter-weight container.
2. Avoid manually lifting or lowering loads to or from the floor. Pallet loads of materials are located at a height that allows workers to lift and lower within their power zone which is between knee height and waist height.
3. Minimize the distances loads are lifted and lowered.
4. Clear spaces to improve access to materials or products being handled. Easy access allows workers to get closer and reduces reaching, bending, and twisting.

5. Adjust work schedules, work pace, or work practices. Provide recovery time (e.g., short rest breaks).
6. Plan the work flow to eliminate unnecessary lifts.
7. Alternate lifting tasks for workers with non-lifting tasks to reduce the frequency of lifting and the amount of time on perform lifting tasks.

Acknowledgments This work was supported by Institute of Occupational Safety and Health of Republic of China. The authors also gratefully acknowledge the helpful comments and suggestions of the reviewers, which have improved the presentation.

References

Bongers PM, de Winter CR, Kompier MA, Hildebrandt VH (1993) Psychosocial factors at work and musculoskeletal disease. Scand J Work, Environ Health 19:297–312
Dempsey PG, Burdorf A, Webster BS (1997) The influence of personal variables on work-related low-back disorders and implications for future research. J Occup Environ Med 39:748–759
Macfarlane GJ, Thomas E, Papageorgiou AC, Croft PR, Jayson MIV, Silman AJ (1997) Employment and physical work activities as predictors of future low back pain. Spine 22:1143–1149
Marras WS, Granata KP, Davis KG, Allread WG, Jorgensen MJ (1999) Effects of box features on spine loading during warehouse order selecting. Ergonomics 42(7):980–996
Waters T, Putz-Anderson V, Barron S, Fine L (1993) National Institute for Occupa-tional Safety and Health, Big Bear Grocery Warehouse, Columbus, Ohio, HETA 91-405-2340
Waters T, Putz-Anderson V, Barron S, Fine L (1995) National Institute for Occupa-tional Safety and Health, Kroger Grocery Warehouse, Nashville, Tennessee, HETA 93-0920-2548
Waters TR (1993) Work workplace factors and trunk motion in grocery selector tasks. In: Proceedings of the human factors and ergonomics society 37th annual meeting. Human Factors and Ergonomics Society, Santa Monica, pp 654–658
Xu Y, Bach E, Orhede E (1997) Work environment and low back pain: the influence of occupational activities. Occup Environ Med 54:741–745

Green Supply Chain Management (GSCM) in an Industrial Estate: A Case Study of Karawang Industrial Estate, Indonesia

Katlea Fitriani

Abstract Increasing the pollution drive government and people consider about environment, therefore green issues become the hot topic lately, especially industrial sector. There are two reasons drive the industries to consider for applying green aspect in their supply chain, which are regulation, Financial and supply chain pressure. This paper intends to introduce the GSCM pressure, describe the GSCM practices, performance in Karawang city, Indonesia and compare the result with previous literature. Survey questionnaire is be conducted in which consist of 3 sections and using five-point scale. The questions are be mainly based on literature review and developed to the conceptual circumstance. The factorial analysis is be used to aggregate the factors, then multiple regression analysis is used to know the relation for each variable. The research provides that Karawang has not considered the GSCM pressure yet. Industries are considering carrying out GSCM practice like improving the quality of environment friendly goods. However, those practices are not enough to prevent their pressure. Nevertheless, those practices did not give huge affect to the performance, especially operation performance.

Keywords Green supply chain management · Operations management · Manufacturing industries · GSCM performance · Factor analysis · Multiple regression analysis

1 Introduction

SCM is a relation from supplier, manufacturer, transportation, distributor until consumer. There are many things which effect SCM. Such as consumer need, efficiency, quality and responsiveness (Zelbst et al. 2010) and lately Environment

---

K. Fitriani (✉)
School of Business and Management Institut Teknologi Bandung, Jl Ganesha No 10, Bandung, Indonesia
e-mail: katlea.fitriani@sbm-itb.ac.id

Y.-K. Lin et al. (eds.), *Proceedings of the Institute of Industrial Engineers Asian Conference 2013*, DOI: 10.1007/978-981-4451-98-7_83, © Springer Science+Business Media Singapore 2013

(Green et al. 2012). People also became more aware about environment. People do not only need quality good and service, they also need cleaner production which can produce safe product. Recently, the objective of manufacturers is not about how to minimize cost but also to apply the cleaner production, like waste disposal and decreasing the use of hazardous material.

Indonesia will be the next manufacturer country in the future. Many Factories will be built to fulfill the demand of the customer. This is will be good opportunity for Indonesia economy. However, Indonesia will face some environmental issue regarding this problem. The pressure will push Indonesia to apply GSCM practice. Thus, Indonesia government, especially Ministry of Environment has established PROPER Project to overcome this matter. This project rated the pollution which has been produced each factories in Indonesia. Ministry of Environment will publish the result and the factories must fulfill the minimum rate of allowed pollution (Anon 2011a, b, c, d). In addition, Indonesia president also established president regulation No 61 2011 about decreasing greenhouse gas emissions (RAN-GRK). This regulation made for fulfilling Indonesia commitment to decrease 26 % gas emissions (Anon 2011a, b, c, d). Karawang is regency in West Java, Indonesia. This area is well-known as Industrial area. Many manufacturers are built their factories in there. Thus, it makes Karawang as one of the highest pollution city in Indonesia.

Because of that overview, this paper aims to introduce the GSCM pressure, practice and performance in Karawang Industrial Area, Indonesia. This paper also showed the comparison among GSCM in China and Japan to give deeper understand. Based on that objective, this paper took four hypotheses which are:

H1. Manufacturers in Karawang Industrial area, Indonesia have thought that the pressures are important to their management especially the Indonesia government regulation about environment, so they need to implement GSCM practice.

H2. Manufacturers in Karawang Industrial area, Indonesia have already considered applying GSCM practice especially to improve the quality of environmental friendly goods.

H3. Most of manufacturers in Karawang, Indonesia measure the company's success from their capabilities to improve the product quality.

H4. All GSCM pressures influence Green production and marketing practice the most.

H5. All GSCM practice in Karawang did not affect manufacturers performance.

H6. Manufacturers in Indonesia, especially in Karawang are lag behind than China and Japan for face the environment pressure or to apply GSCM practice.

2 Literature Review

(Sarkis 1999a, b) said that Supply Chain Management (SCM) is a system which consists of in-bound logistic, production, distribution and reverse logistic. Another literature mentioned that SCM is a relation within the raw material supplier, manufacture, transportation, distribution including the financial and information flow of good and service (Wang and Gupta 2011a, b). Planning and controlling the systems are also mentioned in SCM definition (Wang and Gupta 2011a, b). SCM is an integration from supplier supply raw material until deliver good and service to customer. In other word, SCM consist of four factors which are Raw material, Production, Distribution and Customer.

Because of many environmental issues, people started aware about environment. Balancing financial and environment become new purpose of SCM. To make SCM become green, (Rao 2002) integrated of the green purchasing from supplier to consumer, TQM based environment point, zero waste, Life cycle analysis, and green marketing. (Zhu and Sarkis 2004) defined GSCM as a green purchasing from supplier to customer including the reverse logistic in close-loop net-work. In simple way, GSCM has same function as SCM but in GSCM put environmental point to balancing the chain.

3 Method

This paper used questionnaire to collect the data from respondent. The questionnaire is made based on previous literature (Zhu et al. 2005) and modified by the writer. The questionnaire consists of three sections which are GSCM pressure, practice and performance. In first section, there are 12 questions about some pressure that affects to each company to apply GSCM practice. This section is answered using five-point Liker t-scale (1 = Not very important, 2 = Not important, 3 = Not thinking about it, 4 = Important, 5 = Extremely important). Second section consists of 30 questions about GSCM practice that might be applied by some companies. This section is also answered using five-point liker t-scale (1 = not considering, 2 = planning to consider it, 3 = currently considering it, 4 = starting to carried out, 5 = successfully carried out). Last section consists of 20 questions about some performance indicator that might be measured by companies. The questions are answered also using five-point liker t-scale (1 = not very important, 2 = not important, 3 = not thinking about it, 4 = important, 5 = very important). The questionnaire is made using Indonesia language, so to avoid confusing the respondent, each variable from (Zhu et al. 2005) is developed to some questions. The questionnaire also provided briefly explanation about each five-point liker t-scale, so the respondents can distinguish the differences.

Based on research in China (Zhu et al. 2005) and Japan (Zhu et al. 2010) used large companies as their sample. (Barney 1991) said that large manufacturers in

China and Japan have carried out GSCM practice very well. Larger manufacturers have more financial capabilities to implement GSCM practice in their production. Large manufacturers also get the affect of environmental pressure directly, so to meet the standard of environment requirement, they need to be more care about implement GSCM practice (Russo and Fouts 1997). Because of those issues, this paper use large manufacturers in Karawang as the sample. This Paper also will use high pollution industries to show the affect GSCM clearly. High pollution industries are assessed as one of the effect of increasing pollution and waste in Indonesia.

Based on those requirements, the writer took list of manufacturers in Karawang, Indonesia from BPS (Bank of Statistic Data in Indonesia) West Java about the Medium–Large Manufacturers in Karawang. There are several industries, but the writer just used 53 industries which meet the requirement. Those 53 industries are large manufacturers which can be shown from their amount of employee and production cost. The industries also are high pollution industries which can be shown from the type of industries, such as textile, chemical, plastic, paper and spare part industries.

The writer sent the questionnaire using email to 53 industries. The questionnaires are filled by the production manager because they understand more about the production process in the manufacturers. Out of 53 questionnaires sent, 34 questionnaire gave response (64.15 % respond rate). The manufacturers which respond the questionnaires are six textile manufacturers, four Plastic manufacturers, four chemical manufacturers, four paper manufacturers, three Semiconductor manufacturers, two heavy equipment manufacturers, a hair product manufacturer, a food manufacturer and ten spare part manufacturers.

Before using the factor analysis, reliability and validity test need to be held in the questionnaire data. In this paper, the author used the level of cronbach's alpha. The data will be consistent and valid if the level of cronbach's alpha is above 0.7 (Nunnally 1978a, b). In Pressure factor, all 12 variables are reliable and valid. Practice factor has seven variables to be removed to make the data reliable and valid. In performance factors need to remove six variables to make the data reliable and valid.

4 Result and Discussion

The evaluation just compare the mean score each factor which has determined by using factor analysis. First, The GSCM pressure is divided into two factors, which are Standard Regulation Pressure and Supply Chain Pressure. Manufacturers in Karawang, Indonesia have not taken those pressures as important issue which can make those manufacturers implement GSCM practices. It is shown in the mean score of each factor GSCM pressure. The highest mean score is Standard Regulation pressure which only can reach 3,279 (3 = not thinking about it) and the lowest score is Supply chain pressure, which is 3,015. However, based on the

Green Supply Chain Management (GSCM)

result, variable 'standard for exporting product' got the highest mean score which is 3,884 (4 = important). It means industries started to get the pressure from importer to produce environment-friendly product. Therefore, importers establish some standard for the product which will be imported to protect their customers in their country. Based on the result, H1 is rejected because the manufacturers in Karawang have not considered those pressures as booster to implement GSCM practice yet. Furthermore, Manufacture's in Karawang are more caring about the Standard for exporting product than the Indonesia government regulation.

Second, GSCM practice found three factors, which are Green Production and Marketing, Environmental Design and Management, and Internal Green Processing. Manufacturers in Karawang are in the stage to 'considering it now'. Green production and marketing got the highest mean which 2,772 (3 = considering it now), especially for 'improving the quality of environmental friendly goods', which has mean score 3,088 (3 = considering it now). The lowest mean score is 'Internal green processing' which has mean score 2,512 (2 = Planning to consider it). Some manufacturers do not corporate with consumer for designing environment friendly goods. Those manufacturers just developed it by themselves based on the environment standard. It is shown that variable has the lowest mean score. Overall, Manufacturers in Karawang have not carried out GSCM practices, they are still in the level of considerate it. However most manufacturers have considered improve the quality of environmental friendly goods. Thus, H2 is accepted.

Last, the performance variables that company measure is divided into three factors, which are Operation Performance, Environment Performance and Production Performance. The manufacturers in Karawang are taking care more about production performance. The manufacturers rank increasing of production cost as the measurement whether the company can performs better. Increasing of production can affects in two ways, it can make the product price getting higher and also decreasing of the company benefits. Production performance has the highest mean score, 3, 294 (3 = not thinking about that). The lowest mean score is Operation performance 3, 1 (3 = not thinking about that). The operation performance consists of improving the product quality, 3,912 (4 = important). Some of manufacturers still do not consider about that point to measurement of company performance. Based on the over view, H3 is accepted.

The first regression is between GSCM pressure factors as independent variable and GSCM Practice factors as dependent variable. There are three times regressions from two factors pressure which might be had significant correlation with three factors practice. As can be seen in the result of multiple regression analysis, the independent variable affected the dependent variable if the significant coefficient is less than 0.05. The standard regulation and supply chain pressure gave significant correlation through the green production and marketing practice. It is shown from the significant coefficient, which are 0.032 and 0.020. It means that the standard regulation and supply chain pressure has affected 34.6 and 37.8 % to the green production and marketing practice.

In other hand, GSCM pressure did not all affect in the GSCM practice. The result showed that only Supply chain Pressure could affected in the environmental design and management practice. It is shown from the significant coefficient, which is 0.002. The supply Chain pressure affected 51.2 % environmental design and management practice. Otherwise, internal green processing could be affected 36.6 % by regulation pressure. Only regulation pressure has significant correlation to internal green processing practice, 0.035. Based on that result, H5 is accepted.

Then the regressions between GSCM practice as the independent variable and performance as the dependent variable are applied. There are 3 times regressions between three factor GSCM practices which might be had significant correlation with three factors performance. Based on the SPSS result, there are no any significant correlations from the GSCM practices which can affect the operation performance. All the significant coefficients are higher than 0.05. However green production and marketing practice affected 72.7 % to the environment performance with the significant coefficient, 0.00. The environmental design and management has affected 58.2 % to the production performance with the significant coefficient, 0.00. H5 stated that all GSCM practice in Karawang did not affect the performance. Based on the table V result, H5 is rejected. Because some GSCM practice factors could affect the performance of the manufacturers.

This comparison is to seek the level of GSCM practice in Karawang, Indone-sia compare to China (Zhu et al. 2005) and Japan (Zhu et al. 2010). To give clear explanation, the writer put an ISO 14001 issue because ISO 14001 is a system and certification for companies about their environment action. ISO 14001 Certificate is needed to know how far a company's green action. The manufacturer and the supplier should have ISO 14001 certificate. In Karawang, Indonesia, both manufacturers and suppliers are still considering to get it now. It can be shown in the mean score in table II which have same score, 2,941 (3 = Considering it now). Surprisingly, China is in the same level with manufacturers in Karawang Indonesia. China manufacturers and suppliers are still considering to get ISO 14001 certificate. It is shown by the mean score in the previous paper (Zhu et al. 2005), which are 3.36 and 3.18 (3 = considering it now). In other hand, Japan manufacturers have successfully implemented it. Japan suppliers are still starting to get ISO 14001 certificate. The mean score for this point are 4.89 (5 = successfully implemented) and 4.00 (4 = Starting to implemented) (Zhu et al. 2010). The GSCM practices of manufacturers in Karawang, Indonesia are still left far behind if comparing it with the level of GSCM practice in China and especially Japan. Thus H6 is accepted.

5 Conclusion

The awareness of environment issue has not effect so much to manufacturers in Karawang. Although, Karawang is less awareness, it does not mean that Karawang Industrial area does not face any pressure about environment issue. Manufacturers in Karawang face pressure from importers who require some standard to export products to them. This makes Manufacturers must play smart to overcome this

Green Supply Chain Management (GSCM)

issue. If manufacturers in Karawang do not follow the requirement, they must prepare to lost opportunity to get high foreign income. Moreover, Indonesia government regulation must be considered. Recently, government is aggressively doing environment action, like establishing rule about decreasing carbon emission or ministry of the environment's PROPER project. Thus, next 2–3 years, Indonesia will feel the pressure directly.

Deeper research using more samples with broadly area of research can give clearer overview about GSCM in Indonesia. Moreover, the type of manufacturers can be extended, not only Medium-Large manufacturers, but also Small-Medium manufacturers. The comparison between GSCM in Medium–Large manufacturers and Small–Medium manufacturers can be seen.

Acknowledgments The most sincere appreciation dedicated to my parent who always support every time I lost hope. This research was possible thanks to the participation from all company in Karawang, Indonesia. I also want to say thank you to Prof Dermawan Wibisono and Mr Gatot Yudoko for all suggestions and advices.

References

Journal article

Barney J (1991) Firm resources and sustained competitive advantage. J Manage 17(1):99–120

Green KW Jr, Zelbst PJ, Meacham J, Bhadauria VS (2012) Green supply chain management practice: impact on performance. Supply Chain Manage: Int J 17(3):290–305

Rao P (2002) Greening the supply chain: a new initiative in South East Asia. Int J Oper Prod Manage 22(6):632–655

Russo M, Fouts P (1997) A resource based perspective on corporate environmental performance and profitability. Acad Manag J 40(3):534–559

Anon (2011a) Peraturan Presiden Republik Indonesia Nomor 61 Tahun 2011: Rencana Aksi Nasional Penurunan Emisi Gas Rumah Kaca, Jakarta: s.n

Anon (2011b) PROPER. [Online] Available at: http://www.menlh.go.id/proper. Accessed 23 Apr 2013

Nunnally J (1978a) Psycometric theory. McGraw-Hill, New York

Sarkis J (1999a) How green is the supply chain? Practice and research. Graduate School of Management, Clark University, Worcester

Wang H-F, Gupta SM (2011a) Green supllhain management : product life cycle approach. McGraw-Hill, Chicago

Zelbst P, Green KJ, Sower V, Abshire R (2010) Relationships among market orien-tation, JIT, TQM, and agility. Ind Manage Data Syst 110(5):637–658

Zhu Q, Geng Y, Fujita T, Hashimoto S (2010) Green supply chain management in leading manufacturers: case study in Japanese large companies. Manage Res Paper 33(4):380–392

Zhu Q, Sarkis J (2004) Relationships between operational practices and performance among early adopters of green supply chain management practices in Chinese manufacturing enterprises. J Oper Manag 22(3):265–289

Zhu Q, Sarkis J, Geng Y (2005) Green supply chain management in China: pressures, practice and performance. Int J Oper Prod Manage 25(5):449–468

Journal Article Only by DOI

Sarkis J (1999b) How green is the supply chain? Practice and research, Graduate School of Management, Clark University, Worcester

Anon (2011c) Peraturan Presiden Republik Indonesia Nomor 61 Tahun 2011 : Rencana Aksi Nasional Penurunan Emisi Gas Rumah Kaca, Jakarta: s.n

Book and chapter

Nunnally J (1978b) Psycometric theory. McGraw-Hill, New York

Wang H-F, Gupta SM (2011b) Green Supllhain management: product life cycle approach. McGraw-Hill, Chicago

Online document (no DOI available)

Anon (2011d) PROPER. [Online] Available at: http://www.menlh.go.id/proper/. Accessed 23 Apr 2013

Limits The Insured Amount to Reduce Loss?: Use the Group Accident Insurance as an Example

Hsu-Hua Lee, Ming-Yuan Hsu and Chen-Ying Lee

Abstract Property-Liability insurance Industry pays much attention to the loss ratio of accident insurance. From 2005 to 2011, average loss ratio is 45.24 % in Property-Liability insurance industry and 34.31 % in Life insurance Industry. Most important to achieve the maximum benefit for reducing losses based on the minimum cost of underwriting for property-liability insurer's underwriters. The purpose of this research is to examine limit insured amount can reduce the loss or not and whether business quality control by underwriting system. We complied data on group accident insurance from a Property-Liability Company in Taiwan from 2009 to 2010. There are 4,504 samples in group features. We use χ^2 test, test analysis and ordered logistic regression verify the relevance of the insured amount and loss, business quality control effect and the factor impact the loss. The results of the analysis show that the there was a correlation between insured amount and loss, business quality control that effectively reduce medical loss, Variables of AD&D, MR/AD&D and DHI/AD&D have a significant influence for the loss. The result of the analysis not only help underwriter to adjust underwriting guidelines but also reduce loss amount. Moreover, our results have practical implications for the Property-Liability insurance industry in Taiwan.

Keywords Property-liability insurance · Underwriting system · Group accident insurance · Orded logistic regression · Business quality control

---

H.-H. Lee (✉) · M.-Y. Hsu
Graduate Institute of Management Sciences, Tamkang University, No.151, Yingzhuan Rd., Tamsui, New Taipei City 25137, Taiwan, Republic of China
e-mail: Hxl120@gmail.com

M.-Y. Hsu
e-mail: a6827609@ms18.hinet.net

C.-Y. Lee
Department of Insurance and Finance, Chihlee Institute of Technology, No.313, Sec.1,Wunhau Rd., Banciao, Taipei City, Taiwan, Republic of China
e-mail: chenying0207@yahoo.com.tw

Y.-K. Lin et al. (eds.), *Proceedings of the Institute of Industrial Engineers Asian Conference 2013*, DOI: 10.1007/978-981-4451-98-7_84,
© Springer Science+Business Media Singapore 2013

1 Introduction

As to the loss rate of group accident insurance, the average loss rate of group accident insurance for the property-liability insurance industry was 65 %, which was higher than that for life insurance industry (45 %) by 20 %. After the property-liability insurance industry began participating in the accident insurance market, was business quality control (underwriting) sacrificed with the percentages of market leaders and business growth? This is an issue worthy of study.

This study probed into the following three topics: T-1: Do high-insured amount lead to choose high loss? T-2: Does business quality management help lower the loss? T-3: What factors impact the loss? We use χ^2 test to illustrate T-1, use the decision rule to verify T-2 and using Ordered Logistic Regression analysis to analyze the factors of loss. According to the reimbursement data of case property-liability insurance company, the 1606 claim cases with a total claim amount of NT\$57,497,011 in 2009 were significantly reduced to 899 claim cases with a total reimbursement amount of NT\$32,098,480 in 2010. This was the effect of the case property-liability insurance company practicing precise business screening and quality management in 2010.

For previous underwriting research are concentrated in information asymmetry adverse selection and moral hazard phenomenon of something more than the presence or absence authentication. How to resolve this phenomenon is not taking a step forward of the proposed method, this study attempts to introduce a further view. The underwriters deep-rooted notion that, reduce the insured amount will be able to reduce losses, this study will challenge this argument.

The results suggest that the insured amount related to loss and business quality management helps to reduce the loss. MR and DHI/ADD have a positive relationship; MR/ADD has negative effect of losses.

This paper is constructed as follows: The first section of the article is an introduction. The second section, literature review is explore underwriting, asymmetric information and methodology for quality control in underwriting system. The third section describes the methodology to examine the hypotheses from a case insurance company. The results for various analyses are presented following each of these descriptive sections. Finally, conclusions are presented and suggestions are made for future studies.

2 Literature Review

The insurers' risk includes market risk, credit risk, business risk, underwriting risk, liquid risk and event risk (Report of Solvency Working Party 2002). According to the research of Ryan et al. (2001), from 1969 to 1998, in the U.S. there were 683 events where the insurer failed to pay off. The percentage of events with insurance underwriting risk was 42 %. Thus, underwriting risk is the most significant risk source of the insurers. Luthard and Wiening (2006) defined underwriting as the selection of the insured, the setting of coverage, the decision of the conditions of the insured and the examination and decision-making of business. Dorfman (1998) suggested that the purpose of underwriting is to match the actual loss of insurance companies with the expected loss and avoid the adverse selection of risk.

Group insurance means a corporation of at least five persons, which formation is not for the sole purpose of purchasing insurance, takes out insurance in insurance companies to cover the risk of financial loss caused by diseases, injury, or the death and retirement of the group members and their relatives. In order to have a precise research scope, the researcher focused on samples of group accident insurance.

Arrow (1963) first proposed asymmetric information theory. Rothschild and Stiglitz (1976) and Shavell (1979) applied asymmetric information theory to their studies on insurance and suggested that asymmetric information in the insurance market will cause adverse selection and moral hazard. The related research has mainly focused on automobile insurance and various kinds of life insurance. (Li et al. 2004) suggested that these two kinds of insurance have individual attributes that are close to group attributes.

From the perspective of asymmetric information, group accident insurance assessors are certainly a party receiving asymmetric information. In the acquisition of insurance information, salespersons collect the clients' information and pass it to the underwriters as a reference. However, since the information may be incomplete, the underwriters may be inexperienced, or there may be a lack of samples or analysis in the returned statistics, there can be errors in decision making. After taking out insurance, the individuals participating in group accident insurance may change and the risk is dynamic. There will be various errors from the start to the end of the period of insurance.

Rothschild and Stiglitz (1976) suggested that with asymmetric information, the party having more information makes selections based on personal interest and harms the right of the party with less information. This is known as adverse selection. However, De Meza and Webb (2001) proposed advantage selection and suggested that insurance companies' information asymmetry is reflected on the types of risk and risk preference. Individuals who are risk averse will purchase insurance and be concerned about personal risk factors. Thus, the probability of accidents is low and the insurers can acquire benefits above quota. (Wang et al. 2008) suggested that in a well-developed insurance market, asymmetric information will disappear because of a powerful underwriting system.

In studies on the construction of insurance assessment, (Lee and Ting 2004) evaluated the underwriting system of automobile insurance using the Logit model and the GANN model and demonstrated that the evaluation performance of these two models was at least 80 %. Shih (2004) explored the adverse selection and moral hazard of labor with occupational fatality insurance in Taiwan and realized that labor has adverse selection when taking out insurance and there is moral hazard in the application for payment. Among literatures on rate factors, (Chou et al. 2010) used logistic regression to examine the business of personal accident insurance in one property-liability insurance company and demonstrated that age, place of accident, occupation and place significantly influence the factors for the loss ratio.

Regarding the group insurance of one life insurance company. (Chen et al. 2008) used logistic regression to establish the discriminant claim model and explored the factors of claim using regression analysis. It was found that the precision of the discriminant reimbursement rate constructed by logistic regression was 81.7 %.

3 Research Method

3.1 Data and Research method

1. Data

The research data consisted of group insurance policies and the claim of one property-liability insurance company in Taiwan in 2009 and 2010. The regions of were across Taiwan. The number of samples was 2,373 in 2009 and 2,150 in 2010. After the elimination of invalid samples, the number was 2,362 in 2009 and 2,143 in 2010. The explanatory variables included death and dismemberment insured amount (AD&D), medical expense insured amount (MR), the percentage of medical insurance expense amount in death and dismemberment (MR/AD&D), the daily hospitalization indemnity insured amount (DHI), and the percentage of daily hospitalization indemnity insured amount in death and dismemberment (DHI/AD&D). The variables for validating reimbursement included the loss frequency, loss severity and loss ratio. In order to lower the gap among the variables, the variables were classified according to the degree of risk. Operational definitions of variables are shown in Table 1. In addition, the descriptive statistics of the variables are listed in Table 2.

2. Variables

Explanatory variables: Regarding insured amount, the setting of life insured amount should be appropriate. With low life insured amount, the employees will lack a guarantee, while high coverage will cause moral hazard. The setting of group insurance coverage is not based on group members' decisions; instead, it relies on employees' salary, positions, working years or fixed amounts. There should be no adverse selection or moral hazard. However, since there can be errors in the setting of group insurance that are caused by external factors, an examination is necessary. For the insured, those with higher risk will select higher insured amount. However, the underwriters hope that people with higher risk will select lower insured amount. The balance of the two parties will be the insured amount. Although death and dismemberment loss severity is high, the loss frequency is low. Medical insurance loss severity is low and the loss frequency is much higher than that for death and dismemberment. Thus, the insured will lower the amount of death and dismemberment insurance coverage or increase the amount of medical insurance coverage. Underwriters will adopt a percentage of the medical insured amount in the death and dismemberment insured amount in

Table 1 Operational definition of variables

Variable		Operational definition
Explanatory variables	AD&D	$X \le 1{,}000{,}000$, $V = 1$; $1{,}000{,}000 < X \le 3{,}000{,}000$, $V = 2$; $X > 3{,}000{,}000$, $V = 3$
	MR	$X = 0$, $V = 1$; $X \le 20{,}000$, $V = 2$; $X > 20{,}000$, $V = 3$
	MR/AD&D	$X = 0\,\%$, $V = 1$; $0\,\% < X \le 2\,\%$, $V = 2$; $X > 2\,\%$, $V = 3$
	DHI	$X = 0$, $V = 1$; $X \le 1{,}000$, $V = 2$; $X > 1{,}000$, $V = 3$
	DHI/AD&D	$X = 0$, $V = 1$; $0\,\% < X \le 0.001$, $V = 2$; $X > 0.001$, $V = 3$
Explained variables	Loss frequency	$X = 0$, $V = 1$; $X \le 3$, $V = 2$; $X > 3$, $V = 3$
	Loss severity	0, $V = 1$; $0 < X \le 500{,}000$, $V = 2$; $X > 500{,}000$, $V = 3$
	Loss ratio	0, $V = 1$; $0 < X \le 100\,\%$, $V = 2$; $X > 100\,\%$, $V = 3$

Table 2 Descriptive statistics

Variables		2009					2010				
		N	Min	Max	Mean	S.D.	N	Min	Max	Mean	S.D.
Explanatory variables	ADD	2373	1	3	2.02	0.25	2150	1	3	1.37	0.55
	MR	2373	1	3	1.85	0.57	2150	1	3	1.90	0.48
	MR/ADD	2366	1	3	1.77	0.46	2148	1	3	1.87	0.44
	DHI	2373	0	3	2.01	0.62	2150	1	3	1.85	0.52
	DHI/ADD	2366	1	3	1.81	0.40	2148	1	3	1.80	0.45
Explained variables	Loss frequency	2373	0	3	1.27	0.50	2150	1	3	1.21	0.44
	Loss severity	2373	0	3	1.26	0.46	2150	1	3	1.21	0.42
	Loss ratio	2370	1	3	1.32	0.59	2150	1	3	1.22	0.47
Valid N (listwise)				2,366					2,148		

order to find the adverse selection of the insured when taking out insurance. Explained variables: Loss frequency is the number of losses during the insurance period. If a group has high loss frequency, it means that the group neglect to pay attention to its risk management on behalf of the group. For this type of business, the underwriters will be taken out of business or measures to reduce the insured amount. Loss severity, the characteristics of the accident is the loss of low frequency, the magnitude of such losses. For individual business, the magnitude of such losses and can not prove that this business is bad business. But on the whole, the loss of a large range of cases accumulated as the underwriters to develop underwriting guideline and an important underwriting experience. The loss ratio is equal to loss amount/insured amount. To the individual, high loss rate means that the profit contribution of the business of insurance companies is negative. For these businesses, Underwriters had to understand the exact cause. Indeed own business poor quality, it will increase premiums to reflect reasonable price to the customers.

3.2 Research Model

T-1 χ^2 test for independent

$$\chi^2 = \sum_{i=1}^{a} \sum_{j-1}^{b} \frac{(o_{ij} - e_{ij})^2}{e_{ij}}$$

T-2 Decision Rule

Decision	Population	
	H_0 is true	H_0 is false
Accept H_0	Correct decision	False acceptance
Reject H_0	False rejection	Correct decision

T-3 Ordered Logistic Regression was employed to Verify assumptionslog $\left(\frac{P(y \leq j|x)}{1 - P(y \leq j|x)} \right) = \beta_j - \sum_{k=1}^{K} \beta_k x_k$, j $= 1$, low risk ranking; j $= 2$, middle risk ranking; j $= 3$, high risk ranking. x_k, the kth independent variable.

4 Empirical Results and Analysis

T-1: Do high-insured amount lead to choose high loss?
Table 3 shows the relation insured amount and loss. Apart from 2010 AD&D of the P value was larger, and the rest are very small. It means that there is connected between the insured amount and the loss.

Table 3 The relation of insured amount and loss

Loss p-value	Loss frequency		Loss severity		Loss ratio	
Amount	2009	2010	2009	2010	2009	2010
ADD	0.00	0.12	0.00	0.00	0.00	0.08
MR	0.00	0.00	0.00	0.00	0.00	0.00
MR/ADD	0.00	0.00	0.00	0.00	0.00	0.00
DHI	0.00	0.00	0.00	0.00	0.00	0.00
DHI/ADD	0.00	0.00	0.00	0.00	0.00	0.00

Table 4 Decision rule–insured amount and loss

Type of insurance	Type of claim	Low amount high loss		Correction decision		High amount low loss	
		2009 (%)	2010 (%)	2009 (%)	2010 (%)	2009 (%)	2010 (%)
ADD	Claim frequency	3.20	14.65	23.09	56.56	73.70	28.79
	Claim severity	1.81	14.42	24.61	56.84	73.58	28.74
	Loss ratio	7.30	15.02	19.16	56.23	73.54	28.74
MR	Claim frequency	3.23	2.42	42.23	31.91	54.45	65.67
	Claim severity	2.28	1.77	45.39	32.73	54.81	65.86
	Loss ratio	6.92	3.21	38.95	31.07	54.14	65.72
MR/ADD	Claim frequency	3.89	2.42	43.87	32.26	52.24	65.32
	Claim severity	2.41	1.77	45.39	32.73	52.20	65.50
	Loss ratio	7.61	3.21	40.28	31.42	52.11	65.36
DHI	Claim frequency	2.82	3.95	35.31	34.28	61.86	61.77
	Claim severity	1.39	3.26	36.76	34.84	61.85	61.91
	Loss ratio	5.78	4.56	33.67	33.63	60.55	61.81
DHI/ADD	Claim frequency	3.17	4.00	39.86	34.78	56.97	61.22
	Claim severity	1.69	3.26	41.46	35.47	56.85	61.27
	Loss ratio	7.31	4.66	35.84	34.12	56.85	61.22

T-2: Does business quality management help lower the loss?

Table 4 shows the effects of business quality management. Correct rate, only the AD&D has significantly improved, and the rest showed a downward. In particular, MR, MR/AD&D is even more evident. High insured amount and low losses (False rejection), in underwriting too strict. The one hand, allows the company to generate underwriting profits, but because of reduced premium cash inflow. The ratio was significantly reduced have AD&D, significantly improved MR, DHI. Therefore, AD&D, MR and DHI have a considerable impact on the business quality control.

T-3: What factors impact the loss?

1. Validation of loss frequency

Table 5 shows the factor affect loss frequency. As to the effects of business quality management, MR, DHI/ADD had a positive effect and MR/ADD had a negative effect.

Limits the Insured Amount to Reduce Loss?

Table 5 The factor affect loss frequency

Parameter estimates—loss frequency		2009					2010				
		Est.	Std.	Wald	df	Sig.	Est.	Std.	Wald	df	Sig.
Threshold	lossfre = 1	4.80	0.74	41.92	1	0.00	3.16	0.37	72.30	1	0.00
	lossfre = 2	7.43	0.75	97.48	1	0.00	6.14	0.42	215.33	1	0.00
Location	ADD	−0.41	0.25	2.67	1	0.10	−0.09	0.11	0.64	1	0.42
	MR	0.53	0.16	10.77	1	0.00	−0.13	0.34	0.15	1	0.70
	MR/ADD	0.39	0.23	2.95	1	0.09	0.76	0.37	4.27	1	0.04
	DHI	−0.07	0.12	0.31	1	0.58	−0.37	0.27	1.89	1	0.17
	DHI/ADD	1.53	0.28	29.06	1	0.00	0.77	0.31	6.34	1	0.01

Table 6 Factors affect loss severity

Parameter estimates–loss severity		2009					2010				
		Est.	Std.	Wald	df	Sig.	Est.	Std.	Wald	df	Sig.
Threshold	loss severity = 1	4.98	0.74	44.80	1	0.00	3.06	0.37	68.05	1	0.00
	loss severity = 2	8.64	0.77	124.62	1	0.00	7.07	0.49	209.95	1	0.00
Location	ADD	−0.30	0.25	1.38	1	0.24	−0.10	0.11	0.88	1	0.35
	MR	0.44	0.16	7.12	1	0.01	−0.15	0.34	0.19	1	0.66
	MR/ADD	0.47	0.23	4.14	1	0.04	0.76	0.37	4.28	1	0.04
	DHI	−0.06	0.12	0.23	1	0.63	−0.30	0.27	1.29	1	0.26
	DHI/ADD	1.52	0.28	28.60	1	0.00	0.67	0.30	4.86	1	0.03

Table 7 Factor affect loss ratio

Parameter estimates–loss ratio		2009					2010				
		Est.	Std.	Wald	df	Sig.	Est.	Std.	Wald	df	Sig.
Threshold	lossratio = 1	4.82	0.73	43.43	1	0.00	3.07	0.37	69.20	1	0.00
	lossratio = 2	6.43	0.74	76.30	1	0.00	5.51	0.40	192.38	1	0.00
Location	ADD	−0.34	0.25	1.81	1	0.18	−0.10	0.11	0.75	1	0.39
	MR	0.41	0.16	6.51	1	0.01	−0.17	0.34	0.25	1	0.62
	MR/ADD	0.46	0.23	4.07	1	0.04	0.79	0.37	4.62	1	0.03
	DHI	−0.06	0.12	0.28	1	0.60	−0.29	0.27	1.20	1	0.27
	DHI/ADD	1.52	0.28	29.44	1	0.00	0.65	0.30	4.68	1	0.03

2. Validate of loss amount

Table 6 shows the factor affect loss severity. As to the effects of business quality management, MR, DHI/ADD had a positive effect and MR/ADD had a negative effect.

3. Validate of loss ratio

Table 7 shows the factor affect loss ratio. As to the effects of business quality management, MR, DHI/ADD had a positive effect and MR/ADD had a negative effect.

5 Conclusion and Future Studies

Regarding T-1, Can see that the insured amount is indeed related losses, but can not be further evidence of the high insured amount will lead to high losses. Regarding T-2, Further AD&D significantly improved the probability lead to high losses in high insured amount. The business quality management lowers the effect of loss. However, as to T-3, the factors affect loss were MR, MR/ADD, DHI/ADD. This study explores the correlation between insured amount and losses. The research method is including independent test, decision rule and Ordered Logistic Regression, the region is Taiwan and the data includes 2 years. Increasing the research data periods of free premium rate insurance and validating them by other methods will allow the study of this field to be more complete.

References

Chen C-H, Lin M-C, Gong Y-C (2008) An empirical study on the compensation ratio of group insurance. J Risk Manag 10(2):133–155

Chou PL, Yao CH, Shih PH (2010) An analysis on risk factors of personal injury insurance in property-liability insurance industry. J Underwriters 19:29–60

de Meza D, Webb DC (2001) Advantage selection in insurance market. Rand J Econ 32:249–262

Dorfman MS (1998) Introduction to risk management and insurance. Prentice Hall, Upper Saddle River

Lee WC, Ting J (2004) The construction of the physical damage automobile insurance underwriting system: empirical study in Taiwan. Insur Monograp

Li C-S, Chi C, Yeh JH (2004) The inconsistency between asymmetric information theory and empirical evidence in insurance markets. Insur Monogr 20(2):99–112

Report of Solvency Working Party (2002) International Association of Insurance

Rothschild M, Stiglitz J (1976) Equilibrium in competitive insurance markets: an essay on the economics of imperfect Information. Q J Econ 90(4):629–649

Ryan JP, Archer Lock PR, Czernuszewicz AJ, Gillot NR, Hinton PH, Ibeson D, Malde SA, Paul D, Malde SA, Paul D, Shah N (2001) Financial condition assessment. Inst Actuaries Fac Actuaries

Shih Yang (2004) Note on union workers participating in occupational injury insurance: the perspective of adverse selection and moral hazard. Insur Issue Practices 3(1):53–76

Wang JL, Chung C-F, Tzeng LY (2008) An empirical analysis of the effects of increasing deductibles on moral hazard. J Risk Insur 75(3):551–566

Manipulation Errors in Blindfold Pointing Operation for Visual Acuity Screenings

Ying-Yin Huang and Marino Menozzi

Abstract We developed a self-testing device for screening visual acuity, in which patients used a joystick to enter the responses. Since patients kept fixation of the test chart throughout the test, the joystick was operated blindfold. Pointing errors as function of the orientation of the Landolt ring were computed by analyzing records of 457 patients. In 97 % of errors, patients misestimated the orientation of the Landolt ring by 45°, which is the orientation next to the one presented. Mismatches in counter clockwise direction were 3 times more frequent as in clockwise direction. Findings are compared to results recorded in 25 subjects who performed acuity tests by reporting verbally the orientation of presented Landolt rings. In the verbal reporting condition, error rates for orthogonal and diagonal orientations were similar. We suggest considering the limited accuracy of motor response as a major issue in the blindfolded operation of a joystick which is used in combination with a vision screener. Furthermore we suggest a statistical procedure accounting for manipulation errors in an acuity test. The combination of previous findings (Menozzi 2013, 1995) to findings reported here suggests that computerized screening devices enabling self-testing are a reliable and convenient method for large scale vision screenings.

Keywords Vision screening · Blindfolded pointing · Manipulation · Acuity test

---

Y.-Y. Huang (✉)
Department of Management, Technology and Economics and Department of Health Sciences and Technology, Ergonomics of Information Media Research Group, ETH Zurich, Scheuchzerstrasse 7, CH-8092 Zurich, Switzerland
e-mail: yingyinhuang@ethz.ch

M. Menozzi
Department of Health Sciences and Technology, Ergonomics of Information Media Research Group, ETH Zurich, Scheuchzerstrasse 7, CH-8092 Zurich, Switzerland
e-mail: mmenozzi@ethz.ch

Y.-K. Lin et al. (eds.), *Proceedings of the Institute of Industrial Engineers Asian Conference 2013*, DOI: 10.1007/978-981-4451-98-7_85,
© Springer Science+Business Media Singapore 2013

1 Introduction

Medical screenings of physiological functions have become widespread in the occupational health sector. Depending on the country and on the workplace, medical screenings are performed as part of entrance requirements in many different occupations. Also screenings are applied repeatedly during employment in order to detect adverse health effects of workplace hazards. Occupational medical screenings are important workload in personnel involved in occupational health. From a personal communication of physicians of a Swiss pharmaceutical company we have learned that a large size company may perform as many as 10,000–20,000 medical screening tests per year. Since usually the resources allocated to the occupational screening process are of limited nature, the question arises on how screenings can be accomplished within a given time frame and with respecting quality measurement. A frequently adopted strategy to fulfill mentioned requirements is to adopt a two phase screening. In a first phase the total population is screened by means of a coarse test of a short duration. Patients failing the coarse test are sent to a detailed examination which is performed in a second phase. As has been shown by Krueger (1991), a two phase screening strategy may save an important amount of time when compared to a detailed examination of the total population. A further improvement of efficiency in large scale screenings may be achieved by using fully computerized devices enabling patients to perform sophisticated tests by themselves without the need of personnel for administering the tests. As we were able to show for the case of visual screening tests, acceptability in patients taking self-administered tests is high (Menozzi 2013). Furthermore, we were able to show that for the particular case of acuity testing, the standard procedure of acuity testing (ISO 8596:2009 "Ophthalmic optics—Visual acuity testing—Standard optotype and its presentation") can be accomplished in a short amount of time (Menozzi 2013).

The implementation of standard vision screening procedures as a self-testing method necessitates a closer look on the method used to record patients' responses. Input device requiring a visual feedback are disturbing the process of testing since the patient is required to alternate fixation between the visual target used in the test and the input device. A PC mouse may be used in tests requiring a simple "yes" or "no" answer, such as is the case in perimetry testing. As from our experience (Menozzi et al. 2012) the PC mouse can be operated blindfolded, even by untrained patients and either by using two fingers of the same hand or by using two hands. However, it is hard to use a PC mouse in tests requiring a complex response such as the standard visual acuity measurement using Landolt rings of eight different orientations (see Fig. 1, right). After practical and theoretical considerations the use of keyboards, keypads and voice input devices have been considered as not adequate to be used in large scale vision screenings. We therefore considered the joystick (see Fig. 1, left) as the next best choice to accommodate requirements in acuity vision measurement basing on the ISO 8596:2009 standard ("Ophthalmic optics—Visual acuity testing—Standard optotype and its presentation").

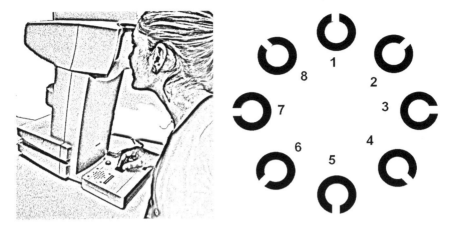

Fig. 1 The picture on the *left* shows a participant gazing into the vision screener and giving a response by means of the joystick included in the answering box which is placed besides the vision screener. For convenience of the photo, the box is placed in front of the screener. On the *right*, Landolt rings in eight different orientations are shown. Used test charts contained five lines, each consisting out of five Landolt rings having the same demand of acuity but varying in orientation

Blindfolded pointing and manipulations have been reported in the scientific literature (Wnuczko and Kennedy 2011; Gentaz et al. 2008). Findings show that blindfolded pointing may deviate from visually guided pointing and that participants perform less accurate in pointing in oblique directions than it is the case for orthogonal directions. However, the accuracy in blindfolded use of a joystick for pointing in eight pre-defined, equidistant orientations remains unclear. The aim of the here reported study was to gain more insight into the accuracy of blindfolded pointing using a joystick. For this purpose we analyzed visual acuity measurements recorded in 457 participants during a health promotion campaign.

In order to separate contributions of the motor system to response errors in blindfold operation of the joystick from response errors originating in the visual system, data recorded in an experiment (unpublished) were analyzed, in which 25 patients gave verbal response on the orientation of the Landolt ring.

2 Method

The accuracy in blindfolded operation of a joystick was investigated using anonymous records of 457 employees working in the railway sector who visited a health promotion campaign which was held in 2006 in Switzerland. Most of the participants (96 %) visiting the health promotion campaign were men and the age in about 45 % of the participants ranged between 41 and 50 y. Further details are given in a previous publication Menozzi (2013). Participants were offered to take an acuity test (Fog. 1) using a computerized self-testing device. The device consisted out of a commercial available vision screener (Rodatest 302) and a purpose built answering box. The answering box included, an analog joystick, a button for starting the test procedure and a loudspeaker. The lever of the joystick had a diameter of 0.8 cm and stick up about 3.5 cm from the top of the answering box. A bearing was inserted inside the housing of the joystick. The bearing served to limit movements of the joystick in the two orthogonal and the two diagonal directions in accordance to the eight orientations of the Landolt rings used in the acuity test.

The vision screener and answering box were connected to a notebook administering the tests in the vision screener and recording the answers of the participants. For taking a test, participants sat in front of the vision screener and pressed their forehead on the appropriate support of the vision screener (Fig. 1, left). After pressing the start button of the answering box instructions for taking the test were given to the participants by means of the loudspeaker. Test charts presented inside the vision screener were viewed though the optic of the vision screener.

Each participant underwent three acuity tests (left eye, right eye, binocular) for far (5 m) vision and one (binocular) acuity test for near (0.4 m) vision. In each acuity test a test chart including five lines of Landolt rings was presented. The acuity demand varied from line to line, starting from decimal acuity 0.5 (logMAR 0.3) in the top line and ending with decimal acuity of 1.25 (logMAR −0.1) in the bottom line. The orientation of the Landolt rings varied pseudo random in accordance to the ISO 8596:2009 standard. Participants were asked to read line by line all rings in the chart, starting from the top left ring and to enter the orientation of each ring by pushing the lever of the joystick in the appropriate direction. Before starting the acuity test, a trial session was performed in order to ascertain the correct use of the joystick. Landolt rings of a low acuity demand (decimal acuity 0.2) were used in the trial session. An acoustic feedback indicated the correct use of the joystick in the trial session.

For each participant a total of 125 responses were recorded. Responses and the correct orientation of each of the presented Landolt rings were stored in a file. The file was imported into an excel data sheet and processed by means of a macro. In a first step, the macro determined acuity achieved in each of the five tests by adopting rules in accordance to the ISO 8596:2009 standard. In a second step, the macro computed the frequency of errors as well as the distribution of sizes of the errors (mismatch in orientation) as function of the orientation of the Landolt ring. In the second step, only responses to Landolt rings having a demand of acuity lower than or equal achieved acuity were considered. This condition is further on referred to as the "pass condition". The second step was repeated considering all responses to Landolt rings having a higher demand of acuity than the achieved one. This condition is further on referred to as the "fail condition". In the "fail condition", Landolt rings are smaller than in the "pass condition".

In order to separate contributions of the motor system to response errors in blindfold operation of the joystick from response errors originating in the visual system, acuity data of a previous experiment (unpublished) were analyzed, in which 25 patients reported verbally the orientation of the Landolt rings. The experimenter recorded the orientation of presented Landolt rings as well as the reported orientation. Acuity measurements were carried monocularly and in accordance to the method described in the ISO 8596:2009 standard. The age of the patient in the previous experiment ranged between 14 y and 57 y and average visual acuity was -0.1 logMAR (decimal acuity 1.2) in eight and -0.2 logMAR (decimal acuity 1.5) in 17 eyes.

3 Results

For responses given to Landolt rings having an acuity demand below the achieved acuity level (the "pass" condition), data sets of 457 participants were processed. For responses given to Landolt rings having an acuity demand above the achieved acuity level (the "fail" condition), data of only 342 participants were processed due to failure of the macro processing the data.

The net diagrams in Fig. 2 report error frequency as function of the orientation of the Landolt ring. The left graph in Fig. 2 has been computed using responses given in the "pass condition". Data reported in the right graph has been computed using responses given to Landolt rings in the "fail condition".

The total number of errors was 6,310 and 1,291 in the "pass condition" and in the "fail condition" respectively. As shown by the graphs in Fig. 2, errors on orthogonal oriented Landolt rings (steps 1, 3, 6 and 7) appear visibly less frequent than errors on diagonal oriented rings. The ratio between the frequency of errors occurring on orthogonal orientations and errors occurring on diagonal orientations is 0.07 in the "pass condition" and 0.26 in the "fail condition". The size of the error is the mismatch between the orientation of the presented Landolt ring and the

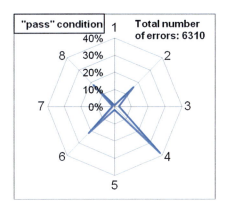 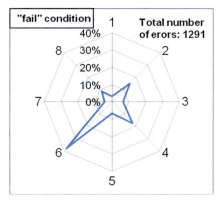

Fig. 2 Error frequency as function of orientation of the Landolt ring (see Fig. 1) for responses given to rings having a demand of acuity below or equal (left graph in Fig. 2) and above (right graph in Fig. 2) achieved acuity

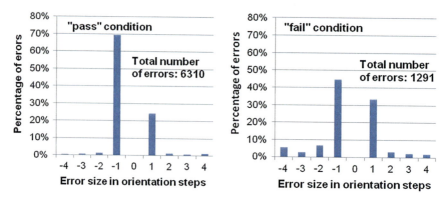

Fig. 3 Distribution of error size in orientation steps. One step corresponds to an error of 45°. Data in the *upper* graph refer to the "pass" condition, i. e. when the acuity demand was lower than achieved acuity. Data in the *bottom* graph report error sizes distribution for acuity demands higher than the achieved acuity

orientation which was pointed by the participant. Since the eight orientations are equidistant distributed, the size of the error varies in steps of 45°. In Fig. 3, the distribution of errors as function mismatch is reported for errors recorded in the "pass condition" (left in Fig. 3) and for errors reported in the "fail condition" (right in Fig. 3). Error size in Fig. 3 is reported in number of steps of orientation mismatch. As can be seen from Fig. 3, mismatch of one step appear much more frequent as larger size of mismatches. In the "pass condition" about 94 % of mismatches were either +1 or −1 step in size. In the "fail condition" mismatches of 1 step appear in 75 % of errors. Also mismatches with negative step size, which are in fact a counter clockwise (CCW) mismatch of the orientation, are more frequent than clockwise (CW) mismatches. The ratio between CCW and CW mismatches is 72/23 in the "pass condition" and 60/40 in the fail condition respectively.

Results of the experiment in which patients' answers were given verbally showed no substantial difference in error rates between errors appearing at orthogonal and errors appearing at diagonal oriented Landolt rings. A total of 398 answers were recorded, 234 for orthogonal and 164 for diagonal oriented Landolt rings. The percentage of errors was 21.8 % in orthogonal oriented rings and 18.9 % in diagonal oriented rings.

4 Discussion

In the experiment with the blindfold operation of the joystick, the errors were more equal distributed across orientation of the Landolt ring in the "fail condition" as was the case for the "pass condition". This is in part due to the reduced visibility of the Landolt rings in the "fail condition". However, in the "fail condition" diagonal errors still appear more pronounced in the diagonal orientation than is the case for the orthogonal orientation. Part of the effect could be due to the varying sensitivity of the visual system with orientation of a target as has been reported for sine wave gratings of different orientations (Watanabe et al. 1968). Possibly other than visual phenomenon may have a major contribution to the variation of errors as function or orientation. For instance, manipulation tasks in an orthogonal direction may be accomplished easier than manipulations in a diagonal direction.

In the "pass condition" 94 % of mismatches are of a size of 1 step, which is 45°. When the ring is less visible, as in the "fail condition", the percentage of mismatches of 1 step in size is reduced to 78 %. For a random response, we would expect the percentage of errors of a size of 1 step to be about 25 %. This finding suggests a limited accuracy in motor response in the blindfolded use of the joystick. Another effect indicating the impact of inaccurate motor response is the asymmetry of mismatches in counter clockwise and in clockwise direction.

Results of the experiment in which patients operated the joystick blindfold show a clear difference in error rates for orthogonal and for diagonal oriented Landolt rings. Such a difference was not found in results of the experiment in which patients reported verbally the orientation of the Landolt rings.

5 Conclusions

Our work aimed to investigate motor response accuracy in blindfolded operation of a joystick which was used as input device in a vision screener. For this purpose we analyzed response errors recorded in acuity tests of 457 participants who reported the orientation of Landolt rings by blindfold operation of a joystick. Possible contributions of the visual system to the inaccuracy in detecting orthogonal and diagonal oriented Landolt rings was investigated using 25 subjects verbally reporting the orientation of presented Landolt rings.

Given above reported findings we suggest to consider limited accuracy in motor response as an important issue in the blindfolded operation of a joystick. Accuracy of motor response for orthogonal directions (up, down, left, right) was found to be clearly better than for diagonal directions. Accuracy in verbal reporting of the orientation of the Landolt ring did not show a significant difference between orthogonal and diagonal orientations.

As for the case of the vision screener, a visual feedback of the joystick action could be displayed in combination with the targets used for the test of visual acuity. Alternatively, one could take into account limited accuracy of the motor responses by altering the evaluation of the response. A mismatch of one step in size could be considered as a correct answer. In order to keep the chance level for passing a given acuity the same as in the ISO 8596:2009 standard "Ophthalmic optics—Visual acuity testing—Standard optotype and its presentation", the stopping criterion must be altered from at least 3 correct readings in 5 presentations to at least 4 correct readings in 5 presentations.

Acknowledgments The authors wish to thank Esther Baumer-Bergande and Urs Hof for data collection and their great help in this project.

References

Gentaz E, Baud-Bovy G, Luyat M (2008) The haptic perception of spatial orientations. Exp Brain Res 187:331–348

Krueger H (1991) Der Betriebsarzt im Spannungsfeld zwischen Arbeitsplatzbegehung und spezieller arbeitsmedizinischer Vorsorgeuntersuchung aus der Sicht eines Arbeitsphysiologen. Zbl Arbeitsmed 41:361–368

Menozzi M (1995) Der Personal Computer im Einsatz beim Screening visueller Funktionen. Klin Monatsbl Augenheilkd 206(5):405–408

Menozzi M (2013) A field report on self-screening visual acuity using a computerized vision screener. Klin Monatsbl Augenheilkd (in press)

Menozzi M, Bauer-Bergande E, Seiffert B (2012) Working towards a test for screening visual skills in a complex visual environment. Ergonomics 55(11):1331–1339

Watanabe A, Mori T, Nagata S, Hiwatashi H (1968) Spatial sine-wave response in the human visual system. Vision Res 8:1245–1263

Wnuczko M, Kennedy JM (2011) Pivots for pointing: Visually-monitored pointing has higher arm elevation than pointing blindfold. J Exp Psychol Hum Percept Perform 37(5):1485–1491

3-Rainbow Domination Number in Graphs

Kung-Jui Pai and Wei-Jai Chiu

Abstract The k-rainbow domination is a location problem in operations research. Give an undirected graph G as the natural model of location problem. We have a set of k colors and assign an arbitrary subset of these colors to each vertex of G. If a vertex which is assigned an empty set, then the union of color set of its neighbors must be k colors. This assignment is called the k-rainbow dominating function, abbreviate as kRDF, of G. The weight of kRDF is the sum of numbers of assigned colors over all vertices of G. The minimum weight of kRDF is defined as the k-rainbow domination number of G. In this paper, we present an exact algorithm and a heuristic algorithm to obtain the 3-rainbow domination number and the weight of 3RDF in graphs, respectively. Then, we test the practical performances of these algorithms, including their run times and solution qualities.

Keywords Domination · k-rainbow domination · Location problems

1 Introduction

Domination problem and its variations have been studied since antiquity. They are natural models for the location problems in operations research. They have been extensively studied in the literature; see Haynes et al. (1998). A *dominating set* of a graph G is a subset D of V(G) such that every vertex not in D is adjacent to some vertex in D. We usually call the vertex in D as *dominating vertex*, and call the

---

K.-J. Pai (✉) · W.-J. Chiu
Department of Industrial Engineering and Management, Ming Chi University
of Technology, New Taipei City, Taiwan
e-mail: poter@mail.mcut.edu.tw

W.-J. Chiu
e-mail: a123750203@yahoo.com.tw

Y.-K. Lin et al. (eds.), *Proceedings of the Institute of Industrial
Engineers Asian Conference 2013*, DOI: 10.1007/978-981-4451-98-7_86,
© Springer Science+Business Media Singapore 2013

vertex not in D as *dominated vertex*. A graph is said to be dominated if any vertex is either a dominating vertex or a dominated vertex. Thus, a *dominated graph* is determined by the number of dominating vertices and the locations where they are placed. It is often desirable to build a fully dominating system with the minimum dominating vertices. The *domination number* $\gamma(G)$ of G is the minimum cardinality of a dominating set of G.

For example, a domination of graph G is shown in Fig. 1. We select vertices V_1, V_6 and V_9 in the dominating set. Vertices V_2, V_3, V_4 and V_5 are dominated by vertex V_1. Vertex V_7 are dominated by vertex V_6, and vertex V_8 are dominated by vertex V_9. Thus, the graph G is dominated and domination number $\gamma(G) = 3$.

The k-rainbow domination was introduced by Brešar et al. (2005), and it is a variant of classical domination. Let $G = (V(G), E(G))$ be a finite, simple and undirected graph, where $V(G)$ and $E(G)$ are the vertex and edge sets of G, respectively. Two vertices u and v are neighbors of each other if $(u, v) \in E(G)$. For a vertex $v \in V(G)$, the open neighborhood $N(v) = \{u \in V(G) \mid (u, v) \in E(G)\}$. Let $C = \{1, 2,\ldots, k\}$ be a set of k colors, and f be a function that assign to each vertex a set of colors chosen from C, that is, f: $V(G) \to P(\{1,\ldots, k\})$. If for each vertex $v \in V(G)$ such that $f(v) = \emptyset$ we have $\cup_{u \in N(v)} f(u) = C$ then f is called a *k-rainbow dominating function* (kRDF) of G. The *weight*, $\omega_k(f)$, of a function f is defined as $\omega_k(f) = \Sigma_{v \in V(G)} |f(v)|$. We call $\omega_k(f)$ as the *number of k-rainbow domination* that is the sum of numbers of assigned colors over all vertices of G. The *k-rainbow domination number*, $\gamma_{rk}(G)$, of G is the minimum weight of a k-rainbow dominating function.

For example, a 3-rainbow domination of graph G is shown in Fig. 2a. We assign a color set $\{1\}$ to V_4, V_6, a color set $\{2\}$ to V_3, V_9 and a color set $\{3\}$ to V_5, V_7. Since the color set of every one of vertices V_1, V_2 and V_8 is empty, the union of color set of their neighbors must be 3 colors. Vertex V_1 has a neighbor V_4 with color set $\{1\}$, a neighbor V_3 with color set $\{2\}$ and a neighbor V_5 with color set $\{3\}$, so Vertices V_1 is dominated. Similarly, Vertices V_2 and V_8 have neighbors with color sets $\{1\}$, $\{2\}$ and $\{3\}$, respectively. Another 3-rainbow domination of graph G is shown in Fig. 2b. We assign a color set $\{1, 2, 3\}$ to V_2, a color set $\{3\}$ to V_4 and a color set $\{1, 2\}$ to V_9. Vertices V_1, V_3, V_6 and V_7 have a neighbor V_2 with a color set $\{1, 2, 3\}$, so they are dominated. Vertices V_5 and V_8 are dominated, since common neighbors V_4 with a color set $\{3\}$ and V_9 with a color set $\{1, 2\}$. Consequently, $\omega_3(f) = 6$ in both Fig. 2a and b. In fact, $\gamma_{r3}(G) = 6$ by brute-force search.

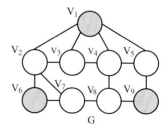

Fig. 1 A domination of graph G. Gray vertices represent dominating vertices and white vertices represent dominated vertices

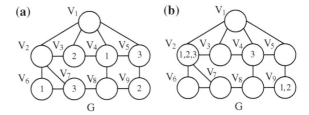

Fig. 2 3-rainbow dominations of G. The vertex with a color set {x, y, z} is labeled by "x, y, z" where y and z can be empty

In this paper, we present an exact algorithm to obtain $\gamma_{r3}(G)$ in graphs. When the number of vertices is fixed and the number of edges is decreasing, the run time of the exact algorithm is rapidly increasing. We design a heuristic algorithm to obtain the weight, $\omega_3(f)$, of 3RDF. Then, we test the practical performances of these algorithms, including their run times and solution qualities.

The remaining sections are organized as follows. In Sect. 2, we give a literature review. In Sect. 3, we present an exact algorithm and a heuristic algorithm for 3-rainbow domination of graph. The experiment results of two algorithms are shown in Sect. 4. Finally, some concluding remarks are given in Sect. 5.

2 Literature Review

The concept of k-rainbow domination of a graph G was introduced by Brešar et al. (2005). They stated that domination represents situations in which each vertex on that is not located by a guard needs to have a guard in a neighboring vertex. Then, they consider a more complex situation where there are different k types of guards. Each vertex that is not located with a guard has in its neighborhood all k types of guards. This relaxation leads to the concept of k-rainbow domination.

Brešar and Šumenjak (2007) showed that the decision version of 2-rainbow domination of graphs is NP-complete even when restricted to chordal graphs (or bipartite graphs). They also gave the exact values of the 2-rainbow domination numbers for paths, cycles and suns. Chang et al. (2010) extended the results of Brešar and Šumenjak (2007). They proved that k-rainbow domination problem is NP-complete even when restricted to chordal graphs or bipartite graphs. They also gave a linear-time algorithm for the k-rainbow domination problem on trees.

Thus, further investigations tended to study bounds on k-rainbow domination number for certain families of graphs, such as trees, generalized Petersen graphs and lexicographic product of graphs. This problem is widely studied in Ali et al. (2011), Brešar et al. (2005, 2008), Brešar and Šumenjak (2007), Chang et al. (2010), Fujita et al. (2012), Meierling et al. (2011), Šumenjak et al. (2012), Tong et al. (2009), Xu (2009), Wu and Rad (2010) and Wu and Xing (2010).

3 Algorithms for 3-Rainbow Domination

All graphs considered here are undirected and simple (i.e., finite, loopless, and without multiple edges). We shall use the following notation and terms in this paper. The degree of a vertex in a graph G is the number of its neighbors in G. A leaf is the vertex with degree 1. A support vertex is the only neighbor of leaf. A strong support vertex is the support vertex with at least 2 leaves. Let $G_1 = (V_1, E_1)$ and $G_2 = (V_2, E_2)$ be two graphs which have disjoint vertex sets V_1 and V_2 and disjoint edge sets E_1 and E_2, respectively. The Cartesian product of G_1 and G_2 is the graph $G_1 \in G_2$ whose set of vertices is the Cartesian product of the sets V_1 and V_2, so $V(G_1 \in G_2) = V_1 \times V_2$. Two vertices of $G_1 \in G_2$, say (v_1, v_2) and (u_1, u_2), are adjacent if and only if $v_1 = u_1$ and $v_2u_2 \in E(G_2)$, or alternatively, if $v_2 = u_2$ and $v_1u_1 \in E(G_1)$.

3.1 An Exact Algorithm

Brešar and Šumenjak (2007) give that the concept of 3-rainbow domination of a graph G coincides with the ordinary domination of the prism $G \in K3,3$, where K_3 stands for a graph with 3 vertices and each pair of vertices is connected by an edge. By brute-force method, the exact algorithm E3RD test all combinations of γ ($G \in K_3$) vertices from $V(G \in K_3)$.

Algorithm E3RD (Exact 3-rainbow domination)

Input: a graph G.
Output: the $\gamma_{r3}(G)$-set S_2.
Step 1. Obtain the graph H from Cartesian product of G with K_3.
Step 2. Solve the dominating set S_1 on the graph H by brute-force method.
Step 3. The graph H can be treated as three copies of the graph G with additional edges.
Step 3.1. For vertex $v \in S_1$,
Step 3.2. If v is in the first copy, then add v with $\{1\}$ in S_2.
Step 3.3. Else If v is in the second copy, then add v with $\{2\}$ in S_2.
Step 3.3. Else add v with $\{3\}$ in S_2.

[+]The time complexity of algorithm E3RD is $O(C_{\gamma(H)}^{|V(H)|})$.

3.2 A Heuristic Algorithm

In order to obtain the number of 3-rainbow domination quickly, we design a heuristic algorithm as follows.

Algorithm H3RD (Heuristic 3-rainbow domination)

Input: a graph.
Output: the set S_1 with a color set $\{1, 2, 3\}$ and the set S_2 with a color set $\{1\}$.
Step 1.1. Let V_d be a strong support vertex in G. If $V_d \neq \emptyset$, go to Step 2.
Step 1.2. Let V_d be a support vertex in G. If $V_d \neq \emptyset$, go to Step 2.
Step 1.3. Let V_d be a vertex with maximum degree in G.
Step 2. Let set $S_1 = S_1 \cup V_d$.
Step 3. Remove V_d and its neighbors from graph G.
Step 4. Repeat from Step 1 until all degrees of vertices in G are less than 2.
Step 5. Let S_2 be the set with remaining vertices in G.
Step 6. Return S_1 with a color set $\{1, 2, 3\}$ and the set S_2 with a color set $\{1\}$.

$^+$The time complexity of algorithm H3RD is $O(|V(G)|^3)$.

4 Experimental Results

In order to observe the results of heuristic algorithm H3RD and exact algorithm E3RD, we have designed a number of experiments. First we design five cases which number of vertices is 16 and numbers of edges are 20, 24, 28, 32 and 36, respectively. For each case we randomly generate 30 connected graphs. The proposed Algorithms E3RD and H3RD were implemented in GNU C programming language. All experiments are performed on a personal computer with 3.4 GHz Intel Core i7-3770 CPU, 8 GB RAM, and running Linux (Fedora Core 17). Two subjects of experiments are to obtain the weight, $\omega_3(f)$, of 3RDF and run times of all graphs. In fact, $\gamma_{r3}(G) = \omega_3(f)$ by Algorithm E3RD.

Once the experiments had been carried out, we used statistics to make sense of the data. Table 1 shows $\omega_3(f)$ and run times of graphs with different number of edges by algorithms E3RD and H3RD. Then, Table 2 shows the difference between $\omega_3(f)$ of H3RD and $\gamma_{r3}(G)$ of E3RD in 30 graphs for each case.

We have the following results.

1. The relations between average $\omega_3(f)$ of 3RDF and the number of edges by two algorithms are shown in Fig. 3. In each algorithm, the average $\omega_3(f)$ of 3RDF is increasing when the number of edges is decreasing. The rum times of algorithm E3RD is increasing so fast when the number of edges is decreasing.
2. As shown in Table 1, the average difference between $\omega_3(f)$ of H3RD and $\gamma_{r3}(G)$ of E3RD is between 1.17 and 1.43 by varied number of edges.
3. As shown in Table 2. In more than 50 % samples, the differences between $\omega_3(f)$ of H3RD and $\gamma_{r3}(G)$ of E3RD are less than or equal to one.
4. In these experiments, the approximate ratio of average $_3(f)$ of H3RD to average $\gamma_{r3}(G)$ of E3RD is less than 1.13.

Table 1 $\omega_3(f)$ and run times of graphs with different number of edges by algorithms E3RD and H3RD

Edge no.	Algorithm E2RD					Algorithm H2RD				
	20	24	28	32	36	20	24	28	32	36
$\omega_3(G)$ (Max.)	12	11	10	9	9	14	13	13	12	11
$\omega_3(G)$ (Min.)	11	8	8	8	7	11	8	9	8	7
$\omega_3(G)$ (Ave.)	11.15	10.19	9.35	8.88	8.37	12.54	11.4	10.54	10.31	9.54
$\omega_3(G)$ (Std.)	0.36	0.8	0.56	0.33	0.56	1.03	1.25	0.99	0.84	0.81
Run time (Ave. seconds)	14,684	4721	1164	398	210	+	+	+	+	+
Run time (Std. seconds)	10,745	2959	599	125	133	+	+	+	+	+

[+] The computing time is less than 1 s

Table 2 The difference between $\omega_3(f)$ of H3RD and $\gamma_{r3}(G)$ of E3RD in 30 graphs for each case

Edge no.	20	24	28	32	36
$\omega_3(f)$ of H3RD = $\gamma_{r3}(G)$ of E3RD	6	11	8	3	6
$\omega_3(f)$ of H3RD = $\gamma_{r3}(G)$ of E3RD + 1	12	9	11	12	16
$\omega_3(f)$ of H3RD = $\gamma_{r3}(G)$ of E3RD + 2	5	4	8	14	6
$\omega_3(f)$ of H3RD = $\gamma_{r3}(G)$ of E3RD + 3	7	6	3	1	2

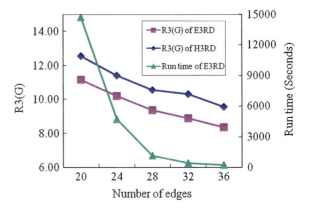

Fig. 3 The relations between average $\omega_3(f)$ of 3RDF and the number of edges by two algorithms

5 Conclusions

By the concept of Brešar and Šumenjak (2007), we present an exact algorithm to obtain $\gamma_{rk}(G)$ in graphs. As shown in experimental results, the run time of exact algorithm is rapidly increasing when number of edges is decreasing. We design a heuristic algorithm to obtain 3-rainbow domination number $\omega_3(f)$. Then, we test the practical performances of these two algorithms, including their run times and solution qualities. As shown in the experimental results, our heuristic algorithm can find 1.15-approximation solutions in graphs with 16 vertices and less than 36 edges. The run time of our heuristic algorithm is less than one second which is much faster than the run time of exact algorithm.

Acknowledgments This work is partially supported by NSC101-221-E-131-039 from the National Science Council, Taiwan. The authors also gratefully acknowledge the helpful comments and suggestions of the reviewers, which have improved the presentation.

References

Ali M, Rahim MT, Zeb M, Ali G (2011) On 2-rainbow domination of some families of graphs. Int J Math Soft Comput 1:47–53

Brešar B, Šumenjak TK (2007) On the 2-rainbow domination in graphs. Discrete Appl Math 155:2394–2400

Brešar B, Henning MA, Rall DF (2005) Paired-domination of Cartesian products of graphs and rainbow domination. Electron Notes Discrete Math 22:233–237

Brešar B, Henning MA, Rall DF (2008) Rainbow domination in graphs. Taiwanese J Math 12:213–225

Chang GJ, Wu J, Zhu X (2010) Rainbow domination on trees. Discrete Appl Math 158:8–12

Fujita S, Furuya M, Magnant C (2012) k-Rainbow domatic numbers. Discrete Appl Math 160:1104–1113

Haynes TW, Hedetniemi ST, Slater PJ (1998) Fundamentals of domination in graphs. Marcel Dekker, NewYork

Meierling D, Sheikholeslami SM, Volkmann L (2011) Nordhaus-Gaddum bounds on the k-rainbow domatic number of a graph. Appl Math Letters 24:1758–1761

Šumenjak TK, Rall DF, Tepeh A (2012) Rainbow domination in the lexicographic product of graphs. Combinatorics arXiv:1210.0514

Tong CL, Lin XH, Yang YS, Lou MQ (2009) 2-rainbow domination of generalized Petersen graphs P(n, 2). Discrete Appl Math 157:1932–1937

Wu Y, Rad NJ (2010) Bounds on the 2-rainbow domination number of graphs. Graphs Combinatorics. doi:10.1007/s00373-012-1158-y

Wu Y, Xing H (2010) Note on 2-rainbow domination and Roman domination in graphs. Appl Math Letter 23:706–709

Xu G (2009) 2-rainbow domination of generalized Petersen graphs P(n, 3). Discrete Appl Math 157:2570–2573

A Semi-Fuzzy AHP Approach to Weigh the Customer Requirements in QFD for Customer-Oriented Product Design

Jiangming Zhou and Nan Tu

Abstract In this paper, a new analytic hierarchy process (AHP) is proposed to determine the importance weights of customer requirements (CRs) in quality function deployment (QFD) for customer-oriented product design. The new approach combines conventional and fuzzy AHP. It takes into account one's uncertainty in comparing different pairwise CRs to improve the imprecise rankings in conventional AHP. By employing semi-fuzzy matrices, it guarantees that the final pairwise comparison matrices based on fuzzy scales are positive reciprocal. The problem of imprecise pairwise comparisons in conventional AHP is ameliorated and more accurate results are provided. Finally, a case study of new sports earphones design is given as an example to illustrate this approach.

Keywords Analytic hierarchy process · Semi-fuzzy · Customer requirements · Importance weights · Quality function deployment · Customer-oriented product design

1 Introduction

The competition in new product development (NPD) has intensified and companies are forced to provide products that meet the fast-changing CRs. For the past decades, various NPD methodologies have been studied for customer-oriented product design and widely used in real cases.

J. Zhou (✉) · N. Tu
Research Center for Modern Logistics, Graduate School at Shenzhen, Tsinghua University, Shenzhen 518055, People's Republic of China
e-mail: jamin.zjm@gmail.com

N. Tu
e-mail: dr.nan.tu@gmail.com

Y.-K. Lin et al. (eds.), *Proceedings of the Institute of Industrial Engineers Asian Conference 2013*, DOI: 10.1007/978-981-4451-98-7_87,
© Springer Science+Business Media Singapore 2013

In order to increase customer satisfaction throughout customer-oriented product design, QFD has been commonly used to integrate CRs into product design (Akao 1990, 1997). The translation process is comprised of six steps (Hsiao 2002). Determining the relative importance of CRs is vital since the target value of DRs could be significantly affected. The conventional QFD employs point scoring scale in weighing CRs (Griffin and Hauser 1993) where human subjective judgments are usually ill-considered and imprecision in weighting CRs may exist.

The AHP, a common tool in solving multiple criteria decision making (MCDM) problems (William 2008), employs pairwise comparison to determine the relative importance to make a more promising decision. Combined with QFD, the methodology has been widely used in product design selection problems: for example, Wang et al. (1998) suggested that CRs and DRs can be regarded as criteria and alternatives in the AHP, Madu et al. (2002) employed AHP to obtain the relative importance weightings of CRs for further study in QFD.

In general, the AHP is comprised of three phases: hierarchy construction, priority analysis and consistency verification. The conventional AHP employs nine-point scale in crisp real numbers as pairwise comparison ratios. Furthermore, fuzzy sets are integrated with AHP to model the uncertainty in human assessment on qualitative attributes (Chan et al. 1999) using such methods as triangular membership functions (Van and Pefrycz 1983) and geometric mean (Buckley 1985). Kwong and Bai (2002, 2003) employed fuzzy comparison matrix to determine the importance weights of CRs in QFD. However, weaknesses exist in previous study:

First, when using the same index of certainty, the differences of one's uncertainty in comparing different CRs are ignored.

Second, when selecting values from the given interval using triangular fuzzy numbers for the whole comparison matrix, the pairwise comparison matrix may be not positive reciprocal.

In this paper, a semi-fuzzy AHP approach is described to provide a more accurate priority analyzing process. First, assessments on CRs are expressed using triangular fuzzy numbers with separate indexes of certainty. Then the pairwise comparison matrix is built with fuzzy numbers to provide interval estimations. When selecting a value from the given interval of an element in the fuzzy comparison matrix, only those with the corresponding crisp real numbers larger than 1 are used, which indicate significant preferences. Subsequently normalized importance weights of CRs at all levels can be calculated applying AHP and synthesization is implemented to get the overall weights of CRs. Finally, the proposed approach has been applied to a new sports earphones design as an illustrative example.

2 The Semi-Fuzzy AHP Approach

The overall procedure of the proposed approach is shown in Fig. 1.

A Semi-Fuzzy AHP Approach

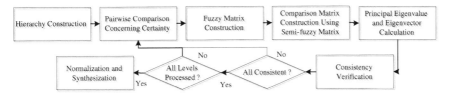

Fig. 1 Overall procedure of the semi-fuzzy AHP approach

2.1 Hierarchy Construction

The first step of applying AHP in weighing the importance of CRs is to structure CRs into different hierarchy levels. The aim is to decompose such a complex MCDM problem into simplified sub-problems for convenience. It has been shown that 7 is a practical bound on the number of elements at the same level, so far as consistency is concerned (Saaty 1980).

2.2 Priority Analysis Based on Semi-Fuzzy AHP

After hierarchy construction, comparisons are carried out on a pairwise basis to evaluate the relative importance. The conventional AHP employs nine-point scale to indicate the degree of preference. To take the uncertainty into account, the fuzzy AHP uses triangular fuzzy numbers, which are characterized as:

$$\forall \alpha \in [0,1], \tilde{x}^\alpha = \begin{cases} [x,(x+2)^\alpha] = [x, x+2-2\alpha], & x = 1 \\ [(x-2)^\alpha, (x+2)^\alpha] = [x-2+2\alpha, x+2-2\alpha], & x = 3, 5, 7 \\ [(x-2)^\alpha, x] = [x-2+2\alpha, x], & x = 9 \end{cases} \quad (1)$$

The level of certainty is indicated by index α varying from 0 to 1, e.g., $\tilde{3}^{0.5}$ indicates that one is 50 % certain that the preference is moderately rather than equally or strongly. The larger the value, the higher degree of certainty. In this research, 0, 0.3, 0.5, 0.8, 1 are used to indicate such degrees as most uncertain, uncertain, neutral, certain and most certain respectively.

The pairwise comparison matrix is constructed based on the relative importance weights using fuzzy numbers. Taking into account the differences of one's uncertainty in comparing different pairwise CRs, the index of certainty α varies.

$$\tilde{a}_{ij}^{\alpha_{ij}} = \left[a_{ijl}^{\alpha_{ij}}, a_{iju}^{\alpha_{ij}}\right] = \begin{cases} [a_{ij}, a_{ij} + 2 - 2\alpha_{ij}], & a_{ij} = 1 \\ [a_{ij} - 2 + 2\alpha_{ij}, a_{ij} + 2 - 2\alpha_{ij}], & a_{ij} = 3, 5, 7 \\ [a_{ij} - 2 + 2\alpha_{ij}, a_{ij}], & a_{ij} = 9 \end{cases} \quad (2)$$

Using fuzzy numbers, interval estimation is given when weighing the relative importance of CRs. In order to prioritize the CRs for further study in NPD, agreements have to be reached when selecting a value from the given interval. The index of optimism μ is employed ($\mu \in [0, 1]$) and defined as linear convex. In this research, fuzzy numbers from the fuzzy comparison matrix are used only when the central value (the corresponding crisp real numbers) are larger than 1, which indicates significant preferences. Thus, the selected value is characterized as:

$$\hat{a}_{ij}^{\alpha_{ij}} = \begin{cases} \mu a_{ijl}^{\alpha_{ij}} + (1 - \mu)a_{iju}^{\alpha_{ij}}, \left(a_{ijl} + a_{iju}\right)/2 \geq 1 \\ 1/\hat{a}_{ji}^{\alpha_{ij}}, \left(a_{ijl} + a_{iju}\right)/2 \geq 1 \end{cases} \tag{3}$$

The final pairwise comparison matrix is characterized as:

$$\hat{A} = \begin{pmatrix} 1 & \hat{a}_{12}^{\alpha_{12}} & \cdots & \hat{a}_{1n}^{\alpha_{1n}} \\ \hat{a}_{21}^{\alpha_{21}} & 1 & \cdots & \hat{a}_{2n}^{\alpha_{2n}} \\ \vdots & \vdots & \ddots & \vdots \\ \hat{a}_{n1}^{\alpha_{n1}} & \hat{a}_{n2}^{\alpha_{n2}} & \cdots & 1 \end{pmatrix} \tag{4}$$

The method of calculating the principle eigenvector is then used to produce the priority vector that indicates the relative importance weights of CRs. The priority vector derived from this method is demonstrated to reproduce itself on a ratio scale under the hierarchy composition principle (Saaty 2003).

To calculate the principal eigenvector, first solve the characteristic equation of matrix $\hat{A}$ as $det\left(\hat{A} - \lambda I\right) = 0$, then substitute the largest eigenvalue into the equation $\hat{A}X = \lambda_m X$ and calculate the vector. The priority vector needs to be normalized and synthesized to get the overall weights at each level.

2.3 Consistency Verification

Since in AHP comparisons are carried out through human judgment, inevitably inconsistency exists ($\tilde{a}_{ij}\tilde{a}_{jk}^T \neq \tilde{a}_{ik}$, e.g., i is preferred to j twice, j is preferred to k twice, but i is preferred to k only three times). It is acceptable whin an extent. Considered as an advantage of the AHP, the verification process is aimed to measure the degree of consistency.

First calculate the consistency index (CI) as $CI = \frac{\lambda_m - n}{n-1}$ where λ_m is the principal eigenvalue and n is the number of elements.

Then compare CI to the random index (RI). It is acceptable if $CI/RI < 0.1$. Otherwise unacceptable inconsistency exists and the comparison matrix should be reviewed and revised. The reference values of RI for different n are shown in Table 1 (Winston 1994).

Table 1 Random index of AHP

n	2	3	4	5	6	7
RI	0	0.58	0.90	1.12	1.24	1.32

3 A Case Study

An example of designing new sports earphones for joggers and hikers is given to demonstrate how the proposed semi-fuzzy AHP approach can be used to weigh the CRs.

3.1 Hierarchy Construction

A three-level hierarchy for the CRs, which were gathered from questionnaires and interviews, was constructed using affinity diagram as shown in Fig. 2. At the top of the hierarchy lies the ultimate design objective, which is sports earphones design. The lower level of the hierarchy are the CRs, which are divided into two levels with three categories so that the number of elements at each level does exceed the upper bound 7.

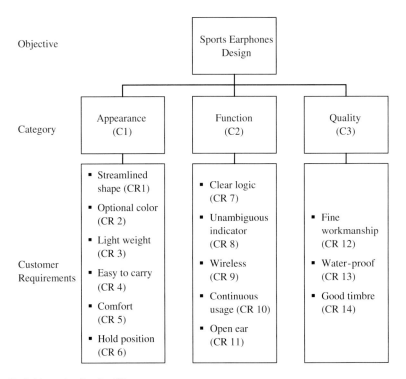

Fig. 2 A hierarchy for the CRs

3.2 Constructing the Comparison Matrix

After the hierarchy structure was constructed, questionnaires were used to gather pairwise comparison preferences of CRs with separate indexes of certainty, which were discussed by the panel of domain and industry experts.

Subsequently the comparison matrices were constructed. Indeed, iterations were carried out to revise the preferences since the consistency ratio might exceed the limit. The comparison matrix of CRs under the category of appearance is shown as an example.

The initial comparison matrix is constructed as:

$$
C1 : \tilde{A}_1 =
\begin{array}{c}
 \\ CR1 \\ CR2 \\ CR3 \\ CR4 \\ CR5 \\ CR6
\end{array}
\begin{array}{cccccc}
CR1 & CR2 & CR3 & CR4 & CR5 & CR6
\end{array}
\begin{pmatrix}
1 & \tilde{3}^{0.5} & \tilde{1}^{0.5} & \tilde{3}^{0.5} & \tilde{1}^{0.3^{-1}} & \tilde{3}^{0.5^{-1}} \\
\tilde{3}^{0.5^{-1}} & 1 & \tilde{1}^{0.5^{-1}} & \tilde{3}^{0.5} & \tilde{5}^{0.5^{-1}} & \tilde{7}^{0.5^{-1}} \\
\tilde{1}^{0.5^{-1}} & \tilde{1}^{0.5} & 1 & \tilde{1}^{0.5} & \tilde{3}^{0.5^{-1}} & \tilde{1}^{0.8^{-1}} \\
\tilde{3}^{0.5^{-1}} & \tilde{3}^{0.5^{-1}} & \tilde{1}^{0.5^{-1}} & 1 & \tilde{5}^{0.5^{-1}} & \tilde{7}^{0.5^{-1}} \\
\tilde{1}^{0.3} & \tilde{5}^{0.5} & \tilde{3}^{0.5} & \tilde{5}^{0.5} & 1 & \tilde{1}^{0.3^{-1}} \\
\tilde{3}^{0.5} & \tilde{7}^{0.5} & \tilde{1}^{0.8} & \tilde{7}^{0.5} & \tilde{1}^{0.3} & 1
\end{pmatrix}
$$

By applying Eq. (2), the fuzzy comparison matrices using interval estimation are characterized as:

$$
C1 : \tilde{A}_1 =
\begin{pmatrix}
1 & [2,4] & [1,2] & [2,4] & [1/2.6,1] & [1/4,1/2] \\
[1/4,1/2] & 1 & [1/2,1] & [2,4] & [1/6,1/4] & [1/8,1/6] \\
[1/2,1] & [1,2] & 1 & [1,2] & [1/4,1/2] & [1/1.4,1] \\
[1/4,1/2] & [1/4,1/2] & [1/2,1] & 1 & [1/6,1/4] & [1/8,1/6] \\
[1,2.6] & [4,6] & [2,4] & [4,6] & 1 & [1/2.6,1] \\
[2,4] & [6,8] & [1,1.4] & [6,8] & [1,2.6] & 1
\end{pmatrix}
$$

By applying Eq. (3) using the index of optimism as 0.95 which indicates a highly optimistic situation, the semi-fuzzy approach construct the final pairwise comparison matrices as:

$$
C1 : \hat{A}_1 =
\begin{pmatrix}
1.0000 & 3.9000 & 1.9500 & 3.9000 & 0.3968 & 0.2564 \\
0.2564 & 1.0000 & 0.5128 & 3.9000 & 0.1695 & 0.1266 \\
0.5128 & 1.9500 & 1.0000 & 1.9500 & 0.2564 & 0.7246 \\
0.2564 & 0.2564 & 0.5128 & 1.0000 & 0.1695 & 0.1266 \\
2.5200 & 5.9000 & 3.9000 & 5.9000 & 1.0000 & 0.3968 \\
3.9000 & 7.9000 & 1.3800 & 7.9000 & 2.5200 & 1.0000
\end{pmatrix}
$$

A Semi-Fuzzy AHP Approach

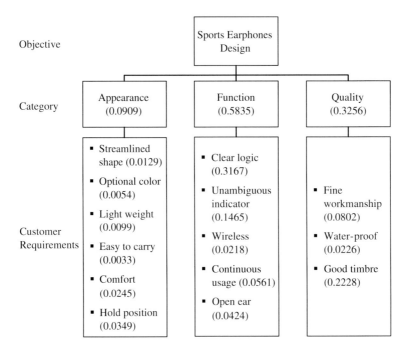

Fig. 3 Importance weights of the CRs

3.3 Prioritizing the CRs and Verification

MATLAB was used to calculate the principal eigenvalue and eigenvector of all the comparison matrices. Afterwards the ratios of CI to RI were calculated to verify the consistency of the matrices separately.

The eigenvectors were later normalized and synthesized to prioritize the overall weights of CRs which are shown in Fig. 3.

4 Conclusions

In this paper, a semi-fuzzy AHP approach is proposed to weigh the CRs for customer-oriented design using QFD, which brings the following benefits as:

First, separate indexes of certainty are used when constructing pairwise comparisons so that the differences of one's uncertainty in comparing different pairwise CRs are taken into account. Thus, more accurate priority rankings of the CRs can be provided.

Second, to construct the final pairwise comparison matrix, fuzzy number is used only when the preference is significant. This guarantees that the matrix is positive reciprocal, which meets the shortfalls using fuzzy AHP.

The proposed semi-fuzzy AHP approach is examined by a real case study of weighing the CRs for a sports earphones design product in Shenzhen, China. The results were later used in QFD to facilitate the decision making in product design.

Acknowledgement This work is partially supported by teams from Vtech Inc. (fake name to hide the identity of the company) in helping with data collection for the new sports earphone design project and providing valuable inputs into this research. The authors also gratefully acknowledge the helpful comments and suggestions of the reviewers, which have improved the presentation.

References

Akao Y (1990) Quality function deployment: integrating customer requirements into product design. Productivity Press, Cambridge

Akao Y (1997) QFD: past, present and future. Proceedings of the international symposium on QFD'97

Buckley JJ (1985) Fuzzy hierarchical analysis. Fuzzy Sets Syst 17:233–247

Chan LK, Kao HP, Ng A, Wu ML (1999) Rating the importance of customer needs in quality deployment by fuzzy and entropy methods. Int J Prod Res 37(11):2499–2518

Griffin K, Hauser JR (1993) The voice of the customer. Mark Sci 12(1):1–27

Hsiao SH (2002) Concurrent design method for developing a new product. Int J Ind Ergon 29:41–45

Kwong CK, Bai H (2002) A fuzzy AHP approach to the determination of importance weights of customer requirements in quality function deployment. J Intell Manuf 13:367–377

Kwong CK, Bai H (2003) Determining the importance weights for the customer requirements in QFD using a fuzzy AHP with an extent analysis approach. IIE Trans 35(7):619–626

Madu CN, Kuei C, Madu IE (2002) A hierarchic metric approach for integration of green issues in manufacturing: a paper recycling application. J Environ Manage 64(3):261–272

Saaty TL (1980) The analytic hierarchy process. McGraw-Hill, New York

Saaty TL (2003) Decision-making with the AHP: why is the principle eigenvector necessary. Eur J Oper Res 145:85–91

Van PJML, Pefrycz W (1983) A fuzzy extension of Saaty's priority theory. Fuzzy Sets Syst 11:229–241

Wang H, Xie M, Goh TN (1998) A comparative study of the prioritization matrix method and the analytic hierarchy process technique in quality function deployment. Total Qual Manag 9(6):412–430

William H (2008) Integrated analytic hierarchy process and its application: a literature review. E J Oper Res 186:211–228

Winston WL (1994) The analytic hierarchy process. Operations research: applications and algorithms. Wadsworth, Belmont, pp 798–806

An Optimization Approach to Integrated Aircraft and Passenger Recovery

F. T. S. Chan, S. H. Chung, J. C. L. Chow and C. S. Wong

Abstract In this paper, the allocation of aircrafts to each rescheduled flight with passengers concerns is considered. The problem consists of a recovered flight schedule within a recovery period, a pool of affected passengers with their initial itineraries, and a fleet of available aircrafts of various configurations. The objective is to route the suitable aircrafts to operate the suitable rescheduled flight legs, and at the same time, generating the corresponding itineraries for affected passengers. This paper proposes a new optimization formulation that integrates the recovery of aircrafts and passengers simultaneously to minimize the sum of passenger delay cost and airline operation cost. With the proposed algorithms, airlines will be able to assign suitable aircrafts to support flight recovery under disruptions within a short time-period, and at the same time reduce passenger delays.

Keywords Aircraft and passenger recovery · Airline scheduling · Disruption management · Fleet assignment · Genetic algorithm

---

F. T. S. Chan (✉) · S. H. Chung · J. C. L. Chow · C. S. Wong
Department of Industrial and Systems Engineering, The Hong Kong Polytechnic University, Hung Hom, Kowloon, Hong Kong
e-mail: f.chan@polyu.edu.hk

S. H. Chung
e-mail: nick.sh.chung@polyu.edu.hk

J. C. L. Chow
e-mail: jenny.chow@polyu.edu.hk

C. S. Wong
e-mail: mfsing@ymail.com

Y.-K. Lin et al. (eds.), *Proceedings of the Institute of Industrial Engineers Asian Conference 2013*, DOI: 10.1007/978-981-4451-98-7_88,
© Springer Science+Business Media Singapore 2013

1 Introduction

As the aviation industry grows more complex and dynamic, effective generation of recovery plans once disruption occurs becomes inevitable for airlines to minimize any potential loses. Resources, including aircrafts and crew members etc., should also be well allocated to optimize the utilization rate during the recovery period, and to minimize the costs associated. In this paper, a model that focuses on integrated aircraft and passenger recovery is presented.

Most work on airline disruption management attempts to schedule aircraft, crew, and passenger recovery in a tractable manner (Filer et al. 2000). Since integrating the recovery of several resources simultaneously is a complicated task, the number of work attempts to integrate a subset of these components is relatively few and new (Kohl et al. 2007; Clausen et al. 2010). Research focusing on passenger recovery is also scarce.

Bratu and Barnhart (2006) described two integrated recovery models by determining whether the disrupted flight legs should be delayed or cancelled. The models were developed on a flight schedule based network with the aim of minimizing airline operation costs and estimated passenger disruption costs. Zhang and Hansen (2008) introduced an integration with other transportation modes to accommodate disrupted passengers in a hub-and-spoke network. An integer programming model was developed to minimize passenger costs caused by flight delays, cancellations, or substitutions with a nonlinear objective function.

A more recent approach by Jafari and Zegordi (2010) introduced an assignment model that recovers disrupted aircraft schedules and passenger itineraries concurrently with a framework of rolling horizon time. The objective of their model is to minimize costs on aircraft recoveries, delays and cancelations. Bisaillon et al. (2011) employed a neighborhood search heuristic in a large-scale to integrate reassignment of fleets, aircraft routings, and passengers to support resumption of regular operations. However, passengers are given low priority in their model. Petersen et al. (2012) presented an optimization approach to solve a fully integrated airline recovery problem. The problem is broken into four sub-problems to recover flight schedule, aircrafts, crews, and passengers within some time horizon. The objective seeks to minimize the total airline operation cost and passenger delay cost.

It is identified that passenger disruptions have rarely been considered or are given low priority in existing airline disruption management literature. In the limited researches that involve passenger considerations, the impact on passengers are usually not being modeled explicitly, in which their delay costs are only approximate. All these operation-centric approaches have led to a fact that passengers often suffer a much greater impact than that of airlines under disruptions. According to a recent report, the direct cost to passengers on flight delay on the U.S. economy in 2007 was US$16.7 billion, and that for airlines were US$8.3 billions only (NEXTOR 2010). In the view of this, an integrated recovery model that is more passenger-centric is therefore proposed, which aims to seek a tradeoff between airline operation and passenger disruption costs.

An Optimization Approach to Integrated Aircraft and Passenger Recovery   731

The remainder of the paper is organized as follow: Sect. 2 gives a description on the airline recovery problem considered in this paper. Focus is put on an aircraft rerouting problem. The proposed model for the integrated aircraft and passenger recovery is formulated in Sect. 3. The operation of the model is also presented. In Sect. 4, some discussions are made on the proposed model and a conclusion is drawn.

2 Problem Description

A *flight schedule* is a set of flights that operated by the airline in a given period of time. A *flight leg* is a non-stop flight from an origin airport to a destination airport. A *Fleet* is a group of aircrafts A that operated as a unit. It may contain aircrafts of more than one model that shares similar configurations. A *route* is the sequence of flight legs assigned to a given aircraft $a \in A$. *Turn-around time* is the time between arrival and departure of aircraft in a rotation.

In this paper, the integrated recovery problem comprises of an aircraft recovery problem and a passenger recovery problem. Given a set of rescheduled flight legs F, individual routings among a single fleet of aircrafts of two models, a_l and a_s, will be assigned to accommodate each rescheduled flight leg $f \in F$. The assignment will base on the number of affected passengers N_p and their itineraries over the recovery period. Passengers who cannot be transferred to the scheduled destination by the end of the recovery period will be transferred to other airlines. The cases of swapping or calling of spare aircraft are allowed. It is assumed the crew base is sufficient enough to cover all modified schedules.

2.1 The Aircraft Rerouting Problem

Once disruption occurs, the disrupted flights corresponded to a single fleet of aircraft bounded in the recovery period (t_0, T) are rescheduled by the airline operation centre. For clear illustration, an example of a repaired flight schedule of a network with 5 airports $(n_1, n_2, ..., n_5)$ is given in Fig. 1. The network is served by a fleet of 2 aircrafts (a_{l1}, a_{s1}) with different seating capacities. Given the rescheduled flight legs $(f_1, f_2, ..., f_{13})$ that would be served by a_{l1} and a_{s1}, and a set of affected passengers $p \in P$ with their initial itineraries, the problem is to construct the best aircrafts routing to utilize the seating capacity to serve as many passengers as possible. Some possible sets of routings are shown in Table 1.

In typical cases, the amount of possible routes can be huge when the recovery period is long enough to cover significantly large number of flight legs. For the illustration above, there can be as many as 14 possible sets of routes for a simple case that consists only of 2 aircrafts with 13 flight legs. In reality, most recovery instances have between 30 and 150 aircrafts and the time horizons can be much longer (Rosenberger et al. 2003). With the cases of operating spare aircrafts also

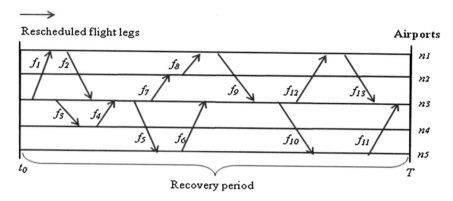

Fig. 1 A time-line network of rescheduled flight legs in a recovery period

Table 1 Some examples of possible routings of aircrafts a_{ll} and a_{sl}

Route 1		Route 2		Route 3	
Operate by aircraft a_{ll}	Operate by aircraft a_{sl}	Operate by aircraft a_{ll}	Operate aircraft a_{sl}	Operate by aircraft a_{ll}	Operate asircraft a_{sl}
f_1	f_3	f_1	f_3	f_3	f_1
f_2	f_4	f_2	f_4	f_4	f_2
f_7	f_5	f_5	f_7	f_5	f_7
f_8	f_6	f_6	f_8	f_6	f_8
f_9	f_{10}	f_{12}	f_9	f_{10}	f_9
f_{12}	f_{11}	f_{13}	f_{10}	f_{11}	f_{12}
f_{13}			f_{11}		f_{13}

being considered, the number of possible aircraft routes can thus be extremely huge. It would be difficult and time consuming to evaluate all possible routings and select the best among them.

In this paper, a comprehensive mathematical formulation that integrates aircraft and passenger rescheduling is presented. The cases of aircraft swapping, ferrying, and spare aircrafts operations, which are seldom being considered in most aircraft rerouting literature are also included.

3 Model Formulation

The objective of the model is to minimize the sum of passenger delay cost and airline operation cost. Passenger delay cost involves a delay cost of arrival time at destination to each passenger (in minute), and an inconvenient cost due to direction to other airlines, which causes a loss of goodwill to the airline. Airline operation cost includes an aircraft operation cost depends on the aircraft model

An Optimization Approach to Integrated Aircraft and Passenger Recovery

(in minute flight time), a cost on swapping, ferrying, or flying spare aircraft, and a compensation cost on meal and drinks to passengers for departure delays over a given limit of time. The parameters common to the proposed model are:

C_{dp} Cost of delay to passenger (per minute)
C_{cp} Inconvenient cost to passengers who are directed to other airlines
C_{pf} Cost of assigning a passenger to flight f
C_{sa} Cost of swapping aircraft a
C_{aa} Cost of operating spare aircraft a
C_{al} Cost of operating aircraft model a_l (per minute)
C_{as} Cost of operating aircraft model a_s (per minute)
C_{mp} Compensation cost to airlines on meals and drinks to passengers with departure delay over a given time limit h_{mp}
H_{ta} Minimum turn-around time for aircraft a
S_{al} Seating capacity of aircraft model a_l
S_{as} Searing capacity of aircraft model a_s
T_{ap} Scheduled arrival time of passenger p
T_{dp} Scheduled departure time of passenger p
N_f Total number of rescheduled flight legs

The decision variables common to the model are:

k_{saf}	$= 1$ if aircraft a of flight f is swapped, and 0 otherwise
k_{fa}	$= 1$ if aircraft a is ferried, and 0 otherwise
k_{aaf}	$= 1$ if flight f is operated by a spare aircraft a, and 0 otherwise
k_{asf}	$= 1$ if flight f is operated by aircraft type a_s, and 0 otherwise
k_{alf}	$= 1$ if flight f is operated by aircraft type a_l, and 0 otherwise
t_{aaf}	$=$ Actual arrival time of aircraft a for recovered flight f
t_{daf}	$=$ Actual departure time of aircraft a for recovered flight f
b_{pf}	$= 1$ if passenger p is being served in flight f, and 0 otherwise
b_{cp}	$= 1$ if passenger p is being directed to other airlines, and 0 otherwise
b_{mp}	$= 1$ if $(b_{pf}\, t_{daf} - T_{dpi}) \geq h_{mp}$, and 0 otherwise
m_{alf}	$= 1$ if flight f is operated by aircraft model a_l, and 0 otherwise
m_{asf}	$= 1$ if flight f is operated by aircraft model a_s, and 0 otherwise
n_{pf}	$=$ Number of passengers being assigned to flight f

The objective function is formulated as follows:

$$\min \sum_{f\in F} \sum_{p\in P} [(t_{aaf} - T_{ap})(b_{pf})C_{dp}]$$

$$+ \sum_{a\in A} \sum_{f\in F} [(t_{daf} - t_{aaf})m_{alf}k_{alf}C_{al}$$

$$+ (t_{daf} - t_{aaf})m_{asf}k_{asf}C_{as} + k_{saf}C_{sa} + k_{aaf}C_{aa}$$

$$+ n_{pf}C_{pf}] + \sum_{p\in P} (b_{mp}C_{mp} + b_{cp}C_{dp})$$

Subject to:

$$m_{alf} + m_{asf} = 1 \, \forall f \in F \tag{1}$$

$$k_{asf}C_{as} + k_{alf}C_{al} \geq n_f \, \forall f \in F \tag{2}$$

$$t_{aaf_{i+1}} + t_{daf_i} \geq H_{ta} \, \forall a \in A, \, f \in F \tag{3}$$

$$b_{fp} + b_{cp} = N_p \, \forall p \in P \tag{4}$$

$$b_{pf} \cdot t_{daf} \geq T_{dp} \, \forall p \in P, \, f \in F \tag{5}$$

$$k_{saf}, \, k_{fa}, \, k_{aaf}, \, k_{asf}, \, k_{alf}, \, b_{cp}, b_{mp}, \, m_{alf}, \, m_{asf} = \{0,1\}, \text{ and}$$
$$t_{aaf}, \, t_{daf} \text{ are REAL, and } n_{pf} \text{ is integer} \tag{6}$$

Constraint (1) ensures all flight legs bounded in the recovery period are assigned to an aircraft of either model a_l or a_s. Constraint (2) is a seat capacity constraint for aircrafts. Constraint (3) guarantees a minimum turnaround time is assigned between flight legs operated by the same aircraft. Constraint (4) ensures all passengers are either being served or redirected to other airlines. Constraint (5) states that no passenger is allowed to depart before the initial scheduled departure time. Finally, constraints in (6) ensures that the decision variables k_{saf}, k_{fa}, k_{aaf}, k_{asf}, k_{alf}, b_{cp}, b_{mp}, m_{alf}, m_{asf} are binary variables, the aircraft departure times (t_{aaf}, t_{daf}) are real, and the number of assigned passengers to a specific flight (n_{pf}) is integer.

3.1 Model Operations

To support effective operation of the model, various forms of information is required. They include the initial fleet schedule before disruption; the initial passengers schedule, including the scheduled departure time, the origin airport, the destination airport, and the scheduled arrival time; the repaired flight schedule correspond to the fleet; the aircrafts combination in the fleet; and the location of each aircraft at the beginning of the recovery period.

Given these, the number of passengers that needed to arrive at a specific airport at a specific timeslot, the number of passengers that scheduled to depart from a specific airport at a specific timeslot, and the number of passengers in each initial scheduled flight can be determined The model would generate possible sets of aircraft routings to cover all rescheduled flight legs based on these and select the optimal set of routings. The proposed framework is modeled in Fig. 2.

The detailed operation process is as follows:

1. Identify the number of passengers whom requirements can be satisfied by traveling on flight f_1
2. Identify the available aircrafts that are able to operate flight f_1
3. Assign an aircraft to flight f_1
4. Check if the aircraft is a swapping or spare aircraft
5. Assign passengers to the flight
6. If the number of passengers exceeds the seating capacity of the assigned aircraft, move the remaining passengers to the next suitable flight
7. If there is no more suitable flight, direct them to other airlines
8. Repeat steps i to vii until all repaired flight legs are covered
9. Direct remaining passengers to another airline
10. Evaluate the generated routing set
11. Repeat steps i to x to get another possible routing set, until a stopping criteria is reached
12. Compare all generated routings and select the best one to implement.

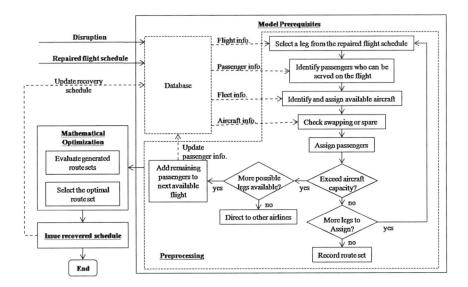

Fig. 2 Proposed model framework on integrated aircraft and passenger recovery

4 Discussions and Conclusions

In a given repaired flight schedule, the possible combination of aircraft routings can be huge. Also, given a high number of decision variables in the proposed model, identifying and evaluating all of them to find an optimized solution would be time consuming. It is recommended to use intelligent search heuristic, such as Genetic Algorithm (GA) or neighborhood search etc., to solve the identified problem to reduce the computation time. In this paper, a framework with the problem model is only provided to give a new research direction on passenger-oriented disruption management. Further investigations can be made to identify the most suitable algorithm in solving the model presented.

With the proposed model being solved, it is believed airlines can be equipped with higher reliability and customer service levels. This in turns increases customer retention rate and confidence of new customers in selecting the airline for air travels. The developed algorithm can further assist airlines in attracting high-value passengers who are sensitive to airline on-time reliability, increasing customer loyalty and satisfactory level, and reducing direct and indirect costs caused by passenger disruptions. These are especially important for airlines as the air travel market grows larger and more competitive.

In conclusion, a new optimization formulation that integrates the recovery of aircrafts and passengers simultaneously is proposed in this paper. The model routes the suitable aircrafts to operate the suitable rescheduled flight legs, and at the same time, generates the corresponding itineraries for affected passengers. The objective is to minimize the costs of passenger delay and airline operation, and to utilize the seating capacity of available aircrafts. After solving the model, airlines will be able to assign suitable aircrafts to support flight recovery under disruptions within a short time-period, reduce passenger delays, and at the same time achieve high customer satisfaction and remain competitive in the market.

Acknowledgments The work described in this paper was substantially supported by The Hong Kong Polytechnic University Research Committee for financial and technical support through an internal grant (Project No. G-UB03). The authors also gratefully acknowledge the helpful comments and suggestions of the reviewers, which have improved the presentation.

References

Bratu S, Barnhart C (2006) Flight operations recovery: new approaches considering passenger recovery. J Sched 9:279–298

Bisaillon S, Cordeau JF, Laporte G, Pasin F (2011) A large neighbourhood search heuristic for the aircraft and passenger recovery problem. Quat J Oper Res 9:139–157

Clausen J, Larsen A, Larsen J, Rezanova NJ (2010) Disruption management in the airline industry—Concepts, models and methods. Comput Oper Res 37:809–821

Filar JA, Manyem P, White K (2000) How airlines and airports recover from schedule perturbations: a survey. Ann Oper Res 108:315–333

Jafari N, Zegordi SH (2010) The airline perturbation problem: considering disrupted passengers. Transp Plann Technol 33:203–220

Kohl N, Larsen A, Larsen J, Ross A, Tiourine S (2007) Airline disruption management—Perspectives, experiences and outlook. J Air Transp Manage 13:149–162

National Center of Excellence for Aviation Operations Research (NEXTOR) (2010) Total delay impact study—A comprehensive assessment of the costs and impacts of flight delay in the United States. National Center of Excellence for Aviation Operations Research. Berkeley, CA

Petersen JD, Solveling G, Clark JP, Johnson EL, Shebalov S (2012) An optimization approach to airline integrated recovery. Trans Sci Artivles in Advance:1–19

Rosenberger JM, Johnson EL, Nemhauser GL (2003) Rerouting aircraft for airline recovery. Transp Sci 37:408–421

Zhang Y, Hansen M (2008) Real-time intermodal substitution: strategy for airline recovery from schedule perturbation and for mitigation of airport congestion. Transp Res Rec 2052:90–99

Minimizing Setup Time from Mold-Lifting Crane in Mold Maintenance Schedule

C. S. Wong, F. T. S. Chan, S. H. Chung and B. Niu

Abstract The integration of production scheduling and maintenance planning has received much attention in the past decade. However, most of the studies only focused on the availability constraint of machines. Other critical resources such as injection molds are usually assumed to operate without breakdown. In fact, the frequency of mold breakdown is even higher than the machine breakdown. It is therefore necessary to consider the availability constraint of injection molds during scheduling. Extending the preliminary study of the production-maintenance scheduling model in (Wong, International J Prod Res 50:5683–5697, 2011), this study aims to solve a new mold maintenance scheduling problem with the consideration of the setup time of using a mold-lifting crane. A Joint Scheduling (JS) approach is proposed to minimize the weighted sum of the makespan and the setup time. The approach is implemented in genetic algorithm to solve five hypothetical problem sets. The results show that the JS approach outperforms the traditional approach.

Keywords Mold maintenance · Production scheduling · Setup time · Genetic algorithm

---

C. S. Wong (✉) · F. T. S. Chan · S. H. Chung
Department of Industrial and Systems Engineering, The Hong Kong Polytechnic University,
Hung Hum, Hong Kong
e-mail: mfsing@ymail.com

F. T. S. Chan
e-mail: f.chan@polyu.edu.hk

S. H. Chung
e-mail: nick.sh.chung@polyu.edu.hk

B. Niu
College of Management, Shenzhen University, Shenzhen 518060, China
e-mail: drniuben@gmail.com

Y.-K. Lin et al. (eds.), *Proceedings of the Institute of Industrial Engineers Asian Conference 2013*, DOI: 10.1007/978-981-4451-98-7_89,
© Springer Science+Business Media Singapore 2013

1 Introduction

Production-maintenance scheduling problem has received much attention both in academia and in industry (Wang 2013; Ebrahimipour et al. 2013; Ben et al. 2011). Fitouhi and Nourelfath (2012) dealt with the production-maintenance scheduling problems considering noncyclical preventive maintenance. Mokhtari et al. (2012) developed a mixed integer nonlinear programming model of the production-maintenance scheduling problems with multiple preventive maintenance services. They solved the problems by a population-based variable neighborhood search (PVNS) algorithm with the objective of minimizing the makespan and the system unavailability. Naderi et al. (2011) integrated periodic preventive maintenance planning and flexible flowshop scheduling. They solved the problem by genetic algorithm and artificial immune system with the objective of makespan minimization. Pan et al. (2010) proposed a single-machine production-maintenance scheduling with the objective of minimizing the maximum weighted tardiness.

In most of the studies, however, the availability of other critical resources such as injection molds is usually ignored. In fact, in many manufacturing firms, injection molds represent a significant share of capital investment (Menges et al. 2001) and it is usually unique for a particular product. If a required mold is not available during production, many operations in the plant will be interrupted. Thus, there is no doubt that keeping all molds in a good condition during production can smoothen the overall manufacturing process. A preliminary study considering mold availability in production scheduling was carried out in Wong et al. (2011). The study proposed to integrate mold maintenance plan with production schedule. A Joint Scheduling (JS) approach was introduced to jointly (but not sequentially) allocate production and maintenance activities in the schedule. With the JS approach, a maintenance activity can be divided into several small-sized activities that may reduce the idle time of machines. It is shown that the JS approach not only improved the makespan, but also ensured all molds in a good working condition. However, with the JS approach, the setup time will increase since a number of small-sized maintenance activities are generated. Mold maintenance requires removal of molds with mold-lifting crane. It is likely that the mold-lifting crane can be the bottleneck of a production system if the molds on the machines are frequently changed. Thus, the frequency of mold changing becomes a critical factor in mold maintenance scheduling. In this connection, this study considers the constraints of a mold-lifting crane in the production-maintenance scheduling problem. In the new problem, we manage to minimize the makespan of jobs and the setup time of using the mold-lifting crane. The weighted sum of both objective values is taken as the combined objective function in the problem. Hypothetical problem sets are generated and solved by genetic algorithm. The JS approach can obtain satisfied results under the tradeoff between the makespan and the setup time from the mold-lifting crane.

2 Problem Description

2.1 Notations

The notations used are summarized as follow:

Index	Descriptions
j	Index of jobs, $j = 1, ..., J$
k	Index of machines, $k = 1, ..., K$
m	Index of injection molds, $m = 1,..., M$
L_j	Processing time of job j
A	Maximum age of injection molds
T	Index of time slots, $t = 1, ...T$
S_j	Starting time of job j
C_j	Completion time of job j
C_{max}	The makespan of all jobs

Decision Variables

X_{jkmt}	= 1, if job j occupies time slot t on machine k with mold m
	= 0, otherwise
Y_{jk}	= 1, if job j is allocated on machine k
	= 0, otherwise
Z_{jm}	= 1, if mold maintenance is performed on mold m after the completion of job j
	= 0, otherwise

2.2 Scheduling Problem Modeling

Objective function:

$$Objective : MIN\{C_{max}\} \tag{1}$$

Equation (1) is the objective function of the problem that minimizing the makespan of all jobs.

Processing time constraint:

$$C_j - S_j = \sum_k Y_{jk}L_j \quad \forall jk \tag{2}$$

$$\sum_{kmt} X_{jkmt} = \sum_k Y_{jk}L_j \quad \forall jkmt \tag{3}$$

742   C. S. Wong et al.

Equation (2) ensures that once a job starts operation, it will be operated without interruption. Equation (3) indicates that the allocated time slot for a job equals to the job processing time.

Job constraint:

$$\sum_k Y_{jk} = 1 \quad \forall jk \tag{4}$$

Equation (4) allows that each job is allocated on one machine only.

Machine capacity constraint:

$$\sum_{jm} X_{jkmt} \leq 1 \quad \forall jkmt \tag{5}$$

Equation (5) defines that one machine can perform one job only at each time unit.

Mold capacity constraint:

$$\sum_{jk} X_{jkmt} \leq 1 \quad \forall jkmt \tag{6}$$

Equation (6) defines that each injection mold can perform one job only at a time unit.

In the production scheduling model, there are J jobs, K injection machines and M injection molds. Each job is required to perform with a specific mold m. Each job spends processing time L_j on the available injection machine.

2.3 Maintenance Modeling

The injection molds in the problem follow an age-dependent maintenance scheme. Mold age is the cumulated operation time of a mold since the performance of the previous mold maintenance. Under the Maximum Age (MA) approach, the maintenance task of a mold will be performed when the mold age reaches the predefined maximum age A. After the maintenance, the mold age will become zero. The next maintenance will be performed when age A is reached again. Under the Joint Scheduling (JS) approach, mold maintenance will be performed according to not only the maximum age but also the maintenance schedule determined by the optimization algorithm.

2.4 Mold-Lifting Crane Modeling

The functions of a mold-lifting crane are to install and remove the injection mold. When a job is required to be operated with Mold m, Mold m must be installed on the machine by the mold-lifting crane before the job starts. If another mold is already installed on the machine, the mold must be removed before installing Mold m. Furthermore, if Mold m is required to perform maintenance, it must be removed from the machine before the maintenance task starts.

3 Genetic Algorithm

In this study, a genetic algorithm is applied for implementing the Joint Scheduling (JS) approach and solving the hypothetical problem sets. It is a well-known fact that genetic algorithms are widely for solving the scheduling problems.

Each gene consists of three digits, representing machine numbers, job numbers, maintenance decision making and domination. For example, there is a gene with the structure of 2–3–1–1. It means that Job 3 is allocated on Machine 2. The third digit is a binary variable. After performing Job 3, mold maintenance will be performed since the digit for maintenance decision making is 1. The last digit for domination is also a binary variable for recording good structure of a chromosome. If the digit is equal to 1, the gene will be recorded and brought to the next generation. This idea was first introduced by Chan et al. (2006).

At the beginning, a pool of chromosomes is generated randomly and evaluated according to a fitness function. The fitness function adopted is one minus the objective value (the weighted sum of the makespan and the setup time) and divided by the sum of the objective values of all chromosomes in the same generation. After that, a roulette wheel selection procedure will be performed to select the fitter chromosomes and form a new chromosome pool. An evolution procedure will then be performed to generate the next generation of the chromosomes. Crossover and mutation operations are included in the evolution procedure. In the crossover operation, the recorded genes in each pair of the chromosomes will be exchanged. In the mutation operation, some genes of the chromosomes will be changed with other parameters randomly. Once the new generation pool is created, the fitness value evaluation and selection procedure will be performed again. The whole evolution procedure will be terminated when the stopping condition is reached.

4 Computational Results

To testify the performance of the Joint Scheduling (JS) approach, five hypothetical problem sets are generated. In the problem sets, there are thirty jobs, three injection machines and five injection molds. The processing time of each job is around 4–25 h. The maximum age of each injection mold is 48 hand the maintenance duration is 10 h. Under the JS approach, mold maintenance can be performed before the maximum age. In that case, the maintenance duration will be

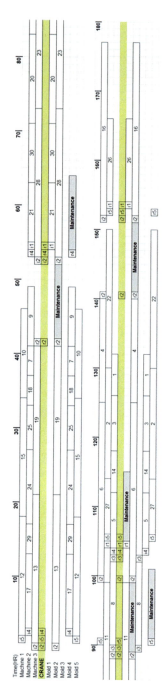

Fig. 1 The schedule from the MA approach optimizing the makespan and the setup time

Minimizing Setup Time from Mold-Lifting Crane 745

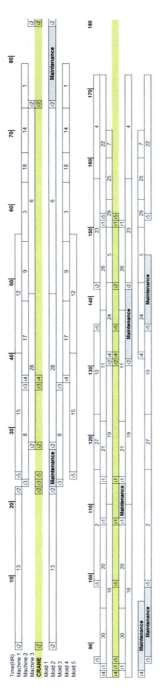

Fig. 2 The schedule from the JS approach optimizing the makespan and the setup time

Table 1 Comparison of the approaches

Problem		MA approach	JS approach	Improvement
Set 1	Function value	74.5	73	2 %
	Makespan	128	125	
	Setup time	21	21	
Set 2	Function value	80	79.5	1 %
	Makespan	135	134	
	Setup time	25	25	
Set 3	Function value	98	96.5	2 %
	Makespan	177	172	
	Setup time	19	21	
Set 4	Function value	101	97	4 %
	Makespan	181	175	
	Setup time	21	19	
Set 5	Function value	107.5	106	1 %
	Makespan	192	189	
	Setup time	23	23	

calculated linearly. The experiment is performed in Visual Basic for Applications (VBA) environment on a personal computer with Intel Core 2 Duo 2.13 GHz CPU.

Figures 1 and 2 are the optimized schedule of the MA and the JS approaches in Problem Set 3. In the figures, the letter "i" means mold installation; the letter "r" means mold removal. For example, "i3" indicates the installation of Mold 2 and "r4" indicates the removal of Mold 4. Table 1 shows the performance of the MA and the JS approaches. Comparing with the MA approach, the JS approach can obtain better function values with 1–4 % improvement. Since the JS approach allows more flexibility in mold maintenance scheduling, the genetic algorithm can decide a more favourable time for mold maintenance.

5 Conclusions

This paper aims to solve a new production-maintenance problem with the consideration of the setup time of a mold-lifting crane. The new problem is identified and demonstrated with the numerical examples. The results show that the proposed Joint Scheduling (JS) approach can achieve better function values than the traditional Maximum Age (MA) approach.

Acknowledgments The work described in this paper was substantially supported by a grant from the Research Grants Council of the Hong Kong Special Administrative Region, China (Project No. PolyU 510410); a research grant from the Hong Kong Polytechnic University (Project No. ZWOW (TCS 147)), and also a grant from The Hong Kong Scholars Program Mainland–Hong Kong Joint Postdoctoral Fellows Program (Project No.: G-YZ24).

References

Ben MA, Sassi M, Gossa M, Harrath Y (2011) Simultaneous scheduling of production and maintenance tasks in the job shop. Int J Prod Res 49(13):3891–3918

Chan FTS, Chung SH, Chan LY, Finke G, Tiwari MK (2006) Solving distributed FMS scheduling problems subject to maintenance: genetic algorithms approach. Robot Comput Integr Manuf 22(5–6):493–504

Ebrahimipour V, Najjarbashi A, Sheikhalishahi M (2013) Multi-objective modeling for preventive maintenance scheduling in a multiple production line. J Intell Manuf. doi:10.1007/s10845-013-0766-6

Fitouhi MC, Nourelfath M (2012) Integrating noncyclical preventive maintenance scheduling and production planning for a single machine. Int J Prod Econ 136(2):344–351

Menges G, Michaeli W, Mohren P (2001) How to make injection molds, 3rd edn. Hanser Publishers, Munich

Mokhtari H, Mozdgir A, Abadi INK (2012) A reliability/availability approach to joint production and maintenance scheduling with multiple preventive maintenance services. Int J Prod Res 50(20):5906–5925

Naderi B, Zandieh M, Aminnayeri M (2011) Incorporating periodic preventive maintenance into flexible flowshop scheduling problems. Appl Soft Comput 11(2):2094–2101

Pan ES, Liao WZ, Xi LF (2010) Single-machine-based production scheduling and integrated preventive maintenance planning. Int J Adv Manuf Technol 50(1–4):365–375

Wang S (2013) Bi-objective optimization for integrated scheduling of single machine with setup times and preventive maintenance planning. Int J Prod Res. doi:10.1080/00207543.2013.765070

Wong CS, Chan FTS, Chung SH (2011) A genetic algorithm approach for production scheduling with mold maintenance consideration. Int J Prod Res 50(20):5683–5697

Differential Evolution Algorithm for Generalized Multi-Depot Vehicle Routing Problem with Pickup and Delivery Requests

Siwaporn Kunnapapdeelert and Voratas Kachitvichyanukul

Abstract This paper presents a Differential Evolution (DE) algorithm for solving generalized multi-depot vehicle routing problem with pickup and delivery requests (GVRP-MDPDR). The GVRP-MDPDR does not require the restricted assumptions of CVRP, VRPSPD, etc. and it contains nearly all characteristics of real world vehicle routing problems. The solution is represented as a multidimensional vector where each dimension is filled with random number and a population of vectors is evolved via the mechanism of differential evolution. A decoding scheme (SD1) is applied to decode the vector into priority of requests and construct the routes of vehicles under the restricted constraints. Five groups of test problem instances, A, B, C, D, and E, with differences geographical data and number of requests are used to evaluate the performance of the algorithm. Each group of instance composes of three different location scenarios of requests: clustered (c), randomly distributed (r), and half-random-half-clustered (rc). The computational results demonstrated that DE algorithm is very competitive when compared to the results obtained by using Particle Swarm Optimization (PSO).

Keywords Metaheuristic · Differential evolution · Pickup and delivery · Vehicle routing problem

---

S. Kunnapapdeelert (✉) · V. Kachitvichyanukul
Industrial and Manufacturing Engineering, School of Engineering and Technology,
Asian Institute of Technology, P.O. Box 4, Klong Luang, Pathumtani 12120, Thailand
e-mail: siwaporn.kunnapapdeelert@ait.ac.th

V. Kachitvichyanukul
e-mail: voratas@ait.ac.th

Y.-K. Lin et al. (eds.), *Proceedings of the Institute of Industrial
Engineers Asian Conference 2013*, DOI: 10.1007/978-981-4451-98-7_90,
© Springer Science+Business Media Singapore 2013

1 Introduction

The generalized multi-depot vehicle routing problem with pickup and delivery requests (GVRP-MDPDR) is the problem for designing the optimal vehicle route under pairing, precedence, vehicle capacity, and time windows constraints. It considered nearly all characteristics of practical vehicle routing problems. It is classified as NP-hard problem which consume high computational time to find the near optimal solutions for large size of problem when solves by exact algorithm. Consequently, various metaheuristics methods such as ant colony algorithm (ACO) (Carabetti et al. 2010; Nanry and Barnes 2000), tabu search (TA) (Cordeau et al. 2001; Wang 2008), genetic algorithm (GA) (Caramia and Onori 2008; Jung and Haghani 2000), and particle swarm optimization (PSO) (Ai and Kachitvichyanukul 2009a, b, c; Geetha et al. 2013; Sombuntham and Kachitvichyanukul 2010) have been developed to find solution in a reasonable time.

Differential evolution (DE) is firstly presented by Storn and Price (1997). The DE algorithm composes of four main process i.e. initialization of vector population, mutation, crossover, and selection. The algorithm has been successfully applied in various fields such as industrial engineering, electrical engineering, and communication. An algorithm based on differential evolution is proposed in this paper.

2 Generalize Multi-Depot Vehicle Routing Problem with Pickup and Delivery Requests

The GVRP-MDPDR can be considered as one of the most practical real world problems in VRP. Some characteristics of real-world practices for VRP can be described as follows.

- *Precedence constraint* is the condition that defines precedence relation of pickup and delivery. Once the request is assigned to a vehicle, location to pickup must be reached before the delivery location.
- *Vehicle constraint* is the limit that a load on a vehicle must not exceed capacity of vehicle.
- *Time window constraint* is the time period for the arrived vehicles to be served in each node. If any vehicle reaches the node too early, it must wait. In contrast, if the vehicle visit the node later than due time, the service will not be allowed for the vehicle.
- *Many requests at locations* refers to each location can have more than one pickup or delivery requests which may have to be delivered to different locations.

The frequently used objective of this problem is to minimize total routing cost which is corresponding to the traveling distance. Input parameters and variable used in the model are listed below follows by mathematical formulation for the problem based on Sombuntham (2010).

Differential Evolution Algorithm for Generalized Multi-Depot Vehicle

751

2.1 Input Parameters

P	set of pickup nodes, $\{1, \ldots, n\}$,
D	set of delivery nodes, $\{n + 1, \ldots, 2n\}$,
N	set of all pickup and delivery nodes, $N = P \cup D$,
H	penalty cost when the request i is not served, $i \in P$,
K	set of all vehicles, $|K| = m$,
C_k	vehicle capacity, k, $k \in K$,
f_k	fix cost of vehicle, k, $k \in K$, if it is used,
g_k	variable cost per a distance unit of vehicle, k, $k \in K$,
τ_k	start node of vehicle, k, $k \in K$,
τ'_k	end node of vehicle, k, $k \in K$,
V	set of all nodes, $V = N \cup \{\tau_1, \ldots, \tau_m\} \cup \{\tau'_1, \ldots, \tau'_m\}$,
A	set of (i, j) which represent an arc from node i to node j, where $j, i \in V$,
$d_{i,j}$	nonnegative distance from node i and node j, $i, j \in N$,
$t_{i,j}$	nonnegative traveling time from node i and node j, $i, j \in N$, travel times satisfy triangle inequality and $t_{i,j} \leq t_{i,l} + t_{l,j}$ for all $j, i, l \in V$,
s_i	fixed service time when visiting node i,
e_i	variable service time per item units of node i,
$[a_i, b_i]$	time windows for node i; a visit to node i can only happen between this time interval,
l_i	quality of goods to be loaded to the vehicle at node i when $i \in P$ and $l_i = -l_{i-n}$ for $i \in D$,
T_k	maximum route time of vehicle k, $k \in K$

In this work, request i is represented by node i and $i + n$, where $i \in P$ and $i + n \in P$. In addition, any node may have same x—y coordinate as the same location could have multiple request.

2.2 Decision variables

The key decision variables for the problem are

x_{ijk} $x_{ijk} = \begin{cases} 1 & \text{if edge between node } i \text{ and } j \text{ is used by vehicle } k, \\ 0 & \text{otherwise} \end{cases}$

S_{ik} nonnegative integer for indicating when vehicle k starts servicing at location i, $i \in V$, $k \in K$,

L_{ik} nonnegative integer that in an upper bound on amount of goods on vehicle k after servicing node i where $i, i \in V$, $k \in K$

z_i $z_i = \begin{cases} 1 & \text{if the request is placerd in te request bank} \\ 0 & \text{otherwise} \end{cases}, i \in P$

A. *Mathematical model*

$$\text{Minimize } \alpha \sum_{k \in K} g_k \sum_{(i,j) \in A} d_{ij} x_{ijk} + \beta \sum_{k \in K} \sum_{j \in P} f_k x_{\tau_k,j,k} + \gamma \sum_{i \in P} H_i z_i = 1 \qquad (1)$$

$$\text{Subject to : } \sum_{k \in K} \sum_{j \in N_k} x_{ijk} + z_i = 1 \qquad \forall i \in P \qquad (2)$$

$$\sum_{j \in V} x_{ijk} - \sum_{j \in V} x_{j,n+i,k} = 0 \qquad \forall k \in K, \forall i \in P \qquad (3)$$

$$\sum_{j \in P \cup \{\tau_k'\}} x_{\tau_k,j,k} = 1 \qquad \forall k \in K \qquad (4)$$

$$\sum_{i \in D \cup \{\tau_k\}} x_{i,\tau_k',k} = 1 \qquad \forall k \in K \qquad (5)$$

$$\sum_{i \in V} x_{ijk} - \sum_{i \in V} x_{j,i,k} = 0 \qquad \forall k \in K, \forall j \in N \qquad (6)$$

$$x_{ijk} = 1 \Rightarrow S_{ik} + s_i + t_{ij} \le S_{jk} \qquad \forall k \in K, \forall (i,j) \in A \qquad (7)$$

$$a_i \le S_{ik} \le b_i \qquad \forall k \in K, \forall i \in V \qquad (8)$$

$$S_{ik} \le S_{n+i,k} \qquad \forall k \in K, \forall i \in P \qquad (9)$$

$$x_{ijk} = 1 \Rightarrow L_{ik} + l_i \le L_{jk} \qquad \forall k \in K, \forall (i,j) \in A \qquad (10)$$

$$L_{ik} \le C_k \qquad \forall k \in K, \forall i \in V \qquad (11)$$

$$L_{\tau_k k} = L_{\tau_k' k} = 0 \qquad \forall k \in K \qquad (12)$$

$$x_{ijk} \in \{0,1\} \qquad \forall k \in K, \forall (i,j) \in A \qquad (13)$$

$$z_i \in \{0,1\} \qquad \forall i \in P \qquad (14)$$

$$S_{ik} \ge 0 \qquad \forall k \in K, \forall i \in V \qquad (15)$$

$$L_{ik} \ge 0 \qquad \forall k \in K, \forall i \in V \qquad (16)$$

$$S_{\tau_k' k} \le T_k \quad \forall k \in K \qquad (17)$$

The objective function is to minimize total distance traveled under various constraints. The first and the second constraints in Eqs. (2) and (3) ensure that both pickup and delivery orders are done with the same vehicle. Equations (4) and (5)

Differential Evolution Algorithm for Generalized Multi-Depot Vehicle 753

make sure that vehicle departs from its start terminal and arrives at its end terminal. Equation (6) guarantees that the consecutive paths between τ_k and τ'_k are constructed for each vehicle $k \in K$. Equations (7) and (8) confirmed that S_{ik} is set accurately along the paths within the given time windows. These constraints also assure that sub tours will not be created. Equation (9) is used to enforce the condition that each pickup must take place before the corresponding delivery. Equations (10)–(12) guarantee that the load variable is precisely set along the path and vehicle capacity constraints are used. Lastly, Eq. (17) limits a maximum route time of each vehicle allowed.

3 Differential Evolution

Differential evolution (DE) algorithm is first proposed by Storn and Price (1997). It has been successfully applied in various scientific and engineering fields (Das and Suganthan 2011). It is a population-based random search that utilizes a population of size N of D-dimensional vectors. Each candidate or target vector in a population is perturbed to form a new vector called a trial vector by mutation and crossover operations as explained below.

For a target vector, $X_{i,g}$, the mutation process is used to create a mutant vector, $V_{i,g}$. In DE, a mutant vector, $V_{i,g}$, at generation g is generated by combining three randomly selected vector from the current population which are mutually exclusive and different from its corresponding mutant vector. The mutant vector can be generated based on $V_{i,g} = X_{r1,g} + F * (X_{r2,g} - X_{r3,g})$ where X_{r1}, X_{r2}, and X_{r3} are the three randomly selected vectors. The scale vector, F, is used for scaling the differential variation $(X_{r2,g} - X_{r3,g})$ and it is a parameter in DE.

The mutant vector, $V_{i,g}$ is then crossover with the target vector, $X_{i,g}$ to create the trial vector, $Z_{i,g}$ to enhance the diversity of the perturbed parameter vectors. In DE algorithm, the two commonly used crossover schemes are the binomial crossover (bin) and the exponential crossover (exp). The crossover probability, C_r, must be specify for controlling the portion of parameter values copied from the mutant vector in each dimension. The smaller C_r value leads to the trial vector, $Z_{i,g}$, that is more similar to the target vector, $X_{i,g}$. In contrast, the higher C_r value leads to the trial vector, $Z_{i,g}$, that is more similar to the mutant vector, $V_{i,g}$.

The selection process is made by selecting the superior vector between the trial vector, $Z_{i,g}$ and the target vector, $X_{i,g}$. If the fitness of the trial vector, $f(Z_{i,g})$ is better than or equal to that of the target vector, $f(X_{i,g})$, the trial vector, $(Z_{i,g})$ will replace the target vector and become part of the next generation $(X_{i,g+1})$. Otherwise, the target vector, $X_{i,g}$ still remains in the population for the next generation.

3.1 Differential Evolution for Solving GVRP-MDPDR

To apply DE algorithm to solve GVRP-MDPDR, the relationship between vector and vehicle route must be defined. The solution representation used in this research is (SD1) from Ai and Kachitvichyanukul (2009a, b, c). Vector in this solution representation is composed of $n + 2\,m$ elements which are divided into two main parts. The first n dimensions of each vector represent priority of the locations. The last $2\,m$ dimensions hold x, y coordinate of the orientation points of the m vehicles. Each vector is decoded into the vehicle route in each iteration. The decoding procedure is divided into three main steps. First, a location priority list is created. The vehicle priority matrix is created next. The last step is to construct the vehicle route. This step can be done by assigning requests based on the outcomes from the first two steps, location priority and vehicle priority.

After the vehicle route is generated based on both customer priority list and vehicle priority matrix, the lowest cost insertion heuristics is applied to find the best insertion position of each customer. However, some unfulfilled requests might still be unassigned in some cases. Neighborhood moves concept from Nanry and Barnes (2000) are then applied for inserting the unfulfilled requests into the vehicle routes. Finally, the attempt for reducing number of vehicle from Sombuntham and Kachitvichyanukul (2010) is also applied to improve the solutions.

4 Experimental Results

Five groups of benchmark instances, A, B, C, D, and E, from Sombuntham and Kunnapapdeelert (2012) are used to test the DE algorithm for solving GVRP-MDPDR. Three different locations scenarios of the request are available in all instances i.e. clustered location (c), random (r), and half-random-half-cluster (rc). Each dataset consists of Cartesian coordinate for representing locations, fix and variable service time, ready and due time. Vehicle information includes vehicle capacity, time limit, fix and variable costs, start and end stations of vehicle. Information of the request such as, quantity, pickup and delivery locations are provided. Additionally, Euclidean distances among locations are also given.

The algorithm is solved for 10 times for each problem instance. Parameters setting of DE algorithm for solving this problem are set as follows. Population size N is set at 200 and number of iteration is 500. Crossover rate, C_r, is linearly increasing from 0 to 1. The scaling factor, F, is randomly change in the range [0, 2] in every iteration. The experimental results of the DE approach for solving GVRP-MDPDR when compared to those of PSO are depicted in Table 1.

The obtained results in Table 1 show that the proposed algorithm outperforms the PSO in terms of solution quality when apply to solve GVRP-MDPDR in most cases. However, it cannot reach the best known solution from PSO in some cases.

Differential Evolution Algorithm for Generalized Multi-Depot Vehicle 755

Table 1 Computational results for solving GVRP-MDPDR via DE and PSO algorithm

Instance	PSO	DE	% deviation
Aac1	864.22	887.59	2.70
Aac2	782.54	903.36	15.44
Aar1	2004.06	2135.32	6.55
Aar2	2373.18	2524.57	6.38
Aarc1	1436.74	1441.76	0.35
Aarc2	1679.34	1741.87	3.72
Bac1	1805.82	1674.92	−7.25
Bac2	1897.47	1833.40	−3.38
Bar1	4196.45	3965.09	−5.51
Bar2	7848.43	7330.13	−6.60
Barc1	2838.10	2646.12	−6.76
Barc2	2838.10	2680.15	−5.57
Cac1	3379.24	3346.96	−0.96
Cac2	4297.98	4296.36	−0.04
Car1	6153.78	5907.51	−4.00
Car2	7643.92	7218.50	−5.57
Carc1	4406.78	4400.51	−0.14
Carc2	4320.85	4229.50	−2.11
Dc1	2715.44	2933.20	8.02
Dc2	2670.12	2777.18	4.01
Dr1	3424.96	2931.97	−14.39
Dr2	3484.56	3057.33	−12.26
Drc1	4388.53	3642.21	−17.01
Drc2	4865.68	4143.95	−14.83
Ec1	3100.78	4220.76	36.12
Ec2	3165.58	4165.22	31.58
Er1	3404.02	3400.22	−0.11
Er2	3260.52	2961.68	−9.17
Erc1	4157.00	4072.59	−2.03
Erc2	6882.52	6184.88	−10.14

This indicates that the search process might require improvement or that the solution representation used in this work cannot reach certain area of the search space.

5 Conclusions

The DE algorithm and solution representation for solving generalized multi-depot vehicle routing problem with pickup and delivery requests (GVRP-MDPDR) are presented in this paper. The results depict that the proposed DE algorithm can provide good quality of solution to GVRP-MDPDR problem. In comparison, the results from the DE algorithm are better than those obtained by PSO for most of the problem instances. The proposed algorithm performs worse than PSO only for dataset A and clustered distributed location instances of some dataset.

References

Ai TJ, Kachitvichyanukul V (2009a) Particle swarm optimization for the vehicle routing problem with time windows. Int J Oper Res 9:519–537

Ai TJ, Kachitvichyanukul V (2009b) Particle swarm optimization and two solution representations for solving the capacitated vehicle routing problem. Comput Ind Eng 56:380–387

Ai TJ, Kachitvichyanukul V (2009c) Particle swarm optimization for the vehicle routing problem with simultaneous pickup and delivery. Comput Oper Res 36:1693–1702

Carabetti EG, de Souza SR, Fraga MCP, Gama PHA (2010) An application of the ant colony system metaheuristic to the vehicle routing problem with pickup and delivery and time windows. Paper presented at 2010 Eleventh Brazilian Symposium on Neural Networks, Centro Fed. de Educ. Technol. De Gerais, Belo Horizonte, Brazil, pp 23–28 Oct 2010

Caramia M, Onori R (2008) Experimenting crossover operators to solve the vehicle routing problem with time windows by genetic algorithm. Int J Oper Res 3:497–514

Cordeau JF, Laporte G, Mercier A (2001) A unified tabu search heuristic for vehicle problem with time windows. J Oper Res Soc 52:928–936

Das S, Suganthan PN (2011) Differential evolution: a survey of the state-of-the-art. IEEE Trans Evol Comput 15:4–31

Geetha S, Poonthalir G, Vanathi PT (2013) Nested particle swarm optimisation for multi-depot vehicle routing problem. Int J Oper Res 16:329–348

Jung S, Haghani A (2000) A genetic algorithm for pickup and delivery problem with time windows. In: Transportation Research Record 1733, Transportation Research Broad 1–7

Nanry WP, Barnes JW (2000) Solving the pickup and delivery problem with time windows using reactive tabu search. Transp Res Part B 34:107–121

Sombuntham P (2010) PSO algorithms for generalized multi-depot vehicle routing problems with pickup and delivery requests. Thesis, Asian Institute of Technology

Sombuntham P, Kachitvichyanukul V (2010) A particle swarm optimization for multi-depot vehicle routing problem with pickup and delivery requests. Paper presented at proceedings of the international multiconference of engineers and computer scientists, Hong Kong (China), pp 17–19

Sombuntham P, Kunnapapdeelert S (2012) Benchmark problem instances for generalized multi-depot vehicle routing problems with pickup and delivery requests. Paper presented at proceedings of the Asia Pacific industrial engineering and management systems conference, Patong Beach, Phuket, Thailand, pp 2–5

Storn R, Price K (1997) Differential evolution—a simple and efficient heuristic for global optimization over continuous spaces. J Global Optim 11:341–359

Wang Y (2008) Study on the model and tabu search algorithm for delivery and pickup vehicle routing problem with times windows. Paper presented at 2008 IEEE international conference on service operations and logistics, and informatics, Beijing, China, pp 12–15

A Robust Policy for the Integrated Single-Vendor Single-Buyer Inventory System in a Supply Chain

Jia-Shian Hu, Pei-Fang Tsai and Ming-Feng Yang

Abstract To find the best production quantity for the vendor and the order quantity for the buyer, the integrated single-vendor single-buyer inventory model is proposed with the aim to minimize the total costs of the entire supply chain. In the Traditional supply chain, the most of integrated inventory is to develop a model that assumes that the input data is deterministic and equal to some nominal values. However, few researches have considered data uncertainty such as in demands, lead times, or even setup/ordering costs. Instead of solving for the optimal solution under the assumption of deterministic demands, here we provide a prescriptive methodology for constructing uncertainty sets within a robust optimization framework for integrated inventory problems with uncertain data. We accomplish this by taking as primitive the decision maker's attitude toward risk. A numerical study and sensitive analysis are conducted to examine the integrated inventory model.

Keywords Robust optimization · Integrated inventory · Supply chain

1 Introduction

Collaboration between buyer and vendor is getting more critical in a supply chain environment. In increasingly competitive and globalized world markets, companies are constantly under pressure to find ways to cut inventory and production

J.-S. Hu · P.-F. Tsai (✉)
Department of Industrial Engineering and Management, National Taipei University of Technology, Taipei, Taiwan, Republic of China
e-mail: ptsai@ntut.edu.tw

M.-F. Yang
Department of Transportation Science, National Taiwan Ocean University, Keelung, Taiwan, Republic of China
e-mail: yang60429@pchome.com.tw

Y.-K. Lin et al. (eds.), *Proceedings of the Institute of Industrial Engineers Asian Conference 2013*, DOI: 10.1007/978-981-4451-98-7_91,
© Springer Science+Business Media Singapore 2013

costs since the annual inventory cost accounts for 10 % of the annual profit on average. To manage the inventory more effectively implies to reduce cost and increase competitive for the company.

In the late 19th century, companies were under the challenges of cost reduction while satisfying customers requires in the timely manner. It led to the success in implementation of just-in-time (JIT) production system, originated from Japan. The objective of JIT manufacturing is to eliminate all wastes by having almost no inventories between operations. Pan and Liao (1989) incorporated JIT concept to the traditional EOQ model. They demonstrated the effect of frequent shipments for a small lot size and total cost. However, small lot-size delivery costs such as shipping, receiving, and inspection costs were ignored on their model. Ramasesh (1990) separated the total ordering cost into the cost of placing a contact order and the cost associated with multiple shipments in small lots.

Instead of focusing on individual inventory control policy, an integrated inventory approach has been proposed to determine optimal order quantity and shipment policy for buyers and vendors simultaneously. In traditional inventory management systems, vendor or buyer adopted an optimal economic order quantity (EOQ) with the consideration of self-interest only. However, this optimal solution approach might not be the best inventory policy for the entire supply chain. Banerjee and Kim(1995) assumed that the optimal inventory policies were determined jointly under the setting of JIT purchasing and JIT manufacturing. Ha and Kim (1997) considered the integration between single-vendor and single-buyer in a JIT environment. Their results showed that this integrated approach might further reduce total costs for both the vendor and the buyer.

Goyal(1977) was first advocated the concept of integrated inventory model for a single supplier–single customer problem. The optimal ordering and production frequencies were decided with the assumption that the ratio between supplier's production cycle time and the customer's order time interval was fixed. This model was further extended by Banerjee (1986), assuming that a vendor had a finite production rate with a lot-for-lot policy. Goyal (1988) relaxed the lot-for-lot policy to achieve lower joint total relevant costs. Goyal and Gupta (1989) showed that the loss of one partner from the buyer-vendor coordination can be compensated by the benefit of other partners. It was suggested that the net benefit should be shared by both parties in some equitable fashion.

Lu (1995) developed a heuristic approach for the one-vendor multi-buyer case, and proved assumption of completing a batch before a shipment is started and explored a model that allowed shipments to take place during production cycle and the delivery quantity to the buyer is known for one-vendor one-buyer problems. Pan and Yang (2002) presented an integrated supplier-purchaser model concentrated on the profit from lead time reduction. Siajadi et al. (2006) presented a single-vendor multiple-buyer integrated inventory model to minimize joint total relevant cost for both vendor and buyer with a multiple shipment policy by determining optimal ordering/production cycle time, and a significant savings in joint total relevant cost is achieved when the total demand rate is close to the production rate.

Uncertainty in the process of globalization is more complex than before. Thus, developing the inventory model of the uncertainty model is inevitable. Most literature on inventory models assumed certainty in demand rates. However, they are usually difficult for the managers to decide the parameters accurately as crisp values in reality. Only limited research discussed the variations in parameters. Here we propose a robust optimization framework for an integrated inventory problem with uncertain demands.

2 A Robust Model

Here a robust optimization framework for an integrated inventory problem was proposed with uncertain demands. The optimization problem with inexact data was investigated by Soyster (1973). A decomposition procedure was then proposed by solving a problem which maximizes the minimum of a finite set of functions over a common domain. The concept of robust optimization was proposed by Bertsimas and Sim (2004), in which the parameters could take values different than the nominal ones. They found that the optimal solution using the nominal data might no longer be optimal or even feasible as several constraints may be violated after the problem was realized.

According to the tradition inventory model, the demand is given and ordering quantity is a decision variable. The total cost is minimized by economic ordering quantity in the model. Now we consider that the demand is happened with uncertainty in the inventory model. In this paper, we assume that the demand is uncertainty, but know that the number of demands will happened in the future. Thus, the different demands produce the total cost functions respectively in the future scenarios. However, by analyzing the demands in the future, we find the worst-case scenario policy in which the solution is the best one should the worst situation is realized. This robust method is as following:

Step 1. To find the minimum total cost, the economic ordering quantity by the possible demand must be calculated. The several ordering quantities are provided the total cost function and find the minimum total cost in the each demand of future scenarios.

Step 2. The ordering quantities which come from possible demand of future scenarios respectively calculate the total cost by the several total cost functions. Therefore, we know what happen the ordering quantity which is the ordering quantity in the specific demand of scenarios apply in the different total cost function, and the order quality may be not the economic ordering quantity in applying total cost function.

Step 3. Following the Step 2, we know the many total costs which are calculated by the economic ordering quantity which is in the specific demand of scenarios. Through the maximin strategy with the total costs find the conservative policy.

2.1 Notations and Assumptions

The notations used in this integrated inventory model are as follows:

D Average demand rate per year for buyer
P Production rate per year
A Ordering cost per order for buyer
S Setup cost per lot
Q Total order/production quantity per cycle time
C_v Unit production cost paid by the vendor
C_b Unit purchase cost paid by the purchaser
r Annual inventory holding cost per dollar invested in stocks

The notations used in this integrated inventory model are as follows:

1. The demand rate is constant and deterministic after realization.
2. The order quantity need not be an integral number of units, and there are no minimum or maximum restrictions on its size.
3. The unit variable cost does not depend on the replenishment quantity; In particular, there are no dis counts in either the unit purchase cost or the unit transportation cost.
4. The cost factors do not change appreciably with time; in particular, inflation is at a low level.
5. The item is treated entirely independently of other items; that is, benefits from joint review or replenishment do not exist or are simply ignored.
6. The replenishment lead time is of zero duration.
7. No shortages are allowed.
8. The entire order quantity is delivered at the same time.
9. The planning horizon is very long. In other words, we assume that all parameters will continue at the same value for a long time.

2.2 Formulations

Based on the above notations and assumptions, the buyer's total relevant cost is the sum of ordering costs and holding costs, which can be calculated as in Eq. (1).

$$TC_p = A\frac{D}{Q} + \frac{Q}{2}C_v r \tag{1}$$

In order to find the minimum cost for this problem, we should take the first partial derivatives of TC_p with Q and the derivatives to zero. The optimal order quantity (or economical ordering quantity) can be evaluated as:

A Robust Policy for the Integrated Single-Vendor Single-Buyer Inventory System 761

$$EOQ = \sqrt{\frac{2AD}{c_p r}} \tag{2}$$

We then consider the vendor's model before introducing the integrated inventory model. The vendor's total relevant cost the sum of set-up costs and holding costs, which can be calculated as in Eq. (3).

$$TC_v = S\frac{D}{Q} + \frac{Q}{2}C_v r\left(1 - \frac{D}{P}\right) \tag{3}$$

As the cost functions for the vendor and the buyers shown above, the joint total cost for the integrated inventory problem is the sum of TC_v and TC_p, as shown in Eq. (4).

$$TC = \frac{D}{Q}(A + S) + \frac{Q}{2}\left[C_v r\left(1 - \frac{D}{P}\right) + C_p r\right] \tag{4}$$

One convenient way to find the minimum is to use the necessary condition that the tangent or slope of the curve is zero at the minimum. Then, we have the optimal order quantity as in Eq. (5).

$$Q = \sqrt{\frac{2D(A + S)}{C_v r\left(1 - \frac{D}{P} + C_p r\right)}} \tag{5}$$

3 Numerical Examples

3.1 EOQ Model

To illustrate the results of the proposed method, consider an EOQ inventory system with data: We consider the possible future demand about D = 500, 1,000 or 1,500 unit/year, A = \$100/order, Cb = \$20/unit, r = 0.2. The optimal ordering quantity (EOQ*) and optimal total cost (TC*) for all possible demands are as shown in Table 1.

We used these three EOQs as the ordering quantity under different demands and calculate the resulting total cost, as shown in Table 2. By the maxi-min principle, we can find the most conservative solution in Table 1, top right point. The solution of EOQ model help people to resist the uncertainty future, letting the user not fall to the worse place. Thus, we chose the ordering quantity 193.65, and the policy is conservative to face future.

To analyze the deviation of total cost to the optimal one, we calculate the absolute difference as in Eq. (6). It means that the costs more close the minimum value when the absolute difference being more small.

762   J.-S. Hu et al.

Table 1 Summary of computation results for EOQ model

D	EOQ*	TC*
500	111.80	223.61
1,000	158.11	316.23
1,500	193.65	387.30

Table 2 Robust method for EOQ model

TC function for demand	Ordering quantity			Max	Min
	111.80	158.11	193.65		
500	223.61	237.17	258.20	258.20	258.20
1,000	335.41	316.23	322.75	335.41	
1,500	447.21	395.28	387.30	447.21	

$$\frac{TC_{(EOQ)} - TC_{(Q)}}{TC_{(EOQ)}} \tag{6}$$

The absolute difference of total costs is showing in Table 3. If we use the robust ordering quantity 431.13 then the average absolute difference 0.058 may not be the best but absolutely not be the worst.

3.2 Integrated Inventory Model

To illustrate the results of the proposed method, consider an integrated inventory system with data: We consider the possible future demand about D = 500, 1,000 or 1,500 unit/year, A = $100/order, Cp = $20/unit, r = 0.2, Cv = $15/unit, P = 32,000 unit/year, S = 400/set-up. The total cost (TC*) including the buyer and vendor cost is calculated from the optimal quality (Q*) in the all possible demands are listed in Table 4.

Table 5, the robust number is showing by the robust method. Thus, the ordering quantity 431.13 is our conservative policy. Then, we calculate the absolute

Table 3 Total Cost of Absolute Difference with TC* in respective total cost function of possible demand for EOQ

TC function for demand	Ordering quantity		
	247.23	350.82	431.13
500	0.000	0.061	0.155
1,000	0.061	0.000	0.021
1,500	0.155	0.021	0.000
Average	0.072	0.027	0.058

Table 4 Summary of computation results for integrated inventory

D	Q*	TC*
500	247.23	1,719.03
1,000	350.82	2,422.87
1,500	431.13	2,957.31

Table 5 Robust method for integrated inventory model

TC function for demand	Ordering quantity			Max	Min
	247.23	350.82	431.13		
500	1,719.03	1,825.38	1,991.75	1,991.75	1,991.75
1,000	2,572.75	2,422.87	2,474.53	2,572.75	
1,500	3,426.48	3,020.37	2,957.31	3,426.48	

Table 6 Total cost of absolute difference with TC* in respective total cost function of possible demand for integrated inventory model

TC function for demand	Ordering quantity		
	247.23	350.82	431.13
500	0.000	0.062	0.159
1,000	0.062	0.000	0.021
1,500	0.159	0.021	0.000
Average	0.074	0.028	0.060

difference similar to Eq. (6) and the results are listed in Table 6. The ordering quantity 431.13 is our conservative decision.

4 Conclusions

Instead of solving for the optimal solution under the assumption of deterministic demands, here we provide a prescriptive methodology for constructing uncertainty sets within a robust optimization framework for integrated inventory problems with uncertain data. From both numerical examples in Sect. 3, it was found that the robust ordering quantity has the lowest average risk among possible demands. Furthermore, decision makers could even weight in additional information on the possibilities among demands to obtain the final solution.

References

Banerjee A (1986) A joint economic-lot-size model for purchaser and vendor. Decis Sci 17(3):292–311

Banerjee A, Kim S-L (1995) An integrated JIT inventory model. Int J Oper Prod Manage 15(9):237–244

Bertsimas D, Sim M (2004) The price of robustness. Oper Res 52(1):35–53

Ha D, Kim S-L (1997) Implementation of JIT purchasing: an integrated approach. Prod Plann Control 8(2):152–157

Goyal SK (1977) An integrated inventory model for a single supplier-single customer problem. Int J Prod Res 15(1):107–111

Goyal SK (1988) A joint economic-lot-size model for purchaser and vendor: a comment. Decis Sci 19(1):236–241

Goyal SK, Gupta YP (1989) Integrated inventory models: the buyer-vendor coordination. Eur J Oper Res 41(3):261–269

Lu L (1995) A one-vendor multi-buyer integrated inventory model. Eur J Oper Res 81(2):312–323

Pan AC, Liao C-J (1989) An inventory model under just-in-time purchasing agreement. Prod Inventory Manage J 30(1):49

Pan JC-H, Yang J-S (2002) A study of an integrated inventory with controllable lead time. Int J Prod Res 40(5):1263–1273

Ramasesh RV (1990) Recasting the traditional inventory model to implement just-in-time purchasing. Prod Inventory Manage J 31(1):71

Siajadi H, Ibrahim RN, Lochert PB (2006) A single-vendor multiple-buyer inventory model with a multiple-shipment policy. Int J Adv Manuf Technol 27(9–10):1030–1037

Soyster AL (1973) Technical note—convex programming with set-inclusive constraints and applications to inexact linear programming. Oper Res 21(5):1154–1157

Cost-Based Design of a Heat Sink Using SVR, Taguchi Quality Loss, and ACO

Chih-Ming Hsu

Abstract This study proposed a cost-based procedure for resolving multi-response parameter design problems using support vector regression (SVR), Taguchi quality loss and ant colony optimization (ACO). A case study aiming to optimize the design of a heat sink applied in a high-power MR16 LED lamp was used to demonstrate the proposed procedure. The experimental results indicated that the proposed procedure can provide highly robust settings of design parameters which can maximize the thermal performance, as well as can minimize the actual material cost of a heat sink. Furthermore, decision makers no longer need to determine the relative weight of each response subjectively. Therefore, the proposed approach in this study can be considered as feasible and effective, and can be popularized to be a useful tool for resolving general multi-response parameter design problems in the real world.

Keywords Heat sink · Support vector regression · Taguchi quality loss · Ant colony optimization · Multi-response parameter design

1 Introduction

The Taguchi method is a well-known traditional approach for tackling multi-response parameter design problems; however, it has not proved to be fully functional for optimizing multiple responses, especially in the case of correlated responses (Sibalija et al. 2011). Therefore, many recent studies have centered on using/integrating techniques from various fields to find the continuous settings of control factors for a multi-response parameter design problem, e.g. Kim and Lin (2000), Kovach and Cho (2008), Sibalija et al. (2011), Bera and Mukherjee (2012),

---

C.-M. Hsu (✉)
Department of Business Administration, Minghsin University of Science and Technology,
1 Hsin-Hsing Road, Hsin-Fong, Hsinchu 304, Taiwan, Republic of China
e-mail: cmhsu@must.edu.tw

Y.-K. Lin et al. (eds.), *Proceedings of the Institute of Industrial Engineers Asian Conference 2013*, DOI: 10.1007/978-981-4451-98-7_92,
© Springer Science+Business Media Singapore 2013

Devi et al. (2012), He et al. (2012), Mukherjee et al. (2012) and Salmasnia et al. (2012).

In the above studies, the estimation of functional relationship between control factors and responses usually relies on the accuracy of second-order polynomials, which are not always suitable (Goethals and Cho 2012). Next, decision makers must specify the relative weight (importance) of each response and/or set the coefficient that manipulate the shape of a desirability function subjectively while transforming multiple responses into a single objective. Furthermore, the manufacturing or material cost of a product is not considered while determining the optimal settings of design/process parameters. To overcome these shortcomings, this study attempts to apply the support vector regression (SVR), Taguchi quality loss and ant colony optimization (ACO) to develop a procedure for resolving a multi-response parameter design problem.

2 Research Methodologies

2.1 Support Vector Regression

Support vector regression (SVR) (Vapnik et al. 1997; Drucker et al. 1997) is the application of support vector machine (SVM) to the case of function approximation or regression. Given a training data $\{X_k, d_k\}_{k=1}^{Q}$ where the input variable $X_k \in \mathbb{R}^n$ is an n-dimensional vector and the output variable $d_k \in \mathbb{R}$ is a real value, SVR approximates a function of the form

$$f(X, W) = \sum_{i}^{m} w_i \varphi_i(X) + w_0 = W^T \Phi(X) + w_0 \tag{1}$$

where w_i is the weight; W is the weight vector; $\varphi_i(X)$ is the feature; $\Phi(X)$ is the feature vector; w_0 is the bias. In order to evaluate the prediction error, Vapnik (1998) introduced the ε-insensitive loss function, defined by

$$L_\varepsilon(d, f(X, W)) = \begin{cases} 0 & \text{if } |d - f(X,W)| \le \varepsilon \\ |d - f(X,W)| - \varepsilon & \text{otherwise} \end{cases}. \tag{2}$$

Therefore, the penalty (loss) can be expressed by

$$d_i - W^T \Phi(X) - w_0 - \varepsilon \le \xi_i, \quad i = 1, \ldots, Q \tag{3}$$

$$W^T \Phi(X) + w_0 - d_i - \varepsilon \le \xi_i', \quad i = 1, \ldots, Q \tag{4}$$

$$\xi_i \ge 0, \quad i = 1, \ldots, Q \tag{5}$$

$$\xi_i' \ge 0, \quad i = 1, \ldots, Q \tag{6}$$

where ξ_i and ξ_i' are non-negative slack variables. The empirical risk minimization problem then can be defined as Vapnik (1995), (1998)

$$\frac{1}{2}||W||^2 + C\left(\sum_{i=1}^{Q}\xi_i + \sum_{i=1}^{Q}\xi_i'\right) \tag{7}$$

subject to the constraints in (3)–(6), where C is a user specified parameter for the trade-off between complexity and losses. To solve the optimization in (7), the Lagrangian in primal variables are constructed and its partial derivatives with respect to the primal variables have to vanish at the saddle point for optimality. Therefore, the simplified dual form then can be obtained, and the optimal weight vectors can be obtained, with the Lagrangian optimization done, as follows

$$\hat{W} = \sum_{i=1}^{Q}(\hat{\lambda}_i - \hat{\lambda}_i')\Phi(X_i) = \sum_{k=1}^{n_s}(\hat{\lambda}_k - \hat{\lambda}_k')\Phi(X_k) \tag{8}$$

where n_s is the number of support vectors. Finally, the optimal bias can be obtained as follows

$$\hat{w}_0 = \frac{1}{n_{us}}\sum_{i=1}^{n_{us}}\left(d_i - \sum_{k=1}^{n_s}\beta_k K(X_k, X_i) - \varepsilon \ \mathrm{sign}(\beta_i)\right) \tag{9}$$

where n_{us} is the number of unbounded support vectors with Lagrangian multipliers satisfying $0 < \lambda_i < C$, and $\beta_i = \hat{\lambda}_i - \hat{\lambda}_i'$.

2.2 Ant Colony Optimization

The ant colony optimization (ACO) approach, which extends to the continuous domain as proposed by Socha (2004), is closest to the spirit of ACO for discrete problems (Blum 2005). Suppose a population with a cardinality of k is used to solve a continuous optimization problem with n dimensions, Socha (2004) used the Gaussian function to estimate the distribution of each member (ant) in the solution population. For the ith dimension, the jth Gaussian function, with the mean μ_j^i and standard deviation σ_j^i, is represented by

$$g_j^i(x) = \frac{1}{\sigma_j^i\sqrt{2\pi}}e^{-\frac{(x-\mu_j^i)^2}{2\sigma_j^{i2}}}, \quad \forall i = 1,\ldots,n, \ \forall j = 1,\ldots,k, \ \forall x \in \mathrm{R} \tag{10}$$

All solutions in the population are ranked based on their fitness with rank 1 for the best solution, and the associated weight of the jth member of the population in the mixture is calculated by Blum (2005):

$$w_j = \frac{1}{qk\sqrt{2\pi}} e^{-\frac{(r-1)^2}{2q^2k^2}}, \quad \forall j = 1, \ldots, k \tag{11}$$

where r is the rank of the jth member and q (> 0) is a parameter of the algorithm. Each ant j must choose one of the Gaussian functions $(g_1^1, g_2^1, \ldots, g_j^1, \ldots, g_k^1)$ for the first dimension, i.e. the first construction step, with the probability (Blum 2005)

$$p_j = \frac{w_j}{\sum_{l=1}^{k} w_l}, \quad \forall j = 1, \ldots, k. \tag{12}$$

For the j^*th Gaussian function in the ith dimension, the mean and standard deviation are set by Blum (2005)

$$\mu_{j^*}^i = x_{j^*}^i, \quad \forall i = 1, \ldots, n, \tag{13}$$

$$\sigma_{j^*}^i = \frac{1}{k} \rho \sum_{j=1}^{k} \sqrt{(x_j^i - x_{j^*}^i)^2}, \quad \forall i = 1, \ldots, n, \tag{14}$$

respectively, where x_j^i is the value of the ith decision variable in solution (ant) j and $\rho \in (0, 1)$ is the parameter which regulates the speed of convergence. For the detailed execution steps of the ant colony optimization for continuous domains, readers can refer to Socha (2004) and Blum (2005).

3 Proposed Cost-Based Parameter Design Procedure

In this study, a cost-based parameter design procedure for resolving multi-response parameter design problems using SVR, Taguchi quality loss and ACO was proposed to overcome the shortcomings in the literatures. The proposed procedure is described as follows:

Step 1: Determine the key quality characteristics of a product and specification limits.
Step 2: Identify and arrange control factors.
Step 3: Identify and arrange noise factors.
Step 4: Conduct each experimental trial and collect data.
Step 5: Build SVR estimation models.
Step 6: Optimize settings of control factors through ACO.
Step 7: Conduct the confirmation experiment.
Step 8: Review the results, re-identify the control factors and repeat Steps 2–7 if the confirmation result is unsatisfactory.

Notably, the overall quality regarding a product under a certain combination of control factors' settings is evaluated through a single minimized objective function calculated by

$$TQL = MC + \sum_{j=1}^{r} AQL_j \qquad (15)$$

where MC is the actual manufacturing or material cost of a product and r is the total number of quality characteristics. In addition, the AQL_j is the average quality loss calculated by

$$AQL_j = \sum_{k=1}^{s} \frac{QL_j^{(k)}}{s} \qquad (16)$$

where s is the total number of combinations of noise factors' evaluation levels and $QL_j^{(k)}$ is the quality loss under a certain combination of control factors' settings along with the kth combination of noise factors' evaluation levels.

4 Case Study on Designing a Heat Sink

Figure 1 is a typical heat sink used in a high-power MR16 LED (light emitting diode) lamp. In order to maximize the thermal performance of a heat sink, the geometric appearance requires an elaborate design as well as the material should be selected carefully. Furthermore, the geometric design affects the volume of a heat sink thus determining its related material cost.

Fig. 1 A typical heat sink for a high-power MR16 LED lamp

Based on the quality improvement objectives, the designers and quality managers of LED lamps determine three key quality characteristics of a heat sink as maximum temperature (y_1), thermal resistance (y_2) and material cost (y_3). The typical upper specification limits for the maximum temperature (y_1) and thermal resistance (y_2) are 85 °C and 3.5 °C/W, respectively, at an ambient air temperature of 25 °C.

Brainstorming with design engineers identified two critical material properties and four main geometric parameters which might significantly affect the three concerned key quality characteristics of a heat sink as control factors including coefficient of thermal radiation (x_1), coefficient of thermal conductivity (x_2), rotation angle (x_3), height of a fin (x_4), width of a fin (x_5), number of fins (x_6). The parameters x_3, x_4 and x_5 are illustrated in Fig. 2. Three types of materials with different coefficients of thermal conductivity and two types of coefficients of thermal conductivity, as summarized in Table 1, were applied in this study. In addition, three experimental levels were set for parameters x_3 to x_6, as summarized in Table 1, in order to estimate their non-linear effects upon the critical quality characteristics. Therefore, a Taguchi $L_{18}(2^1 \times 3^7)$ orthogonal array was selected as the inner array. In addition, a Taguchi $L_9(3^4)$ orthogonal array was used to arrange the noise factors z_1 and z_2 for considering the limitations of machining while making a heat sink as shown in Table 1.

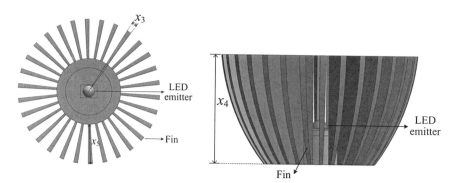

Fig. 2 Illustration of geometric design parameters of a heat sin

Table 1 Experimental settings of control and noise factors

Control factors	x_1	x_2 (W/m·K)	Density (g/cm^3)	Unit cost (NTD/kg)	x_3 (degree)	x_4 (mm)	x_5 (mm)	x_6
Level 1	0.5	170	2.7	120	1.5	18	10.0	20
Level 2	0.9	200	2.8	135	2.5	24	12.5	25
Level 3	–	380	8.9	288	3.5	30	15.0	30
Noise factors	–	–			–	z_1 (mm)	z_2 (mm)	–
Level 1	–	–			–	−0.05	−0.05	–
Level 2	–	–			–	0.00	0.00	–
Level 3	–	–			–	+0.05	+0.05	–

Table 2 Partial experimental trials and collected data

| No. | Control factors | | | | | | Noise factors | | Quality characteristics | | | Average temperature (°C) | Volume (mm^3) |
	x_1	x_2 (W/m·K)	x_3 (degree)	x_4 (mm)	x_5 (mm)	x_6	z_1 (mm)	z_2 (mm)	Maximum temperature (y_1, °C)	Thermal resistance (y_2, °C/W)	Material cost (y_3, NTD)		
1	0.5	170	1.5	18	10.0	20	−0.05	−0.05	69.27	3.5774	0.6686	55.74	2,064
2	0.5	170	1.5	18	10.0	20	−0.05	0.00	69.14	3.5804	0.6713	55.60	2,072
3	0.5	170	1.5	18	10.0	20	−0.05	+0.05	69.01	3.5884	0.6740	55.45	2,080
160	0.9	380	3.5	24	10.0	25	+0.05	−0.05	55.05	3.1633	14.3480	43.10	5,598
161	0.9	380	3.5	24	10.0	25	+0.05	0.00	55.11	3.1585	14.2726	43.17	5,568
162	0.9	380	3.5	24	10.0	25	+0.05	+0.05	55.04	3.1513	14.3571	43.12	5,601

Table 3 Execution results of SVR

Dependent variable	Training MSE ($\times 10^{-5}$)	Test MSE ($\times 10^{-5}$)	Training R^2	Test R^2
Maximum temperature	4.95	5.01	0.9997	0.9999
Average temperature	6.29	4.03	0.9997	0.9995
Volume	7.45	3.66	0.9995	0.9998

The SolidWorks 2010 (http://www.solidworks.com) and ANSYS 13 (http://www.ansys.com) were used to conduct the experiment. Table 2 presents a part of the collected experimental results. The SVR using the LIBSVM (http://www.csie.ntu.edu.tw/~cjlin/libsvm/) was employed to build the estimated mathematical models for describing the dependence of the maximum temperature, average temperature and volume on the design parameters based on the collected data in Table 2. The results are summarized in Table 3.

In order to find the optimal settings of six design parameters for a heat sink, this study applied the ACO algorithm to explore the experimental ranges of design parameters where the SVR models were used to describe the mathematical dependence of the maximum temperature, average temperature and volume on the design parameters, respectively. Notably, the objective function that we want to minimize in ACO was designed by (15). The ACO search procedure was implemented for 10 runs. With the aim of minimizing the objective function's value, the combination of design parameters' settings ($x_1 = 0.9$, $x_2 = 170$, $x_3 = 2.08$, $x_4 = 28.41$, $x_5 = 10$ and $x_5 = 23$) was determined as the optimal solution.

A confirmation experiment was conducted to verify the feasibility and effectiveness of the optimal combination of design parameters' settings. According to the simulation results in Table 4, the maximum temperature (y_1) and thermal resistance (y_2) in all confirmation trials can confirm to their specification limits of 85 °C and 3.5 °C/W, respectively. This implies that the proposed procedure indeed provides an optimal design of a heat sink which can optimize the thermal performance and can minimize the actual material cost of a heat sink.

5 Conclusion

To overcome the shortcomings in the literatures, including insufficient accuracy of second-order polynomials as well as subjective determination of relative weights and shape coefficients, this study applied the support vector regression (SVR), Taguchi quality loss, and ant colony optimization (ACO) to develop a cost-based procedure for resolving a multi-response parameter design problem. The feasibility and effectiveness of the proposed approach were verified through a case study on optimizing the design of a heat sink used in a high-power MR16 LED lamp. The experimental results indicated that the proposed solution procedure can provide highly robust design parameters' settings of a heat sink that can optimize the

Table 4 Summary of confirmation experiment results

No.	x_1	x_2 (W/m·K)	x_3 (degree)	x_4 (mm)	x_5 (mm)	x_6	Maximum temperature (y_1, °C)	Thermal resistance (y_2, °C/Watt)	Material cost (y_3, NTD)	Average temperature (°C)	Volume (mm^3)
1	0.9	170	2.08	28.41	10.00	23	56.41	3.4667	1.3285	42.89	4,100
2	0.9	170	2.08	28.36	9.95	23	56.48	3.4667	1.3217	43.09	4,079
3	0.9	170	2.08	28.36	10.00	23	56.44	3.4667	1.3260	43.01	4,093
4	0.9	170	2.08	28.36	10.05	23	56.40	3.4667	1.3304	42.94	4,106
5	0.9	170	2.08	28.41	9.95	23	56.45	3.4667	1.3241	42.99	4,087
6	0.9	170	2.08	28.41	10.00	23	56.41	3.4667	1.3285	42.98	4,100
7	0.9	170	2.08	28.41	10.05	23	56.37	3.4667	1.3328	42.97	4,114
8	0.9	170	2.08	28.46	9.95	23	56.43	3.4667	1.3265	42.91	4,094
9	0.9	170	2.08	28.46	10.00	23	56.39	3.4667	1.3309	42.98	4,108
10	0.9	170	2.08	28.46	10.05	23	56.35	3.4667	1.3353	42.91	4,121

critical quality characteristics of a heat sink, as well as can minimize the actual material cost.

Acknowledgments The author would like to thank the National Science Council, Taiwan, ROC, for its support of this research under Contract No. NSC 101-2221-E-159-009, and also to thank Raymond Huang for his valuable assistance during this study.

References

Bera S, Mukherjee I (2012) An adaptive penalty function-based maximin desirability index for close tolerance multiple-response optimization problems. Int J Adv Manuf Tech 61(1–4):379–390

Blum C (2005) Ant colony optimization: introduction and recent trends. Phys Life Rev 2(4):353–373

Devi SP, Manivannan S, Rao KS (2012) Comparison of nongradient methods with hybrid Taguchi-based epsilon constraint method for multiobjective optimization of cylindrical fin heat sink. Int J Adv Manuf Tech 63(9–12):1081–1094

Drucker H, Burges CJC, Kaufman L, Smola A, Vapnik V (1997) Support vector regression machines. In: Mozer MC, Jordan MI, Petsche T (eds) Advances in Neural Information Processing Systems, vol 9. MIT Press, Cambridge, MA, pp 155–161

Goethals PL, Cho BR (2012) Extending the desirability function to account for variability measures in univariate and multivariate response experiments. Comput Ind Eng 62(2):457–468

He Z, Zhu PF, Park SH (2012) A robust desirability function method for multi-response surface optimization considering model uncertainty. Eur J Oper Res 221(1):241–247

Kim KJ, Lin DKJ (2000) Simultaneous optimization of mechanical properties of steel by maximizing exponential desirability functions. J Roy Stat Soc C-App 49:311–325

Kovach J, Cho BR (2008) Development of a multidisciplinary-multiresponse robust design optimization model. Eng Optimiz 40(9):805–819

Mukherjee R, Chakraborty S, Samanta S (2012) Selection of wire electrical discharge machining process parameters using non-traditional optimization algorithms. Appl Soft Comput 12(8): 2506–2516

Salmasnia A, Kazemzadeh RB, Tabrizi MM (2012) A novel approach for optimization of correlated multiple responses based on desirability function and fuzzy logics. Neurocomputing 91:56–66

Sibalija TV, Majstorovic VD, Miljkovic ZD (2011) An intelligent approach to robust multi-response process design. Int J Prod Res 49(17):5079–5097

Socha K (2004) ACO for continuous and mixed-variable optimization. Paper presented at the 4th international workshop, Brussels, Belgium, 5–8 Sept 2004

Vapnik V (1995) The nature of statistical learning theory. Springer, New York

Vapnik V (1998) Statistical learning theory. Wiley, New York

Vapnik V, Golowich S, Smola A (1997) Support vector method for function approximation, regression estimation, and signal processing. In: Mozer MC, Jordan MI, Petsche T (eds) Advances in neural information processing systems, vol 9., MIT PressCambridge, MA, pp 281–287